LIVERPOOL JMU LIBRARY

METHODS IN ENZYMOLOGY

Editors-in-Chief

JOHN N. ABELSON and MELVIN I. SIMON
Division of Biology
California Institute of Technology
Pasadena, California

Founding Editors

SIDNEY P. COLOWICK and NATHAN O. KAPLAN

VOLUME FIVE HUNDRED AND THIRTY TWO

METHODS IN ENZYMOLOGY

Research Methods in Biomineralization Science

Edited by

JAMES J. DE YOREO

Pacific Northwest National Laboratory, Richland, WA, USA

AMSTERDAM • BOSTON • HEIDELBERG • LONDON
NEW YORK • OXFORD • PARIS • SAN DIEGO
SAN FRANCISCO • SINGAPORE • SYDNEY • TOKYO

Academic Press is an imprint of Elsevier

Academic Press is an imprint of Elsevier
525 B Street, Suite 1800, San Diego, CA 92101-4495, USA
225 Wyman Street, Waltham, MA 02451, USA
Radarweg 29, PO Box 211, 1000 AE Amsterdam, The Netherlands
The Boulevard, Langford Lane, Kidlington, Oxford, OX5 1GB, UK
32 Jamestown Road, London NW1 7BY, UK

First edition 2013

Copyright © 2013 Elsevier Inc. All Rights Reserved.
For chapters 8 and 10 - Lawrence Livermore National Laboratory/Security, LLC, retains a Government use license under contract no. DE-AC52-07NA27344.

No part of this publication may be reproduced, stored in a retrieval system or transmitted in any form or by any means electronic, mechanical, photocopying, recording or otherwise without the prior written permission of the publisher

Permissions may be sought directly from Elsevier's Science & Technology Rights Department in Oxford, UK: phone (+44) (0) 1865 843830; fax (+44) (0) 1865 853333; email: permissions@elsevier.com. Alternatively you can submit your request online by visiting the Elsevier web site at http://elsevier.com/locate/permissions, and selecting *Obtaining permission to use Elsevier material*

Notice
No responsibility is assumed by the publisher for any injury and/or damage to persons or property as a matter of products liability, negligence or otherwise, or from any use or operation of any methods, products, instructions or ideas contained in the material herein. Because of rapid advances in the medical sciences, in particular, independent verification of diagnoses and drug dosages should be made

> For information on all Academic Press publications
> visit our website at store.elsevier.com

ISBN: 978-0-12-416617-2
ISSN: 0076-6879

Printed and bound in United States of America
13 14 15 16   11 10 9 8 7 6 5 4 3 2 1

CONTENTS

Contributors	xiii
Preface	xvii
Volume in Series	xix

Section I
Investigating Solution Chemistry, Structure, and Nucleation

1. Development of Accurate Force Fields for the Simulation of Biomineralization — 3
Paolo Raiteri, Raffaella Demichelis, and Julian D. Gale

1. Introduction	4
2. Force Field Selection and Refinement	7
3. The Mineral–Aqueous Interface	14
4. Organics in Geochemical Systems	18
5. Summary	19
Acknowledgments	19
References	19

2. The Integration of Ion Potentiometric Measurements with Chemical, Structural, and Morphological Analysis to Follow Mineralization Reactions in Solution — 25
Wouter J.E.M. Habraken

1. Introduction	26
2. Ion Potentiometric Measurements	29
3. Combination of Ion Potentiometric Measurements with other Analysis Techniques	35
4. Comparison with Standard Analyses	41
5. Summary	42
References	42

3. Investigating the Early Stages of Mineral Precipitation by Potentiometric Titration and Analytical Ultracentrifugation — 45
Matthias Kellermeier, Helmut Cölfen, and Denis Gebauer

1. Introduction	46
2. Equipment	48

3. Experimental Procedures	49
4. Data Evaluation	54
5. Conclusions and Outlook	66
References	67

4. Replica Exchange Methods in Biomineral Simulations 71
Adam F. Wallace

1. Introduction	72
2. Monte Carlo Sampling	73
3. Replica Exchange MD	77
4. Application of REMD to Early Stage Mineralization	83
5. Summary	89
References	89

5. SAXS in Inorganic and Bioinspired Research 95
Tomasz M. Stawski and Liane G. Benning

1. Introduction	96
2. General Considerations: How Does SAXS/WAXS Work and When It Is Applicable?	96
3. What Happens in a Solution When Ions Meet and How Do We Quantify it?	99
4. SAXS Data Processing and Interpretation	105
5. Case Studies	111
6. Outlook	122
Acknowledgments	124
References	124

6. *In Situ* Solution Study of Calcium Phosphate Crystallization Kinetics 129
Haihua Pan, Shuqin Jiang, Tianlan Zhang, and Ruikang Tang

1. Introduction	130
2. Solution Preparation and Experimental Considerations	132
3. *In Situ* Experiments	135
4. Data Handling/Processing	138
5. Summary	141
References	141

Section II
Probing Structure and Dynamics at Surfaces and Interfaces

7. Design, Fabrication, and Applications of *In Situ* Fluid Cell TEM — 147
Dongsheng Li, Michael H. Nielsen, and James J. De Yoreo

   1. Introduction — 147
   2. Fluid Cell Design — 150
   3. Cell Fabrication and Assembly — 153
   4. TEM Operation — 154
   5. Effects of the Electron Beam on Reactions in the Fluid Cell — 160
   6. Conclusions — 161
   Acknowledgments — 162
   References — 162

8. X-ray Absorption Spectroscopy for the Structural Investigation of Self-Assembled-Monolayer-Directed Mineralization — 165
Jonathan R.I. Lee, Michael Bagge-Hansen, Trevor M. Willey, Robert W. Meulenberg, Michael H. Nielsen, Ich C. Tran, and Tony van Buuren

   1. Introduction — 166
   2. Theory — 167
   3. Experimental — 172
   4. Analysis — 183
   Acknowledgments — 185
   References — 186

9. Cryo-TEM Analysis of Collagen Fibrillar Structure — 189
Bryan D. Quan and Eli D. Sone

   1. Introduction — 190
   2. Sample Preparation for Cryo-TEM — 193
   3. Image Acquisition and Processing — 195
   4. Image Averaging — 198
   5. Summary and Outlook — 202
   Acknowledgments — 203
   References — 203

Section III
Biomimetic Crystallization Techniques *in vitro*

10. Preparation of Organothiol Self-Assembled Monolayers for Use in Templated Crystallization — 209

Michael H. Nielsen and Jonathan R.I. Lee

1. Introduction — 210
2. Crystallization Approaches Using SAMs as Templates — 211
3. Preparation — 214
4. Why Good SAMs Go Bad — 218
5. SAM Preparation Procedure — 221
6. Summary — 221
Acknowledgments — 222
References — 222

11. Experimental Techniques for the Growth and Characterization of Silica Biomorphs and Silica Gardens — 225

Matthias Kellermeier, Fabian Glaab, Emilio Melero-García, and Juan Manuel García-Ruiz

1. Introduction — 226
2. Silica Biomorphs — 230
3. Chemical Gardens — 240
4. Conclusions and Outlook — 251
Acknowledgments — 252
References — 252

12. Precipitation in Liposomes as a Model for Intracellular Biomineralization — 257

Chantel C. Tester and Derk Joester

1. Introduction — 258
2. Model Systems for Studying the Influence of Confinement on Biomineralization — 260
3. Preparation of Liposomes — 261
4. Precipitation Inside Liposomes — 268
5. Characterization Methods — 270
6. Conclusions — 273
Acknowledgments — 274
References — 274

13. **Polymer-Mediated Growth of Crystals and Mesocrystals**   277

Helmut Cölfen

    1. Introduction   278
    2. Insoluble Polymers   279
    3. Soluble Polymers   286
    4. Conclusion   301
    References   302

14. **Phage Display for the Discovery of Hydroxyapatite-Associated Peptides**   305

Hyo-Eon Jin, Woo-Jae Chung, and Seung-Wuk Lee

    1. Introduction   306
    2. Methods   312
    3. Summary   321
    References   321

Section IV
Protein Structure, Interactions and Function

15. **Quantitatively and Kinetically Identifying Binding Motifs of Amelogenin Proteins to Mineral Crystals Through Biochemical and Spectroscopic Assays**   327

Li Zhu, Peter Hwang, H. Ewa Witkowska, Haichuan Liu, and Wu Li

    1. Introduction   328
    2. Strategy and Rationale   330
    3. Experimental Components and Considerations   331
    4. Experimental Approaches   332
    5. Data Handling and Processing   340
    6. Summary   340
    Acknowledgments   341
    References   341

16. **Using the RosettaSurface Algorithm to Predict Protein Structure at Mineral Surfaces**   343

Michael S. Pacella, Da Chen Emily Koo, Robin A. Thottungal, and Jeffrey J. Gray

    1. Introduction   344
    2. Algorithm Evolution and Prior Results   346
    3. Computational Methods   348

4. Protocol Capture	352
5. Future Challenges	359
Acknowledgments	360
A. Appendix	360
References	363

17. Investigating Protein Function in Biomineralized Tissues Using Molecular Biology Techniques — 367

Christopher E. Killian and Fred H. Wilt

1. Introduction	368
2. Discovery of Proteins Involved in Biomineralization	368
3. Analysis of Function	377
Acknowledgments	385
References	385

Section V
Mapping Biomineral Morphology and Ultrastructure

18. Imaging the Nanostructure of Bone and Dentin Through Small- and Wide-Angle X-Ray Scattering — 391

Silvia Pabisch, Wolfgang Wagermaier, Thomas Zander, Chenghao Li, and Peter Fratzl

1. Introduction	392
2. Experimental Setup	394
3. Overview of Parameters	396
4. Treatment of SAXS Data from Mineralized Tissue	397
5. Treatment of WAXD Data from Mineralized Tissues	405
6. Analysis Tools	408
7. Combination of Scanning SAXS/WAXD with Other Methods	409
8. Conclusions	410
References	412

19. Synchrotron X-Ray Nanomechanical Imaging of Mineralized Fiber Composites — 415

Angelo Karunaratne, Nicholas J. Terrill, and Himadri S. Gupta

1. Introduction	416
2. Molecular and Nanoscale Strains and Stresses in a Prototypical Biomineralized Composite	419
3. Relation between X-Ray Spectra and Nanomechanics	422

4.	Sample Preparation for *In Situ* Mechanical Testing with Synchrotron X-Rays	425
5.	Mechanical Tester Design	435
6.	Synchrotron Setup	441
7.	*In Situ* Experimental Procedure	442
8.	Data Analysis	450
9.	Case Studies	461
10.	Model Interpretation	464
11.	Summary	468
	Acknowledgments	469
	References	469

Section VI
Mapping Mineral Chemistry

20. Application of Total X-Ray Scattering Methods and Pair Distribution Function Analysis for Study of Structure of Biominerals — 477

Richard J. Reeder and F. Marc Michel

1.	Introduction	478
2.	Total Scattering Methodology	479
3.	The PDF	486
4.	Extracting Information from the PDF	486
5.	PDF Studies of Biominerals	492
6.	Summary	496
7.	Links	498
	Acknowledgments	498
	References	498

21. X-Ray Microdiffraction of Biominerals — 501

Nobumichi Tamura and Pupa U.P.A. Gilbert

1.	Introduction	501
2.	Elements of X-Ray (Micro)Diffraction	504
3.	Synchrotron, X-Ray Focusing Optics, and Area Detectors	509
4.	Sample Preparation	514
5.	Powder X-Ray Microdiffraction	515
6.	White-Beam X-Ray Microdiffraction	520
7.	Microbeam Small-Angle X-Ray Scattering	522
8.	Future Developments	526
	References	527

22. FTIR and Raman Studies of Structure and Bonding in Mineral and Organic–Mineral Composites 533
Jinhui Tao

1. Introduction	534
2. FTIR and Raman Spectroscopies Overview	535
3. The Raman Effect	535
4. Measuring Biomimetic Crystals and Biominerals	543
5. Future Directions	553
Acknowledgments	553
References	553

23. A Mixed Flow Reactor Method to Synthesize Amorphous Calcium Carbonate Under Controlled Chemical Conditions 557
Christina R. Blue, J. Donald Rimstidt, and Patricia M. Dove

1. Introduction	558
2. Method	559
3. Characterization	562
4. Results	564
References	567

Author Index	*569*
Subject Index	*597*

CONTRIBUTORS

Michael Bagge-Hansen
Lawrence Livermore National Laboratory, Livermore, California, USA

Liane G. Benning
Cohen Biogeochemistry Laboratory, School of Earth and Environment, University of Leeds, Leeds, United Kingdom

Christina R. Blue
Department of Geosciences, Virginia Polytechnic Institute and State University, Blacksburg, Virginia, USA

Woo-Jae Chung
Department of Bioengineering, University of California; Physical Biosciences Divisions, Lawrence Berkeley National Laboratory, Berkeley, California, USA, and College of Biotechnology and Bioengineering, Sungkyunkwan University, Suwon, South Korea

Helmut Cölfen
Department of Chemistry, Physical Chemistry, University of Konstanz, Konstanz, Germany

James J. De Yoreo
Physical Sciences Division, Pacific Northwest National Laboratory, Richland, Washington, USA

Raffaella Demichelis
Department of Chemistry, Nanochemistry Research Institute, Curtin University, GPO Box U1987, Perth, Western Australia, Australia

Patricia M. Dove
Department of Geosciences, Virginia Polytechnic Institute and State University, Blacksburg, Virginia, USA

Peter Fratzl
Max Planck Institute for Colloids and Interfaces, Potsdam, Germany

Julian D. Gale
Department of Chemistry, Nanochemistry Research Institute, Curtin University, GPO Box U1987, Perth, Western Australia, Australia

Juan Manuel García-Ruiz
Laboratorio de Estudios Cristalográficos, IACT (CSIC-UGR), Av. de las Palmeras 4, Armilla, Spain

Denis Gebauer
Department of Chemistry, Physical Chemistry, University of Konstanz, Konstanz, Germany

Pupa U.P.A. Gilbert
Department of Physics, and Department of Chemistry, University of Wisconsin-Madison, Madison, Wisconsin, USA

Fabian Glaab
Institute of Physical and Theoretical Chemistry, University of Regensburg, Regensburg, Germany

Jeffrey J. Gray
Program in Molecular Biophysics, and Department of Chemical and Biomolecular Engineering, Johns Hopkins University, Baltimore, Maryland, USA

Himadri S. Gupta
Queen Mary University of London, School of Engineering and Material Sciences, London, United Kingdom

Wouter J.E.M. Habraken
Department of Biomaterials, Max Planck Institute for Colloids and Interfaces, Potsdam, Germany

Peter Hwang
Biochemistry & Biophysics Department, School of Medicine, San Francisco, California, USA

Shuqin Jiang
Qiushi Academy for Advanced Studies, and Department of Chemistry, Centre for Biomaterials and Biopathways, Zhejiang University, Hangzhou, China

Hyo-Eon Jin
Department of Bioengineering, University of California, and Physical Biosciences Divisions, Lawrence Berkeley National Laboratory, Berkeley, California, USA

Derk Joester
Northwestern University, Evanston, Illinois, USA

Angelo Karunaratne
Queen Mary University of London, School of Engineering and Material Sciences, London, United Kingdom

Matthias Kellermeier
Department of Chemistry, Physical Chemistry, University of Konstanz, Konstanz, and Institute of Physical and Theoretical Chemistry, University of Regensburg, Regensburg, Germany

Christopher E. Killian
Department of Physics, University of Wisconsin-Madison, Madison, Wisconsin, USA

Da Chen Emily Koo
T. C. Jenkins Department of Biophysics, Johns Hopkins University, Baltimore, Maryland, USA

Jonathan R.I. Lee
Lawrence Livermore National Laboratory, Livermore, California, USA

Seung-Wuk Lee
Department of Bioengineering, University of California, and Physical Biosciences Divisions, Lawrence Berkeley National Laboratory, Berkeley, California, USA

Chenghao Li
Max Planck Institute for Colloids and Interfaces, Potsdam, Germany

Dongsheng Li
Physical Sciences Division, Pacific Northwest National Laboratory, Richland, Washington, USA

Wu Li
Department of Orofacial Sciences, University of California, San Francisco, California, USA

Haichuan Liu
Sandler–Moore Mass Spectrometry Core Facility, University of California, San Francisco, California, USA

Emilio Melero-García
Laboratorio de Estudios Cristalográficos, IACT (CSIC-UGR), Av. de las Palmeras 4, Armilla, Spain

Robert W. Meulenberg
Department of Physics and Astronomy, Laboratory for Surface Science and Technology, University of Maine, Orono, Maine, USA

F. Marc Michel
Department of Geosciences, Virginia Tech, Blacksburg, Virginia, USA

Michael H. Nielsen
Department of Materials Science and Engineering, University of California, and Materials Science Division, Lawrence Berkeley National Lab, Berkeley, California, USA

Silvia Pabisch
Max Planck Institute for Colloids and Interfaces, Potsdam, Germany

Michael S. Pacella
Department of Biomedical Engineering, Johns Hopkins University, Baltimore, Maryland, USA

Haihua Pan
Qiushi Academy for Advanced Studies, Zhejiang University, Hangzhou, China

Bryan D. Quan
Institute of Biomaterials and Biomedical Engineering, University of Toronto, Toronto, Ontario, Canada

Paolo Raiteri
Department of Chemistry, Nanochemistry Research Institute, Curtin University, GPO Box U1987, Perth, Western Australia, Australia

Richard J. Reeder
Department of Geosciences, Stony Brook University, Stony Brook, New York, USA

J. Donald Rimstidt
Department of Geosciences, Virginia Polytechnic Institute and State University, Blacksburg, Virginia, USA

Eli D. Sone
Institute of Biomaterials and Biomedical Engineering; Department of Materials Science and Engineering, and Faculty of Dentistry, University of Toronto, Toronto, Ontario, Canada

Tomasz M. Stawski
Cohen Biogeochemistry Laboratory, School of Earth and Environment, University of Leeds, Leeds, United Kingdom

Nobumichi Tamura
Advanced Light Source, Lawrence Berkeley National Laboratory, Berkeley, California, USA

Ruikang Tang
Department of Chemistry, Centre for Biomaterials and Biopathways, Zhejiang University, Hangzhou, China

Jinhui Tao
Physical Sciences Division, Pacific Northwest National Laboratory, Richland, Washington, USA

Nicholas J. Terrill
Diamond Light Source Ltd., Diamond House, Harwell Science and Innovation Campus, Didcot, Oxfordshire, United Kingdom

Chantel C. Tester
Northwestern University, Evanston, Illinois, USA

Robin A. Thottungal
Program in Molecular Biophysics, Johns Hopkins University, Baltimore, Maryland, USA

Ich C. Tran
Lawrence Livermore National Laboratory, Livermore, California, USA

Tony van Buuren
Lawrence Livermore National Laboratory, Livermore, California, USA

Wolfgang Wagermaier
Max Planck Institute for Colloids and Interfaces, Potsdam, Germany

Adam F. Wallace
Department of Geological Sciences, University of Delaware, Newark, Delaware, USA

Trevor M. Willey
Lawrence Livermore National Laboratory, Livermore, California, USA

Fred H. Wilt
Department of Molecular and Cell Biology, University of California, Berkeley, California, USA

H. Ewa Witkowska
Sandler–Moore Mass Spectrometry Core Facility, University of California, San Francisco, California, USA

Thomas Zander
Max Planck Institute for Colloids and Interfaces, Potsdam, and Helmholtz-Zentrum Geesthacht, Geesthacht, Germany

Tianlan Zhang
Department of Chemical Biology, Peking University School of Pharmaceutical Sciences, Beijing, China

Li Zhu
Department of Orofacial Sciences, University of California, San Francisco, California, USA

PREFACE

The process of biomineralization, the structure of biominerals, and the resulting properties inherently depend on a hierarchy of scales from the molecular to the microscale. Moreover, the very nature of biomineral formation, occurring as it does at the interface between organic and inorganic constituents, places the science of biomineralization at the intersection between physiochemical processes and biological functions. Consequently, unraveling the underlying mechanisms of formation and function demands a wide range of synthetic and analytical techniques that can probe materials and processes over the full range of scales while providing insight into the biology as well as the physics and chemistry of biominerals. In fact, as with most areas of science, much of the progress in the state of knowledge within this field is a result of advances in methodologies that include microscopy, spectroscopy, scattering, molecular biology, computational modeling, and polymer science. The purpose of this volume is to provide researchers in the field of biomineralization science a manual for applying these methods, well armed with the "tricks" one must inevitably master to overcome the technical obstacles associated with any new technique. The collection of techniques described within has been purposely chosen to enable researchers to investigate the scientific questions across the full range of challenges within the field, from biomineral formation to structure, as well as their relationship to function. Although both the methods and the state of knowledge will continue to evolve with time, it is our hope that by documenting the state of the art, that evolution will be accelerated.

JAMES J. DE YOREO
Pacific Northwest National Laboratory,
Richland, WA,
USA

METHODS IN ENZYMOLOGY

VOLUME I. Preparation and Assay of Enzymes
Edited by SIDNEY P. COLOWICK AND NATHAN O. KAPLAN

VOLUME II. Preparation and Assay of Enzymes
Edited by SIDNEY P. COLOWICK AND NATHAN O. KAPLAN

VOLUME III. Preparation and Assay of Substrates
Edited by SIDNEY P. COLOWICK AND NATHAN O. KAPLAN

VOLUME IV. Special Techniques for the Enzymologist
Edited by SIDNEY P. COLOWICK AND NATHAN O. KAPLAN

VOLUME V. Preparation and Assay of Enzymes
Edited by SIDNEY P. COLOWICK AND NATHAN O. KAPLAN

VOLUME VI. Preparation and Assay of Enzymes (*Continued*)
Preparation and Assay of Substrates
Special Techniques
Edited by SIDNEY P. COLOWICK AND NATHAN O. KAPLAN

VOLUME VII. Cumulative Subject Index
Edited by SIDNEY P. COLOWICK AND NATHAN O. KAPLAN

VOLUME VIII. Complex Carbohydrates
Edited by ELIZABETH F. NEUFELD AND VICTOR GINSBURG

VOLUME IX. Carbohydrate Metabolism
Edited by WILLIS A. WOOD

VOLUME X. Oxidation and Phosphorylation
Edited by RONALD W. ESTABROOK AND MAYNARD E. PULLMAN

VOLUME XI. Enzyme Structure
Edited by C. H. W. HIRS

VOLUME XII. Nucleic Acids (Parts A and B)
Edited by LAWRENCE GROSSMAN AND KIVIE MOLDAVE

VOLUME XIII. Citric Acid Cycle
Edited by J. M. LOWENSTEIN

VOLUME XIV. Lipids
Edited by J. M. LOWENSTEIN

VOLUME XV. Steroids and Terpenoids
Edited by RAYMOND B. CLAYTON

VOLUME XVI. Fast Reactions
Edited by KENNETH KUSTIN

VOLUME XVII. Metabolism of Amino Acids and Amines (Parts A and B)
Edited by HERBERT TABOR AND CELIA WHITE TABOR

VOLUME XVIII. Vitamins and Coenzymes (Parts A, B, and C)
Edited by DONALD B. MCCORMICK AND LEMUEL D. WRIGHT

VOLUME XIX. Proteolytic Enzymes
Edited by GERTRUDE E. PERLMANN AND LASZLO LORAND

VOLUME XX. Nucleic Acids and Protein Synthesis (Part C)
Edited by KIVIE MOLDAVE AND LAWRENCE GROSSMAN

VOLUME XXI. Nucleic Acids (Part D)
Edited by LAWRENCE GROSSMAN AND KIVIE MOLDAVE

VOLUME XXII. Enzyme Purification and Related Techniques
Edited by WILLIAM B. JAKOBY

VOLUME XXIII. Photosynthesis (Part A)
Edited by ANTHONY SAN PIETRO

VOLUME XXIV. Photosynthesis and Nitrogen Fixation (Part B)
Edited by ANTHONY SAN PIETRO

VOLUME XXV. Enzyme Structure (Part B)
Edited by C. H. W. HIRS AND SERGE N. TIMASHEFF

VOLUME XXVI. Enzyme Structure (Part C)
Edited by C. H. W. HIRS AND SERGE N. TIMASHEFF

VOLUME XXVII. Enzyme Structure (Part D)
Edited by C. H. W. HIRS AND SERGE N. TIMASHEFF

VOLUME XXVIII. Complex Carbohydrates (Part B)
Edited by VICTOR GINSBURG

VOLUME XXIX. Nucleic Acids and Protein Synthesis (Part E)
Edited by LAWRENCE GROSSMAN AND KIVIE MOLDAVE

VOLUME XXX. Nucleic Acids and Protein Synthesis (Part F)
Edited by KIVIE MOLDAVE AND LAWRENCE GROSSMAN

VOLUME XXXI. Biomembranes (Part A)
Edited by SIDNEY FLEISCHER AND LESTER PACKER

VOLUME XXXII. Biomembranes (Part B)
Edited by SIDNEY FLEISCHER AND LESTER PACKER

VOLUME XXXIII. Cumulative Subject Index Volumes I–XXX
Edited by MARTHA G. DENNIS AND EDWARD A. DENNIS

VOLUME XXXIV. Affinity Techniques (Enzyme Purification: Part B)
Edited by WILLIAM B. JAKOBY AND MEIR WILCHEK

VOLUME XXXV. Lipids (Part B)
Edited by JOHN M. LOWENSTEIN

VOLUME XXXVI. Hormone Action (Part A: Steroid Hormones)
Edited by BERT W. O'MALLEY AND JOEL G. HARDMAN

VOLUME XXXVII. Hormone Action (Part B: Peptide Hormones)
Edited by BERT W. O'MALLEY AND JOEL G. HARDMAN

VOLUME XXXVIII. Hormone Action (Part C: Cyclic Nucleotides)
Edited by JOEL G. HARDMAN AND BERT W. O'MALLEY

VOLUME XXXIX. Hormone Action (Part D: Isolated Cells, Tissues, and Organ Systems)
Edited by JOEL G. HARDMAN AND BERT W. O'MALLEY

VOLUME XL. Hormone Action (Part E: Nuclear Structure and Function)
Edited by BERT W. O'MALLEY AND JOEL G. HARDMAN

VOLUME XLI. Carbohydrate Metabolism (Part B)
Edited by W. A. WOOD

VOLUME XLII. Carbohydrate Metabolism (Part C)
Edited by W. A. WOOD

VOLUME XLIII. Antibiotics
Edited by JOHN H. HASH

VOLUME XLIV. Immobilized Enzymes
Edited by KLAUS MOSBACH

VOLUME XLV. Proteolytic Enzymes (Part B)
Edited by LASZLO LORAND

VOLUME XLVI. Affinity Labeling
Edited by WILLIAM B. JAKOBY AND MEIR WILCHEK

VOLUME XLVII. Enzyme Structure (Part E)
Edited by C. H. W. HIRS AND SERGE N. TIMASHEFF

VOLUME XLVIII. Enzyme Structure (Part F)
Edited by C. H. W. HIRS AND SERGE N. TIMASHEFF

VOLUME XLIX. Enzyme Structure (Part G)
Edited by C. H. W. HIRS AND SERGE N. TIMASHEFF

VOLUME L. Complex Carbohydrates (Part C)
Edited by VICTOR GINSBURG

VOLUME LI. Purine and Pyrimidine Nucleotide Metabolism
Edited by PATRICIA A. HOFFEE AND MARY ELLEN JONES

VOLUME LII. Biomembranes (Part C: Biological Oxidations)
Edited by SIDNEY FLEISCHER AND LESTER PACKER

VOLUME LIII. Biomembranes (Part D: Biological Oxidations)
Edited by SIDNEY FLEISCHER AND LESTER PACKER

VOLUME LIV. Biomembranes (Part E: Biological Oxidations)
Edited by SIDNEY FLEISCHER AND LESTER PACKER

VOLUME LV. Biomembranes (Part F: Bioenergetics)
Edited by SIDNEY FLEISCHER AND LESTER PACKER

VOLUME LVI. Biomembranes (Part G: Bioenergetics)
Edited by SIDNEY FLEISCHER AND LESTER PACKER

VOLUME LVII. Bioluminescence and Chemiluminescence
Edited by MARLENE A. DELUCA

VOLUME LVIII. Cell Culture
Edited by WILLIAM B. JAKOBY AND IRA PASTAN

VOLUME LIX. Nucleic Acids and Protein Synthesis (Part G)
Edited by KIVIE MOLDAVE AND LAWRENCE GROSSMAN

VOLUME LX. Nucleic Acids and Protein Synthesis (Part H)
Edited by KIVIE MOLDAVE AND LAWRENCE GROSSMAN

VOLUME 61. Enzyme Structure (Part H)
Edited by C. H. W. HIRS AND SERGE N. TIMASHEFF

VOLUME 62. Vitamins and Coenzymes (Part D)
Edited by DONALD B. MCCORMICK AND LEMUEL D. WRIGHT

VOLUME 63. Enzyme Kinetics and Mechanism (Part A: Initial Rate and Inhibitor Methods)
Edited by DANIEL L. PURICH

VOLUME 64. Enzyme Kinetics and Mechanism
(Part B: Isotopic Probes and Complex Enzyme Systems)
Edited by DANIEL L. PURICH

VOLUME 65. Nucleic Acids (Part I)
Edited by LAWRENCE GROSSMAN AND KIVIE MOLDAVE

VOLUME 66. Vitamins and Coenzymes (Part E)
Edited by DONALD B. MCCORMICK AND LEMUEL D. WRIGHT

VOLUME 67. Vitamins and Coenzymes (Part F)
Edited by DONALD B. MCCORMICK AND LEMUEL D. WRIGHT

VOLUME 68. Recombinant DNA
Edited by RAY WU

VOLUME 69. Photosynthesis and Nitrogen Fixation (Part C)
Edited by ANTHONY SAN PIETRO

VOLUME 70. Immunochemical Techniques (Part A)
Edited by HELEN VAN VUNAKIS AND JOHN J. LANGONE

VOLUME 71. Lipids (Part C)
Edited by JOHN M. LOWENSTEIN

VOLUME 72. Lipids (Part D)
Edited by JOHN M. LOWENSTEIN

VOLUME 73. Immunochemical Techniques (Part B)
Edited by JOHN J. LANGONE AND HELEN VAN VUNAKIS

VOLUME 74. Immunochemical Techniques (Part C)
Edited by JOHN J. LANGONE AND HELEN VAN VUNAKIS

VOLUME 75. Cumulative Subject Index Volumes XXXI, XXXII, XXXIV–LX
Edited by EDWARD A. DENNIS AND MARTHA G. DENNIS

VOLUME 76. Hemoglobins
Edited by ERALDO ANTONINI, LUIGI ROSSI-BERNARDI, AND EMILIA CHIANCONE

VOLUME 77. Detoxication and Drug Metabolism
Edited by WILLIAM B. JAKOBY

VOLUME 78. Interferons (Part A)
Edited by SIDNEY PESTKA

VOLUME 79. Interferons (Part B)
Edited by SIDNEY PESTKA

VOLUME 80. Proteolytic Enzymes (Part C)
Edited by LASZLO LORAND

VOLUME 81. Biomembranes (Part H: Visual Pigments and Purple Membranes, I)
Edited by LESTER PACKER

VOLUME 82. Structural and Contractile Proteins (Part A: Extracellular Matrix)
Edited by LEON W. CUNNINGHAM AND DIXIE W. FREDERIKSEN

VOLUME 83. Complex Carbohydrates (Part D)
Edited by VICTOR GINSBURG

VOLUME 84. Immunochemical Techniques (Part D: Selected Immunoassays)
Edited by JOHN J. LANGONE AND HELEN VAN VUNAKIS

VOLUME 85. Structural and Contractile Proteins (Part B: The Contractile Apparatus and the Cytoskeleton)
Edited by DIXIE W. FREDERIKSEN AND LEON W. CUNNINGHAM

VOLUME 86. Prostaglandins and Arachidonate Metabolites
Edited by WILLIAM E. M. LANDS AND WILLIAM L. SMITH

VOLUME 87. Enzyme Kinetics and Mechanism (Part C: Intermediates, Stereo-chemistry, and Rate Studies)
Edited by DANIEL L. PURICH

VOLUME 88. Biomembranes (Part I: Visual Pigments and Purple Membranes, II)
Edited by LESTER PACKER

VOLUME 89. Carbohydrate Metabolism (Part D)
Edited by WILLIS A. WOOD

VOLUME 90. Carbohydrate Metabolism (Part E)
Edited by WILLIS A. WOOD

VOLUME 91. Enzyme Structure (Part I)
Edited by C. H. W. HIRS AND SERGE N. TIMASHEFF

VOLUME 92. Immunochemical Techniques (Part E: Monoclonal Antibodies and General Immunoassay Methods)
Edited by JOHN J. LANGONE AND HELEN VAN VUNAKIS

VOLUME 93. Immunochemical Techniques (Part F: Conventional Antibodies, Fc Receptors, and Cytotoxicity)
Edited by JOHN J. LANGONE AND HELEN VAN VUNAKIS

VOLUME 94. Polyamines
Edited by HERBERT TABOR AND CELIA WHITE TABOR

VOLUME 95. Cumulative Subject Index Volumes 61–74, 76–80
Edited by EDWARD A. DENNIS AND MARTHA G. DENNIS

VOLUME 96. Biomembranes [Part J: Membrane Biogenesis: Assembly and Targeting (General Methods; Eukaryotes)]
Edited by SIDNEY FLEISCHER AND BECCA FLEISCHER

VOLUME 97. Biomembranes [Part K: Membrane Biogenesis: Assembly and Targeting (Prokaryotes, Mitochondria, and Chloroplasts)]
Edited by SIDNEY FLEISCHER AND BECCA FLEISCHER

VOLUME 98. Biomembranes (Part L: Membrane Biogenesis: Processing and Recycling)
Edited by SIDNEY FLEISCHER AND BECCA FLEISCHER

VOLUME 99. Hormone Action (Part F: Protein Kinases)
Edited by JACKIE D. CORBIN AND JOEL G. HARDMAN

VOLUME 100. Recombinant DNA (Part B)
Edited by RAY WU, LAWRENCE GROSSMAN, AND KIVIE MOLDAVE

VOLUME 101. Recombinant DNA (Part C)
Edited by RAY WU, LAWRENCE GROSSMAN, AND KIVIE MOLDAVE

VOLUME 102. Hormone Action (Part G: Calmodulin and Calcium-Binding Proteins)
Edited by ANTHONY R. MEANS AND BERT W. O'MALLEY

VOLUME 103. Hormone Action (Part H: Neuroendocrine Peptides)
Edited by P. MICHAEL CONN

VOLUME 104. Enzyme Purification and Related Techniques (Part C)
Edited by WILLIAM B. JAKOBY

VOLUME 105. Oxygen Radicals in Biological Systems
Edited by LESTER PACKER

VOLUME 106. Posttranslational Modifications (Part A)
Edited by FINN WOLD AND KIVIE MOLDAVE

VOLUME 107. Posttranslational Modifications (Part B)
Edited by FINN WOLD AND KIVIE MOLDAVE

VOLUME 108. Immunochemical Techniques (Part G: Separation and Characterization of Lymphoid Cells)
Edited by GIOVANNI DI SABATO, JOHN J. LANGONE, AND HELEN VAN VUNAKIS

VOLUME 109. Hormone Action (Part I: Peptide Hormones)
Edited by LUTZ BIRNBAUMER AND BERT W. O'MALLEY

VOLUME 110. Steroids and Isoprenoids (Part A)
Edited by JOHN H. LAW AND HANS C. RILLING

VOLUME 111. Steroids and Isoprenoids (Part B)
Edited by JOHN H. LAW AND HANS C. RILLING

VOLUME 112. Drug and Enzyme Targeting (Part A)
Edited by KENNETH J. WIDDER AND RALPH GREEN

VOLUME 113. Glutamate, Glutamine, Glutathione, and Related Compounds
Edited by ALTON MEISTER

VOLUME 114. Diffraction Methods for Biological Macromolecules (Part A)
Edited by HAROLD W. WYCKOFF, C. H. W. HIRS, AND SERGE N. TIMASHEFF

VOLUME 115. Diffraction Methods for Biological Macromolecules (Part B)
Edited by HAROLD W. WYCKOFF, C. H. W. HIRS, AND SERGE N. TIMASHEFF

VOLUME 116. Immunochemical Techniques
(Part H: Effectors and Mediators of Lymphoid Cell Functions)
Edited by GIOVANNI DI SABATO, JOHN J. LANGONE, AND HELEN VAN VUNAKIS

VOLUME 117. Enzyme Structure (Part J)
Edited by C. H. W. HIRS AND SERGE N. TIMASHEFF

VOLUME 118. Plant Molecular Biology
Edited by ARTHUR WEISSBACH AND HERBERT WEISSBACH

VOLUME 119. Interferons (Part C)
Edited by SIDNEY PESTKA

VOLUME 120. Cumulative Subject Index Volumes 81–94, 96–101

VOLUME 121. Immunochemical Techniques (Part I: Hybridoma Technology and Monoclonal Antibodies)
Edited by JOHN J. LANGONE AND HELEN VAN VUNAKIS

VOLUME 122. Vitamins and Coenzymes (Part G)
Edited by FRANK CHYTIL AND DONALD B. MCCORMICK

VOLUME 123. Vitamins and Coenzymes (Part H)
Edited by FRANK CHYTIL AND DONALD B. MCCORMICK

VOLUME 124. Hormone Action (Part J: Neuroendocrine Peptides)
Edited by P. MICHAEL CONN

VOLUME 125. Biomembranes (Part M: Transport in Bacteria, Mitochondria, and Chloroplasts: General Approaches and Transport Systems)
Edited by SIDNEY FLEISCHER AND BECCA FLEISCHER

VOLUME 126. Biomembranes (Part N: Transport in Bacteria, Mitochondria, and Chloroplasts: Protonmotive Force)
Edited by SIDNEY FLEISCHER AND BECCA FLEISCHER

VOLUME 127. Biomembranes (Part O: Protons and Water: Structure and Translocation)
Edited by LESTER PACKER

VOLUME 128. Plasma Lipoproteins (Part A: Preparation, Structure, and Molecular Biology)
Edited by JERE P. SEGREST AND JOHN J. ALBERS

VOLUME 129. Plasma Lipoproteins (Part B: Characterization, Cell Biology, and Metabolism)
Edited by JOHN J. ALBERS AND JERE P. SEGREST

VOLUME 130. Enzyme Structure (Part K)
Edited by C. H. W. HIRS AND SERGE N. TIMASHEFF

VOLUME 131. Enzyme Structure (Part L)
Edited by C. H. W. HIRS AND SERGE N. TIMASHEFF

VOLUME 132. Immunochemical Techniques (Part J: Phagocytosis and Cell-Mediated Cytotoxicity)
Edited by GIOVANNI DI SABATO AND JOHANNES EVERSE

VOLUME 133. Bioluminescence and Chemiluminescence (Part B)
Edited by MARLENE DELUCA AND WILLIAM D. MCELROY

VOLUME 134. Structural and Contractile Proteins (Part C: The Contractile Apparatus and the Cytoskeleton)
Edited by RICHARD B. VALLEE

VOLUME 135. Immobilized Enzymes and Cells (Part B)
Edited by KLAUS MOSBACH

VOLUME 136. Immobilized Enzymes and Cells (Part C)
Edited by KLAUS MOSBACH

VOLUME 137. Immobilized Enzymes and Cells (Part D)
Edited by KLAUS MOSBACH

VOLUME 138. Complex Carbohydrates (Part E)
Edited by VICTOR GINSBURG

VOLUME 139. Cellular Regulators (Part A: Calcium- and Calmodulin-Binding Proteins)
Edited by ANTHONY R. MEANS AND P. MICHAEL CONN

VOLUME 140. Cumulative Subject Index Volumes 102–119, 121–134

VOLUME 141. Cellular Regulators (Part B: Calcium and Lipids)
Edited by P. MICHAEL CONN AND ANTHONY R. MEANS

VOLUME 142. Metabolism of Aromatic Amino Acids and Amines
Edited by SEYMOUR KAUFMAN

VOLUME 143. Sulfur and Sulfur Amino Acids
Edited by WILLIAM B. JAKOBY AND OWEN GRIFFITH

VOLUME 144. Structural and Contractile Proteins (Part D: Extracellular Matrix)
Edited by LEON W. CUNNINGHAM

VOLUME 145. Structural and Contractile Proteins (Part E: Extracellular Matrix)
Edited by LEON W. CUNNINGHAM

VOLUME 146. Peptide Growth Factors (Part A)
Edited by DAVID BARNES AND DAVID A. SIRBASKU

VOLUME 147. Peptide Growth Factors (Part B)
Edited by DAVID BARNES AND DAVID A. SIRBASKU

VOLUME 148. Plant Cell Membranes
Edited by LESTER PACKER AND ROLAND DOUCE

VOLUME 149. Drug and Enzyme Targeting (Part B)
Edited by RALPH GREEN AND KENNETH J. WIDDER

VOLUME 150. Immunochemical Techniques (Part K: *In Vitro* Models of B and T Cell Functions and Lymphoid Cell Receptors)
Edited by GIOVANNI DI SABATO

VOLUME 151. Molecular Genetics of Mammalian Cells
Edited by MICHAEL M. GOTTESMAN

VOLUME 152. Guide to Molecular Cloning Techniques
Edited by SHELBY L. BERGER AND ALAN R. KIMMEL

VOLUME 153. Recombinant DNA (Part D)
Edited by RAY WU AND LAWRENCE GROSSMAN

VOLUME 154. Recombinant DNA (Part E)
Edited by RAY WU AND LAWRENCE GROSSMAN

VOLUME 155. Recombinant DNA (Part F)
Edited by RAY WU

VOLUME 156. Biomembranes (Part P: ATP-Driven Pumps and Related Transport: The Na, K-Pump)
Edited by SIDNEY FLEISCHER AND BECCA FLEISCHER

VOLUME 157. Biomembranes (Part Q: ATP-Driven Pumps and Related Transport: Calcium, Proton, and Potassium Pumps)
Edited by SIDNEY FLEISCHER AND BECCA FLEISCHER

VOLUME 158. Metalloproteins (Part A)
Edited by JAMES F. RIORDAN AND BERT L. VALLEE

VOLUME 159. Initiation and Termination of Cyclic Nucleotide Action
Edited by JACKIE D. CORBIN AND ROGER A. JOHNSON

VOLUME 160. Biomass (Part A: Cellulose and Hemicellulose)
Edited by WILLIS A. WOOD AND SCOTT T. KELLOGG

VOLUME 161. Biomass (Part B: Lignin, Pectin, and Chitin)
Edited by WILLIS A. WOOD AND SCOTT T. KELLOGG

VOLUME 162. Immunochemical Techniques (Part L: Chemotaxis and Inflammation)
Edited by GIOVANNI DI SABATO

VOLUME 163. Immunochemical Techniques (Part M: Chemotaxis and Inflammation)
Edited by GIOVANNI DI SABATO

VOLUME 164. Ribosomes
Edited by HARRY F. NOLLER, JR., AND KIVIE MOLDAVE

VOLUME 165. Microbial Toxins: Tools for Enzymology
Edited by SIDNEY HARSHMAN

VOLUME 166. Branched-Chain Amino Acids
Edited by ROBERT HARRIS AND JOHN R. SOKATCH

VOLUME 167. Cyanobacteria
Edited by LESTER PACKER AND ALEXANDER N. GLAZER

VOLUME 168. Hormone Action (Part K: Neuroendocrine Peptides)
Edited by P. MICHAEL CONN

VOLUME 169. Platelets: Receptors, Adhesion, Secretion (Part A)
Edited by JACEK HAWIGER

VOLUME 170. Nucleosomes
Edited by PAUL M. WASSARMAN AND ROGER D. KORNBERG

VOLUME 171. Biomembranes (Part R: Transport Theory: Cells and Model Membranes)
Edited by SIDNEY FLEISCHER AND BECCA FLEISCHER

VOLUME 172. Biomembranes (Part S: Transport: Membrane Isolation and Characterization)
Edited by SIDNEY FLEISCHER AND BECCA FLEISCHER

VOLUME 173. Biomembranes [Part T: Cellular and Subcellular Transport: Eukaryotic (Nonepithelial) Cells]
Edited by SIDNEY FLEISCHER AND BECCA FLEISCHER

VOLUME 174. Biomembranes [Part U: Cellular and Subcellular Transport: Eukaryotic (Nonepithelial) Cells]
Edited by SIDNEY FLEISCHER AND BECCA FLEISCHER

VOLUME 175. Cumulative Subject Index Volumes 135–139, 141–167

VOLUME 176. Nuclear Magnetic Resonance (Part A: Spectral Techniques and Dynamics)
Edited by NORMAN J. OPPENHEIMER AND THOMAS L. JAMES

VOLUME 177. Nuclear Magnetic Resonance (Part B: Structure and Mechanism)
Edited by NORMAN J. OPPENHEIMER AND THOMAS L. JAMES

VOLUME 178. Antibodies, Antigens, and Molecular Mimicry
Edited by JOHN J. LANGONE

VOLUME 179. Complex Carbohydrates (Part F)
Edited by VICTOR GINSBURG

VOLUME 180. RNA Processing (Part A: General Methods)
Edited by JAMES E. DAHLBERG AND JOHN N. ABELSON

VOLUME 181. RNA Processing (Part B: Specific Methods)
Edited by JAMES E. DAHLBERG AND JOHN N. ABELSON

VOLUME 182. Guide to Protein Purification
Edited by MURRAY P. DEUTSCHER

VOLUME 183. Molecular Evolution: Computer Analysis of Protein and Nucleic Acid Sequences
Edited by RUSSELL F. DOOLITTLE

VOLUME 184. Avidin-Biotin Technology
Edited by MEIR WILCHEK AND EDWARD A. BAYER

VOLUME 185. Gene Expression Technology
Edited by DAVID V. GOEDDEL

VOLUME 186. Oxygen Radicals in Biological Systems (Part B: Oxygen Radicals and Antioxidants)
Edited by LESTER PACKER AND ALEXANDER N. GLAZER

VOLUME 187. Arachidonate Related Lipid Mediators
Edited by ROBERT C. MURPHY AND FRANK A. FITZPATRICK

VOLUME 188. Hydrocarbons and Methylotrophy
Edited by MARY E. LIDSTROM

VOLUME 189. Retinoids (Part A: Molecular and Metabolic Aspects)
Edited by LESTER PACKER

VOLUME 190. Retinoids (Part B: Cell Differentiation and Clinical Applications)
Edited by LESTER PACKER

VOLUME 191. Biomembranes (Part V: Cellular and Subcellular Transport: Epithelial Cells)
Edited by SIDNEY FLEISCHER AND BECCA FLEISCHER

VOLUME 192. Biomembranes (Part W: Cellular and Subcellular Transport: Epithelial Cells)
Edited by SIDNEY FLEISCHER AND BECCA FLEISCHER

VOLUME 193. Mass Spectrometry
Edited by JAMES A. MCCLOSKEY

VOLUME 194. Guide to Yeast Genetics and Molecular Biology
Edited by CHRISTINE GUTHRIE AND GERALD R. FINK

VOLUME 195. Adenylyl Cyclase, G Proteins, and Guanylyl Cyclase
Edited by ROGER A. JOHNSON AND JACKIE D. CORBIN

VOLUME 196. Molecular Motors and the Cytoskeleton
Edited by RICHARD B. VALLEE

VOLUME 197. Phospholipases
Edited by EDWARD A. DENNIS

VOLUME 198. Peptide Growth Factors (Part C)
Edited by DAVID BARNES, J. P. MATHER, AND GORDON H. SATO

VOLUME 199. Cumulative Subject Index Volumes 168–174, 176–194

VOLUME 200. Protein Phosphorylation (Part A: Protein Kinases: Assays, Purification, Antibodies, Functional Analysis, Cloning, and Expression)
Edited by TONY HUNTER AND BARTHOLOMEW M. SEFTON

VOLUME 201. Protein Phosphorylation (Part B: Analysis of Protein Phosphorylation, Protein Kinase Inhibitors, and Protein Phosphatases)
Edited by TONY HUNTER AND BARTHOLOMEW M. SEFTON

VOLUME 202. Molecular Design and Modeling: Concepts and Applications (Part A: Proteins, Peptides, and Enzymes)
Edited by JOHN J. LANGONE

VOLUME 203. Molecular Design and Modeling: Concepts and Applications (Part B: Antibodies and Antigens, Nucleic Acids, Polysaccharides, and Drugs)
Edited by JOHN J. LANGONE

VOLUME 204. Bacterial Genetic Systems
Edited by JEFFREY H. MILLER

VOLUME 205. Metallobiochemistry (Part B: Metallothionein and Related Molecules)
Edited by JAMES F. RIORDAN AND BERT L. VALLEE

VOLUME 206. Cytochrome P450
Edited by MICHAEL R. WATERMAN AND ERIC F. JOHNSON

VOLUME 207. Ion Channels
Edited by BERNARDO RUDY AND LINDA E. IVERSON

VOLUME 208. Protein–DNA Interactions
Edited by ROBERT T. SAUER

VOLUME 209. Phospholipid Biosynthesis
Edited by EDWARD A. DENNIS AND DENNIS E. VANCE

VOLUME 210. Numerical Computer Methods
Edited by LUDWIG BRAND AND MICHAEL L. JOHNSON

VOLUME 211. DNA Structures (Part A: Synthesis and Physical Analysis of DNA)
Edited by DAVID M. J. LILLEY AND JAMES E. DAHLBERG

VOLUME 212. DNA Structures (Part B: Chemical and Electrophoretic Analysis of DNA)
Edited by DAVID M. J. LILLEY AND JAMES E. DAHLBERG

VOLUME 213. Carotenoids (Part A: Chemistry, Separation, Quantitation, and Antioxidation)
Edited by LESTER PACKER

VOLUME 214. Carotenoids (Part B: Metabolism, Genetics, and Biosynthesis)
Edited by LESTER PACKER

VOLUME 215. Platelets: Receptors, Adhesion, Secretion (Part B)
Edited by JACEK J. HAWIGER

VOLUME 216. Recombinant DNA (Part G)
Edited by RAY WU

VOLUME 217. Recombinant DNA (Part H)
Edited by RAY WU

VOLUME 218. Recombinant DNA (Part I)
Edited by RAY WU

VOLUME 219. Reconstitution of Intracellular Transport
Edited by JAMES E. ROTHMAN

VOLUME 220. Membrane Fusion Techniques (Part A)
Edited by NEJAT DÜZGÜNEŞ

VOLUME 221. Membrane Fusion Techniques (Part B)
Edited by NEJAT DÜZGÜNEŞ

VOLUME 222. Proteolytic Enzymes in Coagulation, Fibrinolysis, and Complement Activation (Part A: Mammalian Blood Coagulation

Factors and Inhibitors)
Edited by LASZLO LORAND AND KENNETH G. MANN

VOLUME 223. Proteolytic Enzymes in Coagulation, Fibrinolysis, and Complement Activation (Part B: Complement Activation, Fibrinolysis, and Nonmammalian Blood Coagulation Factors)
Edited by LASZLO LORAND AND KENNETH G. MANN

VOLUME 224. Molecular Evolution: Producing the Biochemical Data
Edited by ELIZABETH ANNE ZIMMER, THOMAS J. WHITE, REBECCA L. CANN, AND ALLAN C. WILSON

VOLUME 225. Guide to Techniques in Mouse Development
Edited by PAUL M. WASSARMAN AND MELVIN L. DEPAMPHILIS

VOLUME 226. Metallobiochemistry (Part C: Spectroscopic and Physical Methods for Probing Metal Ion Environments in Metalloenzymes and Metalloproteins)
Edited by JAMES F. RIORDAN AND BERT L. VALLEE

VOLUME 227. Metallobiochemistry (Part D: Physical and Spectroscopic Methods for Probing Metal Ion Environments in Metalloproteins)
Edited by JAMES F. RIORDAN AND BERT L. VALLEE

VOLUME 228. Aqueous Two-Phase Systems
Edited by HARRY WALTER AND GÖTE JOHANSSON

VOLUME 229. Cumulative Subject Index Volumes 195–198, 200–227

VOLUME 230. Guide to Techniques in Glycobiology
Edited by WILLIAM J. LENNARZ AND GERALD W. HART

VOLUME 231. Hemoglobins (Part B: Biochemical and Analytical Methods)
Edited by JOHANNES EVERSE, KIM D. VANDEGRIFF, AND ROBERT M. WINSLOW

VOLUME 232. Hemoglobins (Part C: Biophysical Methods)
Edited by JOHANNES EVERSE, KIM D. VANDEGRIFF, AND ROBERT M. WINSLOW

VOLUME 233. Oxygen Radicals in Biological Systems (Part C)
Edited by LESTER PACKER

VOLUME 234. Oxygen Radicals in Biological Systems (Part D)
Edited by LESTER PACKER

VOLUME 235. Bacterial Pathogenesis (Part A: Identification and Regulation of Virulence Factors)
Edited by VIRGINIA L. CLARK AND PATRIK M. BAVOIL

VOLUME 236. Bacterial Pathogenesis (Part B: Integration of Pathogenic Bacteria with Host Cells)
Edited by VIRGINIA L. CLARK AND PATRIK M. BAVOIL

VOLUME 237. Heterotrimeric G Proteins
Edited by RAVI IYENGAR

VOLUME 238. Heterotrimeric G-Protein Effectors
Edited by RAVI IYENGAR

VOLUME 239. Nuclear Magnetic Resonance (Part C)
Edited by THOMAS L. JAMES AND NORMAN J. OPPENHEIMER

VOLUME 240. Numerical Computer Methods (Part B)
Edited by MICHAEL L. JOHNSON AND LUDWIG BRAND

VOLUME 241. Retroviral Proteases
Edited by LAWRENCE C. KUO AND JULES A. SHAFER

VOLUME 242. Neoglycoconjugates (Part A)
Edited by Y. C. LEE AND REIKO T. LEE

VOLUME 243. Inorganic Microbial Sulfur Metabolism
Edited by HARRY D. PECK, JR., AND JEAN LEGALL

VOLUME 244. Proteolytic Enzymes: Serine and Cysteine Peptidases
Edited by ALAN J. BARRETT

VOLUME 245. Extracellular Matrix Components
Edited by E. RUOSLAHTI AND E. ENGVALL

VOLUME 246. Biochemical Spectroscopy
Edited by KENNETH SAUER

VOLUME 247. Neoglycoconjugates (Part B: Biomedical Applications)
Edited by Y. C. LEE AND REIKO T. LEE

VOLUME 248. Proteolytic Enzymes: Aspartic and Metallo Peptidases
Edited by ALAN J. BARRETT

VOLUME 249. Enzyme Kinetics and Mechanism (Part D: Developments in Enzyme Dynamics)
Edited by DANIEL L. PURICH

VOLUME 250. Lipid Modifications of Proteins
Edited by PATRICK J. CASEY AND JANICE E. BUSS

VOLUME 251. Biothiols (Part A: Monothiols and Dithiols, Protein Thiols, and Thiyl Radicals)
Edited by LESTER PACKER

VOLUME 252. Biothiols (Part B: Glutathione and Thioredoxin; Thiols in Signal Transduction and Gene Regulation)
Edited by LESTER PACKER

VOLUME 253. Adhesion of Microbial Pathogens
Edited by RON J. DOYLE AND ITZHAK OFEK

VOLUME 254. Oncogene Techniques
Edited by PETER K. VOGT AND INDER M. VERMA

VOLUME 255. Small GTPases and Their Regulators (Part A: Ras Family)
Edited by W. E. BALCH, CHANNING J. DER, AND ALAN HALL

VOLUME 256. Small GTPases and Their Regulators (Part B: Rho Family)
Edited by W. E. BALCH, CHANNING J. DER, AND ALAN HALL

VOLUME 257. Small GTPases and Their Regulators (Part C: Proteins Involved in Transport)
Edited by W. E. BALCH, CHANNING J. DER, AND ALAN HALL

VOLUME 258. Redox-Active Amino Acids in Biology
Edited by JUDITH P. KLINMAN

VOLUME 259. Energetics of Biological Macromolecules
Edited by MICHAEL L. JOHNSON AND GARY K. ACKERS

VOLUME 260. Mitochondrial Biogenesis and Genetics (Part A)
Edited by GIUSEPPE M. ATTARDI AND ANNE CHOMYN

VOLUME 261. Nuclear Magnetic Resonance and Nucleic Acids
Edited by THOMAS L. JAMES

VOLUME 262. DNA Replication
Edited by JUDITH L. CAMPBELL

VOLUME 263. Plasma Lipoproteins (Part C: Quantitation)
Edited by WILLIAM A. BRADLEY, SANDRA H. GIANTURCO, AND JERE P. SEGREST

VOLUME 264. Mitochondrial Biogenesis and Genetics (Part B)
Edited by GIUSEPPE M. ATTARDI AND ANNE CHOMYN

VOLUME 265. Cumulative Subject Index Volumes 228, 230–262

VOLUME 266. Computer Methods for Macromolecular Sequence Analysis
Edited by RUSSELL F. DOOLITTLE

VOLUME 267. Combinatorial Chemistry
Edited by JOHN N. ABELSON

VOLUME 268. Nitric Oxide (Part A: Sources and Detection of NO; NO Synthase)
Edited by LESTER PACKER

VOLUME 269. Nitric Oxide (Part B: Physiological and Pathological Processes)
Edited by LESTER PACKER

VOLUME 270. High Resolution Separation and Analysis of Biological Macromolecules (Part A: Fundamentals)
Edited by BARRY L. KARGER AND WILLIAM S. HANCOCK

VOLUME 271. High Resolution Separation and Analysis of Biological Macromolecules (Part B: Applications)
Edited by BARRY L. KARGER AND WILLIAM S. HANCOCK

VOLUME 272. Cytochrome P450 (Part B)
Edited by ERIC F. JOHNSON AND MICHAEL R. WATERMAN

VOLUME 273. RNA Polymerase and Associated Factors (Part A)
Edited by SANKAR ADHYA

VOLUME 274. RNA Polymerase and Associated Factors (Part B)
Edited by SANKAR ADHYA

VOLUME 275. Viral Polymerases and Related Proteins
Edited by LAWRENCE C. KUO, DAVID B. OLSEN, AND STEVEN S. CARROLL

VOLUME 276. Macromolecular Crystallography (Part A)
Edited by CHARLES W. CARTER, JR., AND ROBERT M. SWEET

VOLUME 277. Macromolecular Crystallography (Part B)
Edited by CHARLES W. CARTER, JR., AND ROBERT M. SWEET

VOLUME 278. Fluorescence Spectroscopy
Edited by LUDWIG BRAND AND MICHAEL L. JOHNSON

VOLUME 279. Vitamins and Coenzymes (Part I)
Edited by DONALD B. MCCORMICK, JOHN W. SUTTIE, AND CONRAD WAGNER

VOLUME 280. Vitamins and Coenzymes (Part J)
Edited by DONALD B. MCCORMICK, JOHN W. SUTTIE, AND CONRAD WAGNER

VOLUME 281. Vitamins and Coenzymes (Part K)
Edited by DONALD B. MCCORMICK, JOHN W. SUTTIE, AND CONRAD WAGNER

VOLUME 282. Vitamins and Coenzymes (Part L)
Edited by DONALD B. MCCORMICK, JOHN W. SUTTIE, AND CONRAD WAGNER

VOLUME 283. Cell Cycle Control
Edited by WILLIAM G. DUNPHY

VOLUME 284. Lipases (Part A: Biotechnology)
Edited by BYRON RUBIN AND EDWARD A. DENNIS

VOLUME 285. Cumulative Subject Index Volumes 263, 264, 266–284, 286–289

VOLUME 286. Lipases (Part B: Enzyme Characterization and Utilization)
Edited by BYRON RUBIN AND EDWARD A. DENNIS

VOLUME 287. Chemokines
Edited by RICHARD HORUK

VOLUME 288. Chemokine Receptors
Edited by RICHARD HORUK

VOLUME 289. Solid Phase Peptide Synthesis
Edited by GREGG B. FIELDS

VOLUME 290. Molecular Chaperones
Edited by GEORGE H. LORIMER AND THOMAS BALDWIN

VOLUME 291. Caged Compounds
Edited by GERARD MARRIOTT

VOLUME 292. ABC Transporters: Biochemical, Cellular, and Molecular Aspects
Edited by SURESH V. AMBUDKAR AND MICHAEL M. GOTTESMAN

VOLUME 293. Ion Channels (Part B)
Edited by P. MICHAEL CONN

VOLUME 294. Ion Channels (Part C)
Edited by P. MICHAEL CONN

VOLUME 295. Energetics of Biological Macromolecules (Part B)
Edited by GARY K. ACKERS AND MICHAEL L. JOHNSON

VOLUME 296. Neurotransmitter Transporters
Edited by SUSAN G. AMARA

VOLUME 297. Photosynthesis: Molecular Biology of Energy Capture
Edited by LEE MCINTOSH

VOLUME 298. Molecular Motors and the Cytoskeleton (Part B)
Edited by RICHARD B. VALLEE

VOLUME 299. Oxidants and Antioxidants (Part A)
Edited by LESTER PACKER

VOLUME 300. Oxidants and Antioxidants (Part B)
Edited by LESTER PACKER

VOLUME 301. Nitric Oxide: Biological and Antioxidant Activities (Part C)
Edited by LESTER PACKER

VOLUME 302. Green Fluorescent Protein
Edited by P. MICHAEL CONN

VOLUME 303. cDNA Preparation and Display
Edited by SHERMAN M. WEISSMAN

VOLUME 304. Chromatin
Edited by PAUL M. WASSARMAN AND ALAN P. WOLFFE

VOLUME 305. Bioluminescence and Chemiluminescence (Part C)
Edited by THOMAS O. BALDWIN AND MIRIAM M. ZIEGLER

VOLUME 306. Expression of Recombinant Genes in Eukaryotic Systems
Edited by JOSEPH C. GLORIOSO AND MARTIN C. SCHMIDT

VOLUME 307. Confocal Microscopy
Edited by P. MICHAEL CONN

VOLUME 308. Enzyme Kinetics and Mechanism (Part E: Energetics of Enzyme Catalysis)
Edited by DANIEL L. PURICH AND VERN L. SCHRAMM

VOLUME 309. Amyloid, Prions, and Other Protein Aggregates
Edited by RONALD WETZEL

VOLUME 310. Biofilms
Edited by RON J. DOYLE

VOLUME 311. Sphingolipid Metabolism and Cell Signaling (Part A)
Edited by ALFRED H. MERRILL, JR., AND YUSUF A. HANNUN

VOLUME 312. Sphingolipid Metabolism and Cell Signaling (Part B)
Edited by ALFRED H. MERRILL, JR., AND YUSUF A. HANNUN

VOLUME 313. Antisense Technology
(Part A: General Methods, Methods of Delivery, and RNA Studies)
Edited by M. IAN PHILLIPS

VOLUME 314. Antisense Technology (Part B: Applications)
Edited by M. IAN PHILLIPS

VOLUME 315. Vertebrate Phototransduction and the Visual Cycle
(Part A)
Edited by KRZYSZTOF PALCZEWSKI

VOLUME 316. Vertebrate Phototransduction and the Visual Cycle (Part B)
Edited by KRZYSZTOF PALCZEWSKI

VOLUME 317. RNA–Ligand Interactions (Part A: Structural Biology Methods)
Edited by DANIEL W. CELANDER AND JOHN N. ABELSON

VOLUME 318. RNA–Ligand Interactions (Part B: Molecular Biology Methods)
Edited by DANIEL W. CELANDER AND JOHN N. ABELSON

VOLUME 319. Singlet Oxygen, UV-A, and Ozone
Edited by LESTER PACKER AND HELMUT SIES

VOLUME 320. Cumulative Subject Index Volumes 290–319

VOLUME 321. Numerical Computer Methods (Part C)
Edited by MICHAEL L. JOHNSON AND LUDWIG BRAND

VOLUME 322. Apoptosis
Edited by JOHN C. REED

VOLUME 323. Energetics of Biological Macromolecules (Part C)
Edited by MICHAEL L. JOHNSON AND GARY K. ACKERS

VOLUME 324. Branched-Chain Amino Acids (Part B)
Edited by ROBERT A. HARRIS AND JOHN R. SOKATCH

VOLUME 325. Regulators and Effectors of Small GTPases
(Part D: Rho Family)
Edited by W. E. BALCH, CHANNING J. DER, AND ALAN HALL

VOLUME 326. Applications of Chimeric Genes and Hybrid Proteins
(Part A: Gene Expression and Protein Purification)
Edited by JEREMY THORNER, SCOTT D. EMR, AND JOHN N. ABELSON

VOLUME 327. Applications of Chimeric Genes and Hybrid Proteins (Part B: Cell Biology and Physiology)
Edited by JEREMY THORNER, SCOTT D. EMR, AND JOHN N. ABELSON

VOLUME 328. Applications of Chimeric Genes and Hybrid Proteins (Part C: Protein–Protein Interactions and Genomics)
Edited by JEREMY THORNER, SCOTT D. EMR, AND JOHN N. ABELSON

VOLUME 329. Regulators and Effectors of Small GTPases (Part E: GTPases Involved in Vesicular Traffic)
Edited by W. E. BALCH, CHANNING J. DER, AND ALAN HALL

VOLUME 330. Hyperthermophilic Enzymes (Part A)
Edited by MICHAEL W. W. ADAMS AND ROBERT M. KELLY

VOLUME 331. Hyperthermophilic Enzymes (Part B)
Edited by MICHAEL W. W. ADAMS AND ROBERT M. KELLY

VOLUME 332. Regulators and Effectors of Small GTPases (Part F: Ras Family I)
Edited by W. E. BALCH, CHANNING J. DER, AND ALAN HALL

VOLUME 333. Regulators and Effectors of Small GTPases (Part G: Ras Family II)
Edited by W. E. BALCH, CHANNING J. DER, AND ALAN HALL

VOLUME 334. Hyperthermophilic Enzymes (Part C)
Edited by MICHAEL W. W. ADAMS AND ROBERT M. KELLY

VOLUME 335. Flavonoids and Other Polyphenols
Edited by LESTER PACKER

VOLUME 336. Microbial Growth in Biofilms (Part A: Developmental and Molecular Biological Aspects)
Edited by RON J. DOYLE

VOLUME 337. Microbial Growth in Biofilms (Part B: Special Environments and Physicochemical Aspects)
Edited by RON J. DOYLE

VOLUME 338. Nuclear Magnetic Resonance of Biological Macromolecules (Part A)
Edited by THOMAS L. JAMES, VOLKER DÖTSCH, AND ULI SCHMITZ

VOLUME 339. Nuclear Magnetic Resonance of Biological Macromolecules (Part B)
Edited by THOMAS L. JAMES, VOLKER DÖTSCH, AND ULI SCHMITZ

VOLUME 340. Drug–Nucleic Acid Interactions
Edited by JONATHAN B. CHAIRES AND MICHAEL J. WARING

VOLUME 341. Ribonucleases (Part A)
Edited by ALLEN W. NICHOLSON

VOLUME 342. Ribonucleases (Part B)
Edited by ALLEN W. NICHOLSON

VOLUME 343. G Protein Pathways (Part A: Receptors)
Edited by RAVI IYENGAR AND JOHN D. HILDEBRANDT

VOLUME 344. G Protein Pathways (Part B: G Proteins and Their Regulators)
Edited by RAVI IYENGAR AND JOHN D. HILDEBRANDT

VOLUME 345. G Protein Pathways (Part C: Effector Mechanisms)
Edited by RAVI IYENGAR AND JOHN D. HILDEBRANDT

VOLUME 346. Gene Therapy Methods
Edited by M. IAN PHILLIPS

VOLUME 347. Protein Sensors and Reactive Oxygen Species (Part A: Selenoproteins and Thioredoxin)
Edited by HELMUT SIES AND LESTER PACKER

VOLUME 348. Protein Sensors and Reactive Oxygen Species (Part B: Thiol Enzymes and Proteins)
Edited by HELMUT SIES AND LESTER PACKER

VOLUME 349. Superoxide Dismutase
Edited by LESTER PACKER

VOLUME 350. Guide to Yeast Genetics and Molecular and Cell Biology (Part B)
Edited by CHRISTINE GUTHRIE AND GERALD R. FINK

VOLUME 351. Guide to Yeast Genetics and Molecular and Cell Biology (Part C)
Edited by CHRISTINE GUTHRIE AND GERALD R. FINK

VOLUME 352. Redox Cell Biology and Genetics (Part A)
Edited by CHANDAN K. SEN AND LESTER PACKER

VOLUME 353. Redox Cell Biology and Genetics (Part B)
Edited by CHANDAN K. SEN AND LESTER PACKER

VOLUME 354. Enzyme Kinetics and Mechanisms (Part F: Detection and Characterization of Enzyme Reaction Intermediates)
Edited by DANIEL L. PURICH

VOLUME 355. Cumulative Subject Index Volumes 321–354

VOLUME 356. Laser Capture Microscopy and Microdissection
Edited by P. MICHAEL CONN

VOLUME 357. Cytochrome P450, Part C
Edited by ERIC F. JOHNSON AND MICHAEL R. WATERMAN

VOLUME 358. Bacterial Pathogenesis (Part C: Identification, Regulation, and Function of Virulence Factors)
Edited by VIRGINIA L. CLARK AND PATRIK M. BAVOIL

VOLUME 359. Nitric Oxide (Part D)
Edited by ENRIQUE CADENAS AND LESTER PACKER

VOLUME 360. Biophotonics (Part A)
Edited by GERARD MARRIOTT AND IAN PARKER

VOLUME 361. Biophotonics (Part B)
Edited by GERARD MARRIOTT AND IAN PARKER

VOLUME 362. Recognition of Carbohydrates in Biological Systems (Part A)
Edited by YUAN C. LEE AND REIKO T. LEE

VOLUME 363. Recognition of Carbohydrates in Biological Systems (Part B)
Edited by YUAN C. LEE AND REIKO T. LEE

VOLUME 364. Nuclear Receptors
Edited by DAVID W. RUSSELL AND DAVID J. MANGELSDORF

VOLUME 365. Differentiation of Embryonic Stem Cells
Edited by PAUL M. WASSAUMAN AND GORDON M. KELLER

VOLUME 366. Protein Phosphatases
Edited by SUSANNE KLUMPP AND JOSEF KRIEGLSTEIN

VOLUME 367. Liposomes (Part A)
Edited by NEJAT DÜZGÜNEŞ

VOLUME 368. Macromolecular Crystallography (Part C)
Edited by CHARLES W. CARTER, JR., AND ROBERT M. SWEET

VOLUME 369. Combinational Chemistry (Part B)
Edited by GUILLERMO A. MORALES AND BARRY A. BUNIN

VOLUME 370. RNA Polymerases and Associated Factors (Part C)
Edited by SANKAR L. ADHYA AND SUSAN GARGES

VOLUME 371. RNA Polymerases and Associated Factors (Part D)
Edited by SANKAR L. ADHYA AND SUSAN GARGES

VOLUME 372. Liposomes (Part B)
Edited by NEJAT DÜZGÜNEŞ

VOLUME 373. Liposomes (Part C)
Edited by NEJAT DÜZGÜNEŞ

VOLUME 374. Macromolecular Crystallography (Part D)
Edited by CHARLES W. CARTER, JR., AND ROBERT W. SWEET

VOLUME 375. Chromatin and Chromatin Remodeling Enzymes (Part A)
Edited by C. DAVID ALLIS AND CARL WU

VOLUME 376. Chromatin and Chromatin Remodeling Enzymes (Part B)
Edited by C. DAVID ALLIS AND CARL WU

VOLUME 377. Chromatin and Chromatin Remodeling Enzymes (Part C)
Edited by C. DAVID ALLIS AND CARL WU

VOLUME 378. Quinones and Quinone Enzymes (Part A)
Edited by HELMUT SIES AND LESTER PACKER

VOLUME 379. Energetics of Biological Macromolecules (Part D)
Edited by JO M. HOLT, MICHAEL L. JOHNSON, AND GARY K. ACKERS

VOLUME 380. Energetics of Biological Macromolecules (Part E)
Edited by JO M. HOLT, MICHAEL L. JOHNSON, AND GARY K. ACKERS

VOLUME 381. Oxygen Sensing
Edited by CHANDAN K. SEN AND GREGG L. SEMENZA

VOLUME 382. Quinones and Quinone Enzymes (Part B)
Edited by HELMUT SIES AND LESTER PACKER

VOLUME 383. Numerical Computer Methods (Part D)
Edited by LUDWIG BRAND AND MICHAEL L. JOHNSON

VOLUME 384. Numerical Computer Methods (Part E)
Edited by LUDWIG BRAND AND MICHAEL L. JOHNSON

VOLUME 385. Imaging in Biological Research (Part A)
Edited by P. MICHAEL CONN

VOLUME 386. Imaging in Biological Research (Part B)
Edited by P. MICHAEL CONN

VOLUME 387. Liposomes (Part D)
Edited by NEJAT DÜZGÜNEŞ

VOLUME 388. Protein Engineering
Edited by DAN E. ROBERTSON AND JOSEPH P. NOEL

VOLUME 389. Regulators of G-Protein Signaling (Part A)
Edited by DAVID P. SIDEROVSKI

VOLUME 390. Regulators of G-Protein Signaling (Part B)
Edited by DAVID P. SIDEROVSKI

VOLUME 391. Liposomes (Part E)
Edited by NEJAT DÜZGÜNEŞ

VOLUME 392. RNA Interference
Edited by ENGELKE ROSSI

VOLUME 393. Circadian Rhythms
Edited by MICHAEL W. YOUNG

VOLUME 394. Nuclear Magnetic Resonance of Biological Macromolecules (Part C)
Edited by THOMAS L. JAMES

VOLUME 395. Producing the Biochemical Data (Part B)
Edited by ELIZABETH A. ZIMMER AND ERIC H. ROALSON

VOLUME 396. Nitric Oxide (Part E)
Edited by LESTER PACKER AND ENRIQUE CADENAS

VOLUME 397. Environmental Microbiology
Edited by JARED R. LEADBETTER

VOLUME 398. Ubiquitin and Protein Degradation (Part A)
Edited by RAYMOND J. DESHAIES

VOLUME 399. Ubiquitin and Protein Degradation (Part B)
Edited by RAYMOND J. DESHAIES

VOLUME 400. Phase II Conjugation Enzymes and Transport Systems
Edited by HELMUT SIES AND LESTER PACKER

VOLUME 401. Glutathione Transferases and Gamma Glutamyl Transpeptidases
Edited by HELMUT SIES AND LESTER PACKER

VOLUME 402. Biological Mass Spectrometry
Edited by A. L. BURLINGAME

VOLUME 403. GTPases Regulating Membrane Targeting and Fusion
Edited by WILLIAM E. BALCH, CHANNING J. DER, AND ALAN HALL

VOLUME 404. GTPases Regulating Membrane Dynamics
Edited by WILLIAM E. BALCH, CHANNING J. DER, AND ALAN HALL

VOLUME 405. Mass Spectrometry: Modified Proteins and Glycoconjugates
Edited by A. L. BURLINGAME

VOLUME 406. Regulators and Effectors of Small GTPases: Rho Family
Edited by WILLIAM E. BALCH, CHANNING J. DER, AND ALAN HALL

VOLUME 407. Regulators and Effectors of Small GTPases: Ras Family
Edited by WILLIAM E. BALCH, CHANNING J. DER, AND ALAN HALL

VOLUME 408. DNA Repair (Part A)
Edited by JUDITH L. CAMPBELL AND PAUL MODRICH

VOLUME 409. DNA Repair (Part B)
Edited by JUDITH L. CAMPBELL AND PAUL MODRICH

VOLUME 410. DNA Microarrays (Part A: Array Platforms and Web-Bench Protocols)
Edited by ALAN KIMMEL AND BRIAN OLIVER

VOLUME 411. DNA Microarrays (Part B: Databases and Statistics)
Edited by ALAN KIMMEL AND BRIAN OLIVER

VOLUME 412. Amyloid, Prions, and Other Protein Aggregates (Part B)
Edited by INDU KHETERPAL AND RONALD WETZEL

VOLUME 413. Amyloid, Prions, and Other Protein Aggregates (Part C)
Edited by INDU KHETERPAL AND RONALD WETZEL

VOLUME 414. Measuring Biological Responses with Automated Microscopy
Edited by JAMES INGLESE

VOLUME 415. Glycobiology
Edited by MINORU FUKUDA

VOLUME 416. Glycomics
Edited by MINORU FUKUDA

VOLUME 417. Functional Glycomics
Edited by MINORU FUKUDA

VOLUME 418. Embryonic Stem Cells
Edited by IRINA KLIMANSKAYA AND ROBERT LANZA

VOLUME 419. Adult Stem Cells
Edited by IRINA KLIMANSKAYA AND ROBERT LANZA

VOLUME 420. Stem Cell Tools and Other Experimental Protocols
Edited by IRINA KLIMANSKAYA AND ROBERT LANZA

VOLUME 421. Advanced Bacterial Genetics: Use of Transposons and Phage for Genomic Engineering
Edited by KELLY T. HUGHES

VOLUME 422. Two-Component Signaling Systems, Part A
Edited by MELVIN I. SIMON, BRIAN R. CRANE, AND ALEXANDRINE CRANE

VOLUME 423. Two-Component Signaling Systems, Part B
Edited by MELVIN I. SIMON, BRIAN R. CRANE, AND ALEXANDRINE CRANE

VOLUME 424. RNA Editing
Edited by JONATHA M. GOTT

VOLUME 425. RNA Modification
Edited by JONATHA M. GOTT

VOLUME 426. Integrins
Edited by DAVID CHERESH

VOLUME 427. MicroRNA Methods
Edited by JOHN J. ROSSI

VOLUME 428. Osmosensing and Osmosignaling
Edited by HELMUT SIES AND DIETER HAUSSINGER

VOLUME 429. Translation Initiation: Extract Systems and Molecular Genetics
Edited by JON LORSCH

VOLUME 430. Translation Initiation: Reconstituted Systems and Biophysical Methods
Edited by JON LORSCH

VOLUME 431. Translation Initiation: Cell Biology, High-Throughput and Chemical-Based Approaches
Edited by JON LORSCH

VOLUME 432. Lipidomics and Bioactive Lipids: Mass-Spectrometry–Based Lipid Analysis
Edited by H. ALEX BROWN

VOLUME 433. Lipidomics and Bioactive Lipids: Specialized Analytical Methods and Lipids in Disease
Edited by H. ALEX BROWN

VOLUME 434. Lipidomics and Bioactive Lipids: Lipids and Cell Signaling
Edited by H. ALEX BROWN

VOLUME 435. Oxygen Biology and Hypoxia
Edited by HELMUT SIES AND BERNHARD BRÜNE

VOLUME 436. Globins and Other Nitric Oxide-Reactive Protiens (Part A)
Edited by ROBERT K. POOLE

VOLUME 437. Globins and Other Nitric Oxide-Reactive Protiens (Part B)
Edited by ROBERT K. POOLE

VOLUME 438. Small GTPases in Disease (Part A)
Edited by WILLIAM E. BALCH, CHANNING J. DER, AND ALAN HALL

VOLUME 439. Small GTPases in Disease (Part B)
Edited by WILLIAM E. BALCH, CHANNING J. DER, AND ALAN HALL

VOLUME 440. Nitric Oxide, Part F Oxidative and Nitrosative Stress in Redox Regulation of Cell Signaling
Edited by ENRIQUE CADENAS AND LESTER PACKER

VOLUME 441. Nitric Oxide, Part G Oxidative and Nitrosative Stress in Redox Regulation of Cell Signaling
Edited by ENRIQUE CADENAS AND LESTER PACKER

VOLUME 442. Programmed Cell Death, General Principles for Studying Cell Death (Part A)
Edited by ROYA KHOSRAVI-FAR, ZAHRA ZAKERI, RICHARD A. LOCKSHIN, AND MAURO PIACENTINI

VOLUME 443. Angiogenesis: *In Vitro* Systems
Edited by DAVID A. CHERESH

VOLUME 444. Angiogenesis: *In Vivo* Systems (Part A)
Edited by DAVID A. CHERESH

VOLUME 445. Angiogenesis: *In Vivo* Systems (Part B)
Edited by DAVID A. CHERESH

VOLUME 446. Programmed Cell Death, The Biology and Therapeutic Implications of Cell Death (Part B)
Edited by ROYA KHOSRAVI-FAR, ZAHRA ZAKERI, RICHARD A. LOCKSHIN, AND MAURO PIACENTINI

VOLUME 447. RNA Turnover in Bacteria, Archaea and Organelles
Edited by LYNNE E. MAQUAT AND CECILIA M. ARRAIANO

VOLUME 448. RNA Turnover in Eukaryotes: Nucleases, Pathways and Analysis of mRNA Decay
Edited by LYNNE E. MAQUAT AND MEGERDITCH KILEDJIAN

VOLUME 449. RNA Turnover in Eukaryotes: Analysis of Specialized and Quality Control RNA Decay Pathways
Edited by LYNNE E. MAQUAT AND MEGERDITCH KILEDJIAN

VOLUME 450. Fluorescence Spectroscopy
Edited by LUDWIG BRAND AND MICHAEL L. JOHNSON

VOLUME 451. Autophagy: Lower Eukaryotes and Non-Mammalian Systems (Part A)
Edited by DANIEL J. KLIONSKY

VOLUME 452. Autophagy in Mammalian Systems (Part B)
Edited by DANIEL J. KLIONSKY

VOLUME 453. Autophagy in Disease and Clinical Applications (Part C)
Edited by DANIEL J. KLIONSKY

VOLUME 454. Computer Methods (Part A)
Edited by MICHAEL L. JOHNSON AND LUDWIG BRAND

VOLUME 455. Biothermodynamics (Part A)
Edited by MICHAEL L. JOHNSON, JO M. HOLT, AND GARY K. ACKERS (RETIRED)

VOLUME 456. Mitochondrial Function, Part A: Mitochondrial Electron Transport Complexes and Reactive Oxygen Species
Edited by WILLIAM S. ALLISON AND IMMO E. SCHEFFLER

VOLUME 457. Mitochondrial Function, Part B: Mitochondrial Protein Kinases, Protein Phosphatases and Mitochondrial Diseases
Edited by WILLIAM S. ALLISON AND ANNE N. MURPHY

VOLUME 458. Complex Enzymes in Microbial Natural Product Biosynthesis, Part A: Overview Articles and Peptides
Edited by DAVID A. HOPWOOD

VOLUME 459. Complex Enzymes in Microbial Natural Product Biosynthesis, Part B: Polyketides, Aminocoumarins and Carbohydrates
Edited by DAVID A. HOPWOOD

VOLUME 460. Chemokines, Part A
Edited by TRACY M. HANDEL AND DAMON J. HAMEL

VOLUME 461. Chemokines, Part B
Edited by TRACY M. HANDEL AND DAMON J. HAMEL

VOLUME 462. Non-Natural Amino Acids
Edited by TOM W. MUIR AND JOHN N. ABELSON

VOLUME 463. Guide to Protein Purification, 2nd Edition
Edited by RICHARD R. BURGESS AND MURRAY P. DEUTSCHER

VOLUME 464. Liposomes, Part F
Edited by NEJAT DÜZGÜNEŞ

VOLUME 465. Liposomes, Part G
Edited by NEJAT DÜZGÜNEŞ

VOLUME 466. Biothermodynamics, Part B
Edited by MICHAEL L. JOHNSON, GARY K. ACKERS, AND JO M. HOLT

VOLUME 467. Computer Methods Part B
Edited by MICHAEL L. JOHNSON AND LUDWIG BRAND

VOLUME 468. Biophysical, Chemical, and Functional Probes of RNA Structure, Interactions and Folding: Part A
Edited by DANIEL HERSCHLAG

VOLUME 469. Biophysical, Chemical, and Functional Probes of RNA Structure, Interactions and Folding: Part B
Edited by DANIEL HERSCHLAG

VOLUME 470. Guide to Yeast Genetics: Functional Genomics, Proteomics, and Other Systems Analysis, 2nd Edition
Edited by GERALD FINK, JONATHAN WEISSMAN, AND CHRISTINE GUTHRIE

VOLUME 471. Two-Component Signaling Systems, Part C
Edited by MELVIN I. SIMON, BRIAN R. CRANE, AND ALEXANDRINE CRANE

VOLUME 472. Single Molecule Tools, Part A: Fluorescence Based Approaches
Edited by NILS G. WALTER

VOLUME 473. Thiol Redox Transitions in Cell Signaling, Part A Chemistry and Biochemistry of Low Molecular Weight and Protein Thiols
Edited by ENRIQUE CADENAS AND LESTER PACKER

VOLUME 474. Thiol Redox Transitions in Cell Signaling, Part B Cellular Localization and Signaling
Edited by ENRIQUE CADENAS AND LESTER PACKER

VOLUME 475. Single Molecule Tools, Part B: Super-Resolution, Particle Tracking, Multiparameter, and Force Based Methods
Edited by NILS G. WALTER

VOLUME 476. Guide to Techniques in Mouse Development, Part A Mice, Embryos, and Cells, 2nd Edition
Edited by PAUL M. WASSARMAN AND PHILIPPE M. SORIANO

VOLUME 477. Guide to Techniques in Mouse Development, Part B Mouse Molecular Genetics, 2nd Edition
Edited by PAUL M. WASSARMAN AND PHILIPPE M. SORIANO

VOLUME 478. Glycomics
Edited by MINORU FUKUDA

VOLUME 479. Functional Glycomics
Edited by MINORU FUKUDA

VOLUME 480. Glycobiology
Edited by MINORU FUKUDA

VOLUME 481. Cryo-EM, Part A: Sample Preparation and Data Collection
Edited by GRANT J. JENSEN

VOLUME 482. Cryo-EM, Part B: 3-D Reconstruction
Edited by GRANT J. JENSEN

VOLUME 483. Cryo-EM, Part C: Analyses, Interpretation, and Case Studies
Edited by GRANT J. JENSEN

VOLUME 484. Constitutive Activity in Receptors and Other Proteins, Part A
Edited by P. MICHAEL CONN

VOLUME 485. Constitutive Activity in Receptors and Other Proteins, Part B
Edited by P. MICHAEL CONN

VOLUME 486. Research on Nitrification and Related Processes, Part A
Edited by MARTIN G. KLOTZ

VOLUME 487. Computer Methods, Part C
Edited by MICHAEL L. JOHNSON AND LUDWIG BRAND

VOLUME 488. Biothermodynamics, Part C
Edited by MICHAEL L. JOHNSON, JO M. HOLT, AND GARY K. ACKERS

VOLUME 489. The Unfolded Protein Response and Cellular Stress, Part A
Edited by P. MICHAEL CONN

VOLUME 490. The Unfolded Protein Response and Cellular Stress, Part B
Edited by P. MICHAEL CONN

VOLUME 491. The Unfolded Protein Response and Cellular Stress, Part C
Edited by P. MICHAEL CONN

VOLUME 492. Biothermodynamics, Part D
Edited by MICHAEL L. JOHNSON, JO M. HOLT, AND GARY K. ACKERS

VOLUME 493. Fragment-Based Drug Design Tools,
Practical Approaches, and Examples
Edited by LAWRENCE C. KUO

VOLUME 494. Methods in Methane Metabolism, Part A
Methanogenesis
Edited by AMY C. ROSENZWEIG AND STEPHEN W. RAGSDALE

VOLUME 495. Methods in Methane Metabolism, Part B
Methanotrophy
Edited by AMY C. ROSENZWEIG AND STEPHEN W. RAGSDALE

VOLUME 496. Research on Nitrification and Related Processes, Part B
Edited by MARTIN G. KLOTZ AND LISA Y. STEIN

VOLUME 497. Synthetic Biology, Part A
Methods for Part/Device Characterization and Chassis Engineering
Edited by CHRISTOPHER VOIGT

VOLUME 498. Synthetic Biology, Part B
Computer Aided Design and DNA Assembly
Edited by CHRISTOPHER VOIGT

VOLUME 499. Biology of Serpins
Edited by JAMES C. WHISSTOCK AND PHILLIP I. BIRD

VOLUME 500. Methods in Systems Biology
Edited by DANIEL JAMESON, MALKHEY VERMA, AND HANS V. WESTERHOFF

VOLUME 501. Serpin Structure and Evolution
Edited by JAMES C. WHISSTOCK AND PHILLIP I. BIRD

VOLUME 502. Protein Engineering for Therapeutics, Part A
Edited by K. DANE WITTRUP AND GREGORY L. VERDINE

VOLUME 503. Protein Engineering for Therapeutics, Part B
Edited by K. DANE WITTRUP AND GREGORY L. VERDINE

VOLUME 504. Imaging and Spectroscopic Analysis of Living Cells
Optical and Spectroscopic Techniques
Edited by P. MICHAEL CONN

VOLUME 505. Imaging and Spectroscopic Analysis of Living Cells
Live Cell Imaging of Cellular Elements and Functions
Edited by P. MICHAEL CONN

VOLUME 506. Imaging and Spectroscopic Analysis of Living Cells
Imaging Live Cells in Health and Disease
Edited by P. MICHAEL CONN

VOLUME 507. Gene Transfer Vectors for Clinical Application
Edited by THEODORE FRIEDMANN

VOLUME 508. Nanomedicine
Cancer, Diabetes, and Cardiovascular, Central Nervous System, Pulmonary and Inflammatory Diseases
Edited by NEJAT DÜZGÜNEŞ

VOLUME 509. Nanomedicine
Infectious Diseases, Immunotherapy, Diagnostics, Antifibrotics, Toxicology and Gene Medicine
Edited by NEJAT DÜZGÜNEŞ

VOLUME 510. Cellulases
Edited by HARRY J. GILBERT

VOLUME 511. RNA Helicases
Edited by ECKHARD JANKOWSKY

VOLUME 512. Nucleosomes, Histones & Chromatin, Part A
Edited by CARL WU AND C. DAVID ALLIS

VOLUME 513. Nucleosomes, Histones & Chromatin, Part B
Edited by CARL WU AND C. DAVID ALLIS

VOLUME 514. Ghrelin
Edited by MASAYASU KOJIMA AND KENJI KANGAWA

VOLUME 515. Natural Product Biosynthesis by Microorganisms and Plants, Part A
Edited by DAVID A. HOPWOOD

VOLUME 516. Natural Product Biosynthesis by Microorganisms and Plants, Part B
Edited by DAVID A. HOPWOOD

VOLUME 517. Natural Product Biosynthesis by Microorganisms and Plants, Part C
Edited by DAVID A. HOPWOOD

VOLUME 518. Fluorescence Fluctuation Spectroscopy (FFS), Part A
Edited by SERGEY Y. TETIN

VOLUME 519. Fluorescence Fluctuation Spectroscopy (FFS), Part B
Edited by SERGEY Y. TETIN

VOLUME 520. G Protein Couple Receptors Structure
Edited by P. MICHAEL CONN

VOLUME 521. G Protein Couple Receptors Trafficking and Oligomerization
Edited by P. MICHAEL CONN

VOLUME 522. G Protein Couple Receptors Modeling, Activation, Interactions and Virtual Screening
Edited by P. MICHAEL CONN

VOLUME 523. Methods in Protein Design
Edited by AMY E. KEATING

VOLUME 524. Cilia, Part A
Edited by WALLACE F. MARSHALL

VOLUME 525. Cilia, Part B
Edited by WALLACE F. MARSHALL

VOLUME 526. Hydrogen Peroxide and Cell Signaling, Part A
Edited by ENRIQUE CADENAS AND LESTER PACKER

VOLUME 527. Hydrogen Peroxide and Cell Signaling, Part B
Edited by ENRIQUE CADENAS AND LESTER PACKER

VOLUME 528. Hydrogen Peroxide and Cell Signaling, Part C
Edited by ENRIQUE CADENAS AND LESTER PACKER

VOLUME 529. Laboratory Methods in Enzymology: DNA
Edited by JON LORSCH

VOLUME 530. Laboratory Methods in Enzymology: RNA
Edited by JON LORSCH

VOLUME 531. Microbial Metagenomics, Metatranscriptomics, and Metaproteomics
Edited by EDWARD F. DELONG

VOLUME 532. Research Methods in Biomineralization Science
Edited by JAMES J. DE YOREO

SECTION I

Investigating Solution Chemistry, Structure, and Nucleation

CHAPTER ONE

Development of Accurate Force Fields for the Simulation of Biomineralization

Paolo Raiteri, Raffaella Demichelis, Julian D. Gale[1]
Department of Chemistry, Nanochemistry Research Institute, Curtin University, GPO Box U1987, Perth, Western Australia, Australia
[1]Corresponding author: e-mail address: j.gale@curtin.edu.au

Contents

1. Introduction	4
1.1 Sampling the potential energy landscape	4
1.2 Describing the potential energy landscape	5
1.3 Evaluating the accuracy of the potential energy landscape	6
2. Force Field Selection and Refinement	7
2.1 Water	7
2.2 Minerals	10
2.3 Ions in aqueous solution	13
3. The Mineral–Aqueous Interface	14
3.1 Mineral surfaces	14
3.2 Solubility	15
3.3 Ion pairing	16
4. Organics in Geochemical Systems	18
5. Summary	19
Acknowledgments	19
References	19

Abstract

The existence of an accurate force field (FF) model that reproduces the free-energy landscape is a key prerequisite for the simulation of biomineralization. Here, the stages in the development of such a model are discussed including the quality of the water model, the thermodynamics of polymorphism, and the free energies of solvation for the relevant species. The reliability of FFs can then be benchmarked against quantities such as the free energy of ion pairing in solution, the solubility product, and the structure of the mineral–water interface.

1. INTRODUCTION

Computer simulation is playing an increasingly valuable role, as a complement to experiment, in understanding the atomic details of processes occurring during crystal growth and dissolution. Vital to the success of any such simulation study is the availability of an accurate potential energy surface and the ability to rigorously explore this. Despite the rapid advance of computing resources, this currently requires the use of a force field (FF) description of the energy, rather than potentially more accurate quantum mechanical techniques due to the demanding nature of the latter preventing the use of models with realistic complexity or concentrations. Biomineralization represents the intersection of two communities from mineralogy and the life sciences that have historically distinct approaches to FF derivation. The aim of this article is to establish some criteria against which FFs for biomineralization can be assessed for accuracy and to describe the recent progress achieved in this field.

While there are numerous minerals that can be the product of biomineralization, among the most widely studied are those belonging to the calcium carbonate ($CaCO_3$) family. In the interests of brevity and to avoid dealing only in generalities, this chapter will focus on this specific case. $CaCO_3$ represents a fascinating example, as there are three crystalline polymorphs (calcite, aragonite, and vaterite) that can be produced under different conditions. Furthermore, their nucleation and growth pathway(s) are complex, involving stable prenucleation species (Demichelis, Raiteri, Gale, Quigley, & Gebauer, 2011; Gebauer, Voelkel, & Coelfen, 2008), amorphous $CaCO_3$ with variable water content and polyamorphic structure (Cartwright, Checa, Gale, Gebauer, & Sainz-Díaz, 2012), and liquid precursors (Wolf, Leiterer, Kappl, Emmerling, & Tremel, 2008). Given the rapidly evolving nature of our understanding of $CaCO_3$ formation and the emergence of new ideas from simulation, it is particularly timely to appraise the state of the models that underpin such studies.

1.1. Sampling the potential energy landscape

Processes relevant to biomineralization involve complex potential energy surfaces with a minima count that goes beyond direct enumeration. Statistical sampling of given regions of configuration space can be achieved by two main techniques: Monte Carlo and molecular dynamics (MD) simulations, which rely on a stochastic and a deterministic approach, respectively.

Although the majority of the arguments developed in this chapter apply to both techniques, we will mostly focus on MD, which is arguably the most popular approach of the two. Here atoms are treated as "classical" objects and their motion is described by Newton's equations of motion, which are numerically solved at discrete time intervals (time steps). The time step magnitude depends on the frequency of the fastest atomic vibrations and is usually of the order of 1 fs. Extended simulations are then performed to explore the accessible portion of the phase space and macroscopic properties can be derived by means of thermodynamic averaging. For crystal growth from aqueous solution, there can be barriers that are too great to cross with any substantial probability on the timescales accessible to direct MD. Therefore, techniques for acceleration/enhanced sampling of rare events play a crucial role, as described elsewhere (Bolhuis, Chandler, Dellago, & Geissler, 2002; Kim, Straub, & Keyes, 2007; Laio & Gervasio, 2008; Voter, 1997).

1.2. Describing the potential energy landscape

With increasing computer power, we are now reaching a point where atomistic MD simulations can be used as predictive tools and not simply as an aid to interpret experiment. However, the reliability critically depends on the accuracy of the approximations introduced in the description of the atomic interactions. Here, we can distinguish three main categories: first principles quantum mechanics (QM), semiempirical methods, and FF calculations, in order of decreasing complexity. QM simulations explicitly account for the electrons, make few assumptions, and are potentially the most accurate/transferable. That said, practical incarnations of QM, such as density functional theory, often suffer from errors depending on the approximations made. For example, the density and diffusivity of liquid water can be substantially underestimated if using the generalized gradient approximation (Wang, Román-Pérez, Soler, Artacho, & Fernández-Serra, 2011). The benefit of QM is that such errors are often systematic, and so, failures or deviations from experiment can usually be rationalized. More problematic is that QM is computationally expensive, and with current supercomputers, we can simulate only a few hundred atoms for a real time of a few tens of picoseconds. While this is adequate to describe crystalline materials, complex processes, including biomineralization, are presently beyond reach.

At the other extreme, there are FF simulations where atoms are described as point particles with a mass and a charge that interact via defined functions of the internuclear distances (empirical potentials). For ionic materials, these interactions may consist of just Coulomb's law and a two-body description of short-range repulsion and van der Waals attraction, such as the Lennard-Jones or the Buckingham potentials. The simplicity of this model is partly offset by deriving the parameters to reproduce experimental observables or, where such information is lacking, data from accurate QM. A negative consequence is that any given parameterization may not be transferable outside of the training set for which it was designed and tested. Therefore, it is important that the objectives of an FF model are defined before its development. On the plus side, FFs have a vastly reduced computational cost relative to the QM equivalent and simulations of a few million atoms for several tens of nanoseconds are achievable, with the most ambitious simulations even reaching milliseconds. This allows a significant portion of the accessible phase space to be explored and offers the possibility of achieving statistically significant results even for complex systems.

1.3. Evaluating the accuracy of the potential energy landscape

Although it may appear that a conflicting choice has to be made between accuracy and speed, we contend that if an FF is carefully derived, both objectives can be simultaneously satisfied. In order to achieve this, the key is to focus on the accurate reproduction of *free energies*. In a recent paper (Stack, Gale, & Raiteri, 2013), we propose six key questions to address while developing a reliable FF for aqueous–mineral systems:

1. How accurate are the water density and mobility at the conditions of interest?
2. Does the model predict the correct ground state for the mineral's polymorphs?
3. How accurate is the solvation free energy of the ions in water?
4. Does the exchange rate of water at the cations agree with the experiments?
5. Does the ion pairing free energy compare to the experiments?
6. How accurate is the free energy of dissolution of the mineral (K_{sp})?

Before explaining the importance of these questions and how to address them, it is worth discussing the implications that targeting the free energy has on the fitting procedure. Traditionally, fitting an FF involves determining the parameters that best reproduce a set of experimental quantities (e.g., lattice parameters and elastic constants) and/or the energy of several cluster configurations calculated with some high-level QM technique. Although this is a good start, there is no guarantee that an accurate reproduction of static properties will deliver a correct representation of the underlying free-energy surface. However, targeting the free energy takes the fitting procedure to a higher level of complexity and computational cost. In the following sections, we examine the possible choices of FFs for different components that make up the full complexity of the biomineralization problem.

2. FORCE FIELD SELECTION AND REFINEMENT
2.1. Water

Crucial to any simulation of biomineralization will be an accurate description of water, as the medium in which minerals nucleate. The literature already contains a plethora of different FFs for water (Guillot, 2002), and so a new one for biomineralization is unnecessary. Instead, an existing, well-tested model should be selected based on an evaluation of literature results. At the outset, it should be noted that water is a complex system to simulate as the properties of the liquid/solid phases exhibit a substantial many-body effect relative to the gas-phase molecule, at least partly due to the change in polarity. Therefore, any fixed-charge model is unlikely to be able to accurately capture both extremes. However, the use of variable charge models substantially increases the computational cost and still suffers from numerous limitations. Similarly, the quantum nature of the hydrogen nucleus should be considered. While path-integral simulations have been employed (Marx, Tuckerman, Hutter, & Parrinello, 1999; Wong et al., 2010), most models are able to subsume nuclear quantum effects into the FF parameterization as a first approximation. As resources increase in the future, the transition to path-integral simulations can be readily made as several FFs have been derived with both classical and quantum variants (Paesani, Zhang, Case, Cheatham, & Voth, 2006).

The choice of FF for water is one area that highlights the different philosophies of sections of the mineralogical community versus those from the biomolecular sciences. When modeling solid mineral phases, the use of the shell model to allow for ion dipolar polarizability has long been recognized as important (Dick & Overhauser, 1958). As this community began to study mineral–aqueous interfaces, it was therefore natural to explore whether it was possible to extend the shell model concept to the simulation of water. Here, de Leeuw and Parker (1998) have proposed a model that has been applied to several aqueous–mineral systems, including that of $CaCO_3$

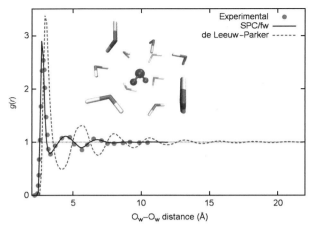

Figure 1.1 Oxygen–oxygen radial distribution function (RDF) of water as measured experimentally (Vaknin, Bu, & Travesset, 2008) and computed according to the de Leeuw and Parker (1998) and SPC/Fw (Wu et al., 2006) FFs. Note the extended range of oscillations out to ∼20 Å for the de Leeuw and Parker model indicative of overstructuring. The inset figure shows a sample configuration of the local coordination environment for a water molecule (blue) by neighboring waters (oxygen in red and hydrogen in white). The integral over $g(r)$ out to the first minimum gives an estimated coordination number for water of 13 for the de Leeuw and Parker model. (See color plate.)

(de Leeuw, Parker, & Harding, 1999). Unfortunately, this model has several major flaws. First, the density of water is too high by more than 20% and the local structure is too ordered with respect to experiment (see Fig. 1.1). Second, and most crucially, it has been observed that with this model, pure liquid water freezes at room temperature. Recently, Wolthers, di Tommaso, Du, and de Leeuw (2012) have argued that the reported freezing of this shell model (van Maaren & van der Spoel, 2001) was due to the small supercell size employed and presented data suggesting that the liquid phase is stable at 300 K. However, this argument fails to allow for the dependence of the rate of nucleation on the simulation size and so the water may be in a supercooled metastable state.

In order to absolutely resolve the previously mentioned issue, we have performed moving interface simulations for this system (Broughton, Gilmer, & Jackson, 1982; Harafuji, Tsuchiya, & Kawamura, 2004; Karim & Haymet, 1988). Here, a simulation is run in which a surface of ice is placed in contact with liquid water at a given temperature (see

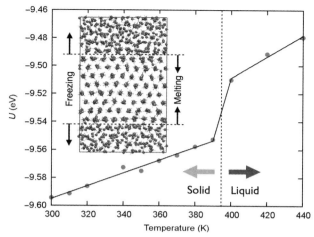

Figure 1.2 Moving interface simulation between ice IX (001) and liquid water. Graph shows the average potential energy per water molecule calculated during a 100 ps simulation as a function of temperature; the first 50 ps were considered as equilibration and discarded. The discontinuity corresponds to the melting point of ice IX. Given this is a metastable phase at these conditions, it therefore represents a lower bound to the melting point. Positive deviations from the potential energy versus temperature curve, such as at 340 K, can be due to defects in the ice structure that are trapped during growth. (See color plate.)

Fig. 1.2). The presence of an existing surface removes homogeneous nucleation as a rate-limiting step. By varying the temperature, the approximate melting point for the given FF can be determined. Here, we have used ice IX as the solid since it has a right-angled unit cell (the initial structure is cubic, though this relaxes to become orthorhombic with the de Leeuw and Parker model) and taken the (001) surface. Despite considering an ice phase that is not the ground state for ambient conditions, which maximizes the chance of stabilizing the liquid phase, after only 17 ps of simulation at 300 K, the entire liquid has frozen. In Fig. 1.2, we show the variation of potential energy from the moving boundary simulations versus temperature. This proves that this water model has a melting temperature in the range of 390–400 K. On a final technical point, many of the simulations performed using this model utilized the code DL_POLY. Here, there is a fundamental issue in the finite mass algorithm: The shell degrees of freedom are not thermostated, and so, the user has to hope that no energy transfer occurs; the impact of any deviation is hard to predict.

Within the biomolecular community, many water models have been extensively validated, and in light of the previous text, it appears advisable to take a more established FF. We have explored four literature models with different characteristics. Initially, SWM4-NDP was used to explore an alternative polarizable water FF (Bruneval, Donadio, & Parrinello, 2007; Lamoureux, Harder, Vorobyov, Roux, & MacKerell, 2006). However, the computational expense was a limitation. The next step was to use TIP4-Ew as a rigid water model (Horn et al., 2004), allowing the use of an increased time step (Raiteri, Gale, Quigley, & Rodger, 2010). Although this should improve the speed, the complexity of handling rigid bodies largely negates this. Finally, we have settled on the SPC/Fw (Wu, Tepper, & Voth, 2006) as being appropriate for large-scale simulations within this field (Demichelis et al., 2011; Raiteri & Gale, 2010). The positive characteristics of this choice, aside from efficiency, are that it offers a good description of water density, self-diffusion, heat of vaporization, and dielectric constant (Wu et al., 2006), which determines the range over which the solute ions interact. Conversely, the downsides are that it performs less well when transferred to ice or isolated molecules. However, the primary concern here is the liquid phase and so these issues are less relevant.

Besides the previously mentioned three conventional models, we have also examined using a reactive FF based on ReaxFF (Gale, Raiteri, & Van Duin, 2011; Van Duin, Dasgupta, Lorant, & Goddard, 2001). The benefit is that proton transfer events are captured since there is no need to predefine the connectivity. Due to the complexity and greater computational cost, the determination of the thermodynamic characteristics of this model is still in progress.

2.2. Minerals

There has been an extensive history of deriving FF models for mineral phases, including $CaCO_3$. Although the initial focus was often to calculate the lattice energy, this evolved into the derivation of FFs primarily designed to reproduce structural and mechanical properties, including vibrations (Dove, Winkler, Leslie, Harris, & Salje, 1992; Jackson & Price, 1992; Pavese, Catti, Price, & Jackson, 1992). For $CaCO_3$, the thermodynamics of polymorphism was rarely considered. However, the relative stabilities of calcite, aragonite, and vaterite are crucial to any simulation of biomineralization. Therefore, the focus of recent work has been to derive FFs for $CaCO_3$ capable of reproducing the free-energy differences between the crystalline phases. Here, the structural and mechanical properties are not ignored; they just carry a reduced weight in the fit. A crucial feature of our method is to use the relaxed fitting approach (Gale, 1996) in which the structure is fully optimized at every point of the FF refinement such that the structural changes are used as observables, rather than the forces. This ensures that the Hessian is well behaved and means that fitted properties, including the thermodynamics, are computed for the actual relaxed structure.

Some of the history of $CaCO_3$ FFs and resulting models has been discussed elsewhere (Raiteri et al., 2010). In Table 1.1, we present the results of several models for the properties of calcite (Archer et al., 2003; Bruneval et al., 2007; Demichelis et al., 2011; Dove et al., 1992; Freeman et al., 2007; Jackson & Price, 1992; Pavese et al., 1992; Raiteri & Gale, 2010; Ricci, Spijker, Stellacci, Molinari, & Voïtchovsky, 2013; Rohl, Wright, & Gale, 2003; Xiao, Edwards, & Grater, 2011) and the stability of the polymorphs. Here, ΔG values were computed by free-energy minimization using the ZSISA approximation (Gale, 1998; Taylor, Barrera, Allan, & Barron, 1997) within quasi-harmonic lattice dynamics with the phonons sampled using a shrinking factor of 6. Here, the contribution of the zero-point energy (ZPE) was excluded. Given that the objective is to apply the FFs in MD, the ZPE contribution must be ignored due to the classical treatment of the nuclei within standard MD. For some of the FFs, either calcite or aragonite was found to contain imaginary phonon modes for the experimental space group and so the symmetry had to be lowered to obtain a stable structure. These instabilities correspond to rotations of the carbonate groups.

The results in Table 1.1 demonstrate that the majority of FFs are unable to describe the delicate thermodynamic balance between calcite and aragonite. Several of the models give lower-symmetry structures for either aragonite or calcite; when this is the case, the thermodynamics were computed for these distorted, but stable, structures. In addition, many of the FFs were unable to complete a free minimization at 298 K due to a premature breakdown of the quasi-harmonic approximation from mode softening. By design, our recent FFs reproduce the calcite–aragonite free-energy difference, though even here, the balance between enthalpic and entropic contributions is not perfect. Consequently, the FFs are only likely to be accurate for temperatures close to ambient conditions, which are the conditions most relevant to biomineralization. For comparison, the free-energy difference is also given between calcite and vaterite in Table 1.1. Here, there are a number of caveats though. First, structural models for vaterite are still evolving

Table 1.1 Results of literature FFs for the properties of calcite

Force field	Phonon-unstable?	V (Å^3)	K (GPa)	ΔG_{c-a} (kJ/mol)	ΔG_{c-v} (kJ/mol)
Dove et al. (1992)	A	124.15 (+1.2%)	64.5	−8.48	−3.71
Jackson and Price (1992)		121.96 (−0.6%)	95.9	+2.32*	+2.4*
Pavese et al. (1992) Rigid ion model		121.69 (−0.8%)	79.7	−5.09*	−5.1
Pavese et al. (1992) Shell model		121.43 (−1.0%)	79.9	+2.02	−5.8
Freeman et al. (2007)	A	116.65 (−4.9%)	83.1	−6.96*	−6.8
Rohl et al. (2003)		122.27 (−0.4%)	81.0	−3.80	−4.2
Archer et al. (2003)		121.71 (−0.8%)	84.4	+0.16*	−8.3*
Bruneval et al. (2007)		118.62 (−3.3%)	82.4	+3.32*	−3.5*
Xiao et al. (2011)		128.00 (+4.3%)	90.0	+10.5*	−3.0*
Ricci et al. (2013)	C	98.31 (−19.9%)	201.1	−43.1	+14.0
Raiteri et al. (2010)		120.29 (−2.0%)	87.7	−0.83	−5.6
Demichelis et al. (2011)		120.46 (−1.8%)	90.4	−0.91	−5.5
Experiment		122.69	67/73	−0.82	−3.2

The volume of the optimized primitive unit cell (V) with percentage error in parenthesis, the bulk modulus (K), and the free-energy differences at 298 K between calcite and aragonite (ΔG_{c-a}), and calcite and vaterite (ΔG_{c-v}) (a negative value implies that calcite is the more stable phase) are shown. Here, the orthorhombic structure is used for vaterite (Meyer, 1959). Where the experimental structure is phonon-unstable, this is indicated by A or C, which stands for aragonite or calcite, respectively. An asterisk indicates that the free-energy minimization failed due to breakdown of the quasi-harmonic approximation below 298 K and so the single point value is given instead, based on the internal energy-optimized structure. Experimental thermodynamic data are from Plummer and Busenberg (1982).

and *ab initio* QM has led to several new low-symmetry configurations (Demichelis, Raiteri, Gale, & Dovesi, 2012). However, none of the FFs can describe these; thus, a less stable high-symmetry model for vaterite has to be used. Second, since vaterite is a disordered material, there should be a correction for the configurational entropy. Allowing for both of these factors, the calculated FF free-energy difference should exceed the experimental value. Overall, most FFs perform better for the calcite–vaterite thermodynamics than for aragonite.

As an alternative to computing ΔG between calcite and aragonite using lattice dynamics, we have also estimated this quantity directly from MD by connecting both phases to the Einstein crystal (Frenkel & Ladd, 1984), or Einstein molecule (Noya, Conde, & Vega, 2008) in the case of carbonate, using a modified version of LAMMPS (Plimpton, 1995). Here, we find that the computed ΔG is equivalent to that from lattice dynamics to within the statistical uncertainty of the simulations. Given that free-energy perturbation (FEP) to the Einstein solid is much more complex and computationally demanding than the use of lattice dynamics, the latter technique is preferable for determining the low-temperature thermodynamics where anharmonicity is less significant.

2.3. Ions in aqueous solution

As proposed by Aqvist (1990), another important quantity that needs to be captured in order to correctly study the dissolution and growth of minerals is the solvation free energy of the ions, that is, ΔG between the solvated state and the ion in vacuum. Although some care has to be paid to the definition of the reference state, experimental measurements for this quantity are readily available for many atomic and molecular ions. From a computational point of view, the solvation free energy corresponds to taking an isolated ion and fully immersing it in the simulation cell. This can be done by using either FEP (Zwanzig, 1954) or thermodynamic integration (TI) (Kirkwood, 1935). The idea behind these techniques is that the solute–solvent interactions are switched on (or off) either in a stepwise (FEP) or a continuous (TI) fashion; more details can be found elsewhere (Chipot & Pohorille, 2007; Darve, 2007).

As an illustrative example, we can take the case of Ca^{2+} for which we have tested both of the previously mentioned techniques. First, it is worth discussing the experimental values available in the literature, which seem to have changed significantly over time (Raiteri et al., 2010). This was mostly due to an improvement in the experimental measurements that greatly affected the chosen value for the solvation free energy of $H^+_{(aq)}$, a fundamental quantity for the thermodynamic cycle that is used to determine the solvation free energy of the ions. Therefore, we decided to target the most recent number and to validate our choice by looking at other quantities related to the strength of ion solvation, such as the water residence time and coordination number. For these calculations, we have applied FEP using a local version of DL_POLY 2.19 (Raiteri & Gale, 2010; Smith, 2006) and with TI, available in LAMMPS (Eike & Maginn, 2006). Although these calculations are relative expensive, with today's supercomputers, it is feasible to test several different sets of parameters in a few days/weeks and then to refine against experiment. In order to make a meaningful comparison between theory and experiment, there are a few correction terms that need to be accounted for (Hummer, Pratt, & Garcia, 1997; Kastenholz & Huenenberger, 2006a, 2006b).

Besides fitting the solvation free energy of the ions, it is also important to capture the dynamics of water around the ions, which determines the kinetics of desolvation of the particles and therefore contributes to the rates of processes. Experimentally, this is not easy to measure, especially for weakly solvated, large cations, where the water exchange rate is close to the instrument precision. For $Ca^{2+}_{(aq)}$, there are experimental reports of oxygen exchange times that range from <0.1 up to 3 ns (Atkinson, Emara, & Fernandez-Prini, 1974; Helm & Merbach, 1999; Ohtaki & Radnai, 1993). While the computed value for our latest model (0.23 ns) lies within this range, the experimental uncertainty makes it difficult to judge the true accuracy of the FF.

3. THE MINERAL–AQUEOUS INTERFACE

Having considered the components of the FFs for the key interactions in the preceding text, it is now possible to examine the simulation outcomes for the combined system. This is most readily achieved via several quantities for which experimental data exist, namely, the mineral surface interface with aqueous solution, the solubility, and the formation of ion pairs in solution.

3.1. Mineral surfaces

The calcite–water interface represents an ideal case for study. From an experimental perspective, calcite offers extended regions of relatively clean basal (104) surface, making it suitable for study by both x-ray reflectivity and atomic force microscopy. Experimental work from the group of Fenter and collaborators has demonstrated that there are two well-defined layers of water above the calcite (104) surface, with some evidence for perhaps a third more weakly ordered layer beyond this (Fenter & Sturchio, 2004; Geissbuhler et al., 2004). Numerous simulation studies have examined this case with different models, and nearly all claim good agreement (Kerisit & Parker, 2004b; Kerisit, Parker, & Harding, 2003; Ricci et al., 2013). All FFs seem to exhibit ordering of water at this interface with two or more layers, so that this feature does not provide a discerning evaluation of simulation models. However, recent work (Fenter, Kerisit, Raiteri, & Gale, 2013) has shown that quantitative direct comparison against the x-ray reflectivity is possible, showing that none of the available FFs tested so far quantitatively agree with the experimental data to within the level of precision of the measurement. However, such deviations tend to be dominated by the error in the relaxations of the calcite structure at the interface, more than the water structure.

3.2. Solubility

Arguably the most important characteristic of a mineral when it comes to crystal growth behavior is its solubility. Despite this, almost all simulations of crystallization have avoided benchmarking their FF against this quantity. As will be shown in the succeeding text, an error in the solubility not only changes the quantitative values obtained but can also significantly alter the qualitative observations. One of the reasons the solubility calculation has become the "elephant in the room" is because direct evaluation by MD is a complex task. As nucleation and growth rates usually exceed the timescales accessible to simulation, achieving direct equilibrium between the solid and ions in solution is largely untenable. Therefore, ΔG between ions in solution and the solid must be obtained indirectly. In our work, we have adopted a procedure that readily allows the solubility to be estimated. Using the hydration free energy of the ions, the Born–Haber cycle can be completed by evaluating ΔG between the ions in the gas phase and the solid state. Here, quasi-harmonic lattice dynamics can provide the required vibrational free energy of the solid and isolated carbonate group, while the translational and rotational free energies of the gaseous ions/molecules can be computed from standard statistical mechanics. There are two more subtle points relating to the vibrational calculation: first, the ZPE is excluded to mimic what we would obtain from a direct MD simulation, as previously noted. Second, lattice dynamics correctly allows for vibrational quantization, which MD does not, leading to potential differences in the free energy of dissolution. In practice, we find that for calcite at 298 K, the difference between the classical and quantized free energy is a small perturbation on the solubility relative to the variation between different FFs.

To ensure that our latest generation of FFs accurately describes the solubility of calcite, ΔG between the solid and the ions in solution is explicitly included in the fitting procedure. For our latest model, the dissolution free energy of calcite is calculated to be +46.3 kJ/mol at 298 K (the experimental value is +48.4 kJ/mol based on a log K_{sp} of -8.48 (Plummer & Busenberg, 1982)). It should be noted that many other literature models for aqueous $CaCO_3$ systems have ion hydration free energies that are in error by more than 100 kJ/mol, which equates to errors in the solubility of greater than 17 orders of magnitude! To illustrate that this is not just a quantitative problem, let us consider the free-energy profile for adsorption of Ca^{2+} at the calcite (104)–water interface. The first simulations of this indicated that binding of this cation at the calcite surface was effectively isoenergetic with the ion in solution (Kerisit & Parker, 2004a). Hence, one would predict that the surface of calcite would be coated with a significant concentration of adsorbed ions. However, when this calculation was revisited with free-energy-calibrated FFs, it was found that Ca^{2+} is repelled by the calcite (104)–water interface, indicating that the surface should be relatively clean (Raiteri & Gale, 2010; Raiteri et al., 2010), as is observed experimentally (Heberling et al., 2011).

3.3. Ion pairing

Another important quantity that can be used to validate/develop FFs for minerals in aqueous environments is the pairing free energy of the constituent ions. Although experimentally this is not an easy task, values of the equilibrium constant ($\Delta G = -k_B T \ln K$) for the formation of aqueous ion pairs between the most common ions with both inorganic and organic anions have been successfully obtained (Busenberg & Plummer, 1986, 1989; Covington & Danish, 2009; Daniele, Foti, Gianguzza, Prenesti, & Sammartano, 2008). However, most of the experimental techniques are blind to the actual structure of the complex, as they cannot distinguish between contact- and solvent-separated ion pairs or if multi-ion complexes are formed. While for $CaCO_3$ the former should not be an issue, there is growing evidence that complex liquid-like ionic aggregates can be present in solution (Demichelis et al., 2011; Gebauer et al., 2008), which may give rise to an equilibrium constant that is an average of multiple different structures and then complicate the comparison with the theoretically calculated pairing free energy.

From a computational point of view, the pairing free energy can be obtained by several techniques. In the past few years, we have successfully used umbrella sampling (Torrie & Valleau, 1977), steered MD (Jarzynski, 1997), and metadynamics (Laio & Parrinello, 2002) for various ion pairs to develop FFs. All the aforementioned techniques are based on the choice of a suitable reaction coordinate, which naturally can be the cation–anion distance, and if the simulations are carefully conducted until convergence is reached, they produce equivalent results. The results for the ion pairing free energy in aqueous solution are given in Fig. 1.3 for our three most recent $CaCO_3$ FFs. All three models are qualitatively similar in that the contact ion pair has two minima corresponding to bi- and monodentate coordination of calcium by carbonate, with the latter being more stable. Quantitatively, the older rigid molecule FF (Raiteri et al., 2010) gives a free energy of binding that is almost twice that of the two later flexible models. Experimentally, the free energy of this ion pair has been determined to be −18.5 kJ/mol (Plummer & Busenberg, 1982), which is in good agreement with the flexible FFs, suggesting that these should be preferred for simulations of ion association in solution.

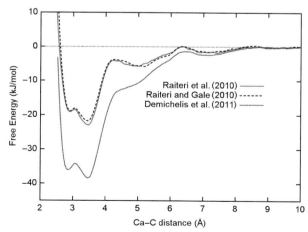

Figure 1.3 Free-energy profiles for the ion pairing of Ca^{2+} and CO_3^{2-} in aqueous solution as a function of the Ca–C distance for three different force fields. The free energy is taken relative to that of the separated ions. (See color plate.)

4. ORGANICS IN GEOCHEMICAL SYSTEMS

So far, our discussion has focused on ensuring an accurate description of the mineral–water thermodynamics in order to provide accurate predictions of crystal growth mechanisms in aqueous solution. The key challenge is now to include biomolecules so that their influence can be understood too. As noted previously, there is already an extensive range of biomolecular FFs available, the only issue being how to link them to an accurate model for the mineral phase. Most biomolecular FFs already include the interactions that are needed for metal cations, such as Ca^{2+}, and carboxylate groups since these species also occur *in vivo*. Therefore, it might be expected that there is little requirement for FF derivation.

We have investigated the parameterization of an FF for three different organics that bind to calcium via carboxylate groups, namely, acetate, aspartate, and citrate (Raiteri et al., 2012). Starting from the CHARMM FF (Vanommeslaeghe et al., 2010), the same procedure was followed, where possible, as per the generation of the $CaCO_3$ model. The procedure is somewhat complicated by the lack of experimental data for some quantities, such as the solvation free energy. Here, additional information can be provided through the use of QM data. Similarly, bulk solubility of anhydrous calcium salts is often unavailable. Instead, the fitting of the calcium–carboxylate interaction parameters can be guided by the ion pairing free energies in solution.

Three important conclusions can be reached from the limited benchmarking of organic–$CaCO_3$ interaction thermodynamics so far:

1. The use of "off-the-shelf" FF parameters or simple combination rules alone to determine the interactions between carboxylate groups and both water, plus calcium ions, is likely to be inadequate. For thermodynamic accuracy, specific fitting of the interaction parameters will usually be required.
2. The association between organics with a single or even two geometrically constrained carboxylate groups and $CaCO_3$ is generally weak. For example, the free energy of binding to a $CaCO_3^{(0)}$ ion pair is of the order of ambient thermal energy for acetate and aspartate, with only citrate achieving more significant binding. This is reasonable since molecules such as polyaspartate (Elhadj, de Yoreo, Hoyer, & Dove, 2006) can interact via multiple functional groups to achieve stronger coordination. If binding were too strong, then biomolecules would ultimately suppress crystallization by irreversibly blocking active sites for growth.

3. Association between organics and $CaCO_3$ is stronger for prenucleation species, rather than those occurring postnucleation. None of the three considered molecules appear to bind with sufficient strength to displace the ordered first aqueous layer over the calcite (104) surface. Hence, any attachment must occur via defects where the water structure is already disrupted.

5. SUMMARY

In this chapter, we have attempted to briefly describe the principles underlying the derivation of an accurate FF for the biomineralization of $CaCO_3$ from aqueous solution. Although the focus has been on only one specific system, the approach is general and could be equally applied to other minerals. The focus of the FF parameterization has been on obtaining accurate thermodynamics and more specifically on free energies. While this means that other properties, such as mechanical behavior, may be inferior to other models, this is a necessary compromise when the aim is to study solution-based processes. In the future, more sophisticated FFs could be derived that remove the need to compromise, but at present, reducing the computational cost is paramount so that extensive sampling of configuration space can be achieved.

ACKNOWLEDGMENTS

The authors would like to acknowledge funding from the ARC (DP0986999) and Curtin University and the provision of computer time from iVEC@Murdoch, iVEC@UWA, and National Computational Infrastructure.

REFERENCES

Aqvist, J. (1990). Ion-water interaction potentials derived from free energy perturbation simulations. *Journal of Physical Chemistry, 94*(21), 8021–8024.
Archer, T., Birse, S., Dove, M. T., Redfern, S., Gale, J. D., & Cygan, R. T. (2003). An interatomic potential model for carbonates allowing for polarization effects. *Physics and Chemistry of Minerals, 30*(7), 416–424.
Atkinson, G., Emara, M. M., & Fernandez-Prini, R. (1974). Ultrasonic absorption in aqueous solutions of calcium acetate and other bivalent metal acetates. *Journal of Physical Chemistry, 78*(19), 1913–1917.
Bolhuis, P. G., Chandler, D., Dellago, C., & Geissler, P. (2002). Transition path sampling: Throwing ropes over rough mountain passes, in the dark. *Annual Review of Physical Chemistry, 53*, 291–318.
Broughton, J. Q., Gilmer, G. H., & Jackson, K. A. (1982). Crystallization rates of a Lennard-Jones liquid. *Physical Review Letters, 49*(20), 1496–1500.
Bruneval, F., Donadio, D., & Parrinello, M. (2007). Molecular dynamics study of the solvation of calcium carbonate in water. *Journal of Physical Chemistry B, 111*, 12219–12227.

Busenberg, E., & Plummer, N. L. (1986). The solubility of $BaCO_3$ (cr)(witherite) in CO_2-H_2O solutions between 0 and 90 °C, evaluation of the association constants of $BaHCO^{+3}$(aq) and $BaCO_0^3$ (aq) between 5 and 80 °C, and a preliminary evaluation of the thermodynamic properties of Ba^{2+}(aq). *Geochimica et Cosmochimica Acta, 50*, 2225–2233.

Busenberg, E., & Plummer, N. L. (1989). Thermodynamics of magnesian calcite solid-solutions at 25 C and 1 atm total pressure. *Geochimica et Cosmochimica Acta, 53*(6), 1189–1208.

Cartwright, J. H. E., Checa, A. G., Gale, J. D., Gebauer, D., & Sainz-Díaz, C. I. (2012). Calcium carbonate polyamorphism and its role in biomineralization: How many amorphous calcium carbonates are there? *Angewandte Chemie International Edition, 51*(48), 11960–11970.

Chipot, C., & Pohorille, A. (2007). Calculating Free energy Differences Using Pertubation Theory. In C. Chipot & A. Pohorille (Eds.), *Free Energy Calculations* (pp. 33–72). Berlin: Springer-Verlag.

Covington, A. K., & Danish, E. Y. (2009). Measurement of magnesium stability constants of biologically relevant ligands by simultaneous use of pH and ion-selective electrodes. *Journal of Solution Chemistry, 38*(11), 1449–1462.

Daniele, P. G., Foti, C., Gianguzza, A., Prenesti, E., & Sammartano, S. (2008). Weak alkali and alkaline earth metal complexes of low molecular weight ligands in aqueous solution. *Coordination Chemistry Reviews, 252*(10), 1093–1107.

Darve, E. (2007). Thermodynamic integration Using Constrained and Unconstrained Dynamics. In C. Chipot & A. Pohorille (Eds.), *Free Energy Calculations* (pp. 33–72). Berlin: Springer-Verlag.

de Leeuw, N. H., & Parker, S. C. (1998). Molecular-dynamics simulation of MgO surfaces in liquid water using a shell-model potential for water. *Physical Review B, 58*(20), 13901–13908.

de Leeuw, N. H., Parker, S. C., & Harding, J. H. (1999). Molecular dynamics simulation of crystal dissolution from calcite steps. *Physical Review B, 60*(19), 13792–13799.

Demichelis, R., Raiteri, P., Gale, J. D., & Dovesi, R. (2012). A new structural model for disorder in vaterite from first-principles calculations. *CrystEngComm, 14*(1), 44.

Demichelis, R., Raiteri, P., Gale, J. D., Quigley, D., & Gebauer, D. (2011). Stable prenucleation mineral clusters are liquid-like ionic polymers. *Nature Communications, 2*, 590.

Dick, B. G., Jr., & Overhauser, A. W. (1958). Theory of the dielectric constants of alkali halide crystals. *Physical Review, 112*(1), 90.

Dove, M. T., Winkler, B., Leslie, M., Harris, M. J., & Salje, E. K. (1992). A new interatomic potential model for calcite: Applications to lattice dynamics studies, phase transition, and isotope fractionation. *American Mineralogist, 77*(3–4), 244–250, Mineralogical Society of America.

Eike, D. M., & Maginn, E. J. (2006). Atomistic simulation of solid-liquid coexistence for molecular systems: Application to triazole and benzene. *Journal of Chemical Physics, 124*(16), 164503.

Elhadj, S., de Yoreo, J. J., Hoyer, J., & Dove, P. M. (2006). Role of molecular charge and hydrophilicity in regulating the kinetics of crystal growth. *Proceedings of the National Academy of Sciences of the United States of America, 103*(51), 19237–19242.

Fenter, P., Kerisit, S., Raiteri, P., & Gale, J. D. (2013). Is the calcite-water interface understood? Direct comparisons of molecular dynamics simulations with specular X-ray reflectivity data. *Journal of Physical Chemistry C, 117*(10), 5028–5042.

Fenter, P., & Sturchio, N. C. (2004). Mineral-water interfacial structures revealed by synchrotron X-ray scattering. *Progress in Surface Science, 77*(5–8), 171–258.

Freeman, C. L., Harding, J. H., Cooke, D. J., Elliott, J. A., Lardge, J. S., & Duffy, D. M. (2007). New forcefields for modeling biomineralization processes. *Journal of Physical Chemistry C, 111*, 11943–11951.

Frenkel, D., & Ladd, A. J. C. (1984). New Monte Carlo method to compute the free energy of arbitrary solids. Application to the fcc and hcp phases of hard spheres. *Journal of Chemical Physics, 81*(7), 3188.

Gale, J. D. (1996). Empirical potential derivation for ionic materials. *Philosophical Magazine Part B, 73*(1), 3–19.

Gale, J. D. (1998). Analytical free energy minimization of silica polymorphs. *Journal of Physical Chemistry B, 102*(28), 5423–5431, ACS Publications.

Gale, J. D., Raiteri, P., & Van Duin, A. C. T. (2011). A reactive force field for aqueous-calcium carbonate systems. *Physical Chemistry Chemical Physics, 13*(37), 16666–16679.

Gebauer, D., Voelkel, A., & Coelfen, H. (2008). Stable prenucleation calcium carbonate clusters. *Science, 322*(5909), 1819–1822.

Geissbuhler, P., Fenter, P., DiMasi, E., Srajer, G., Sorensen, L., & Sturchio, N. C. (2004). Three-dimensional structure of the calcite-water interface by surface X-ray scattering. *Surface Science, 573*(2), 191–203.

Guillot, B. (2002). A reappraisal of what we have learnt during three decades of computer simulations on water. *Journal of Molecular Liquids, 101*(1–3), 219–260.

Harafuji, K., Tsuchiya, T., & Kawamura, K. (2004). Molecular dynamics simulation for evaluating melting point of wurtzite-type GaN crystal. *Journal of Applied Physics, 96*(5), 2501.

Heberling, F., Trainor, T. P., Lützenkirchen, J., Eng, P., Denecke, M. A., & Bosbach, D. (2011). Structure and reactivity of the calcite-water interface. *Journal of Colloid and Interface Science, 354*, 843–857.

Helm, L., & Merbach, A. E. (1999). Water exchange on metal ions: Experiments and simulations. *Coordination Chemistry Reviews, 187*(1), 151–181.

Horn, H., Swope, W., Pitera, J., Madura, J., Dick, T., Hura, G., et al. (2004). Development of an improved four-site water model for biomolecular simulations: TIP4P-Ew. *Journal of Chemical Physics, 120*(20), 9665–9678.

Hummer, G., Pratt, L. R., & Garcia, A. (1997). Ion sizes and finite-size corrections for ionic-solvation free energies. *Journal of Chemical Physics, 107*(21), 9275–9277.

Jackson, R. A., & Price, G. D. (1992). A transferable interatomic potential for calcium carbonate. *Molecular Simulation, 9*(2), 175–177.

Jarzynski, C. (1997). Nonequilibrium equality for free energy differences. *Physical Review Letters, 78*(14), 2690–2693.

Karim, O. A., & Haymet, A. D. J. (1988). The ice/water interface: A molecular dynamics simulation study. *Journal of Chemical Physics, 89*(11), 6889.

Kastenholz, M. A., & Huenenberger, P. H. (2006a). Computation of methodology-independent ionic solvation free energies from molecular simulations. I. The electrostatic potential in molecular liquids. *Journal of Chemical Physics, 124*(12), 124106.

Kastenholz, M. A., & Huenenberger, P. H. (2006b). Computation of methodology-independent ionic solvation free energies from molecular simulations. II. The hydration free energy of the sodium cation. *Journal of Chemical Physics, 124*(22), 224501.

Kerisit, S., & Parker, S. C. (2004a). Free energy of adsorption of water and calcium on the [10 1 4] calcite surface. *Chemical Communications, 1*, 52–53.

Kerisit, S., & Parker, S. C. (2004b). Free energy of adsorption of water and metal ions on the [1014] calcite surface. *Journal of the American Chemical Society, 126*(32), 10152–10161.

Kerisit, S., Parker, S. C., & Harding, J. H. (2003). Atomistic simulation of the dissociative adsorption of water on calcite surfaces. *Journal of Physical Chemistry B, 107*(31), 7676–7682.

Kim, J., Straub, J. E., & Keyes, T. (2007). Statistical temperature molecular dynamics: Application to coarse-grained β-barrel-forming protein models. *Journal of Chemical Physics, 126*(13), 135101.

Kirkwood, J. G. (1935). Statistical mechanics of fluid mixtures. *Journal of Chemical Physics, 3*(5), 300.

Laio, A., & Gervasio, F. L. (2008). Metadynamics: A method to simulate rare events and reconstruct the free energy in biophysics, chemistry and material science. *Reports on Progress in Physics, 71*(12), 126601.

Laio, A., & Parrinello, M. (2002). Escaping free-energy minima. *Proceedings of the National Academy of Sciences of the United States of America, 99*(20), 12562–12566.

Lamoureux, G., Harder, E., Vorobyov, I., Roux, B., & MacKerell, A. D., Jr. (2006). A polarizable model of water for molecular dynamics simulations of biomolecules. *Chemical Physics Letters, 418*(1–3), 245–249.

Marx, D., Tuckerman, M., Hutter, J., & Parrinello, M. (1999). The nature of the hydrated excess proton in water. *Nature, 397*(6720), 601–604.

Meyer, H. J. (1959). Über vaterit und seine struktur. *Angewandte Chemie International Edition, 71*(21), 678–679.

Noya, E. G., Conde, M. M., & Vega, C. (2008). Computing the free energy of molecular solids by the Einstein molecule approach: Ices XIII and XIV, hard-dumbbells and a patchy model of proteins. *Journal of Chemical Physics, 129*(10), 104704.

Ohtaki, H., & Radnai, T. (1993). Structure and dynamics of hydrated ions. *Chemical Reviews, 93*(3), 1157–1204.

Paesani, F., Zhang, W., Case, D. A., Cheatham, T. E., & Voth, G. A. (2006). An accurate and simple quantum model for liquid water. *Journal of Chemical Physics, 125*(18), 184507.

Pavese, A., Catti, M., Price, G. D., & Jackson, R. A. (1992). Interatomic potentials for $CaCO_3$ polymorphs (calcite and aragonite), fitted to elastic and vibrational data. *Physics and Chemistry of Minerals, 19*(2), 80–87, Springer.

Plimpton, S. (1995). Fast parallel algorithms for short-range molecular dynamics. *Journal of Computational Physics, 117*(1), 1–19, Citeseer.

Plummer, L., & Busenberg, E. (1982). The solubilities of calcite, aragonite and vaterite in H_2O-CO_2 solutions between 0 and 90 °C, and an evaluation of the aqueous model for the system $CaCO_3$-CO_2-H_2O. *Geochimica et Cosmochimica Acta, 46*(6), 1011–1040.

Raiteri, P., Demichelis, R., Gale, J. D., Kellermeier, M., Gebauer, D., Quigley, D., et al. (2012). Exploring the influence of organic species on pre- and post-nucleation calcium carbonate. *Faraday Discussions, 159*, 61–85.

Raiteri, P., & Gale, J. D. (2010). Water is the key to nonclassical nucleation of amorphous calcium carbonate. *Journal of the American Chemical Society, 132*(49), 17623–17634.

Raiteri, P., Gale, J. D., Quigley, D., & Rodger, P. M. (2010). Derivation of an accurate force-field for simulating the growth of calcium carbonate from aqueous solution: A new model for the calcite-water interface. *Journal of Physical Chemistry C, 114*(13), 5997–6010.

Ricci, M., Spijker, P., Stellacci, F., Molinari, J.-F., & Voïtchovsky, K. (2013). Direct visualization of single ions in the stern layer of calcite. *Langmuir, 29*(7), 2207–2216.

Rohl, A., Wright, K., & Gale, J. D. (2003). Evidence from surface phonons for the (2×1) reconstruction of the (10-14) surface of calcite from computer simulation. *American Mineralogist, 88*(5–6), 921–925.

Smith, W. (2006). DL_POLY 2.19. *Molecular Simulation, 32*, 933–1121.

Stack, A. G., Gale, J. D., & Raiteri, P. (2013). Virtual probes of mineral-water interfaces: The more flops, the better!. *Elements, 9*, 211–216.

Taylor, M. B., Barrera, G. D., Allan, N. L., & Barron, T. (1997). Free-energy derivatives and structure optimization within quasiharmonic lattice dynamics. *Physical Review B, 56*(22), 14380.

Torrie, G., & Valleau, J. (1977). Nonphysical sampling distributions in Monte Carlo free-energy estimation: Umbrella sampling. *Journal of Computational Physics, 23*(2), 187–199.

Vaknin, D., Bu, W., & Travesset, A. (2008). Extracting the pair distribution function of liquids and liquid-vapor surfaces by grazing incidence X-ray diffraction mode. *Journal of Chemical Physics, 129*(4), 044504.

Van Duin, A. C. T., Dasgupta, S., Lorant, F., & Goddard, W. A., III. (2001). ReaxFF: A reactive force field for hydrocarbons. *Journal of Physical Chemistry A, 105*(41), 9396–9409.

van Maaren, P., & van der Spoel, D. (2001). Molecular dynamics simulations of water with novel shell-model potentials. *Journal of Physical Chemistry B, 105*(13), 2618–2626.

Vanommeslaeghe, K., Hatcher, E., Acharya, C., Kundu, S., Zhong, S., Shim, J., et al. (2010). CHARMM general force field: A force field for drug-like molecules compatible with the CHARMM all-atom additive biological force fields. *Journal of Computational Chemistry, 31*, 671–690.

Voter, A. (1997). A method for accelerating the molecular dynamics simulation of infrequent events. *Journal of Chemical Physics, 106*(11), 4665–4677.

Wang, J., Román-Pérez, G., Soler, J. M., Artacho, E., & Fernández-Serra, M. V. (2011). Density, structure, and dynamics of water: The effect of van der Waals interactions. *Journal of Chemical Physics, 134*(2), 024516.

Wolf, S. E., Leiterer, J., Kappl, M., Emmerling, F., & Tremel, W. (2008). Early homogenous amorphous precursor stages of calcium carbonate and subsequent crystal growth in levitated droplets. *Journal of the American Chemical Society, 130*(37), 12342–12347.

Wolthers, M., di Tommaso, D., Du, Z., & de Leeuw, N. H. (2012). Calcite surface structure and reactivity: Molecular dynamics simulations and macroscopic surface modelling of the calcite–water interface. *Physical Chemistry Chemical Physics, 14*(43), 15145.

Wong, K. F., Sonnenberg, J. L., Paesani, F., Yamamoto, T., Vanicek, J., Zhang, W., et al. (2010). Proton transfer studied using a combined Ab initio reactive potential energy surface with quantum path integral methodology. *Journal of Chemical Theory and Computation, 6*(9), 2566–2580.

Wu, Y., Tepper, H., & Voth, G. A. (2006). Flexible simple point-charge water model with improved liquid-state properties. *Journal of Chemical Physics, 124*(2), 024503.

Xiao, S., Edwards, S. A., & Grater, F. (2011). A new transferable forcefield for simulating the mechanics of $CaCO_3$ crystals. *Journal of Physical Chemistry C, 115*, 20067–20075.

Zwanzig, R. (1954). High-temperature equation of state by a perturbation method. I. Nonpolar gases. *Journal of Chemical Physics, 22*(8), 1420.

CHAPTER TWO

The Integration of Ion Potentiometric Measurements with Chemical, Structural, and Morphological Analysis to Follow Mineralization Reactions in Solution

Wouter J.E.M. Habraken[1]

Department of Biomaterials, Max Planck Institute for Colloids and Interfaces, Potsdam, Germany
[1]Corresponding author: e-mail address: wouter.habraken@mpikg.mpg.de

Contents

1. Introduction	26
1.1 Mineralization reactions	27
2. Ion Potentiometric Measurements	29
2.1 Concentration/pH measurements to follow reaction kinetics	30
2.2 Using ion potentiometric measurements to determine the solubility of a bulk species	33
3. Combination of Ion Potentiometric Measurements with other Analysis Techniques	35
3.1 Light scattering/zeta-potential	36
3.2 FTIR/Raman spectroscopy	37
3.3 Synchrotron SAXS and WAXS	38
3.4 Cryo-TEM	39
3.5 Alternative techniques	40
4. Comparison with Standard Analyses	41
5. Summary	42
References	42

Abstract

The solution crystallization of biominerals like calcium carbonate or calcium phosphate is a process that requires a high level of control over reaction kinetics. Ion potentiometric measurements are a way to follow and control reaction kinetics by measuring changes in pH and ion concentration, also allowing quantification of chemical compositions and solubility characteristics. By combining these measurements with various analysis techniques, one can acquire a complete spectrum of chemical, structural, and morphological data, even on metastable precursors. Therefore, in this chapter, the use of potentiometric measurements in standard solution crystallization experiments is described and the integration of these measurements with microscopy, spectroscopic and scattering analyses in such a way that the control over the reaction kinetics is maintained.

1. INTRODUCTION

To understand the formation of biological mineral specimens, showing complex morphologies or remarkable stable amorphous phases, a lot of comparative crystallization studies with calcium carbonate or calcium phosphate have been performed, often in the presence of an additive like an organic scaffold, protein, polypeptide, small molecule, or inorganic impurity that is expected to direct the mineralization toward the properties of the investigated biomineral (Nudelman et al., 2010; Shechter et al., 2008). To understand the mechanisms behind solution crystallization in a more general context, the formation of both sparingly soluble calcium salts has been investigated in detail over decades, rendering insights with respect to nucleation mechanisms and formation of amorphous (precursor) phases (Chistofferensen, Christoffersen, Kibalczyc, & Andersen, 1989; Gebauer, Völkel, & Cölfen, 2008). Recently, it has been discovered that the solution crystallization of calcium phosphate involves the assembly and subsequent chemical transformation of ion association complexes via a range of metastable precursor phases (Habraken et al., 2013), a process that obviously requires a high degree of control over the reaction kinetics. As a result of such a mechanism, the outcome of many mineralization experiments seems to be unpredictable and highly dependent on parameters like humidity, CO_2 concentrations, stirring speed, vessel size or dimension, and purity of used chemicals often not described in the methods section of most papers.

Additionally, to monitor the evolving mineral during these experiments, an array of techniques has been used like quantitative analysis on the composition of the reaction solution (Chistofferensen et al., 1989; Gebauer et al., 2008; Habraken et al., 2013; Meyer & Eanes, 1978), calculations on solubility or equilibrium behavior of the mineral precursors (Gebauer et al., 2008; Habraken et al., 2013; Meyer & Eanes, 1978), and qualitative analysis on chemistry, structure, and morphology (Chistofferensen et al., 1989; Habraken et al., 2013; Kazanci, Fratzl, Klaushofer, & Paschalis, 2006; Nudelman et al., 2010). As every analysis technique has its own drawbacks and often requires a different sampling method, integration of these techniques is not straightforward, especially as standard sampling techniques often require operations like drying, filtration, or centrifugation where the control over reaction kinetics is lost. On the other hand, to fully understand what is occurring during a mineralization reaction, the use of multiple techniques covering different aspects of the growing mineral is a necessity, though by doing so, one might compare minerals at different stages of the reaction.

Ion potentiometric measurements can be used to follow the kinetics of a reaction inside the reaction solution by measuring the pH or ion (Ca^{2+}) concentration, thereby also offering a way to calculate changes in chemistry and solubility or equilibrium behavior of evolving species. Upon knowing and controlling the reaction kinetics, one can use the ion potentiometric measurements as a standard for other techniques that often require sampling of the reaction medium. Therefore, this chapter focuses on the integration of ion potentiometric measurements with more or less standard chemical, structural, and morphological investigations and shows how these techniques can be combined to retrieve trustworthy data on something as untenable as solution crystallization.

1.1. Mineralization reactions

To study the formation of minerals in solution, in general, two different types of experiments can be employed, that is, a reaction where upon mixing of two reactants, an instant (super)saturation with respect to a bulk phase is generated or a reaction where one reactant is slowly added to an excess of the other (Fig. 2.1).

1.1.1 Instant (super)saturation
By varying the concentration from slightly saturated with respect to the crystalline bulk phase (i.e., apatite and calcite) to highly supersaturated, this method is the most common way to also form metastable phases like amorphous calcium carbonate (ACC) or amorphous calcium phosphate (ACP). Depending on the extent of (super)saturation with respect to the first precipitated phase, there is often a lag time between mixing both stock solutions to actual precipitation (Termine & Posner, 1970), the so-called prenucleation stage where the prenucleation species (ions, complexes, or clusters of complexes) assemble to eventually form a nucleus during time (Habraken et al., 2013). This lag time is an indication for the kinetics after nucleation, though parameters that increase the lag time, such as a decreasing Ca/P ratio or increasing pH in the case of calcium phosphate, might increase

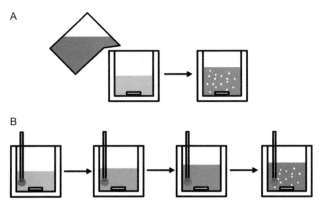

Figure 2.1 Mineralization reactions: (A) instant (super)saturation and (B) gradual increase in concentration; blue and pink refer in both cases to the separate cation (Ca^{2+}) and anion (carbonate and phosphate) stock solutions. Preferably, the cation solution is added to the anion solution, which because of its buffering capacity prevents large fluctuations in pH to occur. (For interpretation of the references to color in this figure legend, the reader is referred to the online version of this chapter.)

the kinetics for any subsequent reaction (Termine, Peckauskas, & Posner, 1970). The facile preparation method, which enables a convenient upscaling of reaction volume, is also an advantage, though for a good reproducibility, the mixing procedure between both stock solutions should be kept constant.

1.1.2 Gradual increase in concentration

The controlled (constant) addition of one reactant into an excess of the other by titration (Gebauer et al., 2008; Verch, Gebauer, Antonietti, & Cölfen, 2011) optimally should lead to a high control of reaction kinetics, where gradually, saturation with respect to the bulk phase is obtained. However, as it takes a long time before a bulk phase precipitates, it is sensitive to small changes in physicochemical environment during this time like stirring speed, temperature, and CO_2 diffusion or outgassing, which can lead to relatively large differences in precipitation time. Furthermore, as one is gradually inducing supersaturation, the formation of metastable phases that need high supersaturation is less likely. However, by changing reaction parameters (like pH) in the case of $CaCO_3$, the formation of two different types of ACC was reported using this method (Gebauer et al., 2008). For $CaCO_3$ crystallization, specialized methods have been developed that rely on the gradual diffusion of CO_2 vapor (i.e., $(NH_4)_2CO_3$) into a calcium stock using a closed desiccator (Politi, Mahamid, Goldberg, Weiner, & Addadi, 2007) or by outgassing of CO_2 from a saturated $Ca(HCO_3)_2$ solution (Kitano method, Kitano, 1962). Both experiments are ideal for the formation of crystals at a surface or underneath a monolayer; however, as both diffusion and outgassing of CO_2 occur at the gas–water interface, this is where the crystals nucleate and grow, making it impossible to perform most of the bulk physicochemical analysis (including potentiometric measurements) described in the following chapters.

Finally, because phase transitions in many systems, including calcium carbonate and calcium phosphate, are accompanied by a sudden drop in pH, for both type of experiments (instant addition or gradual), one can choose to fix the pH at a certain value by titrating NaOH/HCl or using a buffer (see also Section 2.1).

2. ION POTENTIOMETRIC MEASUREMENTS

To measure the concentration of ions in solution, use can be made of an ion-selective electrode (ISE) of which the pH electrode is the most commonly used version, but nowadays, electrodes for multiple mono- and divalent cations and anions like Na^+, K^+, F^-, I^-, Ca^{2+}, and Pb^{2+} are available. The basis of these electrode measurements is an ion-selective membrane, which allows only one type of ion to diffuse into the electrode, thereby causing a potential difference between the indicator electrode and a reference electrode according to Nernst law (Eq. 2.1; Hunter, 1994):

$$E = E_{ind} - E_{ref} = E^0 - \frac{RT}{nF} \ln a_{ion} = E^0 - \frac{RT}{nF} \ln \left(\frac{\gamma_{ion} \cdot c_{ion}}{c_0} \right) \quad (2.1)$$

where E is the measured voltage (mV), E_{ind} is the voltage of indicator electrode (mV), E_{ref} is the voltage of reference electrode (mV), E^0 is a constant characteristic for a given cell (mV), R is the gas constant (8.3144 J/K), T is the absolute temperature (K), F is the Faraday's constant, n is the valency of the ions, a_{ion} is the activity of the ions, γ_{ion} is the activity coefficient, c_{ion} is the concentration of the ions (M), and c_0 is the standard concentration (=1 M). The activity coefficient translates measured activities into concentrations and is dependent on the total ionic strength (I) of the solution and the type of ion (valency and atomic weight) and temperature. Furthermore, the equation in the preceding text is a simplified version, not taking into account interference of other ions on the potentiometric measurements. A correct calibration procedure preceding the measurement is therefore crucial for quantitative investigations, of which the implications together with activity coefficient theory will be discussed in the next section (2.1).

2.1. Concentration/pH measurements to follow reaction kinetics

Even when the ISEs are used as just an indication of reaction kinetics, showing changes in pH or calcium concentration at a specific time point due to a nucleation event or phase transition, a quantitative measurement of the pH is necessary to somehow relate the obtained reaction kinetics with literature data. This requires a calibration procedure, which in the case of a pH measurement is a standardized procedure where by the use of calibration standards of pH 4.00, 7.00, and 10.00, a calibration line is automatically generated (mV vs. pH $(=-\log[a_{H^+}])$), often even showing the level of accuracy of the measurement.

A measurement using, for example, a Ca^{2+}-ISE is not so dissimilar; however, these electrodes are often not as ion specific and have a higher detection limit. Furthermore, where most of the time for a pH measurement it is enough to know the order of magnitude of H^+ ion activity in the solution, to distinguish between the solubility of, for example, aragonite and calcite at 25 °C using a Ca^{2+}-ISE, we need to distinguish a Ca^{2+} activity between 7.7×10^{-5} (aragonite) and 5.8×10^{-5} (calcite), a difference in ion activity that is at the limit of accuracy for any normal pH meter ($-\log[7.7 \times 10^{-5}] = 4.24$, where $-\log[5.8 \times 10^{-5}] = 4.11$).

Calibration experiments for ISEs can be performed by titrating a constant rate of an ion solution containing a certain fixed concentration of the ion of interest to a solution where optimally the pH is kept constant. Similar to the pH-ISE, as a response on the increasing ion concentration, the ISE will measure a signal (in mV) that is linear with the logarithm of the activity of the measured ion (see Fig. 2.2). However, this is not always the case for the whole concentration range; instead, there is often a less linear behavior at low or high values. Therefore, to minimize the error in the measurement, the middle part of the curve that actually shows the linear behavior should correspond to the expected concentration range of the experiment. Furthermore, to enable an easy conversion of the calibration experiment to the actual mineralization experiment, in practice, two parameters are of major importance, the use of activity coefficients and interaction with CO_2 from the air.

2.1.1 Activity coefficients

As activities are measured instead of concentrations, in the calibration experiment, one needs to convert added concentrations of ions into activities and in the actual experiment measured activities into concentrations by use of the activity coefficient (γ). Mostly depending on the total ionic strength used

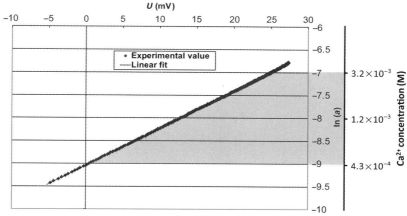

Figure 2.2 Calibration curve for a Ca^{2+}-ISE based on Eq. (2.1), showing the electrode potential (mV) as a function of the activity (ln (a)) and corresponding Ca^{2+} concentration. The filled area indicates the concentration region in which the experiment should be preferentially performed. However, as in this case the ionic strength is kept constant at ~0.2 M by the addition of excess NaCl, even at the low and high concentration ranges, an acceptable linear response is observed. As can be deducted from Table 2.1, therefore, the Ca^{2+} concentration has to be converted into activity using the Davies-extended Debye–Hückel equation, rendering an activity coefficient that is approximately constant throughout the whole calibration curve ($\gamma_{Ca^{2+}} = 0.288 - 0.285$). (For color version of this figure, the reader is referred to the online version of this chapter.)

($I = \frac{1}{2}\sum_{i=1}^{n} c_i z_i^2$, with c_i being the concentration and z_i the charge (Burgot, 2012; Hunter, 1994)), there are different regimes for calculating the activity coefficient as explained in Table 2.1, where only at very low ionic strengths, the difference between concentration and activity is negligible. While with mineralization experiments this is never the case, an easy way to directly convert measured activities into concentrations is by adding a high amount of spectator ions (like NaCl) to the solution, assuring that the overall ionic strength does not change significantly during the mineralization reaction as compared to the calibration experiment. For quantitative calculations, though, even small changes in concentration can be of importance, especially when two or three valency ions like Ca^{2+}, CO_3^{2-}, HPO_4^{2-}, and PO_4^{3-} are present.

2.1.2 Interaction with CO_2

For mineralization reactions at high pH (8 and higher) and actually always in the case of carbonate-containing minerals ($CaCO_3$ and $MgCO_3$), CO_2 coming from the air could be a disturbing factor influencing the kinetics

Table 2.1 Converting activities into concentrations by use of activity coefficients at different applicability conditions

Equation	Applicability conditions
$\gamma_i = 1 \left(a_i = \frac{c_i}{c_0}, \text{with } c_0 = 1 \right)$	Ionic strengths <0.1 mM
$\log(\gamma_i) = -A z_i^2 \sqrt{I}$ (Debye–Hückel)	Ionic strengths up to 10 mM
$\log(\gamma_i) = -A z_i^2 \left(\frac{\sqrt{I}}{1+Ba\sqrt{I}} \right)$ (extended Debye–Hückel)	Ionic strengths up to 0.1 M
$\log(\gamma_i) = -A z_i^2 \left(\frac{\sqrt{I}}{1+Ba\sqrt{I}} - cI \right)$ (Davies-type-extended Debye–Hückel)	Ionic strengths up to 1 M, most used for calculations of divalent/trivalent ions also at lower concentrations, values of c vary depending on type of ion used

Here, A = Debye–Hückel constant (5.1×10^{-1} at 25 °C), typical values for Ba are 1.0 or 1.5, and typical values for c are 0.1–0.3 (Burgot, 2012).

Table 2.2 Theoretical activity of HCO_3^- and CO_3^{2-} by CO_2 diffusion from the air at the indicated pH and $p_{CO_2} = 3.14 \times 10^{-5}$ atm, $k_H = 29.41$ l atm/mol, $K_{a1,app} = 4.6 \times 10^{-7}$, $K_{a2} = 4.7 \times 10^{-11}$ (Haynes, 2013)

	Theoretical activities due to CO_2 diffusion	
pH	HCO_3^-	CO_3^{2-}
7.0	4.9×10^{-5}	2.3×10^{-8}
8.0	4.9×10^{-4}	2.3×10^{-6}
9.0	4.9×10^{-3}	2.3×10^{-4}
10.0	4.9×10^{-2}	2.3×10^{-2}
11.0	4.9×10^{-1}	2.3

These values could be achieved by pH = constant titration (using NaOH) in the open air; however, they are not corrected for the formation of $Na_2CO_3/NaHCO_3$.

of the reaction and the outcome of the potentiometric measurements. When performing a mineralization reaction in the open air at high pH, amounts of CO_2 possibly incorporated into the solution can be enormous (Table 2.2), as the absorbed CO_2 from the air will convert into HCO_3^- or CO_3^{2-} according to the equilibria described in Fig. 2.3.

At these high pHs, the reaction and calibration should be performed in a closed vessel, optimally purged by nitrogen or argon to prevent the little amount of air inside the vessel to contribute to the carbonate concentration inside the reaction solution. As in all cases the pH will drop upon diffusion of

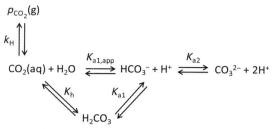

Figure 2.3 Carbonic acid equilibria. Here, p_{CO_2} is the partial pressure of CO_2 in the air, k_H is the Henry's law constant, K_h is the hydration equilibrium constant, K_{a1} and K_{a2} are the first and second dissociation constants of H_2CO_3, and $K_{a1,app}$ is the apparent first dissociation constant. As in practice $CO_2(aq) \gg H_2CO_3$, one normally calculates the concentrations of HCO_3^- and CO_3^{2-} by this apparent dissociation constant, although using the route via H_2CO_3 leads to the same values.

CO_2, a reference experiment should be performed where the pH of the solution of one of the reactants (normally the carbonate or phosphate buffer) is followed over a longer period of time to exclude any significant diffusion of CO_2 into the solution. One should be aware that due to their high pH, NaOH solutions, which are often used to keep the pH stable, are easily affected by CO_2 diffusion.

As most ISEs are pH-sensitive, constant pH titration (by NaOH/HCl) or the use of a buffer can be recommended for both calibration experiments and reaction, fixing the pH at a certain value. For this purpose, an overdose of phosphate or carbonate stock could be used. However, if this is not desirable, one could think of using one of the standard organic buffers like Tris or HEPES. One should take into account that also buffers contribute to the total ionic strength of the solution, and due to their pH dependency, calculation of this contribution is not so straightforward and requires insight into the equilibrium behavior of the buffers (pK_a).

2.2. Using ion potentiometric measurements to determine the solubility of a bulk species

Knowing concentrations of free ions in the solution and amounts of released protons, by making charge and mass balances over the whole mineralization reaction, one could derive a model to calculate the chemistry of the reacted species at each time point, as has recently been done for calcium phosphate (Habraken et al., 2013). However, the potentiometric measurements can also be used to determine the solubility constant of a (known) species, a parameter that is crucial in explaining kinetic pathways in mineralization reactions. For a certain compound, this parameter is not always fixed but depends on the size and morphology of the precipitated bulk phase (see, e.g., Chow & Sun, 2004) or the presence of both inorganic and organic impurities at the surface or incorporated into the bulk of the mineral.

The solubility (K_{sp}) of a bulk phase like calcite is expressed as Eq. (2.2):

$$K_{sp} = [Ca^{2+}]^*[CO_3^{2-}] \qquad (2.2)$$

Here, both $[Ca^{2+}]$ and $[CO_3^{2-}]$ are activities and the concentration ratios do not have to reflect the 1:1 stoichiometry of the bulk mineral but can have any value as long as this equality is fulfilled.

1. So when the activity of free calcium is known from the ion potentiometric measurement, to calculate the activity of $[CO_3^{2-}]$ and after that the solubility constant (K_{sp}), one first needs to know with how much carbonate the calcium reacted to form the calcite. For this, the measured $[Ca^{2+}]$ activity has to be converted into concentrations and accordingly (for ease of explanation) into moles using the activity coefficient and volume of the reaction solution.
2. Accordingly, this can be subtracted from the total amount of Ca^{2+} (in moles) added to the system, rendering the total amount of Ca^{2+} into the precipitated mineral and the total amount of CO_3^{2-} in the precipitate (as there is a ratio of 1:1).
3. By subtracting the amount of CO_3^{2-} (in moles) incorporated into the mineral from the total amount of carbonate added, one can calculate the total amount (and thereby concentration) of carbonate species (at pH >7, total amount of carbonate $\cong CO_3^{2-} + HCO_3^-$) still present inside the solution after mineral formation.
4. Knowing the pH of the solution, by use of the equilibrium constant(s) between the carbonate species in solution, their activity ratio can be calculated (i.e., $pH = pK_a + \log[CO_3^{2-}]/[HCO_3^-]$).
5. Finally, by use of the activity coefficient for each species, by combining results from step 3+4, the activity of individual species (like $[CO_3^{2-}]$) and the solubility constant can be deducted.

In the literature, the solubility (K_{sp}) is often expressed in concentrations instead of activities, which is only correct when ionic strengths are very low (Table 2.1). Of course, solubility constants are also temperature- and pressure-dependent, and while most reactions are performed at ambient pressure, the K_{sp} should always be accompanied with the temperature at which the solubility is determined. Finally, if ion pairs are expected to be present in the medium as well, this will increase the value for the solubility calculated in this manner. However, this is normally only a small fraction of the total amount of bound ions (up to a few percentage with calcium phosphates (Habraken et al., 2013)). Furthermore, using equilibrium constants for these ion pairs, which are often given in the literature (i.e., Chughtai, Marshall, & Nancollas, 1968), one is able to correct for this.

3. COMBINATION OF ION POTENTIOMETRIC MEASUREMENTS WITH OTHER ANALYSIS TECHNIQUES

Upon knowing and controlling the kinetics of a mineralization reaction, this information gives a time window that can be used for various other analysis techniques that often require sampling of the reaction solution and precipitate (Fig. 2.4). Depending on the time used for each specific analysis technique, one is able to delay reaction kinetics in such a way that the

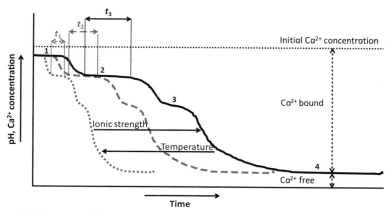

Figure 2.4 The use of ion potentiometric measurements to control reaction kinetics. The diagram corresponds to a multistep calcium phosphate mineralization reaction as can be observed in the literature (Chistofferensen et al., 1989; Habraken et al., 2013) with 1 being the prenucleation stage, 2+3 the metastable intermediates, and 4 the final bulk phase. Reaction kinetics can be slowed down by decreasing the temperature and increasing the ionic strength; note that the height of the different plateaus may change as a result of that. As indicated in the graph, the acquisition time for 2 hereby increases from t_1, t_2, to t_3. For stage 4, amounts of free and bound calcium are indicated by arrows, from which it can be deducted that during such a multistage mineralization experiment, the solubility (K_{sp}) of the successive mineral phases (2–4) decreases. (For color version of this figure, the reader is referred to the online version of this chapter.)

composition of the solution should not change during the analysis time. Convenient ways to slow down the kinetics is by performing the reaction (or even sampling) at lower temperatures (sparingly soluble salts like calcium carbonates and calcium phosphates have a reverse solubility), adding a high concentration of spectator ions (i.e., NaCl) or decreasing effective stirring. To assure similar kinetics throughout the reaction medium, for the mineralization reaction, some sort of stirring is recommended. As most analyses on solute samples are performed on static solutions, kinetics will be slowed down, giving even a larger time window for the measurements. In the case of a multistage mineralization reaction, as often seen with calcium phosphate, sampling should be done preferably at the plateaus where both pH and Ca^{2+} concentrations show the smallest change.

To be certain that the reaction kinetics do not increase upon sampling, one should avoid any preparation that includes concentrating the reaction solution, like drying, centrifugation, or filtration, but measure the native reaction solution directly after removing the sample volume. Pumping a small part of the reaction solution into a measurement cell, as is possible with some spectrometry or scattering analyses, is also an option; however, here, also changes in flow rate (pressure changes) can induce significant differences in reaction kinetics and cause problems with clogging of reactants upon sedimentation. Furthermore, care should be taken with high-energy radiation like X-rays, electrons, and lasers, as by locally irradiating the sample, reaction kinetics can also increase. Despite these restrictions, still, many techniques of comparative analysis of the morphology, structure, and chemistry of the evolving mineral can be applied. These will be further described in this section.

3.1. Light scattering/zeta-potential

In the case of evolving crystal precursors before/upon nucleation, following the average particle size in solution and the amount of particles (count rate) using dynamic light scattering (DLS) can provide valuable information on the mechanism of nucleation (Onuma & Ito, 1998). For example, when particles aggregate upon nucleation, this should lead to a sudden increase in particle size, often followed by a decreasing count rate and rapid sedimentation. This is in contrast to a system with a fixed amount of nuclei, which continuously grow or densify, where only gradual increases in size and count rate are expected. Particle size measurements using DLS are based on differences in Brownian motion between different particle sizes and are most sensitive to particles in the intermediate size ranges (around 100 nm (Hunter, 1994)). Studies on 1–10-nm-sized prenucleation species require specialized equipment and above all no interference of large (dust) species. Conversely, species that are larger (from 1 to 10 μm) in general have a very low mobility and therefore require longer measurement times. Next to DLS (time-resolved), static light scattering (SLS) can provide additional information on particle mass and radius of gyration, as has been done recently on evolving ACC particles in solution (Liu, Rieger, & Huber, 2008).

If a suspension of particles with a certain size is present at any time point of the reaction, it might be worth performing a zeta-potential measurement. The zeta-potential is defined as the potential difference between the dispersion medium and the stationary layer of fluid attached to the dispersed particle. It is measured by putting a potential difference over the sample solution, where the velocity of the particle migration to one of the electrodes is proportional to its zeta-potential and calculated using either the more general von Smoluchowski equation or the Hückel equation (Hunter, 1994) derived for nanometer-sized particles. The value of the zeta-potential can be used to investigate the stability of a colloidal suspension of particles, where in general zeta-potentials higher than $|\pm 25|$ mV are regarded as electrostatically stable. Without the addition of a polyelectrolyte (Morgan et al., 2008), this is seldom the case in a mineralization reaction; however, even when particles aggregate, the overall sign of the zeta-potential can tell something about the charge of individual particles in solution (Habraken et al., 2013). Sample volumes for DLS/SLS measurements can be relatively small (starting from 200 μl), whereas the standard zeta-potential cuvette acquires a sample volume of about 2 ml.

3.2. FTIR/Raman spectroscopy

Though Fourier-transform infrared (FTIR) spectroscopy is more often used to determine the mineral-specific phosphate or carbonate peak positions (i.e., vibrations at a specific wavelength) of extracted crystals from biological samples or powdered synthetic material, with the use of some small or larger adaptations to any standard FTIR equipment, it can also be used to follow mineralization reactions in solution. At a first glance, due to its high sensitivity to water, FTIR does not seem to be the most optimal technique for solution reactions. However, using a flow cell, where the distance between the IR-transparent plates is kept constant, one is able to successfully subtract the water background, giving high-resolution signals at positions where water does not contribute in the IR spectrum (from 900 to 1500 cm^{-1} and 1700 to 2600 cm^{-1}). Here, the range of the "blind regions" is also dependent on the thickness of the measured water layer.

An even more convenient technique is by the use of a horizontal attenuated reflectance (ATR) device (Habraken et al., 2013). This equipment consists of a large rectangular ATR crystal placed in the bottom of a well, which can be completely filled with fluid. Varying the angle of incidence, the IR radiation is reflected multiple times through the crystal/solution interface before reaching the detector, therefore screening the 100–200 µm layer touching the crystal. The penetration depth here is also dependent on the wavelength of the IR source used. This method enables not only an optimal subtraction of the aqueous background but also a fast measurement with a relative high amount of signal and an easy means of cleaning the crystal between different measurements. For mineralization experiments in aqueous solution, the standard ZnSe would be the crystal of choice, though pH should not drop below pH 5.0 to prevent dissolving the crystal and the subsequent formation of the very toxic H_2Se gas. Quantitative calculations of amount of mineral present cannot be deduced, as upon sedimentation of the mineral, the amount of signal will increase.

Raman spectroscopy is far less sensitive for water than IR, which simplifies both sample handling and background subtraction, and therefore is more commonly used for analyzing samples dispersed in liquids using various types of sample holders like capillaries, cuvettes, glass vials, or simple microscope slides. In the literature, numerous examples of Raman spectroscopy on aqueous solutions of ions (including carbonate and phosphate) have been performed (Févotte, 2002; Pye & Rudolph, 2003; Rudolph, Irmer, & Königsberger, 2008), showing that although the amount of Raman scattering events is low with respect to IR absorption, intensity should be high enough to follow association and nucleation of ions inside a solution (up to 10–20 mg/l (Cunningham, Goldberg, & Weiner, 1977)). A possible disadvantage, however, is the high energy of the laser used, heating up the reaction solution during the measurement. Therefore, an optimum should be found in achieving the highest amount of signal (varying the concentration of the reactants) without evoking any unwanted side effects. However, also without the use of specialized Raman equipment, performing analysis on mineral precipitate that is still dispersed in the reaction solution is easily done.

3.3. Synchrotron SAXS and WAXS

To investigate the evolving structure of a mineral in solution by X-ray diffraction, difficulties have to be overcome like the small difference in electron density between the mineral and the medium, the overall low concentration of mineral in solution, and the long measurement times normally required to attain enough signal. Small- and/or wide-angle X-ray scattering (SAXS or WAXS) studies on *in situ*-grown minerals are therefore preferably performed using synchrotron radiation, thereby seriously decreasing measurement times and increasing resolution.

The setups used here are either a flow system, continuously pumping part of the reaction volume into a sample chamber, or introducing part of the reaction fluid into a capillary (Bots, Benning, Rodriguez-Blanco, Roncal-Herrero, & Shaw, 2012; Habraken et al., 2013). While the SAXS data give additional information about the shape and size of evolving particles, refining results from DLS, WAXS would be the preferred technique to investigate the appearance of long-range order inside the solution, as this will concomitantly lead to an increase in electron density. Due to the high brilliance of the beam, reference experiments should always be performed, focusing on sudden (heat-induced) changes in the reaction medium as a response to the chosen sampling time. Furthermore, both techniques can be used to verify morphological and or structural data obtained by microscopic techniques (like cryogenic transmission electron microscopy (cryo-TEM), discussed in Section 3.4) on a bulk scale.

3.4. Cryo-TEM

To investigate the very early stages of mineralization, cryo-TEM has proven to be a very useful tool, not only for morphological characterization but also to determine the long- or short-range order (electron diffraction), chemical environment (energy-dispersive X-ray spectroscopy, electron energy loss spectroscopy, and scanning transmission electron microscopy), and growth kinetics (fractal analysis on cryo-tomograms) (Dey, de With, & Sommerdijk, 2009; Habraken et al., 2013). The technique is based on sampling a small amount of reaction solution onto a porous TEM grid, which after blotting of excess water and rapid cooling in liquid ethane ($-180\ °C$) is vitrified in a state of amorphous ice. This whole process takes about 20 s (sampling, blotting, and freezing), assuring a good control over the reaction kinetics.

Cryo-TEM allows us to visualize the mineralization reaction to the nanometer scale in which nucleation of a mineral phase really occurs. Cryo-TEM analysis resulted in the first visualization of prenucleation species and gave insight into the nucleation of calcium carbonate, calcium phosphate, and even iron oxides under various conditions (Baumgartner et al., 2013; Dey et al., 2010; Habraken et al., 2013; Nudelman et al., 2010; Pouget et al., 2009). The additional analyses cryo-TEM offers, although sometimes destructive in nature as they require higher electron dosages, give a complete array to structurally and chemically characterize the visualized morphology. However, for an adequate translation to the more quantitative ion potentiometric measurements, cryo-TEM should always be accompanied by earlier described techniques like DLS, Raman, FTIR, and/or diffraction data, which can verify that the processes occurring on the nanoscale represent the bulk of the solution.

Finally, initial investigations using *in situ* TEM equipped with a fluid cell were performed to follow calcium carbonate formation in solution (Nielsen, Lee, Hu, Hanc, & De Yoreo, 2012), thereby possibly overcoming the last drawbacks of cryo-TEM, which are related to the sampling of small amounts of reaction solution. However, integration of *in situ* TEM with conventional bulk analysis seems less straightforward due to the same size restrictions (thickness of investigated area) that are accompanied with any TEM analysis.

3.5. Alternative techniques

Although the techniques described in the previous chapters are enough to fully characterize a growing mineral in solution, additional techniques might be useful depending on the exact reaction conditions and sample extraction procedure.

An *in situ* analysis that has been used for many decades on mineralizing solutions is calorimetry, where by analyzing temperature changes in the reaction solution, one is able to determine the exo- or endothermic nature of a phase transition (Bewernitz, Gebauer, Long, Cölfen, & Gower, 2012; Kibalczyc, Zielenkiewicz, & Zielenkiewicz, 1988).

Solution NMR (H, C, or P) has been used to measure and chemically identify solution carbonate or phosphate ions (Bewernitz et al., 2012; Yesinowski & Benedict, 1983) or a stable colloidal suspension of apatite particles (Yesinowski, 1981); however, until recently (Bewernitz et al., 2012), attempts to measure prenucleation species with NMR (complexes and clusters) were less successful due to the low amount of material present and, in the case of species like ACC or ACP, a rapid sedimentation and possible line broadening.

Cryo-scanning electron microscopy (cryo-SEM) combined with freeze-fracture is a strong tool in characterizing biological mineral specimens (Mahamid et al., 2010); however, the low concentrations of mineral present inside the reaction solution (especially in the early stages) make this analysis very time-consuming. In the case of a mineral precipitate, cryo-SEM would work much better, as unlike cryo-TEM, it is able to characterize the large precipitate particles. In such a case, one could also think of scooping up the material on a flat surface and, by keeping the material moist, measuring it using environmental SEM or atomic force microscopy. However, reaction kinetics for such experiments are difficult to control due to changes in reaction environment with the original reaction solution, mainly caused by evaporation and/or increased interaction with CO_2 from the air.

Of course, when particles grow to micrometer size, they can be observed with light microscopy, where additional use of crossed polarizers can distinguish between amorphous and crystalline samples (Schenk et al., 2012), at least when the crystals are not isometric like NaCl.

In the case of a large precipitate that immediately settles when stirring of the reaction solution is stopped, one could think of analyzing the supernatant by atomic adsorption spectrometry or inductively coupled plasma–optical emission spectrometry to chemically identify the total amounts of Ca or P. However, here, one cannot discriminate between ions and nano-sized (solution) species that might still be present. Furthermore, as explained before, collection of low-density intermediates like ACC or ACP by filtration or centrifugation must be avoided in all cases.

4. COMPARISON WITH STANDARD ANALYSES

Not all of the described techniques are readily available in every lab, and some of them (i.e., synchrotron SAXS/WAXS and cryo-TEM) are even scarce on a global scale. As a first trial, analysis of extracted or dried samples therefore might be unavoidable and could even facilitate the way to a more sophisticated analysis. While even with all the artifacts present, these experiments have been able to identify amorphous (intermediate) phases for calcium carbonate and calcium phosphate, each with their own distinct chemical and structural characteristics, irrespective of whether they really resemble the mineral present in the reaction medium. Once a dried sample is obtained though, other difficulties arise as, upon contact with humidity from the air, metastable (amorphous) phases can transform quite rapidly, creating very high local levels of supersaturation. To increase the lifetime of such samples, they preferably should be kept in an inert atmosphere and at a low temperature. Because one can only delay reaction kinetics by doing so, it is anyhow advisable to characterize the dried sample within a limited time window.

5. SUMMARY

To monitor and control the kinetics of a mineralization reaction, ion potentiometric measurements are a strong tool, which also can be used to determine the chemistry of evolving species and solubility of precipitated bulk phases. The technique can be combined with multiple analysis techniques on morphology, chemistry, or structure like DLS, Raman and FTIR spectroscopy, synchrotron SAXS/WAXS, and cryo-TEM that require sampling of the reaction medium without additional sample preparation. Only by knowing and controlling the kinetics, especially in the case of metastable intermediates like ACC or ACP, can one be certain that the mineral analyzed still represents the sample taken at a specific time point.

REFERENCES

Baumgartner, J., Dey, A., Bomans, P. H. H., Le Coadou, C., Fratzl, P., Sommerdijk, N. A. J. M., et al. (2013). Nucleation and growth of magnetite from solution. *Nature Materials, 12*, 310–314.

Bewernitz, M. A., Gebauer, D., Long, J., Cölfen, H., & Gower, L. B. (2012). A metastable liquid precursor phase of calcium carbonate and its interactions with polyaspartate. *Faraday Discussions, 159*, 291–312.

Bots, P., Benning, L. G., Rodriguez-Blanco, J., Roncal-Herrero, T., & Shaw, S. (2012). Mechanistic insights into the crystallization of amorphous calcium carbonate (ACC). *Crystal Growth & Design, 12*, 3806–3814.

Burgot, J. (2012). *Ionic equilibria in analytical chemistry*. New York: Springer.

Chistofferensen, J., Christoffersen, M. R., Kibalczyc, W., & Andersen, F. A. (1989). A contribution to the understanding of the formation of calcium phosphates. *Journal of Crystal Growth, 94*, 767–777.

Chow, L. C., & Sun, L. (2004). Properties of nanostructured hydroxyapatite prepared by a spray drying technique. *Journal of Research of the National Institute of Standards and Technology, 109*(6), 543–551.

Chughtai, A., Marshall, R., & Nancollas, G. H. (1968). Complexes in calcium phosphate solutions. *Journal of Physical Chemistry, 72*(1), 208–211.

Cunningham, K. M., Goldberg, M. C., & Weiner, E. R. (1977). Investigation of detection limits for solutes in water measured by laser Raman spectroscopy. *Analytical Chemistry, 49*(1), 70–75.

Dey, A., Bomans, P. H. H., Muller, F. A., Will, J., Frederik, P. M., de With, G., et al. (2010). The role of prenucleation clusters in surface-induced calcium phosphate crystallization. *Nature Materials, 9*(12), 1010–1014.

Dey, A., de With, G., & Sommerdijk, N. A. J. M. (2009). In situ techniques in biomimetic mineralization studies of calcium carbonate. *Chemical Society Reviews, 39*(2), 397–409.

Févotte, G. (2002). New perspectives for the on-line monitoring of pharmaceutical crystallization processes using in situ infrared spectroscopy. *International Journal of Pharmaceutics, 241*(2), 263–278.

Gebauer, D., Völkel, A., & Cölfen, H. (2008). Stable prenucleation calcium carbonate clusters. *Science, 322*, 1819–1822.

Habraken, W. J. E. M., Tao, J., Brylka, L. J., Friedrich, H., Bertinetti, L., Schenk, A. S., et al. (2013). Ion-association complexes unite classical and non-classical theories for the biomimetic nucleation of calcium phosphate. *Nature Communications*, 4(1507), 1–12.

Haynes, W. M. (2013). *CRC handbook of chemistry and physics, Internet version* (93rd ed.). Boca Raton: CRC Press/Taylor & Francis.

Hunter, R. J. (1994). *Introduction to modern colloid science* (1st ed.). New York: Oxford University Press, pp. 44–45, 207, 217, 238–241.

Kazanci, M., Fratzl, P., Klaushofer, K., & Paschalis, E. P. (2006). Complementary information on in vitro conversion of amorphous (precursor) calcium phosphate to hydroxyapatite from Raman microspectroscopy and wide-angle X-ray scattering. *Calcified Tissue International*, 79, 354–359.

Kibalczyc, W., Zielenkiewicz, A., & Zielenkiewicz, W. (1988). Calorimetric investigations of calcium phosphate precipitation in relation to solution composition and temperature. *Thermochimica Acta*, 131, 47–55.

Kitano, Y. (1962). The behavior of various inorganic ions in the separation of calcium carbonate from a bicarbonate solution. *Bulletin of the Chemical Society of Japan*, 35(12), 1973–1980.

Liu, J., Rieger, J., & Huber, K. (2008). Analysis of the nucleation and growth of amorphous $CaCO_3$ by means of time-resolved static light scattering. *Langmuir*, 24, 8262–8271.

Mahamid, J., Aichmayer, B., Shimoni, E., Ziblat, R., Li, C. H., Siegel, S., et al. (2010). Mapping amorphous calcium phosphate transformation into crystalline mineral from the cell to the bone in zebrafish fin rays. *Proceedings of the National Academy of Sciences of the United States of America*, 107, 6316–6321.

Meyer, J. L., & Eanes, E. D. (1978). A thermodynamic analysis of the amorphous to crystalline calcium phosphate transformation. *Calcified Tissue Research*, 25, 59–68.

Morgan, T. T., Muddana, H. S., Altinoglu, E. I., Rouse, S. M., Tabakovic, A., Tabouillot, T., et al. (2008). Encapsulation of organic molecules in calcium phosphate nanocomposite particles for intracellular imaging and drug delivery. *Nano Letters*, 8(12), 4108–4115.

Nielsen, M. H., Lee, J. R. I., Hu, Q., Hanc, T. Y., & De Yoreo, J. J. (2012). Structural evolution, formation pathways and energetic controls during template-directed nucleation of $CaCO_3$. *Faraday Discussions*, 159, 105–121.

Nudelman, F., Pieterse, K., George, A., Bomans, P. H. H., Friedrich, H., Brylka, L. J., et al. (2010). The role of collagen in bone apatite formation in the presence of hydroxyapatite nucleation inhibitors. *Nature Materials*, 9, 1004–1009.

Onuma, K., & Ito, A. (1998). Cluster growth model for hydroxyapatite. *Chemistry of Materials*, 10(11), 3335–3346.

Politi, Y., Mahamid, J., Goldberg, H., Weiner, S., & Addadi, L. (2007). Asprich mollusk shell protein: In vitro experiments aimed at elucidating function in $CaCO_3$ crystallization. *CrystEngComm*, 9, 1171–1177.

Pouget, E. M., Bomans, P. H. H., Goos, J. A. C. M., Frederik, P. M., de With, G., & Sommerdijk, N. A. J. M. (2009). The initial stages of template-controlled $CaCO_3$ formation revealed by CryoTEM. *Science*, 323(5920), 1455–1458.

Pye, C. C., & Rudolph, W. W. (2003). An ab initio, infrared, and Raman investigation of phosphate ion hydration. *Journal of Physical Chemistry A*, 107, 8746–8755.

Rudolph, W. W., Irmer, G., & Königsberger, E. (2008). Speciation studies in aqueous $HCO_3^--CO_3^{2-}$ solutions. A combined Raman spectroscopic and thermodynamic study. *Dalton Transactions*, 7, 900–908.

Schenk, A. S., Zope, H., Kim, Y., Kros, A., Sommerdijk, N. A. J. M., & Meldrum, F. C. (2012). Polymer-induced liquid precursor (PILP) phases of calcium carbonate formed in the presence of synthetic acidic polypeptides-relevance to biomineralization. *Faraday Discussions*, 159, 327–344.

Shechter, A., Glazer, L., Cheled, S., Mor, E., Weil, S., Berman, A., et al. (2008). A gastrolith protein serving a dual role in the formation of an amorphous mineral containing extracellular matrix. *Proceedings of the National Academy of Sciences of the United States of America, 105*(20), 7129–7134.

Termine, J. D., Peckauskas, R. A., & Posner, A. S. (1970). Calcium phosphate formation in vitro II. Effects of environment on amorphous-crystalline transformation. *Archives of Biochemistry and Biophysics, 140*, 318–325.

Termine, J. D., & Posner, A. S. (1970). Calcium phosphate formation in vitro I. Factors affecting initial phase separation. *Archives of Biochemistry and Biophysics, 140*, 307–317.

Verch, A., Gebauer, D., Antonietti, M., & Cölfen, H. (2011). How to control the scaling of $CaCO_3$: A "fingerprinting technique" to classify additives. *Physical Chemistry Chemical Physics, 13*, 16811–16820.

Yesinowski, J. P. (1981). High-resolution NMR spectroscopy of solids and surface-adsorbed species in colloidal suspension: ^{31}P-NMR spectra of hydroxyapatite and diphosphonates. *Journal of the American Chemical Society, 103*, 6266–6267.

Yesinowski, J. P., & Benedict, J. J. (1983). ^{31}P NMR as a spectroscopic monitor of the spontaneous precipitation of calcium phosphates. *Calcified Tissue International, 35*(3), 284–286.

CHAPTER THREE

Investigating the Early Stages of Mineral Precipitation by Potentiometric Titration and Analytical Ultracentrifugation

Matthias Kellermeier, Helmut Cölfen, Denis Gebauer[1]
Department of Chemistry, Physical Chemistry, University of Konstanz, Konstanz, Germany
[1]Corresponding author: e-mail address: denis.gebauer@uni-konstanz.de

Contents

1. Introduction	46
2. Equipment	48
2.1 Titration setup	48
2.2 Analytical ultracentrifuge	49
3. Experimental Procedures	49
3.1 Calibration of electrodes and sensors	49
3.2 Crystallization assay: Potentiometry, conductivity, and pH titration	51
3.3 Sample preparation for AUC measurements	53
4. Data Evaluation	54
4.1 Treatment of primary data from the titration setup	54
4.2 Advanced analyses: The multiple-binding equilibrium	57
4.3 Cluster detection and characterization in AUC	61
5. Conclusions and Outlook	66
References	67

Abstract

Despite the importance of crystallization for various areas of research, our understanding of the early stages of the mineral precipitation from solution and of the actual mechanism of nucleation is still rather limited. Indeed, detailed insights into the processes underlying nucleation may enable a systematic development of novel strategies for controlling mineralization, which is highly relevant for fields ranging from materials chemistry to medicine. In this work, we describe experimental aspects of a quantitative assay, which relies on pH titrations combined with *in situ* metal ion potentiometry and conductivity measurements. The assay has originally been designed to study the crystallization of calcium carbonate, one of the most abundant biominerals. However, the developed procedures can also be readily applied to any compound containing cations for which ion-selective electrodes are available. Besides the possibility to quantitatively assess ion association prior to nucleation and to directly determine thermodynamic solubility products of precipitated phases, the main advantage of the crystallization assay is the unambiguous identification of the different stages of precipitation (i.e., prenucleation, nucleation, and early postnucleation) and the characterization of the multiple effects of additives. Furthermore, the experiments permit targeted access to distinct precursor species and intermediate stages, which thus can be analyzed by additional methods such as cryo-electron microscopy or analytical ultracentrifugation (AUC). Regarding ion association in solution, AUC detects entities significantly larger than simple ion pairs, so-called prenucleation clusters. Sedimentation coefficient values and distributions obtained for the calcium carbonate system are discussed in light of recent insights into the structural nature of prenucleation clusters.

1. INTRODUCTION

Crystallization phenomena are ubiquitous in daily life and bear fundamental importance for a broad variety of scientific disciplines, ranging from materials chemistry and crystallography over pharmacy and medicine to industrial processing. Still, much effort is devoted to study biomineralization processes (Lowenstam & Weiner, 1989), as the obtained insight may be utilized in bio-inspired approaches to novel functional materials with potential applications in diverse fields (Fratzl & Weiner, 2010; Sommerdijk & Cölfen, 2010; Weiner & Addadi, 2011). In recent years, there is increasing evidence that the classical textbook perspective (Mullin, 2001) on nucleation and growth of minerals fails to explain a number of crystallization phenomena and thus seems to have rather limited relevance, at least in some systems (Banfield, Welch, Zhang, Ebert, & Penn, 2000; Cölfen & Antonietti, 2008; Gebauer & Cölfen, 2011; Li et al., 2012; Meldrum & Cölfen, 2008). It is now widely accepted that mineralization sequences according to Ostwald's rule of stages is a common phenomena; that is, the initially precipitated phase is not necessarily the most stable polymorph under the given conditions and, in particular, that amorphous precipitates are common precursors in the crystallization of a number of minerals (Weiner, Mahamid, Politi, Ma, & Addadi, 2009). In fact, such intermediates appear to play a key role in the directed self-assembly of crystalline structures with complex morphologies, both in biological and bio-inspired environments (Addadi, Raz, & Weiner, 2003; Gower, 2008). Regarding the onset of precipitation, classical theories assume that the first nuclei in solution are formed via stochastic collisions of atoms, ions, or molecules (depending on the type of crystal). Eventually, this leads to the occurrence of a metastable cluster—the critical nucleus—which subsequently grows to a macroscopic crystal via continuous addition of these monomers (De Yoreo & Vekilov, 2003). As opposed to that, so-called nonclassical concepts have been reported, where the species considered to be central to mineral nucleation and growth are significantly larger than the basic atomic or molecular constituents and may range from stable solute clusters to nanoparticle building units (Cölfen & Antonietti, 2008; Gebauer & Cölfen, 2011).

For the study of nucleation phenomena, induction time statistics have often been employed (Izmailov, Myerson, & Arnold, 1999); however, in most cases, this approach cannot shed light on the fundamental processes that underlie nucleation by principle (Davey, Schroeder, & ter Horst, 2013). As pointed out by the latter authors, it is crucial to investigate solute association in order to obtain a novel—molecular—perspective on nucleation. In the 1970s, it was recognized that any change in intensive parameters such as pH can sensibly affect nucleation and growth; therefore, the constant-composition method was established so as to allow for truly quantitative physicochemical analyses (Tomson & Nancollas, 1978). Inspired by this technique, we have designed a crystallization assay in which supersaturation is slowly generated under otherwise constant (or explicitly known and controlled) conditions, while the concentrations of relevant species are continuously monitored. Corresponding measurements have led to the discovery and characterization of so-called prenucleation clusters in solutions of calcium carbonate (Gebauer, Völkel, & Cölfen, 2008) and moreover have shown that amorphous intermediates with distinct short-range structures are generated at different pH levels in the absence of any additives (Gebauer et al., 2010, 2008). By using the developed methodology, it is possible to quantitatively assess ion association in solution, to trace the nucleation event and its kinetics, and to examine the nature of initially formed nanoparticles of amorphous calcium carbonate (ACC) and their subsequent transformation toward more stable (crystalline) polymorphs. Beyond that, the role and effect of certain additives during the distinct stages of precipitation can be evaluated (Gebauer, Cölfen, Verch, & Antonietti, 2009) and, for instance, used to optimize industrial products like scale inhibitors. With respect to fundamental analyses of early precursor phases, the crystallization assay offers the advantage that it clearly indicates, at all times, the stage in which a given sample currently exists (i.e., still prenucleation, near or at nucleation, or already postnucleation). This sets a valuable basis for in-depth studies of these distinct stages, for example, by means of cryo-electron microscopy (Dey, de With, & Sommerdijk, 2010) or analytical ultracentrifugation (AUC) (Planken & Cölfen, 2010), which have turned out to be especially powerful techniques in this context.

Herein, we detail the basic equipment for this crystallization assay, give an outline of experimental procedures and evaluation routines, and finally exemplify how AUC can be used to investigate solute association prior to nucleation. Recent advancements and crucial points in data analysis and interpretation are critically discussed. We hope that this contribution will foster further research into the role of ion association during the early stages of crystallization, which may lead to a better understanding of nucleation mechanisms in general. In the following, we will focus on experiments conducted with calcium carbonate as a model system but note that the presented methodology can be directly transferred to other calcium minerals like phosphates, oxalates, or sulfates as well. Furthermore, with suitable ion-selective electrodes (commercial or self-made), a vast amount of other materials is analytically accessible in an analogous manner.

2. EQUIPMENT
2.1. Titration setup

Our measurements are performed using a computer-controlled titration system supplied by Metrohm (Filderstadt, Germany), with corresponding commercial software (Tiamo™, current version: 2.3). The principal setup consists of a titration instrument (Titrando 905, Metrohm No. 2.905.0020) that controls two dosing devices (Dosino 800, Metrohm No. 2.800.0010), which in turn operate two 807 Dosing Units (Metrohm No. 6.3032.120, one each for addition of $CaCl_2$ and $NaOH$ solution; see succeeding text). The whole assembly is run by an independent power supply, which decouples the electric circuit of the instruments from that of the laboratory. The dosing units are equipped with 2 mL glass cylinders as internal reservoirs, which allow titrant solutions to be dispensed in volume steps down to 200 nL, thereby rendering very fine and precise titrations possible. Different pH electrodes are employed depending on actual needs and requirements, such as fast response times (Metrohm Flat-Membrane Electrode, No. 6.0256.100), robustness at elevated temperatures (Metrohm Unitrode, No. 6.0258.010), small sample volumes (Metrohm Biotrode, No. 6.0224.100), or routine applications (Metrohm Microelectrode, No. 6.0234.100). In order to measure free concentrations of Ca^{2+} during titration, we use calcium ion-selective electrodes (Ca^{2+}-ISE, Metrohm No. 6.0508.110 (half-cell) or 6.0510.100 (combined sensor)) and, if required, tap the potential of the internal Ag/AgCl reference system of the pH electrode.

The setup can be extended to also monitor the conductivity of the studied solutions by a conductometric cell (Metrohm No. 6.0910.120 or 6.0915.130) connected via a corresponding module (Conductivity Module 856, Metrohm No. 2.856.0010), while the turbidity of the samples can be traced by means of an appropriate optical sensor (Optrode, Metrohm No. 6.1115.000; this option will however not be discussed further hereinafter). For quantitative analyses, experiments are carried out in commercial double-walled titration vessels (Metrohm Nos. 6.1418.220 and 6.1414.010), which are fed by oil from an attached thermostat, so that the temperature of the receiver solution (measured by an immersed Pt100 probe) is maintained at a preset value in the range of 10–90 °C (usually 25 °C). The vessels are closed and largely decoupled from the atmosphere, in order to minimize potential artifacts that may arise from in-diffusion of atmospheric CO_2 or evaporation of the solutions, especially at higher temperatures.

2.2. Analytical ultracentrifuge

We use a Beckman-Coulter XL-I ultracentrifuge equipped with Rayleigh interference optics for detecting sedimenting species via spatiotemporal changes in refractive index. Measurements are performed at 25 °C and a rotor speed of 60,000 RPM in self-made 12 mm 2.5° titanium double-sector centerpieces, with at least 8 h duration per experiment. Resulting primary data are processed and evaluated using the SEDFIT program, in which $c(s)$ as well as Lamm equation modeling is performed (Schuck, 2000, 2013).

3. EXPERIMENTAL PROCEDURES

3.1. Calibration of electrodes and sensors

3.1.1 Ca^{2+}-ISE

When sparingly soluble minerals like calcium carbonate are investigated, calibration of the Ca^{2+}-ISE can be conducted in water, because ionic strengths in the actual measurements are rather low (≤ 20 mM) and can be neglected to a good approximation. Indeed, very recent work has quantitatively confirmed earlier qualitative claims (Gebauer & Cölfen, 2011) that contributions originating from ionic activity just slightly exceed the range of typical experimental error and that ideal treatment does not affect as-determined physicochemical parameters (binding constants and free energies, and solubility products) to any significant extent (Kellermeier,

Picker, Kempter, Cölfen, & Gebauer, 2013). Thus, for calibration, 10 mM calcium chloride solution is dosed at a constant rate of 10 µL/min into 50 mL water (Milli-Q quality), which has been adjusted to the pH of the subsequent crystallization experiment by adding aliquots of 10 mM NaOH. During titration, the receiver solution is continuously stirred and showered with a gentle stream of water-saturated nitrogen to avoid CO_2 in-diffusion. Simultaneously, any decrease in pH due to addition of the slightly acidic Ca^{2+} solution is automatically counterbalanced by titration with 10 mM NaOH. The Ca^{2+} potential is recorded every 10 s and, since all added volumes are known at any time, can be related to the actual calcium concentration via a Nernstian approach according to

$$U(Ca^{2+}) = U_0 + \frac{RT}{2F} \ln[c(Ca^{2+})] \quad (3.1)$$

where $U(Ca^{2+})$ is the measured potential, U_0 the electrode intercept, R the gas constant, T the temperature, and F the Faraday constant. In principle, calibration is carried out mainly to determine the electrode intercept U_0, since the slope is given as $m = RT/2F$ in case of Nernstian behavior. However, for data analysis, we always use the values obtained from calibration, in order to account for any possible deviations. Then, the potentials measured in the subsequent crystallization assays can be converted into concentrations by using Eq. (3.1) and the as-obtained values for U_0 and m, under the assumption of ideality (Kellermeier et al., 2013). We normally calibrate Ca^{2+}-selective electrodes at least once per week.

On the other hand, when the analyzed solutions exhibit noticeably higher ionic strengths ($I > \sim 25$ mM)—for example, due to the presence of charged additives or when more soluble minerals are investigated—the error made by ideal treatment (and calibration in water) becomes significant, and the ISE should preferably be calibrated in ionic strength-adjusted environments. This can be realized by titrating $CaCl_2$ into solutions containing suitable amounts of sodium chloride as a background electrolyte, meant to emulate the ionic strength in the samples. In this way, activity effects that arise from Coulomb interactions are accounted for in calibration (they affect primarily the electrode intercept U_0), and resulting U_0^{NaCl} and m values allow for directly translating experimental potentials into actual free Ca^{2+} concentrations (Kellermeier et al., 2013). In turn, when accurate thermodynamic parameters are to be derived from the titration data, these concentrations have to be converted into activities. The required activity coefficients can be either directly obtained from a comparison of ISE calibrations in water and NaCl solutions (Kellermeier et al., 2013) or calculated with the aid of theoretical expressions that are suitable for the given range of ionic strengths (such as extended Debye–Hückel or Davies equations) (Robinson & Stokes, 2002).

3.1.2 pH electrodes
Calibration of the pH electrodes is achieved by means of a three-point routine utilizing commercial buffers from Mettler Toledo (pH 4.00, 7.00, and 9.21). Biweekly calibration has turned out to be sufficient.

3.1.3 Conductivity probes
For accurate conductivity measurements, the cell constant of the employed probe needs to be determined precisely. In general, the transport of electricity in salt solutions is characterized by Ohm's electrical resistance (R_Ω), which can be expressed as the product of the (shape-independent) specific resistance ρ (in Ωm) and the cell constant Z (in m^{-1}), which accounts for the particular geometry of the experimental setup. Specific conductivity (χ, in S/m) and resistance are directly related by $\chi = 1/\rho$. In order to determine the cell constant, R_Ω is measured in a fixed geometrical setting (as defined by the construction of the sensor) for an electrolyte with known conductivity. For the latter purpose, we use potassium chloride standard solutions supplied by Merck (CertiPUR, either 0.01 M (1.41 mS/cm) or 0.1 M (12.88 mS/cm), depending on the estimated range of χ in the samples). In this way, measured resistance values can be converted into normalized specific conductivities, which then may be used to calculate ion concentrations in an alternative approach.

3.2. Crystallization assay: Potentiometry, conductivity, and pH titration

A schematic illustration of the experimental setup is shown in Fig. 3.1A. In the crystallization assays, 10 mM calcium chloride solution is titrated at a rate of 10 µL/min into 50 mL 10 mM sodium (bi)carbonate buffer at a given preset pH, applying the same routine as detailed earlier for the ISE calibration. Calcium potentials, pH, and conductivity values are read from the immersed probes in intervals between 2 and 10 s. Carbonate buffers can be readily prepared by mixing appropriate volumes of 10 mM NaHCO$_3$ and 10 mM Na$_2$CO$_3$, until the desired pH is reached (the overall carbonate concentration thus remains 10 mM). The fraction of CO$_3^{2-}$ ions is very low (<0.05%) at neutral pH values and increases from ca. 4% to 25% as the pH is raised from 9 to 10. Therefore, association of Ca^{2+} and CO$_3^{2-}$ in solution

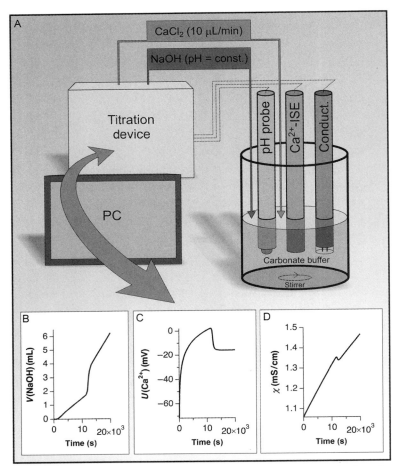

Figure 3.1 Schematic overview of the employed titration setup (A) with plots showing the time developments of typical primary data (B–D, raw data). The titration device is controlled by a computer (PC) and operates two dosing units for the continuous addition of calcium chloride and sodium hydroxide solution (the latter being used for keeping the pH constant by automatic countertitration). Three sensors are connected: a pH probe, a calcium ion-selective electrode (Ca^{2+}-ISE), and a conductivity sensor (Conduct.). During addition of $CaCl_2$, the volume of NaOH necessary to maintain the preset pH value (V(NaOH) in B), the calcium potential (U(Ca^{2+}) in C), and the specific conductivity (χ in D) are recorded. Upon nucleation of $CaCO_3$, all curves respond by a distinct change.

prior to nucleation has been examined in the latter range (Gebauer et al., 2008), while experiments at near-neutral pH have evidenced that binding of HCO_3^- to Ca^{2+} occurs only upon nucleation of a liquid intermediate (and *not* significantly before nucleation) (Bewernitz, Gebauer, Long, Cölfen, & Gower, 2012).

In the course of $CaCl_2$ addition, the pH of the (bi)carbonate solution has to be kept constant by countertitration of 10 mM NaOH, as in the case of the calibrations. However, this effect is much more pronounced in the buffer, because binding of carbonate ions in $CaCO_3$ species (before, during, and after nucleation) leads to the effective removal of a base or, in other words, to a decrease in pH. The volume of sodium hydroxide required to maintain a constant pH level can be used to calculate the absolute amount of carbonate bound at any given stage (Gebauer et al., 2008), as detailed later. We note that NaOH solutions are not titer-stable and hence should be prepared freshly for quantitative analyses. As already indicated earlier, this pH titration is performed automatically by the computer-controlled system. A fundamental prerequisite for accurate pH titrations is that the electrode potential has stabilized before Ca^{2+} addition is started (which typically takes about 1–2 min), so as to avoid overtitrations. Moreover, the set point of the titration needs to be precisely adjusted to the actual pH value of the buffer. Although pH measurements generally have an absolute error in the range of 0.05 units, our setup displays three digits in order to facilitate fine-tuning. Typically, the measured pH of the buffer agrees with the preset value within 0.01–0.03 units. The set point for the titration is then adjusted to ca. 0.005 pH units below the measured current value of the buffer, before initiating the addition of $CaCl_2$. Apart from the set point itself, the success of the automatic countertitrations relies on three further parameters (in Tiamo), namely, the dynamic pH range, the minimum addition rate, and maximum addition rate. For our particular setting, optimal results are obtained for a dynamic range of 0.1 pH units, a minimum rate of 5 μL/min, and a maximum rate of 2 mL/min.

3.3. Sample preparation for AUC measurements

In order to analyze solute association by means of AUC, aliquots are drawn from the crystallization assay at distinct stages prior to nucleation. The nucleation process is concurrently indicated by a steep increase in the volume of NaOH needed to maintain the pH (Fig. 3.1B), a drop in the amount of free Ca^{2+} detected by the ISE (Fig. 3.1C), as well as a kink in the development of the measured conductivity (Fig. 3.1D). Hence, it is possible to directly assess which state is actually analyzed, that is, at which level of under- or supersaturation the samples were in fact drawn. Aliquots of the solutions are immediately filled into the AUC cells and subsequently analyzed in so-called sedimentation velocity experiments (Demeler, 2005), which yield—as a primary result—the sedimentation coefficient (s) distribution for a predefined number of distinct components (see succeeding text). However, also, Diffusion coefficients (*D*) are also accessible from the data by analyzing the time-dependent broadening of sedimenting boundaries. Detailed experimental procedures—concerning, for example, the appropriate filling of cells, calibrations of radial position and angular velocity, or balancing of the rotor—are described in the literature (Mächtle & Börger, 2006; Ralston, 1993).

4. DATA EVALUATION

4.1. Treatment of primary data from the titration setup

First, it is important to note that ion-selective electrodes only detect single and free—that is, noncomplexed and unbound—ions in their hydrated state. Thus, the recorded Ca^{2+} potential (Fig. 3.1C) reflects the concentration (when ideality is assumed; otherwise, it reflects the activity) of free calcium ions, $c_{free}(Ca^{2+})$, which is accessible via Eq. (3.1). In turn, since the dosed amount of calcium, that is, the total concentration of Ca^{2+} present in the system $c_{added}(Ca^{2+})$, is known at all times, the concentration of bound calcium, $c_{bound}(Ca^{2+})$, can readily be calculated according to

$$c_{bound}(Ca^{2+}) = c_{added}(Ca^{2+}) - c_{free}(Ca^{2+}) \qquad (3.2)$$

Second, the volume of NaOH needed to keep the pH constant (Fig. 3.1B) correlates with the amount of base removed from the buffer. With the known buffer composition at the given pH (i.e., the fractions of HCO_3^- and CO_3^{2-} in equilibrium), the amount of added NaOH can be directly converted into absolute numbers of bound carbonate ions (Gebauer et al., 2008). This treatment neglects association between calcium and bicarbonate ions, which, however, is a good approximation at pH levels above ca. pH 8.5 (Kellermeier et al., 2013). Countertitration with NaOH accounts for the removal of diprotic base (CO_3^{2-}) as well as for protons generated by dissociation of bicarbonate ions ($HCO_3^- \rightarrow CO_3^{2-} + H^+$) as the equilibrium carbonate/bicarbonate ratio at the given pH is restored. On that basis, the amount of bound carbonate, $n_{bound}(CO_3^{2-})$, can be calculated from the titrated volume of sodium hydroxide solution, $V(NaOH)$, with its concentration, $c(NaOH)$, and the pH-dependent fractions of carbonate, $\lambda(CO_3^{2-})_{pH}$, and bicarbonate ions, $\lambda(HCO_3^-)_{pH}$, in the buffer by the following equation (Gebauer et al., 2008):

$$n_{bound}(CO_3^{2-}) = c(NaOH) \cdot V(NaOH) \cdot \left[2 \cdot \lambda(CO_3^{2-})_{pH} + \lambda(HCO_3^{-})_{pH}\right]^{-1}$$
(3.3)

For calcium carbonate, such analyses evidence that equal amounts of CO_3^{2-} and Ca^{2+} ions are bound throughout the different stages within experimental accuracy and, hence, that any associated prenucleation species are neutral in charge on average (Gebauer et al., 2008). Similar expressions can be derived for the binding of phosphate species during calcium phosphate precipitation. In turn, for (very) weak bases like sulfate or oxalate, there will be only minor effects on pH upon ion association, and anion binding cannot be quantified via pH titration for corresponding minerals.

Third, the measured conductivity values (Fig. 3.1D) result from the contributions of all cations and anions in the system, which can be modeled using Kohlrausch's law of independent ion migration for the case of infinite dilution (ideal conditions) or a combination of Fuoss–Onsager equations for nonideal behavior at higher ionic strengths (Robinson & Stokes, 2002). Hence, concentrations of free and bound Ca^{2+} and CO_3^{2-} ions can be calculated from the conductivity data and applied to test values obtained from calcium potential measurements and pH titrations. On the one hand, this allows for confirming the validity of certain assumptions made, for example, that calcium bicarbonate association can be neglected at the investigated pH levels. On the other hand, when combined with ISE data, the conductivity measurements provide an alternative means to assess anion binding in situations where pH titrations do not give corresponding information (i.e., for weakly basic anions).

Practically, we use spreadsheet analyses to process primary data sets, which are imported as tables together with calibration values as well as initial concentrations and volumes, in order to derive relevant secondary quantities (such as bound amounts of Ca^{2+} and CO_3^{2-}) at any time during the titration. Commonly, each experiment is performed at least in triplicate, so as to ensure reproducibility and estimate experimental errors simultaneously. Figure 3.2 shows a scheme depicting the typical development of detected free Ca^{2+} during the early stages of $CaCO_3$ precipitation (dashed line), as obtained by evaluation of the primary data displayed in Fig. 3.1C. Initially, the measured amount of free calcium increases linearly until a maximum is reached, which corresponds to a critical stage where phase separation occurs. At this point, solid particles are nucleated and the calcium level rapidly drops to a plateau that reflects the solubility of

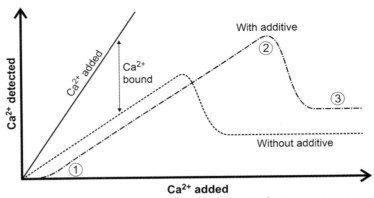

Figure 3.2 Qualitative representation of the amount of Ca^{2+} detected during titration of $CaCl_2$ into dilute carbonate buffer in the absence (dashed line) and presence (dashed dotted line) of an additive. In both cases, significantly less Ca^{2+} is measured than actually added, reflecting the fraction of calcium bound in each of the different stages. The amount of free Ca^{2+} first increases linearly and then drops to a level corresponding to the solubility of the phase formed upon nucleation. From changes in distinct features of the curve, the influence of crystallization additives can be directly identified and characterized. The hypothetical additive depicted in the plot shows the following effects: (1) complexation of calcium ions, (2) inhibition of nucleation, and (3) stabilization of a more soluble phase after nucleation.

the precipitated phase (usually amorphous $CaCO_3$ under the given conditions). In the presence of additives, the different features of the curves can change markedly, allowing the various possible roles of crystallization additives to be quantified and categorized (Gebauer, Cölfen, et al., 2009; Verch, Gebauer, Antonietti, & Cölfen, 2011). Figure 3.2 (dashed dotted line) exemplifies a case where an additive is able to (i) complex Ca^{2+} ions (apparent from a delayed increase in free calcium), (ii) inhibit nucleation (causing a shift of maximum in the curve toward larger amounts of added calcium and, thus, higher supersaturation), and (iii) induce the formation of a less stable solid phase (reflected by an increase in solubility). Indeed, other effects like destabilization of ion associates in the prenucleation regime or growth inhibition of precipitated particles can also be quantitatively characterized by such titrations (both not being explicitly shown in Fig. 3.2).

Finally, the data enable direct calculation of actual free ion products ($IP = c_{\text{free}}(Ca^{2+}) \cdot c_{\text{free}}(CO_3^{2-})$) during any of the distinct stages of precipitation probed by the experiments. In principle, corresponding values result immediately from thorough analyses of primary data, as the calcium concentration is measured by the ion-selective electrode, while the carbonate concentration can be independently obtained from the pH titration and/or conductivity data. However, since 1:1 binding of the ions has been demonstrated for the $CaCO_3$ system (cf. preceding text), it is practically sufficient to measure free concentrations of Ca^{2+} with an ISE and derive those of CO_3^{2-} via straightforward mass balance considerations; that is, subtract the number of bound calcium from the total amount of carbonate present and restore the bicarbonate/carbonate equilibrium under the new conditions according to

$$c_{free}(CO_3^{2-}) = \lambda(CO_3^{2-})_{pH} \cdot [n_{total}(HCO_3^-/CO_3^{2-}) - n_{bound}(Ca^{2+})] \cdot V_{total}^{-1}$$

(3.4)

In this way, the titrations readily give access to thermodynamic solubility products of solid phases detected after nucleation (Kellermeier et al., 2013) and thus allow for the identification and characterization of any occurring metastable intermediates, both in the absence (Gebauer et al., 2010) and presence of additives (Gebauer, Cölfen, et al., 2009; Gebauer, Verch, Börner, & Cölfen, 2009; Picker, Kellermeier, Seto, Gebauer, & Cölfen, 2012; Verch et al., 2011).

4.2. Advanced analyses: The multiple-binding equilibrium

The temporal profiles traced for the concentrations of the relevant ions using titration as well as potentiometric and conductometric experimentation can further be used to derive thermodynamic parameters for ion associates existing in solution prior to nucleation, based on different models. First of all, it is crucial to emphasize that any successful description of experimental results by a certain assumed model cannot prove the validity of this model as such, because any theory always poses distinct basic axioms. A given physico-chemical model that fits actual data should be critically assessed whether it bears the ability to make predictions while admitting that there might always be other models with greater explanatory power. On the other hand, if the chosen model is incompatible with experimental observations, the only safe conclusion to be drawn is that it relies on insufficient (not necessarily erroneous) assumptions.

In the case of calcium carbonate, association in solution has traditionally been discussed in the framework of ion pairing (i.e., $Ca^{2+}_{(aq)} + CO_3^{2-}_{(aq)} \rightarrow [CaCO_3]^0_{(aq)}$), although this concept is still under debate, perhaps owing to the rather large variance in reported equilibrium constants (Gal, Bollinger, Tolosa, & Gache, 1996). One argument being put forward to support the formation of simple ion pairs (rather than larger associated species) relies on the finding that calcium binding profiles show linear behavior prior to nucleation (cf. Fig. 3.2), as outlined, for instance, by Moore and Verine (1981). Thereby, the thermodynamic equilibrium underlying association between n calcium and n carbonate ions is formulated as

$$n\text{Ca}^{2+} + n\text{CO}_3^{2-} \rightleftharpoons [\text{CaCO}_3]_n \qquad (3.5)$$

If there is a large excess of carbonate species present in the system, changes in the concentration of CO_3^{2-} may be neglected during titration (which, in a rough approximation, is true for our experiments), and the association constant (K_A) linked to the equilibrium in Eq. (3.5) can be written as follows:

$$\frac{c([\text{CaCO}_3]_n)}{c_{\text{free}}^n(\text{Ca}^{2+})} = K_A \cdot c_{\text{free}}^n(\text{CO}_3^{2-}) \approx \text{constant} \qquad (3.6)$$

Linear binding profiles at constant CaCl_2 addition rates imply that the ratio of free and bound calcium (the latter being equal to n times the concentration of $[\text{CaCO}_3]_n$) remains constant throughout the experiment (or, respectively, during the entire prenucleation stage). With regard to Eq. (3.6), this is obviously only possible if $n = 1$ (i.e., ion pair formation). Although this speciation model can successfully describe measured data for $n = 1$, it does not allow to infer that ion pairing is the only explanation for the observed binding behavior, because any conclusion drawn on the basis of an applied model is strictly confined to the underlying assumptions—such as, in this case, to represent cluster formation as an elementary reaction between a number of constituent ions in a single step according to Eq. (3.5).

In an alternative approach, we may consider the existence of larger ion associates that assemble (and disintegrate) via multiple steps and a series of coupled binding equilibria. The resulting (much more complex) system of equations can be significantly simplified by assuming that all binding events on the way to larger ion associates are independent and equal (Gebauer et al., 2008). At first sight, this may seem to be a drastic simplification, which, however, is sustained by results of extensive computer simulations (Demichelis, Raiteri, Gale, Quigley, & Gebauer, 2011; Raiteri & Gale, 2010). Under this particular assumption, binding of ions in ion pairs and higher associates cannot be distinguished anymore from a macroscopic point of view, as the equilibrium described by Eq. (3.5) then has to be reformulated according to

$$Ca^{2+} + CO_3^{2-} \rightleftharpoons [CaCO_3]_{Cluster} \tag{3.7}$$

where $[CaCO_3]_{Cluster}$ represents an ion pair within a cluster. It is directly evident that, if the chosen model applies, the formation of clusters containing more than one $CaCO_3$ formula unit (actually any arbitrary number) can also be reconciled with linear calcium binding profiles—the expression obtained by this model for the association constant via the law of mass action is mathematically the same as that for the ion pair model discussed earlier.

In order to assess the thermodynamics of cluster formation, we use a so-called multiple-binding model (Demichelis et al., 2011; Gebauer et al., 2008) that was first introduced for the quantitative analysis of protein–ligand interactions (Scatchard, 1949). It relies on the basic assumption mentioned earlier (equal and independent binding events with identical equilibrium constants K) and formally considers the existence of several binding sites for calcium ions on one carbonate ion, as illustrated by the following reaction scheme:

$$CO_3^{2-} \underset{K}{\overset{+Ca^{2+}}{\rightleftharpoons}} CaCO_3 \underset{K}{\overset{+Ca^{2+}}{\rightleftharpoons}} [Ca_2CO_3]^{2+} \underset{K}{\overset{+Ca^{2+}}{\rightleftharpoons}} \cdots \tag{3.8}$$

It has to be emphasized that multiple coordination of Ca^{2+} around a central CO_3^{2-} in a microscopic perspective does not contradict the macroscopically observed 1:1 binding of both ions in prenucleation associates, because the number of binding sites for calcium on carbonate ions, x, does not reflect the stoichiometry of the clusters (Gebauer et al., 2008) and coordination of carbonate on calcium is not necessarily limited to one CO_3^{2-} per Ca^{2+}, either. Based on the binding scheme indicated by Eq. (3.8), it can be shown that

$$\frac{n_{bound}(Ca^{2+})}{n_{bound}(CO_3^{2-}) + n_{free}(CO_3^{2-})} = x \cdot \frac{K \cdot c_{free}(Ca^{2+})}{1 + K \cdot c_{free}(Ca^{2+})} \tag{3.9}$$

where the unknown microscopic parameters x and K are expressed by a combination of experimentally accessible values, namely, the free and bound amounts of calcium and/or carbonate ions (n_{free} and n_{bound}, respectively) as well as the free calcium concentration (c_{free}). Corresponding fits provide evidence that the measured data are well described by the relation given in Eq. (3.9) and thus demonstrate that linear Ca^{2+} profiles are perfectly compatible with the model of multiple binding and the formation of associates larger than simple ion pairs. Evaluation of fitted data directly yields the number of binding sites for calcium on a carbonate ion (x) and the microscopic equilibrium constant (K). These parameters can then be used to derive the actual calcium carbonate coordination within the clusters, $N_{Calcium}$, via (Demichelis et al., 2011)

$$N_{Calcium} = x \cdot \frac{n_{bound}(Ca^{2+}) + n_{free}(Ca^{2+})}{n_{bound}(Ca^{2+})} \qquad (3.10)$$

as well as their macroscopic thermodynamic stability (ΔG) according to (Gebauer et al., 2008)

$$\Delta G = -RT\ln(x \cdot K) \qquad (3.11)$$

In the same way as the treatment according to Eq. (3.5) cannot fundamentally exclude that association may proceed beyond the ion pair, successful data analysis on the basis of multiple-binding equilibria does not prove the existence of prenucleation clusters as such, either. However, we can now turn to discuss the explanatory power of this model. First, it may be regarded as the simplest approach to theoretically describe the formation of large (≥ 1 nm) ion associates in solution, which have been detected by means of AUC (see succeeding text) and cryo-TEM (Kellermeier et al., 2012; Pouget et al., 2009), while still being in agreement with experimentally observed ion binding profiles. Second, the thermodynamic parameters resulting from the model indicate that the stability of the associated species varies with the pH of the system (Gebauer et al., 2008), which is hard to explain when admitting ion pairs only. In turn, changes in the internal structure of larger clusters may well account for these variations in stability. Furthermore, when considering that nucleation might occur through cluster–cluster aggregation (Kellermeier et al., 2012; Pouget et al., 2009), the model can rationalize why a more stable form of ACC is initially precipitated at lower than at higher pH values (Cartwright, Checa, Gale, Gebauer, & Sainz-Díaz, 2012; Gebauer & Cölfen, 2011; Gebauer et al., 2010). This indicates that the clusters exhibit some kind of encoded proto-structure that is conveyed into the emerging solid phase during nucleation. Last but not least, multiple-binding analyses suggest that, on average, two calcium ions are bound per carbonate, and vice versa, throughout the entire cluster (Demichelis et al., 2011). This mean calcium carbonate coordination number hints at linear chains or rings of polymerized ion pairs (i.e., $\cdots Ca^{2+} - CO_3^{2-} - Ca^{2+} - CO_3^{2-} \cdots$, rather than $[CaCO_3]^0$) and is consistent with dynamic structures observed in computer simulations of aqueous $CaCO_3$ solutions (Demichelis et al., 2011). These so-called dynamically ordered liquid-like oxyanion polymers (DOLLOPs) were proposed to represent the actual structural form of $CaCO_3$ prenucleation clusters.

4.3. Cluster detection and characterization in AUC

Figure 3.3 illustrates the principal setup of the sedimentation velocity experiments used to trace ion clusters in prenucleation $CaCO_3$ solutions. The analytical ultracentrifuge operates at very high rotational speeds that generate centrifugal accelerations in the range of $\sim 280{,}000 \times g$ ($\omega^2 r$, where ω is the

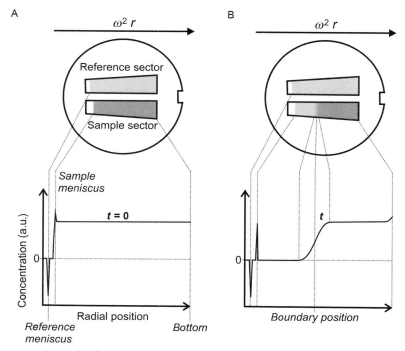

Figure 3.3 Principle of sedimentation velocity experiments in analytical ultracentrifugation. Sample and reference solutions are filled into sector-shaped titanium centerpieces (top) and then become exposed to a centrifugal field $\omega^2 r$ (ω, angular velocity; r, radial position). (A) Before the beginning of the experiment ($t = 0$), the solution to be analyzed is homogeneously distributed in the sample cell, and corresponding concentration profiles (bottom) indicate the position of the sample and reference meniscuses as sharp peaks (which are artifacts induced by the air/liquid interface). (B) After a certain time t, species present in the sample have sedimented to a certain degree, and the position of the sedimenting boundary can be determined from the measured concentration profile (bottom). Diffusion of the sedimenting species leads to progressive broadening of the boundary with time, as indicated.

angular velocity and r the distance from the center of rotation). The measurement cells are sector-shaped (so as to avoid convection caused by collisions of sedimenting species with the wall of the cells) and aligned in the path of analytical optics, which allow for the determination of concentration profiles in the sample cell relative to a reference, typically water (Fig. 3.3A and B, top). Generally, different methods can be employed for monitoring the concentrations of species and their spatiotemporal development, such as UV–vis spectroscopy or interferometry. In case of the nonlight-absorbing sedimenting ions and ion associates discussed here, the only detectable difference with respect to the solvent is the refractive index (n) and corresponding increments, which can be detected by means of a Rayleigh interferometer.

At the beginning of the experiment ($t=0$), the solution to be analyzed is continuously distributed in the sample cell, and the optics merely detect the meniscus of both sample and reference (Fig. 3.3A). After a certain time t, the solutes (or particles) of interest have sedimented due to the applied centrifugal field (Fig. 3.3B). By tracking the position of the sedimenting boundary as a function of time (i.e., experiment duration), the sedimentation velocity (v) can be obtained and used to derive the sedimentation coefficient ($s=v\omega^{-2}r^{-1}$), which is typically given in units of Svedberg (1 S $= 10^{-13}$ s). In practice, data analyses are performed with the SEDFIT software (Schuck, 2000, 2013), where the measured time-dependent sedimentation profiles are evaluated on the basis of the Lamm equation, which describes transport processes occurring due to concurrent sedimentation and diffusion in the experiments (note that diffusion leads to a broadening of the sedimenting boundary, which becomes more and more pronounced with time, cf. Fig. 3.3B).

Because the sedimentation of single ions is very slow and close to the detection limit of the technique, the uncertainty of corresponding s-values is rather high (estimated to ca. 0.3 S) (Gebauer et al., 2008). Moreover, it is crucial to realize that there is a large excess of spectator ions in the buffer (in particular Na^+ and HCO_3^-, as well as the chloride ions introduced with Ca^{2+} during titration), which do not participate in cluster formation. In fact, calcium only makes up a fraction of about 1% of all present ions in typical samples (Gebauer et al., 2008). Depending on the pH, 30–70% of these calcium ions are bound in the prenucleation stage so that, compared to all free ions, the concentration of clusters is certainly low. To address this issue, the SEDFIT evaluation routines have been tested by simulations concerning whether they can give correct sedimentation coefficients for this extreme situation; it was found that the applied procedure can distinguish a fraction of 1 wt% clusters next to an excess of 99% spectator ions (Gebauer et al., 2008), which thus is sufficient to characterize ion association in the studied systems.

Conventionally, AUC data are fitted to the Lamm equation utilizing a model that assumes noninteracting and monodisperse species and yields sedimentation as well as diffusion coefficients for a preset number of components (≤ 4). Each data set is evaluated for one, two, three, and four different species. Then, the best fit is chosen based on the quality of the SEDFIT parameters. When AUC analyses are carried out with samples drawn from the $CaCO_3$ crystallization assay before nucleation, clusters with $s = 1.4 \pm 0.8$ S are found with good statistical significance next to a majority of sedimenting ions ($s = 0.11 \pm 0.05$ S) (Gebauer et al., 2008). The clusters are reliably detected about as soon as the solution becomes supersaturated with respect to the initially precipitated phase, whereas their concentration seems to be too low in the undersaturated regime (we note, however, that ion association occurs in the same manner also below the saturation limit, where the presence of clusters has been confirmed by cryo-TEM (Pouget et al., 2009)). A second, larger cluster species ($s = 5 \pm 1$ S) is observed close to the point of nucleation (yet with lower statistical relevance), while a third and even bigger population ($s \approx 7$–10 S) could occasionally be traced in solutions after nucleation (Gebauer et al., 2008).

In addition to the sedimentation coefficients, the SEDFIT routine also yields corresponding diffusion coefficients (D) for each of the distinct species, which can be derived from the broadening of the sedimenting boundary and converted into hydrodynamic diameters (d_H) utilizing the Stokes–Einstein equation:

$$d_H = \frac{k_B T}{3\pi \eta D} \qquad (3.12)$$

where k_B is the Boltzmann constant, T the absolute temperature, and η the solvent viscosity. Since the sedimenting species are very small, sizes estimated on the basis of the diffusion coefficient should be more reliable than values obtained from sedimentation coefficients. However, while the free ions can be considered monodisperse within experimental accuracy, the clusters may well be polydisperse and, if so, the derived diffusion coefficients would be too high, because broadening of the boundary is in this case not caused by diffusion alone, but contains unknown contributions from the polydispersity of the analyzed species (i.e., its s-distribution). In this regard, diffusion coefficients determined for the clusters by Lamm modeling have to be regarded as upper limits, whereas diameters calculated from D via the Stokes–Einstein relation consequently represent lower limits (cf. Eq. 3.12). For the smallest cluster species (i.e., those reliably detected in supersaturated solutions with $s=1.4$ S), Eq. (3.12) yields a diameter of 0.9 ± 0.2 nm, while 1.8 ± 0.5 and ~ 4 nm are calculated for the larger clusters observed close to and after nucleation, respectively (Gebauer et al., 2008).

Alternatively, the hydrodynamic diameter can also be directly obtained from the measured sedimentation coefficient by assuming spherically shaped clusters and applying the following equation:

$$d_{\mathrm{H}} = \sqrt{\frac{18\eta s}{\rho_{\mathrm{Cluster}} - \rho_{\mathrm{S}}}} \qquad (3.13)$$

where ρ_{Cluster} and ρ_{S} are the densities of the clusters and the solvent, respectively. Here, the major problem is that the density of the clusters is not known; in a first approximation, one may use the value of ACC ($\rho_{\mathrm{ACC}}=1.48$ g/mL (Cölfen & Völkel, 2006)), which is supposed to be the phase that nucleates through aggregation of the clusters. This gives sizes of 2.1 ± 0.6, 3.9 ± 0.4, and ca. 5 nm for the different cluster species discussed earlier (Gebauer et al., 2008). If the density of the clusters is lower than that of ACC (which is well possible as the solute clusters are expected to be more hydrated than nucleated ACC particles), corresponding diameters become larger (cf. Eq. 3.13). Thus, s-based cluster sizes must also be considered as lower limits.

Knowledge of the cluster size and density furthermore allows for an estimation of the number of calcium and carbonate ions combined in single clusters (Gebauer et al., 2008). For example, a spherical cluster of 2 nm in diameter and with the density of ACC would contain about 35 formula units of calcium carbonate. Even though this may be a rough approximation, it is evident that the species detected next to single ions by means of AUC must be significantly larger, on average, than simple ion pairs. In our opinion, a realistic estimate is that typical clusters comprise some tens of ions, plus an as yet unknown amount of hydration water.

While the earlier considerations essentially reflect the AUC evaluation made for $CaCO_3$ clusters in the original work (Gebauer et al., 2008), we can now turn to discuss the results in light of recent insight gained by computer simulations (Demichelis et al., 2011), which suggest that prenucleation

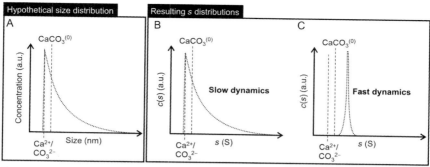

Figure 3.4 Hypothetical distributions for the size (A) and the diffusion-corrected sedimentation coefficient (B, C) of prenucleation clusters in case of slow (B) and fast dynamics (C) with respect to the duration of the experiment. In solution at room temperature, the dynamics of ion association are expected to be orders of magnitude faster than the typical duration of an AUC measurement (thus excluding scenario B). Therefore, AUC data reflect an average response and, depending on the chosen model, the mean values found for s will be significantly larger than those of single ions (C, model of prenucleation clusters). Note that spectator ions are neglected in the scheme and that peak integrals are not to scale.

clusters are in fact highly dynamic, chain-like polymers of alternating Ca^{2+} and CO_3^{2-} ions (DOLLOPs). In general, the size distribution expected for such dynamic polymers is rather broad (Fig. 3.4A), perhaps in analogy to the outcome of classical polycondensation reactions (Flory, 1936). In contrast to that, diffusion-corrected sedimentation coefficient distributions obtained by AUC are very narrow, thus pointing toward fairly monodisperse species at first glance (Fig. 3.4C) (Gebauer et al., 2008). However, it must be taken into account that structural rearrangements of DOLLOPs occur on timescales of molecular processes in solution, that is, within hundreds of picoseconds (Demichelis et al., 2011), and that the clusters decompose and reform at rates that are multiple orders of magnitude faster than the duration of a typical AUC experiment (which takes several hours). Hence, it is impossible to resolve the polydispersity of these species in AUC (as hypothetically done in Fig. 3.4B), owing to their dynamics, and the values measured for s and D therefore reflect average states. In this regard, the apparent narrow size distribution indicated by AUC is perfectly consistent with the speciation envisaged in the DOLLOP concept, which has been proposed to explain the structure of $CaCO_3$ prenucleation clusters. Moreover, since the species involved in the formation of DOLLOPs (i.e., single Ca^{2+} and CO_3^{2-}, ion pairs, and dimers of ion pairs) are all connected

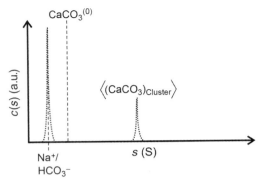

Figure 3.5 Schematic representation of typical s-distributions determined experimentally by means of AUC (integrals not to scale). The peak at low s originates from spectator ions in the buffer (primarily sodium and bicarbonate but also chloride ions), whereas the signal at higher s reflects the average of all associated calcium carbonate species that are interconnected by fast equilibria (i.e., the mean cluster size, as indicated by angle brackets).

by very fast equilibria, the mean size seen by AUC will be an average of all these states, weighted according to their relative abundance in the system. In other words, if there were only single ions and ion pairs, sedimentation and diffusion coefficients would reflect the average of these species and, hence, the determined sizes and s-values would lie in between those of single ions and ion pairs. As this is not the case and the resulting hydrodynamic diameters are distinctly larger (i.e., a scenario as depicted in Fig. 3.5), we conclude that ion association in aqueous $CaCO_3$ solutions proceeds beyond the ion pair, leading to the formation of prenucleation clusters. It further becomes clear that the actual population of single ions detected in the AUC experiments comprises only the spectator ions mentioned earlier (Na^+ and HCO_3^- in first place) and not Ca^{2+} and CO_3^{2-} ions, which participate in the DOLLOP equilibria. The fact that larger clusters can be detected next to smaller clusters close to nucleation may indicate that the latter show different dynamics and are not connected to the smaller DOLLOPs by fast equilibria. This notion is also supported by computer simulations, which evidence a distinct change of DOLLOP dynamics above a certain critical size (Cartwright et al., 2012).

5. CONCLUSIONS AND OUTLOOK

We have described a crystallization assay that provides a straightforward means to assess ion association before, during, and after the nucleation of minerals in a quantitative manner. It further allows for the identification of different precursor and intermediate stages, both in the absence and presence of additives, and thus renders them analytically accessible. Though not explicitly addressed here, the collected data can also be directly used to characterize precipitation kinetics and, when studies are carried out at distinct temperatures, permit even deeper insights into the thermodynamics of ion binding and nucleation as such. By combining the crystallization assay with a high-resolution technique like AUC, ion associates existing in solution prior to nucleation can be investigated in detail. For calcium carbonate, this methodology shows that ions assemble into clusters that are significantly larger than ion pairs, while size distributions obtained from the AUC measurements agree with the dynamic structural form (DOLLOP) suggested for these prenucleation clusters on the basis of computer simulations. AUC is an absolute technique that, in our opinion, will turn out to be valuable also for analyses of interactions between crystallization additives and prenucleation species.

Experiments as those discussed in this work are crucial to gain a more profound understanding of the early stages of mineral precipitation, which is urgently needed and will have fundamental implications for diverse applications across various fields of research. Quantitative crystallization assays may contribute to achieve this by tracing distinct phases and intermediates on the way to final stable phases, which then can be studied by a range of analytical techniques including not only those introduced here but also others such as light scattering or *in situ* IR and Raman spectroscopy.

REFERENCES

Addadi, L., Raz, S., & Weiner, S. (2003). Taking advantage of disorder: Amorphous calcium carbonate and its roles in biomineralization. *Advanced Materials*, *15*, 959–970.

Banfield, J. F., Welch, S. A., Zhang, H. Z., Ebert, T. T., & Penn, R. L. (2000). Aggregation-based crystal growth and microstructure development in natural iron oxyhydroxide biomineralization products. *Science*, *289*, 751–754.

Bewernitz, M. A., Gebauer, D., Long, J. R., Cölfen, H., & Gower, L. B. (2012). A metastable liquid precursor phase of calcium carbonate and its interactions with polyaspartate. *Faraday Discussions*, *159*, 291–312.

Cartwright, J. H. E., Checa, A. G., Gale, J. D., Gebauer, D., & Sainz-Díaz, C. I. (2012). Calcium carbonate polymorphism and its role in biomineralization: How many amorphous calcium carbonates are there? *Angewandte Chemie, International Edition*, *51*, 11960–11970.

Cölfen, H., & Antonietti, M. (2008). *Mesocrystals and nonclassical crystallization*. Chichester: John Wiley & Sons, Ltd.

Cölfen, H., & Völkel, A. (2006). Application of the density variation method on calcium carbonate nanoparticles. *Progress in Colloid and Polymer Science*, *131*, 126–128.

Davey, R. J., Schroeder, S. L. M., & ter Horst, J. H. (2013). Nucleation of organic crystals—A molecular perspective. *Angewandte Chemie, International Edition*, *52*, 2166–2179.

Demeler, B. (2005). UltraScan—A comprehensive data analysis package for analytical ultracentrifugation experiments. In *Modern analytical ultracentrifugation: Techniques and methods* (pp. 210–229). Cambridge, UK: The Royal Society of Chemistry.

Demichelis, R., Raiteri, P., Gale, J. D., Quigley, D., & Gebauer, D. (2011). Stable prenucleation mineral clusters are liquid-like ionic polymers. *Nature Communications*, *2*, 590.
Dey, A., de With, G., & Sommerdijk, N. A. J. M. (2010). In situ techniques in biomimetic mineralization studies of calcium carbonate. *Chemical Society Reviews*, *39*, 397.
De Yoreo, J. J., & Vekilov, P. G. (2003). Principles of crystal nucleation and growth. *Reviews in Mineralogy and Geochemistry*, *54*, 57–93.
Flory, P. J. (1936). Molecular size distribution in linear condensation polymers. *Journal of the American Chemical Society*, *58*, 1877–1885.
Fratzl, P., & Weiner, S. (2010). Bio-inspired materials—Mining the old literature for new ideas. *Advanced Materials*, *22*, 4547–4550.
Gal, J.-Y., Bollinger, J.-C., Tolosa, H., & Gache, N. (1996). Calcium carbonate solubility: A reappraisal of scale formation and inhibition. *Talanta*, *43*, 1497–1509.
Gebauer, D., & Cölfen, H. (2011). Prenucleation clusters and non-classical nucleation. *Nano Today*, *6*, 564–584.
Gebauer, D., Cölfen, H., Verch, A., & Antonietti, M. (2009). The multiple roles of additives in $CaCO_3$ crystallization: A quantitative case study. *Advanced Materials*, *21*, 435–439.
Gebauer, D., Gunawidjaja, P. N., Ko, J. Y. P., Bacsik, Z., Aziz, B., Liu, L. J., et al. (2010). Proto-calcite and proto-vaterite in amorphous calcium carbonates. *Angewandte Chemie, International Edition*, *49*, 8889–8891.
Gebauer, D., Verch, A., Börner, H. G., & Cölfen, H. (2009). Influence of selected artificial peptides on calcium carbonate precipitation—A quantitative study. *Crystal Growth & Design*, *9*, 2398–2403.
Gebauer, D., Völkel, A., & Cölfen, H. (2008). Stable prenucleation calcium carbonate clusters. *Science*, *322*, 1819–1822.
Gower, L. B. (2008). Biomimetic model systems for investigating the amorphous precursor pathway and its role in biomineralization. *Chemical Reviews*, *108*, 4551–4627.
Izmailov, A. F., Myerson, A. S., & Arnold, S. (1999). A statistical understanding of nucleation. *Journal of Crystal Growth*, *196*, 234–242.
Kellermeier, M., Gebauer, D., Melero-García, E., Drechsler, M., Talmon, Y., Kienle, L., et al. (2012). Colloidal stabilization of calcium carbonate prenucleation clusters with silica. *Advanced Functional Materials*, *22*, 4301–4311.
Kellermeier, M., Picker, A., Kempter, A., Cölfen, H., & Gebauer, D. (2013). A straightforward treatment of activity in aqueous $CaCO_3$ and its consequences for nucleation theory (submitted for publication).
Li, D., Nielsen, M. H., Lee, J. R. I., Frandsen, C., Banfield, J. F., & De Yoreo, J. J. (2012). Direction-specific interactions control crystal growth by oriented attachment. *Science*, *336*, 1014–1018.
Lowenstam, H., & Weiner, S. (1989). *On biomineralization*. New York: Oxford University Press.
Mächtle, W., & Börger, L. (2006). *Analytical ultracentrifugation of polymers and nanoparticles*. Berlin/New York: Springer.
Meldrum, F. C., & Cölfen, H. (2008). Controlling mineral morphologies and structures in biological and synthetic systems. *Chemical Reviews*, *108*, 4332–4432.
Moore, E. W., & Verine, H. J. (1981). Pancreatic calcification: Formation constants of $CaHCO_3^+$ and $CaCO_3^{(0)}$ complexes determined with Ca^{2+} electrode. *American Journal of Physiology—Gastrointestinal and Liver Physiology*, *241*, G182–G190.
Mullin, J. (2001). *Crystallization* (4th ed.). Oxford: Butterworth–Heinemann.
Picker, A., Kellermeier, M., Seto, J., Gebauer, D., & Cölfen, H. (2012). The multiple effects of amino acids on the early stages of calcium carbonate crystallization. *Zeitschrift für Kristallographie—Crystalline Materials*, *227*, 744–757.

Planken, K. L., & Cölfen, H. (2010). Analytical ultracentrifugation of colloids. *Nanoscale*, *2*, 1849–1869.

Pouget, E. M., Bomans, P. H. H., Goos, J. A. C. M., Frederik, P. M., de With, G., & Sommerdijk, N. A. J. M. (2009). The initial stages of template-controlled $CaCO_3$ formation revealed by cryo-TEM. *Science*, *323*, 1455–1458.

Raiteri, P., & Gale, J. D. (2010). Water is the key to nonclassical nucleation of amorphous calcium carbonate. *Journal of the American Chemical Society*, *132*, 17623–17634.

Ralston, G. (1993). *Introduction to analytical ultracentrifugation*. Fullerton, CA: Beckman Instruments.

Robinson, R. A., & Stokes, R. H. (2002). *Electrolyte solutions* (2nd rev. ed.). Mineola, NY: Dover Publications.

Scatchard, G. (1949). The attractions of proteins for small molecules and ions. *Annals of the New York Academy of Sciences*, *51*, 660–672.

Schuck, P. (2000). Size-distribution analysis of macromolecules by sedimentation velocity ultracentrifugation and Lamm equation modeling. *Biophysical Journal*, *78*, 1606–1619.

Schuck, P. (2013). *SEDFIT*. http://www.analyticalultracentrifugation.com.

Sommerdijk, N. A. J. M., & Cölfen, H. (2010). Lessons from nature—Biomimetic approaches to minerals with complex structures. *MRS Bulletin*, *35*, 116–121.

Tomson, M. B., & Nancollas, G. H. (1978). Mineralization kinetics: A constant composition approach. *Science*, *200*, 1059–1060.

Verch, A., Gebauer, D., Antonietti, M., & Cölfen, H. (2011). How to control the scaling of $CaCO_3$: A "fingerprinting technique" to classify additives. *Physical Chemistry Chemical Physics*, *13*, 16811–16820.

Weiner, S., & Addadi, L. (2011). Crystallization pathways in biomineralization. *Annual Review of Materials Research*, *41*, 21–40.

Weiner, S., Mahamid, J., Politi, Y., Ma, Y., & Addadi, L. (2009). Overview of the amorphous precursor phase strategy in biomineralization. *Frontiers of Materials Science in China*, *3*, 104–108.

CHAPTER FOUR

Replica Exchange Methods in Biomineral Simulations

Adam F. Wallace[1]
Department of Geological Sciences, University of Delaware, Newark, Delaware, USA
[1]Corresponding author: e-mail address: afw@udel.edu

Contents

1. Introduction — 72
2. Monte Carlo Sampling — 73
    2.1 The Metropolis method — 73
    2.2 Parallel tempering — 74
3. Replica Exchange MD — 77
    3.1 Choice of replica temperatures — 78
    3.2 Optimal exchange frequency — 79
    3.3 Overcoming system size limitations — 79
4. Application of REMD to Early Stage Mineralization — 83
    4.1 Modified Kawska–Zahn method of cluster growth — 83
    4.2 Aggregation-based model of amorphous $CaCO_3$ — 85
    4.3 Thermodynamic properties — 87
5. Summary — 89
References — 89

Abstract

Replica exchange methods are widely used for the purposes of accelerating the conformational sampling of small organic molecules and biopolymers and, to a lesser extent, to explore structural transformations in small Lennard-Jones clusters. Though the general ability of such approaches to enhance the sampling efficiency of both inorganic and organic systems makes replica exchange methods ideal candidates for simulations of biominerals and biomineralization, inherent limitations have largely restricted their applicability to small system sizes and/or short timescales, and their potential in this area has not yet been thoroughly explored. This chapter provides an introduction to the standard replica exchange molecular dynamics method and presents more advanced algorithmic variants, which are designed to improve the efficiency of the replica exchange procedure in large solvent-dominated systems.

1. INTRODUCTION

Atomistic simulations are potentially powerful tools for investigating the materials properties of biominerals and the processes by which they form. However, due to time- and length-scale limitations inherent to such methods, many aspects of the biomineralization process are either at the limit or beyond the reach of standard simulation techniques, and overlap with experiment is difficult to obtain. Electronic structure methods are often too computationally intensive to be tractable for low-frequency events that involve the collective rearrangement of many particles; such processes must be treated with methods that use simplified representations of the particle–particle interactions to reduce computational overhead and extend the simulation-accessible timescale. Even classical molecular dynamics (MD), which is the method of choice for studying the chemical evolution of complex systems, probes relatively short periods of time (typically up to ~ 100 ns in reasonably sized explicitly solvated systems), such that rare events like nucleation and protein folding generally remain significant computational challenges.

Although the accessibility and speed of high-performance computing resources continues to grow, facilitating ever larger and longer simulations, it is unlikely that hardware advances alone can bridge the gap between simulation and experiment. Therefore, complimentary development of algorithmic solutions that improve system sampling by making more efficient use of existing computational resources is also necessary. Here, we present a primer on one such class of algorithms, replica exchange methods, which are widely used to accelerate the exploration of phase space. Despite the many benefits of the replica exchange procedure, the efficiency of the standard method decreases significantly as the number of particles in the system increases, which places an upper limit on the system size for which its use is practical. Consequently, recent developments are focused on approaches that expand the applicability of the replica exchange methodology to larger systems, particularly those that are solvent-dominated. These emergent approaches are discussed in the following sections within the context of studies designed to enhance the conformational sampling of biopolymers and hydrated calcium carbonate cluster phases in aqueous solution.

2. MONTE CARLO SAMPLING

2.1. The Metropolis method

Monte Carlo sampling is the underlying engine that drives the replica exchange process. These methods comprise a large class of numerical algorithms that generate stochastic trajectories within a given statistical mechanical ensemble. The expectation value of a given property, A, which is a function of the positions, r^N, of all N particles in the system, is then taken as the mean value or ensemble average, $\langle A \rangle$, of that property computed at each of the discrete microstates sampled by the trajectory:

$$\langle A \rangle \approx \frac{1}{N_{\text{MC}}} \sum_{i=1}^{N_{\text{MC}}} A(r_i^N) \qquad (4.1)$$

Typical choices include the canonical (NVT) and isothermal–isobaric ensembles (NPT) (N=number of particles, V=volume, P=pressure, and T=temperature) though there are many other possibilities. Regardless, any viable Monte Carlo procedure must sample the Boltzmann distribution and uphold the principle of detailed balance (microscopic reversibility). This is accomplished by accepting Monte Carlo moves subject to a conditional probability rule. The most commonly applied procedure is that initially proposed by Metropolis (Metropolis, Rosenbluth, Rosenbluth, Teller, & Teller, 1953):

$$P(i \rightarrow j) = \min \begin{cases} 1, & \text{for } \Delta \leq 0 \\ e^{-\Delta}, & \text{for } \Delta > 0 \end{cases} \qquad (4.2)$$

where $P(i \rightarrow j)$ is the probability of accepting a move from the current system configuration, i, to a trial configuration, j, which differ from one another only by the displacement of a randomly selected particle. In the canonical ensemble, $\Delta = \beta(E_j - E_i)$, where E_i and E_j are the potential energies of configurations i and j, respectively, and $\beta = (k_B T)^{-1}$. In the event that the trial configuration energy is less than or equal to that of the current configuration, $\Delta \leq 0$, and the trial move is always accepted. However, if the trial configuration energy is higher than E_i, the move is accepted only if a random number generated from a uniform distribution on the interval [0,1] is less than the value of $e^{-\Delta}$. Provided that the sampling procedure is *ergodic* (i.e. every point in configuration space is accessible from any other point), Monte Carlo sampling can be coupled with statistical mechanics to obtain the equilibrium thermodynamic properties of a system.

2.2. Parallel tempering

Monte Carlo methods are advantageous in the sense that they generate random configurations with appropriate statistical weights; they are also highly adaptable and readily extendable to various ensembles and novel types of moves. The trajectories are not required to satisfy any equations of motion, and therefore, nonphysical jumps between microstates are permissible, which can greatly enhance sampling efficiency. Even so, for many problems of interest, basic sampling approaches are not adequate. This is typically the case when energy minima are separated by significant barriers. In such situations, the simulation may become trapped in its basin of origin such that it is only able to explore a subregion of the energy landscape. There are now many methods that are designed to enable barrier-crossing events. Many of these are biased non-Boltzmann type sampling procedures such as umbrella sampling (Torrie & Valleau, 1974) and metadynamics (Ensing, De Vivo, Liu, Moore, & Klein, 2006; Iannuzzi, Laio, & Parrinello, 2003; Laio & Gervasio, 2008; Laio & Parrinello, 2002; Laio, Rodriguez-Fortea, Gervasio, Ceccarelli, & Parrinello, 2005; Quigley & Rodger, 2009). While such approaches have their advantages, they typically require the user to specify certain degrees of freedom, or collective variables, to which the bias is applied, and the choice of the collective variables may influence how the system evolves between states.

The parallel tempering Monte Carlo sampling scheme (Earl & Deem, 2005; Freeman, 2000; Frenkel & Smit, 2002; Geyer & Thompson, 1995; Swendsen & Wang, 1987) is a fundamentally different approach whereby exploration of the energy landscape is enhanced, in principle, without introducing a bias to the system. In standard parallel tempering, several non-interacting copies (or replicas) of the system are all initiated at once and run in parallel, with each replica in the series equilibrated at progressively higher-temperature conditions or with different Hamiltonians, such that there is overlap in the potential energy distributions of the neighboring replicas. Periodically during the simulation, the potential energies of the various replicas are compared; when a given replica encounters a state that is lower in energy than its neighbor, the two systems are allowed to swap states according to the Metropolis acceptance criterion specified in the preceding text (Eq. 4.2), where

$$\Delta = \beta_i[U_i(x_j) - U_i(x_i)] + \beta_j[U_j(x_i) - U_j(x_j)] \quad (4.3)$$

Here, β_i represents the thermodynamic temperature of replica i and $U_i(x_j)$ is the value of the potential energy function of replica i operating on the current configuration, x_j, of replica j. For variants of parallel tempering that enhance the sampling of the system by differentiating between replicas on the basis of their respective Hamiltonians rather than their temperatures, it is typical for all replicas to be thermostated at the same value (i.e., $\beta_i = \beta_j$) and Δ becomes

$$\beta[U_i(x_j) - U_i(x_i) + U_j(x_i) - U_j(x_j)] \quad (4.4)$$

However, for conventional parallel tempering, the potential energies of all the replicas are evaluated with the same Hamiltonian (i.e., $U_i = U_j$), and Δ takes the following form:

$$\Delta = (\beta_i - \beta_j)[U(x_j) - U(x_i)] \quad (4.5)$$

With the convention that $i < j$, if the potential energy of configuration x_j is less than or equal to that of configuration x_i, the swap is always accepted. However, in the event that $U(x_j) > U(x_i)$, the swap may still be accepted provided that $e^{-\Delta}$ is greater than a random number generated on the interval [0, 1]. This procedure ensures that the simulation does not get trapped in a local energy minimum and that the lowest energy configurations are continually promoted towards lower temperature replicas as they are encountered. In practice, swaps are usually only attempted between neighboring replicas because the probability of accepting a swap between nonadjacent pairs diminishes rapidly.

For a simple example of parallel tempering and how it enhances the exploration of energy landscapes, refer to Fig. 4.1. In Fig. 4.1A, a one-dimensional energy surface, $U(x)$, is presented that displays several degenerate local energy minima that are separated by barriers that increase in magnitude as x tends towards higher values. First, consider the behavior of a particle operating on $U(x)$ at several increasing temperatures from T_1 to T_5. Also, let the initial position of the particle at each temperature be the minimum centered at $x = -1.25$. If an ordinary Monte Carlo simulation is performed in the canonical ensemble with $T = T_1$, the particle is unable to overcome many of the barriers on the landscape, and according to the probability distribution, $P(x)$, shown in Fig. 4.1B, resides only in the two leftmost wells throughout the run. However, if $T = T_5$, the particle easily traverses

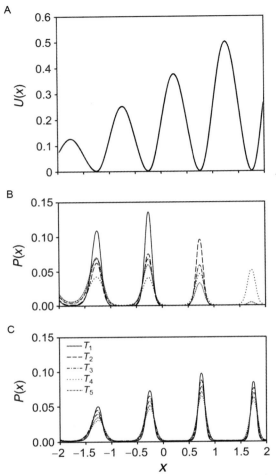

Figure 4.1 Parallel tempering/replica exchange sampling of a single particle on a one-dimensional energy landscape, $U(x)$, as depicted in (A). The value of the potential goes to infinity at the plot boundaries so that the system is confined to the region shown. (B) Probability distributions compiled for five standard Monte Carlo sampling runs with temperatures (T_1 through T_5) that all originate in the potential well centered at $x = -1.25$. (C) Probability distributions compiled from analogous Monte Carlo runs with replica exchange. In contrast to the standard runs, all the trajectories sample each of the potential wells.

the full potential. The parallel tempering method harnesses the ability of the higher-temperature simulations to easily traverse barriers to overcome the sampling deficiency that exists at lower temperatures. Figure 4.1C shows the corresponding probability distributions that are obtained at each temperature from parallel tempering. In contrast to those obtained from standard sampling, the distributions show that all the potential wells are sampled at every temperature. Moreover, the probability of finding the particle in any one of the energy wells is approximately the same because all the wells are at the same energy level, and since parallel tempering ensures proper sampling of the Boltzmann distribution, the system properties determined from ensemble averages taken at each temperature are also valid. If the sampling is thorough enough, it is even possible to obtain the energy landscape directly from any one of the finite temperature probability distributions (i.e., $A(x) = -\beta^{-1} \ln P(x)$), particularly the high-temperature ones; however, at lower temperatures, $P(x)$ is more sparse in the high-energy regions of the potential because the particle avoids the barriers entirely by hopping between the local minima.

3. REPLICA EXCHANGE MD

MD simulations are generally preferable to Monte Carlo methods in complex condensed matter systems because the high probability of particle overlaps results in diminishing acceptance probabilities. Moreover, in MD, the positions and momenta of each particle are updated at each timestep. However, because particles in MD simulations have both kinetic and potential energy, whereas particles in Monte Carlo have only potential energy, the MD-based version of the parallel tempering algorithm is a bit more challenging to implement. Sugita and Okamoto (1999) showed in their derivation of the conditional exchange probability for replica exchange molecular dynamics (REMD) (same as Eq. 4.5) that the kinetic energy components of the replica Hamiltonians cancel, however, in order to ensure that the average kinetic energy, $\langle U(p^N) \rangle_T$, of each temperature replica remains constant:

$$\langle U(p^N) \rangle_T = \frac{3}{2} \beta^{-1} N \qquad (4.6)$$

it is still necessary to rescale the particle momenta when a configuration swap is accepted. Conventionally, the individual particle momenta are rescaled as

$$p^{[i]'} = p^{[i]} \sqrt{\frac{T_n}{T_m}} \quad p^{[j]'} = p^{[j]} \sqrt{\frac{T_m}{T_n}} \qquad (4.7)$$

where $p^{[i]}$ denotes the momentum of a particle in replica i at temperature m and $p^{[i]'}$ is the scaled momentum of that particle after the configuration is swapped with replica j at temperature n. Alternatively, all the particle momenta in a given configuration can also be randomly reassigned at the new temperature; however, there is some indication that the sampling efficiency may be negatively impacted by such a procedure (Cooke & Schmidler, 2008; Rosta & Hummer, 2009; Sindhikara, Emerson, & Roitberg, 2010). Wang, Zhu, Li, and Hansmann (2011) recently suggested a slightly modified procedure for scaling the velocities that may be more suitable for the use of hybrid treatment of solvation effects during exchange steps (see Section 3.3).

3.1. Choice of replica temperatures

In order for REMD to be effective, replicas must swap frequently enough that the individual trajectories can diffuse over the entire temperature distribution. The acceptance probability associated with exchanging replicas depends on the temperature interval between the adjacent replicas and the width of the potential energy distributions at each temperature (i.e., on the order of the typical energy fluctuation in the system). If the replicas are separated by too large a temperature interval such that the potential energy distributions do not overlap, the acceptance probability drops to zero and there is no advantage in running parallel simulations. Conversely, if the replica temperatures are spaced too closely such that there is substantial overlap in the energy distributions of the neighboring replicas, swaps are accepted at a high rate, but the efficiency of the REMD approach decreases as resources must be allocated to simulate more replicas than necessary. Ideally, as few replicas are chosen as is necessary to span the temperature distribution, the total number of replicas required is proportional to $1/\Delta T$ and increases as $\sqrt{N}$ (Earl & Deem, 2005; Kone & Kofke, 2005; Periole & Mark, 2007; Sugita & Okamoto, 1999). The temperature range, ΔT, should be specified so that the energy barrier opposing the process of interest can be surmounted at the highest temperature (Sindhikara, 2010). Typically, temperatures are chosen from an exponential distribution, $T_i = T_0 e^{ki}$ (with k and i as adjustable parameters), which helps ensure that the exchange probability is roughly consistent between all replica pairs; however, some amount of adjustment is generally required (Nymeyer, Gnanakaran, & García, 2004; Sugita & Okamoto, 1999). Moreover, when performing REMD simulations, it is important to use a canonical thermostat (i.e., Langevin, Andersen, or Nosé–Hoover) as noncanonical thermostats have been shown to generate incorrect results (Sindhikara, 2010).

3.2. Optimal exchange frequency

The optimal exchange attempt frequency continues to be a source of debate in the REMD community. The conventional view is that exchanges should be attempted as frequently as possible provided that the system has adequate time to relax, as measured by the potential energy autocorrelation time, before subsequent exchanges are attempted (Abraham & Gready, 2008). If swap attempts are too frequent, then the replicas may remain highly correlated. Periole and Mark (2007) tested various REMD methods and exchange times on the equilibration of β-heptapeptide in explicit solvent. They concluded that equilibration was most efficient when the exchange frequency was just small enough to avoid correlation effects but that even smaller attempt rates provided no additional advantages. Abraham and Gready (2008) likewise concluded that the exchange frequency should be no faster than once every picosecond using a typical timestep of 1–2 fs. However, more recent work (Sindhikara, 2010; Sindhikara, Meng, & Roitberg, 2008) suggests that exchanges should be attempted as frequently as possible (every 10–100 integration steps) provided that the exchanges are done "properly," meaning that velocity rescaling is performed as discussed earlier and a canonical thermostat is used.

3.3. Overcoming system size limitations

Unfortunately, the total value of the Hamiltonian increases with the number of degrees of freedom in the system such that obtaining reasonable exchange probabilities requires a large number of closely spaced replicas. Naturally, this places a practical limit on the size of systems where the application of temperature REMD is a tractable solution. This is especially problematic for solvated systems because the solvent–solvent interactions dominate the potential energy. Due to the system size limitations described in the previous section, most REMD simulations in the literature are focused on coarse-grained systems with reduced degrees of freedom and small organic molecules, peptides, and proteins in the presence of implicit solvation fields and/or a small number of explicit solvent molecules (Chaudhury, Olson, Tawa, Wallqvist, & Lee, 2012; Chebaro, Dong, Laghaei, Derreumaux, & Mousseau, 2009; Jiang & Roux, 2010; Nymeyer et al., 2004; Okur et al., 2006; Dashti, Meng, & Roitberg, 2012; Sugita & Okamoto, 1999; Zhang, Wu, & Duan, 2005).

However, there are also recent efforts aimed at extending the applicability of replica exchange methods to larger explicitly solvated systems. One such approach is replica exchange with solute tempering (REST) (Liu, Kim, Friesner, & Berne, 2005; Terakawa, Kameda, & Takada, 2010). In REST, temperature exchanges are performed using a temperature-scaled potential energy function that is designed to nullify the contribution of the solvent–solvent interactions in the evaluation of the exchange probability. The potential energy of a given configuration is decomposed into contributions from the solute (S), solute–water (SW), and water–water (WW) interactions:

$$U_0(x) = U_S(x) + U_{SW}(x) + U_{WW}(x) \qquad (4.8)$$

As the temperature of the ith replica increases away from the target temperature (β_0), the potential energy is scaled as follows:

$$U_i(x) = U_S(x) + \left[\frac{\beta_0 + \beta_i}{2\beta_i}\right] U_{SW}(x) + \left[\frac{\beta_0}{\beta_i}\right] U_{WW}(x) \qquad (4.9)$$

such that the solvent–solvent interaction term cancels out of the resulting expression for the exchange probability (obtained by substituting Eq. 4.9 into Eq. 4.3):

$$\Delta = \left(\beta_i - \beta_j\right)\left[\left(U_S(x_j) + \frac{1}{2}U_{SW}(x_j)\right) - \left(U_S(x_i) + \frac{1}{2}U_{SW}(x_i)\right)\right] \qquad (4.10)$$

The consequence of eliminating $U_{WW}(x)$ from the previously mentioned expression is that solvent–solvent interactions that traditionally cause poor scaling with respect to system size no longer contribute to declining acceptance probabilities in explicitly solvated systems. For a given system size, REST acceptance probabilities are significantly larger than traditional REMD and far fewer replicas are needed to span the temperature distribution (Liu et al., 2005) (Fig. 4.2).

The number of replicas may be similarly reduced by using an alternative representation of the solvent. Implicit solvation models are particularly attractive for REMD simulations; however, in instances where results can be compared against explicitly solvated REMD results, significant quantitative differences have been noted (Chaudhury et al., 2012; Okur et al., 2006). Okur et al. (2006) showed using alanine peptides that the proper structural ensemble could be reproduced for the alanine dipeptide, but not for the tetrapeptide. To improve the agreement of implicit solvent REMD results with those obtained from the explicitly solvated systems without increasing the total number of replicas required to sample the system, Okur proposed

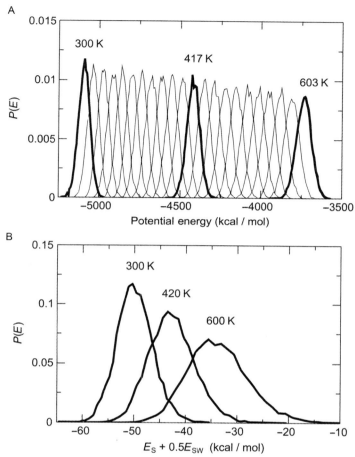

Figure 4.2 Potential energy distributions obtained from replica exchange molecular dynamics simulations of the alanine dipeptide. (A) Distributions obtained from standard REMD. (B) Distributions obtained from replica exchange with solute tempering (REST). This approach extends the applicability of replica exchange methods to larger explicitly solvated systems by eliminating the solvent–solvent interactions from the evaluation of the exchange probability (see main text). Using the effective potential energy (solute potential energy + 1/2 (solute–water potential energy)) enables the full temperature range to be spanned by 3 replicas, rather than 22 as needed by standard REMD. Figure reproduced with permission from Liu et al. (2005).

the use of a hybrid method that can substantially reduce the quantitative errors due to sampling under solvent-deficient conditions. In this approach, explicit solvation is maintained during MD steps; however, during Monte Carlo moves, the evaluation of the exchange probability is determined by reevaluating the energy of the solute in implicit solvent. Mu, Yang, and Xu (2007) quantified the conformational ensemble sampled by standard temperature REMD simulations for three pentapeptides and compared the results to those obtained using the hybrid approach with a Poisson–Boltzmann (PB) solvation model. As seen in Fig. 4.3, the hybrid results are comparable to those obtained from explicit solvent REMD but required one-third the number of replicas. Okur et al. and Chudhury et al. used the generalized

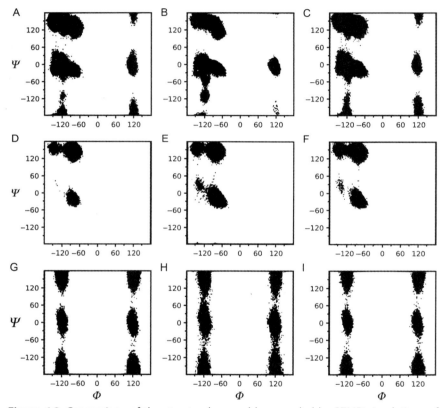

Figure 4.3 Comparison of the structural ensembles sampled by REMD simulations of three pentapeptides, met-enkephalin (first row), alanine (second row), and glycene (third row), whose conformational states are described by the backbone torsion angles, Φ and Ψ. The first column presents results obtained from standard REMD with explicit solvent and 24 temperature replicas. The second and third columns display results from hybrid REMD simulations in which a Poisson–Boltzmann implicit solvation model was used during the evaluation of the exchange probabilities. The hybrid simulations both operated with eight temperature replicas but used somewhat different van der Waals radii. *Figure reproduced with permission from Mu et al. (2007).*

Born approximation rather than a full PB solver to compute implicit solvation energies. Both studies found that the simulation results are sensitive to the amount of explicit water molecules included along with the solute during Monte Carlo moves. Okur showed that very accurate results (compared to explicit solvent REMD) could be attained if 1–2 layers of explicit solvent molecules are maintained about the solute as the energy is calculated in the implicit solvent field.

4. APPLICATION OF REMD TO EARLY STAGE MINERALIZATION

Most applications of REMD in the literature focus on small organic molecules and biopolymers (Mitsutake, Sugita, & Okamoto, 2001; Pitera & Swope, 2003). However, there is a history of using parallel tempering Monte Carlo to explore structural transitions in Lennard-Jones clusters (Mandelshtam, Frantsuzov, & Calvo, 2005; Sharapov & Mandelshtam, 2007). Therefore, the REMD method appears to be well suited to probe the formation and structural character of mineral clusters and the influence of biopolymers on them. However, to date, there is only a single study (Wallace et al., 2013), which has studied a biomineral phase with REMD techniques. In Wallace et al., a hybrid temperature-based REMD approach is used to probe the growth and conformational dynamics of hydrated calcium carbonate clusters up to ∼2 nm in diameter. This size range encompasses the full size distribution of the "prenucleation" clusters that have been observed in the $CaCO_3$–H_2O system by cryo-TEM (Gebauer & Coelfen, 2011; Gebauer, Voelkel, & Coelfen, 2008) and analytical ultracentrifugation (Gebauer et al., 2008; Pouget et al., 2009). It also overlaps with the size of anhydrous mineral clusters sampled by metadynamics (Quigley, Freeman, Harding, & Rodger, 2011; Quigley & Rodger, 2008). However, the approach put forth by Wallace et al. is distinguished by its ability to explore the phase space available to the clusters while simultaneously allowing water to be incorporated or expelled during growth.

4.1. Modified Kawska–Zahn method of cluster growth

In Wallace et al., hydrated calcium carbonate mineral clusters were grown following a modified version of a procedure first suggested by Kawska and Zahn (Kawska, Brickmann, Kniep, Hochrein, & Zahn, 2006). The original approach utilized an iterative three-step cycle to simulate the early stages of nanoparticle growth; these are (i) ion addition to a fixed cluster, (ii) solvent relaxation about the fixed cluster, and (iii) combined relaxation of both cluster and solvent. Starting from an initial cluster composition (in this case, one $CaCO_3^°$ ion pair), the first step of the process commences with the addition of an additional growth unit to the initial cluster. This was performed by placing an additional ion pair randomly on a sphere centered on the cluster center of mass. Keeping the coordinates of the cluster fixed, the interaction force between the ion pair and the initial cluster was minimized, directing the new growth unit to dock at the surface of the cluster. In previous implementations of this approach, the addition of new growth units was performed in the absence of solvent; however, due to the importance of water in the structure of amorphous calcium carbonate (ACC) (and other biominerals), Wallace et al. modified this procedure such that all solvent molecules within 4.5 Å of any of the cluster's constituent ions were maintained during ion pair docking. This important modification allowed solvent molecules to be incorporated into the clusters during growth. In the second step of the sequence, the hydrated cluster (with the new growth unit attached) was immersed in a large volume of preequilibrated solvent. The cluster geometry then remained fixed while the solvent was further relaxed about the cluster. In the third step of the process, the solvent and the cluster were both relaxed. In the original implementation of this stage, a high-temperature MD simulation was employed to hasten the relaxation of the system and to accelerate the sampling of cluster configurations. Wallace et al. replaced the high-temperature MD simulation with REMD. Eight temperature replicas (1:300, 2:310, 3:320, 4:330, 5:340, 6:360, 7:380, and 8:400 K) were run for 0.5 ns at each cluster size. Six trajectories were constructed in this way so that each cluster size was subject to 3 ns of REMD simulation in total. The simulations were performed at constant volume so that liquid water was stable at the pressure and temperature conditions of all replicates. During the REMD simulations, the exchange probability was evaluated for all even (2–3, 4–5, and 6–7) and odd (1–2, 3–4, 5–6, and 7–8) replica pairs in an alternating fashion. The simulations were performed in the presence of explicit solvent; however, the exchange probability was evaluated using a hybrid explicit/implicit solvent model (Chaudhury et al., 2012; Okur et al., 2006). Prior to each evaluation of the exchange probability, the number of explicit solvent molecules whose center of mass resided within 4.5 Å of each cluster's constituent ions was tallied and the highest solvent occupancy number, n_{occ}, was recorded. Solvent molecules were then removed around each cluster, leaving n_{occ} nearest neighbor solvent molecules in association with each cluster. This procedure ensured that all the clusters were fully hydrated and the same number of solvent molecules was present for the potential energy comparison. The potential energy of the explicitly solvated clusters was then computed with the general utility lattice program (Gale, 1997) using the COSMO (Gale & Rohl, 2007; Klamt & Schuurmann, 1993) continuum solvation model to account for long-range effects. Here, the atomic radii were fitted to reproduce the free energy of solvation of the Ca^{2+} and CO_3^{2-} ions given by the explicit force field and the solvation free energy of the H_2O molecule. Following the conclusion of the REMD simulation, the cluster configuration in the 300 K replica was used as the starting configuration for the next growth cycle. Snapshots from the REMD simulations are shown in Fig. 4.4A.

4.2. Aggregation-based model of amorphous $CaCO_3$

Among the many varied experimental observations concerning the early stages of calcium carbonate crystallization, the structure of ACC is by far the most consistent. At ambient conditions, the pair distribution function (PDF) obtained from total X-ray scattering consistently displays the same reproducible features in both synthetic and biogenic ACC (Goodwin et al., 2010; Michel et al., 2008; Reeder et al., 2013). At elevated pressures, there is now convincing evidence that a partially reversible phase transition takes place as ACC adopts a somewhat denser configuration (Fernandez-Martinez, Kalkan, Clark, & Waychunas, 2013). However, because the PDF structure is so consistent, it can be used to benchmark simulation results.

To construct a model of ACC, Wallace et al. selected one of the REMD-grown mineral clusters whose diameter was comparable to the coherent X-ray scattering length in ACC. In their approach, the initial aggregate structure was obtained by packing randomly oriented copies of that cluster into a solvent-filled cell. Subsequently, the calcium to water ratio was slowly adjusted to match that of synthetic ACC ($Ca/H_2O \sim 1.0$) over a series of constant pressure MD simulations. At each iteration, a few solvent molecules are preferentially removed from the solvent-rich volumes between the clusters. The net effect of this procedure was that the clusters were driven to coalesce without disturbing the structural water molecules that are intrinsic to their character. Once this process was complete, the individual PDFs, $g_{ij}(r)$, between all atom pairs were calculated as follows:

$$g_{ij}(r) = \frac{n_{ij}(r)}{4\pi r^2 \rho_i dr} \quad (4.11)$$

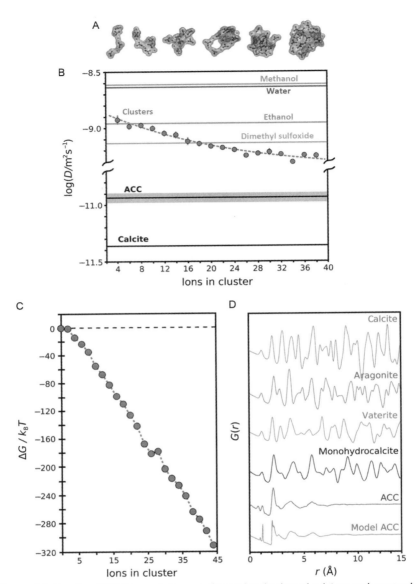

Figure 4.4 Results of REMD simulations performed on hydrated calcium carbonate clusters. (A) Snapshots taken at various stages of cluster growth. (B) The diffusivity of the calcium ions is plotted as a function of size and compared against calculated values for calcite and hydrous ACC. The significantly greater diffusivity of the ions in the cluster phase along with the similarity with the self-diffusivities of several common solvents suggests that the clusters are in a dense liquid phase. (C) The free energy landscape underlying phase separation and $[Ca^{2+}] = [CO_3^{2-}] = 15$ mmol/L. The energy landscape is consistent with liquid–liquid separation and spinodal decomposition. (D) Pair distribution functions for several crystalline and amorphous phases of calcium carbonate as measured by total X-ray scattering. The model ACC structure constructed as per the process described in the text by Wallace et al. (2013) is shown for comparison. *Reproduced with permission from Wallace et al. (2013). (See color plate.)*

where $n_{ij}(r)$ is the number of atoms of type j located within the radial distances r and $r+dr$ from an atom of type i and ρ_i is the total number density of atom type i. To facilitate the comparison of the theoretical results with the experimental PDF, the individual $g_{ij}(r)$ functions for all m atom pairs were summed up and weighted (Egami & Billinge, 2003; Fernandez-Martinez et al., 2010) (ρ_0 is the total atom number density):

$$G(r) = 4\pi\rho_0 \sum_{i,j}^{m} w_{ij}\left[g_{ij}(r) - 1\right] \quad (4.12)$$

The weighting factors, w_{ij}, are defined as follows, where c_i and c_j are the concentrations of atom types i and j, respectively, and f_i and f_j are their corresponding atomic form factors evaluated when the scattering vector $q=0$ (under this condition, the atomic form factors are equal to the atomic number):

$$w_{ij} = \frac{c_i\, c_j\, f_i\, f_j}{\left(\sum_{i}^{m} c_i f_i\right)^2} \quad (4.13)$$

The model ACC structure produced by this procedure faithfully reproduces all the salient aspects of the experimental PDF (Fig. 4.4D). Although the theoretical PDF is not as exact a match as those for candidate structures produced by reverse Monte Carlo (RMC) modeling (Goodwin et al., 2010; McGreevy & Howe, 1992), the procedure outlined in the preceding text has the advantage that it is guaranteed to produce a structure that occupies a local minimum on the energy landscape. The results of MD simulations by Singer, Yazaydin, Kirkpatrick, and Bowers (2012) show that the RMC-derived structure of ACC (Goodwin et al., 2010) is unstable.

4.3. Thermodynamic properties

It is standard practice in REMD simulations to compile probability distributions with respect to given order parameters that may then be used to determine thermodynamic properties. For example, were we to run a REMD simulation on a single $CaCO_3^{\,\circ}$ ion pair, we would find that the calcium ion oscillates between two bound states (monodentate and bidentate coordination by the carbonate ion). For the sake of simplicity, if we ignore the possibility of unbound states, then a sufficiently long REMD simulation will provide the relative probability that the ion pair exists in one state or the other. If a probability distribution is compiled with respect to a metric that can distinguish between the two states, one may determine the relative free energies between them. This type of approach was also outlined in the example in Section 2.2.

While the earlier approach works well, in many complex systems, selection of an appropriate order parameter with which to project the free energy landscape is not always as obvious as in the previously mentioned examples. Therefore, use of free energy methods that do not rely on order parameters can be advantageous. Demichelis, Raiteri, Gale, Quigley, and Gebauer (2011) performed standard MD simulations in concentrated calcium carbonate and bicarbonate solutions. From the concentrations of cluster species in solution, they were able to determine equilibrium constants for certain reactions including

$$CaCO_3^\circ + (CaCO_3^\circ)_n \rightarrow (CaCO_3^\circ)_{n+1} \quad \log K = -21.7 \quad (4.14)$$

which indicates that the free energy landscape for forming uncharged clusters is entirely downhill.

Wallace et al. calculated the free energy as a function of cluster size using an absolute two-phase thermodynamic (2PT) model (Huang, Pascal, Goddard, Maiti, & Lin, 2011; Lin, Blanco, & Goddard, 2003; Lin, Maiti, & Goddard, 2010) and obtained comparable results (Fig. 4.4C).

The 2PT method yields thermodynamic properties through compilation, decomposition, and integration of the phonon density of states (DOS). The DOS may also be used to quantify the diffusive characteristics of a system, which are proportional to its zeroth frequency, $S(0)$ (Huang et al., 2011; Lin et al., 2003, 2010; Pascal, Lin, & Goddard, 2011):

$$D = S(0)\left(\frac{k_B T}{12mN}\right) \quad (4.15)$$

Wallace et al. used the previous expression to determine the diffusivity of calcium ions within calcium carbonate cluster species; here, $N=$ total number of calcium ions, $m=$ particle mass, and $T=$ temperature. Because calcium ions have only translational degrees of freedom, the total DOS corresponding to the calcium ion population was calculated as

$$S(v) = \frac{1}{k_B T} \sum_{l=1}^{N} \sum_{k=1}^{3} \lim_{\tau \to \infty} \frac{m_l}{\tau} \left| \int_{-\tau}^{\tau} v(t) e^{-i2\pi v t} dt \right|^2 \quad (4.16)$$

where m_l is the mass of atom l and $v_l^k(t)$ is the velocity of atom l in the k direction (k is one of three Cartesian basis vectors, $x=1$, $y=2$, and $z=3$) at time t. The calcium ion diffusivities determined in this manner were smoothly trending and decreased with increasing cluster size (Fig. 4.4B), slowly approaching a constant value that was significantly more diffusive than calculated for solid phases of $CaCO_3$ (ACC and calcite). Based on this observation, and the similarity of the ion diffusivities with the self-diffusivities of several common solvents, Wallace et al. argued, as did Demichelis et al. (2011), that the clusters occupy a liquid-like if not bona fide liquid state.

5. SUMMARY

Although simulations are commonly utilized in molecular-level science, significant challenges still limit their application to many problems of interest. The gap between experimental and simulation-accessible timescales/length scales is especially difficult to bridge. Replica exchange methods are designed to improve the sampling efficiency of standard techniques and are well suited to the simulation of biomineral phases; however, to date, these methods have not been widely utilized within the biomineralization community. This chapter provides an introduction to replica exchange techniques, their origins, implementation details, and ongoing development. Particular focus is given to methods that aim to extend the applicability of the method to large solvent-dominated systems.

REFERENCES

Abraham, M. J., & Gready, J. E. (2008). Ensuring efficiency of replica exchange molecular dynamics simulations. *Journal of Chemical Theory and Computation, 4*, 1119–1128.

Chaudhury, S., Olson, M. A., Tawa, G., Wallqvist, A., & Lee, M. S. (2012). Efficient conformational sampling in explicit solvent using a hybrid replica exchange molecular dynamics method. *Journal of Chemical Theory and Computation, 8*(2), 677–687. http://dx.doi.org/10.1021/ct200529b.

Chebaro, Y., Dong, X., Laghaei, R., Derreumaux, P., & Mousseau, N. (2009). Replica exchange molecular dynamics simulations of coarse-grained proteins in implicit solvent. *The Journal of Physical Chemistry B, 113*(1), 267–274. http://dx.doi.org/10.1021/jp805309e.

Cooke, B., & Schmidler, S. C. (2008). Preserving the Boltzmann ensemble in replica-exchange molecular dynamics. *The Journal of Chemical Physics, 129*(16), 164112. http://dx.doi.org/10.1063/1.2989802.

Dashti, D. S., Meng, Y., & Roitberg, A. E. (2012). pH-replica exchange molecular dynamics in proteins using a discrete protonation method. *The Journal of Physical Chemistry B, 116*(30), 8805–8811. http://dx.doi.org/10.1021/jp303385x.

Demichelis, R., Raiteri, P., Gale, J. D., Quigley, D., & Gebauer, D. (2011). Stable prenucleation mineral clusters are liquid-like ionic polymers. *Nature Communications*, *2*(590), 1–6. http://dx.doi.org/10.1038/ncomms1604.

Earl, D. J., & Deem, M. W. (2005). Parallel tempering: Theory, applications, and new perspectives. *Physical Chemistry Chemical Physics*, *7*(23), 3910–3916.

Egami, T., & Billinge, S. J. L. (2003). The method of total scattering and atomic pair distribution function analysis. Chapter 3, pages 55–99. In *Underneath the Bragg peaks: Structural analysis of complex materials. Pergamon materials series*, Vol. 7. Amsterdam: Elsevier.

Ensing, B., De Vivo, M., Liu, Z., Moore, P., & Klein, M. L. (2006). Metadynamics as a tool for exploring free energy landscapes of chemical reactions. *Accounts of Chemical Research*, *39*(2), 73–81. http://dx.doi.org/10.1021/ar040198i.

Fernandez-Martinez, A., Kalkan, B., Clark, S. M., & Waychunas, G. A. (2013). Pressure-induced polymorphism and formation of "aragonitic" amorphous calcium carbonate. *Angewandte Chemie International Edition*, *52*(32), 8354–8357. http://dx.doi.org/10.1002/anie.201302974.

Fernandez-Martinez, A., Timon, V., Roman-Ross, G., Cuello, G. J., Daniels, J. E., & Ayora, C. (2010). The structure of schwertmannite, a nanocrystalline iron oxyhydroxysulfate. *American Mineralogist*, *95*(8–9), 1312–1322. http://dx.doi.org/10.2138/am.2010.3446.

Freeman, D. L. (2000). Phase changes in 38-atom Lennard-Jones clusters. I. A parallel tempering study in the canonical ensemble. *Chemical Physics*, *112*(23), 10340–10349.

Frenkel, D., & Smit, B. (2002). *Understanding molecular simulation: From algorithms to applications* (2nd ed.). London: Academic Press, p. 638.

Gale, Julian D. (1997). GULP: A computer program for the symmetry-adapted simulation of solids. *Journal of the Chemical Society, Faraday Transactions*, *93*(4), 629–637. http://dx.doi.org/10.1039/A606455H.

Gale, J. D., & Rohl, A. L. (2007). An efficient technique for the prediction of solvent-dependent morphology: The COSMIC method. *Molecular Simulation*, *33*(15), 1237–1246. http://dx.doi.org/10.1080/08927020701713902.

Gebauer, D., & Coelfen, H. (2011). Prenucleation clusters and non-classical nucleation. *Nano Today*, *6*(6), 564–584. http://dx.doi.org/10.1016/j.nantod.2011.10.005.

Gebauer, D., Voelkel, A., & Coelfen, H. (2008). Stable prenucleation calcium carbonate clusters. *Science*, *322*(5909), 1819–1822.

Geyer, C. J., & Thompson, E. A. (1995). Annealing Markov chain Monte Carlo with applications to ancestral inference. *Journal of the American Statistical Association*, *90*(431), 909–920.

Goodwin, A. L., Michel, F. M., Phillips, B. L., Keen, D. A., Dove, M. T., & Reeder, R. J. (2010). Nanoporous structure and medium-range order in synthetic amorphous calcium carbonate. *Chemistry of Materials*, *22*(10), 3197–3205.

Huang, S.-N., Pascal, T. A., Goddard, W. A., Maiti, P. K., & Lin, S.-T. (2011). Absolute entropy and energy of carbon dioxide using the two-phase thermodynamic model. *Journal of Chemical Theory and Computation*, *7*(6), 1893–1901. http://dx.doi.org/10.1021/ct200211b.

Iannuzzi, M., Laio, A., & Parrinello, M. (2003). Efficient exploration of reactive potential energy surfaces using Car-parrinello molecular dynamics. *Physical Review Letters*, *90*(23), 238302. http://dx.doi.org/10.1103/PhysRevLett.90.238302.

Jiang, W., & Roux, B. (2010). Free energy perturbation Hamiltonian replica-exchange molecular dynamics (FEP/H-REMD) for absolute ligand binding free energy calculations. *Journal of Chemical Theory and Computation*, *6*(9), 2559–2565. http://dx.doi.org/10.1021/ct1001768.

Kawska, A., Brickmann, J., Kniep, R., Hochrein, O., & Zahn, D. (2006). An atomistic simulation scheme for modeling crystal formation from solution. *The Journal of Chemical Physics, 124*(2), 24513–24517.

Klamt, A., & Schuurmann, G. (1993). COSMO: A new approach to dielectric screening in solvents with explicit expressions for the screening energy and its gradient. *Journal of the Chemical Society, Perkin Transactions, 2*(5), 799–805. http://dx.doi.org/10.1039/P29930000799.

Kone, A., & Kofke, D. A. (2005). Selection of temperature intervals for parallel-tempering simulations. *The Journal of Chemical Physics, 122*(20), 206101. http://dx.doi.org/10.1063/1.1917749.

Laio, A., & Gervasio, F. L. (2008). Metadynamics: A method to simulate rare events and reconstruct the free energy in biophysics, chemistry and material science. *Reports on Progress in Physics, 71*(12), 126601. http://dx.doi.org/10.1088/0034-4885/71/12/126601.

Laio, A., & Parrinello, M. (2002). Escaping free-energy minima. *Proceedings of the National Academy of Sciences of the United States of America, 99*(20), 12562–12566. http://dx.doi.org/10.1073/pnas.202427399.

Laio, A., Rodriguez-Fortea, A., Gervasio, F. L., Ceccarelli, M., & Parrinello, M. (2005). Assessing the accuracy of metadynamics. *The Journal of Physical Chemistry B, 109*(14), 6714–6721. http://dx.doi.org/10.1021/jp045424k.

Lin, S.-T., Blanco, M., & Goddard, W. A. (2003). The two-phase model for calculating thermodynamic properties of liquids from molecular dynamics: Validation for the phase diagram of Lennard-Jones fluids. *The Journal of Chemical Physics, 119*(22), 11792. http://dx.doi.org/10.1063/1.1624057.

Lin, S.-T., Maiti, P. K., & Goddard, W. A. (2010). Two-phase thermodynamic model for efficient and accurate absolute entropy of water from molecular dynamics simulations. *The Journal of Physical Chemistry B, 114*(24), 8191–8198. http://dx.doi.org/10.1021/jp103120q.

Liu, P., Kim, B., Friesner, R. A., & Berne, B. J. (2005). Replica exchange with solute tempering: A method for sampling biological systems in explicit water. *Proceedings of the National Academy of Sciences of the United States of America, 102*(39), 13749–13754. http://dx.doi.org/10.1073/pnas.0506346102.

Mandelshtam, V. A., Frantsuzov, P. A., & Calvo, F. (2005). Structural transitions and melting in LJ74-78 Lennard-Jones clusters from adaptive exchange Monte Carlo simulations†. *The Journal of Physical Chemistry A, 110*(16), 5326–5332. http://dx.doi.org/10.1021/jp055839l.

McGreevy, R. L., & Howe, M. A. (1992). RMC: Modeling disordered structures. *Annual Review of Material Science, 22*, 217–242.

Metropolis, N., Rosenbluth, A. W., Rosenbluth, M. N., Teller, A. H., & Teller, E. (1953). Equation of state calculations by fast computing machines. *The Journal of Chemical Physics, 21*(6), 1087. http://dx.doi.org/10.1063/1.1699114.

Michel, F. M., MacDonald, J., Feng, J., Phillips, B. L., Ehm, L., Tarabrella, C., et al. (2008). Structural characteristics of synthetic amorphous calcium carbonate. *Chemistry of Materials, 20*(14), 4720–4728. http://dx.doi.org/10.1021/cm800324v.

Mitsutake, A., Sugita, Y., & Okamoto, Y. (2001). Generalized-ensemble algorithms for molecular simulations of biopolymers. *Peptide Science, 60*(2), 96–123. http://dx.doi.org/10.1002/1097-0282(2001)60:2<96::AID-BIP1007>3.0.CO;2-F.

Mu, Y., Yang, Y., & Xu, W. (2007). Hybrid Hamiltonian replica exchange molecular dynamics simulation method employing the Poisson-Boltzmann model. *The Journal of Chemical Physics, 127*(8), 084119. http://dx.doi.org/10.1063/1.2772264.

Nymeyer, H., Gnanakaran, S., & García, A. E. (2004). Atomic simulations of protein folding, using the replica exchange algorithm. *Methods in Enzymology, 383*(2000), 119–149. http://dx.doi.org/10.1016/S0076-6879(04)83006-4.

Okur, A., Wickstrom, L., Layten, M., Geney, R., Song, K., Hornak, V., et al. (2006). Improved efficiency of replica exchange simulations through Use of a hybrid explicit/implicit solvation model. *Journal of Chemical Theory and Computation, 2*(2), 420–433. http://dx.doi.org/10.1021/ct050196z.

Pascal, T. A., Lin, S.-T., & Goddard, W. A., III. (2011). Thermodynamics of liquids: Standard molar entropies and heat capacities of common solvents from 2PT molecular dynamics. *Physical Chemistry Chemical Physics, 13*(1), 169–181. http://dx.doi.org/10.1039/C0CP01549K.

Periole, X., & Mark, A. E. (2007). Convergence and sampling efficiency in replica exchange simulations of peptide folding in explicit solvent. *The Journal of Chemical Physics, 126*(1), 014903. http://dx.doi.org/10.1063/1.2404954.

Pitera, J. W., & Swope, W. (2003). Understanding folding and design: Replica-exchange simulations of "Trp-cage" miniproteins. *Proceedings of the National Academy of Sciences of the United States of America, 100*(13), 7587–7592.

Pouget, E. M., Bomans, P. H. H., Goos, J. A. C. M., Frederik, P. M., de With, G., & Sommerdijk, N. A. J. M. (2009). The initial stages of template-controlled $CaCO_3$ formation revealed by Cryo-TEM. *Science, 323*(5920), 1455–1458.

Quigley, D., Freeman, C. L., Harding, J. H., & Rodger, P. M. (2011). Sampling the structure of calcium carbonate nanoparticles with metadynamics. *Journal of Chemical Physics, 134*, 044703. http://dx.doi.org/10.1063/1.3530288.

Quigley, D., & Rodger, P. M. (2008). Free energy and structure of calcium carbonate nanoparticles during early stages of crystallization. *The Journal of Chemical Physics, 128*(22), 221101. http://dx.doi.org/10.1063/1.2940322.

Quigley, D., & Rodger, P. M. (2009). A metadynamics-based approach to sampling crystallisation events. *Molecular Simulation, 35*(7), 613–623. http://dx.doi.org/10.1080/08927020802647280.

Reeder, R. J., Tang, Y., Schmidt, M. P., Kubista, L. M., Cowan, D. F., & Phillips, B. L. (2013). Characterization of structure in biogenic amorphous calcium carbonate: Pair distribution function and nuclear magnetic resonance studies of lobster gastrolith. *Crystal Growth & Design, 13*(5), 1905–1914. http://dx.doi.org/10.1021/cg301653s.

Rosta, E., & Hummer, G. (2009). Error and efficiency of replica exchange molecular dynamics simulations. *The Journal of Chemical Physics, 131*(16), 165102. http://dx.doi.org/10.1063/1.3249608.

Sharapov, V. A., & Mandelshtam, V. A. (2007). Solid−solid structural transformations in Lennard-Jones clusters: Accurate simulations versus the harmonic superposition approximation†. *The Journal of Physical Chemistry A, 111*(41), 10284–10291. http://dx.doi.org/10.1021/jp072929c.

Sindhikara, D. J. (2010). Exchange often and properly in replica exchange molecular dynamics. *Journal of Chemical Theory and Computation, 6*(9), 2804–2808. http://dx.doi.org/10.1021/ct100281c.

Sindhikara, D. J., Emerson, D. J., & Roitberg, A. E. (2010). Exchange often and properly in replica exchange molecular dynamics. *Journal of Chemical Theory and Computation, 6*(9), 2804–2808.

Sindhikara, D., Meng, Y., & Roitberg, A. E. (2008). Exchange frequency in replica exchange molecular dynamics. *The Journal of Chemical Physics, 128*(2), 024103. http://dx.doi.org/10.1063/1.2816560.

Singer, J. W., Yazaydin, A. O., Kirkpatrick, R. J., & Bowers, G. M. (2012). Structure and transformation of amorphous calcium carbonate: A solid-state ^{43}Ca NMR and computational molecular dynamics investigation. *Chemistry of Materials, 24*(10), 1828–1836.

Sugita, Y., & Okamoto, Y. (1999). Replica-exchange molecular dynamics method for protein folding. *Chemical Physics Letters*, *314*(1–2), 141–151. http://dx.doi.org/10.1016/S0009-2614(99)01123-9.

Swendsen, R. H., & Wang, J.-S. (1987). Nonuniversal critical dynamics in Monte Carlo simulations. *Physical Review Letters*, *58*(2), 86–88.

Terakawa, T., Kameda, T., & Takada, S. (2010). On easy implementation of a variant of the replica exchange with solute tempering in GROMACS. *Journal of Computational Chemistry*, *32*(7), 1228–1234. http://dx.doi.org/10.1002/jcc.

Torrie, G. M., & Valleau, J. P. (1974). Monte Carlo free energy estimates using non-Boltzmann sampling: Application to the sub-critical Lennard-Jones fluid. *Chemical Physics Letters*, *28*(4), 578–581.

Wallace, A. F., Hedges, L. O., Fernandez-Martinez, A., Raiteri, P., Waychunas, G. A., Gale, J. D., et al. (2013). Microscopic evidence for liquid-liquid separation in supersaturated $CaCO_3$ solutions. *Science*, *341*, 885–889.

Wang, J., Zhu, W., Li, G., & Hansmann, U. H. E. (2011). Velocity-scaling optimized replica exchange molecular dynamics of proteins in a hybrid explicit/implicit solvent. *The Journal of Chemical Physics*, *135*(8), 084115. http://dx.doi.org/10.1063/1.3624401.

Zhang, W., Wu, C., & Duan, Y. (2005). Convergence of replica exchange molecular dynamics. *The Journal of Chemical Physics*, *123*(15), 154105. http://dx.doi.org/10.1063/1.2056540.

CHAPTER FIVE

SAXS in Inorganic and Bioinspired Research

Tomasz M. Stawski[1], Liane G. Benning

Cohen Biogeochemistry Laboratory, School of Earth and Environment, University of Leeds, Leeds, United Kingdom
[1]Corresponding author: e-mail address: t.m.stawski@leeds.ac.uk

Contents

1. Introduction 96
2. General Considerations: How Does SAXS/WAXS Work and When It Is Applicable? 96
3. What Happens in a Solution When Ions Meet and How Do We Quantify it? 99
   3.1 Formation of inorganic materials from solutions 99
   3.2 The suitability of synchrotron-based scattering methods 103
4. SAXS Data Processing and Interpretation 105
   4.1 Data treatment 106
   4.2 Data interpretation 106
5. Case Studies 111
   5.1 Formation and growth of amorphous $CaCO_3$ and its crystallization 111
   5.2 Nucleation and growth of silica nanoparticles 115
   5.3 Evolution of the colloidal precursors for $BaTiO_3$ 119
6. Outlook 122
Acknowledgments 124
References 124

Abstract

In situ and time-resolved structural information about emergent microstructures that progressively develop during the formation of inorganic or biologically mediated solid phases from solution is fundamental for understanding of the mechanisms driving complex precipitation reactions, for example, during biomineralization. In this brief chapter, we present the use of small- and wide-angle X-ray scattering (SAXS and WAXS) techniques and show how SAXS can be used to gather structural information on the nanoscale properties of the *de novo*-forming entities. We base the discussion on several worked examples of inorganic materials such as calcium carbonate, silica, and perovskite-type titanates.

1. INTRODUCTION

The making of bonds during the formation of inorganic or biologically mediated solid phases from solution usually follows a series of complex steps and pathways that most often depend on molecular-level reactions occurring at fluid–surface interfaces. *In situ* and time-resolved structural information about size, shape, crystallinity, and any mutual interaction between emergent microstructures that progressively develop during such processes provides us with the crucial building blocks to further our fundamental understanding of the kinetics, energetics, and mechanisms driving complex phase-forming reactions. The fact that the emergent species are often structurally poorly ordered, nanoparticulate and in many cases unstable dictates that ideally they must be characterized not just at length scales <100 nm but also as *in situ* as possible to avoid any artifacts due to sample handling, etc. Therefore, solution-based X-ray scattering methods are one of the most effective tools in studying nanostructured materials as they form in fluid media. In this brief chapter, we will discuss specifically the use of (ideally simultaneous) small- and wide-angle X-ray scattering (SAXS and WAXS) techniques and show how SAXS can be used to gather structural information on the nanoscale properties of the *de novo*-forming entities, whereas WAXS (basically X-ray diffraction) provides us with the atomic length-scale characterization information of the same reaction.

2. GENERAL CONSIDERATIONS: HOW DOES SAXS/WAXS WORK AND WHEN IT IS APPLICABLE?

Let us consider a sample that is either particulate or bicontinuous, that is, nonparticulate in nature but that has nanometer-range electronic density variations. Such variations could originate from homogenous colloidal particles of electronic density ρ that are dispersed in a matrix of a constant, but different, electronic density ρ_0 (e.g., a solvent). Equally, such variations could be due to presence of pores in a constant electronic density matrix or the coexistence of two continuous phases of different electronic densities but separated at a nanometer scale within a sample body (e.g., bicontinuous emulsions; Craievich, 2002; Glatter & Kratky, 1982; Guinier & Fournet, 1955; Stribeck, 2007). In all these cases, the difference in average electron density, $\Delta\rho$, leads to a "two-electron density system." $\Delta\rho$ will hereafter be referred to as the scattering contrast and $\Delta\rho \neq 0$ is a necessary condition for scattering to occur.

Applying this to SAXS experiments and combining it with a monochromatic collimated X-ray beam that passes through a sample that exhibits a scattering contrast, $\Delta\rho$, leads to the recording of the elastically scattered photons at very small angles. This is most often done using a 2D plate detector (Fig. 5.1A).

Overall, both small- and wide-angle scattering measurements can be performed either with conventional laboratory X-ray lamps or with synchrotron X-ray radiation. Depending on the measured angular ranges and light wavelengths, a SAXS pattern allows a series of structural parameters about the solid entities in a sample to be evaluated (e.g., size, size distribution, geometric shape, and agglomeration with dimensions typically 0.1–<500 nm; Bras, 1998; Craievich, 2002; Rieckel, Burghammer, & Müller, 2000; Stribeck, 2007). The resulting recorded scattering intensity is conventionally plotted as the function of the modulus of the scattering vector **q** (nm^{-1}), which is related to the scattering angle 2θ and the wavelength λ (nm) of the incident beam via (Eq. 5.1)

$$q = (4\pi/\lambda)\sin\theta \tag{5.1}$$

where q is the modulus (magnitude) of **q**.

A representative 2D scattering pattern of a sample containing agglomerated particles in solution is shown in Fig. 5.1B. It is worth mentioning that in order to extract the maximum information about the in-sample-contained structures, it is paramount that the used q-range must be adjusted before each SAXS experiment to match as best as possible the expected range of sizes of the investigated structural features. This is the necessary practical condition for a successful SAXS experiment. This condition is important because if the minimum expected particle size is denoted by D_{min} and the maximum by D_{max}, this leads to the q-range needing to be set to fit $q_{min} = \pi/D_{max} < q < q_{max} = \pi/D_{min}$ (Glatter & Kratky, 1982). For a q-range defined in such a way, the scattering data will contain the full information for all particles in the size range between D_{min} and D_{max}. However, if a scattering pattern contains information from semiperiodic (e.g., correlated or *pseudo*crystalline) structures, the q-range is also dependent on the maximum and minimum interparticle distances, d_{max} and d_{min}, with $q_{min} = 2\pi/d_{max} < q < q_{max} = 2\pi/d_{min}$, as defined by the Bragg equation. These parameters are obviously dependent not only on the wavelength of the incident X-ray beam but also on the sample-to-detector distance, the area of the 2D detector, and the position of the beam on the detector (beam-stop position; e.g., placing a beam stop in the corner of the detector instead of the

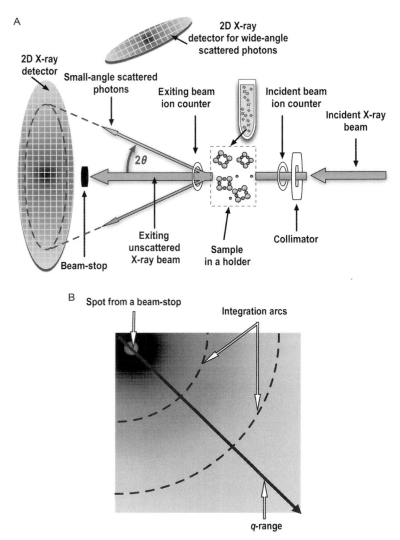

Figure 5.1 (A) Scheme of the typical SAXS/WAXS experiment; (B) example of a scattering pattern from a sample containing agglomerated particles in solution recorded with a 2D SAXS detector. Some of the elements relevant for the data-reduction steps explained in Section 4 are marked in the figure. (See color plate.)

middle allows for more than doubling of the measured q-range; Fig. 5.1B). The actual measured q-range in absolute units (nm^{-1} or Å^{-1}) with respect to pixel/channel positions on any detector needs to be established, each time, by means of a standard, that is, a chemical compound of known structure exhibiting Bragg diffraction peaks in the measured low-angular range. For instance, silver behenate ($d_{max} = 5.8380(3)$ nm; Huang, Toraya, Blanton, & Wu, 1993) or wet rat tail collagen ($d_{max}=$ ca. 67 nm; Fratzl et al., 1997) is most often used for this purpose depending on the investigated q-range.

SAXS instruments are often complemented with WAXS detectors (Fig. 5.1A). These record those photons scattered at higher angles, hence, extending the q-range to higher magnitudes of the scattering vector. The investigated system is targeting not just poorly ordered nanostructures but possibly also particles with emerging and developing crystalline or nanocrystalline properties. These higher-angle scatterers would produce a diffraction pattern, which will be recorded in the WAXS regime. Such patterns complement the SAXS data and give us quantitative information about the internal structure of forming ordered entities.

3. WHAT HAPPENS IN A SOLUTION WHEN IONS MEET AND HOW DO WE QUANTIFY IT?

In this section, we very briefly review how amorphous and crystalline inorganic materials are known to form from solutions. It is not our aim to discuss these complex processes in detail nor will we address the thermodynamics or detailed kinetic aspects related to such reactions, but we will focus merely on a simplified structural description and only partially address basic kinetic aspects. This introduction is necessary for a better understanding of not only the advantages but also limitations of SAXS for solution-based particle nucleation and growth investigations and to indicate where complementary and/or alternative methods should be/need to be used to derive quantitative data about such reactions. In every scientific study, it should be always clear that no single method alone ever provides a conclusive answer to any scientific question and that in all cases multiple and complementary analytic and experimental approaches need to be employed to achieve a rigorous scientific result.

3.1. Formation of inorganic materials from solutions

The growth of any solid material from solution is essentially an evolutionary process in which at different stages of growth, species of various shapes, size, and internal structure form and further convert into a final product (amorphous colloidal gel, crystal, etc.).

In the case of nucleation and crystallization phenomena occurring in solutions, relevant but not exclusive to (bio)mineralization, two descriptions are used: the classical nucleation theory (CNT) often applied universally as a starting point for further discussions (Becker & Döring, 1935; Benning & Waychunas, 2007; Volmer & Weber, 1925) and the relatively novel prenucleation concept developed and confirmed for calcium carbonate and calcium phosphate (Dey et al., 2010; Gebauer & Cölfen, 2011; Gebauer, Völkel, & Cölfen, 2008; Fig. 5.2A). In both cases, the formation of a new entity starts with a change in supersaturation of a reacting solution of dissolved species that leads progressively, fast or slow, to the coalescence of a first "solid" entity. From this initial entity, the reaction can follow many pathways, but ultimately, the reaction chain completes with the formation of a stable crystal. From the structural point of view, CNT assumes that elementary nuclei form in the supersaturated solution and that this proceeds further through attachment of basic monomers (atoms, ions, and molecules) to this newly formed entity (Benning & Waychunas, 2007). Structurally, this newly formed elementary nucleus is indistinguishable from that of the intended bulk material. However, these smaller nuclei are most often thermodynamically unstable and dissolve and thus the material grows from nuclei of a certain critical size (critical nucleus), leading eventually to the formation of the so-called primary particle. CNT has been used universally in the last few decades, and it is still applicable in many systems. For example, just recently, CNT has been confirmed for the nucleation and growth of magnetite crystals (Fe_3O_4), which have been shown to form via a basic-pH coprecipitation route, describable within CNT (Baumgartner et al., 2013). On the other hand, in the case of $CaCO_3$, Gebauer and co-workers (Gebauer & Cölfen, 2011; Gebauer et al., 2008) proposed an alternative mechanism (Fig. 5.2A) that implies that within a supersaturated solution, the first entities that form are stable prenucleation clusters. Only in subsequent stages do these prenucleation clusters aggregate into larger amorphous particles, which can further agglomerate, coalesce, and eventually covert via various pathways into a crystalline phase. This also implies that any species forming before the nucleation and growth of the crystal must be structurally different from the final material, yet they suggest that likely there is a structural "memory" effect that affects these reactions. For example, in the $CaCO_3$ system, depending on physicochemical conditions at which the experiments are carried out, the prenucleation clusters, and the subsequently formed amorphous calcium carbonate (ACC), seem to have a structural "memory" that leads to ACC that is structurally more alike with the final crystalline entity that it will transform into (e.g., ACC with a vaterite-, calcite-, or aragonite-like protostructure; Bots, Benning, Rodriguez-Blanco,

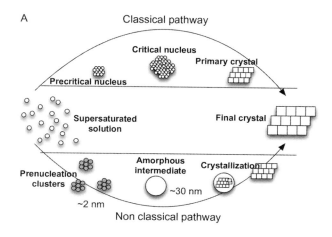

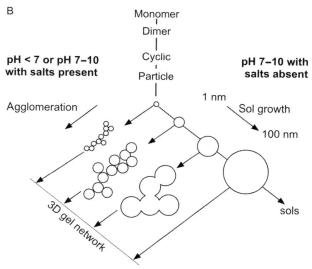

Figure 5.2 (A) Classical and nonclassical pathways describing growth of the crystalline materials from solutions. (B) Scheme representing the evolution of aqueous silica system from monomers to developed sols and gels, in relation with reaction conditions. Panel (A) is based on Meldrum and Sear (2008) and Gebauer and Cölfen (2011), and panel (B) is based on Iler (1979). (For color version of this figure, the reader is referred to the online version of this chapter.)

Roncai-Herrero, & Shaw, 2012; Gebauer & Cölfen, 2011). These matters are still in debate and discussion as the reactions are fairly complex, but scattering can help elucidate some of the still open questions. Naturally, these descriptions of complex reactions have been in this chapter highly simplified. Many reviews that discuss the pathways of the formation of ACC in pure or biomimetic/biomineralization scenarios in detail are available (e.g., Gower, 2008; Meldrum & Cölfen, 2008). However, what is important for this discussion is the fact that on the pathway between the solution stage and the final (most often crystalline) stable end product, emergent discrete nanostructures form. The morphology and rate of development of these nanostructures depends on physical and chemical parameters of the evolving system, and any small modification in any of the parameters at any stage in the process has invariably nonnegligible implications for any of the consequent steps. For such emerging properties, SAXS is an ideal technique as it can provide information at fast timescales and also at the spatial scale needed to observe these entities (provided that enough particles are present in the X-ray-illuminated volume of fluid).

Another example of emerging entities from supersaturated aqueous solutions is the evolution of aqueous silica from solution. The SiO_2 system is ideal to illustrate the connection between reaction conditions and development of solid amorphous silica nanoparticles (Fig. 5.2B; Brinker & Scherer, 1990; Iler, 1979). In aqueous solution and depending on conditions, silica, $Si(OH)_4$, can polymerize via various steps: (1) $Si(OH)_4$ monomers coalesce into particles through dimer-cluster and cyclic-cluster stages; (2) in acidic solutions or in the presence of flocculating salts, sub-10 nm particles aggregate into networks, which form a gel; or (3) in basic solutions with salts absent, the growth of homogenous sol particles occurs, with particle sizes reaching 100 nm (Fig. 5.2B; Brinker & Scherer, 1990; Iler, 1979). These systems have been widely investigated mainly due to not only the technological, geochemical, and biological significance of silica but also the impeded reactivity of silicon-carrying precursors in comparison with transition metal analogues, providing with a good study model system for metal oxide colloid science in general (Brinker & Scherer, 1990; Wright & Sommerdijk, 2001). Aqueous and nanoparticulate silica has many technological applications and is industrially directly obtained in its amorphous form most often from alkoxide precursors. The use of metal–organic starting materials makes it simpler to link the properties of sols with the properties of expected solid final products. For such studies, the use of SAXS is ideal because in the alkoxide silica system, homogenous and most often relatively monodispersed particles form from solution which is favorable in many industrial applications. However, such organically produced SiO_2 particles have little to say about how silica, the most abundant compound in the Earth's crust, forms in a natural—nonindustrial—setting. We will discuss later (Section 5.2) how this has been the focus of recent studies (Tobler & Benning, 2013; Tobler, Shaw, & Benning, 2009) that employed synchrotron-based X-ray scattering to quantify the nucleation and growth of silica nanoparticles from inorganic solutions, mimicking natural processes.

3.2. The suitability of synchrotron-based scattering methods

Synchrotron-based SAXS in particular and WAXS in some cases are powerful tools enabling extraction of structural information about emergent properties of new phases forming from and in solutions because of the following:

- From its very principle requiring the occurrence of the scattering contrast, scattering methods are sensitive to the presence of particulates suspended in solution. Furthermore, ideally, the growing phases have relatively high electron density in comparison with the solvent, and thus, the scattering contrast is high.
- Scattering methods enable *in situ* and time-resolved measurements of an evolving system in which particles are in a constantly changing quasiequilibrium with the reacting medium; in some cases, measurements under other physical conditions can also be achieved, for example, nonequilibrium physical drying (Stawski, Veldhuis, Castricum, et al., 2011; Stawski et al., 2011b).
- The emergent species range in size from <1 nm to a few micrometer, with length scales at the most crucial prenucleation, nucleation, and growth phases contained at sub-200 nm. This is the length scale ideally adjusted for X-ray scattering experiments.
- From the kinetic point of view, any reaction that is followed using scattering needs to ideally occur at resolvable timescales; these can be sub-100 ms to 1 min and/or be up to many hours (Bots et al., 2012). For instance, as mentioned earlier, crystalline $CaCO_3$ phases form via ACC precursors (Meldrum & Cölfen, 2008). ACC precipitates from supersaturated solutions at ambient conditions as nanoparticles within seconds (Rodriguez-Blanco, Shaw, & Benning, 2008). *In situ* and time-resolved scattering and diffraction studies (Bots et al., 2012; Rodriguez-Blanco, Shaw, & Benning, 2011) showed that in purely inorganic systems, ACC transforms to crystalline $CaCO_3$ polymorphs within minutes to hours and that nucleation, growth, and transformation mechanisms and kinetics depend on variations in saturation states, temperature, pH, or the presence or absence of foreign ions (Loste, Wilson, Seshadri, & Medrum, 2003; Meldrum &

Cölfen, 2008). However, so far, the extremely low concentration of prenucleation clusters in a supersaturated calcium carbonate solution and the invariably small electron density contrasts between these clusters and the reacting solutions have precluded the use of scattering methods and the characterization of the clusters *in situ*. The data about their existence are primarily derived from titration and analytic centrifugation studies (Gebauer & Cölfen, 2011; Gebauer et al., 2008).

- In SAXS experiments, only a small fraction of the incident X-ray beam is scattered by the sample; however, the required time resolution and scattering intensity are achievable with the use of the high brightness and high flux of synchrotron-based SAXS. Modern pixel array detectors, such as Dectris Pilatus series (Mueller, Wang, & Schulze-Briese, 2012), may be able to provide the time resolution necessary to monitor cluster formation and all transformations reactions over the desired range of length scales, with high-quality data now possible to be acquired in sub-50 ms time-resolved snapshots. On the other hand, time-resolved experiments on systems of low electron density contrast cannot be practically carried out with SAXS instruments using conventional X-ray sources (laboratory SAXS). The required data acquisition time per snapshot with acceptable signal-to-noise ratio reaches many minutes or hours, depending on the scattering contrast and the size of nanostructures. However, it should be pointed out that the laboratory SAXS instruments are also improved constantly in terms of X-ray collimation quality and detector sensitivity and resolution. For instance, recently, the formation of gold clusters (high-scattering contrast system) was followed with time resolution of sub-50 ms using a conventional, laboratory-based SAXS system (Polte et al., 2010).
- The development of crystalline structures of an emergent inorganic nanostructure is not measureable within the q-range recorded by SAXS, unless the developing ordered morphologies show diffraction peaks at low angles like, for example, metal–organic transition phases (Stawski, Veldhuis, Castricum, et al., 2011). This is clearly a hardware limitation related to the very physical area of the detector, but this can be partially overcome by using an auxiliary WAXS detector.
- WAXS is a diffraction method at very high q, which is ideally suited to extract information about the internal structure of the emergent phases, kinetics and energetics of their growth, and basic phase identification. This is valid provided that the phases are crystalline and that the number of particles within the scattering volume is sufficient for the diffraction

pattern to be recorded with any reasonable signal-to-noise ratio. WAXS methods have been used in the last two decades intensively in many inorganic systems to quantify the crystallization mechanisms, kinetics, and energetics of various systems with important biological and/or geochemical implications. For example, in highly anoxic systems, the solution-based crystallization of preformed, poorly ordered iron sulfide (mackinawite, nominal FeS) has been shown to, depending on redox conditions and additives, follow various pathways (Cahill, Benning, Barnes, & Parise, 2000). Under highly reducing conditions, mackinawite transforms first to a highly magnetic and Fe^{2+}/Fe^{3+}-bearing intermediate, greigite (Fe_3S_4, Hunger & Benning, 2007), which ultimately recrystallized and leads to the formation of the geologically stable pyrite (FeS_2). Similarly, *in situ* and time-resolved WAXS combined with electrochemical measurements have been recently used (Ahmed et al., 2010) to quantify in detail the formation pathways and mechanisms of green rust sulfate, a layered double-hydroxide phase with the capacity to reduce a range of inorganic and organic species. Finally, combining *in situ* crystallization with element-partitioning measurements in aqueous solutions, the formation and intertransformation of important iron-bearing phases (ferrihydrite to goethite and hematite, Davidson, Shaw, & Benning, 2008 or ferrihydrite to magnetite, Sumoondur, Shaw, Ahmed, & Benning, 2008) in the absence or presence of various toxic compounds (Vu, Shaw, Brinza, & Benning, 2010) have been quantified using WAXS and other *in situ* and time-resolved X-ray diffraction methods. Elucidating the mechanisms, kinetics, and energetics of these reactions helps understand the fate of toxic compounds in various natural and anthropogenically modified environments and this knowledge plays a crucial role in guiding remediation efforts in contaminated settings. Thus, it is clear that WAXS in combination with SAXS is ideally suited to mark both the onset of transition between the solution and the amorphous stages and the transitions between the amorphous and crystalline stages in any reaction and that the *in situ* and real-time capabilities combined with special experimental settings that they offer can and are more and more used to better understand a plethora of industrial and natural processes.

4. SAXS DATA PROCESSING AND INTERPRETATION

In this section, we present the very basics of scattering data processing and interpretation. We do not explicitly describe WAXS data interpretation since diffraction methods are well established and covered in the literature. However, what we summarize here is only the basic theory of scattering and as such this brief description should be treated accordingly.

4.1. Data treatment

If we consider the case of particles in a solution, the recorded scattering pattern is isotropic because nanometer-range electronic density variations are randomly oriented with respect to each other and the incident beam. In such a case, a 2D scattering pattern is transformed to a 1D scattering curve in a data-reduction step (Craievich, 2002; Glatter & Kratky, 1982; Guinier & Fournet, 1955; Stribeck, 2007). The as-measured intensity values need to be corrected for detector alinearities, fluctuating X-ray beam intensity using the exiting beam ion counter (Fig. 5.2A)—and background scattering. The scattering curve of interest is subsequently obtained by integration of the measured intensities along the arcs having an origin at $q=0$ (center of the beam spot on the detector, hence, $q=0$ is extrapolated from the calibration data), having radii corresponding to consequent discrete q values and lengths defined by the shape of the detector (Fig. 5.2B) (Pauw, 2013).

4.2. Data interpretation

In the case of particles forming from a supersaturated solution, SAXS can be used to derive information about the size and shape of individual particles, their interactions, and their polydispersity, number density (concentration), or molecular weight. However, because these parameters are in fact mutually coupled, only certain combinations of structural elements can be obtained from a single system measured at a given q-range. That means that either certain simplifications concerning the structure need to be made or the sample has to be prepared and measured in a way that enables the exclusion of some of the "unwanted" factors (e.g., interparticle interactions can be avoided by using dilute systems). Similarly, missing data on specific parameters (e.g., shape of particles, polydispersity, and aggregation) must be obtained by other spectroscopic (e.g., vibrational, X-ray absorption, and nuclear magnetic resonance) or imaging methods (e.g., high-resolution scanning or transmission electron microscopy). These limitations result from the very physics of scattering, as will be described further later.

In general, the scattering amplitude $F(\mathbf{q})$ from an individual scattering primary particle that is embedded in a homogenous matrix can be approximated by Eq. (5.2) (Feigin & Svergun, 1987; Glatter & Kratky, 1982;

Guinier & Fournet, 1955; Vachette & Svergun, 2000). In this equation, it is assumed that in a given direction, only the scattering contrast $\Delta\rho(\mathbf{r})$ would be contributing to the scattering intensity. This direction is defined by the scattering vector $\mathbf{q}$ representing the electron density dependent on the position vector $\mathbf{r}$:

$$F(\mathbf{q}) = \int_V \Delta\rho(\mathbf{r})e^{-i\mathbf{q}\cdot\mathbf{r}}d\mathbf{r} \qquad (5.2)$$

Hence, the scattered intensity of a single object, I_1, in the direction defined by a scattering vector is expressed by Eq. (5.3), where F^* is the complex conjugate of F:

$$I_1(\mathbf{q}) = F(\mathbf{q})F^*(\mathbf{q}) = \int_V\int_V \Delta\rho(\mathbf{r})\Delta\rho(\mathbf{r}')e^{-i\mathbf{q}\cdot(\mathbf{r}-\mathbf{r}')}d\mathbf{r}d\mathbf{r}' \qquad (5.3)$$

For our purposes, we only consider (i) an isotropic system (hence describable by scalars q and r) of identical particles and a (ii) system showing no long-distance order (it is "dilute"). For a collection of identical non-interacting particles of particle volume V, number N embedded in the homogenous matrix so that scattering contrast is $\Delta\rho$, the total scattering intensity $I(q)$ can be expressed as

$$I(q) = N\cdot I_1(q) = N(\Delta\rho)^2 V \cdot 4\pi \int_0^{D_{max}} r^2\gamma(r)\frac{\sin(qr)}{qr}dr \qquad (5.4)$$

where D_{max} denotes a maximum dimension within the scattering object and $\gamma(r)$ is known as the correlation function (Feigin & Svergun, 1987; Glatter, 1977; Glatter & Kratky, 1982; Guinier & Fournet, 1955; Vachette & Svergun, 2000). From this equation, we can derive the fact that the scattering intensity at $q=0$—termed $I(0)$—must be equal to $N(\Delta\rho)^2 V^2$.

To simplify matters, more commonly, instead of the correlation function, $p(r) = \gamma(r)r^2$ is used. This is referred to as a pair (distance) distribution function and $p(r)$ is related to the electron density contrast through Eq. (5.5).

$$p(r) = \gamma(r)r^2 = r^2 \left\langle \int_V \Delta\rho(\mathbf{r})\Delta\rho(\mathbf{r}') \right\rangle_\Omega \qquad (5.5)$$

where $\langle\rangle_\Omega$ denotes spherical averaging and Ω is a solid angle in the reciprocal space. $p(r)$ contains information on the complex shape and the structure of the scattering object. For many cases, it can be retrieved from the scattering intensity, as, for example, in the case of scattering from dilute solutions of proteins (Feigin & Svergun, 1987; Putnam, Hammel, Hura, & Tainer, 2007; Vachette & Svergun, 2000). Overall, the value for $p(r)$ relates to $I(q)$ through an inverse Fourier transform:

$$p(r) = \frac{1}{2\pi^2} \int_0^\infty qrI(q)\sin(qr)\,dq \qquad (5.6)$$

By finding $p(r)$, one can extract structural information from scattering objects of complicated shapes, because the pair distribution function (PDF) plot is essentially a histogram of all distances within a particle up to D_{max}. This is valid provided that the considered system is monodisperse and dilute, as is often the case in proteins in solution (naturally depending on sample preparation). Equation (5.6) implies that $I(q)$ is measured at the infinite q-range and furthermore assumes a continuous nature of $I(q)$. However, these two conditions cannot be fulfilled in practice, and a direct Fourier transform of measured intensities often yields unreliable results (Glatter & Kratky, 1982). Instead, $p(r)$ is usually obtained through an indirect Fourier transform method (Glatter, 1977, 1981; Glatter & Kratky, 1982) that is usually implemented via software packages such as GNOM (Svergun, 1992) or GIFT (Brunner-Popela & Glatter, 1997).

It has to be noted, however, that $F(q)$ is also referred to as the amplitude of the scattering form factor, $P(q) = F^2(q)$. In the case of many known and classical geometric shapes (spheres, shells, cylindrical objects, Gaussian chains, etc.), the mathematical expressions for the form factors $P(q)$ are relatively easy to derive analytically and are available readily in the literature (Feigin & Svergun, 1987; Pedersen, 1997). For example, scattering from a collection of noninteracting, monodisperse homogenous spheres of radii R is expressed by Eq. (5.7), which is derived directly from Eq. (5.3), assuming an isotropic system and assuming $\Delta\rho(\mathbf{r}) = \Delta\rho$ for $\mathbf{r}$ inside the particle and $\Delta\rho(\mathbf{r}) = 0$ for $\mathbf{r}$ outside the particle:

$$I(q) = N(\Delta\rho)^2 V^2 P(q) = N(\Delta\rho)^2 V^2 \left(3\frac{\sin(qR) - qR\cos(qR)}{q^3 R^3}\right)^2 \qquad (5.7)$$

However, for changes in shape or other morphological parameters during a reaction, this is often far from trivial. We can apply certain approximations when evaluating SAXS data in order to obtain information about sizes and basic structural characteristics of the investigated species. For instance, for a dilute system of scattering particulate features, a shape-independent radius of gyration, R_g, can be found through applying the Guinier approximation to the form factor, $P(q)$ (Guinier & Fournet, 1955). This is only valid, however, for $qR_g < 1$. Then, the scattering intensity is given by

$$I(q) = N(\Delta\rho)^2 V^2 \exp\left(\frac{-q^2 R_g^2}{3}\right) \tag{5.8}$$

Provided that the particles are sufficiently monodisperse, it is important to note that Eq. (5.8) is independent of the actual shape of the scattering features in the applicable q-range. In addition, a further generalization of the Guinier approximation to highly elongated and plate-like particles allows us to discriminate these geometries and obtain basic information on some of the characteristic dimensions, for example, cross-sectional radius of gyration of elongated particles (Feigin & Svergun, 1987; Glatter & Kratky, 1982; Guinier & Fournet, 1955).

If we aim to also include interparticle correlations in our scattering data evaluations, we can do that by taking into account Eqs. (5.1)–(5.5) and denoting that $P(q) = F^2(q)$. Thus, we can introduce interparticle spatial correlations by implementing a structure factor under the assumption that these correlations have an isotropic character:

$$I(q) = N(\Delta\rho)^2 V^2 P(q) S(q) \tag{5.9}$$

In this equation, $S(q)$ represents a structure factor term describing the mutual arrangement of primary particles of a given form factor, $P(q)$, within the measured sample volume of nondiluted systems. $S(q)$ takes into account any scattering interferences from different species due to any interactions between particles. Such interactions include Coulomb repulsion–attraction phenomena, aggregation and clustering effects, or other spatial correlations within the system. Hence, $S(q)$ provides additional geometric information, and this can be used indirectly to extract information about the dynamic processes and physicochemical interactions between the particles. In analogy to Eq. (5.4), we can express the structure factor $S(q)$ of an isotropic system through a correlation function. For the sake of clarity, this correlation function is denoted as $g(r)$ in Eq. (5.10) (Glatter & Kratky, 1982; Guinier & Fournet, 1955; Pedersen, 1997; Squires, 1978):

$$S(q) = 1 + 4\pi n \int_0^\infty [g(r) - 1] r^2 \frac{\sin(qr)}{qr} dr \tag{5.10}$$

where n denotes the number density of particles.

For dilute system of noninteracting particles, $S(q) = 1$. Please note that in the formalism, $\gamma(r)$ discussed in Eqs. (5.4) and (5.5) is related to the internal structure of the individual scattering primary particles, whereas the $g(r)$ introduced earlier in Eq. (5.10) is related to the arrangement of the primary particles in space. Similarly to $\gamma(r)$ (and $p(r)$), $g(r)$ can under certain circumstances be extracted from the scattering data by reverse Fourier transform. Furthermore, one may include polydispersity, assuming different types of distributions, variable types of averaging (intensity, number, and weight), and assumptions about whether polydispersity is related to primary particles or structures formed by primary particles (e.g., whether it is related to $P(q)$ or $S(q)$; Pedersen, 1997).

It should now be clear from the description earlier that the evolution of inorganic solution-based nanostructures in the prenucleation phase and during the various stages of nucleation, aggregation, and/or growth/crystallization involves the formation of particles of certain sizes and shapes and that these change with time. Furthermore, any of the various stages in a whole reaction can involve processes of agglomeration and coalescence. Hence, the direct extraction of values for either $p(r)$ or $g(r)$ from any set of scattering data is difficult and possible only in a limited number of cases because of the following:

1. In order to extract complete information on the form factor, $P(q)$ (and $p(r)$), the scattering system must be monodisperse and dilute so that either $S(q) = 1$ or $S(q)$ must be explicitly known from other measurements or alternatively convincingly assumed or deduced (Brunner-Popela & Glatter, 1997). In order to account for polydispersity of $P(q)$, and extract information about the distribution, one needs to know (or assume) the form of monodisperse $P(q)$.
2. In order to extract high-resolution information about the structure factor $S(q)$ (and $g(r)$), the form factor must be known or assumed, or for the considered q-range, the interparticle interactions need to occur on a length scale large enough to assume $P(q) = 1$. In order to account for polydispersity of $S(q)$ and extract information about the distribution, one needs to know (or assume) the form of monodisperse $S(q)$.
3. Importantly, both $P(q)$ and $S(q)$ contain relevant information characterizing an evolving system. For instance, dilution to the point where $S(q) = 1$ would mean practically quenching the process and altering the investigated system. Therefore, in the case of complex multistep-formed materials, the behavior of $I(q)$ is usually modeled assuming that $P(q)$ can be represented through simple geometric shapes (Glatter & Kratky, 1982; Pedersen, 1997). Similarly, this applies to $S(q)$, which is

usually modeled within specifically proposed structural arrangements and interactions (hard spheres, sticky hard spheres, fractal aggregates, etc.; Freltoft, Kjems, & Sinha, 1986; Kinning & Thomas, 1984; Pedersen, 1997; Sorensen, 2001; Teixeira, 1988).

A more detailed description and overview of the mathematical functions used to express various form and structure factors can be found in Pedersen (1997). Once derived, the so-obtained mathematical models can then be fitted to experimental data leading to an evaluation of parameter values that will characterize the investigated sample. Clearly, such an approach is often limited by the availability of existing models and their physical validity. Nevertheless, it has proven to be useful in providing information on the microstructure of such inorganic systems. To point a few, we mention here silica (Besselink, Stawski, Castricum, & ten Elshof, 2013; Freltoft et al., 1986; Teixeira, 1988), titania (Jalava et al., 2000; Kamiyama, Mikami, & Suzuki, 1992), zirconia (Riello, Minesso, Craievich, & Benedetti, 2003; Stawski et al., 2012), colloidal precursors for perovskite-type ceramics (Stawski et al., 2011a,2011b, 2012), sodium hydrous sulfate (Brand, Scarlett, Grey, Knott, & Kirby, 2013), or calcium carbonate (Bolze et al., 2002; Pontoni, Bolze, Dingenouts, Narayanan, & Ballauff, 2003).

Finally, it has to be mentioned that after the arduous stages of data processing and handling, the next step of data interpretation constitutes an equally nontrivial element when it comes to SAXS measurements. Additionally, because conventional SAXS is an electron density contrast-based method, it does not allow extracting chemical (compositional) information. Therefore, other auxiliary methods must be used to support findings obtained by SAXS. These would include, among others, element-specific anomalous (resonant) A-SAXS; small-angle neutron scattering and contrast matching; wide-angle X-ray scattering, that is, diffraction (WAXS) and atomic PDF analysis; X-ray absorption spectroscopy; various vibrational spectroscopic analyses; and natural hard and soft X-ray and electron microscopic approaches combined with related electron energy loss and energy dispersive spectroscopies and electron diffraction.

5. CASE STUDIES

5.1. Formation and growth of amorphous CaCO$_3$ and its crystallization

For the research community working in the field of biomineralization, investigations about the mechanisms controlling the precipitation of CaCO$_3$

from supersaturated solutions constitute a key aspect of research. However, it was only in the last 10 years, once time-resolved SAXS/WAXS was successfully applied for the first time to study these processes (Bolze et al., 2002), and that advances could be made. In the considered experiments, calcium carbonate was precipitated from aqueous solution of $CaCl_2 \cdot 2H_2O$ and Na_2CO_3 mixed using a stopped-flow device. By combining synchrotron-based SAXS data acquisition with a fast stopped-flow injection and mixing system, Bolze et al. (2002) were able to achieve time resolution of sub-100 ms and this allowed them to follow the nucleation and growth reaction through the changes in SAXS patterns for many minutes (Fig. 5.3). They

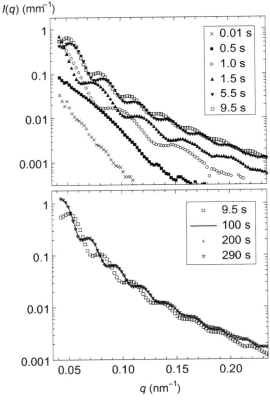

Figure 5.3 Time-resolved SAXS data measured after rapidly mixing aqueous solutions of $CaCl_2$ and Na_2CO_3 in a stopped-flow apparatus. After a short induction period of less than 500 ms, the scattering patterns exhibit marked oscillations. Particle growth is evidenced from the increase in the oscillation frequency and the concomitant increase of the intensity. *Printed with permission from Bolze et al. (2002). Copyright 2002 American Chemical Society.*

found that the developing particles were describable within the polydisperse spherical form factor of very narrow distribution but that they had diameters up to ca. 270 nm and the polydispersity was decreasing with the increasing size of the spheres. Their results also showed that the forming particles were amorphous in nature, which was further confirmed (Pontoni et al., 2003) by the lack of diffraction peaks in the simultaneously collected WAXS patterns. The formed ACC particles were found to remain "stable" in solution as isolated nonagglomerated entities for the measured time of 5 min. This conclusion was derived from data collected over a q-range extended to low values of the scattering vector, where no increase in $I(q)$ was observed and the assumption that such an increase would normally be attributable to the presence of agglomerates. Furthermore, the authors showed that after a short nucleation period, the number density of the growing spheres remained constant. From the evaluation of the absolute scattering intensities, the particle mass density was determined to be ca. 1.62 g/cm^3, which was lower than the density of the crystalline polymorphs of $CaCO_3$. By using time-resolved SAXS/WAXS, the transformation of ACC into microcrystals was observed in supersaturated solutions *in situ*. It was possible to detect the onset of particle dissolution by changes in the frequency of the oscillations in the scattered intensity typical of the spherical form factor (Fig. 5.4A), followed by the growth of crystalline $CaCO_3$ demonstrating itself as clear single diffraction peaks in the 2D WAXS pattern (Fig. 5.4B). By these means, it was deduced that crystalline polymorphs of $CaCO_3$ were formed from ACC via dissolution and subsequent heterogeneous nucleation of the crystals. Hence, the authors did not observe any solid–solid transition.

However, this latter transition from ACC to stable crystalline $CaCO_3$ phases has more recently been reevaluated also by simultaneous SAXS and WAXS (Bots et al., 2012; Rodriguez-Blanco, Bots, Roncal-Herrero, Shaw, & Benning, 2012; Rodriguez-Blanco et al., 2011). Using higher supersaturations compared to Bolze et al. (2002), Bots et al. (2012) showed that the crystallization of pure ACC to vaterite (μ-$CaCO_3$) proceeded via multiple stages but that the disappearance of the precipitated ACC particles, which were only 35–40 nm in diameter, occurred in less than 90 s (Fig. 5.5). These initial ACC nanoparticles dehydrated and dissolved and concomitantly vaterite formed via a nucleation-controlled mechanism. The data clearly showed that already after 2 min, vaterite nanocrystals, initially only ~9 nm in diameter, formed. In the next ~15 min, in a secondary stage, these individual nanocrystals grew to micron-sized spherulites. Noteworthy

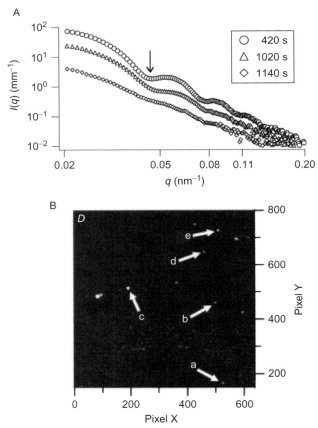

Figure 5.4 (A) SAXS curves after mixing reagent solutions for CaCO$_3$. The arrow indicates the first minimum of the spherical form factor. Its shift toward larger scattering angles indicates dissolution of the amorphous particles. The particles become more polydisperse, as suggested by the less pronounced oscillations in the scattering curve. The overall intensity decrease is due to particle dissolution and sedimentation. (B) WAXS pattern showing diffraction spots that can be assigned to the various crystalline metamorphs of calcium carbonate. The spots indicated by arrows correspond to the following Bragg reflections: (a) calcite (012), (b) vaterite (004), (c) aragonite (032), (d) aragonite (310), and (e) aragonite (302). *Printed with permission from Pontoni et al. (2003). Copyright 2003 American Chemical Society.*

however is the fact that each of these spherulites in turn was made up of individual nanoparticles that reached ~60 nm in size. This intermediate stage of crystallization was rather short and in a subsequent stage the vaterite transformed to the stable calcite. This final reaction stage can, depending on temperature, last from a few to many hours (Rodriguez-Blanco et al., 2011) and the vaterite spherulites transform to calcite, via a slower dissolution and reprecipitation mechanism. Interestingly, any change in parameters will also lead to a dramatic change in the crystallization pathway and mechanism. For example, the addition of magnesium leads to the direct transformation of ACC to calcite with no vaterite intermediate (Rodriguez-Blanco et al., 2012), while the addition of organics (e.g., ethanol, Sand, Rodriguez-Blanco, Makovicky, Benning, & Stipp, 2012) will stabilize aragonite instead of calcite (Fig. 5.5).

5.2. Nucleation and growth of silica nanoparticles

The mechanisms controlling silica precipitation are essential to the understanding of natural processes involving this material such as biosilicification (Konhauser, Jones, Phoenix, Ferris, & Renaut, 2004) but are also relevant to industrial applications when desired microstructure of silica is of crucial importance (Besselink et al., 2013; Brinker & Scherer, 1990; Iler, 1979). As a matter of fact, silica is probably one of the most intensively investigated by SAXS inorganic systems. Major concepts of clustering and aggregation, for example, mass-fractal aggregates (Freltoft et al., 1986; Teixeira, 1988), were experimentally derived and coined on the basis of silica-based systems. Nevertheless, in-depth studies about the nature and evolution of silica are in part still in their infancy. This is despite the fact that we desperately need this information in particular with a view towards improving our understanding of how the intricate biological shapes are created by living organisms (e.g., diatoms) as they control the global Si cycle in marine settings. Similarly, we need to understand how silica precipitates in geothermal systems and why and how to possibly prevent the clogging of pipes of heat exchangers in geothermal power plants in order to increase energy production.

In this worked example based on Tobler et al. (2009) and Tobler and Benning (2013), the nucleation and growth of silica nanoparticles was followed in aqueous solutions with initial silica concentrations of 640–1600 ppm, at different ionic strengths and at different temperatures. Silica polymerization and silica nanoparticle formation were induced either by adjusting the high-pH precursor solution to 7 or by inducing polymerization through an increase in supersaturation due to a fast temperature drop from 230 to 30–60 °C. SAXS scattering curves were collected for up to 3 h at a time resolution of 5 min/pattern. This was primarily because the electron density contrast between silica and aqueous solution was very low and longer time lengths were needed to improve a signal-to-noise ratio.

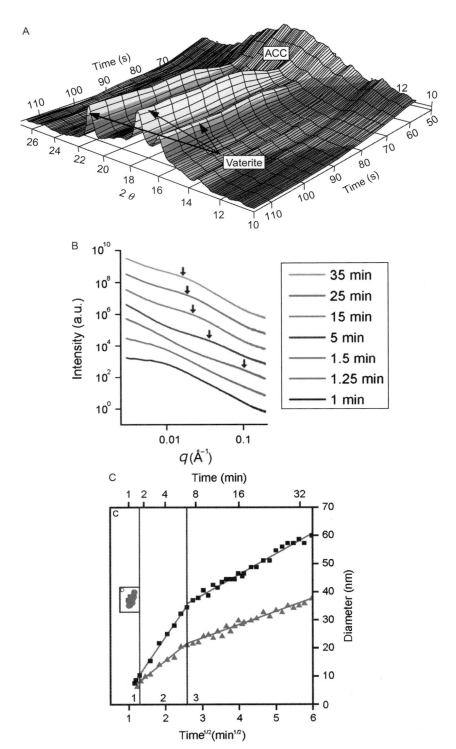

Figure 5.5—Cont'd

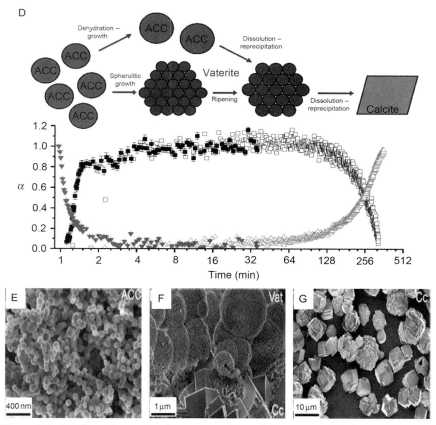

Figure 5.5 (A) Three-dimensional representations of the time-resolved WAXS patterns from the pure ACC experiment (time is plotted on a base 2 log scale for clarity); (B) stacked time series of selected SAXS patterns from the pure ACC experiment, with the legend showing time in minutes and the arrows illustrating the position of the peaks caused by the scattering from the growing vaterite crystallites. (C) ACC nanoparticle and vaterite crystallite sizes derived from the SAXS data versus time (on a $t^{1/2}$ scale) for the pure ACC and an ACC doped with SO_4 experiments. (D) Schematic representation of the multistage ACC → vaterite → calcite crystallization pathway. (E–G) Electron microscope microphotographs of solids quenched throughout a full ACC to calcite transformation reaction. *Printed with permission from Bots et al. (2012). Copyright 2012 American Chemical Society.* (See color plate.)

In Fig. 5.6A, a typical time-resolved SAXS curve from an experiment with one of the experimental solutions of SiO_2 is presented. The evolving system could practically, at all times, be treated as dilute, and the primary particles grew, but the interparticle interactions, such as aggregation, were practically negligible. These conclusions were drawn from the scattering curves where

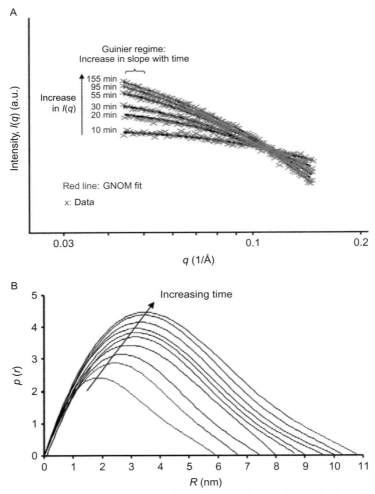

Figure 5.6 (A) $I(q)$ as a function of time for 640 ppm SiO_2. Solid lines depict GNOM fits; (B) pair distribution function plots, $p(r)$, of scattered silica nanoparticles as a function of r and time evaluated with GNOM. *Printed with permission from Tobler et al. (2009). Copyright 2009 Elsevier.* (For color version of this figure, the reader is referred to the online version of this chapter.)

two dominant features were present: (i) the overall increase in scattering intensity, $I(q)$ with time and (ii) the progressive shift of the scattering curves towards lower q in time. The increase in $I(q)$ was attributed to a change in electron density contrast between the matrix and the newly formed particles and a change in the total scattering volume of particles (i.e., increase in particle volume or number), whereas the shift of the scattering curves towards lower q indicated an increase in the particles' sizes with time. Therefore, the cited example constitutes the case when $p(r)$ could be extracted from the scattering curves using the indirect Fourier transform methods, implemented, for example, in GNOM (Svergun, 1992). An example of a time-resolved distance distribution function, $p(r)$, for a polymerizing solution with 1600 ppm SiO_2 is shown in Fig. 5.6B. Over time, the $p(r)$ plot showed an increase in both the area under the curve and a shift in the apex of the curve indicating an increase in particle size. In the case of perfect, monodispersed spheres, the shape of each individual $p(r)$ curve should be Gaussian. Despite the $p(r)$ being slightly right-skewed, the near-Gaussian shape of the $p(r)$ curves obtained in this study suggested relatively low polydispersity of the system, although the actual distributions cannot be evaluated from these measurements due to the limited q-range. Hence, it was concluded that the observed tail could be induced by the presence of a few aggregates or it could also indicate the presence of some degree of polydispersity. The authors noted that the overall shape of the $p(r)$ curves did not differ between experiments (i.e., over the studied silica concentration, ionic strength, or temperature conditions), suggesting that the shape of the growing particles did not change between experiments. Electron microscopic evaluations using both conventional and cryoelectron microscopy verified the SAXS data and confirmed the spherical and hydrous structure of the forming silica nanoparticles. Information obtained from the SAXS patterns was further used to extract kinetic information about the process by fitting relevant kinetic models to the data. The results revealed that once polymerization was induced, a fast decrease in monomeric silica in solution was accompanied by a simultaneous increase in scattering intensity through the formation of silica nanoparticles with a ∼1–2 nm diameter. With time, these grew to reach a final size of about 7–8 nm. They showed that the nucleation and growth of silica nanoparticles from supersaturated solutions follows a three-stage process that proceeds through the homogenous and instantaneous nucleation of these initial nanoparticles followed by a 3D surface-controlled particle growth following a first-order reaction kinetics and in a latter step through Ostwald ripening and particle aggregation.

5.3. Evolution of the colloidal precursors for BaTiO$_3$

This example addresses the evolution of colloidal amorphous precursors for barium titanate also followed by time-resolved SAXS (Stawski et al., 2011a,2011b). BaTiO$_3$ is not a naturally occurring material, and it is derived from alkoxides–carboxylate sol–gel process. This resembles in numerous aspects the previously described and well-studied similarly alkoxide-based formation of amorphous silica and again illustrates well the advantages and limitations of scattering methods. The $BaTiO_3$ precursor system is based on two components mixed together: titanium(IV) alkoxide dissolved in alcohol and barium acetate dissolved in acetic acid. Upon hydrolysis, initiated by water addition, sol particles of Ti(IV)oxo(hydroxo)acetate formed through condensation and further agglomeration resulting in an amorphous gel. The gelation process was attributed to the evolution of Ti-based phase and the formation of amorphous titania (TiO_x). Based on the SAXS and other complementary data sets (e.g., rheological characterization), the process could be derived into three growth stages: (i) formation of spherical primary Ti-(hydroxo)oxo clusters with an inorganic core diameter $2r_0 = 0.9$ nm and an outer organic ligand shell of ~ 0.45 nm thickness, (ii) agglomeration of these primary particles into mainly mass-fractal-like agglomerates but with a minor component of correlated semiordered agglomerates, (iii) and cluster–cluster aggregation of structures from the previous phase into a 3D macroscopic gel of low density. The morphology and the kinetics of formation of the gel depended on initial hydrolysis conditions and the entire gelation process took between 0.5 and 3 h. The time-resolved scattering curves obtained from the *in situ* evolving system are presented in Fig. 5.7A.

The previously mentioned structural information was based on a scattering model developed to describe the precursor solution in terms of a mixture of mass-fractal-like agglomerates and structures with internal correlations, that is, semiordered agglomerates of similarly sized primary spherical particles. Both types of structures were composed of the same type of particles with a radius r_0 and a spherical form factor, $P(q)$. The fractal-like branched oligomeric structures were described using the mass-fractal structure $S_F(q)$ as defined in Freltoft et al. (1986), Sorensen (2001), and Teixeira (1988). The parameter D, the fractal dimension, can have values between 1 and 3, and this value relates the number of primary particles n of radius r_0 to its radius of gyration R_g. Therefore, a mass-fractal agglomerate fulfills the condition (Sorensen, 2001):

$$n \propto \left(\frac{R_g}{r_0}\right)^D \quad (5.11)$$

The second type of entities in solution was semiordered agglomerates of scatterers with similar size. These were described best by the hard-sphere

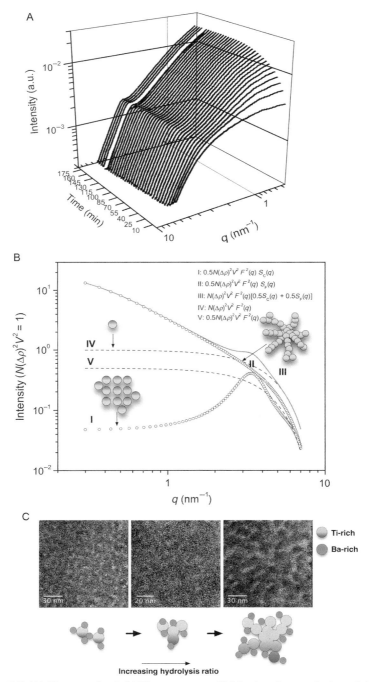

Figure 5.7 (A) Time-resolved SAXS patterns at 60 °C, showing evolution of barium titanate precursor sols of initial 0.5 mol/dm^3 concentration with hydrolysis ratio [H$_2$O]/[Ti] = 5.6. (B) Simulated SAXS curves using Eq. (5.12), with $N(\Delta\rho)^2 V^2$ set to 1.

(*Continued*)

Percus–Yevick structure factor $S_C(q)$ (Kinning & Thomas, 1984; Percus & Yevick, 1958) and depend on the hard-sphere radius R_{HS} (correlation radius) of the randomly packed spherical particles and their volume fraction v. The degree of ordering of such particles, located at a distance $2R_{HS}$ from each other, is expressed by v. To account for the fact that the relative contributions to scattering of the fractal and semiordered agglomerates in solution vary with time, a weighting factor $0 < \varepsilon < 1$ was introduced. This described the contribution of the fractal-like structures to the total scattering intensity. The scattering intensity from the sol upon gelation as the function of time was thus given by

$$I(q,t) = N(\Delta\rho)^2 P(q,t)\{\varepsilon(t)S_F(q,t) + [1 - \varepsilon(t)]S_C(q,t)\} \quad (5.12)$$

A few examples of scattering curves that were simulated using Eq. (5.12) are presented in Fig. 5.7B. The SAXS data as such did not indicate whether the emergent nanostructures were Ti- or Ba-related. This problem was overcome by juxtapositioning the results from SAXS with those from electron energy loss spectroscopy, (EELS) and energy-filtered transmission electron microscopy obtained from dried gel samples produced using various hydrolysis ratios (Fig. 5.7C). The resulting nanoscale resolution elemental maps allowed the authors demonstrate that processing of multicomponent amorphous precursors for barium titanate leads to spatially nonhomogenous structures resulting from the growth of amorphous titania only.

6. OUTLOOK

The previously described worked examples showing applications of SAXS in inorganic and bioinspired research and some of the examples contained in the references in this chapter cover only a miniscule fraction of SAXS knowledge base relevant to investigations related to the formation processes of inorganic phases from solution.

Figure 5.5—Cont'd (i) $I(q)$ of semiordered agglomerates with $r_0 = 0.5$ nm, $v = 0.3$, $2R_{HS} = 1.8$ nm, and $\varepsilon = 0.5$. (ii) $I(q)$ of fractal agglomerates with $r_0 = 0.5$ nm, $D = 1.8$, $R_g = 4.76$ nm, and $\varepsilon = 0.5$. (iii) $I(q)$ of linear combination of structures i and ii. (iv) $I(q)$ of sphere form factor of $r_0 = 0.5$ nm. (v) $I(q)$ of sphere form factor of $r_0 = 0.5$ nm and $\varepsilon = 0.5$. (C) EELS mappings of Ba (red) and Ti (green) of as-dried BTO films with increasing hydrolysis ratios. *Panels (A) and (B) are printed with permission from Stawski et al. (2011a) and panel (C) is printed with permission from Stawski et al. (2011b). Copyright 2011 American Chemical Society.* (See color plate.)

The advances in detector technology in terms of acquisition accuracy, speed, and signal-to-noise levels now allow us to follow a process faster or at higher spatial resolution. This leads unfailingly to a dramatic increase in the amount of data acquired in each experiment. In addition, the higher quality of data that can now be achieved allows us to derive much more information from each data set. However, inconsistencies and limitations in data reduction and limited theoretical understanding of how we should or can process and/or model the resulting data sets are in many cases the bottleneck in our ability to quantitatively assess even simple solution-based nucleation and growth reactions. This is caused by the fact that existing theories of scattering do not handle well simultaneous contributions from different form factors, structure factor, and polydispersity. On the other hand, it should be stated that scattering methods are far more used and applied in fields such as polymer physics or protein structural biology. It is not an exaggeration if one says that the progress in our understanding of scattering, as well as development of data interpretation methods and theories, is due to the challenges in these two broad fields of interest. The inorganic scattering world in comparison is still in its infancy, yet advances in our theoretical understanding of nanoparticle interactions and how scattering data should be interpreted will lead to further advancement in the next decade. As long as scattering models developed in the other fields can be transferred to the solution-based nucleation and growth phenomena in inorganic systems, further progress in the data interpretation methods will also be essential. For example, one such advance is the use of reverse Monte Carlo modeling especially suited for disordered systems. Such approaches open up new possibilities in SAXS data interpretation in combination with WAXS and XAS (McGreevy & Pusztai, 1988).

In fact, electron microscopy is the most powerful ally of SAXS. For example, by using cryoquench TEM, one can obtain information from different stages of the synthesis reaction by cryoquenching of samples, which allows to look at various time-resolved "snapshots" of the reaction. With this approach, samples in their liquid state are rapidly (<1 s) cryoquenched and flash-frozen and subsequently imaged in cryomode at high resolution. Such cryoimaging of developing structures is a major first step in understanding reactions at high resolution and these are highly complementary to any SAXS or WAXS data sets. However, one of the most exciting recent developments is a liquid-cell system for TEM/STEM, where the self-assembly of the forming structures can be followed *in situ* (Li et al., 2012; Liao, Cul, Whitelam, & Zheng, 2012). In conventional electron microscopy, samples are usually exposed to very high vacuum, completely removing the solvent environment, while in cryoquenched systems, a continuous process cannot be followed. Rapid drying or cryoquenching often leads to artifacts, and this hinders the direct data correlation with the other measurements in which the solvent environment is preserved (i.e., SAXS/WAXS). Using a liquid-cell system in conjunction with the high-resolution imaging allows us to characterize the dynamics and progress of reactions at the nanoscale but with the samples still in the fluid phase. In such a TEM fluid cell, a droplet of the reacting liquid sample is sealed between two electron-transparent membranes. This allows for quantitative and qualitative image and spectral analyses of the nucleation, crystallization, self-assembly, and agglomeration processes of inorganic phases at an unprecedented resolution and in a most direct manner. This way, a direct link between *in situ* TEM imaging and scattering data from *in situ* synchrotron-based SAXS/WAXS is warranted.

ACKNOWLEDGMENTS

This work was made possible by a Marie Curie grant from the European Commission in the framework of the MINSC ITN (Initial Training Research network), project number 290040.

REFERENCES

Ahmed, I. A. M., Benning, L. G., Kakonyi, G., Sumoondur, A., Terrill, N. J., & Shaw, S. (2010). The formation of green rust sulfate: Situ and time-resolved scattering and electrochemistry. *Langmuir, 26*(9), 6593–6603.

Baumgartner, J., Dey, A., Bomans, P. H. H., Le Coadou, C., Fratzl, P., Sommerdijk, N. A. J. M., et al. (2013). Nucleation and growth of magnetite from solution. *Nature Materials, 12*, 310–314.

Becker, D., & Döring, W. (1935). The kinetic treatment of nuclear formation in supersaturated vapors. *Annalen der Physik, 24*, 719–752.

Benning, L. G., & Waychunas, G. (2007). Nucleation, growth, and aggregation of mineral phases: Mechanisms and kinetic controls. In S. L. Brantley, J. D. Kubicki, & A. F. White (Eds.), *Kinetics of water-rock interaction*. New York: Springer-Verlag, chapter 7.

Besselink, R., Stawski, T. M., Castricum, H. L., & ten Elshof, J. E. (2013). Evolution of microstructure in mixed niobia-hybrid silica thin films from sol–gel precursors. *Journal of Colloid and Interface Science, 404*, 24–35.

Bolze, J., Peng, B., Dingenouts, N., Panine, P., Narayanan, T., & Ballauff, M. (2002). Formation and growth of amorphous colloidal $CaCO_3$ precursor particles as detected by time-resolved SAXS. *Langmuir, 18*, 8364–8369.

Bots, P., Benning, L. G., Rodriguez-Blanco, J.-D., Roncai-Herrero, T., & Shaw, S. (2012). Mechanistic insights into the crystallization of amorphous calcium carbonate (ACC). *Crystal Growth and Design, 12*(7), 3806–3814.

Brand, H. E. A., Scarlett, N. V. Y., Grey, I. E., Knott, R. B., & Kirby, N. (2013). In situ SAXS studies of the formation of sodium jarosite. *Journal of Synchrotron Radiation, 20*(4), 626–634.

Bras, W. (1998). An SAX/WAXS beamline at the ESRF and future experiments. *Journal of Macromolecular Science, Part B: Physics, 37*(4), 557–565.

Brinker, C. J., & Scherer, G. W. (1990). *Sol-gel: The physics and chemistry of sol-gel processing*. London: Academic Press.

Brunner-Popela, J., & Glatter, O. (1997). Small-angle scattering of interacting particles. I. Basic principles of a global evaluation technique. *Journal of Applied Crystallography, 30*(4), 431–442.

Cahill, C. L., Benning, L. G., Barnes, H. L., & Parise, J. B. (2000). In situ time-resolved X-ray diffraction of iron sulfides during hydrothermal pyrite growth. *Chemical Geology, 167*(1–2), 53–63.

Craievich, A. F. (2002). Synchrotron SAXS studies of nanostructured materials and colloidal solutions. A review. *Materials Research, 5*(1), 1–11.

Davidson, L. E., Shaw, S., & Benning, L. G. (2008). The kinetics and mechanisms of schwertmannite transformation to goethite and hematite under alkaline conditions. *American Mineralogist, 93*, 1326–1337.

Dey, A., Bomans, P. H. H., Müller, F. A., Will, J., Frederik, P. M., de With, G., et al. (2010). The role of prenucleation clusters in surface-induced calcium phosphate crystallization. *Nature Materials, 9*, 1010–1014.

Feigin, L. A., & Svergun, D. I. (1987). *Structure analysis by small-angle X-ray and neutron scattering*. New York and London: Plenum Press.

Fratzl, P., Misof, K., Zizak, I., Rapp, G., Amenitsch, H., & Bernstorff, S. (1997). Fibrillar structure and mechanical properties of collagen. *Journal of Structural Biology, 122*(1–2), 119–122.

Freltoft, T., Kjems, J. K., & Sinha, S. K. (1986). Power-law correlations and finite-size effects in silica particle aggregates studied by small-angle neutron scattering. *Physical Review B, 33*, 269–275.

Gebauer, D., & Cölfen, H. (2011). Prenucleation clusters and non-classical nucleation. *Nano Today, 6*, 564–584.

Gebauer, D., Völkel, A., & Cölfen, H. (2008). Stable prenucleation calcium carbonate clusters. *Science, 322*(5909), 1819–1822.

Glatter, O. (1977). A new method for the evaluation of small-angle scattering data. *Journal of Applied Crystallography, 10*(5), 415–421.

Glatter, O. (1981). Convolution square root of band-limited symmetrical functions and its application to small-angle scattering data. *Journal of Applied Crystallography, 12*(2), 101–108.

Glatter, O., & Kratky, O. (Eds.), (1982). *Small angle X-ray scattering*. London: Academic Press.

Gower, L. B. (2008). Biomimetic model systems for investigating the amorphous precursor pathway and its role in biomineralization. *Chemical Reviews, 108*, 4551–4627.

Guinier, A., & Fournet, G. (1955). *Small angle scattering of X-rays*. New York: John Wiley & Sons Inc.

Huang, T. C., Toraya, H., Blanton, T. N., & Wu, Y. (1993). X-ray powder diffraction analysis of silver behenate, a possible low-angle diffraction standard. *Journal of Applied Crystallography, 26*(2), 180–184.

Hunger, S., & Benning, L. G. (2007). Greigite: The intermediate phase on the pyrite formation pathway. *Geochemical Transactions, 8*(1), 1–20.

Iler, R. K. (1979). *The chemistry of silica*. New York: John Wiley & Sons.

Jalava, J.-H., Hiltunen, E., Kähkönen, H., Erkkilä, H., Härmä, H., & Taavitsainen, V.-M. (2000). Structural investigation of hydrous titanium dioxide precipitates and their formation by small-angle X-ray scattering. *Industrial & Engineering Chemistry Research, 39*, 349–361.

Kamiyama, T., Mikami, M., & Suzuki, K. (1992). A SAXS study of the gelation process of silicon and titanium alkoxides. *Journal of Non-Crystalline Solids, 150*, 157–162.

Kinning, D. J., & Thomas, E. L. (1984). Hard-sphere interactions between spherical domains in diblock copolymers. *Macromolecules, 17*, 1712–1718.

Konhauser, K. O., Jones, B., Phoenix, V. R., Ferris, F. G., & Renaut, R. W. (2004). The microbial role in hot spring silicification. *Ambio, 33*, 552–558.

Li, D., Nielsen, M. H., Lee, J. R. I., Frandsen, C., Banfield, J. F., & De Yoreo, J. J. (2012). Direction-specific interactions control crystal growth attachment. *Science, 336*, 1014–1018.

Liao, H.-G., Cul, L., Whitelam, S., & Zheng, H. (2012). Real-time imaging of Pt_3Fe nanorod growth in solution. *Science, 336*, 1011–1014.

Loste, E., Wilson, R. M., Seshadri, R., & Medrum, F. C. (2003). The role of magnesium in stabilizing amorphous calcium carbonate and controlling calcite morphologies. *Journal of Crystal Growth, 254*, 206–218.

McGreevy, R. L., & Pusztai, L. (1988). Reverse Monte Carlo simulation: A new technique for the determination of disordered structures. *Molecular Simulation, 1*(6), 359–367.

Meldrum, F. C., & Cölfen, H. (2008). Controlling mineral morphologies and structures in biological and synthetic system. *Chemical Reviews, 108*, 4332–4432.

Meldrum, F. C., & Sear, R. P. (2008). Now you see them. *Science, 322*, 1802–1803.

Mueller, M., Wang, M., & Schulze-Briese, C. (2012). Optimal fine φ-slicing for single-photon-counting pixel detectors. *Acta Crystallographica Section D: Biological Crystallography, D68*, 42–56.

Pauw, B. R. (2013). Everything SAXS: small-angle scattering pattern collection and correction. *Journal of Physics: Condensed Matter, 25*, 383201 (24pp).

Pedersen, J. K. (1997). Analysis of small-angle scattering data from colloids and polymer solutions: Modelling and least-squares fitting. *Advances in Colloid and Interface Science, 70*, 171–210.

Percus, J. K., & Yevick, G. J. (1958). Analysis of classical statistical mechanics by means of collective coordinates. *Physical Review, 110*, 1–13.

Polte, J., Erler, R., Thünermann, A. F., Sokolov, S., Ahner, T. T., Rademannm, K., et al. (2010). Nucleation and growth of gold nanoparticles studied via in situ small angle X-ray scattering at millisecond time resolution. *ACS Nano, 4*(2), 1076–1082.

Pontoni, D., Bolze, J., Dingenouts, N., Narayanan, T., & Ballauff, M. (2003). Crystallization of calcium carbonate observed in-situ by combined small- and wide-angle X-ray scattering. *The Journal of Physical Chemistry, B, 107*, 5123–5125.

Putnam, C. D., Hammel, M., Hura, G. L., & Tainer, J. A. (2007). X-ray solution scattering (SAXS) combined with crystallography and computation: Defining accurate macromolecular structures, conformations and assemblies in solution. *Quarterly Reviews of Biophysics, 40*(3), 191–285.

Rieckel, C., Burghammer, M., & Müller, M. (2000). Microbeam small-angle scattering experiments and their combination with microdiffraction. *Journal of Applied Crystallography, 33*, 421–423.

Riello, P., Minesso, A., Craievich, A., & Benedetti, A. (2003). Synchrotron SAXS study of the mechanisms of aggregation of sulfate zirconia sols. *The Journal of Physical Chemistry B, 107*(15), 3390–3399.

Rodriguez-Blanco, J.-D., Bots, P., Roncal-Herrero, T., Shaw, S., & Benning, L. G. (2012). The role of pH and Mg on the stability and crystallization of amorphous calcium carbonate. *Journal of Alloys and Compounds, 536*(S1), S477–S479.

Rodriguez-Blanco, J.-D., Shaw, S., & Benning, L. G. (2008). How to make stable ACC: Protocol and structural characterization. *Mineralogical Magazine, 72*(1), 283–286.

Rodriguez-Blanco, J.-D., Shaw, S., & Benning, L. G. (2011). The kinetics and mechanisms of amorphous calcium carbonate (ACC) crystallization to calcite, via vaterite. *Nanoscale, 3*, 265–271.

Sand, K. K., Rodriguez-Blanco, J.-D., Makovicky, E., Benning, L. G., & Stipp, S. S. L. (2012). Crystallization of $CaCO3$ in water/alcohol mixtures: Spherulitic growth, polymorph stabilization and morphology change. *Crystal Growth and Design, 12*(2), 842–853.

Sorensen, C. M. (2001). Light scattering by fractal aggregates: A review. *Aerosol Science and Technology*, *35*, 648–687.

Squires, G. L. (1978). *Introduction to the theory of thermal neutron scattering*. Cambridge: Cambridge University Press.

Stawski, T. M., Besselink, R., Veldhuis, S. A., Castricum, H. L., Blank, D. H. A., & ten Elshof, J. E. (2012). Time-resolved small angle X-ray scattering study of sol-gel precursor solutions of lead zirconate titanate and zirconia. *Journal of Colloid and Interface Science*, *369*, 184–192.

Stawski, T. M., Veldhuis, S. A., Besselink, R., Castricum, H. L., Portale, G., Blank, D. H. A., et al. (2011a). Nanoscale structure evolution in alkoxide-carboxylate sol-gel precursor solutions of barium titanate. *The Journal of Physical Chemistry C*, *115*(42), 20449–20459, Correction to nanoscale structure evolution in alkoxide-carboxylate sol-gel precursor solutions of barium titanate. The Journal of Physical Chemistry C, 115, 24028-24028.

Stawski, T. M., Veldhuis, S. A., Besselink, R., Castricum, H. L., Portale, G., Blank, D. H. A., et al. (2011b). Nanostructure development in alkoxide-carboxylate-derived precursor films of barium titanate. *The Journal of Physical Chemistry C*, *116*(1), 425–434.

Stawski, T. M., Veldhuis, S. A., Castricum, H. L., Keim, E. G., Eeckhaut, G., Bras, W., et al. (2011). Development of nanoscale inhomogeneities during drying of sol-gel derived amorphous lead zirconate titanate precursor thin films. *Langmuir*, *27*(17), 11081–11089.

Stribeck, N. (2007). *X-ray scattering of soft matter*. Berlin, Heidelberg, New York: Springer.

Sumoondur, A., Shaw, S., Ahmed, I., & Benning, L. G. (2008). Green rust a precursor for magnetite: An in situ synchrotron based study. *Mineralogical Magazine*, *72*(1), 201–204.

Svergun, D. I. (1992). Determination of the regularization parameter in indirect-transform methods using perceptual criteria. *Journal of Applied Crystallography*, *25*(4), 495–503.

Teixeira, J. (1988). Small-angle scattering by fractal systems. *Journal of Applied Crystallography*, *21*(6), 781–785.

Tobler, D. J., & Benning, L. G. (2013). The in situ and time resolved nucleation and growth of silica nanoparticles under simulated geothermal conditions. *Geochimica et Cosmochimica Acta*, *144*, 156–168.

Tobler, D. J., Shaw, S., & Benning, L. G. (2009). Quantification of initial steps of nucleation and growth of silica nanoparticles. An in-situ SAXS and DLS study. *Geochimica et Cosmochimica Acta*, *73*(18), 5377–5393.

Vachette, P., & Svergun, D. I. (2000). Small-angle X-ray scattering by solutions of biological macromolecules. In E. Fanchon, G. Geissler, J.-L. Hodeau, J.-R. Regnard, & P. A. Timmins (Eds.), *Structure and dynamics of biomolecules*.New York: Oxford University Press, Chapter 11.

Volmer, M., & Weber, A. (1925). Keimbildung in übersättigten Gebilden. *Zeitschrift für Physikalische Chemie*, *119*, 277–301.

Vu, H. P., Shaw, S., Brinza, L., & Benning, L. G. (2010). Crystallization of hematite (alpha-Fe_2O_3) under alkaline condition: The effects of Pb. *Crystal Growth and Design*, *10*(4), 1544–1551.

Wright, J. D., & Sommerdijk, N. A. J. M. (2001). *Sol-gel materials chemistry and applications. Advanced chemistry texts*Boca Raton: CRC Press.

CHAPTER SIX

In Situ Solution Study of Calcium Phosphate Crystallization Kinetics

Haihua Pan*, Shuqin Jiang*,†, Tianlan Zhang‡, Ruikang Tang†,1

*Qiushi Academy for Advanced Studies, Zhejiang University, Hangzhou, China
†Department of Chemistry, Centre for Biomaterials and Biopathways, Zhejiang University, Hangzhou, China
‡Department of Chemical Biology, Peking University School of Pharmaceutical Sciences, Beijing, China
1Corresponding author: e-mail address: rtang@zju.edu.cn

Contents

1. Introduction 130
2. Solution Preparation and Experimental Considerations 132
   2.1 Stock solution preparation 132
   2.2 Working solution preparation 133
   2.3 Carbon dioxide 133
   2.4 Solution crystallization 134
3. *In Situ* Experiments 135
   3.1 Constant composition 135
   3.2 pH meter 137
   3.3 Stopped-flow spectrophotometer 137
4. Data Handling/Processing 138
   4.1 Constant composition 138
   4.2 pH curve 139
   4.3 Stopped-flow spectrophotometer 140
5. Summary 141
References 141

Abstract

Calcium phosphate minerals are the main inorganic component of the bones and teeth. It is of fundamental importance to discover the effect of biomolecules on kinetics of nucleation and crystal growth of calcium phosphate crystals, which shed light on the understanding of the mechanism of biomineralization. Here, we introduce some general solution-based *in situ* detection methods, including analyses of chemical composition and pH and stopped-flow spectrophotometry, to study the crystallization kinetics of calcium phosphate in solution. We present the details of experimental components and considerations, such as stock solution preparation and preservation, working solution preparation, and the protocols for mixing and solution crystallization. The factors that might influence the crystallization process such as temperature control, stirring and flow rate, ionic strength, ionic species ratio, and foreign particles are also discussed here. Finally, we describe the protocols for each method and the processing of experimental data to extract the kinetics of nucleation and crystal growth. The advantages and disadvantages for each method are summarized in this chapter.

1. INTRODUCTION

Biomineralization is mineralization that happens in biological environments in which an organic matrix or soluble biomolecules, along with biological-induced local environments, facilitate the crystallization of minerals and control their morphologies and locations of nucleation (Lowenstam & Weiner, 1989; Mann, 2001). Biomolecules are generally considered to play important roles in controlling the kinetics of the biomineralization process (Boskey, 2003; George & Veis, 2008; Suzuki et al., 2009; Veis, 2004). To a better understanding of biomineralization process, the kinetics of biomimetic crystallizations are extensively investigated (Borkiewicz, Rakovan, & Cahill, 2010; Hu et al., 2012; Jiang, Liu, Zhang, & Li, 2005; Koutsoukos, Amjad, Tomson, & Nancollas, 1980; Nancollas & Wu, 2000; Termine & Eanes, 1972; Tomson & Nancollas, 1978; Uskokovic, Li, & Habelitz, 2011; Wallace, DeYoreo, & Dove, 2009; Wang, Ma, & Liu, 2009; Yang et al., 2011). Among biomineral phases, calcium phosphate has attracted much attention due to implications as to its role in bone and tooth formation (Mathew & Takagi, 2001). For calcium phosphate crystals, hydroxyapatite [$Ca_{10}(PO_4)_6(OH)_2$, HAP], octacalcium phosphate [$Ca_8H_2(PO_4)_6 \cdot 5H_2O$, OCP], brushite ($CaHPO_4 \cdot 2H_2O$, DCPD), and tricalcium phosphate ($Ca_3(PO_4)_2$, TCP) have been extensively investigated.

In principle, crystallization can be divided into two important processes: nucleation and crystal growth (De Yoreo & Vekilov, 2003; Mersmann, 2001; Mullin, 2001). The supersaturation, S, is the thermodynamic driving force for crystallization, which is given by

$$S = \frac{IP}{K_{sp}} \qquad (6.1)$$

where IP is the ionic activity product and K_{sp} is the solubility product. However, homogeneous nucleation does not always occur spontaneously under supersaturated conditions; instead, a metastable solution persists during an induction time, t_{ind}, before the nucleation, which in principle is determined by the nucleation barrier and kinetic factors (Liu, 2001; Liu & De Yoreo, 2004).

Crystal growth rate laws are dependent upon the specific choice of growth model (Beckmann, 2013; Lacmann, Herden, & Mayer, 1999; Nielsen, 1964), which include birth and spread, spiral growth, adhesive growth, surface diffusion and dislocation, and bulk diffusion. For sparingly soluble calcium phosphate salts, the rate of growth, J, can be empirically given as (Koutsoukos et al., 1980; Wang & Nancollas, 2008)

$$J = k\sigma^n \qquad (6.2)$$

$$\sigma = \left(\frac{\text{IP}}{K_{sp}}\right)^{1/\nu} - 1 \qquad (6.3)$$

where k is a rate constant; n is the effective reaction order, which is dependent on the growth model; σ is the relative supersaturation; and ν is the number of ions in a formula unit of growing phase.

To obtain the kinetics of crystallization, one can either track the evolution of mineral phase or monitor the concentrations of ionic species in solution, which gives the consumption of nutrients during crystallization. Methods of *ex situ* characterization by transmission electron microscopy (Pan, Liu, Tang, & Xu, 2010; Tao et al., 2008; Tsuji, Onuma, Yamamoto, Iijima, & Shiba, 2008), X-ray diffraction (XRD) (Tao, Zhou, Zhang, Xu, & Tang, 2009; Tsuji et al., 2008), Raman (Tao et al., 2009; Tsuji et al., 2008; Wang, Kim, Stephens, Meldrum, & Christenson, 2012) and Fourier-transformed infrared (Beniash, Metzler, Lam, & Gilbert, 2009; Tao et al., 2009) have the advantage of verifying the phase of minerals and in semiquantifying their compositions during crystallization. However, it is still a challenge to extract samples without disturbing the process of crystallization. Besides, the *ex situ* characterization is time-consuming and the temporal resolution is low. So, *in situ* characterization of crystallization kinetics has attached much attention. Techniques include solution-based methods such as constant composition (CC) (Amjad, Koutsoukos, Tomson, & Nancollas, 1979; Koutsoukos et al., 1980; Nancollas & Wu, 2000; Tang, Henneman, & Nancollas, 2003; Tomson & Nancollas, 1978; Wang & Nancollas, 2009), pH meter (Meyer & Eanes, 1978; Ofir, Govrin-Lippman, Garti, & Furedi-Milhofer, 2004; Wang, Liao, et al., 2009; Yang et al., 2011), stopped-flow spectrophotometer (Wang, Liao, et al., 2009; Wang, Ma, et al., 2009), ion-selective electrode (Gebauer,

Völkel, & Cölfen, 2008), and electrolyte conductivity (Klepetsanis & Koutsoukos, 1998) and *in situ* mineral characterization by atomic force microscopy (Stephenson et al., 2008; Wallace et al., 2009), Raman (Ramírez-Rodríguez, Delgado-López, & Gómez-Morales, 2013), XRD (Pina, Torres, Goetz-Neunhoeffer, Neubauer, & Ferreira, 2010; Song, Feng, & Wang, 2007), turbidity (Dorozhkina & Dorozhkin, 2002; Pan et al., 2010; Tay, Pashley, Rueggeberg, Loushine, & Weller, 2007; Wang et al., 2012), and quartz crystal microbalance (Tong et al., 2004; Yang, Si, Zeng, Zhang, & Dai, 2008). In this chapter, we will focus on the details of using CC, monitoring of pH, and stopped-flow spectrophotometry to measure the kinetics of calcium phosphate crystallization.

2. SOLUTION PREPARATION AND EXPERIMENTAL CONSIDERATIONS

2.1. Stock solution preparation

Because of the convenience of measuring volumes by adjustable pipettes as compared with weighting chemical reagents using a balance, using stock solutions can save much time and improve the accuracy in preparing working solutions. However, care should be taken to check the accuracy of the pipette before use. (After years of use, pipettes can become inaccurate with the relative error becoming larger than 3%.) Once prepared, the concentration of stock solutions should be calibrated against standard reagents if needed (in case the salt is prone to losing waters of crystallization or adsorbing water).

Another issue to be addressed is the preservation of stock solutions. Phosphate solutions are liable to induce bacterial growth, so they cannot be stored at room temperature for days unless the solution is sterile. Usually, solutions must be stored under refrigeration to prevent bacterial growth. However, when they were taken out of the refrigerator, phosphate stock solutions should be kept at room temperature for some time before opening the vessel. On the one hand, phosphate stock solutions may precipitate when cooled, so some time is needed to completely dissolve such precipitates at room temperature. On the other hand, water vapor from the atmosphere is readily condensed on the inner wall of cooled containers, so stock solutions can become diluted.

After preparation, all stock solutions should be sealed and be stored under proper conditions. All stock solution should be refreshed every month. However, in the case of bicarbonate and carbonate stock solutions that are highly super- or undersaturated, they should be freshly prepared each time to prevent the concentration from shifting due to the release or absorption of CO_2 vapor from the atmosphere.

2.2. Working solution preparation

The stock solutions are diluted and mixed to get working solutions with designated compositions and concentrations. When preparing supersaturated solution, special protocols must be applied to prevent precipitation. For simulated body fluid solutions (Kokubo & Takadama, 2006), please refer to Kokubo and Takadama (2006) for details (in Appendix). In brief, most water should be first added into the container and then add the relatively soluble salts (KCl, NaCl, and $NaHCO_3$) and then phosphate, $MgCl_2$, $CaCl_2$, Na_2SO_4, and buffer (Tris). When the solution contains sufficient amounts of calcium and phosphate, adjusting pH by using concentrated base should be avoided because, near the droplet of base, the local supersaturation can become quite high, triggering precipitation of calcium phosphate. To avoid this, one protocol is to prepare calcium (containing $CaCl_2$ and $MgCl_2$) and phosphate solutions (containing $NaHCO_3$, Na_2SO_4, Tris, etc.), separately, and preadjust the pH for each solution. After mixing calcium and phosphate solution, check the pH of the final solution. By trial and error, one can get a working solution with the desired pH and supersaturation.

Mixing is an important issue in preparing supersaturated solution. There are several ways to mix two solutions:
1. Adding one solution into another while stirring
2. Mixing two solutions via a tube into a container
3. Rapidly injecting of one solution into another and pipetting the sample up and down several times

Supersaturated solutions are unstable, and they are sensitive to the protocols of mixing. For a given system, mixing protocols should be kept the same (e.g., the rate of addition and injection, the same type of pipette, tip, stirring rate, and the size of tube) for all solutions.

For nucleation kinetics studies, the supersaturated solutions should be freshly prepared. For crystal growth kinetics, they should be used within the induction period and refreshed periodically.

2.3. Carbon dioxide

As noted earlier, aqueous solutions can absorb CO_2 from the atmosphere, resulting in changes in solution pH and introducing bicarbonate ions into the solutions. Therefore, CO_2 must be driven out from system by bubbling with water-saturated nitrogen gas prior to use or during experiments.

2.4. Solution crystallization

There are several factors that may affect crystal nucleation and growth:
1. Temperature.

    As crystallization rate is a function of temperature, a thermostat is needed to keep the solution at constant temperature during crystallization.
2. Volume.

    As absolute nucleation rates are a function of the volume of mother liquor, all volumes of supersaturated solutions should be kept the same for any set of experiments on a given system.
3. Ionic strength.

    The ionic strength (IS) not only influences the activity of ions in solution but also affects the electrolyte double layer on the crystal interface, which can have great impact on the incorporation of ions into the crystal lattice (i.e., crystal growth) and the aggregation of ionic species (i.e., nucleation or formation of growth units for crystal growth). In practice, the IS for a given system should be fixed.
4. Ionic species ratio.

    The nucleation event can be treated as chain-like reactions that combine ionic species into a big cluster or an aggregate. Similarly, crystal growth proceeds by incorporation of ionic species into the crystal lattice. The ratio of ionic species with respect to the stoichiometry of crystal unit should have an impact on the efficiency of reactions. Thus, in practice, the ratio of ionic species should be kept the same for a given system.
5. Stirring or flow rate.

    In studies of nucleation kinetics, stirring is usually not recommended for highly supersaturated solutions, because it frequently induces nucleation (by introduction of foreign particles or creation of new high energy surfaces when the stirrer impacts the container, shear-induced nucleation, or increases in the fluctuation of the local environment that can trigger nucleation). When stirring is necessary, the stirring rate and stirrer size should be fixed.

    When the rate of crystal growth is limited by diffusion or supply of nutrient to the crystal surface, the stirring rate or flowing rate can influence the apparent kinetics of crystallization. Therefore, the rate of stirring or flow rate should be large enough to ensure that the crystal growth is surface-controlled, which means it is controlled by the growth kinetics of the crystal itself (i.e., the incorporation process).

6. Foreign particles, bubbles, or crystal seeds.

Because foreign particles or bubbles may induce heterogeneous nucleation, all working solutions—except for ammonia or bicarbonate solutions—should be degassed and filtered through 0.22 μm or 20 nm pore-size filters prior to use. However, for some supersaturated solutions, filtration can also trigger the precipitation, presumably due to shear-induced nucleation. In such cases, one should filter either the solvent or each reagent solution before preparing supersaturated solutions.

The amount and the surface quality (number density of dislocations, facets, surface areas, etc.) of crystal seeds can influence the overall growth rate. The same batch of crystal seeds should be introduced into the system to obtain more consistent result and increase the reproducibility. The total surface area should be known to get the crystal growth rate (per unit area).

7. Reproducibility.

Due to the sensitivity of crystallization to various factors, each experiment should be repeated at least three times.

3. *IN SITU* EXPERIMENTS

3.1. Constant composition

For free-drift crystallization processes, the nutrient ions in mother liquor are consumed by crystal growth. In such a situation, the supersaturation drifts downward, which, in turn, affects the rate of crystal growth. With the CC method (Koutsoukos et al., 1980; Tomson & Nancollas, 1978), the composition of mother liquor is maintained by titrating nutrient ions into it during crystallization. Thus, the flux of nutrient ions to or from the crystallites is identical to the titration rate. The advantage of this method is that it can measure even a very small crystal growth rate and the stoichiometry of the solid phases undergoing growth can be determined. Titrant addition is potentiometrically controlled by the error signal between the electrode of the working solution and the reference potential, which is initially set to the value of the solution. The titrants are designed to have a stoichiometry matching that of the growing phase while considering the dilution of the reaction solutions due to addition of multiple titrants. Thus, a constant thermodynamic driving force for crystal growth is maintained during an experiment.

As calcium phosphate precipitates, phosphate ions (PO_4^{3-}) are removed from a solution. This will disrupt the previous balance between different types of phosphate ions in a solution:

$$H_3PO_4 \rightleftharpoons H^+ + H_2PO_4^- \rightleftharpoons H^+ + HPO_4^{2-} \rightleftharpoons H^+ + PO_4^{3-} \quad (6.4)$$

From Le Chatelier's principle, upon removal of PO_4^{3-}, the reactions will move to the right sides of Eq. (6.4) and reestablish a new balance. So, protons are released as calcium phosphate precipitates; this is proved by experiments (Koutsoukos et al., 1980; Meyer & Eanes, 1978; Wang, Liao, et al., 2009). In this regard, the pH will also need to be maintained by titration. If the stoichiometry of the precipitant is known, the titrants can be designed according to the equation of chemical equilibrium. Take the HAP system as an example:

$$10CaCl_2 + 6KH_2PO_4 + 14KOH = Ca_{10}(PO_4)_6(OH)_2 (HAP) \\ \downarrow + 12H_2O + 20KCl \quad (6.5)$$

The titrants can be designed as (1) Ca solution, $CaCl_2$ and KCl, and (2) P solution, KH_2PO_4, KOH, and KCl. The concentration ratios are set according to the stoichiometry of the reaction (Eq. 6.5): $C_{Ca}/C_P = 10/6$; $C_{Ca}/C_{OH} = 10/14$. The IS is reduced upon addition of titrants, so more electrolyte (KCl) should be added to maintain the IS: $C_{KCl} = IS - 2C_{Ca}$. In this way, by a pH-stat, CC of mother liquor can be maintained. In practice, a pH drop of approximately 0.003 is set to trigger the addition of titrants. Note, Ca- and P-containing solutions should be loaded in two mechanically coupled burettes to keep the stoichiometry of titration constant.

In some cases, the stoichiometry of the crystallization reaction is unknown or it will change at different stages of crystallization. The addition of Ca-containing, P-containing, and base solutions should be controlled by suitable ion-selective electrodes and a pH electrode, separately.

To trigger crystal growth, crystal seeds are introduced. Because the overall crystal growth rate is proportional to the total area of crystal seeds, the specific surface area (SSA) and total amount of seed material must be determined, usually by BET (Koutsoukos et al., 1980), prior to use.

Here is the protocol for CC:
1. Prepare and calibrate stock solutions: $CaCl_2$, KH_2PO_4, and NaCl.
2. Prepare titrants: (1) $CaCl_2$; (2) $KH_2PO_4 + KOH$.

3. Calibrate pH electrode.
4. Prepare supersaturated calcium phosphate solutions.
5. Set the parameters for maintaining CC (titration step; trigger threshold).
6. Wait for quasi-equilibrium of the solution, as indicated by stabilization of the electrode potential.
7. Introduce known amount of crystal seed to initiate crystal growth.
8. Automatic titration will be triggered as the crystals grow; the volumes of titrants added as a function of time are recorded.
9. Wash the containers and tubes with dilute HCl and water, respectively, three times.

3.2. pH meter

Because the precipitation or crystal growth of calcium phosphate crystals is always accompanied with a change in the pH, a pH meter can be used to monitor crystallization (see Eq. 6.5). However, for highly supersaturated solutions, as the mixing of calcium and phosphate solution proceeds, amorphous calcium phosphate (ACP) can be formed before crystallization. So, the observed drop in pH is actually the transformation of ACP to HAP:

$$Ca_9(PO_4)_6 + Ca^{2+} + 2H_2O = Ca_{10}(PO_4)_6(OH)_2(HAP) \downarrow + 2H^+ \quad (6.6)$$

Thus, with this method, the nucleation kinetics can be investigated either for calcium phosphate solutions or for ACP-mediated crystallization.

Here is the protocol for pH measurement:
1. Calibrated pH electrode using standard buffer solutions.
2. Thermostat calcium and phosphate solutions at designated temperature for 15 min.
3. Mix calcium and phosphate solutions at designated pH and temperature.
4. Record the pH value as a function of time during crystallization.
5. After each experiment, rinse the electrode and containers with $0.1\ M$ HCl and water, respectively, two times to remove any residue from precipitation.

3.3. Stopped-flow spectrophotometer

Upon mixing of calcium and phosphate solutions, the formation of ACP is so fast (within seconds) that a pH meter cannot capture the entire process. By the introduction of an acid–base indicator, the pH of solution can instead be measured by UV–vis spectrophotometry. By combining a stopped-flow method with spectrophotometry, one can track changes in solution pH that take place on millisecond-to-minute timescale during the mixing of calcium and phosphate solutions. Here, we describe the use of bromothymol blue (BTB, $C_{27}H_{28}Br_2O_5S$) as a pH indicator. In practice, the ratio for the absorptions at 418 and 618 nm is correlated for pH. But in calcium phosphate solutions, light will also be scattered by particles as precipitation of calcium phosphate occurs, and this will influence the detection of absorption (A) from solution. (Actually, the measured datum is extinction (E), which equals absorption (A) plus scattering (S), $E = A + S$.) A reference curve should be measured for calcium phosphate solutions without pH indicator to determine the magnitude of the scattering term.

Here is the protocol for this measurement:
1. Double distil water degas in vacuum for 40 min.
2. Prepare calcium phosphate stock solutions: 36 mM $CaCl_2$ and 21.6 mM NaH_2PO_4.
3. Prepare 8 mM BTB stock solution: put 25 mg BTB in 0.5 mL 10 mM NaOH, and dilute to 50 mL.
4. Calibrate the pH curve: Prepare 30 µM BTB solution with pH 6.2, 6.6, 7.0, and 7.4 (adjusted with 1 M NaOH), and measure the absorptions at 418 and 618 nm, respectively.
5. Prepare working solutions for calcium solutions (8.0 mM $CaCl_2$, pH 7.40, adjusted with 10 mM NaOH) and phosphate solutions (4.8 mM NaH_2PO_4 with and without indicator, pH 7.40), respectively. Filter with 20 nm film and degas in vacuum for 40 min before use.
6. Rapidly mix calcium and phosphate solutions with BTB by stopped-flow spectrometer to detect the extinction spectra of the working solutions (degas water as reference).
7. Measure the extinction spectra of control solutions (without BTB) (degas water as reference).
8. Wash the cuvette, flow path, and mixers with 0.1 M HCl, 5 mM NaH_2PO_4, and water, respectively, two times.

4. DATA HANDLING/PROCESSING

4.1. Constant composition

1. Crystallization rate.
    The molar rates of reaction (J) can be calculated from the volume of added titrants using Eq. (6.7).

$$J = \frac{C_{\text{eff}}}{S_A m} \frac{dV}{dt} \qquad (6.7)$$

where dV/dt is the titrant curve gradient, C_{eff} is the equivalent number of moles precipitated in a liter of added titrant, S_A is the SSA of the seed crystals, and m is the weight of minerals. Note the mass and SSA of minerals will change as crystallization. So, during the crystal growth, slurry samples need to be withdrawn periodically to characterize SSA and check the phase and morphology. The crystal growth rate can be determined by the initial (first 5–15 min) slope of titration curve or by linear fit of the surface area-scaled titration curve against time.

2. Nucleation rate.

   For nucleation kinetics, during the induction period, there is no titration. The starting point of titration is the induction time for nucleation.

3. Reaction order.

   Linear fit of the crystal growth rate ($\ln J$ or $\log J$) against relative supersaturation ($\ln \sigma$ or $\log(IP^{1/v} - K_{sp}^{1/v})$) can be applied to determine the effective reaction order (n) (cf. Eq. 6.2).

4.2. pH curve

In pH curves (Fig. 6.1), the crystallization process can be divided into three stages. In stage I, the solution pH is stable and this can be considered to be the induction period. In stage II, the fast drop of pH is observed, suggesting the occurrence of nucleation and crystallization. In stage III, the pH level off,

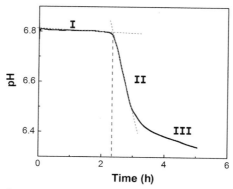

Figure 6.1 pH curve for calcium phosphate solution. The induction time is determined by the intersection of the tangent with the initially flat baseline level of pH (stage I) and that for the subsequent fast pH drop in stage II. (For color version of this figure, the reader is referred to the online version of this chapter.)

and it is postcrystallization. The induction times are determined by the intersections of tangents over the initially flat levels of pH (stage I) and over the subsequent fast pH drop during the precipitation of HAP (stage II) (cf. Fig. 6.1).

4.3. Stopped-flow spectrophotometer

1. UV–vis spectrums of pH indicator.

    Figure 6.2 shows the UV–vis spectrums of 30 μM BTB at pH 6.2–7.4. There are two absorption peaks, one at 418 nm (produced by acid form) and another at 618 nm (produced by base form). The ratio of the absorptions, A_{418} and A_{618}, is correlated to pH, which is given by

$$\mathrm{pH} = a - b\log\left(\frac{A_{418}}{A_{618}}\right) \qquad (6.8)$$

   By linear fit of pH against $\log(A_{418}/A_{618})$, we get pH $= 7.00 - 1.05 \log(A_{418}/A_{618})$, $R^2 = 0.9994$.

2. Real-time spectrums for stopped-flow spectrophotometer.

    Record the UV–vis spectrums data as a function of time for calcium phosphate solutions containing 30 μM BTB and without BTB (control), respectively. Get the data for the extinctions at 418 and 618 nm (E_{418} and E_{618}), respectively. Discount the effect of scattering by subtracting the data for control experiment, that is, $A_{418,618} = E_{418,\ 618}$ (BTB) −

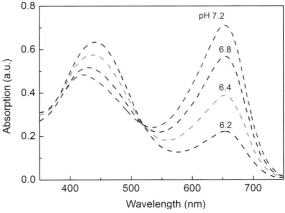

Figure 6.2 UV–vis spectrums of 30 μM BTB at pH 6.2–7.4. The absorption peak at 418 nm is produced by acid formation and that at 618 nm is produced by base formation. (See color plate.)

$E_{418,618}$ (control). By converting the value of $\log(A_{418}/A_{618})$ to pH (Eq. 6.8), the fast change of pH with reaction time can be detected.

5. SUMMARY

This chapter presented general solution-chemistry-based methods for *in situ* kinetics studies of calcium phosphate system. These protocols are adaptable to analyze the kinetics of many solution crystallization systems. The advantage of CC method is to detect qualitatively the flux of nutrients ions to and from the overall crystallites in solution. The reaction order of the crystallization detected by CC methods indicated the possible model of crystallization, and the promotion or inhibiting of biomolecules can be precisely determined by CC method (Dorozhkin & Dorozhkina, 2007; O'Young et al., 2011; Wang & Nancollas, 2009). The free-drift pH curve is a convenient and general method to detect the nucleation kinetics for the crystallization systems that is with pH change, which is especially suitable for calcium phosphate system. The influence of biomolecules or additives on the nucleation of calcium phosphate crystals can be clearly demonstrated by pH curve (Ofir et al., 2004; Yang et al., 2011). The fast reactions during the initial processes of crystallization can be determined by stopped-flow spectrophotometer, which is recommended to investigate the important process of prenucleation stages and ACP formation. With this in mind, we would encourage the use of these general methods for kinetics of biomimetic mineralization of calcium phosphate systems in both fundamental and applied research.

REFERENCES

Amjad, Z., Koutsoukos, P., Tomson, M. B., & Nancollas, G. H. (1979). The growth of hydroxyapatite from solution. A new constant composition method. *Journal of Dental Research, 58*(4), 1431–1432.

Beckmann, W. (2013). Mechanisms of crystallization. *Crystallization* (pp. 7–33). Weinheim: Wiley-VCH Verlag GmbH & Co. KGaA.

Beniash, E., Metzler, R. A., Lam, R. S. K., & Gilbert, P. U. P. A. (2009). Transient amorphous calcium phosphate in forming enamel. *Journal of Structural Biology, 166*(2), 133–143.

Borkiewicz, O., Rakovan, J., & Cahill, C. L. (2010). Time-resolved in situ studies of apatite formation in aqueous solutions. *American Mineralogist, 95*(8–9), 1224–1236.

Boskey, A. L. (2003). Mineral analysis provides insights into the mechanism of biomineralization. *Calcified Tissue International, 72*(5), 533–536.

De Yoreo, J. J., & Vekilov, P. G. (2003). Principles of crystal nucleation and growth. *Reviews in Mineralogy and Geochemistry, 54*(1), 57.

Dorozhkin, S. V., & Dorozhkina, E. I. (2007). Crystallization from a milk-based revised simulated body fluid. *Biomedical Materials, 2*(2), 87–92.

Dorozhkina, E. I., & Dorozhkin, S. V. (2002). Application of the turbidity measurements to study in situ crystallization of calcium phosphates. *Colloids and Surfaces A: Physicochemical and Engineering Aspects, 203*(1–3), 237–244.
Gebauer, D., Völkel, A., & Cölfen, H. (2008). Stable prenucleation calcium carbonate clusters. *Science, 322*(5909), 1819–1822.
George, A., & Veis, A. (2008). Phosphorylated proteins and control over apatite nucleation, crystal growth, and inhibition. *Chemical Reviews, 108*(11), 4670–4693.
Hu, Q., Nielsen, M. H., Freeman, C., Hamm, L. M., Tao, J., & Lee, J. R. (2012). Concluding remarks: The thermodynamics of calcite nucleation at organic interfaces: Classical vs. non-classical pathways. *Faraday Discussions, 159*, 509–523.
Jiang, H. D., Liu, X. Y., Zhang, G., & Li, Y. (2005). Kinetics and template nucleation of self-assembled hydroxyapatite nanocrystallites by chondroitin sulfate. *Journal of Biological Chemistry, 280*(51), 42061–42066.
Klepetsanis, P. G., & Koutsoukos, P. G. (1998). Kinetics of calcium sulfate formation in aqueous media: Effect of organophosphorus compounds. *Journal of Crystal Growth, 193*(1–2), 156–163.
Kokubo, T., & Takadama, H. (2006). How useful is SBF in predicting in vivo bone bioactivity? *Biomaterials, 27*(15), 2907–2915.
Koutsoukos, P., Amjad, Z., Tomson, M., & Nancollas, G. (1980). Crystallization of calcium phosphates. A constant composition study. *Journal of the American Chemical Society, 102*(5), 1553–1557.
Lacmann, R., Herden, A., & Mayer, C. (1999). Kinetics of nucleation and crystal growth. *Chemical Engineering and Technology, 22*(4), 279–289.
Liu, X. (2001). Generic mechanism of heterogeneous nucleation and molecular interfacial effects. In K. Sato, K. Nakajima, & Y. Furukawa (Eds.), *Advances in crystal growth research* (pp. 42–61). Amsterdam: Elsevier Science B.V.
Liu, X., & De Yoreo, J. (2004). *Nanoscale structure and assembly at solid-fluid interfaces: Interfacial structures versus dynamics*. London: Springer.
Lowenstam, H., & Weiner, S. (1989). *On biomineralization*. USA: Oxford University Press.
Mann, S. (2001). *Biomineralization: Principles and concepts in bioinorganic materials chemistry*. USA: Oxford University Press.
Mathew, M., & Takagi, S. (2001). Structures of biological minerals in dental research. *Journal of Research—National Institute of Standards and Technology, 106*(6), 1035–1044.
Mersmann, A. (2001). *Crystallization technology handbook*. New York: CRC Press.
Meyer, J. L., & Eanes, E. D. (1978). A thermodynamic analysis of the amorphous to crystalline calcium phosphate transformation. *Calcified Tissue International, 25*(1), 59–68.
Mullin, J. W. (2001). *Crystallization*. Oxford: Butterworth-Heinemann.
Nancollas, G. H., & Wu, W. (2000). Biomineralization mechanisms: A kinetics and interfacial energy approach. *Journal of Crystal Growth, 211*(1–4), 137–142.
Nielsen, A. E. (1964). *Kinetics of precipitation*. Oxford: Pergamon Press(distributed in the Western Hemisphere by Macmillan, New York).
Ofir, P., Govrin-Lippman, R., Garti, N., & Furedi-Milhofer, H. (2004). The influence of polyelectrolytes on the formation and phase transformation of amorphous calcium phosphate. *Crystal Growth & Design, 4*(1), 177–183.
O'Young, J., Liao, Y. Y., Xiao, Y. Z., Jalkanen, J., Lajoie, G., & Karttunen, M. (2011). Matrix gla protein inhibits ectopic calcification by a direct interaction with hydroxyapatite crystals. *Journal of the American Chemical Society, 133*(45), 18406–18412.
Pan, H., Liu, X., Tang, R., & Xu, H. (2010). Mystery of the transformation from amorphous calcium phosphate to hydroxyapatite. *Chemical Communications, 46*(39), 7415–7417.
Pina, S., Torres, P. M., Goetz-Neunhoeffer, F., Neubauer, J., & Ferreira, J. M. F. (2010). Newly developed Sr-substituted α-TCP bone cements. *Acta Biomaterialia, 6*(3), 928–935.

Ramírez-Rodríguez, G. B., Delgado-López, J. M., & Gómez-Morales, J. (2013). Evolution of calcium phosphate precipitation in hanging drop vapor diffusion by in situ Raman microspectroscopy. *CrystEngComm, 15*, 2206–2212.

Song, Y., Feng, Z., & Wang, T. (2007). In situ study on the curing process of calcium phosphate bone cement. *Journal of Materials Science Materials in Medicine, 18*(6), 1185–1193.

Stephenson, A. E., DeYoreo, J. J., Wu, L., Wu, K. J., Hoyer, J., & Dove, P. M. (2008). Peptides enhance magnesium signature in calcite: Insights into origins of vital effects. *Science, 322*(5902), 724–727.

Suzuki, M., Saruwatari, K., Kogure, T., Yamamoto, Y., Nishimura, T., & Kato, T. (2009). An acidic matrix protein, Pif, is a key macromolecule for nacre formation. *Science, 325*(5946), 1388–1390.

Tang, R., Henneman, Z. J., & Nancollas, G. H. (2003). Constant composition kinetics study of carbonated apatite dissolution. *Journal of Crystal Growth, 249*(3–4), 614–624.

Tao, J., Pan, H., Wang, J., Wu, J., Wang, B., & Xu, X. (2008). Evolution of amorphous calcium phosphate to hydroxyapatite probed by gold nanoparticles. *Journal of Physical Chemistry C, 112*(38), 14929–14933.

Tao, J., Zhou, D., Zhang, Z., Xu, X., & Tang, R. (2009). Magnesium-aspartate-based crystallization switch inspired from shell molt of crustacean. *Proceedings of the National Academy of Sciences, 106*(52), 22096–22101.

Tay, F. R., Pashley, D. H., Rueggeberg, F. A., Loushine, R. J., & Weller, R. N. (2007). Calcium phosphate phase transformation produced by the interaction of the portland cement component of white mineral trioxide aggregate with a phosphate-containing fluid. *Journal of Endodontics, 33*(11), 1347–1351.

Termine, J. D., & Eanes, E. D. (1972). Comparative chemistry of amorphous and apatitic calcium phosphate preparations. *Calcified Tissue International, 10*(1), 171–197.

Tomson, M. B., & Nancollas, G. H. (1978). Mineralization kinetics: A constant composition approach. *Science (New York, NY), 200*(4345), 1059–1060.

Tong, H., Ma, W., Wang, L., Wan, P., Hu, J., & Cao, L. (2004). Control over the crystal phase, shape, size and aggregation of calcium carbonate via a l-aspartic acid inducing process. *Biomaterials, 25*(17), 3923–3929.

Tsuji, T., Onuma, K., Yamamoto, A., Iijima, M., & Shiba, K. (2008). Direct transformation from amorphous to crystalline calcium phosphate facilitated by motif-programmed artificial proteins. *Proceedings of the National Academy of Sciences, 105*(44), 16866–16870.

Uskokovic, V., Li, W., & Habelitz, S. (2011). Amelogenin as a promoter of nucleation and crystal growth of apatite. *Journal of Crystal Growth, 316*(1), 106–117.

Veis, A. (2004). Biomineralization. *Science, 305*(5683), 480, 480.

Wallace, A. F., DeYoreo, J. J., & Dove, P. M. (2009). Kinetics of silica nucleation on carboxyl- and amine-terminated surfaces: Insights for biomineralization. *Journal of the American Chemical Society, 131*(14), 5244–5250.

Wang, Y. W., Kim, Y. Y., Stephens, C. J., Meldrum, F. C., & Christenson, H. K. (2012). In situ study of the precipitation and crystallization of amorphous calcium carbonate (ACC). *Crystal Growth & Design, 12*(3), 1212–1217.

Wang, C.-G., Liao, J.-W., Gou, B.-D., Huang, J., Tang, R.-K., & Tao, J.-H. (2009). Crystallization at multiple sites inside particles of amorphous calcium phosphate. *Crystal Growth & Design, 9*(6), 2620–2626.

Wang, Z. Q., Ma, G. B., & Liu, X. Y. (2009). Will fluoride toughen or weaken our teeth? Understandings based on nucleation, morphology, and structural assembly. *The Journal of Physical Chemistry B, 113*(51), 16393–16399.

Wang, L., & Nancollas, G. H. (2008). Calcium orthophosphates: Crystallization and dissolution. *Chemical Reviews, 108*(11), 4628–4669.

Wang, L., & Nancollas, G. H. (2009). Pathways to biomineralization and biodemineralization of calcium phosphates: The thermodynamic and kinetic controls. *Dalton Transactions*, (15), 2665–2672.

Yang, Z., Si, S., Zeng, X., Zhang, C., & Dai, H. (2008). Mechanism and kinetics of apatite formation on nanocrystalline TiO2 coatings: A quartz crystal microbalance study. *Acta Biomaterialia*, *4*(3), 560–568.

Yang, X., Xie, B., Wang, L., Qin, Y., Henneman, Z. J., & Nancollas, G. H. (2011). Influence of magnesium ions and amino acids on the nucleation and growth of hydroxyapatite. *CrystEngComm*, *13*(4), 1153–1158.

SECTION II

Probing Structure and Dynamics at Surfaces and Interfaces

CHAPTER SEVEN

Design, Fabrication, and Applications of *In Situ* Fluid Cell TEM

Dongsheng Li[*,1], Michael H. Nielsen[†,‡], James J. De Yoreo[*,1]
[*]Physical Sciences Division, Pacific Northwest National Laboratory, Richland, Washington, USA
[†]Department of Materials Science and Engineering, University of California, Berkeley, California, USA
[‡]Materials Science Division, Lawrence Berkeley National Lab, Berkeley, California, USA
[1]Corresponding author: e-mail address: dongsheng.li2@pnnl.gov; James.DeYoreo@pnnl.gov

Contents

1. Introduction 147
2. Fluid Cell Design 150
3. Cell Fabrication and Assembly 153
4. TEM Operation 154
5. Effects of the Electron Beam on Reactions in the Fluid Cell 160
6. Conclusions 161
Acknowledgments 162
References 162

Abstract

In situ fluid cell TEM is a powerful new tool for understanding dynamic processes during liquid phase chemical reactions, including mineral formation. This technique, which operates in the high vacuum of a TEM chamber, provides information on crystal structure, phase, morphology, size, aggregation/segregation, and crystal growth mechanisms in real time. *In situ* TEM records both crystal structure and morphology at spatial resolutions down to the atomic level with high temporal resolution of up to 10^{-6} s per image, giving it distinct advantages over other *in situ* techniques such as optical microscopy, AFM, or X-ray scattering or diffraction. This chapter addresses the design, fabrication, and assembly of TEM fluid cells and applications of fluid cell TEM to understanding mechanisms of mineralization.

1. INTRODUCTION

In situ techniques are useful for understanding the kinetics and mechanisms of chemical reactions, crystal nucleation and growth, development of morphology, structural transitions, changes in chemical composition and electronic structure, and the formation of defect structures such as vacancies, dislocations, and kinks. *In situ* experiments provide information that may be missed via *ex situ* observations, such as a transient step in a reaction process. Moreover, it is often difficult or impossible to infer mechanistic information about material formation processes simply from the final composition, structure, or morphology. Thus, the inherent lack of temporal resolution in *ex situ* experiments limits their contribution towards understanding dynamic processes in materials. As a consequence, many characterization tools, including optical microscopy, atomic force microscopy (AFM), scanning tunneling microscopy (STM), X-ray photoelectron spectroscopy (XPS), X-ray diffraction, and transmission electron microscopy (TEM), have been modified to deliver *in situ* monitoring of chemical reactions, crystal growth, mechanical properties, phase transformation, and other dynamic processes (Brennan, Fuoss, Kahn, & Kisker, 1990; Hu et al., 2012; Land, DeYoreo, & Lee, 1997; Teng, Dove, Orme, & De Yoreo, 1998; Todorov, Martins, & Viana, 2013; Tsuchiya, Taniwatari, Uomi, Kawano, & Ono, 1993).

Each of these *in situ* methods has its advantages and disadvantages. *In situ* optical microscopy and spectroscopy provide information on particle size, composition, phase, and morphology during chemical reactions with a resolution in the range of tens to hundreds of nanometer. *In situ* AFM is useful for imaging particle size and morphological changes in liquids at single digit nanometer lateral resolution and provides 3D surface profiles during reactions with subangstrom resolution. Thus, *in situ* AFM has been a valuable tool for investigating the nucleation and growth of biomineral phases. However, AFM has three drawbacks. First, conventional AFM requires a minute or two to acquire an image, though the recent development of high-speed AFM promises to reduce that time to a second or less per image in both air and fluids (Ando et al., 2001; Schitter et al., 2007), albeit with some loss of resolution. The second drawback of AFM is that it generally cannot provide crystal structure or phase information. Finally, the processes of interest must occur on a fixed substrate, which can either influence or be incompatible with the process of interest.

In situ STM can probe particle and domain sizes and morphology with 0.1 nm lateral resolution and 0.01 nm height resolution while providing information on electronic structure, but it cannot be used on insulating materials, which largely eliminates its utility in biomineralization studies. *In situ* XPS directly measures changes in elemental composition, chemical state, and electronic state of the elements in the upper 1–10 nm of the surface of materials during reactions. However, XPS has traditionally required ultrahigh vacuum (UHV) conditions, and although there are now liquid cells for XPS, the technique has been mostly used for solid-state reactions and has seen only a few applications to studies of biomineralization (Lee, Han, Willey, Nielsen, et al., 2013; Lee, Han, Willey, Wang, et al., 2007). *In situ* X-ray diffraction is a fast and accurate tool for identifying the structure, phases, and sizes of crystals but requires a large sample volume and provides little information on morphology or small-scale variations in structure.

In situ TEM has a number of attributes that make it particularly useful for investigating mineralization. First, it not only provides information on crystal size, phase, and morphology but also can probe transient and dynamic processes like nucleation and growth, morphological evolution, aggregation, segregation, and phase transformation with atomic spatial resolution at video rates. Moreover, the development of dynamic TEM in which the normal thermionic or field emission source is replaced by a photoemission source promises to deliver a temporal resolution of $\sim 10^{-6}$ s (Armstrong et al., 2007).

Unlike AFM, TEM operation requires UHV conditions (10^{-8} to 10^{-6} Torr). Thus, early applications were carried out in vacuum. In 1986–1998, *in situ* TEM was used to observe the atomic mechanisms of phase transformation associated with dislocations and stacking faults (Parker, Sigmon, & Sinclair, 1986), interphase boundary dynamics (Howe et al., 1998), nanocrystallite nucleation, and coarsening (Li-Chi & Risbud, 1994) in real time at elevated temperatures by video recording of images. In 1998, Ross from IBM introduced gas phase to *in situ* TEM experiments and reported real-time imaging of Ge island growth in a UHV TEM equipped with chemical vapor deposition (Ross, Tersoff, Reuter, Legoues, & Tromp, 1998).

The extension of TEM to liquid environments came in 2003 when Ross reported a real-time TEM study of the growth of Cu clusters in liquid phase (Williamson, Tromp, Vereecken, Hull, & Ross, 2003). In this work, a sealed fluid cell compatible with UHV was made from Si wafers. The cell had a thin layer of liquid sandwiched between two ultrathin Si_3N_4 membranes, through which the electron beam passed. This fluid cell was later modified to include electrochemical control (Radisic, Ross, & Searson, 2006; Radisic, Vereecken, Hannon, Searson, & Ross, 2006). Imaging of cells (de Jonge, Peckys, Kremers, & Piston, 2009) and, later, protein structures (Evans et al., 2012; Mirsaidov, Zheng, Casana, & Matsudaira, 2012) was reported using similar cell designs. Atomic resolution TEM imaging in fluid cells of this type was reported for the study of oriented attachment between ferrihydrite (Li, Nielsen, Lee, Frandsen, Banfield, & De Yoreo, 2012) and Pt_3Fe (Liao, Cui, Whitelam, & Zheng, 2012) nanoparticles. In 2012, a new type of liquid cell based on entrapment of a very thin-liquid film between layers of graphene was introduced (Jong Min et al., 2012). This cell also provided atomic-level resolution imaging. However, image resolution in both types of TEM fluid cells is primarily limited by the thickness of the liquid layer, and, to date, precise control over that thickness has been difficult. Therefore, atomic resolution has not been obtained reproducibly.

Since development of the first liquid cell for TEM, cell designs have been modified to include electrochemical and heating control, as well as single-channel and two-channel flow. With the growing demand for *in situ* TEM, many manufactures have developed a wide range of TEM holders for various purposes, such as heating, electrical biasing, magnetizing, and fluid flow. Here, we will discuss the design and fabrication of the basic fluid cell made from Si wafers, the addition of electrochemical control and heating, and the application to studies of mineralization reactions.

2. FLUID CELL DESIGN

Most silicon-based fluid cell designs are based on Ross's work at IBM. A schematic side-view diagram of the basic design for an experimental cell used by Li et al. (2012) and Nielsen, Lee, Hu, Han, and De Yoreo (2012) is presented in Fig. 7.1. Each cell is hermetically sealed to isolate the solution from the high vacuum environment of the TEM chamber. The cell is mainly constructed by gluing together two pieces of Si (Fig. 7.1A). The cell contains two solution reservoirs (Fig. 7.1G), which are composed of hollow Si towers (Fig. 7.1D) sealed by glass caps (Fig. 7.1E) to provide large volumes of excess solution that ensure the hydration of the imaging area. In between the two reservoirs, there is an electron transparent window (Fig. 7.1F) constructed from two Si_3N_4 membranes separated by a spacer (Fig. 7.1C), which provides a gap for a thin layer of solution. In principle, the distance between the two Si_3N_4 membranes is determined by the thickness of the spacer. This spacer is typically 200–500 nm in thickness but can be increased to accommodate larger biological samples or decreased to achieve the highest possible resolution. However, in reality, the solution thickness under the electron beam can be as large as a many micrometers due to an outward bowing

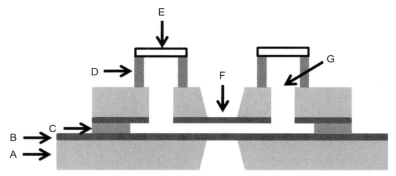

Figure 7.1 Schematic cross section of the TEM fluid cell. Components: A, Si wafer with; B, 50–100 nm Si_3N_4 coating; C, 200–500 nm metal or Si_3N_4 spacer; D, Si tower; E, glass cap; F, electron transparent Si_3N_4 window; G, solution reservoir (Li et al., 2012). (For color version of this figure, the reader is referred to the online version of this chapter.)

of the Si_3N_4 membranes or very close to zero due to an inward bowing of the membranes.

This basic fluid cell design can be modified to provide electrochemical and temperature control to initiate crystal nucleation or other reactions in the microscope. The system consists of a hermetically sealed cell on a custom TEM stage containing electrical feed-throughs for connection to external electronics and is compatible with commercial electron microscopes. Electrochemical control is enabled by incorporation of three electrodes: the reference, working, and counter electrodes as shown in Fig. 7.2a (A, B, and C, respectively). The external portion of the working electrode extends to the edge of the Si bottom chip, which is connected via the electrode feed-through, and the internal portion extends over the electron transparent Si_3N_4 window. This design ensures that electrochemically induced processes of interest occur within the field of view of the microscope. In some cases, the effects of an applied potential can be reversed by inverting or removing the voltage, which enables one to reversibly grow and dissolve an inorganic phase and thus image nucleation repeatedly in a single experiment (Radisic, Ross, & Searson, 2006; Radisic, Vereecken, Hannon, et al., 2006; Radisic, Vereecken, Searson, & Ross, 2006). The Au film can also be used for studies of biological materials by providing a platform upon which to deposit self-assembled monolayers of organothiol monomers, which can be used to immobilize biomolecules.

For cell heating and temperature control, a platinum resistive heater 100 μm wide by 100–150 nm thick was deposited on the backside of the

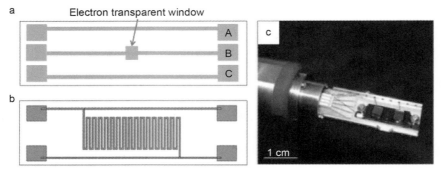

Figure 7.2 Sketches of basic design for electrochemical and thermal control and a photo of the fluid cell holder and fluid cell. (a) Patterning of Au electrodes for electrochemical control showing: reference electrode (A), working electrode (B), and counter electrode (C); (b) Pt resistor pattern for thermal control; (c) photograph of electrochemical cell on custom-made fluid stage made by Hummingbird Scientific (Nielsen et al., 2012). (For color version of this figure, the reader is referred to the online version of this chapter.)

bottom silicon wafer (Fig. 7.2b). Platinum, copper, and nickel are known to have a unique and repeatable resistance versus temperature relationship over an element-dependent temperature range. Platinum was chosen due to its chemical inertness in air and because it has the most stable resistance–temperature relationship over the largest temperature range. Application of a voltage across the Pt resister results in a temperature rise due to Joule heating. By measuring the change of the resistance while applying this voltage, the temperature can be calculated, monitored, and therefore controlled. Thus, the Pt resistive element serves as both heater and thermometer, though it is important to note that, because the Pt film is on the outside of the lower wafer, it may not give an accurate reading of the temperature in the liquid between the membranes should there be substantial heating by the beam.

The resistance and temperature follow a linear relationship between 0 and 100 °C given by:

$$R(t) \approx R(0)(1 + \alpha T) \tag{7.1}$$

where $R(T)$ is the resistance at temperature T in °C, $R(0)$ is the resistance at 0 °C, and α is a constant called the temperature coefficient. For pure Pt, α equals 0.003925 1°C between 0 and 100 °C. Depending on the Pt purity, the value of α can vary. The value of α for the Pt resistor of the cell is calibrated via a benchtop experiment. This Pt heater design enables temperature control between room temperature and about 100 °C, which is a temperature range well suited to the study of reactions in aqueous solutions.

Because each cell is constructed for an individual set of experiments, additional functionality can be included via modification of the core design. For example, choice of the thickness of the cell for a given model of TEM determines whether chemical mapping via energy dispersive X-ray spectroscopy (EDS) and electron energy loss spectroscopy are possible.

Recently, an alternative design based on graphene liquid cells (GLCs) was introduced (Jong Min et al., 2012). A graphene layer is grown on a copper foil substrate via chemical vapor deposition and then directly transferred onto a gold TEM mesh with amorphous carbon film support. The reaction solution is pipetted directly onto two graphene-coated TEM grids facing in opposite directions. Upon wetting, one of the graphene membranes detaches from its associated TEM grid. Due to van der Waals interaction between graphene sheets, the liquid droplets of various thicknesses from 6 to 200 nm can be securely trapped between the double-membrane pocket. While this cell design may give thin liquid layers more consistently, reaction volumes are highly constrained and, unlike, Si-based fluid cells where routine lithographic patterning, Si etching, e-beam evaporating, and liftoff processes make incorporating temperature, electrochemical, flow, or other environmental controls fairly straightforward (Li et al., 2012; Yu, Liu, & Yang, 2013) expansion of the GLC design presents a formidable challenge.

3. CELL FABRICATION AND ASSEMBLY

Fluid cells are fabricated using 4 in. silicon wafers 300 μm in thickness. Low-stress silicon nitride membranes 50–100 nm thick are grown on both sides of silicon wafers (e.g., B in Fig. 7.1). Lithographic patterning and KOH etching of these wafers are then used to fabricate arrays of top and bottom halves—or "chips"—of the fluid cell (e.g., A in Fig. 7.1), the solution reservoir access ports on the top Si chips (port below D in Fig. 7.1), and the free-standing Si_3N_4 membranes that provide the electron transparent windows (e.g., F in Fig. 7.1) on the top and bottom chips. The window size is set as desired. The original design utilized windows that were 200 μm × 1000 μm, where the larger dimension was incorporated to accommodate line of sight from the window to an EDS detector. However, due to substantial issues with bowing, the window dimensions were reduced to 50 μm by 50 μm. An alternative design with windows that are narrower, but longer, for example, 25 μm × 100 μm, and oriented in orthogonal directions on the two wafers would further stiffen the windows while increasing the ease of alignment during assembly.

The Si towers are fabricated from Si wafers by lithographic patterning and deep dry Si etching. The lower wafer is coated with a rectangular ring of Si_3N_4 or a metal that is inert to the reactants to be used in the experiments in order to provide a gap between the upper and lower chips that accommodates the solution layer (C in Fig. 7.1). This spacer is deposited by e-beam evaporation or plasma-enhanced chemical vapor deposition. Different thicknesses of the spacer can be chosen according to experimental needs. The Au electrodes and Pt resistive heaters used for electrochemical and temperature control, as described earlier, are deposited onto the bottom Si chips by lithographic patterning and e-beam evaporative deposition followed by a liftoff process.

The bottom and top chips of the liquid cell are plasma cleaned for 1–3 min to make the Si_3N_4 membrane window hydrophilic for aqueous reaction solutions and then glued together using epoxy (M-Bond 610, SPI supplies Inc.) around the outer edges of the wafers after alignment of the bottom and top windows. The glue is cured at a temperature of 100–150 °C, which is reached using a slow ramp upward of approximately 20 °C/h from room temperature. The towers that serve as solution reservoirs are aligned with the reservoir access ports on the top chip and glued on using the same epoxy. Others have reported sealing static fluid cells by plasma-activated wafer bonding of the bottom and top chips and by using O-rings for the solution reservoirs (Grogan & Bau, 2010).

4. TEM OPERATION

To perform an experiment, an aliquot of a few microliters of solution is loaded into one of the two reservoirs. After the solution is drawn between the two Si_3N_4 membranes via capillary action, the second reservoir is also filled with the solution. The two reservoirs are then covered with glass caps (E in Fig. 7.1) and sealed using UV curable glue (Norland Opticure 63).

The sealed cell-containing reaction solution is then mounted onto the TEM sample holder (a portion of a custom-made holder from Hummingbird Scientific is shown in Fig. 7.2c). The holder is put into a prepump stage to test for leakage before being put into the TEM. In our research, imaging is conducted with a JEOL 2100F TEM at an accelerating voltage of 200 kV. The overall thickness of the cell exceeds the thickness of normal foils or grids.

However, in this microscope, the holder still can be tilted by ±20° in the alpha direction. Video is recorded with the VirtualDub software from the live images in Digital Micrograph (Gatan). We have carried out fluid cell TEM experiments with this combination of cell, holder, and microscope using a variety of solution-based systems, in which the chemical reactions were trigged by electron beam energy, electrochemical control, and cell heating.

The electron beam in TEM is known to carry enough energy to damage a sample through heating, electrostatic charging, ionization damage, displacement damage, and hydrocarbon contamination (de Jonge et al., 2009). An electron beam of 200–300 keV has enough energy to trigger some chemical reactions (Li et al., 2012) or reduce metal ions to metal (Zheng, Smith, Young-wook, Kisielowski, Dahmen, & Alivisatos, 2009). Beam-induced heating is due to inelastic scattering between the incoming electrons and particles. The energy transferred in the process ends up as heat, resulting in a locally higher temperature within the specimen. Although the temperature of the sample holder rises only a few K under the electron beam, depending on the material's thermal conductivity, it is not unusual for the electron beam to melt common thin metal samples under high electron fluxes (Egerton, Li, & Malac, 2004).

While beam-induced effects are a problem in many TEM experiments at high incident beam currents, with *in situ* TEM experiments, we have found we can take advantage of beam-induced initiation of reactions. For example, in the system iron (III) chloride in a potassium dihydrogen phosphate solution, we showed that by exposing the solution to an electron beam for a period of a few seconds to minutes, previously precipitated akaganeite nanorods dissolved and ferrihydrite nanoparticles were produced in their place (Li et al., 2012). We were then able to examine the dynamics of ferrihydrite nanoparticle interaction and attachment.

Figure 7.3A presents a sequence of *in situ* TEM images at atomic resolution showing the details of the ferrihydrite nanoparticle attachment process. This work highlights the unique advantage of *in situ* TEM in terms of spatial and temporal resolution. Here, it provided information about the dynamics of the direction-specific attachment process—a process that takes place within a few seconds—at lattice resolution. A similar study was performed on metallic nanoparticles using GLCs. As shown in Fig. 7.3B, the thinness of the GLCs enabled atomic resolution of Pt nanoparticle fusion.

For both of the GLC- and Si chip-based cells, precise control over the thickness of the fluid layer is critical for reproducible imaging. We noted earlier that the spacers used to fabricate the silicon-based cells are

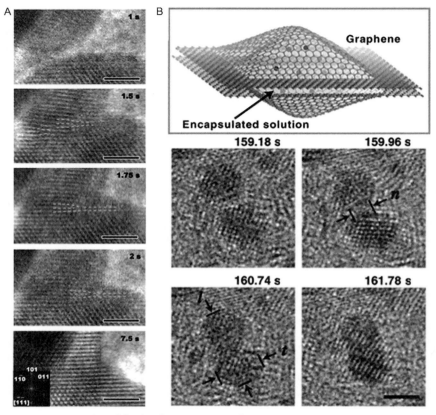

Figure 7.3 Examples of fluid cell TEM images of nanoparticle aggregation and attachment from two different fluid cell designs and an illustration of a graphene liquid cell (GLC). (A) Sequence of *in situ* TEM images with atomic resolution showing the details of the ferrihydrite nanoparticle attachment process. The dashed lines indicate edge dislocations formed at the moment of attachment. These rapidly translate laterally across the boundary and are expelled, leaving behind a defect-free interface. (B) An illustration (top right) of a GLC encapsulating growth solution and TEM stills (bottom right) from a movie of Pt nanocrystal growth via coalescence (Jong Min et al., 2012). All scale bars are 2 nm.

200–500 nm in thickness. Based on our experience and that of others, atomic resolution cannot be obtained unless the fluid thickness is about 100 nm or less. Yet the images in Fig. 7.3 reveal the atomic details of the attachment process. This apparent contradiction is likely due to the fact that the even a slight bowing of the Si_3N_4 membranes can substantially alter the spacing between the windows. In this case, the atomic resolution indicates an inward bending of the two windows. In contrast, we find that when the windows bend outward from each other, the liquid layer under the electron beam can be up to a couple of microns in thickness, resulting in poor resolution. A number of design and fabrication features have been implemented to producing thinner liquid layers. Patterns of SiO_2 nanopillars were fabricated and deposited on the bottom chip to define the minimum thickness of the liquid layer (Grogan & Bau, 2010). A plasma wafer-bonding method was used to create a liquid layer of ~100 nm (Creemer et al., 2008). However, even with these approaches, the outward bending of the Si_3N_4 membrane often makes the upper limit of the thickness difficult to control. Consequently, efforts to develop procedures for reproducibly fabricating Si-based cells with ultrathin windows and precisely defined gap sizes below 100 nm are ongoing and appear to offer the greatest promise for widespread application of the technique.

Electrochemical control can be very useful in studies of mineralization, because it can be used to alter the pH of the solution, which typically results in changes in mineral solubility. When an electrical bias (~1.2 V) is applied to the working electrode of a fluid cell filled with an aqueous solution, the reduction of dissolved molecular oxygen in the solution produces hydroxide ions at the metal/liquid interface according to the following reactions (Tlili, Benamor, Gabrielli, Perrot, & Tribollet, 2003):

$$O_2 + 2H_2O + 2e^- \rightarrow H_2O_2 + 2OH^-$$

$$H_2O_2 + 2e^- \rightarrow 2OH^-$$

$$\overline{}$$

$$O_2 + 2H_2O + 4e^- \rightarrow 4OH^-$$

This reaction then increases the pH locally near the surface of the electrode. In a solution of a mineral phase like $CaCO_3$, this in turn results in a local decrease in solubility. As long as the calcium and carbonate concentrations are chosen appropriately, the solution can begin in an undersaturated state and become supersaturated only when the voltage is turned on and then only near the electrode. By placing the electrode on the window, this electrochemical trigger ensures that nucleation occurs where and when it can be imaged. We have demonstrated that this approach works in benchtop tests (Nielsen et al., 2012); however, we have not yet implemented this in the TEM. Moreover, we have found that even in the absence of a voltage, the electron beam itself induces nucleation and growth of $CaCO_3$ on the

Au electrode, giving a clear indication that the beam causes dissociation of H_2O.

Other applications of electrochemical control over reactions during *in situ* TEM have been reported. Here, we consider two examples. The first comes from the Ross group's research on electrochemical nucleation and growth of Cu nanoparticles on Au (Radisic, Vereecken, Searson, et al., 2006) (Fig. 7.4A–E). The high spatial and temporal resolution of *in situ*

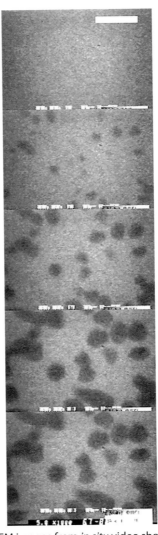

Figure 7.4 Sequences of TEM images from *in situ* video showing electrochemical nucleation and growth of Cu nanoclusters on Au (Radisic, Vereecken, Hannon, et al., 2006).

TEM enabled collection of data on the formation and growth of individual Cu nanoclusters, leading to a significant revision of conventional models. The improved understanding provided a more quantitative approach to the electrochemical fabrication of nanoscale structures.

The second example comes from the work of Huang et al. (2010) who used this approach to study the failure mechanism during battery charging and discharging by driving electrochemical lithiation of a single SnO_2 nanowire electrode (Huang et al., 2010). Figure 7.5 presents a time-lapse sequence showing the morphological evolution of the nanowire during charging at −3.5 V against a $LiCoO_2$ cathode (Huang et al., 2010). Upon charging, a reaction front along the nanowire caused the nanowire to swell, elongate, and spiral. Because lithiation-induced volume expansion,

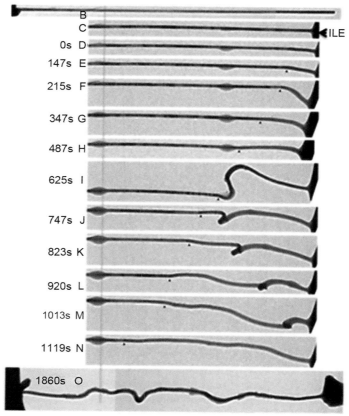

Figure 7.5 Sequences of TEM images from *in situ* video showing morphological evolution of a SnO_2 nanowire anode during charging (Huang et al., 2010). (For color version of this figure, the reader is referred to the online version of this chapter.)

plasticity, and pulverization of electrode materials are the key mechanical factors that affect the performance and lifetime of high-capacity anodes in lithium-ion batteries, real-time observations of these mechanical effects should provide key mechanistic insights for the design of advanced batteries.

These examples demonstrate the power of *in situ* electrochemical control for understanding reaction, nucleation, and growth kinetics in a wide range of materials, including insulating crystals, thin metal films, and semiconducting nanowires.

Numerous chemical reactions cannot be triggered electrochemically or by the e-beam, but can be successfully driven thermally. Heating control has been used for some time in the study of solid-state phases and phase transformations (Parker et al., 1986), interphase boundary dynamics (Howe et al., 1998), and nanocrystallite nucleation and coarsening in vapor (Li-Chi & Risbud, 1994). The biggest obstacle to the use of controlled heating with liquid cells is leakage due to increased pressure inside the cell. Even though the cell and heater detailed earlier can easily reach 100 °C, in practice, we have found that the temperature must be kept below 60 °C when filled with an aqueous solution in order to avoid leakage out of the cell and into the TEM column. Research into different methods for sealing the cell is under way to extend this upper limit on temperature.

5. EFFECTS OF THE ELECTRON BEAM ON REACTIONS IN THE FLUID CELL

Whether the electron beam serves as an aid to studying reactions *in situ* or is only a source for imaging, the potential for deleterious effects is substantial. Most of the effects seen with dry samples (Egerton et al., 2004) are also possible in solution. These include heating, electrostatic charging, ionization damage (radiolysis), displacement damage, sputtering, and hydrocarbon contamination. The extent of radiation damage is proportional to the electron dose, regardless of beam diameter. However, the magnitude of heating and sputtering are likely to be greatly reduced by the presence of the surrounding fluid. On the other hand, the interaction of the high-energy electron beam with the water (if in aqueous solution) will lead to production of free radicals and radiolysis of water produces H, H_2, OH, H_2O_2, and hydrated electrons (Garrett et al., 2005). The presence of these radiolysis products and their effect on both pH and solute speciation probably explains the many observations of nucleation, growth, and dissolution induced by the beam. In heating experiments, these effects may be enhanced due to the high temperature. In each case, the only strategy used to date to minimize the damage is to reduce the electron dose by either limiting the time of exposure to the beam through beam shuttering or lowering of the beam current.

6. CONCLUSIONS

With the development of *in situ* TEM techniques, real-time imaging of various kinds of chemical reactions, morphologic development, crystal growth, and chemical and electronic structure development in gas-, liquid-, and solid-state phases is feasible (de Jonge & Ross, 2011). The advantage of *in situ* TEM over other *in situ* techniques, such as optical microscopy, AFM, or X-ray scattering or diffraction, is that it records both structure and morphology at spatial resolutions down to the atomic level with high temporal resolution of up to 10^{-6} s per image. Thus, fluid cell *in situ* TEM is a powerful new tool for understanding dynamic processes during liquid phase chemical reactions, including mineral formation.

Si- and graphene-based liquid cells have been developed and used to study the mechanisms of nanocrystal nucleation and growth, as well as their aggregation via oriented attachment. The thickness of the liquid layer is the key factor in determining the spatial resolution, with reported thicknesses of Si-based fluid cells ranging from near zero up to a couple of micrometers and those of GLCs ranging from 6 to 200 nm. For both the GLC- and Si-based cells, precise control over that thickness has been difficult. While GLCs provide a narrow thickness range, the advantage of Si-based fluid cells is that incorporating temperature, electrochemical, flow, or other environmental controls is fairly straightforward through routine Si wafer fabrication processes.

Electrochemical control has been developed to study electrochemically induced nucleation and growth, as well as failure mechanisms during electrode charging and discharging. Heating control is, in principle, straightforward to implement for liquid phase reactions that require high temperature. However, in practice, leakage has been a problem during heating of silicon-based cells due to the method by which cells are sealed. Modifications of the cell design and methods of sealing are under study to avoid leakage.

Reproducible control of the liquid layer thickness in the 10–100 nm range is critical for maximizing the impact of this technique in applications where atomic resolution is important, such as the study of crystal structure development during nucleation. Development of methods for fabricating

Si-based cells with ultrathin windows and precisely defined gap sizes in this range is currently under way.

ACKNOWLEDGMENTS

This research was supported by the US Department of Energy, Office of Basic Energy Sciences (OBES), and by LBNL under contract no. DE-AC02-05CH11231. Development of the TEM fluid cell was supported by the OBES, Division of Chemical, Biological and Geological Sciences; analysis of iron oxide formation was supported by the OBES, Division of Materials Science and Engineering; and cell fabrication and TEM analysis were performed at the Molecular Foundry, LLNL, which is supported by the OBES, Scientific User Facilities Division. M. H. N. acknowledges government support under and awarded by the Department of Defense, the Air Force Office of Scientific Research, and a National Defense Science and Engineering Graduate Fellowship, 32 CFR 168a.

REFERENCES

Ando, T., Kodera, N., Takai, E., Maruyama, D., Saito, K., & Toda, A. (2001). A high-speed atomic force microscope for studying biological macromolecules. *Proceedings of the National Academy of Sciences of the United States of America*, 98(22), 12468–12472. http://dx.doi.org/10.1073/pnas.211400898.

Armstrong, M. R., Boyden, K., Browning, N. D., Campbell, G. H., Colvin, J. D., DeHope, W. J., et al. (2007). Practical considerations for high spatial and temporal resolution dynamic transmission electron microscopy. *Ultramicroscopy*, 107(4–5), 356–367. http://dx.doi.org/10.1016/j.ultramic.2006.09.005.

Brennan, S., Fuoss, P. H., Kahn, J. L., & Kisker, D. W. (1990). Experimental considerations for in situ X-ray-scattering analysis of OMVPE growth. *Nuclear Instruments and Methods in Physics Research Section A: Accelerators, Spectrometers, Detectors and Associated Equipment*, 291(1–2), 86–92. http://dx.doi.org/10.1016/0168-9002(90)90038-8.

Creemer, J. F., Helveg, S., Hoveling, G. H., Ullmann, S., Molenbroek, A. M., Sarro, P. M., et al. (2008). Atomic-scale electron microscopy at ambient pressure. *Ultramicroscopy*, 108(9), 993–998. http://dx.doi.org/10.1016/j.ultramic.2008.04.014.

de Jonge, N., Peckys, D. B., Kremers, G. J., & Piston, D. W. (2009). Electron microscopy of whole cells in liquid with nanometer resolution. *Proceedings of the National Academy of Sciences of the United States of America*, 106(7), 2159–2164. http://dx.doi.org/10.1073/pnas.0809567106.

de Jonge, N., & Ross, F. M. (2011). Electron microscopy of specimens in liquid. *Nature Nanotechnology*, 6(11), 695–704. http://dx.doi.org/10.1038/nnano.2011.161.

Egerton, R. F., Li, P., & Malac, M. (2004). Radiation damage in the TEM and SEM. *Micron*, 35(6), 399–409. http://dx.doi.org/10.1016/j.micron.2004.02.003.

Evans, J. E., Jungjohann, K. L., Wong, P. C. K., Chiu, P.-L., Dutrow, G. H., Arslan, I., et al. (2012). Visualizing macromolecular complexes with in situ liquid scanning transmission electron microscopy. *Micron*, 43(11), 1085–1090. http://dx.doi.org/10.1016/j.micron.2012.01.018.

Garrett, B. C., Dixon, D. A., Camaioni, D. M., Chipman, D. M., Johnson, M. A., Jonah, C. D., et al. (2005). Role of water in electron-initiated processes and radical chemistry: Issues and scientific advances. *Chemical Reviews*, 105(1), 355–389. http://dx.doi.org/10.1021/cr030453x.

Grogan, J. M., & Bau, H. H. (2010). The nanoaquarium: A platform for in situ transmission electron microscopy in liquid media. *Journal of Microelectromechanical Systems*, *19*(4), 885–894. http://dx.doi.org/10.1109/jmems.2010.2051321.

Howe, J. M., Murray, T. M., Moore, K. T., Csontos, A. A., Tsai, M. M., Garg, A., et al. (1998). Understanding interphase boundary dynamics by in situ high-resolution and energy-filtering transmission electron microscopy and real-time image simulation. *Microscopy and Microanalysis*, *4*(3), 235–247.

Hu, Q., Nielsen, M. H., Freeman, C. L., Hamm, L. M., Tao, J., Lee, J. R. I., et al. (2012). The thermodynamics of calcite nucleation at organic interfaces: Classical vs. non-classical pathways. *Faraday Discussions*, *159*, 509–523. http://dx.doi.org/10.1039/c2fd20124k.

Huang, J. Y., Zhong, L., Wang, C. M., Sullivan, J. P., Xu, W., Zhang, L. Q., et al. (2010). In situ observation of the electrochemical lithiation of a single SnO2 nanowire electrode. *Science*, *330*(6010), 1515–1520. http://dx.doi.org/10.1126/science.1195628.

Jong Min, Y., Jungwon, P., Ercius, P., Kwanpyo, K., Hellebusch, D. J., Crommie, M. F., et al. (2012). High-resolution EM of colloidal nanocrystal growth using graphene liquid cells. *Science*, *335*(6077), 61–64. http://dx.doi.org/10.1126/science.1217654.

Land, T. A., DeYoreo, J. J., & Lee, J. D. (1997). An in-situ AFM investigation of canavalin crystallization kinetics. *Surface Science*, *384*(1–3), 136–155. http://dx.doi.org/10.1016/s0039-6028(97)00187-8.

Lee, J. R. I., Han, T. Y.-J., Willey, T. M., Nielsen, M. H., Klivansky, L. M., Liu, Y., et al. (2013). Cooperative reorganization of mineral and template during directed nucleation of calcium carbonate. *Journal of Physical Chemistry C* http://dx.doi.org/10.1021/jp400279f.

Lee, J. R. I., Han, T. Y.-J., Willey, T. M., Wang, D., Meulenberg, R. W., Nilsson, J., et al. (2007). Structural development of mercaptophenol self-assembled monolayers and the overlying mineral phase during templated CaCO3 crystallization from a transient amorphous film. *Journal of the American Chemical Society*, *129*(34), 10370–10381. http://dx.doi.org/10.1021/ja071535w.

Li, D., Nielsen, M. H., Lee, J. R. I., Frandsen, C., Banfield, J. F., & De Yoreo, J. J. (2012). Direction-specific interactions control crystal growth by oriented attachment. *Science*, *336*(6084), 1014–1018. http://dx.doi.org/10.1126/science.1219643.

Liao, H.-G., Cui, L., Whitelam, S., & Zheng, H. (2012). Real-time imaging of Pt3Fe nanorod growth in solution. *Science*, *336*(6084), 1011–1014. http://dx.doi.org/10.1126/science.1219185.

Li-Chi, L., & Risbud, S. H. (1994). Real-time hot-stage high-voltage transmission electron microscopy precipitation of CdS nanocrystals in glasses: Experiment and theoretical analysis. *Journal of Applied Physics*, *76*(8), 4576–45804580.

Mirsaidov, U. M., Zheng, H., Casana, Y., & Matsudaira, P. (2012). Imaging protein structure in water at 2.7 nm resolution by transmission electron microscopy. *Biophysical Journal*, *102*(4), L15–L17. http://dx.doi.org/10.1016/j.bpj.2012.01.009.

Nielsen, M. H., Lee, J. R. I., Hu, Q., Han, T. Y.-J., & De Yoreo, J. J. (2012). Structural evolution, formation pathways and energetic controls during template-directed nucleation of CaCO3. *Faraday Discussions*, *159*, 105–121. http://dx.doi.org/10.1039/c2fd20050c.

Parker, M. A., Sigmon, T. W., & Sinclair, R. (1986). In-situ high resolution transmission electron microscopy of dynamic events during the amorphous to crystalline phase transformation in silicon. In L. W. Hobbs, K. W. Westmacott, & D. B. Williams (Eds.), *Materials problem solving with the transmission electron microscope.Materials research society symposium proceedings*. Materials Research Society, Warrendale, PA, *62*, 311–322.

Radisic, A., Ross, F. M., & Searson, P. C. (2006). In situ study of the growth kinetics of individual island electrodeposition of copper. *The Journal of Physical Chemistry B*, *110*(15), 7862–7868. http://dx.doi.org/10.1021/jp057549a.

Radisic, A., Vereecken, P. M., Hannon, J. B., Searson, P. C., & Ross, F. M. (2006). Quantifying electrochemical nucleation and growth of nanoscale clusters using real-time kinetic data. *Nano Letters*, *6*(2), 238–242. http://dx.doi.org/10.1021/nl052175i.

Radisic, A., Vereecken, P. M., Searson, P. C., & Ross, F. M. (2006). The morphology and nucleation kinetics of copper islands during electrodeposition. *Surface Science*, *600*(9), 1817–1826. http://dx.doi.org/10.1016/j.susc.2006.02.025.

Ross, F. M., Tersoff, J., Reuter, M., Legoues, F. K., & Tromp, R. M. (1998). In situ transmission electron microscopy observations of the formation of self-assembled Ge islands on Si. *Microscopy Research and Technique*, *42*(4), 281–294. http://dx.doi.org/10.1002/(sici)1097-0029(19980915)42:4<281::aid-jemt7>3.0.co;2-t.

Schitter, G., Astroem, K. J., DeMartini, B. E., Thurner, P. J., Turner, K. L., & Hansma, P. K. (2007). Design and modeling of a high-speed AFM-scanner. *IEEE Transactions on Control Systems Technology*, *15*(5), 906–915. http://dx.doi.org/10.1109/tcst.2007.902953.

Teng, H. H., Dove, P. M., Orme, C. A., & De Yoreo, J. J. (1998). Thermodynamics of calcite growth: Baseline for understanding biomineral formation. *Science*, *282*(5389), 724–727. http://dx.doi.org/10.1126/science.282.5389.724.

Tlili, M. M., Benamor, M., Gabrielli, C., Perrot, H., & Tribollet, B. (2003). Influence of the interfacial pH on electrochemical CaCO3 precipitation. *Journal of the Electrochemical Society*, *150*(11), C765–C771. http://dx.doi.org/10.1149/1.1613294.

Todorov, L. V., Martins, C. I., & Viana, J. C. (2013). In situ WAXS/SAXS structural evolution study during uniaxial stretching of poly(ethylene therephthalate) nanocomposites in solid state: Poly(ethylene therephthalate)/montmorillonite nanocomposites. *Journal of Applied Polymer Science*, *128*(5), 2884–2895. http://dx.doi.org/10.1002/app.38368.

Tsuchiya, T., Taniwatari, T., Uomi, K., Kawano, T., & Ono, Y. (1993). In-situ X-ray monitoring of metalorganic vapor-phase epitaxy. *Japanese Journal of Applied Physics Part 1-Regular Papers Short Notes & Review Papers*, *32*(10), 4652–4655. http://dx.doi.org/10.1143/jjap.32.4652.

Williamson, M. J., Tromp, R. M., Vereecken, P. M., Hull, R., & Ross, F. M. (2003). Dynamic microscopy of nanoscale cluster growth at the solid-liquid interface. *Nature Materials*, *2*(8), 532–536. http://dx.doi.org/10.1038/nmat944.

Yu, X.-Y., Liu, B., & Yang, L. (2013). Imaging liquids using microfluidic cells. *Microfluidics and Nanofluidics*, *15*, 1–20. http://dx.doi.org/10.1007/s10404-013-1199-4.

Zheng, H., Smith, R. K., Young-wook, J., Kisielowski, C., Dahmen, U., & Alivisatos, A. P. (2009). Observation of single colloidal platinum nanocrystal growth trajectories. *Science*, *324*(5932), 1309–1312. http://dx.doi.org/10.1126/science.1172104.

CHAPTER EIGHT

X-ray Absorption Spectroscopy for the Structural Investigation of Self-Assembled-Monolayer-Directed Mineralization

Jonathan R.I. Lee[*,1], Michael Bagge-Hansen[*], Trevor M. Willey[*], Robert W. Meulenberg[†], Michael H. Nielsen[‡], Ich C. Tran[*], Tony van Buuren[*]

[*]Lawrence Livermore National Laboratory, Livermore, California, USA
[†]Department of Physics and Astronomy, Laboratory for Surface Science and Technology, University of Maine, Orono, Maine, USA
[‡]Department of Materials Science and Engineering, University of California, Berkeley, California, USA
[1]Corresponding author: e-mail address: lee204@llnl.gov

Contents

1. Introduction	166
2. Theory	167
3. Experimental	172
3.1 Establishing the degree of linear polarization and energy calibration of the beamline	172
3.2 XAS measurements of SAM/crystal	173
4. Analysis	183
Acknowledgments	185
References	186

Abstract

Self-assembled monolayers (SAMs) of organothiol molecules prepared on noble metal substrates are known to exert considerable influence over biomineral nucleation and growth and, as such, offer model templates for investigation of the processes of directed biomineralization. Identifying the structural evolution of SAM/crystal systems is essential for a more comprehensive understanding of the mechanisms by which organic monolayers mediate mineral growth. X-ray absorption spectroscopy (XAS) provides the attractive ability to study SAM structure at critical stages throughout the processes of crystallization in SAM/mineral systems. Here, we discuss important theoretical and experimental considerations for designing and implementing XAS studies of SAM/mineral systems.

Methods in Enzymology, Volume 532
ISSN 0076-6879
http://dx.doi.org/10.1016/B978-0-12-416617-2.00008-4

© 2013 Elsevier Inc.
All rights reserved.

165

1. INTRODUCTION

In this chapter, we discuss the use of synchrotron-based X-ray absorption spectroscopy (XAS) for the structural investigation of directed mineralization self-assembled monolayer (SAM) templates. SAMs make attractive platforms for investigating the processes of directed mineralization because of their ability to direct mineral nucleation and growth (Sommerdijk & de With, 2008), combined with their straightforward preparation and simple, yet well-ordered, structures (Ulman, 1996). Prior studies of SAM/mineral systems indicate that the organic monolayers induce biomineral nucleation and growth on distinct crystallographic planes with a high degree of specificity (Aizenberg, Black, & Whitesides, 1999; Han & Aizenberg, 2003; Sommerdijk & de With, 2008). Furthermore, the orientation of the crystals is sensitive to subtle changes in the SAM composition, such as the end-group functionality and the chain length of the organic thiol molecules that compose the monolayer. Mechanistic insight obtained from the investigation of mineral nucleation and growth on idealized organic films, such as SAMs, is of fundamental importance in enhancing our understanding of the considerably more complex natural processes of organic-matrix-directed biomineralization. Meanwhile, identifying the means by which the organic monolayers influence mineral growth could hold the key to developing advanced strategies for synthesizing designer materials of controlled structure and composition (Sommerdijk & de With, 2008).

Characterization of the evolution in structure during directed mineralization, with a particular emphasis on the interfacial structure, is a critical component in deriving new mechanistic understanding; determining the structure of the interface and how it evolves during nucleation and growth can yield valuable information regarding the interactions between the organic molecules in the SAM and the mineral phase. As such, it is essential to implement techniques that probe the structure of the SAM buried beneath the mineral and, by extension, the structure of the interface between the two phases (Fig. 8.1). XAS is an established technique for the characterization of SAMs (Grunze, 1993; Hahner, Woll, Buck, & Grunze, 1993; Zharnikov & Grunze, 2001) that, in contrast to many alternative techniques, such as scanned probe microscopies, provides the capability to study the monolayer structure even when buried beneath a condensed mineral phase. In particular, XAS enables the assignment of bond/functional group orientations within the organothiol molecules that

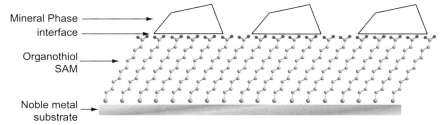

Figure 8.1 Schematic (not to scale) of a SAM/mineral sample viewed in cross-section. The SAM displayed is composed of mercaptoundecanoic acid molecules, for which carbon, oxygen, and sulfur atoms are denoted by gray, red, and yellow spheres, respectively. (For interpretation of the references to color in this figure legend, the reader is referred to the online version of this chapter.)

compose the SAMs. Identifying the bond/functional group orientations at various stages during the crystallization process, such as immediately after SAM preparation, following exposure of the SAM to the growth solution and after nucleation and growth of the crystal, can be extremely instructive in characterizing the nature of interactions between the monolayer and mineral. Recent XAS studies of SAM/crystal systems indicate, for example, that flexibility in the monolayer structure is crucial for directed mineralization as part of a cooperative process between the organic and inorganic phases (Lee et al., 2013, 2007).

Although XAS is a powerful tool for investigating directed mineralization on SAMs, the technique has been underutilized to date in the field. This is, in part, due to the need for access to the high X-ray flux and the ability to tune the X-ray photon energy provided by synchrotron sources and the challenges associated with the design and implementation of experiments on synchrotron beamlines, specifically the soft X-ray beamlines required for the XAS study of SAMs. Thus, the aim of this chapter is to address the important components of experimental design required for successful XAS measurements of SAM/mineral systems on soft X-ray beamlines and to identify the more common challenges one can encounter, along with approaches used to mitigate these challenges.

2. THEORY

In an XAS measurement, the absorption of X-ray photons is recorded as the photon energy is scanned through the electronic core-level binding energy of a specific element within the sample of interest. As the incident

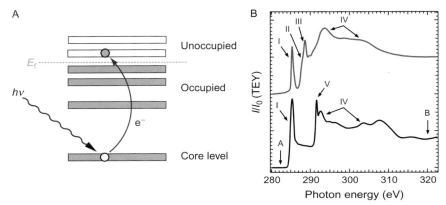

Figure 8.2 (A) Energy-level diagram of the X-ray absorption process and (B) representative XAS spectra recorded in the TEY mode for HOPG (black trace, collected at beamline 8.0 of the Advanced Light Source) and a 4-mercaptodecyl benzoic acid (MDBA) SAM on an Au(111) substrate (red trace, collected at beamline 8.2 of the Stanford Synchrotron Radiation Lightsource). The experimental spectra display a range of distinct resonances one can encounter in carbon K-edge XAS data including (I) π^* resonance for HOPG and aryl ring of the MDBA, (II) C–H σ^*/Rydberg-like R* resonances, (III) π^* resonance for the carboxyl group of MDBA, (IV) C—C σ^* resonances within the HOPG and MDBA, and (V) a σ-exciton feature observed for HOPG. (For color version of this figure, the reader is referred to the online version of this chapter.)

photon energy is increased, an absorption onset will be observed corresponding to the photoexcitation of core-level electrons into the lowest-energy unoccupied electronic states associated with the atom (Fig. 8.2A). Subsequent increases in the incident photon energy will be accompanied by resonances due to transitions from the core level into higher-energy unoccupied states, up to a limit of $\sim$30–100 eV above the absorption onset. Significantly, the dipole selection rule of $\Delta l = \pm 1$ must be fulfilled for an electron transition to occur via photon absorption and for a resonance to be observed in the XAS. The region of the XAS spectrum between the absorption edge and the $\sim$30–100 eV limit above it is typically referred to as the near-edge absorption fine structure (NEXAFS) for measurements in the soft X-ray regime and provides the main focus for structural studies of SAMs. Beyond the NEXAFS regime, one reaches the extended X-ray absorption fine structure (EXAFS) region of the XAS spectrum, in which photoelectron scattering by neighboring atoms results in small oscillations in absorption defined by the local environment of the absorbing atom. Structural investigations of SAMs via measurement of the EXAFS are impractical, however, because the photoelectron backscattering by neighboring low-Z elements is weak and the EXAFS region is often obscured by other X-ray absorption edges that are close in energy.

For low-Z elements within the organothiol SAMs, such as carbon, nitrogen, and oxygen (with K-edges at ~288, ~399, and ~543 eV, respectively), the unoccupied electronic density of electronic states probed with NEXAFS is primarily composed of the antibonding orbitals (e.g., π^* and σ^* orbitals) associated with covalent bonding to neighboring atoms (Fig. 8.2B). Subtle changes in the local bonding environment, including the inductive effects of electronegative elements, result in chemical shifts between these resonances (e.g., the π^* resonances for graphite and a carboxyl group are separated by ~3.3 eV). Therefore, one can isolate resonances for specific bonds/functional groups. One will also reach the ionization potential/binding energy within the NEXAFS regime, which manifests as a steplike feature beginning at the minimum energy required to promote a core-level electron above the vacuum level/into the continuum.

A comprehensive discussion of the theoretical basis for obtaining orientational information from the XAS data for monolayer films lies beyond the scope of this chapter but is the subject of numerous excellent treatments available in the literature (Stöhr, 1992; Stohr & Outka, 1987). It is instructive, however, to briefly introduce the key equations that define the necessary XAS data for deriving molecular orientation in organothiol SAMs prepared on noble metal substrates. The discussion herein will only address K-edge XAS (excitation from the 1s state) because measurements at the K-edge are essential for study of the second period elements that predominate in organothiol molecules. In addition, K-edge XAS measurements of third period elements, such as the sulfur of a sulfate or mercaptan head group, are also readily used for the structural investigation of SAMs and, due to the spherical symmetry of the 1s orbital, require less complicated analysis than measurements at the L_3 and L_2 edges.

XAS enables a quantitative assignment of orbital orientation via analysis of the angular dependence of the associated resonance intensity. For organothiol SAMs, one can consider the intensity of an XAS resonance to be proportional to the dot product of the direction of maximum probability density (collinear with the transition dipole moment) and the electric field vector of the incident X-ray beam. Therefore, any bonds/orbitals with well-defined orientations will exhibit changes in the resonance intensity as a function of the angle of incidence between the X-ray beam and sample, provided that the beam is linearly polarized (Fig. 8.3). The light provided on the majority of soft X-ray beamlines is highly linearly polarized in the plane of

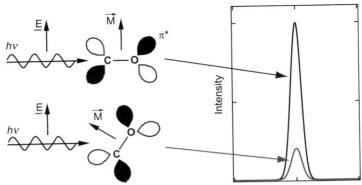

Figure 8.3 Simplified schematic illustrating the effects of the transition dipole moment orientation with respect to the electric vector of the incident X-ray beam on the resonance strength. (For color version of this figure, the reader is referred to the online version of this chapter.)

the storage ring, which makes the use of synchrotron radiation advantageous for XAS studies of SAM structure. The degree of linear polarization, P, is defined by (Stöhr, 1992; Stohr & Outka, 1987)

$$P = \frac{E_p^2}{E_s^2 + E_p^2} \quad (8.1)$$

where E_p and E_s are the electric field components of the beam in the plane of the storage ring and perpendicular to this plane, respectively. A simple method for determining P on an experimental beamline is presented later in this chapter.

An established approach to deriving the orientation of a specific bond/orbital is to model the transition dipole moment as (i) a vector for single bonds or molecular orbitals (Hahner et al., 1991), such as the carbon–carbon σ^* orbital for an alkyl chain or the π^* orbital of a carboxyl group, or (ii) a collection of resonances in a given plane, such as the C–H σ^*/Rydberg-like R^* resonances of an alkyl chain (Bagus et al., 1996). If one assumes threefold azimuthal symmetry for the substrate (appropriate for the Au(111) and Ag(111) substrates used in SAM preparation) and that the beam spot is larger than the domain size (typically $\sim 1\, \text{mm}^2$ vs 100s of nm^2, respectively), the resonance intensity for the vector case, I_v, is expressed as (Stöhr, 1992; Stohr & Outka, 1987)

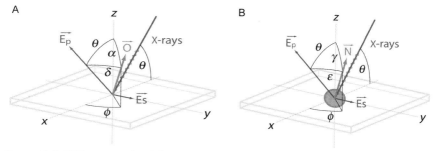

Figure 8.4 XAS geometries (after (Stöhr, 1992)) for (A) vector-like transition dipole moments and (B) multiple transition dipole moments that lie in a plane. **O** corresponds to the transition dipole moment vector of a specific orbital, while **N** corresponds to the vector that lies normal to the plane containing multiple transition dipole moments. (For color version of this figure, the reader is referred to the online version of this chapter.)

$$I_v(\theta,\alpha) \propto \frac{1}{3}P\left[1+\frac{1}{2}(3\cos^2\alpha-1)(3\cos^2\theta-1)\right]+\frac{1}{2}(1-P)\sin^2\alpha \quad (8.2)$$

where α corresponds to the angle between the transition dipole moment vector and the surface normal and θ represents the incident angle between the incident radiation and the sample surface ($\leq 90°$) (Fig. 8.4A). Meanwhile, the resonance intensity for a series of resonances whose transition dipole moment is defined by a plane is expressed as (Stöhr, 1992; Stohr & Outka, 1987)

$$I_p(\theta,\gamma) \propto \frac{2}{3}P\left[1-\frac{1}{4}(3\cos^2\theta-1)(3\cos^2\gamma-1)\right] \\ +\frac{1}{2}(1-P)(1+\cos^2\gamma) \quad (8.3)$$

where γ corresponds to the angle between the normal to the plane of the transition dipole moment and the normal to the sample surface (Fig. 8.4B). A key feature of Eqs. (8.2) and (8.3) is that one can remove the proportionality by taking the ratio of spectra recorded at multiple angles of incidence. If one leaves the intensities as functions of cosine squared, Eqs. (8.2) and (8.3) become (Willey et al., 2004)

$$\frac{I_v(\Theta_i,A)}{I_v(\Theta_j,A)} = \frac{P(3A-1)\Theta_i - A + 1}{P(3A-1)\Theta_j - A + 1} \quad (8.4)$$

$$\frac{I_p(\Theta_i,\Gamma)}{I_p(\Theta_j,\Gamma)} = \frac{P(3\Gamma-1)\Theta_i - \Gamma - 1}{P(3\Gamma-1)\Theta_j - \Gamma - 1} \quad (8.5)$$

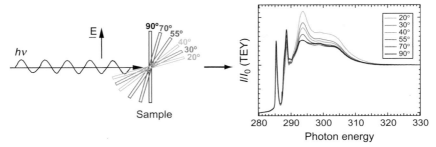

Figure 8.5 Schematic of the geometric configuration for angular-dependent XAS measurements on a planar SAM/crystal sample and an experimental dataset collected at the carbon K-edge for an MDBA SAM on Au(111) using this geometry. Note that the experimental spectra reveal significant angular dependence in the σ^* resonances at ~294 and 304 eV but comparatively little angular dependence in the π^* resonances of the aryl ring (285.4 eV) and carboxyl end-group (~289 eV). (See color plate.)

where $\Theta = \cos^2\theta$, $A = \cos^2\alpha$, and $\Gamma = \cos^2\gamma$. Equations (8.4) and (8.5) are both linear in Θ_i and, importantly, a linear regression can be obtained for all of the spectra recorded at Θ_i versus each individual spectrum recorded at Θ_j. For a series of spectra collected at N different angles, linear regression provides N slopes and N offsets ($N-1$ offsets if data are collected at normal incidence) that each yield α or γ. A value of α or γ is then returned from the average of $2N$ angles, with an error in precision obtained from their standard deviation.

From an experimental perspective, this discussion illustrates that a series of spectra must be recorded over a range of angles of incidence (Fig. 8.5) if one aims to derive information about molecular orientation within a SAM.

3. EXPERIMENTAL

3.1. Establishing the degree of linear polarization and energy calibration of the beamline

As Section 2 illustrates, determining P for a given soft X-ray beamline is a crucial step in XAS studies of SAM structure (Watts & Ade, 2008) and should be conducted once during each beamtime. This is readily achieved via analysis of the π^* resonance of a freshly cleaved sample of highly oriented pyrolytic graphite (HOPG). Since there is a low angular spread between the sp^2 carbon sheets of HOPG (<1°), the transition dipole moment of the π^*

resonance is aligned with the surface normal, which corresponds to $\alpha=0$. As a function of E_p and θ, the resonance intensity for an orbital with a transition dipole moment modeled by a vector reduces to

$$I_v(E_p,\theta) \propto E_p^2 \cos^2\theta \tag{8.6}$$

Thus, the acquisition of spectra that provide the intensity of the π^* resonance as a function of the incident angle enables one to measure the squared amplitudes of the electric vector in the plane of the storage ring and perpendicular to this plane.

In practice, most beamlines offer the capability to rotate a vertically mounted sample from normal to grazing incidence with the major electric field component, E_p, of the X-ray beam in the plane of incidence, while the minor electric field component, E_s, lies in the plane of the sample and does not contribute to the π^* resonance. In contrast, few beamlines enable a comparable degree of rotation with E_s in the plane of incidence and E_p in the plane of the sample. This limitation is overcome experimentally by mounting an additional sample(s) on a wedge at close to grazing incidence (e.g., 30°) with the sample holder normal to the incident radiation, as displayed in Fig. 8.6. Plotting the π^* resonance intensity as a function of the angle of incidence for (i) E_p and (ii) E_s will result in two sinusoidal curves (Fig. 8.6). The intensities of these curves correspond to the relative intensities of E_p^2 and E_s^2 and, therefore, one can determine P using Eq. (8.1).

HOPG also provides an excellent standard for calibrating the energy of the experimental beamline prior to carbon K-edge XAS measurements because the π^* resonance is reported to arise at 285.38 eV (Batson, 1993) (Fig. 8.2B).

3.2. XAS measurements of SAM/crystal
3.2.1 Modes of detection
Several modes of detection can be used for the XAS investigation of SAM structure, which offer distinct advantages and are separated into two categories: the electron yield (EY) and fluorescence yield (FY) modes. We note that XAS measurements in the traditional transmission mode are not possible due to extensive attenuation of the incident beam by the SAM/mineral samples; the inelastic mean free path of soft X-rays in solids (≤ 1–2 μm) is more than two orders of magnitude smaller than the typical sample thickness and precludes the transmission of sufficient beam intensity for meaningful measurements. All EY and FY detection modes provide indirect measurements

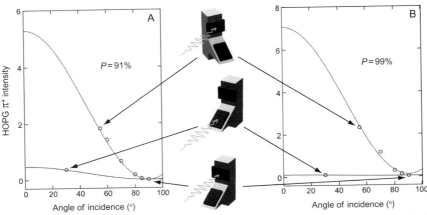

Figure 8.6 HOPG π^* resonance intensity as a function of angle of incidence recorded during distinct experiments on beamline 8.2 of the Stanford Synchrotron Radiation Lightsource under (A) SPEAR II and (B) SPEAR III. The upper data points in each plot correspond to variation in the angle of incidence with E_p in the plane of incidence and E_s in the plane of the sample, that is, rotation about the vertical axis. Following Eq. (8.6), a $\cos^2\theta$ fit to the experimental data is displayed in red. The lower trace (blue) corresponds to a $\cos^2\theta$ fit to two data points recorded with E_s in the plane of incidence and E_p in the plane of the HOPG—the first for normal incidence and the second for a sample mounted on a wedge at 30° grazing incidence. The relative intensities of the two $\cos^2\theta$ fits are used to derive the degree of linear polarization of the X-ray beam using Eq. (8.1). The contrasting relative intensities of the $\cos^2\theta$ fits displayed in (A) and (B) are indicative of different degrees of linear polarization during the two experiments, which are calculated to be $P=91\%$ and 99%, respectively. Schematics of the HOPG sample and holder are provided to illustrate the experimental configuration for specific data points. (For interpretation of the references to color in this figure legend, the reader is referred to the online version of this chapter.)

of the absorption coefficient because they monitor the decay of the photoexcited atom of interest. Nonetheless, the signal obtained from each EY and FY method is directly proportional to the intensity of absorption for low-Z elements present in organothiol molecules.

All EY modes are valuable for the measurement of SAM structure because they are inherently surface-sensitive. For beamlines that provide an electron energy analyzer, one can obtain particularly surface-sensitive information via measurement of absorption signal in the Auger electron yield (AEY) mode. Electrons emitted from the sample during the nonradiative Auger decay have characteristic and element-specific kinetic energies (Fig. 8.7). As such, if one sets the electron energy analyzer to the appropriate fixed energy for a specific Auger electron, the signal intensity

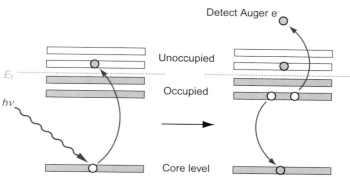

Figure 8.7 Auger yield energy-level diagram. The AEY XAS signal is obtained by detecting Auger electrons ejected during decay of an excited atom, which have characteristic energies for specific elements. (For color version of this figure, the reader is referred to the online version of this chapter.)

as a function of incident energy provides an AEY XAS spectrum. We note that the Auger electrons emitted from the prominent elements within the organothiol molecules have short mean free paths (≤ 16 Å for C, N, and O (Ohara et al., 1999)) and, therefore, are highly attenuated by the monolayer itself, which dictates that a substantial component of the AEY signal arises from atoms close to the SAM surface, that is, within the functional groups that interact with the mineral phase.

Measurements in the total electron yield (TEY) mode are readily conducted on almost all soft X-ray beamlines, without the need for an electron energy analyzer. The decay of an excited atom, generated via absorption of an X-ray photon, results in emission of either an Auger electron or a fluorescence photon. Inelastic scattering of either the Auger electron or fluorescence photon creates a cascade of excited electrons leaving the sample (Fig. 8.8). The associated drain current is large enough to be measured with a picoammeter, and since the current into the sample is proportional to the absorption, its evolution as a function of incident photon energy yields an XAS spectrum. Given that the attenuation length of the fluorescence photons (and, by extension, the secondary electron cascade) significantly exceeds the mean free path of the Auger electrons, the TEY signal offers lower surface sensitivity than the AEY mode (mean free path of escaping electrons is ≥ 35 Å (Ohara et al., 1999)). Nonetheless, the TEY mode retains the elemental specificity required for studies of the SAM structure and is extremely attractive because TEY data can be collected simultaneously to any AEY or FY measurement. TEY XAS also requires only a simple

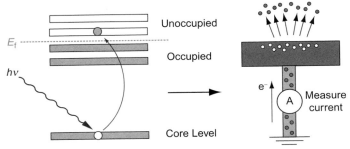

Figure 8.8 Diagram of the TEY mode. Inelastic scattering of Auger electrons and reabsorption of fluorescence photons emitted during decay of the photoexcited atom result in an electron cascade. This cascade is detected as the current into the sample and the variation in the current as a function of energy provides the TEY XAS spectrum. (For color version of this figure, the reader is referred to the online version of this chapter.)

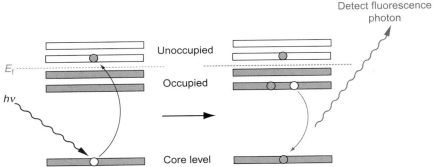

Figure 8.9 Fluorescence yield energy-level diagram. The detected intensity of the fluorescence photons emitted during decay of the excited state as a function of energy yields the FY XAS spectrum. (For color version of this figure, the reader is referred to the online version of this chapter.)

experimental setup: the sample must be mounted on a conductive holder that is both electrically floating and connected to a picoammeter via a shielded wire.

FY XAS data are obtained by measuring the intensity of fluorescence photons emitted during the decay of atoms excited by absorption of the incident X-ray photons (Fig. 8.9). The comparatively long attenuation length for the fluorescence photons (≥ 500 nm through a mineral layer at the carbon K-edge) versus photoelectrons makes the FY mode a complementary and often preferable approach to EY methods for studying the structure of SAMs composed of long-chain organothiol molecules or those buried beneath a condensed inorganic layer in SAM/crystal samples. FY detectors with the capability for energy discrimination (e.g., Si drift detectors) enable data collection in the partial fluorescence yield mode, which is significant because one can preferentially detect fluorescence photons with characteristic energies associated with the decay of specific elements. Even so, detectors without the capability for energy discrimination are prevalent on soft X-ray beamlines, because the total fluorescence yield data they provide remain representative of the absorption coefficient.

3.2.2 Normalization of the incident flux

Invariably, the intensity of the incident beam will vary as a function of the photon energy, which will cause associated variations in the raw XAS signal. The energy-dependent intensity variations observed in the incident beam arise from several sources, including photon absorption by organic contaminants adsorbed on the surface of the beamline optics or subtle differences in the optimization of the optical configuration as a function of photon energy. To account for these effects, it is crucial to normalize the XAS signal to the incident flux, I_0, which should be measured simultaneously with the XAS signal. Typically, I_0 is measured in the TEY mode for an Au grid (or thin, metallized polymer film) located in the beam path immediately upstream of the experimental sample. The absorption of organic contaminants on this grid over time will also lead to signal distortion, which can be remedied by recoating the grid with Au prior to any XAS measurements.

In the event that recoating a contaminated I_0 grid is not possible, a "double normalization" approach can be used to account for variations in the incident flux. The "double normalization" protocol requires the collection of an XAS spectrum, I_{dn}, for an experimental control that closely approximates the experimental samples without containing the specific material of interest (for the purposes of studying SAM/crystal systems, an ideal control is a bare metal substrate). One can then eliminate the effects of the contaminated I_0 by dividing the normalized sample signal, I_s, by the normalized control signal according to

$$I_{dn} = \frac{\left(\frac{I_s}{I_{0s}}\right)}{\left(\frac{I_c}{I_{0c}}\right)} \tag{8.7}$$

where I_c is the signal for the control sample and I_{0s} and I_{0c} are the I_0 signals for the sample and control, respectively. In principle, applying Eq. (8.7) will yield a representative XAS spectrum for the SAM or SAM/crystal layer, provided the control sample is itself free of any surface contamination. Equation (8.7) necessitates that conditions are identical for the sample and control scans, and therefore, great care must be taken to ensure that the beamline configuration and scan parameters remain constant throughout data collection.

3.2.3 Scope of experiment and data optimization

As the discussion presented in Section 2 indicates, obtaining the molecular orientation within a SAM necessitates the collection of XAS data at multiple angles of incidence, preferably over as wide a range of angles of incidence as possible. At the minimum, this dictates that measurements are conducted at three angles for each individual sample: (1) normal incidence between the axis of the incident beam and sample, (2) as close to grazing incidence as possible within the constraints of the experimental beamline, and (3) an intermediate angle that allows for meaningful linear regression analysis. Nonetheless, it is valuable to consider the following during the design and execution of any XAS studies of SAM/crystal systems:

1. *Spectra recorded at grazing incidence are prone to errors in measurement:* As the sample approaches grazing incidence, the projection of the surface perpendicular to the incident beam decreases laterally until it approaches the dimensions of the beam spot. This situation can prove problematic because any lateral drift of the beam during an XAS scan will cause photons to leak off the sample, thereby introducing inconsistencies in the experimental data and potentially rendering an entire dataset unusable. To mitigate this effect, one can record data at multiple angles approaching grazing incidence (e.g., 20°, 30°, and 40°) to follow trends in the XAS resonance intensities, which allows obviously erroneous spectra to be omitted from analysis. If possible, one should also collect multiple spectra at each angle for improved statistics.
2. *Angular dependence (linear dichroism) for a specific resonance disappears for a randomly oriented bonds/functional groups or for orbitals that have a defined orientation close to $arcsin[(2/3)^{1/2}]$ ($\sim 54.7°$), often referred to as the magic angle:* As a consequence, it is valuable to use 55° as one of the intermediate angles of incidence used during data collection.

Rotation of the experimental sample should always be accompanied by realignment with respect to the electron energy analyzer or FY detector to ensure that the sample is in the detector "sweet spot" for maximum signal. Ideally, the sample alignment should be optimized at a fixed incident photon energy corresponding to the resonance of most interest for structural studies of the SAM. The TEY signal is an invaluable diagnostic during alignment because the current will drop to ~50% of the maximum when the center of the photon beam is at the sample edge, thereby defining the limits of the sample surface.

The acquisition parameters used in each experiment will be defined by the specific sample of study, the element of interest, and the capabilities of the beamline (flux, energy resolution, etc.). Even so, it is crucial that each spectrum begins at an energy below the absorption edge, where there will be no contribution to the signal, and ends at an energy where all of the electrons excited during photoabsorption will be emitted to the continuum (e.g., ~280 and 320 eV, respectively, for measurements at the carbon K-edge as displayed in Fig. 8.2). The difference between these two limits, often referred to as the "edge step", is proportional to the amount of the element of interest in the beam and is used for the purposes of normalization. This normalization is crucial for angular-dependent measurements of SAMs because the footprint of the X-ray beam on the sample surface and, by extension, the amount of organothiol molecules in the beam increase as a function of the angle of incidence.

Collecting with a high density of points across the edge and near-edge region of the XAS spectrum is advantageous for resolving peaks associated with molecular orientation within the SAM, because it minimizes error during fitting and allows one to deconvolve different resonances within the near-edge fine structure. The energy step size chosen depends on beamline resolution and natural linewidth. Generally, a 0.05 eV step size provides high-quality data in the edge and near-edge region, while lower resolutions (perhaps 0.2–0.5 eV) are sufficient at higher energies.

3.2.4 Identifying and mitigating X-ray-induced beam damage

The organothiol molecules within SAMs are susceptible to beam damage, primarily via bond breaking due to the absorption of either X-ray photons or low-energy photoelectrons emitted from the underlying metal substrate or neighboring molecules. The onset of beam damage is immediate but meaningful XAS spectra are readily collected if the extent of the SAM degradation is limited throughout data collection. The rate of beam damage is beamline-dependent because it is proportional to the X-ray flux per unit area delivered to the sample surface. As such, it is important to establish an exposure limit for the specific beamline configuration used during the XAS measurements, beyond which further irradiation induces unacceptable degradation of the experimental sample.

A simple approach for addressing the exposure limit is to record successive carbon K-edge XAS spectra on the same region of a SAM prepared from a simple alkanethiol, such as dodecanethiol, and to follow the evolution of the carbon–carbon π^* resonance at $\sim$285.4 eV as a function of time (Fig. 8.10). This provides an excellent system diagnostic because a pristine alkanethiol SAM does not contain any double bonds, whereas carbon–carbon double bonds are known to form during the processes of irradiation-induced degradation of the monolayer (Zharnikov, Geyer, Golzhauser, Frey, & Grunze, 1999; Zharnikov & Grunze, 2002). Hence, the appearance and subsequent increase in intensity of a carbon–carbon π^* feature for an alkanethiol SAM is indicative of beam damage. As a general rule, the presence of a carbon–carbon π^* resonance in the XAS spectrum of an alkanethiol SAM signifies that the degree of beam damage is too great for further analysis. Additional information is required to establish acceptable exposure limits. At the minimum, one should aim to collect two successive

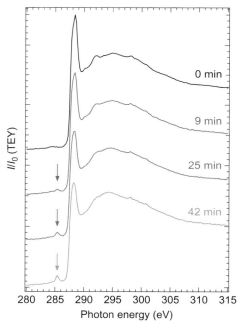

Figure 8.10 A series of carbon K-edge XAS spectra recorded for a hexadecanethiol SAM on Au(111) at normal incidence after varying lengths of exposure to the X-ray beam. Note the appearance of a π^* resonance at 285.4 eV with increased exposure that is indicative of damage to the SAM induced by irradiation. (For color version of this figure, the reader is referred to the online version of this chapter.)

XAS spectra that do not exhibit any appreciable changes (i.e., a π^* resonance does not appear for the saturated alkanethiol SAM). Meeting this condition ensures that limited sample degradation occurs during the time frame for the first XAS spectrum.

In the event that the damage-related π^* feature is observed in the second, or even first, XAS spectrum recorded for the alkanethiol SAM standard, several approaches can be adopted to mitigate sample degradation:

1. *Alter the data collection parameters:* Reducing the point density, data collection time per point, and energy range (length) of the XAS scan will all have the effect of reducing the scan duration and, therefore, beam exposure and damage.
2. *Defocus the X-ray beam:* Most soft X-ray beamlines have tightly focused beam spots (≤ 1 mm^2) to enable the measurement of small samples/features or provide high photon flux for studies of low concentration samples. Defocusing the beam will increase the size of the beam spot and reduce the flux per unit area at the SAM surface.
3. *Attenuate the incident beam via insertion of a thin film, typically Al, into the beam path prior to the I_0 grid.*

Assuming all concerns regarding beam damage have been addressed, it is good practice to follow two protocols during the XAS measurements, irrespective of the beamline used to conduct the experiments:

1. Do not collect more than one spectrum on each region of the experimental sample.
2. Limit the scan duration to the minimum required for good S/N.

3.2.5 Sample preparation and composition for XAS studies

Excellent descriptions of methods for preparing high-quality SAMs and amorphous or crystalline biominerals on the surface of these monolayers abound in the literature (Arnold, Azzam, Terfort, & Woll, 2002; Love, Estroff, Kriebel, Nuzzo, & Whitesides, 2005), and therefore, sample synthesis will not be discussed in this manuscript. Nonetheless, it is important to address the composition, structure, and dimensions of the experimental samples necessary for a meaningful XAS study, as well as the postsynthesis protocols required before transfer into the ultrahigh vacuum (UHV) environment of an experimental beamline.

3.2.5.1 Sample preparation

Ideally, all experimental samples should be prepared at the synchrotron source and inserted into the load–lock/introduction chamber of the beamline as soon as possible after synthesis to minimize ambient sample degradation prior to measurement. SAMs left under ambient conditions are subject to rapid degradation due to oxidation of the mercaptan head groups via reaction with ambient ozone (Willey et al., 2005). If immediate transfer into the beamline end station is not possible, samples should be stored under inert atmosphere (e.g., dry argon).

UHV compatibility is an essential requirement of all samples because soft X-ray beamlines almost universally operate at $\leq 10^{-9}$ torr, which presents a significant experimental challenge for the samples that have been immersed in ionic aqueous solution, for example, during biomineral growth. The removal of water molecules retained by these samples can require extended pumping times to reach UHV conditions and the associated loss of valuable synchrotron beamtime. As such, all samples should be dried in a diffuse stream of dry nitrogen or argon prior to transfer into the load–lock to remove as much water/solvent as possible.

3.2.5.2 Sample composition

In general, it is desirable to prepare the largest possible samples that are compatible with the experimental beamline used to conduct the XAS measurements for two important reasons:

1. Increasing the width of the sample allows the collection of XAS data at angles closer to grazing incidence.
2. Larger samples offer improved statistics: The effects of beam damage necessitate a minimum sample size for conducting beam alignment, a full series of XAS measurements (between grazing and normal incidence) at the carbon (and, potentially, oxygen, nitrogen, and sulfur) edge without using the same region of the surface twice. It is preferable, however, to maximize the sample size in order that one can repeat XAS measurements at all angles of incidence to verify experimental reproducibility for a single sample and to improve statistics.

Meanwhile, the composition of the SAM/mineral samples must be carefully controlled to ensure the collection of meaningful data. Since the XAS signal is a superposition of all of environments irradiated by the incident beam, a high surface coverage (preferably $\geq 67\%$) of the mineral is required to ensure that the spectrum is representative of the monolayer structure at the SAM/mineral interface. At low coverages, the mineral-free regions of the SAM provide a large enough contribution to the total XAS signal to preclude an unambiguous assignment of organothiol orientation beneath the mineral phase. One must also limit the thickness of the inorganic phase. Even crystals with moderate thickness (as little as a few nanometers) can prevent photoelectrons from escaping from the sample to vacuum, which has two undesirable effects: (i) charging of the insulating mineral, which can lead to distortion of the experimental data, and (ii) the inability to conduct any measurements in the EY mode. In principle, FY modes of detection allow the measurement of samples supporting crystals with thicknesses in excess of 500 nm, but with reduced S/N because the mineral will attenuate both the incident beam and photons ejected from the sample. An added consideration for FY studies of crystallization on SAMs is that the signal from elements within the mineral (e.g., carbon atoms within the carbonate ion for carbon K-edge XAS) can contribute to, and potentially dominate, the XAS spectrum if the mineral layer is too thick.

4. ANALYSIS

Following the preliminary steps of energy calibration and normalization of the experimental data to the incident flux (using I_0 or the double normalization protocol described previously) and the absorption "edge step," it is often necessary to fit the spectra to extract the resonance intensities required for analysis of the monolayer orientation. This fitting is imperative for deconvolving overlapping resonances, to account for any angular-independent background (which includes the step-function-like absorption edge near the ionization potential) and to extract accurate energies and intensities for these resonances. The extent and complexity of the fitting procedure remains at the discretion of the individual, and numerous commercial programs provide the capabilities for tailoring a fitting protocol to the requirements of a specific experimental dataset. Nonetheless, one can save time during analysis and avoid significant error in fitting the experimental data by considering the following: since only the angle of incidence changes for a single dataset on a given sample, the energy of a specific resonance and all sources of peak broadening (such as the core–hole lifetime and instrumental broadening) should not vary between spectra recorded at different angles. As such, one must ensure that the full width at half maximum and energy of a specific resonance are the same for the fit to each spectrum; only the amplitude of each resonance should change as a function of the angle of incidence.

We note that the experimental geometry available on most experimental beamlines limits the information available from the resonance intensities to a single degree of freedom, because one can only probe a sample in a single orientation with a single beam of X-ray photons at any given time. More explicitly, linear regression analysis using Eqs. (8.4) and (8.5) yields a single parameter: either the tilt angle of the transition dipole moment vector or the plane containing the transition dipole moment, respectively, relative to the surface normal of the sample. These angles are extremely informative about the structure of the SAM but do not enable the assignment of a unique solution for the orientation of functional groups and/or molecules that are described by more than one degree of freedom. For example, assigning the specific orientation of an aromatic ring requires the assignment of two parameters, such as the tilt (colatitudal, θ) and twist (dihedral, φ) angles of the ring (Fig. 8.11A). Although it is not possible to deconvolve two interdependent parameters of this kind from analysis of the resonance intensities alone, a consideration of steric factors enables one to derive a manifold of physically viable combinations of the two parameters that is consistent with the experimental XAS. A thorough description of the process for

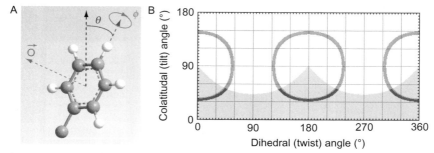

Figure 8.11 (A) Schematic to indicate the tilt (colatitudal, θ) and twist (dihedral, φ) angles that describe the orientation of a phenyl ring. Carbon and hydrogen atoms are denoted by gray and white spheres, respectively. The black arrow is coincident with the surface normal of the metal substrate and the green arrow corresponds to the transition dipole moment, which lies perpendicular to the plane of the ring. (B) A graph indicating the permissible end-group orientations within an MDBA SAM on Au(111) as a subset of all possible orientations. The combinations of colatitudal and tilt angles for the aromatic end-group of MDBA obtained via linear regression analysis are displayed as a red annulus and correspond to the XAS data presented in Fig. 8.2B. Meanwhile, the sterically permissible and forbidden end-group orientations are displayed as gray and white regions, respectively. Thus, overlap between the red annulus and the gray region corresponds to a viable combination of colatitudal and dihedral angles for the MDBA end groups. The blue annulus is a guide for the eye and corresponds to an end-group orientation exactly at the magic angle, ~54.7° (or a random distribution of end-group orientations). Conventions for defining the colatitudal and dihedral angles are available in the literature (Lee et al., 2013). (See color plate.)

addressing steric considerations is available in the literature (Willey et al., 2009), but the key steps are summarized as follows:

1. Calculate theoretical tilt angles for the transition dipole moment vector or plane containing the transition dipole moment for all possible orientations of the bond(s)/functional group(s) in question.
2. Compare theoretical XAS intensities for all possible orientations with those obtained experimentally—close agreement between the theoretical and experimental intensities indicates a possible bond/functional group orientation.
3. Assess the subset of orientations obtained via comparison of the theoretical and experimental resonance intensities for physical viability through steric considerations. For example, an orientation that causes the bond/functional group to be coincident with a neighboring molecule within the SAM or the underlying substrate would be forbidden due to steric constraints.

In general, the manifold of possible orientations that are both (i) consistent with the XAS resonance intensities and (ii) physically viable is very limited and provides an excellent indication of the bond/functional group orientation (Fig. 8.11B).

The element specificity of XAS enables the collection of XAS data for different elements within a single functional group (such as the C and O of a carboxyl moiety), which provides a valuable diagnostic of the quality of the experimental data and/or the analysis of the resonance intensities. For a bond of well-defined orientation between two different elements in the functional group, linear regression analysis of the XAS spectra recorded for the two different elements must yield the same orientation, to within experimental error.

ACKNOWLEDGMENTS

This work was performed under the auspices of the US DoE, Office of Basic Energy Sciences (OBES), and Division of Materials Sciences and Engineering by Lawrence Livermore National Laboratory under Contract No. DE-AC52-07NA27344. M. H. N. acknowledges government support under and awarded by the Department of Defense, the Air Force Office of Scientific Research, and National Defense Science and Engineering Graduate Fellowship, 32 CFR 168a. The Advanced Light Source is supported by the Director, Office of Science, OBES, of the US DoE under Contract No. DE-AC02-05CH11231. Portions of this research were carried out at the Stanford Synchrotron Radiation Lightsource, a Directorate of SLAC National Accelerator Laboratory and an Office of Science User Facility operated for the US DoE, Office of Science by Stanford University.

REFERENCES

Aizenberg, J., Black, A. J., & Whitesides, G. H. (1999). Oriented growth of calcite controlled by self-assembled monolayers of functionalized alkanethiols supported on gold and silver. *Journal of the American Chemical Society, 121*(18), 4500–4509. http://dx.doi.org/10.1021/ja984254k.

Arnold, R., Azzam, W., Terfort, A., & Woll, C. (2002). Preparation, modification, and crystallinity of aliphatic and aromatic carboxylic acid terminated self-assembled monolayers. *Langmuir, 18*(10), 3980–3992. http://dx.doi.org/10.1021/la0117000.

Bagus, P. S., Weiss, K., Schertel, A., Woll, C., Braun, W., Hellwig, C., et al. (1996). Identification of transitions into Rydberg states in the X-ray absorption spectra of condensed long-chain alkanes. *Chemical Physics Letters, 248*(3–4), 129–135. http://dx.doi.org/10.1016/0009-2614(95)01315-6.

Batson, P. E. (1993). Carbon-1S near-edge-absorption fine-structure in graphite. *Physical Review B, 48*(4), 2608–2610. http://dx.doi.org/10.1103/PhysRevB.48.2608.

Grunze, M. (1993). Preparation and characterization of self-assembled organic films on solid substrates. *Physica Scripta, T49B*, 711–717. http://dx.doi.org/10.1088/0031-8949/1993/t49b/056.

Hahner, G., Kinzler, M., Woll, C., Grunze, M., Scheller, M. K., & Cederbaum, L. S. (1991). Near edge X-ray-absorption fine-structure determination of alkyl-chain orientation—Breakdown of the building-block scheme. *Physical Review Letters, 67*(7), 851–854. http://dx.doi.org/10.1103/PhysRevLett.67.851.

Hahner, G., Woll, C., Buck, M., & Grunze, M. (1993). Investigation of intermediate steps in the self-assembly of n-alkanethiols on gold surfaces by soft X-ray spectroscopy. *Langmuir, 9*(8), 1955–1958.

Han, Y. J., & Aizenberg, J. (2003). Face-selective nucleation of calcite on self-assembled monolayers of alkanethiols: Effect of the parity of the alkyl chain. *Angewandte Chemie International Edition, 42*(31), 3668–3670. http://dx.doi.org/10.1002/anie.200351655.

Lee, J. R. I., Han, T. Y.-J., Willey, T. M., Nielsen, M. H., Klivansky, L. M., Liu, Y., et al. (2013). Cooperative reorganization of mineral and template during directed nucleation of calcium carbonate. *Journal of Physical Chemistry C, 117*, 11076–11085. http://dx.doi.org/10.1021/jp400279f.

Lee, J. R. I., Han, T. Y.-J., Willey, T. M., Wang, D., Meulenberg, R. W., Nilsson, J., et al. (2007). Structural development of mercaptophenol self-assembled monolayers and the overlying mineral phase during templated CaCO3 crystallization from a transient amorphous film. *Journal of the American Chemical Society, 129*(34), 10370–10381. http://dx.doi.org/10.1021/ja071535w.

Love, J. C., Estroff, L. A., Kriebel, J. K., Nuzzo, R. G., & Whitesides, G. M. (2005). Self-assembled monolayers of thiolates on metals as a form of nanotechnology. *Chemical Reviews, 105*(4), 1103–1169. http://dx.doi.org/10.1021/cr0300789.

Ohara, H., Yamamoto, Y., Kajikawa, K., Ishii, H., Seki, K., & Ouchi, Y. (1999). Effective escape depth of photoelectrons for hydrocarbon films in total electron yield measurement at the C K-edge. *Journal of Synchrotron Radiation, 6*, 803–804. http://dx.doi.org/10.1107/s0909049599004033.

Sommerdijk, N., & de With, G. (2008). Biomimetic CaCO3 mineralization using designer molecules and interfaces. *Chemical Reviews, 108*(11), 4499–4550. http://dx.doi.org/10.1021/cr078259o.

Stöhr, J. (1992). *NEXAFS spectroscopy*. Berlin, Heidelberg: Springer Verlag.

Stohr, J., & Outka, D. A. (1987). Determination of molecular orientations on surfaces from the angular-dependence of near-edge X-ray-absorption fine-structure spectra. *Physical Review B, 36*(15), 7891–7905. http://dx.doi.org/10.1103/PhysRevB.36.7891.

Ulman, A. (1996). Formation and structure of self-assembled monolayers. *Chemical Reviews*, *96*(4), 1533–1554. http://dx.doi.org/10.1021/cr9502357.

Watts, B., & Ade, H. (2008). A simple method for determining linear polarization and energy calibration of focused soft X-ray beams. *Journal of Electron Spectroscopy and Related Phenomena*, *162*(2), 49–55. http://dx.doi.org/10.1016/j.elspec.2007.08.008.

Willey, T. M., Lee, J. R. I., Fabbri, J. D., Wang, D., Nielsen, M. H., Randel, J. C., et al. (2009). Determining orientational structure of diamondoid thiols attached to silver using near-edge X-ray absorption fine structure spectroscopy. *Journal of Electron Spectroscopy and Related Phenomena*, *172*(1–3), 69–77. http://dx.doi.org/10.1016/j.elspec.2009.03.011.

Willey, T. M., Vance, A. L., van Buuren, T., Bostedt, C., Nelson, A. J., Terminello, L. J., et al. (2004). Chemically transformable configurations of mercaptohexadecanoic acid self-assembled monolayers adsorbed on Au(111). *Langmuir*, *20*(7), 2746–2752. http://dx.doi.org/10.1021/la036073o.

Willey, T. M., Vance, A. L., van Buuren, T., Bostedt, C., Terminello, L. J., & Fadley, C. S. (2005). Rapid degradation of alkanethiol-based self-assembled monolayers on gold in ambient laboratory conditions. *Surface Science*, *576*(1–3), 188–196. http://dx.doi.org/10.1016/j.susc.2004.12.022.

Zharnikov, M., Geyer, W., Golzhauser, A., Frey, S., & Grunze, M. (1999). Modification of alkanethiolate monolayers on Au-substrate by low energy electron irradiation: Alkyl chains and the S/Au interface. *Physical Chemistry Chemical Physics*, *1*(13), 3163–3171. http://dx.doi.org/10.1039/a902013f.

Zharnikov, M., & Grunze, M. (2001). Spectroscopic characterization of thiol-derived self-assembling monolayers. *Journal of Physics—Condensed Matter*, *13*(49), 11333–11365. http://dx.doi.org/10.1088/0953-8984/13/49/314.

Zharnikov, M., & Grunze, M. (2002). Modification of thiol-derived self-assembling monolayers by electron and X-ray irradiation: Scientific and lithographic aspects. *Journal of Vacuum Science & Technology B*, *20*(5), 1793–1807. http://dx.doi.org/10.1116/1.1514665.

CHAPTER NINE

Cryo-TEM Analysis of Collagen Fibrillar Structure

Bryan D. Quan[*], Eli D. Sone[*,†,‡,1]

[*]Institute of Biomaterials and Biomedical Engineering, University of Toronto, Toronto, Ontario, Canada
[†]Department of Materials Science and Engineering, University of Toronto, Toronto, Ontario, Canada
[‡]Faculty of Dentistry, University of Toronto, Toronto, Ontario, Canada
[1]Corresponding author: e-mail address: eli.sone@utoronto.ca

Contents

1. Introduction — 190
   1.1 Collagen fibrillar structure — 191
   1.2 Techniques for research into collagen structure — 192
2. Sample Preparation for Cryo-TEM — 193
   2.1 Solutions — 193
   2.2 Dental and periodontal tissues — 194
   2.3 Soft tissues — 194
   2.4 Cryosectioning — 194
   2.5 Section vitrification — 195
3. Image Acquisition and Processing — 195
   3.1 Cryo-TEM — 195
   3.2 Interpretation of Cryo-TEM micrographs — 196
   3.3 Analysis of whole micrographs — 196
4. Image Averaging — 198
   4.1 Image selection — 198
   4.2 Image alignment — 200
   4.3 Measurement of fibril periodicity from averaged images — 201
5. Summary and Outlook — 202
Acknowledgments — 203
References — 203

Abstract

Fibrillar collagens are important structural proteins and are known to be closely associated with mineral in the case of mineralized tissues. However, the precise role of collagen in the mineralization process remains unclear, and the evaluation of structural differences in collagen from mineralized and nonmineralized tissues may be instructive in this regard. Here, we review the use of cryo-transmission electron microscopy to investigate the axial structure of collagen fibrils in tissue sections from both mineralizing and nonmineralizing tissues. By examining collagen fibrillar structure in an unstained frozen-hydrated state, it is possible to avoid artifacts normally associated with staining and dehydration that are required for conventional TEM. We describe both sample preparation and image analysis with emphasis on the particular challenges of using image averaging techniques, which can be used to overcome the low signal-to-noise ratio that is inherent in this technique. Detailed banding patterns can be obtained from averaged images, and these can be analyzed to obtain quantitative information on fibril periodicity and structure.

1. INTRODUCTION

Type I collagen (COL I) is the most abundant protein in vertebrates, making up about 25% of proteins in mammals, and is a major component of both mineralized (bone, dentin, and cementum) and unmineralized (tendon, ligament, and dermis) connective tissues. In mineralized connective tissues, fibrils of COL I are embedded with crystals of carbonated apatite, a calcium phosphate mineral. The crystal phase is oriented with its c-axis parallel to the long axis of the collagen fibril and has been shown to be initially localized to the gap region of collagen fibrils in bone (Landis, Hodgens, Arena, Song, & McEwen, 1996) and mineralizing turkey leg tendon (Landis, Song, Leith, McEwen, & McEwen, 1993). The precise location and orientation of apatite crystals within collagen fibrils point to the importance of collagen in the mineralization process. Indeed, mutations in the amino acid sequence of collagen, as in osteogenesis imperfecta, can seriously impair the formation, orientation, and organization of the apatite crystals, leading to increased brittleness of the bone (Fratzl, Paris, Klaushofer, & Landis, 1996). Moreover, it has been shown that significant changes to collagen structure in predentin precede collagen mineralization (Beniash, Traub, Veis, & Weiner, 2000). In terms of collagen biochemistry, differences between lysine hydroxylation and cross-linking of mineralizing and nonmineralizing tissues are well known (Eyre & Wu, 2005). However, the precise role of collagen in the mineralization process is still not well understood. Early *in vitro* work in this area suggested that collagen fibrils could themselves nucleate mineral phases (Glimcher, 1959), but more recent emphasis has been on the role of noncollagenous proteins in mediating collagen–mineral interactions (for a recent review, see Nudelman, Lausch, Sommerdijk, & Sone, 2013). These two aspects of the mineralizing system are not mutually exclusive as variations in collagen composition, posttranslational modification, and structure may modulate collagen–protein interactions. Hence, there remains a need to characterize collagen structure in mineralizing and nonmineralizing tissues to better understand the role of collagen in biological mineralization.

1.1. Collagen fibrillar structure

Collagen molecules (tropocollagen) consist of three left-handed polyproline II-like helices wound together to form a right-handed triple helix. Hodge and Petruska (1963) deduced from transmission electron microscopic (TEM) observations the ¼-stagger arrangement of tropocollagen in COL I fibrils (Fig. 9.1): aligned rodlike tropocollagen molecules, ∼300 nm long and ∼1.5 nm in diameter, are axially offset from each other by their D-period, a multiple of ∼67 nm, and pack laterally to form fibrils. Since the length of a collagen molecule is not an integer multiple of the offset distance, the remainder is made up by a gap between axially adjacent molecules. This generates a gap/overlap structure where for every five molecules in the overlap region, there are four in the gap region. TEM and X-ray diffraction (XRD) of COL I fibrils indicate that the gap region is ∼27 nm and the overlap region ∼40 nm, or ∼0.4 and ∼0.6 of the banding period (Claffey, 1977; Hodge & Petruska, 1963; Marchini, Morocutti, Castellani, Leonardi, & Ruggeri, 1983).

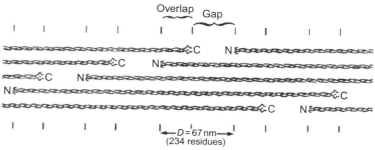

Figure 9.1 Schematic of the 1/4-stagger arrangement of collagen fibrils. ∼300 nm long tropocollagen molecules are aligned approximately parallel to the fibrillar axis. Lateral neighbors are offset axially by ∼67 nm, generating a ∼67 nm fibrillar period. A gap between the ends of axially successive molecules accounts for the apparent mismatch between the length of tropocollagen molecules and the period of fibrillar collagen, resulting in the characteristic gap/overlap structure where for every five molecules in the overlap region, there are four in the gap region. *Reprinted from Starborg et al. (2008), Copyright 2008, with permission from Elsevier.*

1.2. Techniques for research into collagen structure

Collagen structure has been probed by spectroscopic methods including infrared (Carden and Morris, 2000), Raman (Carden and Morris, 2000), and nuclear magnetic resonance (Zernia and Huster, 2008) and scattering methods including optical microscopes (Whittaker and Canham, 1991), X-ray (Marchini, Morocutti, Ruggeri, Koch, Bigi, and Roveri, 1986) and neutron scattering (Bonar, Lees, and Mook, 1985), and electron microscopy (EM) (Marchini et al., 1986). While significant information has been gleaned from spectroscopy and light microscopy, methods with high spatial resolution are needed to probe collagen structure at the fibrillar level, particularly for native collagens. This has given rise to a dependence on EM. However, while conventional EM is convenient for analyzing collagen structure due to direct interpretation of images, there are drawbacks to sample preparation; dehydration is well known to induce shrinkage of the collagen D-period, and polymer embedding necessitates heavy-metal staining to provide contrast but emphasizes chemical environment over structure.

Atomic force microscopy (AFM) is popular in collagen research since it can be performed in aqueous conditions with minimally prepared samples and has been applied to the hydrated mineralized collagen of dentin (Habelitz, Balooch, Marshall, Balooch, & Marshall, 2002) and bone (Sasaki et al., 2002), as well as nonmineralizing tissues such as dermis (Erickson, Fang, Wallace, Orr, Les, and Holl, 2012). Transmission EM of dehydrated negatively stained collagen (Bos et al., 2001) and scanning EM of freeze-fracture replicas (Meadows, Holmes, Gilpin, & Kadler, 2000) have also been used to elucidate topographic features of native collagen fibrils. One of the most significant recent advances in collagen research is the XRD crystal structure of hydrated COL I in native rat tail tendon (RTT) (Orgel, Irving, Miller, & Wess, 2006), which provides a relatively low-resolution (5.16 Å × 11.1 Å) 3-dimensional model of collagen structure. Native COL I fibrils from tissues other than RTT are not as well described due to the inherent disorder both of these tissues and of collagen fibrils. Therefore, while periodic packing of collagen fibrils from a variety of tissues including the ligament (Gathercole, Porter, & Scully, 1987), skin (Gathercole et al., 1987), cornea (Meek, Elliott, Hughes, & Nave, 1983), and bone (Almer & Stock, 2005) has been characterized by XRD, significantly less is known about the structure of fibrillar collagen in these and other tissues.

Here, we describe the use of uncorrected cryo-TEM images of fixed, frozen-hydrated tissue sections to evaluate the axial ultrastructure of collagen fibrils from mineralized and nonmineralized collagenous tissues. The advantages of this technique are that samples are maintained in a hydrated state throughout the sectioning and microscopy and heavy-metal staining is not necessary, which allows examination of local collagen structure directly from density of organic material within the fibril. Banding analysis has been performed from cryo-TEM images of reconstituted collagen fibrils (Nudelman et al., 2010), and several previous studies (Sabanay, Arad, Weiner, & Geiger, 1991; Weiner, Arad, Sabanay, & Traub, 1997; and Beniash et al., 2000) examined fixed collagenous tissues using cryo-TEM. One of these (Beniash et al., 2000) investigated in detail the banding pattern of collagen fibrils, using averaged images of predentin collagen to demonstrate structural changes in collagen fibrils prior to mineralization. This underused technique is well placed to provide a relatively high-resolution link between conventional TEM, AFM, SEM, and XRD analyses of native fibrillar collagen structure.

2. SAMPLE PREPARATION FOR CRYO-TEM

Two criteria must be met to investigate the structure of collagen in mineralized tissues by TEM: collagen must be visible in the TEM, and collagen structure must be adequately preserved. Mineralized tissues are typically demineralized in a neutral solution of ethylenediaminetetraacetic acid (EDTA) to eliminate the dominant contribution of inorganic components to image formation, while glutaraldehyde (GA) fixation is used to maintain ultrastructural integrity. We employ the cryosectioning method of Tokuyasu (1973) combined with the method of thin section vitrification employed by Sabanay et al. (1991) to investigate mineralized and nonmineralized collagenous tissues as described later. This hybrid approach allows the production of vitreous hydrated samples without the limitations on sample size and vitrification depth imposed by high-pressure freezing or slam freezing, respectively (McDonald & Auer, 2006).

2.1. Solutions

Buffered fixative (BF). Dulbecco's phosphate-buffered saline (DPBS) pH 7.4 with 0.8% paraformaldehyde (PFA) and 0.2% GA

Demineralizing fixative (DF). DPBS pH 7.4 with 9.3% EDTA, 0.2% PFA, and 0.05% GA

2.3 M Sucrose solution (SS). 78.729 g sucrose dissolved in ultrapure (18.2 MΩ/cm) water made up to 100 mL

Uranyl acetate (UA) solution. 2% uranyl acetate dissolved in ultrapure (18.2 MΩ/cm) water

2.2. Dental and periodontal tissues

CD1 mice (Charles River, QC, CA) are sacrificed by cervical dislocation according to an approved animal protocol, and their mandibles are immediately dissected out and immersed in fresh BF on ice. Excess soft tissue is subsequently removed from the bone surface and each hemimandible segment fixed in approximately 2.5 mL fresh BF at 4 °C for 24 h with shaking. Hemimandible segments are then demineralized at 4 °C with shaking in approximately 2.5 mL DF, changed daily for 7 days.

2.3. Soft tissues

Skin from male CD1 mice and tail tendon from male Wistar rats is harvested immediately after sacrifice and immersed in BF on ice, then fixed with shaking for 24 h at 4 °C. Tissue samples are stored in BF at 4 °C until they are prepared for cryosectioning.

2.4. Cryosectioning

1. Cryoprotect tissue samples by incubating a small (~3 mm × 3 mm × 8 mm) block of tissue in ~3 mL (or at least 10 × the tissue volume) of SS 2 h at 4 °C with shaking, replace SS, and incubate overnight at 4 °C. Once the sample sinks, it is ready for cryosectioning. Tissue blocks are subsequently stored at 4 °C until mounted for cryosectioning and in liquid nitrogen (LN) thereafter.
2. Mount SS-infused sample on a specimen carrier and freeze in LN or by introducing to the nitrogen environment of the EM UC6-NT ultramicrotome EM FC6 cryochamber (Leica Microsystems GmbH, Wetzlar, DE) at −80 °C.
3. Trim the specimen using a diamond cryotrimming tool (Diatome, USA, Hatfield, PA), so the block face is approximately 200 μm × 300 μm.
4. Cool the cryochamber to −120 °C.
5. Cut 100 nm thick sections using a diamond CryoImmuno knife (Diatome, USA, Hatfield, PA) with a sectioning speed of ~1 mm/s.

6. Pick up ribbon of sections on a droplet of aqueous 2.3 M sucrose using a loop, then remove sections from the cryochamber and thaw the droplet at room temperature.
7. Deposit sections on 600 mesh nickel or lacey carbon-supported 200 mesh nickel TEM grids (Electron Microscopy Sciences, Hatfield, PA) by touching the droplet of 2.3 M sucrose to the grid.
8. Wash sucrose from the tissue section by floating the TEM grid, section side down, on 50 µL droplets of ultrapure (18.4 MΩ/cm) water three times for 10 min each.

2.5. Section vitrification

We typically vitrify our sucrose-free cryosections within 2 h of sectioning, as described in the succeeding text:

1. A grid is picked up from a droplet of ultrapure water using locking tweezers and mounted on the central axis of a Vitrobot Mark IV (FEI, Hillsboro, OR) equilibrated to 100% humidity and 22 °C for at least 20 min prior to use.
2. Excess water is automatically removed from the grid by blotting with filter paper simultaneously from both sides within the climate-controlled chamber.
3. The grid is plunged into liquid ethane, freezing the water fast enough to vitrify the layer of water on the grid.
4. To avoid condensation and crystallization of ice on the vitrified sample, the grid is transferred to a cryo grid box under an atmosphere of dry nitrogen at LN temperature and stored in LN until use, typically ~24 h.

3. IMAGE ACQUISITION AND PROCESSING

3.1. Cryo-TEM

We perform cryo-TEM using a model 626-DH cryotransfer holder (Gatan, Pleasanton, CA) cooled to approximately −179 °C in a Tecnai 20 TEM (FEI, Hillsboro, OR) with a LaB_6 filament (Electron Microscopy Sciences, Hatfield, PA) operating at 200 keV, spot size ~5, and −3.00 µm defocus. Images are collected at 25,000× magnification in low-dose mode using an AMT1600 side-mount camera (Advanced Microscopy Techniques, Woburn, MA). We use exposures in the range of 30–60 electrons/Å^2 to limit beam damage (Kudryashev, Castano-Diez, & Stahlberg, 2012).

3.2. Interpretation of Cryo-TEM micrographs

Using these imaging conditions, micrographs have a spatial sampling frequency of $1/1.15$ nm^{-1}. According to the Nyquist–Shannon sampling theorem (Shannon, 1949), this limits our resolution to no better than 2.30 nm. However, the contrast transfer function (CTF) for these conditions is predicted by ctfExplorer to have its first node at about $1/3$ nm^{-1} (Sidorov, 2002), which corresponds to a resolution of $\sim$3 nm. This is a typical resolution for biological samples (Massover, 2011), partially related to sectioning damage as demonstrated by comparing the resolution of polymer-embedded protein crystals in XRD and TEM (Sader, Reedy, Popp, Lucaveche, & Trinick, 2007; Weaver, McDowall, Oliver, & Deisenhofer, 1992). We do not employ CTF correction since we are unable to resolve higher-frequency contributions to the image, and we adopt a direct interpretation of the correspondence between micrographs and variations in the density of organic material in the tissue sections.

3.3. Analysis of whole micrographs

In low-dose cryo-TEM micrographs of COL I tissues (Fig. 9.2), collagen fibrils are the dominant feature, densely packing the field of view. Along the length of each collagen fibril, broad axially alternating light and dark stripes are apparent, corresponding to gap and overlap regions, respectively. Since COL I is the major component of these tissues, constituting some 60–86% tendon by dry weight (Kannus, 2000), and 90% of the organic component of dentin (Goldberg & Takagi, 1993; Linde & Robins, 1988) and cementum (Goncalves et al., 2005), it can be expected to dominate micrographs of collagenous tissues. This is clearly the case given the abundance of banded fibrils observed in a variety of tissues, and so we can estimate tissue-specific collagen D-periods from power spectra of micrographs (Fig. 9.2).

Power spectra from images of collagenous tissues consist of series of lines oriented perpendicular to the collagen fibril axes, representing the average frequency of collagen banding and its harmonics. They exhibit a strong peak near $1/67$ nm^{-1}, which corresponds to the collagen D-period. Harmonics of this period extend out to at least the 5th order depending on the organization of the tissue and the collagen fibrils, and in the case of RTT, the 12th harmonic is apparent. Each 2-dimensional power spectrum is condensed to one dimension (1-D) by radially integrating over an angular range concomitant with the arc traced by the frequency components. 1-D power spectra of

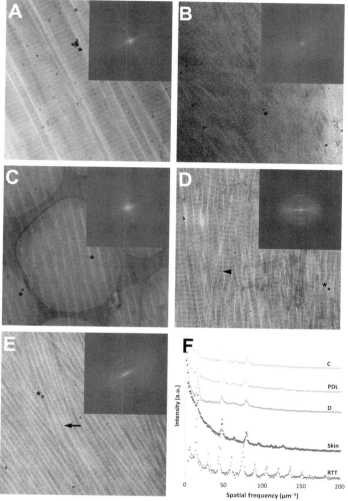

Figure 9.2 2.3 μm × 2.3 μm areas of selected collagenous tissues (rat tail tendon (A), mouse cementum (B), mouse skin (C), mouse dentin (D), mouse periodontal ligament (E)). Suprafibrillar architecture varies widely by tissue, RTT, skin, and PDL having a largely parallel arrangement with some apparent curvature (arrow in E), and dentin and cementum a more tortuous appearance with fibrils passing over/under one another and into/out of the section plane (arrow head in D). Inset: power spectrum of the associated image. Averages of 10 power spectra for each tissue area (F) clearly exhibit strong frequency components up to the fifth harmonic of the D-period, though the 12th harmonic is apparent in RTT due to the high degree of organization. Some ice contamination is visible (asterisks). (For color version of this figure, the reader is referred to the online version of this chapter.)

the same tissue are then averaged. Using the position of the fifth harmonic, we estimate the D-period in RTT and skin as 67.2 and 63.5 nm, respectively, which are in reasonable agreement with XRD results for hydrated and unfixed RTT and mouse skin of 67.8 nm (Orgel et al., 2006) and 65 nm (Stinson & Sweeny, 1980), respectively. The collagen period in cementum, dentin, and PDL is 63.1, 63.7, and 64.6 nm, respectively, by this method. However, in these micrographs, some collagen fibrils become indistinct and end as they pass beyond the plane of the section, while others are curved. These sorts of distortions contribute to the power spectrum and can skew the period estimate downward, so this method only provides a lower limit on periodicity, particularly for tissues with complex fibril organization.

4. IMAGE AVERAGING

Contrast is low for unstained biological samples in TEM, and this situation is exacerbated because the signal-to-noise ratio of low-dose micrographs is inherently low. To improve contrast and facilitate analysis of banding structure, we average images of collagen fibrils from a given tissue.

4.1. Image selection

When averaging images, a standard assumption is that the objects under consideration are identical. While this is usually reasonable in single particle reconstruction, where a single protein or complex is under investigation, this is patently untrue for collagen fibrils. TEM observation of collagen fibrils indicates that, while roughly circular in cross section, collagen fibrils of a single tissue have a distribution of diameters (Parry, Barnes, & Craig, 1978). Besides, a significant proportion of fibrils are disrupted by sectioning. It is therefore not possible to reconstruct the three-dimensional structure of collagen fibrils from single images of different fibrils. Moreover, a resolution limit of ~3 nm precludes analysis of the lateral structure of collagen fibrils, which typically have a lateral spacing of ~1.3 nm. However, the axial structure of collagen fibrils is regular and lends itself to image averaging techniques if some care is taken to select appropriate fibrils. Several studies (Beniash et al., 2000; Chapman, 1974; Chapman, Tzaphlidou, Meek, & Kadler, 1990; Tzaphlidou, 2005; Zervakis, Gkoumplias, & Tzaphlidou, 2005) make use of this to treat collagen banding as a 1-D structure.

Two issues we consider when selecting fibrils for image averaging are deformation of the collagen fibril and orientation. As seen in Fig. 9.2, in-plane curvature is evident in fibrils of PDL, while fibrils of dentin appear

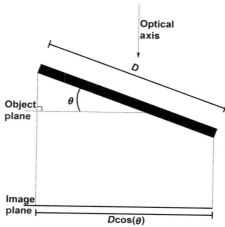

Figure 9.3 Projection of a collagen D-period (heavy line) onto the image plane showing that misalignment of fibrils relative to the object/image planes compresses the apparent D-period by a factor of $\cos(\theta)$.

to pass over/under one another. In addition to complicating image alignment by affecting the relative orientation of individual bands, curvature-induced bending strains can cause variations in periodicity. Misorientation of the fibril axis relative to the object plane will cause an apparent compression of the D-period by a factor of $\cos(\theta)$ (Fig. 9.3). Therefore, to limit variations in the axial banding of selected images, we impose the following criteria:

- Fibrillar banding must be clearly visible.
- Fibrils must appear straight.
- Fibrils should persist in the section for ≥ 1 μm.

Assuming that the fibril under consideration is perfectly straight and has a diameter less than the ~100 nm thickness of the section, the final criterion limits angular misalignment to less than 5.7° (arcsin (100 nm/1000 nm)) or a D-period contraction of about 0.5%. The prior two criteria ensure that only relatively undistorted fibrils with distinct banding structure are selected for further analysis. Using WEB (Health Research Inc., Albany, NY) to visualize micrographs and record coordinates of specific regions of interest, we choose segments of collagen fibrils that meet the selection criteria as demonstrated in Fig. 9.4; a representative area is selected from the larger RTT fibril, and the fibril axis is oriented vertically. The ends of the selected area are padded with the average grayscale value to facilitate axial alignment of images.

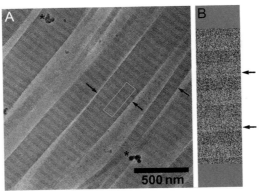

Figure 9.4 A subfibrillar image (rectangle) is selected from a micrograph of RTT (A). The only asymmetric band of type I collagen fibrils (arrows) is clearly visible in the gap region. The selected area is oriented vertically and padded at the top and bottom with its average grayscale value (B) to facilitate alignment with other images. Note that it is more difficult to identify the asymmetric band outside of the context of the whole fibril. (For color version of this figure, the reader is referred to the online version of this chapter.)

4.2. Image alignment

Due to asymmetry of the tropocollagen molecule and a tendency for these molecules to align N-telopeptide to C-telopeptide, fibrils of COL I are polar (Kadler, Holmes, Trotter, & Chapman, 1996). This can be seen clearly in the asymmetric banding of heavy-metal-stained collagen fibrils or in cryo-TEM of unstained collagen (Fig. 9.4); a dark band is visible in the gap region, offset from the center, demonstrating polarity of the collagen fibrils. This polarity must be taken into account when averaging together images of collagen fibrils, otherwise a symmetric banding pattern may be erroneously generated, as described later.

We initially implemented in Spider (Health Research Inc., Albany, NY) a script that orients fibrils vertically by maximizing the axial mean-squared deviation of grayscale values and then aligns and averages the fibril images using a reference-free shift align operation. We found, however, that this alignment operation is not very well able to identify the polarity of collagen fibrils, resulting in a symmetric averaged fibril image (Fig. 9.5). Thus, while the fibrillar D-period is preserved, the banding structure is misrepresented and differences between gap and overlap regions are obliterated. We attribute the difficulty in maintaining fibril polarity in automated alignment to the low signal-to-noise ratio and the apparent symmetry of the most prominent features of the collagen fibril, namely the gap and overlap regions.

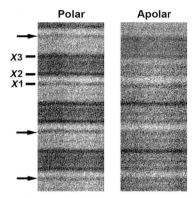

Figure 9.5 Average of 103 RTT fibril images where fibril polarity was identified manually and correctly displays asymmetric banding (polar), and where polarity was identified automatically by autocorrelation, erroneously generating symmetric (apolar) banding. In the polar image, gap and overlap are clearly identifiable as light and dark regions, respectively, and a dark band in the gap region (arrows) indicates fibril polarity, whereas the apolar image maintains the D-period but merges features of the gap and overlap region.

To preserve fibril polarity, we included a user input step to manually indicate the polarity and angular orientation of each fibril by selecting two points coaxial with the area of interest, the order of this selection indicating polarity. This simplified the alignment and averaging process and improved consistency of the average.

The resultant average asymmetric banding pattern in COL I fibrils (Fig. 9.5) consists of broad gap and overlap regions, about 40 and 27 nm wide, respectively. At the transitions between gap and overlap regions, and within the gap (light) region, dark bands are clearly visible. We label these bands X1, X2, and X3 according to the convention of Marchini et al. (1983), who observed analogous ridges on freeze-etched collagen fibrils. X1 is the asymmetric band, its nearest neighbor, X2, presumably corresponds to the region of the N-telopeptide, and X3 to the C-telopeptide.

4.3. Measurement of fibril periodicity from averaged images

To obtain an initial estimate of fibril periodicity, we apply a Fourier transform to individual images and use the position of the fifth harmonic in the resulting power spectrum to estimate periodicity. By this method, we find that the variation in periodicity is fairly broad and the average periodicity is somewhat lower than accepted values. We presume that these variations in fibril periodicity are induced by misalignment (see 3.3). We, therefore, determine the maximum fibril period by imposing further constraints on data selection. We do this very simply by dividing the fibril images into two approximately equally sized subsets based on their periodicity and selecting the larger D-period subset.

To obtain independent measurements from averaged collagen fibrils, which benefit from improved signal-to-noise ratio, we produce averages from five randomly selected subsets of the retained images. We prepare 1-D power spectra for these averages and use the position of the fifth harmonic to estimate the D-period. Using this method, we determine the periodicity of mouse skin and RTT collagen to be 64.8 ± 0.5 nm and 68.0 ± 0.2 nm, respectively, which is in very good agreement with XRD measurements of hydrated RTT (67.8 nm) (Orgel et al., 2006) and skin (65 nm) (Stinson & Sweeny, 1980).

To evaluate gap and overlap size from the averages of image subsets, we generate 1-D banding profiles by integrating across the fibrils to determine the average axial variation in grayscale values. We measure gap and overlap sizes directly from the 1-D banding profiles, arbitrarily defining the boundary between gap and overlap regions as the point midway between the maximum and minimum values at each gap/overlap transition. Though reported gap/overlap sizes will vary depending on how boundaries are chosen, tissue-specific differences in gap/overlap size should be relatively independent of this. Interestingly, we find that the overlap size is the same in RTT (31.9 ± 0.7 nm) and skin (31.5 ± 0.7 nm) collagen, whereas the gap of RTT (36.4 ± 0.9 nm) is significantly larger than the gap of skin (33.3 ± 1.0 nm). This method can be readily applied to demineralized connective tissues to obtain accurate measurements of fibril periodicity and gap/overlap size.

5. SUMMARY AND OUTLOOK

This chapter presents a method for averaging cryo-TEM images of collagen fibrils. Having confirmed the validity of our method by comparing periodicity measurements to results from XRD, we report measurements for gap and overlap sizes in hydrated collagen fibrils of RTT and skin. Though similar parameters have previously been estimated from TEM, they are almost entirely from dehydrated samples, which limits the biological relevance of these structural parameters.

Cryo-TEM and related techniques like cryo-electron tomography promise more biologically relevant structural information of collagenous tissues with quantitative measurements. Their application to normal and pathological mineralizing tissues, and in particular early stages of mineralization, will be invaluable in elucidating the role of collagen structure in mediating intrafibrillar mineralization, and determining collagen–mineral interactions.

ACKNOWLEDGMENTS

We thank Doug Holmyard and Bob Temkin of the Advanced Bioimaging Centre at Mount Sinai Hospital, Toronto, for technical advice. This work was supported by a National Sciences and Engineering Research Council of Canada (NSERC) Discovery grant to E. D. S. We are grateful for NSERC Postgraduate Scholarships and an Ontario Graduate Scholarship to B. D. Q.

The authors declare no conflict of interest.

REFERENCES

Almer, J. D., & Stock, S. R. (2005). Internal strains and stresses measured in cortical bone via high-energy X-ray diffraction. *Journal of Structural Biology*, *152*, 14–27.

Beniash, E., Traub, W., Veis, A., & Weiner, S. (2000). A transmission electron microscope study using vitrified ice sections of predentin: Structural changes in the dentin collagenous matrix prior to mineralization. *Journal of Structural Biology*, *132*, 212–225.

Bonar, L. C., Lees, S., & Mook, H. A. (1985). Neutron diffraction studies of collagen in fully mineralized bone. *Journal of Molecular Biology*, *181*, 265–270.

Bos, K. J., Holmes, D. F., Kadler, K. E., McLeod, D., Morris, N. P., & Bishop, P. N. (2001). Axial structure of the heterotypic collagen fibrils of vitreous humour and cartilage. *Journal of Molecular Biology*, *306*, 1011–1022.

Carden, A., & Morris, M. D. (2000). Application of vibrational spectroscopy to the study of mineralized tissues. *Journal of Biomedical Optics*, *5*(3), 259–268.

Chapman, J. A. (1974). The staining pattern of collagen fibrils: I. an analysis of electron micrographs. *Connective Tissue Research*, *2*, 137–150.

Chapman, J. A., Tzaphlidou, M., Meek, K. M., & Kadler, K. E. (1990). The collagen fibril— A model system for studying the staining and fixation of a protein. *Electron Microscopy Reviews*, *3*, 143–182.

Claffey, W. (1977). Interpretation of the small-angle X-ray diffraction of collagen in view of the primary structure of the $\alpha1$ chain. *Biophysical Journal*, *19*, 63–70.

Erickson, B., Fang, M., Wallace, J. M., Orr, B. G., Les, C. M., & Holl, M. M. B. (2012). Nanoscale structure of type I collagen fibrils: Quantitative measurement of D-spacing. *Biotechnology Journal*, *8*, 117–126.

Eyre, D. R., & Wu, J.-J. (2005). Collagen cross-links. *Topics in Current Chemistry*, *247*, 207–229.

Fratzl, P., Paris, O., Klaushofer, K., & Landis, W. J. (1996). Bone mineralization in an osteogenesis imperfecta mouse model studied by small-angle x-ray scattering. *Journal of Clinical Investigation*, *97*, 396–402.

Gathercole, L. J., Porter, S., & Scully, C. (1987). Axial periodicity in periodontal collagens. Human periodontal ligament and gingival connective tissue collagen fibers possess a dermis-like D-period. *Journal of Periodontal Research*, *22*, 408–411.

Glimcher, M. J. (1959). Molecular biology of mineralized tissues with particular reference to bone. *Reviews of Modern Physics*, *31*(2), 359–419.

Goldberg, M., & Takagi, M. (1993). Dentine proteoglycans: Composition, ultrastructure and functions. *Histochemical Journal*, 25, 781–806.

Goncalves, P. F., Sallum, E. A., Sallum, A. W., Casati, M. Z., Toledo, S., & Nociti, F. H., Jr. (2005). Dental cementum reviewed: Development, structure, composition, regeneration and potential functions. *Brazilian Journal of Oral Sciences*, 4(12), 651–658.

Habelitz, S., Balooch, M., Marshall, S. J., Balooch, G., & Marshall, G. W., Jr. (2002). In situ atomic force microscopy of partially demineralized human dentin collagen fibrils. *Journal of Structural Biology*, 138, 227–236.

Hodge, A. J., & Petruska, J. A. (1963). Recent studies with the electron microscope on ordered aggregates of the tropocollagen macromolecule. In G. N. Ramachandran (Ed.), *Aspects of protein structure* (pp. 289–300). New York: Academic Press.

Kadler, K. E., Holmes, D. F., Trotter, J. A., & Chapman, J. A. (1996). Collagen fibril formation. *Biochemical Journal*, 316, 1–11.

Kannus, P. (2000). Structure of the tendon connective tissue. *Scandinavian Journal of Medical & Science in Sports*, 10, 312–320.

Kudryashev, M., Castano-Diez, D., & Stahlberg, H. (2012). Limiting factors in single particle cryo electron tomography. *Computational and Structural Biotechnology Journal*, 1(2), 1–6.

Landis, W. J., Hodgens, K. J., Arena, J., Song, M. J., & McEwen, B. F. (1996). Structural relations between collagen and mineral in bone as determined by high voltage electron microscopic tomography. *Microscopy Research and Technique*, 33(2), 192–202.

Landis, W. J., Song, M. J., Leith, A., McEwen, L., & McEwen, B. F. (1993). Mineral and organic matrix interaction in normally calcifying tendon visualized in three dimensions by high-voltage electron microscopic tomography and graphic image reconstruction. *Journal of Structural Biology*, 110, 39–54.

Linde, A., & Robins, S. P. (1988). Quantitative assessment of collagen crosslinks in dissected predentin and dentin. *Collagen and Related Research*, 8, 443–450.

Marchini, M., Morocutti, M., Castellani, P. P., Leonardi, L., & Ruggeri, A. (1983). The banding pattern of rat tail tendon freeze-etched collagen fibrils. *Connective Tissue Research*, 11, 175–184.

Marchini, M., Morocutti, M., Ruggeri, A., Koch, M. H. J., Bigi, A., & Roveri, B. (1986). Differences in the fibril structure of corneal and tendon collagen. An electron microscopy and X-ray diffraction investigation. *Connective Tissue Research*, 15, 269–281.

Massover, W. H. (2011). New and unconventional approaches for advancing resolution in biological transmission electron microscopy by improving macromolecular specimen preparation and preservation. *Micron*, 42, 141–151.

McDonald, K. L., & Auer, M. (2006). High-pressure freezing, cellular tomography, and structural cell biology. *BioTechniques*, 41(2), 137–143.

Meadows, R. S., Holmes, D. F., Gilpin, C. J., & Kadler, K. E. (2000). Electron cryomicroscopy of fibrillar collagens. In C. H. Streuli & M. E. Grant (Eds.), *Extracellular matrix protocols: 139. Methods in molecular biology* (pp. 95–109). Totowa, NJ: Humana Press.

Meek, K. M., Elliott, G. F., Hughes, R. A., & Nave, C. (1983). The axial electron density in collagen fibrils from human corneal stroma. *Current Eye Research*, 2(7), 471–477.

Nudelman, F., Lausch, A. J., Sommerdijk, N. A. J. M., & Sone, E. D. (2013). In vitro models of collagen biomineralization. *Journal of Structural Biology*, 183(2), 258–269.

Nudelman, F., Pieterse, K., George, A., Bomans, P. H. H., Friedrich, H., Brylka, L. J., et al. (2010). The role of collagen in bone apatite formation in the presence of hydroxyapatite nucleation inhibitors. *Nature Materials*, 9, 1004–1009.

Orgel, J. P. R. O., Irving, T. C., Miller, A., & Wess, T. J. (2006). Microfibrillar structure of type I collagen in situ. *Proceedings of the National Academy of Sciences*, 103(24), 9001–9005.

Parry, D. A. D., Barnes, G. R. G., & Craig, A. S. (1978). A comparison of the size distribution of collagen fibrils in connective tissues as a function of age and a possible relation between fibril size distribution and mechanical properties. *Proceedings of the Royal Society B: Biological Sciences, 203*, 305–321.

Sabanay, I., Arad, T., Weiner, S., & Geiger, B. (1991). Study of vitrified, unstained frozen tissue sections by cryoimmunoelectron microscopy. *Journal of Cell Science, 100*, 227–236.

Sader, K., Reedy, M., Popp, D., Lucaveche, C., & Trinick, J. (2007). Measuring the resolution of uncompressed plastic sections cut using an oscillating knife ultramicrotome. *Journal of Structural Biology, 159*, 29–35.

Sasaki, N., Tagami, A., Goto, T., Taniguchi, M., Nakata, M., & Hikichi, K. (2002). Atomic force microscopic studies on the structure of bovine femoral cortical bone at the collagen fibril-mineral level. *Journal of Materials Science Materials in Medicine, 13*(3), 333–337.

Shannon, C. E. (1949). Communication in the presence of noise. *Proceedings of the Institute of Radio Engineers, 37*(1), 10–21.

Sidorov, M. (2002). ctfExplorer: Interactive software for 1d and 2d calculation and visualization of TEM phase contrast transfer function. *Microscopy and Microanalysis, 8*(Suppl. 2), 1572–1573.

Starborg, T., Lu, Yinhui, Meadows, R. S., Kadler, K. E., & Holmes, D. F. (2008). Electron microscopy in cell-matrix research. *Methods, 45*(1), 53–64.

Stinson, R. H., & Sweeny, P. R. (1980). Skin collagen has an unusual D-spacing. *Biochimica et Biophysica Acta, 621*(1), 158–161.

Tokuyasu, K. T. (1973). A technique for ultracryotomy of cell suspensions and tissues. *The Journal of Cell Biology, 57*, 551–565.

Tzaphlidou, M. (2005). The role of collagen in bone structure: An image processing approach. *Micron, 36*, 593–601.

Weaver, A. J., McDowall, A. W., Oliver, D. B., & Deisenhofer, J. (1992). Electron microscopy of thin-sectioned three-dimensional crystals of SecA protein from Escherichia coli: Structure in projection at 40 A resolution. *Journal of Structural Biology, 109*, 87–96.

Weiner, S., Arad, T., Sabanay, I., & Traub, W. (1997). Rotated plywood structure of primary lamellar bone in the rat: Orientations of the collagen fibril arrays. *Bone, 20*(6), 509–514.

Whittaker, P., & Canham, P. B. (1991). Demonstration of quantitative fabric analysis of tendon collagen using two-dimensional polarized light microscopy. *Matrix, 11*, 59–62.

Zernia, G., & Huster, D. (2008). Investigation of collagen dynamics by solid-state NMR spectroscopy. In G. A. Webb (Ed.), *Modern magnetic resonance* (pp. 87–92). Dordrecht, Netherlands: Springer.

Zervakis, M., Gkoumplias, V., & Tzaphlidou, M. (2005). Analysis of fibrous proteins from electron microscopy images. *Medical Engineering & Physics, 27*, 655–667.

SECTION III

Biomimetic Crystallization Techniques *in vitro*

CHAPTER TEN

Preparation of Organothiol Self-Assembled Monolayers for Use in Templated Crystallization

Michael H. Nielsen[*,†,1], Jonathan R.I. Lee[‡]

[*]Department of Materials Science and Engineering, University of California, Berkeley, California, USA
[†]Materials Science Division, Lawrence Berkeley National Lab, Berkeley, California, USA
[‡]Lawrence Livermore National Laboratory, Livermore, California, USA
[1]Corresponding author: e-mail address: mhnielsen@lbl.gov

Contents

1. Introduction	210
2. Crystallization Approaches Using SAMs as Templates	211
2.1 CO_2-in	211
2.2 Kitano	213
2.3 Fixed composition	213
3. Preparation	214
3.1 Substrate preparation	214
3.2 Monolayer preparation	215
3.3 Assessing monolayer quality	217
4. Why Good SAMs Go Bad	218
4.1 Contamination	219
4.2 Thiol solution	220
4.3 Film degradation	220
5. SAM Preparation Procedure	221
6. Summary	221
Acknowledgments	222
References	222

Abstract

Organothiol self-assembled monolayers (SAMs) have garnered much interest as templates for oriented crystallization of biominerals. While, on the surface, SAM preparation appears to be straightforward, there are many subtleties that may yield films that lack the desired effect on the mineral component in subsequent use for templated mineralization. Herein, we discuss literature that uses organothiol SAMs to understand various principles in biomineralization, to motivate the following discussion of preparation procedures and pitfalls that may arise while working with SAMs. We provide a range of parameters for each element of a SAM-forming process, which have been shown in the literature to produce monolayers suitable for mineralization experiments, and close with a step-by-step procedure, based on findings in the cited literature, that yields functional SAMs with very high fidelity.

1. INTRODUCTION

Many studies in the past 15 years have used organothiol SAMs as templates in biomimetic approaches to understanding core concepts in biomineralization. These SAMs have been shown to preferentially nucleate calcite on specific faces depending on a combination of the SAM monomer chemistry and the underlying substrate. This templating behavior has made SAMs an attractive platform with which to investigate the roles that macromolecular complexes and surfaces may play in the formation and growth of biominerals. As such, there is much in the literature on organothiol SAM preparation and subsequent characterization. Many of these studies that were conducted up to 2005 can be found discussed in the excellent review by Love, Estroff, Kriebel, Nuzzo, & Whitesides (2005). The much smaller body of work that uses SAMs as templates for biomimetic crystallization can be found reviewed by Sommerdijk and de With (2008), along with descriptions of other organic matrices used as templates (e.g., Langmuir monolayers). However, even with this large body of literature on organothiol SAMs, many researchers find there to be gaps in the reported information such that consistent and reproducible preparation of highly ordered SAMs may be a difficult endeavor. In this chapter, we will present a collection existent in the literature and provide additional details from our extensive work with organothiol SAMs to mitigate the preparation pitfalls that may arise. We first provide an overview of the major work that has been undertaken in utilizing SAMs to investigate issues in biomineral formation. Following this, we will discuss the spectrum of materials thus far used as substrates for SAM formation in such studies and continue on to a discussion of the range of conditions under which SAMs are formed on the underlying substrates. The next section of this chapter will discuss a few characterization techniques for assessing the quality of SAMs and the ramifications they have on experimental approach. At the end of the chapter, we provide a detailed step-by-step procedure that we have found to be highly successful in reproducibly forming SAMs for use in a wide variety of experimental techniques.

2. CRYSTALLIZATION APPROACHES USING SAMs AS TEMPLATES

2.1. CO$_2$-in

The single most prevalent approach to calcite crystallization using SAM is the "CO$_2$-in" method, shown schematically in Fig. 10.1A. In this technique, a SAM-bearing substrate is placed in a calcium-bearing solution, the open container of which is placed in a sealed environment (e.g., desiccator) along with an open container of ammonium carbonate. The decomposition of ammonium carbonate produces carbon dioxide gas, which is in part taken up by the calcium-bearing solution as the partial pressure of CO$_{2(g)}$ increases. Calcium carbonate supersaturation continuously rises in the

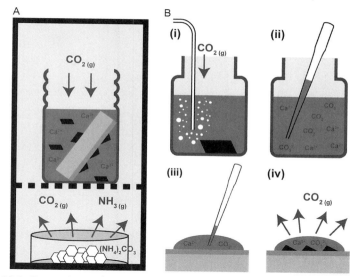

Figure 10.1 Comparison of (A) CO$_2$-in diffusion and (B) Kitano methods. The CO$_2$-in method (A) uses the decomposition of (NH$_4$)$_2$CO$_3$ to continually increase the partial pressure of CO$_{2(g)}$ in an enclosure. This drives up the concentrations of carbonate species in the calcium-bearing solution that contains the SAM-covered substrate. The substrate is placed facedown to prevent homogeneously formed solids from settling on the SAM. The Kitano (B) method (i) bubbles CO$_{2(g)}$ into a solution of water and solid calcite, dissolving the calcite to produce (ii) a solution rich in Ca^{2+} and CO$_2$ that is filtered to remove any remaining solid. (iii) An aliquot of this solution is placed onto a SAM-bearing substrate. (iv) CO$_{2(g)}$ diffusion out of the Kitano solution continuously raises the calcium carbonate supersaturation and leads to mineralization. (For color version of this figure, the reader is referred to the online version of this chapter.)

SAM-containing solution, and mineralization is described by the following chemical reactions:

$$(NH)_4CO_{3(s)} \rightarrow 2NH_{3(g)} + CO_{2(g)} + H_2O \qquad (10.1)$$

$$CO_2 + Ca^{2+} + H_2O \rightarrow CaCO_{3(s)} + 2H^+ \qquad (10.2)$$

$$2NH_3 + 2H^+ \rightarrow 2NH_4^+ \qquad (10.3)$$

Due to the continuous evolution of gases, this system is problematic in terms of accurately describing the solution state during the nucleation process, although recent efforts have been undertaken to make the CO_2-in approach more rigorous (Ihli, Bots, Kulak, Benning, & Meldrum, 2013). Even so, a surprising number of results have come from using this procedure. Utilizing the CO_2-in method, Aizenberg and colleagues conducted a series of systematic studies on SAMs, varying surface chemistry, monomer alkane chain length, and underlying substrate, to produce SAM-specific, highly oriented nucleation of calcite crystals in a pure calcium carbonate system (Aizenberg, Black, & Whitesides, 1999; Han & Aizenberg, 2003; Kwak, DiMasi, Han, Aizenberg, & Kuzmenko, 2005) and in the presence of Mg^{2+} (Han, Wysocki, Thanawala, Siegrist, & Aizenberg, 2005; Kwak et al., 2005). Such studies showed the importance of the following combined features of SAMs to oriented crystallization: surface functional group (–X), orientation of –X, and packing geometry of –X. These results also highlight the necessity of well-formed and highly reproducible SAMs for use in templated crystallization. A SAM lacking a significant degree of uniformity across the surface does not provide the controlled surface required for exerting a high degree of influence over nucleation.

In addition to working with single-species monolayers, Aizenberg and colleagues have shown that patterned SAMs can further control calcite nucleation (Aizenberg, 2000; Aizenberg, Black, & Whitesides, 1998; Aizenberg et al., 1999). Furthermore, when the experimental system first forms and stabilizes amorphous calcium carbonate (ACC), a small localized SAM can serve as the nucleation site for large-scale, single crystals with complex architectures (Aizenberg, Muller, Grazul, & Hamann, 2003). Han and Aizenberg also took advantage of the observation that certain SAMs stabilized ACC to create a reservoir from which to controllably nucleate calcite in arbitrary patterns on a second SAM (Han & Aizenberg, 2008). The Estroff group demonstrated that a hybrid system composed of a SAM covered by a hydrogel can control both nucleation and growth of calcite (Li & Estroff, 2007).

Lee and colleagues used the CO_2-in method to investigate the structural relationship between the SAM and the mineral phase, through the use of synchrotron-based X-ray spectroscopies (Lee et al., 2007, 2013; Nielsen, Lee, Hu, Han, & De Yoreo, 2012). These studies suggest that there is a dynamic relationship between the two phases in the system and that one should, in fact, consider the process of surface-mediated nucleation on SAMs as a cotemplating effect whereby the structures of each phase evolve throughout the nucleation process to achieve an optimal structural relationship. Lastly, the CO_2-in method has been used to suggest differences in mineralization pathways on differently functionalized SAMs (Hu et al., 2012).

2.2. Kitano

Another approach that has been used in the past in the calcium carbonate system is the Kitano method (Kitano, 1962), shown schematically in Fig. 10.1B. The approach is centered upon bubbling CO_2 into a mixture of solid $CaCO_3$ in pure water to drive the following reaction to the right:

$$2HCO_{3(aq)}^- + Ca_{(aq)}^{2+} \rightleftarrows CaCO_{3(s)} + CO_{2(g)} + H_2O_{(l)} \qquad (10.4)$$

When the solution is filtered to remove the solid $CaCO_3$, this produces a solution that becomes increasingly supersaturated with respect to calcium carbonate mineralization as the artificially high level of CO_2 diffuses out of the solution and eventually leads to crystallization. Travaille used this approach, placing SAMs in the Kitano solution, to show a unidirectional 1:1 lattice relationship between the in-plane orientation of the nucleation calcite and the underlying Au film (Travaille et al., 2002, 2003). As with the work of Aizenberg and Han detailed earlier, these results suggest the need for a SAM preparation that is consistent and produces highly ordered surfaces.

2.3. Fixed composition

The third experimental approach that has been reported for crystallizing biominerals on SAMs is to mix precursor salt solutions at fixed ratios. This method allows for experiments under tightly controlled solution conditions, where the supersaturation at the onset of mineralization is well known. One may either take an aliquot of the mixed solution for a stagnant volume experiment or continuously flow the solutions so as to ensure that nucleation on the SAM is reaction-limited rather than mass transport-limited. Travaille used the stagnant volume approach to estimate initial thermodynamic driving forces of SAM-templated calcite nucleation (Travaille, Steijven, Meekes, & van Kempen, 2005). Wallace (Wallace, De Yoreo, & Dove,

2009) and Hu (Hu et al., 2012) used the continuous flow setup to measure nucleation rates of silica and calcite, respectively, on SAMs. This allowed quantitative investigations of kinetic and thermodynamic barriers for these minerals on different SAM surfaces. Hu et al. also used the stagnant volume approach to study nucleation pathways and showed evidence that suggests different pathways of calcite nucleation on chemically distinct SAM surfaces.

3. PREPARATION

The preceding section describes a number of experimental approaches for using SAMs as templates in crystallization. The tacit assumption throughout that section is that the SAM is a highly ordered, near-uniform structure that presents a well-defined, repeating chemical arrangement with which the solvated ions and incipient mineral phase can interact. This assumption, however, is only as good as the template preparation. In this section, we describe the many approaches taken to prepare the underlying substrate and the SAM and discuss methods by which one may assess a SAM's viability for use in mineralization experiments.

3.1. Substrate preparation

The literature abounds with a multitude of preparations for substrates on which to organize organothiol SAMs. We take as examples those studies of SAM-templated biomineral crystallization detailed in the earlier section. In each case, a metal film is deposited on an underlying substrate. This substrate is usually a Si(100) wafer, either *n*- or *p*-type (Aizenberg et al., 1999, 2003; Han et al., 2005; Han & Aizenberg, 2008; Hu et al., 2012; Kwak et al., 2005; Li & Estroff, 2007), although mica (Travaille et al., 2002; Wallace et al., 2009), bare glass slides (Aizenberg, 2000), and photoresist-covered glass slides (Aizenberg et al., 2003) have also been used as substrates. In order to promote adhesion of the noble metal film and prevent delamination in aqueous environments, an adhesion layer of Ti or Cr with a thickness between 2.0 and 5.0 nm has been first evaporated onto the underlying substrate, as detailed in many of the previous references in this section. The metal film, with a thickness ranging between 15 and 500 nm, is then evaporated onto the adhesion layer. Au is the standard film material, although Ag and Pd have also been used by Han and Aizenberg. Typically, the films are composed of a single metal, although patterned films of different metals have also been used by Han and Aizenberg to great effect. The (111) structure and grain size of Au can be improved through heated evaporation and annealing at temperature (Travaille et al., 2002).

Lee et al. used a hydrogen flame annealing method to clean the Au surface of any adsorbed species and improve the (111) structure of the film immediately prior to use in forming the organothiol template (Lee et al., 2007). Wallace et al., by contrast, prepared ultraflat Au(111) surfaces through the template-stripping procedure (Hegner, Wagner, & Semenza, 1993) of transferring evaporated Au on mica to another substrate (e.g., Si). This latter technique has the advantage of providing large, atomically flat gold terraces, a characteristic that was necessary for observing the formation of small (<10 nm) silica nuclei (Wallace et al., 2009).

From this section, one can clearly see that there are a wide range of suitable substrates on which to organize a SAM. While the noble metal surface is a common feature in the aforementioned approaches, the underlying substrate, metal thickness, surface roughness, and surface preparation are examples of SAM substrate characteristics that can vary widely and still provide an amenable surface for SAM organization and subsequent biomineral crystallization.

3.2. Monolayer preparation

As with the underlying substrates, there are many reported conditions for preparing organothiol monolayers on the underlying noble metal films. Although the literature regarding organothiol SAMs is quite extensive, we will again restrict our examples of approaches to monolayer formation to the studies outlined in the aforementioned section on experimental uses of SAMs in biomineral systems. The chemistries in the following examples are not identical, varying primarily in carbon chain length and functional group at the nonthiol terminus.

To prepare a single-species SAM, most studies utilize an ethanolic solution, irrespective of surface functional group, with thiol concentrations ranging between 1 and 30 mM, placing the substrates in those solutions for 1–24 h, and rinsing with ethanol, with (Li & Estroff, 2007; Travaille et al., 2002) or without (Aizenberg, 2000; Aizenberg et al., 1999; Han et al., 2005; Han & Aizenberg, 2008; Kwak et al., 2005) an additional water rinse. Long soak times in the thiol solutions allow the SAM monomers to reorganize themselves into increasingly well-packed structures and more upright configurations following adsorption onto the substrate. The additional water rinse was intended to remove unbound thiols and eliminate the presence of bilayer patches on the surface.

Lee (Lee et al., 2007, 2013) and Hu (Hu et al., 2012) employed different conditions for SAM formation of different surface group functionalities, based on the results of Arnold and Willey (Arnold, Azzam, Terfort, & Wöll, 2002; Willey et al., 2004). These results demonstrated that carboxyl-terminated SAM monomers can readily form bilayers across a significant portion of the surface or otherwise form dimers that decrease the degree to which the monolayer is well ordered, as a result of hydrogen bonding between the –COOH functionalities on proximate monomers. As such, Arnold and Willey suggested using a small amount of acetic acid to form and rinse SAMs—as noted earlier, others have used an additional water rinse to achieve the same purpose for –COOH SAMs formed in ethanol. Furthermore, Arnold recommended using lower thiol concentrations to minimize the ability of the monomers to form dimers and to promote monolayer coverage. Willey showed by spectroscopic means that the amount of unbound thiols decreased with solution concentration, with the maximum amount of unbound thiol estimated to be around 5% at 1 mM and decreasing to roughly 3.5% at 1 μM. Lee et al. (2007) and Hu et al. (2012) thus prepared –OH SAMs by immersing substrates in 1–2 mM ethanolic thiol solutions and –COOH SAMs by placing substrates in 1–2 mM thiol solutions with a solvent composed of 95:5 ethanol:acetic acid. Each type of SAM was rinsed with its respective solvent and dried in a diffuse nitrogen stream immediately prior to use.

The following is a brief presentation of slight modifications to the basic procedure, as well as other approaches, used by the aforementioned groups for SAM formation. Gently warming the SAM solution has been used to form SAMs at relatively low concentrations and short soaking periods (Travaille et al., 2003). For mixed monolayers, Aizenberg sequentially soaked patterned substrates in two separate thiol solutions for different soaking times to control the extent of thiol exchange (Aizenberg et al., 1998). PDMS stamps inked with thiols have been used for microcontact printing (Kumar, Abbott, Kim, Biebuyck, & Whitesides, 1995) specific patterns of one thiol onto the surface, followed by a short soak in a second thiol solution to fill in uncontacted regions (Aizenberg et al., 1998, 1999; Wallace et al., 2009). Aizenberg used a similar approach by inking an atomic force microscope (AFM) probe with thiols to deposit a small region of thiol on the surface. The remainder of the surface was covered through soaking in a mixed thiol solution (Aizenberg et al., 2003).

Again it can be seen that a wide range of approaches to monolayer formation can be taken—from thiol concentration to assembly time to rinse solvent(s)—with each reported to produce monolayers suitable for crystallization. There is a paucity of discussion, however, as to the robustness of a given approach and the degree to which each produces consistent results. We emphasize the results of Arnold et al. (2002) and Willey et al. (2004) that detail general suggestions for reproducible approaches to forming well-ordered monolayers. Lower concentrations, down to sub-mM, tend to form better monolayers by mitigating the extent to which dimerization can occur. The second is that a solvent mixture can be used as both a SAM solution and a rinse for hydrogen bond forming surface chemistries to prevent dimer formation and to better rinse off dimers or adsorbed species on the monolayer, leading to better monolayer quality. In the following section, we will suggest an easy, qualitative method by which one may quickly ascertain the suitability of a substrate that has been soaked in a thiol solution for use as a SAM template for biomineral crystallization.

3.3. Assessing monolayer quality

Various groups have utilized many methods to characterize the quality of organothiol SAMs, discussed in the aforementioned reviews. Most of these techniques are not suitable, however, for a general assessment of a prepared SAM's suitability for use in experiment. Nevertheless, the results of a few approaches do provide useful information in determining whether or not a usable SAM is present on a given substrate. Assuming that the substrate size is large enough to see the spreading of a small droplet of an aqueous solution, the wettability of the surface can provide a very qualitative sense as to whether or not the SAM has the correct surface functionality. Looking at some of the literature on contact angle measurements for variously functionalized organothiol SAMs (Bain et al., 1989; Bain & Whitesides, 1988; Laibinis, Fox, Folkers, & Whitesides, 1991; Laibinis & Whitesides, 1992) in conjunction with calcite crystallization studies on SAMs in which contact angle measurements were made (Aizenberg et al., 1999; Lee et al., 2007, 2013), it can be readily seen that surfaces that promote oriented nucleation have polar, and thus hydrophilic, functional groups at the surface. Conversely, hydrophobic surfaces suppress nucleation, and calcite crystals that are observed on those regions are not preferentially oriented. The utility of these results is that after a SAM-covered substrate is removed from the thiol solution, rinsed, and dried, the general quality and viability of that SAM for use in templated crystallization can be tested by dropping a small volume of water or aqueous solution onto the surface. If the liquid easily

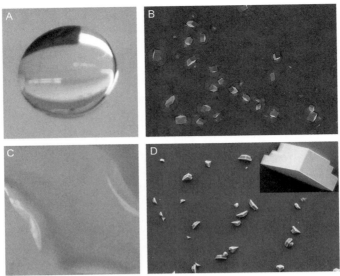

Figure 10.2 Optical (A and C) and SEM (B and D) comparisons of poorly ordered (A and B) and well-ordered (C and D) MHA SAMs. (A) A drop of a Ca^{2+} solution beads up on a poorly ordered MHA film, whereas on a well-ordered monolayer (C), the solution readily spreads across the surface. (B) Calcite mineralization on MHA films lacking a high degree of order yields calcite crystals with a wide range of orientations, while crystallization on highly ordered MHA (D) yields primarily (012) calcite crystals. Scale bars are 2 mm, 20 μm, and 1 μm for optical, SEM, and inset SEM images, respectively. (For color version of this figure, the reader is referred to the online version of this chapter.)

spreads across the surface, this is a good indication that the SAM has been formed as expected, with the polar headgroups at or near the surface of the SAM. In contrast, if the liquid beads up on the surface, it suggests that something has gone wrong in the film formation and the sample should not be used. A comparison of the results of this approach for well versus poorly formed SAMs are shown for carboxyl-terminated monolayers in Fig. 10.2, along with SEM images of the end products of subsequent calcite crystallization on those substrates.

4. WHY GOOD SAMs GO BAD

Despite the seemingly straightforward preparation of SAMs, the formation of well-ordered films useful for crystallization can go wrong in a number of ways. The purpose of this section is to discuss the various aspects of SAM preparation, both the obvious and perhaps the more subtle, that may prevent the formation of the desired film.

4.1. Contamination

There are many potential sources of contamination, from the tools used to handle the metal substrates to the substrates and chemicals themselves. In particular, contamination from tweezers can be particularly insidious. Prior to use handling the substrates, tweezers should be thoroughly rinsed with acetone and 200 proof ethanol and dried with a diffuse nitrogen stream. It can sometimes be helpful to wipe the tweezers with an acetone-wetted cloth instead of a simple rinse, depending on the condition of the tweezers. Furthermore, when the substrates are removed from the thiol solution to be rinsed and dried, any potential residue on the tweezers can wash off onto the SAM and degrade the monolayer. To avoid this, the rinsing solvent(s) should be rinsed across the film surface towards the tweezers, and likewise, the nitrogen stream should be directed across the surface towards the tweezers.

The metal substrates can also become contaminated if left to sit in ambient laboratory conditions. As such, many groups use freshly evaporated metal films for SAM substrates. We have found that cleaning metal films that have sat in the lab for longer periods of time can also yield suitable substrates, provided the metal films have no substantial surface defects such as scratches. Cleaning the metal film can be accomplished through flame annealing the film, as referenced earlier, through plasma cleaning, or through the use of Piranha solution.

In making the monolayer, we have also found solvent purity, ethanol in particular, to sometimes pose an issue. When making thiol solutions, we use 200 proof ethanol stored in glass containers. As water can absorb into the alcohol over time, if we notice that SAMs are not forming as expected, and the other potential causes are eliminated, we switch to a new source of ethanol.

With all of the potential sources of contamination described earlier, one might think that thiol purity might be a significant factor in film quality. In our experience, however, we have not observed differences in films produced from thiols of 90% purity versus thiols of 99% purity. Love et al. suggest that SAM structures are not adversely affected by <5% disulfide impurities in thiol solutions (Love et al., 2003). Nonetheless, higher purity thiol powders minimize potential SAM defects, all other factors being equal.

4.2. Thiol solution

Some thiols, in particular those with carboxyl functionalities, do not readily disperse in the solvent solutions used in SAM preparation. This can result in suboptimal SAMs, sometimes to the degree that they are unusable. If the thiol is stored under refrigeration, warming the powder up to room temperature prior to adding it to the solvent can help the thiol more readily disperse. Furthermore, we have found that thiol freshness can affect dispersion into solution. Thiols more recently purchased disperse quicker than those that have sat in the lab for months or years. Excessively aged thiols may not disperse entirely, tend to be problematic in SAM formation, and should be replaced. Lastly, placing the container of the thiol and solvent in an ultrasonic bath for a few minutes can aid in thiol dispersion.

4.3. Film degradation

A study by Willey et al. used X-ray spectroscopy to look at the degradation of SAM quality over time (Willey et al., 2005). The results of this study show that SAMs exposed to ambient laboratory conditions over time rapidly degrade and have a much lower degree of orientational order via ozone-induced oxidation of the thiolate. Films exposed to air, but placed in small vials and capped, retained much of the quality of the originally prepared films, but do degrade to a small degree. These results suggest that SAMs should be minimally exposed to air between preparation and crystallization experiments. Furthermore if the experimental design requires the SAM to sit in air for any prolonged amount of time, it should be placed in a small, sealed container to best conserve monolayer quality. If possible, storing the SAM in an inert environment (e.g., under a nitrogen blanket) leads to less degradation of SAM order than exposure to air. Imaging SAM surfaces with an AFM is an example of where this may play an important role in SAM quality. After the substrate is removed from the thiol solution and rinsed, it must be secured with epoxy onto a chuck for use in the AFM. Epoxy cure times vary from minutes to more than an hour. During this time, the SAM can be covered with a droplet of a calcium-bearing solution, in order to prevent SAM disorganization and oxidation. Another approach would be to store the substrate and chuck under an inert atmosphere, if the epoxy does not require oxygen to cure.

5. SAM PREPARATION PROCEDURE

This discussion of SAM preparation would be incomplete without an example of monolayer preparation. Following is our recommended approach to preparing a single-species SAM that templates calcite nucleation on a uniform metal surface. This approach incorporates all of the elements of the discussion in the preceding text.

1. Evaporate (or clean) a metal film of desired thickness on an underlying substrate (e.g., Si(100)), preferably with a Cr/Ti adhesion layer.
2. Clean tweezers and other tools used to handle the substrate with acetone and 200 proof ethanol.
3. Cleave the film-bearing substrate into sizes suitable to the experiment.
4. Optically inspect the substrate piece for gross surface defects (e.g., scratches).
5. Disperse thiol powder (at room temperature) in an appropriate solvent solution (using 200 proof ethanol) to a concentration of 0.1–2.0 mM.
6. Sonicate the thiol solution if necessary to aid in dispersion.
7. Using cleaned tweezers, place the substrate face up in thiol solution and allow it to sit for 18–24 h.
8. Remove the substrate from the thiol solution and rinse with 3–5 mL thiol-less solvent solution, washing the solvent across the substrate towards the tweezers.
9. Dry the substrate in a diffuse nitrogen stream, blowing the nitrogen towards the tweezers to mitigate flow of solvent back onto the substrate surface.
10. Place a droplet of pure water or calcium-bearing solution onto the surface to check wettability. If the water/solution beads up, discard the substrate.
11. Immediately place in a crystallization solution, if possible. If not, store in a small enclosed container with the surface covered by the above water/solution.

6. SUMMARY

In this chapter, we have provided a discussion of the use of organothiol SAMs in biomimetic crystallization. We discussed a wide range of experimental approaches that have been attempted and the types of information gained from using SAMs as crystallization templates. Variations in preparation procedures for the organic films and the underlying substrates were presented, as well as a quick and qualitative method for ascertaining the quality of a prepared SAM. Following was a presentation of the myriad potential causes for SAMs to not behave as expected during experimental use and a generic procedure for SAM preparation that we have found to be highly robust.

ACKNOWLEDGMENTS

This work was supported by the Department of Defense, Air Force Office of Scientific Research, and National Defense Science and Engineering Graduate (NDSEG) Fellowship, 32 CFR 168a. Portions of this work were performed at LLNL under the auspices of the U.S. DoE by LLNL under Contract DE-AC52-07NA27344.

REFERENCES

Aizenberg, J. (2000). Patterned crystallisation on self-assembled monolayers with integrated regions of disorder. *Journal of the Chemical Society Dalton Transactions, 21*, 3963–3968.

Aizenberg, J., Black, A. J., & Whitesides, G. M. (1998). Controlling local disorder in self-assembled monolayers by patterning the topography of their metallic supports. *Nature, 394*, 868–871.

Aizenberg, J., Black, A. J., & Whitesides, G. H. (1999). Oriented growth of calcite controlled by self-assembled monolayers of functionalized alkanethiols supported on gold and silver. *Journal of the American Chemical Society, 121*, 4500–4509.

Aizenberg, J., Muller, D. A., Grazul, J. L., & Hamann, D. R. (2003). Direct fabrication of large micropatterned single crystals. *Science, 299*, 1205–1208.

Arnold, R., Azzam, W., Terfort, A., & Wöll, C. (2002). Preparation, modification, and crystallinity of aliphatic and aromatic carboxylic acid terminated self-assembled monolayers. *Langmuir, 18*, 3980–3992.

Bain, C. D., Troughton, E. B., Tao, Y. T., Evall, J., Whitesides, G. M., & Nuzzo, R. G. (1989). Formation of monolayer films by the spontaneous assembly of organic thiols from solution onto gold. *Journal of the American Chemical Society, 111*, 321.

Bain, C. D., & Whitesides, G. M. (1988). Correlations between wettability and structure in monolayers of alkanethiols adsorbed on gold. *Journal of the American Chemical Society, 110*, 3665–3666.

Han, Y. J., & Aizenberg, J. (2003). Face-selective nucleation of calcite on self-assembled monolayers of alkanethiols: Effect of the parity of the alkyl chain. *Angewandte Chemie, International Edition, 42*, 3668–3670.

Han, T. Y. J., & Aizenberg, J. (2008). Calcium carbonate storage in amorphous form and its template-induced crystallization. *Chemistry of Materials, 20*, 1064–1068.

Han, Y. J., Wysocki, L. M., Thanawala, M. S., Siegrist, T., & Aizenberg, J. (2005). Template-dependent morphogenesis of oriented calcite crystals in the presence of magnesium ions. *Angewandte Chemie, International Edition, 44*, 2386–2390.

Hegner, M., Wagner, P., & Semenza, G. (1993). Ultralarge atomically flat template-stripped Au surfaces for scanning probe microscopy. *Surfaces Science, 291*, 39–46.

Hu, Q., Nielsen, M. H., Freeman, C. L., Hamm, L. M., Tao, J., Lee, J. R. I., et al. (2012). The thermodynamics of calcite nucleation at organic interfaces: Classical vs. non-classical pathways. *Faraday Discussions, 159*, 509–523.

Ihli, J., Bots, P., Kulak, A., Benning, L. G., & Meldrum, F. C. (2013). Elucidating mechanisms of diffusion-based calcium carbonate synthesis leads to controlled mesocrystal formation. *Advanced Functional Materials*, *23*, 1965–1973. http://dx.doi.org/10.1002/adfm.201201742.

Kitano, Y. (1962). The behavior of various inorganic ions in the separation of calcium carbonate from a bicarbonate solution. *Bulletin of the Chemical Society of Japan*, *35*, 1980–1985.

Kumar, A., Abbott, N. L., Kim, E., Biebuyck, H. A., & Whitesides, G. M. (1995). Patterned self-assembled monolayers and mesoscale phenomena. *Accounts of Chemical Research*, *28*, 219–226.

Kwak, S. Y., DiMasi, E., Han, Y. J., Aizenberg, J., & Kuzmenko, I. (2005). Orientation and Mg incorporation of calcite grown on functionalized self-assembled monolayers: A synchrotron X-ray study. *Crystal Growth & Design*, *5*, 2139–2145.

Laibinis, P. E., Fox, M. A., Folkers, J. P., & Whitesides, G. M. (1991). Comparisons of self-assembled monolayers on silver and gold—Mixed monolayers derived from HS(CH2) 21X and HS(CH2)10Y (X,Y = CH3, CH2OH) have similar properties. *Langmuir*, *7*, 3167–3173.

Laibinis, P. E., & Whitesides, G. M. (1992). Omega-terminated alkanethiolate monolayers on surfaces of copper, silver, and gold have similar wettabilites. *Journal of the American Chemical Society*, *114*, 1990–1995.

Lee, J. R. I., Han, T. Y. J., Willey, T. M., Nielsen, M. H., Klivansky, L. M., Liu, Y., et al. (2013). Cooperative reorganization of mineral and template during directed nucleation of calcium carbonate. *Journal of Physical Chemistry C*, *117*, 11076–11085. http://dx.doi.org/10.1021/jp400279f.

Lee, J. R. I., Han, T. Y. J., Willey, T. M., Wang, D., Meulenberg, R. W., Nilsson, J., et al. (2007). Structural development of mercaptophenol self-assembled monolayers and the overlying mineral phase during templated CaCO3 crystallization from a transient amorphous film. *Journal of the American Chemical Society*, *129*, 10370–10381.

Li, H., & Estroff, L. A. (2007). Hydrogels coupled with self-assembled monolayers: An in vitro matrix to study calcite biomineralization. *Journal of the American Chemical Society*, *129*, 5480–5483.

Love, J. C., Estroff, L. A., Kriebel, J. K., Nuzzo, R. G., & Whitesides, G. M. (2005). Self-assembled monolayers of thiolates on metals as a form of nanotechnology. *Chemical Reviews*, *105*, 1103–1169.

Love, J. C., Wolfe, D. B., Haasch, R., Chabinyc, M. L., Paul, K. E., Whitesides, G. M., et al. (2003). Formation and structure of self-assembled monolayers of alkanethiolates on palladium. *Journal of the American Chemical Society*, *125*, 2597–2609.

Nielsen, M. H., Lee, J. R. I., Hu, Q., Han, T. Y. J., & De Yoreo, J. J. (2012). Structural evolution, formation pathways and energetic controls during template-directed nucleation of CaCO3. *Faraday Discussuions*, *159*, 105–121.

Sommerdijk, N., & de With, G. (2008). Biomimetic CaCO3 mineralization using designer molecules and interfaces. *Chemical Reviews*, *108*, 4499–4550.

Travaille, A. M., Donners, J., Gerritsen, J. W., Sommerdijk, N., Nolte, R. J. M., & van Kempen, H. (2002). Aligned growth of calcite crystals on a self-assembled monolayer. *Advanced Materials*, *14*, 492–495.

Travaille, A. M., Kaptijn, L., Verwer, P., Hulsken, B., Elemans, J., Nolte, R. J. M., et al. (2003). Highly oriented self-assembled monolayers as templates for epitaxial calcite growth. *Journal of the American Chemical Society*, *125*, 11571–11577.

Travaille, A. M., Steijven, E. G. A., Meekes, H., & van Kempen, H. (2005). Thermodynamics of epitaxial calcite nucleation on self-assembled monolayers. *The Journal of Physical Chemistry B*, *109*, 5618–5626.

Wallace, A. F., De Yoreo, J. J., & Dove, P. M. (2009). Kinetics of silica nucleation on carboxyl- and amine-terminated surfaces: Insights for biomineralization. *Journal of the American Chemical Society*, *131*, 5244–5250.

Willey, T. M., Vance, A. L., van Buuren, T., Bostedt, C., Nelson, A. J., Terminello, L. J., et al. (2004). Chemically transformable configurations of mercaptohexadecanoic acid self-assembled monolayers adsorbed on Au(111). *Langmuir, 20*, 2746–2752.

Willey, T. M., Vance, A. L., van Buuren, T., Bostedt, C., Terminello, L. J., & Fadley, C. S. (2005). Rapid degradation of alkanethiol-based self-assembled monolayers on gold in ambient laboratory conditions. *Surface Science, 576*, 188–196.

CHAPTER ELEVEN

Experimental Techniques for the Growth and Characterization of Silica Biomorphs and Silica Gardens

Matthias Kellermeier[*,†,1], **Fabian Glaab**[†], **Emilio Melero-García**[‡], **Juan Manuel García-Ruiz**[‡,1]

[*]Department of Chemistry, Physical Chemistry, University of Konstanz, Konstanz, Germany
[†]Institute of Physical and Theoretical Chemistry, University of Regensburg, Regensburg, Germany
[‡]Laboratorio de Estudios Cristalográficos, IACT (CSIC-UGR), Av. de las Palmeras 4, Armilla, Spain
[1]Corresponding authors: e-mail address: matthias.kellermeier@uni-konstanz.de; jmgruiz@ugr.es

Contents

1. Introduction — 226
2. Silica Biomorphs — 230
    2.1 Growth from solutions — 230
    2.2 Growth from gels — 233
    2.3 Morphological and structural *ex situ* analyses — 236
    2.4 *In situ* monitoring of the growth process — 238
    2.5 Probing concentrations and chemistry in solution — 239
3. Chemical Gardens — 240
    3.1 Classical preparation method — 241
    3.2 Liquid injection method — 244
    3.3 Shaping silica gardens by magnetic and electric fields — 246
    3.4 Isolation and characterization of silica gardens — 246
    3.5 Tracing dynamic diffusion and precipitation processes in silica gardens — 248
4. Conclusions and Outlook — 251
Acknowledgments — 252
References — 252

Abstract

Silica biomorphs and silica gardens are canonical examples of precipitation phenomena yielding self-assembled nanocrystalline composite materials with outstanding properties in terms of morphology and texture. Both types of structures form spontaneously in alkaline environments and rely on simple, and essentially similar, chemistry. However, the underlying growth processes are very sensitive to a range of experimental parameters, distinct preparation procedures, and external conditions. In this chapter, we report detailed protocols for the synthesis of these extraordinary biomimetic materials and identify critical aspects as well as advantages and disadvantages of different approaches. Furthermore, modifications of established standard procedures are reviewed and discussed with respect to their benefit for the control over morphogenesis and the reproducibility of the experiments in both cases. Finally, we describe currently used techniques for the characterization of these fascinating structures and devise promising ways to analyze their growth behavior and formation mechanisms *in situ* and as a function of time.

1. INTRODUCTION

During biomineralization, inorganic materials such as carbonates or silica are precipitated by living organisms as solid frameworks of virtually unlimited complexity and diversity (Addadi & Weiner, 1998; Behrens & Bäuerlein, 2009; Cusack & Freer, 2008). In most cases, the resulting mineral structures display morphologies that do not bear any obvious relation to the symmetry of the underlying crystal phase. Thus, biominerals often exhibit sinuous shapes with smooth curvatures (Thompson, 2004), which clearly distinguish them from their mostly faceted counterparts formed in nonliving, geologic settings (Garcia-Ruiz, Villasuso, Ayora, Canals, & Otalora, 2007). In addition, these biogenic architectures usually feature superior performance in terms of mechanical, optical, or magnetic properties (tailored by evolution for the respective needs of the organism) (Faivre & Schüler, 2008; Sun & Bhushan, 2012), as compared to the current state of the art with respect to synthetic materials. Therefore, numerous attempts have been made to mimic biological mineralization in the laboratory, with the aim to design novel synthetic materials with enhanced properties (Fratzl & Weiner, 2010; Meldrum & Cölfen, 2008; Nudelman & Sommerdijk, 2012). The vast majority of these biomimetic approaches employs organic additives and/or template matrices to direct crystallization of a crystalline inorganic phase, in analogy to the role of specialized organic interfaces and dissolved macromolecules in the natural process.

An alternative promising route to the fabrication of advanced materials relies on self-organization, that is, chemical cocktails that upon precipitation produce higher-order structures with textures and morphology similar to biominerals. In this context, one quite remarkable case is the formation of so-called silica biomorphs, unusual crystal aggregates that develop upon slow precipitation of alkaline-earth carbonates into silica-rich media at elevated pH (Garcia-Ruiz & Amoros, 1981; Kellermeier, Cölfen, &

Garcia-Ruiz, 2012). Interestingly, despite their purely inorganic origin (hybrids of carbonate and silica), biomorphs show elaborate morphologies with smooth curvature, like twisted ribbons as well as coral-like and flower-like patterns (see Fig. 11.1A), all being very much reminiscent of biological (i.e., living) forms (Garcia-Ruiz, Carnerup, Christy, Welham, & Hyde, 2002; Garcia-Ruiz, Hyde, et al., 2003). Beyond that, the precipitates consist of numerous individual carbonate nanocrystals, which are mutually aligned and follow a specific long-range order on the mesoscale (Hyde, Carnerup, Larsson, Christy, & Garcia-Ruiz, 2004), while each of the building units as well as the whole aggregate is sheathed by certain amounts of comineralized silica. Thus, silica appears to play the structure-directing role usually performed by complex organic compounds (Garcia-Ruiz, Melero-Garcia, & Hyde, 2009; Kellermeier, Cölfen, & Garcia-Ruiz, 2012; Kellermeier, Melero-Garcia, Glaab, et al., 2012). In light of these textural characteristics, biomorphs can be regarded as hierarchically structured nanocomposites—a feature that is also found in many biominerals (Beniash, 2011; Seto et al., 2012).

Another prominent example for the self-assembly of inorganic matter into biomimetic architectures is the formation of silica or, more generally speaking, chemical gardens. These peculiar materials are familiar to almost any (chemistry) student, and their preparation serves as an often chosen demonstration experiment in school classes (Hazlehurst, 1941). Upon reaction of solid metal salts with a solution of sodium silicate, hollow tubular structures are generated that look like trees or distinct aquatic plants (Fig. 11.1B). The tubes are composed of a mixture of amorphous (metal) silicate and partially crystalline metal hydroxide or oxide, depending on the chosen cation (Balköse et al., 2002; Glaab, Kellermeier, Kunz, Morallon, & Garcia-Ruiz, 2012; Pagano, Thouvenel-Romans, & Steinbock, 2007; Parmar et al., 2009). Indeed, chemical garden-like structures have remained the focus of scientific interest during the last century (Coatman, Thomas, & Double, 1980) due to their potential use as catalyst materials (Collins, Mokaya, & Klinowski, 1999; Pagano, Bansagi, & Steinbock, 2008) and/or microreactors (Maselko & Strizhak, 2004); their relevance for important processes like the hydration of Portland cement (Birchall, Howard, & Bailey, 1978; Double & Hellawell, 1976; Sorensen, 1981), steel corrosion and ferrotube formation (Butler & Ison, 1958; Stone & Goldstein, 2004), or the generation of deep-sea hydrothermal vents (Boyce, Coleman, & Russell, 1983; Kelley, 2005); and recurrent claims of a relation to the origin of life (Leduc, 1911; Russell, Daniel, Hall, & Sherringham, 1994). With

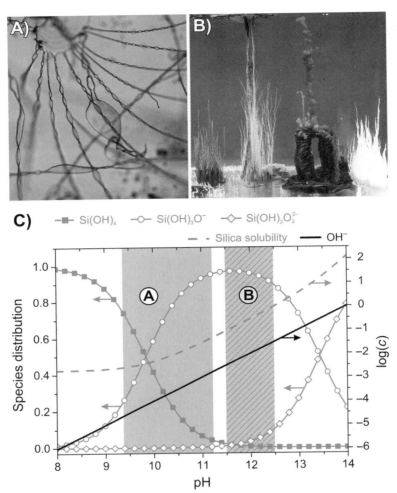

Figure 11.1 Biomimetic self-assembly in inorganic environments. (A) Silica biomorphs grown by slow crystallization of barium carbonate in a silica gel with a starting pH of 10.5. The dominant morphologies are twisted ribbons and flat sheets, which show optical birefringence when viewed between crossed polarizers. (B) Silica gardens prepared by addition of concentrated silica sol (pH ~ 11.6) to crystals of (from left to right) $CoCl_2$ (blue), $FeCl_2$ (green), $FeCl_3$ (brown-orange), and $CaCl_2$ (white). The precipitates consist of multiple hollow tubes with different thickness and shape. (C) Plot of the fractions of monomeric silica species (symbols), the total solubility of silica (dashed line), and the hydroxide concentration (full line) in solution as a function of pH. The shaded areas represent pH ranges in which the two types of structures were reported to form (Coatman et al., 1980; Eiblmeier, Kellermeier, Rengstl, et al., 2013; Melero-Garcia et al., 2009). At the moderately alkaline pH required for the growth of biomorphs (ca. 9–11), dissolved silica species carry a relatively low number of charges and are prone to condensation

regard to biomineralization, silica gardens constitute a remarkable example of spontaneous compartmentalization: during their formation, two solutions with substantially dissimilar chemical compositions and strong differences in pH are separated by a membrane (i.e., the tube wall). Recently, it has been demonstrated that these gradients in conditions lead to considerable electrochemical potentials, which persist over periods of up to days (Glaab et al., 2012). This raises fascinating questions about whether such spontaneously generated potentials could be used to drive chemical reactions and, what is more, if related processes may have contributed to the creation of complex organic molecules in prebiotic chemistry. Indeed, geochemical environments considered to be plausible for the formation of silica gardens on early Earth were also likely to contain significant amounts of reduced carbon, as a consequence of serpentinization and other mineral redox processes (Sleep, Meibom, Fridriksson, Coleman, & Bird, 2004), thus further fueling the idea of chemical gardens having acted as prebiotic batteries.

When comparing silica biomorphs and silica gardens, it becomes evident that both kinds of material arise from basically the same chemistry (i.e., metal salts and silica), but nonetheless, the underlying mechanisms of self-assembly, as well as the texture, morphology, and composition of the resulting structures, are fundamentally distinct. One of the main differences is that in biomorphs, the metal ions crystallize as carbonates and silica is only coprecipitated in minor fractions, whereas in chemical gardens, the cations directly react with silicate and hydroxide ions to yield metal silicate hydrates. This behavior originates to a large extent from the different pH at which the structures are formed, as explained in Fig. 11.1C. Another main factor is the actual silica concentration used in the experiments, which is drastically higher in the case of chemical gardens (typically 1–2 M) than during growth of biomorphs (lower mM range) (Coatman et al., 1980; Eiblmeier, Kellermeier, Rengstl, Garcia-Ruiz, & Kunz, 2013). This propels metal silicate precipitation in silica gardens and, vice versa, rationalizes why the cations prefer crystallization as carbonates in biomorphs.

---

reactions (Iler, 1979). This lowers their solubility and leads to the formation of amorphous silica next to carbonate. At the more basic conditions needed for the formation of silica gardens (>11.5), deprotonation of silicate species is enhanced and the solubility of silica increases, so that direct precipitation of SiO_2 is suppressed and the charged silicate anions rather react with the metal cations. In parallel, the higher concentrations of OH^- ions favor the formation of metal hydroxides and oxides. *(A) Reproduced with permission from Kellermeier, Cölfen, and Garcia-Ruiz (2012). Copyright 2012 John Wiley & Sons. (See color plate.)*

In the present contribution, we describe straightforward experimental procedures to grow silica biomorphs and chemical gardens in the laboratory. Moreover, we discuss methods for the characterization of the materials and outline recent progress made toward advanced *in situ* analyses of their growth behavior. For further information on theoretical and analytical aspects of the formation mechanism, the interested reader is referred to a recent review on silica biomorphs (Kellermeier, Cölfen, & Garcia-Ruiz, 2012) and corresponding studies on silica gardens (Cartwright, Garcia-Ruiz, Novella, & Otalora, 2002; Coatman et al., 1980; Collins, Zhou, Mackay, & Klinowski, 1998; Glaab et al., 2012; Pagano et al., 2008; Pantaleone et al., 2009).

2. SILICA BIOMORPHS

Unlike the complexity of the resulting structures, the preparation of biomorphs is quite straightforward and does not require any elaborate equipment and/or special conditions. In general, these biomimetic aggregates form spontaneously upon slow crystallization of alkaline-earth carbonates (typically $BaCO_3$) in silica-containing media at high pH (Kellermeier, Cölfen, & Garcia-Ruiz, 2012). This can be realized by two distinct methods, namely, in dilute silica sols and in silica gels, whereby in both cases, one of the relevant ions (Ba^{2+} and CO_3^{2-}) is usually introduced by diffusion. These different approaches are described in the following, before addressing experimental techniques by which the structure and composition of biomorphs can be investigated.

2.1. Growth from solutions

Using dilute aqueous solutions is clearly the easiest way to obtain silica biomorphs (see Fig. 11.2 for a schematic drawing of the experimental procedure). To that end, an appropriate silica sol has to be prepared first, either by dissolving solid sodium metasilicate in water or by diluting commercially available, concentrated sodium silicate stock solutions (so-called water glass). In most previous studies (e.g., Eiblmeier, Kellermeier, Rengstl, et al., 2013; Garcia-Ruiz, 1998; Garcia-Ruiz, Hyde, et al., 2003, 2009; Hyde et al., 2004; Kellermeier, Cölfen, & Garcia-Ruiz, 2012; Kellermeier, Eiblmeier, Melero-Garcia, et al., 2012; Kellermeier, Melero-Garcia, Glaab, et al., 2012), the silica concentration of the sol was chosen to be 15–20 mM, which can, for instance, be achieved by mixing 1 mL of water glass (containing

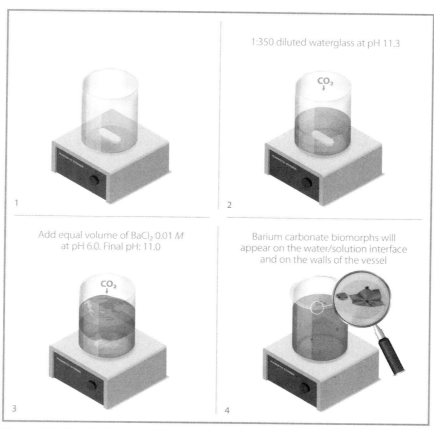

Figure 11.2 Sketch of the synthesis of biomorphs in dilute silica solutions. See text for explanations. (See color plate.)

~12.5 wt% Si, as supplied by Sigma–Aldrich) with 349 mL water (giving 16.8 mM SiO$_2$ in the sol). For the sake of reproducibility, fresh silica stock should be employed to avoid aging effects due to silica polycondensation, and, ideally, the dilute sol should be purged with nitrogen to drive out preabsorbed carbon dioxide. In a second step, the pH of the silica solution has to be adjusted to sufficiently alkaline values, typically by adding 1 mL 0.1 M NaOH to 20 mL of the as-prepared sol, which then has a pH of about 11.3 (Panel 2 in Fig. 11.2). Finally, to initiate the growth process, equal volumes of the silica mixture and BaCl$_2$ solution are combined to yield samples that effectively contain about 8.4 mM SiO$_2$ and 8.9 mM Na$^+$, which corresponds to a Na/Si molar ratio of ~1.06. The concentration of BaCl$_2$ can be varied over a quite large range (ca. 5–100 mM) without noticeable differences in the resulting structures (Hyde et al., 2004). We normally use 10 mM BaCl$_2$ solutions to obtain a final reaction mixture with 5 mM Ba^{2+} and a pH of about 11.0 (Panel 3 in Fig. 11.2).

Slow crystallization of barium carbonate then occurs upon exposure of the system to the atmosphere at ambient temperature and under quiescent conditions. With time, carbon dioxide diffuses from the air into the alkaline solutions, where it is converted to bicarbonate and carbonate ions ($CO_2 + H_2O \rightarrow H_2CO_3 \rightarrow HCO_3^- + H^+ \rightarrow CO_3^{2-} + 2H^+$), so that the supersaturation of BaCO$_3$ increases and ultimately triggers precipitation. In principle, growth of biomorphs by this method can be carried out in virtually any reaction container, as long as sufficient contact to the atmosphere is provided. However, the period required for self-assembly to be completed, as well as to some extent also the growth rate of individual biomorphs, depends intimately on the surface-to-volume ratio of the samples, which determines the flux of CO$_2$ into the system (Kellermeier, Melero-Garcia, Glaab, et al., 2012). We found that commercial polystyrene, flat-bottom multiwell plates are very suitable vessels for the synthesis of well-developed biomorphs, as they allow direct observation of the samples in an optical microscope and give fairly reproducible results. Therefore, either 24- or 6-well dishes (2.0 and 9.6 cm^2 growth area per pit) can be used, whereby each well should be filled with 2 mL or 10 mL, respectively, of the reaction mixture. Afterward, the wells are covered with the lid of the dish, which avoids rapid evaporation and contamination of the solutions, while still permitting uptake of CO$_2$ from the atmosphere.

Under these circumstances, the formation of mature silica biomorphs takes around 8–10 h (Kellermeier, Melero-Garcia, Glaab, et al., 2012). First precipitates are observed after about 2 h, floating on the surface of the samples (cf. Panel 4 in Fig. 11.2). Eventually, the fully grown crystal aggregates can be visually discerned as discrete spots scattered across the walls of the vessel or as networks of material at the solution surface. Along with the biomorphs, there is usually also a certain amount of amorphous silica present at the end of experiments, which coagulated independently of carbonate crystallization (owing to a decrease in the bulk pH during growth) and occurs as floccules or fluffy carpet on the bottom of the wells. In order to isolate the desired aggregates, the mother liquor is withdrawn and quickly replaced by water. The silica floccules are then dispersed in the supernatant by pumping with a pipette, and are finally removed. This washing step is repeated at least three times. Ultimately, ethanol is added to arrest any ongoing processes, such as superficial dissolution of the carbonate material. Extensive rinsing with ethanol and subsequent drying in air eventually gives clean and stable silica biomorphs (Kellermeier, Melero-Garcia, Glaab, et al., 2012).

We note that there are a number of modifications to this protocol reported in the literature, primarily concerning the concentrations of the reagents, the pH of the initial solution, and the silica source (Bittarello & Aquilano, 2007; Bittarello, Massaro, & Aquilano, 2009; Eiblmeier, Kellermeier, Rengstl, et al., 2013; Garcia-Ruiz, 1998; Hyde et al., 2004; Kellermeier, Cölfen, & Garcia-Ruiz, 2012; Voinescu et al., 2007). Indeed, the entire space of parameters affecting morphogenesis has still not been completely explored, but it seems clear that biomorphs can form under quite distinct conditions; for instance, structures identical to those obtained from our experiments were observed when both the pH and the silica concentration were drastically increased (Bittarello & Aquilano, 2007). However, there are also certain boundaries outside of which self-assembly cannot take place, such as a lower limit in pH (Eiblmeier, Kellermeier, Rengstl, et al., 2013; Garcia-Ruiz, 1998) and a minimum required silica content (Bittarello et al., 2009; Garcia-Ruiz, 1998; Hyde et al., 2004). One variation in the synthesis protocol that is worth mentioning refers to the actual source of carbonate: instead of using atmospheric CO_2, one can also introduce the carbonate by directly adding either $NaHCO_3$ (Bittarello et al., 2009) or Na_2CO_3 (Eiblmeier, Kellermeier, Deng, et al., 2013; Kellermeier, Melero-Garcia, Kunz, & Garcia-Ruiz, 2012) to the mother solution. In the latter case, it was found that, due to the high initial supersaturation thus imposed, amorphous $BaCO_3$ is generated immediately and becomes stabilized by a coating of silica (Eiblmeier, Kellermeier, Deng, et al., 2013). Upon subsequent equilibration, this transient precursor phase was redissolved and acted as a depot, gradually releasing $BaCO_3$ units for growth. In this way, carbonate crystallization could proceed slowly over hours and resulted in ultrastructures virtually indistinguishable from counterparts produced in conventional diffusion-based assays.

2.2. Growth from gels

Though being experimentally more demanding, the synthesis of biomorphs in gels is a valuable method for studying these materials (Bittarello & Aquilano, 2007; Garcia-Ruiz, 1985; Garcia-Ruiz et al., 2009; Imai, Terada, Miura, & Yamabi, 2002; Melero-Garcia, Santisteban-Bailon, &

Garcia-Ruiz, 2009; Terada, Yamabi, & Imai, 2003), especially when it comes to *in situ* investigations of the growth behavior. Here, the first problem is to prepare a stable and homogeneous silica hydrogel at a predefined bulk pH, as this parameter has proven to be crucial for the outcome of silica-controlled carbonate crystallization in both gels and solutions. Silica gels are most commonly obtained by lowering the pH of a sol, which will induce enhanced protonation of siliceous species ($\equiv Si - O^- + H^+ \rightarrow \equiv Si - OH$) and trigger condensation reactions ($\equiv Si - OH + HO - Si \equiv \rightarrow \equiv Si - O - Si \equiv + H_2O$) (Iler, 1979). At sufficiently high concentrations, this will result in extensive cross-linking and the formation of a continuous silica matrix. However, predicting the pH of the final gel is *per se* not trivial, because after initial acidification, the pH increases during gelation as acidic silanol (Si—OH) groups are consumed in the course of polycondensation (Voinescu et al., 2007). This issue was addressed in detail in a recent work, where the volume of hydrochloric acid added to a given volume of silica sol was correlated with the pH measured after completed gelation (Melero-Garcia et al., 2009). In essence, this and other studies (Bittarello & Aquilano, 2007; Garcia-Ruiz et al., 2009; Imai et al., 2002) have led to the conclusion that growth in gels is most efficient when a hydrogel with a starting pH of about 10.5 and a silica concentration of 0.45–0.50 M is used. This can be accomplished by diluting commercial silica stock at a ratio of 1:10 (v/v) with water and adding 3.25 mL of 1 M HCl to 10 mL of this solution under vigorous stirring (Melero-Garcia et al., 2009). When left to stand under quiescent conditions, the initially clear mixture turns turbid within few hours and ultimately transforms into a rigid gel. The gelation process can be considered completed after 5 days latest. Notably, for a gel body with spatially uniform pH to be obtained, the ~0.6 M silica solution should be equilibrated for about one month prior to acidification, in order to allow for silica particles and oligomers (present in the concentrated stock) to be entirely dissolved in the sol. Otherwise, there will be a slight gradient in pH from the top to the bottom, owing to sedimentation of the colloids in the used vessel (Melero-Garcia et al., 2009).

During gelation, the alkaline sol is covered loosely (e.g., by parafilm), so that it still has access to the atmosphere. Thus, significant amounts of CO_2 are taken up by the system on its way to the final gel state, resulting in a carbonate-containing silica matrix. Therefore, $BaCO_3$ precipitation occurs when Ba^{2+} ions are introduced into the gel. This can readily be done by placing a layer of concentrated $BaCl_2$ solution (0.25–0.50 M) on top of the gel body. Diffusion of Ba^{2+} across the gel–solution interface will then gradually increase the supersaturation of carbonate and decrease the pH, with both parameters being intimate functions of time and location in the gel (Melero-Garcia et al., 2009). Correspondingly, a range of different morphologies are found in distinct regions of the gel, with most developed crystal aggregates usually occurring at greater distances from the interface, where smoothly varying conditions are provided and ordered growth is favored. In parallel, counterdiffusion of silicate species and OH^- ions into the supernatant metal salt reservoir raises its pH (initially ca. 5.5), such that carbonate crystallization will eventually take place there, too, yielding structures similar to those formed in a regular solution synthesis.

As in the case of the experiments in solution, arbitrary vessels can be employed to grow biomorphs from gels, although very small gel bodies and nonunidirectional in-diffusion of $BaCl_2$ should be avoided. Generally, it is desirable to fabricate layers of gel that are thin enough to enable direct imaging of the embedded aggregates with the aid of a microscope. To that end, syntheses can be performed in glass capillaries of 1–2 mm thickness, ideally with rectangular geometries, which are charged with freshly acidified sol to about half of their height by means of a syringe and thin needles. After completed gelation, the second half is filled with 0.5 M $BaCl_2$ while ensuring sufficient contact between solution and gel (Kellermeier, Melero-Garcia, Glaab, et al., 2012). In an alternative approach, we have manufactured custom-designed crystallization cassettes that facilitate observation of much larger areas (Fig. 11.3). For this purpose, two identical rectangular plates (typically about 10 × 4 cm) of thin glass are used to sandwich an appropriately cut frame of rubber (1–2 mm thick), which previously has been covered with grease on both sides to seal the contact area with the plates. Further, three needles are inserted between the rubber and the plates, as depicted in Fig. 11.3 (Panel 1 in Fig. 11.3).

The whole assembly is finally fixed by applying a number of clamps around the rim, pressing together the glass plates and the rubber. After flushing the interior volume with nitrogen (Panel 2 in Fig. 11.3), the acidified silica sol is introduced to the bottom part of the as-prepared cell via the lower needle (Panel 3 in Fig. 11.3). Gelation is then allowed to proceed with the cassette left to stand up, while CO_2 diffuses into the system through the two remaining needles (cf. Panel 3 in Fig. 11.3). After gel formation is completed, the $BaCl_2$ solution is injected into the upper part of the cell (Panel 4 in Fig. 11.3), and growth begins. We note that the capillaries and the cassettes are sealed against the atmosphere after preparation (Panel 5

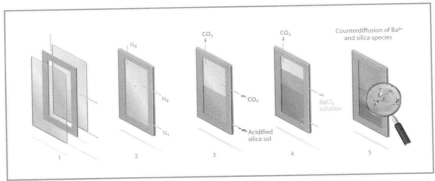

Figure 11.3 Schematic representation for the preparation of biomorphs in silica gels, utilizing custom-designed crystallization cassettes. The yellow lines indicate needles through which reagents can successively be injected. For further explanations, see text. (For interpretation of the references to color in this figure legend, the reader is referred to the online version of this chapter.)

in Fig. 11.3), so that crystallization relies solely on the carbonate predissolved in the silica sol. In both cases, precipitation of $BaCO_3$ occurs within minutes to few hours nearby the gel–solution interface, whereas the formation of well-developed biomorphs in regions ≥ 1 cm down in the gel takes between 8 h and several days (Kellermeier, Melero-Garcia, Glaab, et al., 2012; Melero-Garcia et al., 2009).

2.3. Morphological and structural *ex situ* analyses

Crystal aggregates resulting from the described syntheses are usually large enough (≥ 10 μm) to be routinely investigated by means of optical microscopy. This can be done directly in the well plates (solution method) or the capillaries/cassettes containing the gel samples as obtained (cf. Fig. 11.3). Alternatively, in the solution case, small glass coverslips can be placed on the bottom of the wells and used as growth substrates (note that biomorphs grow indistinctively on a range of different surfaces, including plastic, glass, and metal substrates). Aggregates formed thereon are washed by simple rinsing with water and ethanol and can subsequently be imaged in the microscope without any further manipulation. Likewise, the substrates (favorably pieces of conducting foil in this case) can be directly mounted on sample holders for scanning electron microscopy (SEM), thus excluding any damage to the aggregates during transfer. Successful isolation of biomorphs formed in gels is much more difficult, as the precipitates first have to be removed from the surrounding silica matrix. This can be achieved by using a fine brush and microtools to carefully detach the gel step-by-step under an optical microscope. However, this procedure requires some experience and it is often not possible to completely get rid of gel-borne silica on the surface of the aggregates, hence clearly illustrating one of the main drawbacks of the gel method.

For transmission electron microscopy (TEM), intact biomorphs are generally too thick to be directly examined. To solve this problem, one can, for example, crush the aggregates and grind them in an organic solvent like butanol to gain a fine dispersion, which then can be evaporated on a TEM grid (Kellermeier, Melero-Garcia, Glaab, et al., 2012). An obvious disadvantage of this procedure is that the particles on the grids are fragments of the original structure. This can be overcome by preparing thin sections of the crystal assemblies. For this purpose, suspensions of biomorphs are incubated with epoxy resin and cured at elevated temperatures. This gives stable blocks that can be cut into slices of about 70–100 nm in thickness by means of an ultramicrotome. In the TEM, these slices display a projection of the crystallites and their local arrangement in the native state, providing unique insight into the internal texture (Kellermeier, Melero-Garcia, Glaab, et al., 2012). In both TEM and SEM, the composition of biomorphs can further be analyzed by energy-dispersive X-ray (EDX) spectroscopy, which gives relative contents of barium carbonate and silica for individual building units (TEM) or larger regions of the entire assembly (SEM).

A very illustrative way to demonstrate the composite nature of silica biomorphs is to perform selective dissolution studies (Garcia-Ruiz et al., 2002, 2003; Imai et al., 2002; Kellermeier et al., 2009; Terada et al., 2003). For example, when aggregates consisting of an inner carbonate-rich core and an outer silica skin are treated with dilute acid, then the carbonate part will dissolve and leave a hollow silica framework (a so-called ghost) that is insoluble under these conditions. Such leaching experiments can readily be done by placing a small drop of ca. $0.03\ M$ acetic acid onto a biomorph fixed on a coverslip and observing the gradual disappearance of the carbonate domains, as well as the emerging ghost, under an optical microscope (a movie illustrating this process is shown in Supplementary Video S1, http://dx.doi.org/10.1016/B978-0-12-416617-2.00011-4). In turn, it is also possible to selectively remove the siliceous component, for example, by immersing the precipitates overnight in $1\ M$ sodium hydroxide. This will lay bare the carbonate core and often facilitates studies of the nanocrystals and their mutual arrangement in the SEM. In particular, it also helps to clean gel-grown aggregates from any associated remainders of the surrounding silica matrix.

For techniques that require larger amounts of sample (as, for instance, powder XRD or IR spectroscopy), solution-based syntheses have proven to be the method of choice, as isolation is fast and formed aggregates can be almost quantitatively recovered. However, it is important to note that the number of biomorphs generated in a typical batch is rather low and, to our knowledge, cannot be scaled up noticeably without changing their structure. In fact, from 10 mL initial reaction mixture, 1–2 mg of well-developed biomorphs can at most be retrieved (corresponding to a yield of 10–20% with respect to the amount of Ba^{2+} provided) (Kellermeier, Melero-Garcia, Glaab, et al., 2012). Therefore, in order to obtain practicable quantities of material, growth has to be carried out in multiple wells and the resulting precipitates need to be collected manually. To that end, the biomorphs adhering to the vessel walls are loosened by gently pushing with a brush, so that they accumulate at the bottom of the well. By suspension in ethanol, the aggregates can then be removed from each pit and are eventually combined to dry in air.

2.4. *In situ* monitoring of the growth process

Directly observing the formation of biomorphs in their native growth environment is a most promising way to learn more about the mechanisms underlying the self-assembly of these peculiar materials. This is where the gel method has its main strengths, as it allows most elaborate architectures to be reproducibly synthesized and can be carried out in thin layers permitting sufficient transmission of light. Thus, it is possible to watch biomorphs grow just by placing samples under an optical microscope and recording time-lapse series of images, typically at intervals ranging from 10 s to 1 min (Garcia-Ruiz et al., 2003, 2009). These may then be composed into a movie, as exemplified by Supplementary Video S2, http://dx.doi.org/10.1016/B978-0-12-416617-2.00011-4. The major challenge here is to find suitable aggregates that still are in an early stage of development and will finally evolve into shapes of interest. Generally, best results are obtained with biomorphs growing in the middle part of the gel, that is, at a distance of about 1–2 cm from the gel–solution interface, where interesting structures usually evolve between 8 and 24 h after addition of $BaCl_2$ (Melero-Garcia et al., 2009). The resulting data enable detailed studies of the phenomenological growth behavior at the micron scale and furthermore provide quantitative information on the kinetics of the growth process (Garcia-Ruiz et al., 2009; Garcia-Ruiz & Moreno, 1997; Kellermeier, Melero-Garcia, Glaab, et al., 2012). High-quality movies of biomorphs growing in solutions can be produced with confocal laser scanning microscopes, which are able to specifically focus planes nearby the bottom of the wells, where the aggregates are known to form (Kellermeier, Eiblmeier, Melero-Garcia, et al., 2012). This is also possible with inverted optical microscopes.

2.5. Probing concentrations and chemistry in solution

While imaging techniques like those discussed in the previous section can grant a unique view on structure evolution, they are restricted in a sense that only individual biomorphs are sampled and that any traced behavior may not be of general validity. Therefore, it is important to collect data from the bulk of the growth medium during mineralization as complementary information. This is more convenient in solutions than in gels, due to the easier handling and because drastic concentration gradients are not expected in this case. Considering the composition of biomorphs, there are evidently three parameters that appear worth characterizing in a time-dependent manner: the concentration of dissolved Ba^{2+}, the fractions of dissolved and precipitated silica, and the actual free amount of carbonate ions in the solution.

The most reliable way to follow changes in the concentration of Ba^{2+} in the mother solution is to measure the intensity of fluorescence induced by irradiation with X-rays, which is a noninvasive *in situ* technique that does not affect the growth process (Kellermeier, Melero-Garcia, Glaab, et al., 2012). Since biomorphs generally form attached to interfaces in solution syntheses (and not in the bulk), the fluorescence emitted from the supernatant is directly proportional to the concentration of dissolved barium. As a main shortcoming of this approach, synchrotron radiation is required in order to get sufficient signal from the dilute systems. In turn, the experimental setup is quite straightforward, as one of the commonly used polystyrene wells can directly be placed in the beam at a suitable position relative to the fluorescence detector. Concentration data obtained from such measurements show a linear decrease of dissolved Ba^{2+} with time; this behavior is perfectly in line with the constant growth rates determined by video microscopy (Eiblmeier, Kellermeier, Rengstl, et al., 2013; Kellermeier, Melero-Garcia, Glaab, et al., 2012).

The silica concentration in solution can be monitored by the so-called molybdosilicate method, in which dissolved silicate is reacted with ammonium heptamolybdate to give a yellow complex ($H_4SiO_4 \cdot 12MoO_3 \cdot xH_2O$), whose concentration can be determined spectrophotometrically at 410 nm. However, these experiments have to be carried out in a discontinuous fashion, that is, by quenching and subsequent *ex situ* analysis. In particular, the mother solutions of growing biomorphs are withdrawn after different times, passed through membrane filters to remove colloidal silica, and are then rapidly converted with the molybdate reagent and HCl to achieve a final pH of around 1.3 (see Kellermeier, Melero-Garcia, Glaab, et al., 2012 for a detailed recipe). Under these conditions, only low-molecular-weight silicate species (predominantly the mono- and the dimer) transform at significant rates into the dye, so that the resulting values reflect concentrations of "reactive" silica in solution, which do not include higher oligomers and polymers. Studies based on this methodology have shed novel light on the mode and degree of silica incorporation into the emerging crystal aggregates at distinct stages of growth (Kellermeier, Melero-Garcia, Glaab, et al., 2012).

Finally, the fractions of dissolved carbonate and bicarbonate ions can be assessed by tracing the temporal development of the bulk pH in the growth medium. This possibility is due to the fact that diffusion of CO_2 into the alkaline mixtures and its subsequent conversion to HCO_3^- and CO_3^{2-} is accompanied by a release of protons, which progressively reduces the pH. Therefore, observed changes in pH can be translated into apparent numbers of protons generated due to CO_2 uptake, from which actual concentrations of HCO_3^- and CO_3^{2-} can be calculated based on the known dissociation constants of carbonic acid, taking into account the amount of carbonate removed via $BaCO_3$ precipitation (known from simultaneous Ba^{2+} measurements) (Kellermeier, Melero-Garcia, Glaab, et al., 2012). Thus, simply by immersing a pH probe into the samples and continuously recording values during crystallization, time-dependent carbonate concentration profiles are obtained that, when combined with corresponding *in situ* Ba^{2+} data, give access to actual levels of supersaturation in the bulk. These can then be assigned to characteristic structures found under the respective circumstances and hence help to explain their formation, as has been done for range of different conditions in a recent study (Eiblmeier, Kellermeier, Rengstl, et al., 2013).

3. CHEMICAL GARDENS

In this section, we describe experimental procedures applied for the preparation and characterization of silica gardens. First, we will summarize classical approaches to the synthesis of these materials and briefly discuss variations in their composition. Subsequently, advanced methods for the growth of more tailored silica gardens will be introduced, before specifying suitable experimental techniques for both *ex situ* and *in situ* analyses. Finally, we present a new methodology that allows well-defined macroscopic silica gardens to be prepared in a reproducible manner and enables direct monitoring of dynamic chemical processes occurring in these simple precipitation systems.

3.1. Classical preparation method

The easiest way to produce chemical gardens is to place crystals of suitable metal salts into a beaker and pour alkaline silica sol on them (or, conversely, submerge the salt crystals into a large volume of the sol). This will induce the self-assembly of forms as those displayed in Fig. 11.1B (see Supplementary Video S3, http://dx.doi.org/10.1016/B978-0-12-416617-2.00011-4, for a movie of the growth process). Indeed, almost any metal salt can be used to grow such structures, provided that it contains cations that will precipitate upon reaction with hydroxide and/or silicate anions (which is essentially true for all multivalent cations). Typical cations employed for this purpose include Co^{2+}, $Fe^{2+/3+}$, Cu^{2+}, Ni^{2+}, Zn^{2+}, Mn^{2+}, Al^{3+}, and Ca^{2+}, in salts with counterions such as Cl^-, NO_3^-, or SO_4^{2-} (Balköse et al., 2002; Cartwright, Escribano, Khokhlov, & Sainz-Diaz, 2011; Cartwright, Escribano, & Sainz-Diaz, 2011; Cartwright et al., 2002; Coatman et al., 1980; Collins et al., 1998; Duan, Kitamura, Uechi, Katsuki, & Tanimoto, 2005; Parmar et al., 2009). The silica sol needed to prepare chemical gardens can readily be obtained by diluting commercial water glass (~6.25 M "SiO_2"$_{(aq)}$). In fact, the concentration of the sol (and with it, the pH) is the only crucial parameter for the morphological evolution, and an optimum structuring effect is generally observed for a certain range of silica concentrations, which may vary depending on the type of metal salt. For instance, in the case of cobalt chloride, well-developed silica gardens have been generated with 1:4 and 1:8 dilutions (v/v) of water glass, whereas toward both lower and higher silica content, either the structures became more and more ill-defined or growth did not occur at all (Coatman et al., 1980).

The choice of the cation determines the color of the resulting precipitates (cf. Fig. 11.1A), whereas corresponding morphologies do not show any systematic differences. Usually, silica gardens consist of hollow tubes that may extend over several decimeters in length, while their diameter is limited to a few millimeters and their walls are only some tens of microns in width

(Cartwright, Escribano, & Sainz-Diaz, 2011; Coatman et al., 1980; Parmar et al., 2009). These tubes grow more or less vertically from the immersed metal salt crystal, by a mechanism involving forced (osmosis) and free convection (buoyancy) as well as chemical reaction (precipitation), as illustrated schematically in Fig. 11.4 (Cartwright et al., 2002; Novella, 2000): upon addition of silica solution to solid metal salt (Panel 1), a film of hydrated metal silicate is immediately formed over the surface of the crystals

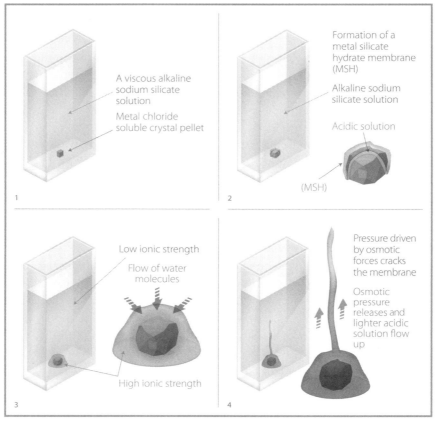

Figure 11.4 Sketch depicting the formation of silica gardens in a classical setup, where sodium silicate solution is added to random crystals or pellets of solid metal salt (1). Growth of vertical tubes occurs by the following successive events: initial dissolution of metal salt and precipitation of a metal silicate membrane (2), inflow of water through the membrane caused by osmosis (3), rupture of the membrane and ejection of concentrated metal salt solution, which rises vertically due to buoyancy and becomes solidified into a tube by instant precipitation of metal hydroxide and silica (4). (For a color version of this figure, the reader is referred to the online version of this chapter.)

(Panel 2). Then, osmosis causes water to flow through this barrier and dissolve the enclosed solid salt (Panel 3). This increases the internal pressure and leads to swelling of the flexible membrane, until a critical stage is reached and rupture occurs. At this point, the outer layer breaks, and a jet of concentrated metal salt solution is ejected into the surrounding medium (Panel 4). Owing to buoyancy, the lighter acidic salt solution ascends vertically into the heavier alkaline silica sol and becomes rapidly solidified due to simultaneous precipitation of metal hydroxide and/or silicate at the interface between the chemically distinct environments. In this way, a wall is generated around the initially liquid jet, producing the observed capillary-like tubules. The whole process can be directly visualized by Mach–Zehnder interferometry, as shown in Supplementary Video S4, http://dx.doi.org/10.1016/B978-0-12-416617-2.00011-4.

In contrast to variations of the cation, the nature of the anion has initially received much less attention, although it is well known for a long time that silicate can be replaced by species such as borates or oxalates without fundamentally changing the precipitation behavior (Hazlehurst, 1941). In the past years, interest in chemical gardens based on compounds other than silica has been fuelled, as they may give straightforward access to functional materials with specific structures. Corresponding work was not only focused on exploring other oxyanions like aluminates (Coatman et al., 1980; Collins et al., 1999; Collins, Mokaya, & Klinowski, 1999), carbonates (Maselko & Strizhak, 2004), oxalates (Baker et al., 2009), or phosphates (Maselko, Geldenhuys, Miller, & Atwood, 2003; Toth, Horvath, Smith, McMahan, & Maselko, 2007) but also included distinct inorganic species such as hexacyanoferrates, which were reacted with Fe^{3+} to give Prussian blue-type precipitates (Bormashenko, Bormashenko, Stanevsky, & Pogreb, 2006; Coatman et al., 1980). Recently, the concept of chemical gardens was even extended to organic–inorganic composite systems based on polyoxometalates and polyaromatic cations (Ritchie et al., 2009). Procedures applied for the preparation of these modified materials are generally similar to those described earlier for their siliceous counterparts, with greater or lesser modifications as detailed in the respective literature.

A first step toward an improved reproducibility of the experiments is to use pressed metal salt pellets, rather than irregular crystals (Baker et al., 2009; Cartwright, Escribano, & Sainz-Diaz, 2011; Glaab et al., 2012; Maselko et al., 2003; Maselko & Strizhak, 2004). This will affect morphogenesis for two reasons. On the one hand, the rate of salt dissolution is lower for pellets than for random microcrystals, so that the kinetics of membrane precipitation will change. On the other hand, commercially available crystals often contain significant amounts of air enclosed in voids, which become released upon dissolution and may form a gas bubble that sometimes guides tube formation in the following (see Section 3.2). This effect can be excluded in experiments with pressed pellets, especially if the tablets have been further degassed in vacuum prior to use. To prepare the pellets, a predefined mass of metal salt is ground in either an agate mortar or an electric mill for periods of at least one minute (ideally in steps of 20 s to avoid overheating) (J. Maselko, personal communication). For this purpose, salts with the highest possible amount of hydration water should be employed, in order to minimize uncontrolled uptake of the partially strongly hygroscopic substances during grinding. Finally, the homogenized powder is pressed into a tablet with uniform dimensions by means of custom-designed pellet makers at controlled pressures of typically a few bars.

3.2. Liquid injection method

Instead of using solid crystals, the metal salt can also be introduced into the silica sol by injecting concentrated solutions. This technique was developed by Steinbock and coworkers (Pagano, Thouvenel-Romans, & Steinbock, 2007; Thouvenel-Romans & Steinbock, 2003) to gain a better control over the formation of the tubular structures. In a typical experiment (Fig. 11.5A), a glass nozzle (inner diameter, 1 mm) is inserted vertically at the bottom of a glass cylinder containing ca. 1 M silica sol. The metal solution (e.g., $CuSO_4$, $c = 0.1$–0.5 M) is injected through the nozzle using a peristaltic (or syringe) pump at rates ranging from 1 to 20 mL/h. It is important to fill the nozzle and the tubing between the pump and the nozzle first with metal salt solution, before adding the silica sol to the glass cylinder. Otherwise, uncontrolled precipitation at the outlet may cause clogging of the nozzle; however, any solid deposits in the nozzle are usually removed once the injection is started (L. Roszol, personal communication). In analogy to the classical preparation method, liquid injection leads to the formation of a continuous buoyant jet of solution (Fig. 11.5B), which is progressively mineralized to yield capillaries of several hundreds of microns in width (independent of the actual nozzle diameter). As a main advantage of this technique, growth occurs in a highly controlled manner under these conditions and affords well-defined tubules with reproducible dimensions, while avoiding random processes like branching or intergrowth.

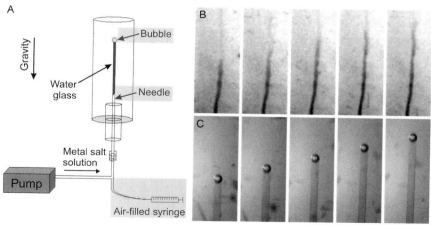

Figure 11.5 Growth of silica gardens by means of liquid injection. (A) Schematic drawing of the experimental setup. Gray-shaded parts are modifications required for the controlled introduction of a structure-guiding air bubble. (B) Time-lapse series of pictures showing the formation of a tubular structure by injection of 0.25 M cupric sulfate solution into 1 M sodium silicate sol (field of view (per panel), 14.0×4.1 mm^2). (C) Image sequence illustrating how a manually generated air bubble directs vertical growth of a perfectly straight tube upon injection of 0.5 M CuSO$_4$ into 1 M silicate solution (field of view, 7.3×3.0 mm^2). *(A) The sketch is based on a drawing in Pagano (2008). Adapted with permission of the author. (B) Reproduced with permission from the Royal Society of Chemistry from Pagano, Thouvenel-Romans, and Steinbock (2007). (C) Reproduced with permission from the Royal Society of Chemistry from Thouvenel-Romans et al. (2005).* (For a color version of this figure, the reader is referred to the online version of this chapter.)

Even greater control can be gained via a templating approach, in which an *ad hoc* generated gas bubble directs the flow of the metal salt solution and thus determines the shape of the emerging tubes (Pagano, 2008; Pagano et al., 2008; Roszol & Steinbock, 2011; Thouvenel-Romans, Pagano, & Steinbock, 2005). To that end, a needle is inserted into the nozzle and connected to an air-filled syringe (cf. gray-shaded parts in Fig. 11.5A). Salt solution and air are injected at the same time using T-shaped tubing, whereby the tip of the needle should slightly protrude from the top of the nozzle. While injecting the metal salt solution, air bubbles (typical volumes of 0.1–1 μL) are generated manually at regular intervals, until one of them gets pinned to the forming tube. The bubble then rises vertically in the silica sol, due to buoyancy, and guides the metal salt jet to give perfectly straight hollow tubules (Fig. 11.5C). The tube diameter scales in a linear fashion with the size of the bubble and hence can be tuned rather precisely within a range of 100–600 μm (Thouvenel-Romans et al., 2005). Further parameters able to affect the size and shape of the tube include the concentrations of both the metal salt solution and the silica sol: while increasing amounts of metal ions may favor budded patterns over straight tubules, higher silica contents in the sol can yield intricate twisted ribbonlike forms (Thouvenel-Romans & Steinbock, 2003). Finally, it is worth mentioning that regular silica garden tubes can also be produced in a reversed setup, that is, when silica sol is pumped from the top into an excess of metal salt solution (Pagano, 2008; Pagano, Bansagi, & Steinbock, 2007). In this case, the tubules grow along the downward stream of the heavier silicate solution and exhibit roughly similar diameters and morphologies.

3.3. Shaping silica gardens by magnetic and electric fields

Another elegant means to guide the flow of salt solution and thus control the shape of silica gardens is to perform growth under the influence of an external magnetic field. In this way, tubes with helical morphologies were prepared using distinct metal salts (e.g., $ZnSO_4$ or $MgCl_2$) in vessels that were exposed to a vertical field of ca. 15 T (Duan et al., 2005; Uechi, Katsuki, Dunin-Barkovskiy, & Tanimoto, 2004). Chiral structures were obtained regardless of the magnetism of the metal ions, and the pitch of the helices was found to decrease with increasing field strength (generally ranging from ca. 5 to 25 mm). Moreover, their handedness was not arbitrary, but could be tuned by inverting the field direction, which caused a switch from left- to right-handed (or vice versa). These observations were explained on the basis of hydrodynamic considerations and the role of the Lorentz force in directing the movement of the involved ions (Duan et al., 2005). In an alternative approach, the growth behavior of chemical gardens was regulated by means of electric fields (Ritchie et al., 2009), which again affect the flow of ions and, in some cases, can prompt forming tubules to run through multiple successive 180° turns upon changing the field polarity.

3.4. Isolation and characterization of silica gardens

In order to analyze the obtained tubular structures by conventional *ex situ* techniques, they first have to be removed from their growth environment without substantial damage. To that end, the tubes are carefully detached from any remnants of seed crystal or pellet (if applicable) and finally are withdrawn from their mother liquor using a pair of tweezers. In the case of the liquid injection method, the precipitates can also be harvested by enclosing them in a glass tube and placing a finger on the top of the tube, which allows transferring the chemical garden from the silica sol into distilled water (L. Roszol, personal communication). Typically, the tubular precipitates are rinsed several times with water to displace residual metal salt and silicate solution from the surface. In a final washing step, water can be replaced by ethanol, so as to avoid extensive dissolution of solid material and facilitate drying in air. In general, silica garden tubes are very fragile and thus have to be handled with great care. Nonetheless, manipulations are possible with the aid of a thin brush or needle or by using electrostatic forces with, for example, a charged plastic pipette tip.

The dry tube walls may then be ground to give a homogeneous powder suitable for bulk analyses by XRD as well as IR, Raman, NMR, and XPS spectroscopy or porosity measurements based on adsorption isotherms (Balköse et al., 2002; Cartwright, Escribano, Khokhlov, et al., 2011; Cartwright, Escribano, & Sainz-Diaz, 2011; Collins, Mann, et al., 1999; Collins, Mokaya, & Klinowski, 1999; Pagano, Thouvenel-Romans, & Steinbock, 2007; Parmar et al., 2009). Alternatively, in order to examine the microstructure of the walls in their native state, the intact tubes can be broken into manageable pieces, which subsequently are fixed onto SEM stubs for morphological studies or on X-ray diffraction holders for textural analyses. By varying the orientation of the membrane on the stub, the inner and outer surface of the tube walls as well as their cross section can be studied separately by SEM and EDX. In this way, it was found that the tube walls show a gradient in their composition, with silica-rich domains characterizing the outer side of the wall and an excess of metal hydroxide/oxides covering the inner surface (Balköse et al., 2002; Cartwright, Escribano, Khokhlov, et al., 2011; Cartwright, Escribano, & Sainz-Diaz, 2011; Collins, Zhou, & Klinowski, 1999; Pagano et al., 2008; Pagano, Thouvenel-Romans, & Steinbock, 2007; Parmar et al., 2009). In order to complement this information, the overall content of relevant elements can be independently determined by dissolving the tube wall in acid and measuring the concentrations of the different species with the aid of atomic emission spectroscopy (AES) (Balköse et al., 2002). Finally, the morphology and crystallinity of nano- and micron scale subunits constituting silica garden membranes can be investigated in detail by means of TEM and electron diffraction, after successful transfer of small fragments onto suitable grids (Collins, Zhou, & Klinowski, 1999; Pagano et al., 2008; Pagano, Thouvenel-Romans, & Steinbock, 2007; Parmar et al., 2009).

Despite the insight gained into the nature of chemical gardens via *ex situ* methods, a more promising way to study the formation of these structures is to follow the growth process *in situ*. Most frequently, this has been achieved by acquiring time-lapse pictures of the emerging structures, using standard photographic equipment. Such experiments should preferably be carried out in flat rectangular cells (similar to those shown in Fig. 11.3) to improve the quality of the images. On that basis, the morphological evolution of the system can be monitored and quantified by measuring time-dependent tube dimensions. This sheds light on precipitation kinetics and allows distinct growth regimes (e.g., jetting or budding) to be identified (Bormashenko et al., 2006; Cartwright, Escribano, & Sainz-Diaz, 2011; Pagano, Bansagi, & Steinbock, 2007; Thouvenel-Romans et al., 2005; Thouvenel-Romans & Steinbock, 2003). Another interesting class of experiments is dedicated to the role of osmotic pressures and membrane rupture events during the formation of silica gardens. For example, the osmotically forced inflow of water from the silica sol into the inner metal salt compartment was investigated by placing a metal silicate-based membrane in between the two solutions and tracing the volume change on the metal salt side (Coatman et al., 1980). In turn, the circumstances leading to membrane rupture were studied by pumping metal salt solution into silica sol and measuring the pressure at the inlet as a function of time (Pantaleone et al., 2009). Both experiments showed that the self-assembly of silica gardens is driven by consecutive cycles of osmotic pressure increase and relief.

3.5. Tracing dynamic diffusion and precipitation processes in silica gardens

Beyond simple monitoring of the evolution of these fascinating structures, a recurrent main goal has been to track time-dependent variations in chemical conditions during growth and aging of silica gardens *in situ*. However, a limiting factor in this context is the diameter of the tubes, which usually reaches some hundreds of microns at most (Pagano et al., 2008). Therefore, the inner metal salt compartment cannot be easily accessed, rendering direct analyses of solution chemistry difficult. This problem was solved in a recent study, where a new experimental setup was developed that allows single macroscopic tubes to be obtained in a very reproducible fashion (Glaab et al., 2012). The method relies on the use of uniform pressed salt pellets and the controlled and slow addition of silica sol to these pellets, as illustrated by Fig. 11.6A (see Supplementary Movie S5, http://dx.doi.org/10.1016/B978-0-12-416617-2.00011-4, for a video of the growth process). Under

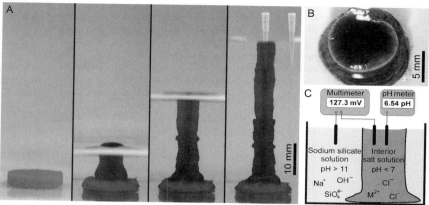

Figure 11.6 New preparation method for silica gardens. (A) Image sequence showing the formation of a macroscopic tube from a pellet of CoCl$_2$·6H$_2$O by controlled dosing of silica sol. Pipette tips indicate that samples can easily be drawn from both the inner solution enclosed by the membrane and the outer reservoir surrounding it. (B) Top view of the tube, demonstrating that it has an open end. (C) Schematic representation of the experimental setup used for characterizing the temporal evolution of the macroscopic silica garden system. Relevant species in the inner and outer solution are indicated together with the devices used for measuring potential differences and pH. *Reproduced with permission from Glaab et al. (2012). Copyright 2012 John Wiley & Sons.* (See color plate.)

suitable conditions, this procedure affords tubes with lengths up to 5 cm and diameters of around 1 cm. Moreover, the tubes have an open end on top rather than being closed (Fig. 11.6B), so that the solutions on both sides of the membrane can readily be sampled (as indicated by the pipette tips in Fig. 11.6A).

For the preparation of such large silica garden tubes, ca. 0.5 g of dried metal salt (typically CoCl$_2$·6H$_2$O or CoI$_2$·H$_2$O) is first ground to a fine powder, which is subsequently transferred to a hydraulic press and converted to a tablet of about 13 mm in diameter and 2 mm in height by applying a pressure of 3.5 bar for several minutes. The pellet is then fixed at the bottom of a beaker (total volume of 120 mL) by using a piece of double-sided adhesive tape. To initiate growth, 10 mL of silica sol (1:4 dilution of water glass) is added within around 10 s over the rim of the beaker through a needle attached to a syringe. This leads to gradual dissolution of the metal salt pellet, which soon becomes visibly covered by a layer of hydrated silicate. At this point, further silica sol is added, now however by means of an automated dosing device that dispenses the sol at a constant rate of 2 mL/min. In the following, the tube grows continuously, until it has reached its maximum length after about 15 min. Similar structures can be obtained in the same way for other metal salts like $FeCl_2$ or $FeCl_3$ (Glaab, 2011), although addition rates have to be adjusted if the cation is changed (generally 1–10 mL/min). It is worth mentioning that after dosing of silica sol has been completed and macroscopic growth essentially ceased, continued inflow of water can cause an overflow of metal salt solution at the open top of the tube, leading to uncontrolled outgrowths along the air–liquid interface. This unwanted process can be avoided by removing small volumes of the inner solution after completed addition, giving uniform vertical tubes as shown in Fig. 11.6A. The as-formed precipitates are relatively stable and can be isolated into dry state without destroying their ultrastructure. For this purpose, the outer silica sol is withdrawn with a syringe, whereas the remaining inner metal salt solution can be carefully sucked out of the tube utilizing a thin needle. After washing with water and ethanol, the dried membranes can be characterized as described in the previous section.

The major advantage of this new synthesis method is that it offers the possibility to monitor ongoing diffusion and precipitation processes by measuring distinct physicochemical parameters as a function of aging time after completed tube preparation (Fig. 11.6C). For instance, the concentrations of ions dissolved in the inner and outer solution can be determined by drawing aliquots from the two reservoirs and analyzing them by means of AES (Glaab et al., 2012). As the amount of liquid inside the tube is rather limited (usually about 0.2–0.5 mL), only small volumes (e.g., 10 µL) should be taken, whereas larger samples (100 µL) can be drawn from the outer sol. Subsequently, the samples are diluted with 10 mL water and characterized by AES. This yields precise concentrations for all relevant elements in both compartments at any given time. In essence, these studies have revealed that the development of silica gardens is not at all terminated once macroscopic tube growth is completed and that ion diffusion and precipitation remain active over time frames as long as tens of hours, due to the enormous concentration gradients established during initial separation of the two compartments. Moreover, the collected concentration data provide clear evidence that the walls of silica gardens cannot be considered semipermeable membranes, as commonly believed, but rather combine properties of both diaphragms and membranes, which allow multiple ionic species to diffuse through them with time (Glaab et al., 2012).

Instead of discontinuous sample drawing as required for AES measurements, the evolution of silica gardens can also be directly followed *in situ* by introducing suitable sensors into the inner and/or outer solution. Owing to the large diameter of the synthesized tubes and its open end, it is, for example, possible to immerse a pH microprobe in the metal salt reservoir and continuously collect data (cf. Fig. 11.6C). Again, valuable information can be derived concerning the progress of precipitation and diffusion, in particular because the pH determines the solubility of both metal hydroxides/oxides and silica (Glaab et al., 2012). Another parameter worth to be investigated in light of the drastic concentration gradients across the tube wall is the electrochemical potential difference between the two compartments. Such measurements merely require two platinum stick electrodes, which are submerged in the solutions inside and outside the membrane (cf. Fig. 11.6C). By connecting the electrodes to a conventional multimeter, potentials can directly be recorded in a time-resolved manner. Corresponding data indicate that the compartmentalization occurring during the early stages of silica garden formation generates considerable electrochemical potential differences with initial values as high as 100–200 mV (Barge et al., 2012; Glaab et al., 2012). The measured voltage then decreases with time in several steps, which immediately correlate with the progress of ion diffusion and precipitation. Overall, these findings indicate that silica gardens are complex systems that run through a cascade of dynamic coupled processes before ultimately returning to equilibrium. During this period, significant potential differences are created spontaneously and maintained over hours, which might be promising for the use of silica gardens as self-catalyzed chemical reactors and/or batteries (Glaab, 2011).

4. CONCLUSIONS AND OUTLOOK

Silica biomorphs and chemical gardens represent excellent laboratory model systems for the study of self-organization phenomena in biomineralization and related disciplines, because they grow spontaneously and by the interaction of very simple, cheap, and readily available compounds. However, the formation of these extraordinary structures requires certain conditions to be met, which have been described in detail in this chapter. Furthermore, we have reviewed procedures for the preparation of these materials, both classical and advanced in terms of control over morphogenesis, and discussed multiple ways to analyze the resulting precipitates.

In our opinion, it is crucial to perform further time-dependent *in situ* experiments as those discussed here to gain a more profound picture of the morphological and structural evolution, as well as the chemistry and kinetics of precipitation, in both silica gardens and biomorphs. For example, online imaging of the growth processes should not be limited to conventional light microscopy, but extended to other methods with higher resolution, such as X-ray microscopy (Rieger, Thieme, & Schmidt, 2000) or even TEM (Li et al., 2012), which are capable of tracing the development of the structures at the nanoscale. Another potentially interesting *in situ* technique is microfocus X-ray diffraction, which can deliver detailed textural information from localized regions and permit time-resolved scans of the progress of crystallization (Kellermeier, Cölfen, & Garcia-Ruiz, 2012). Regarding the chemistry and dynamics of the systems, it would be highly desirable to find further ways to monitor variations of conditions over time and *in situ*. One approach might be to use ion-selective electrodes for continuous measurements of metal ion concentrations, ideally with miniaturized probes that can be directly immersed into the tubes of silica gardens.

Despite many years of research, much remains to be discovered about biomorphs and chemical gardens. We hope that this contribution will foster research in this area and that the preparation and characterization methods summarized in this chapter will help to enable a deeper understanding of these peculiar materials in the future.

ACKNOWLEDGMENTS

The authors thank Jerzy Maselko (University of Alaska), Laszlo Roszol and Oliver Steinbock (both Florida State University) for providing detailed information on the use of solid pellets and the liquid injection method for the preparation of silica gardens. We are further grateful to Werner Kunz and Josef Eiblmeier (both University of Regensburg), Stephen Hyde (Australian National University), Luis David Patiño (LEC Granada), Emilia Morallon (University of Alicante), Luis Gago (University of Vigo), and Helmut Cölfen (University of Konstanz) for the fruitful discussions and their participation in some of the experiments. Financial support by BASF SE (M. K.) and the MICINN projects "Factoría de Cristalización," CSD2006-00015, Consolider-Ingenio 2010, and CGL2010-16882 (J. M. G. R.) is greatly appreciated.

REFERENCES

Addadi, L., & Weiner, S. (1998). Control and design principles in biological mineralization. *Angewandte Chemie, International Edition, 31*, 153–169.
Baker, A., Toth, A., Horvath, D., Walkush, J., Ali, A. S., Morgan, W., et al. (2009). Precipitation pattern formation in the copper(II) oxalate system with gravity flow and axial symmetry. *The Journal of Physical Chemistry, A, 113*, 8243–8248.
Balköse, D., Özkan, F., Köktürk, U., Ulutan, S., Ülkü, S., & Nisli, G. (2002). Characterization of hollow chemical garden fibers from metal salts and water glass. *Journal of Sol-Gel Science and Technology, 23*, 253–263.

Barge, L. M., Doloboff, I. J., White, L. M., Stucky, G. D., Russell, M. J., & Kanik, I. (2012). Characterization of iron-phosphate-silicate chemical garden structures. *Langmuir, 28,* 3714–3721.

Behrens, P., & Bäuerlein, E. (2009). *Handbook of biomineralization.* Weinheim: Wiley-VCH.

Beniash, E. (2011). Biominerals—Hierarchical nanocomposites: The example of bone. *Wiley Interdisciplinary Reviews Nanomedicine and Nanobiotechnology, 3,* 47–69.

Birchall, J. D., Howard, A. J., & Bailey, J. E. (1978). On the hydration of Portland cement. *Proceedings of the Royal Society A, 360,* 445–453.

Bittarello, E., & Aquilano, D. (2007). Self-assembled nanocrystals of barium carbonate in biomineral-like structures. *European Journal of Mineralogy, 19,* 345–351.

Bittarello, E., Massaro, F. R., & Aquilano, D. (2009). The epitaxial role of silica groups in promoting the formation of silica/carbonate biomorphs: A first hypothesis. *Journal of Crystal Growth, 312,* 402–412.

Bormashenko, E., Bormashenko, Y., Stanevsky, O., & Pogreb, R. (2006). Evolution of chemical gardens in aqueous solutions of polymers. *Chemical Physics Letters, 417,* 341–344.

Boyce, A. J., Coleman, M. L., & Russell, M. J. (1983). Formation of fossil hydrothermal chimneys and mounds from Silvermines, Ireland. *Nature, 306,* 545–550.

Butler, G., & Ison, H. C. K. (1958). An unusual form of corrosion product. *Nature, 182,* 1229–1230.

Cartwright, J. H. E., Escribano, B., Khokhlov, S., & Sainz-Diaz, C. I. (2011). Chemical gardens from silicates and cations of group 2: A comparative study of composition, morphology and microstructure. *Physical Chemistry Chemical Physics, 13,* 1030–1036.

Cartwright, J. H. E., Escribano, B., & Sainz-Diaz, C. I. (2011). Chemical-garden formation, morphology, and composition. I. Effect of the nature of the cations. *Langmuir, 27,* 3286–3293.

Cartwright, J. H. E., Garcia-Ruiz, J. M., Novella, M. L., & Otalora, F. (2002). Formation of chemical gardens. *Journal of Colloid and Interface Science, 256,* 351–359.

Coatman, R. D., Thomas, N. L., & Double, D. D. (1980). Studies of the growth of "silicate gardens" and related phenomena. *Journal of Materials Science, 15,* 2017–2026.

Collins, C., Mann, G., Hoppe, E., Duggal, T., Barr, T. L., & Klinowski, J. (1999). NMR and ESCA studies of the "silica garden" Bronsted acid catalyst. *Physical Chemistry Chemical Physics, 1,* 3685–3687.

Collins, C., Mokaya, R., & Klinowski, J. (1999). The "silica garden" as a Bronsted acid catalyst. *Physical Chemistry Chemical Physics, 1,* 4669–4672.

Collins, C., Zhou, W., & Klinowski, J. (1999). A unique structure of $Cu_2(OH)_3 \cdot NH_3$ crystals in the 'silica garden' and their degradation under electron beam irradiation. *Chemical Physics Letters, 306,* 145–148.

Collins, C., Zhou, W., Mackay, A. L., & Klinowski, J. (1998). The 'silica garden': A hierarchical nanostructure. *Chemical Physics Letters, 286,* 88–92.

Cusack, M., & Freer, A. (2008). Biomineralization: Elemental and organic influence in carbonate systems. *Chemical Reviews, 108,* 4433–4454.

Double, D. D., & Hellawell, A. (1976). The hydration of Portland cement. *Nature, 261,* 486–488.

Duan, W., Kitamura, S., Uechi, I., Katsuki, A., & Tanimoto, Y. (2005). Three-dimensional morphological chirality induction using high magnetic fields in membrane tubes prepared by a silicate garden reaction. *The Journal of Physical Chemistry. B, 109,* 13445–13450.

Eiblmeier, J., Kellermeier, M., Deng, M., Kienle, L., Garcia-Ruiz, J. M., & Kunz, W. (2013). Bottom-up self-assembly of amorphous core-shell-shell nanoparticles and biomimetic crystal forms in inorganic silica-carbonate systems. *Chemistry of Materials, 25,* 1842–1851.

Eiblmeier, J., Kellermeier, M., Rengstl, D., Garcia-Ruiz, J. M., & Kunz, W. (2013). Effect of bulk pH and supersaturation on the growth behavior of silica biomorphs in alkaline solutions. *Crystal Engineering Communications, 15*, 43–53.

Faivre, D., & Schüler, D. (2008). Magnetotactic bacteria and magnetosomes. *Chemical Reviews, 108*, 4875–4898.

Fratzl, P., & Weiner, S. (2010). Bio-inspired materials—Mining the old literature for new ideas. *Advanced Materials, 22*, 4547–4550.

Garcia-Ruiz, J. M. (1985). On the formation of induced morphology crystal aggregates. *Journal of Crystal Growth, 73*, 251–262.

Garcia-Ruiz, J. M. (1998). Carbonate precipitation into alkaline silica-rich environments. *Geology, 26*, 843–846.

Garcia-Ruiz, J. M., & Amoros, J. L. (1981). Morphological aspects of some symmetrical crystal aggregates grown by silica gel technique. *Journal of Crystal Growth, 55*, 379–383.

Garcia-Ruiz, J. M., Carnerup, A., Christy, A. G., Welham, N. J., & Hyde, S. T. (2002). Morphology: An ambiguous indicator of biogenicity. *Astrobiology, 2*, 353–369.

Garcia-Ruiz, J. M., Hyde, S. T., Carnerup, A. M., Christy, A. G., Van Kranendonk, M. J., & Welham, N. J. (2003). Self-assembled silica-carbonate structures and detection of ancient microfossils. *Science, 302*, 1194–1197.

Garcia-Ruiz, J. M., Melero-Garcia, E., & Hyde, S. T. (2009). Morphogenesis of self-assembled nanocrystalline materials of barium carbonate and silica. *Science, 323*, 362–365.

Garcia-Ruiz, J. M., & Moreno, A. (1997). Growth behaviour of twisted ribbons of barium carbonate/silica self-assembled ceramics. *Anales de Quimica International Edition, 93*, 1–2.

Garcia-Ruiz, J. M., Villasuso, R., Ayora, C., Canals, A., & Otalora, F. (2007). Formation of natural gypsum megacrystals in Naica, Mexico. *Geology, 35*, 327–330.

Glaab, F. (2011). In-situ examination of diffusion and precipitation processes during the evolution of chemical garden systems. PhD thesis, University of Regensburg.

Glaab, F., Kellermeier, M., Kunz, W., Morallon, E., & Garcia-Ruiz, J. M. (2012). Formation and evolution of chemical gradients and potential differences across self-assembling inorganic membranes. *Angewandte Chemie, International Edition, 124*, 4393–4397.

Hazlehurst, T. H. (1941). Structural precipitates: The silicate garden type. *Journal of Chemical Education, 18*, 286–289.

Hyde, S. T., Carnerup, A. M., Larsson, A. K., Christy, A. G., & Garcia-Ruiz, J. M. (2004). Self-assembly of carbonate-silica colloids: Between living and non-living form. *Physica A, 339*, 24–33.

Iler, R. K. (1979). *The chemistry of silica. Solubility, polymerization, colloid and surface properties, and biochemistry*. New York: John Wiley & Sons.

Imai, H., Terada, T., Miura, T., & Yamabi, S. (2002). Self-organized formation of porous aragonite with silicate. *Journal of Crystal Growth, 244*, 200–205.

Kellermeier, M., Cölfen, H., & Garcia-Ruiz, J. M. (2012). Silica biomorphs: Complex hybrid materials from "sand and chalk" *European Journal of Inorganic Chemistry, 32*, 5123–5144.

Kellermeier, M., Eiblmeier, J., Melero-Garcia, E., Pretzl, M., Fery, A., & Kunz, W. (2012). Evolution and control of complex curved form in simple inorganic precipitation systems. *Crystal Growth and Design, 12*, 3647–3655.

Kellermeier, M., Glaab, F., Carnerup, A. M., Drechsler, M., Gossler, B., Hyde, S. T., et al. (2009). Additive-induced morphological tuning of self-assembled silica-barium carbonate crystal aggregates. *Journal of Crystal Growth, 311*, 2530–2541.

Kellermeier, M., Melero-Garcia, E., Glaab, F., Eiblmeier, J., Kienle, L., Rachel, R., et al. (2012). Growth behavior and kinetics of self-assembled silica-carbonate biomorphs. *Chemistry—A European Journal, 18*, 2272–2282.

Kellermeier, M., Melero-Garcia, E., Kunz, W., & Garcia-Ruiz, J. M. (2012). The ability of silica to induce biomimetic crystallization of calcium carbonate. *Advances in Chemical Physics, 151*, 277–307.

Kelley, D. S. (2005). From the mantle to microbes: The Lost City hydrothermal field. *Oceanography, 18*, 32–45.

Leduc, S. (1911). *The mechanism of life*. London: Rebman.

Li, D., Nielsen, M. H., Lee, J. R. I., Frandsen, C., Banfield, J. F., & De Yoreo, J. J. (2012). Direction-specific interactions control crystal growth by oriented attachment. *Science, 336*, 1014–1018.

Maselko, J., Geldenhuys, A., Miller, J., & Atwood, D. (2003). Self-construction of complex forms in a simple chemical system. *Chemical Physics Letters, 373*, 563–567.

Maselko, J., & Strizhak, P. (2004). Spontaneous formation of cellular chemical system that sustains itself far from thermodynamic equilibrium. *The Journal of Physical Chemistry B, 108*, 4937–4939.

Meldrum, F. C., & Cölfen, H. (2008). Controlling mineral morphologies and structures in biological and synthetic systems. *Chemical Reviews, 108*, 4332–4432.

Melero-Garcia, E., Santisteban-Bailon, R., & Garcia-Ruiz, J. M. (2009). Role of bulk pH witherite biomorph growth in silica gels. *Crystal Growth and Design, 9*, 4730–4734.

Novella, M. L. (2000). Caracterización por interferometría Mach-Zehnder de procesos de crecimiento de cristales a partir de soluciones. PhD Thesis, University of Granada.

Nudelman, F., & Sommerdijk, N. A. J. M. (2012). Biomineralization as an inspiration for materials chemistry. *Angewandte Chemie, International Edition, 51*, 6582–6596.

Pagano, J. J. (2008). Growth dynamics and composition of tubular structures in a reaction-precipitation system. PhD thesis, Florida State University.

Pagano, J. J., Bansagi, T., & Steinbock, O. (2007). Tube formation in reverse silica gardens. *The Journal of Physical Chemistry C, 111*, 9324–9329.

Pagano, J. J., Bansagi, T., & Steinbock, O. (2008). Bubble-templated and flow-controlled synthesis of macroscopic silica tubes supporting zinc oxide nanostructures. *Angewandte Chemie, International Edition, 47*, 9900–9903.

Pagano, J. J., Thouvenel-Romans, S., & Steinbock, O. (2007). Compositional analysis of copper-silica precipitation tubes. *Physical Chemistry Chemical Physics, 9*, 110–116.

Pantaleone, J., Toth, A., Horvath, D., RoseFigura, L., Morgan, W., & Maselko, J. (2009). Pressure oscillations in a chemical garden. *Physical Review E, 79*, 056221–056228.

Parmar, K., Chaturvedi, H. T., Akhtar, M. W., Chakravarty, S., Das, S. K., Pramanik, A., et al. (2009). Characterization of cobalt precipitation tube synthesized through "silica garden" route. *Materials Characterization, 60*, 863–868.

Rieger, J., Thieme, J., & Schmidt, C. (2000). Study of precipitation reactions by X-ray microscopy: $CaCO_3$ precipitation and the effect of polycarboxylates. *Langmuir, 16*, 8300–8305.

Ritchie, C., Cooper, G. J. T., Song, Y. F., Streb, C., Yin, H., Parenty, A. D. C., et al. (2009). Spontaneous assembly and real-time growth of micrometre-scale tubular structures from polyoxometalate-based inorganic solids. *Nature Chemistry, 1*, 47–52.

Roszol, L., & Steinbock, O. (2011). Controlling the wall thickness and composition of hollow precipitation tubes. *Physical Chemistry Chemical Physics, 13*, 20100–20103.

Russell, M. J., Daniel, R. M., Hall, A. J., & Sherringham, J. A. (1994). A hydrothermally precipitated catalytic iron sulphide membrane as a first step toward life. *Journal of Molecular Evolution, 39*, 231–243.

Seto, J., Ma, Y., Davis, S. A., Meldrum, F. C., Gourrier, A., Kim, Y. Y., et al. (2012). Structure-property relationships of a biological mesocrystal in the adult sea urchin spine. *Proceedings of the National Academy of Sciences of the United States of America, 109*, 3699–3704.

Sleep, N. H., Meibom, A., Fridriksson, T., Coleman, R. G., & Bird, D. K. (2004). H2-rich fluids from serpentinization: Geochemical and biotic implications. *Proceedings of the National Academy of Sciences of the United States of America, 101*, 12818–12823.

Sorensen, T. S. (1981). A theory of osmotic instabilities of a moving semipermeable membrane: Preliminary model for the initial stages of silicate garden formation and of Portland cement hydration. *Journal of Colloid and Interface Science, 79*, 192–208.

Stone, D. A., & Goldstein, R. E. (2004). Tubular precipitation and redox gradients on a bubbling template. *Proceedings of the National Academy of Sciences of the United States of America, 101*, 11537–11541.

Sun, J., & Bhushan, B. (2012). Hierarchical structure and mechanical properties of nacre: A review. *RSC Advances, 2*, 7617–7632.

Terada, T., Yamabi, S., & Imai, H. (2003). Formation process of sheets and helical forms consisting of strontium carbonate fibrous crystals with silicate. *Journal of Crystal Growth, 253*, 435–444.

Thompson, D. W. (2004). *On growth and form*. Cambridge: Cambridge University Press.

Thouvenel-Romans, S., Pagano, J. J., & Steinbock, O. (2005). Bubble guidance of tubular growth in reaction-precipitation systems. *Physical Chemistry Chemical Physics, 7*, 2610–2615.

Thouvenel-Romans, S., & Steinbock, O. (2003). Oscillatory growth of silica tubes in chemical gardens. *Journal of American Chemical Society, 125*, 4338–4341.

Toth, A., Horvath, D., Smith, R., McMahan, J. R., & Maselko, J. (2007). Phase diagram of precipitation morphologies in the Cu^{2+}-PO_4^{3-} system. *The Journal of Physical Chemistry C, 111*, 14762–14767.

Uechi, I., Katsuki, A., Dunin-Barkovskiy, L., & Tanimoto, Y. (2004). 3D-Morphological chirality induction in zinc silicate membrane tube using a high magnetic field. *The Journal of Physical Chemistry. B, 108*, 2527–2530.

Voinescu, A. E., Kellermeier, M., Carnerup, A. M., Larsson, A. K., Touraud, D., Hyde, S. T., et al. (2007). Co-precipitation of silica and alkaline-earth carbonates using TEOS as silica source. *Journal of Crystal Growth, 306*, 152–158.

CHAPTER TWELVE

Precipitation in Liposomes as a Model for Intracellular Biomineralization

Chantel C. Tester, Derk Joester[1]
Northwestern University, Evanston, Illinois, USA
[1]Corresponding author: e-mail address: d-joester@northwestern.edu

Contents

1. Introduction — 258
2. Model Systems for Studying the Influence of Confinement on Biomineralization — 260
3. Preparation of Liposomes — 261
   3.1 Lipid selection — 262
   3.2 Preparation of liposomes by extrusion or sonication — 263
   3.3 Preparation of giant liposomes — 265
   3.4 Encapsulation efficiency — 267
   3.5 Exchange of ions in the rehydration medium — 267
4. Precipitation Inside Liposomes — 268
5. Characterization Methods — 270
6. Conclusions — 273
Acknowledgments — 274
References — 274

Abstract

Liposomes present a versatile platform to model intracellular, biologically controlled mineralization. Perhaps, most importantly, precipitation in the confinement of liposomes excludes heterogeneous nucleators that facilitate formation of the thermodynamically most stable crystalline phase in bulk. This provides access to metastable amorphous precursors even in the absence of other additives that interact strongly with the mineral and is fundamental to the capability of cells to prevent spurious nucleation and to select a specific polymorph. Herein, we summarize methods to prepare liposomes from the nanometer to micron length scale and review strategies to carry out precipitation reactions of iron oxide, calcium carbonate, and calcium phosphate in the confinement of such liposomes. In addition, we discuss methods to characterize the morphology, structure, and growth kinetics of crystalline and amorphous precipitates, with particular emphasis on *in situ* characterization approaches.

1. INTRODUCTION

A widespread strategy to assert control over the nucleation, shape, and location of biominerals in mineralized tissues is to initiate precipitation in "privileged environments" (Lowenstam & Weiner, 1989). Intracellular compartmentalization occurs from the nanometer to micrometer length scale, using phospholipid-bilayer membranes that act as selective diffusion barriers. These membrane-delimited compartments are part of the endomembrane system of the cell and can be involved in accumulation, storage, detoxification, and transport of ions or may serve as the final mineralization site. Examples include barium sulfate precipitation in the terminal vacuoles of desmids (Krejci, Finney, Vogt, & Joester, 2011; Krejci, Wasserman, et al., 2011), formation of highly hierarchically structured valves in the silica deposition vesicle in diatoms (Coradin & Lopez, 2003), and uptake and transport of mineral building blocks in seawater vacuoles in foraminifera (Bentov, Brownlee, & Erez, 2009). A more complex form of extracellular compartmentalization is observed, for instance, in mollusk shells, where the privileged environment is created by the cells of the mantle epithelium on one side and the acellular periostracum or previously deposited mineral on the other (Addadi, Joester, Nudelman, & Weiner, 2006).

In its simplest form, the privileged environment need not be more complex than a phospholipid-bilayer vesicle that concentrates ions for deposition at the site of mineral growth (Weiner & Addadi, 2011). However, within some intracellular mineral vesicles, the ions are stored not in aqueous solution, but in the form of a metastable amorphous precursor. For example, during spiculogenesis in the sea urchin embryo, vesicles loaded with amorphous calcium carbonate (ACC) are found in the cells adjacent to regions of active spicule growth (Beniash, Addadi, & Weiner, 1999). Membrane-bound amorphous calcium phosphate (ACP) is present in the cells responsible for the growth and development of zebrafish fins and mouse calvarial bone (Mahamid et al., 2011). Crystallization occurs only after the mineral-bearing vesicles are secreted into the extracellular collagenous matrix. While the mechanism of stabilization of these amorphous precursors is not well understood, it is thought that the phospholipid membrane may prevent crystallization by confinement or direct contact with the mineral. It follows that exocytosis of the amorphous mineral and removal of the phospholipid membrane may serve as a trigger for crystallization.

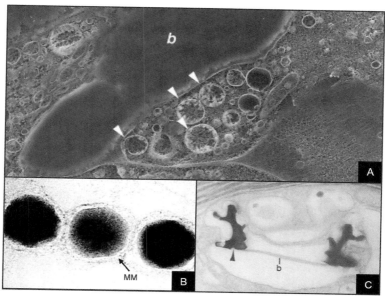

Figure 12.1 (A) Cryo-SEM image of a cryo-sectioned calvarial bone displaying multiple intracellular mineral-containing vesicles (white arrowheads) in cells adjacent to mineralizing bone (b, white) (Mahamid et al., 2011). *Reproduced with permission from Elsevier.* (B) TEM image of a sectioned magnetotactic bacteria containing magnetite crystals growing within the vesicles of the magnetosome membrane (MM) (Abreu et al., 2013; Faivre & Schuler, 2008). *Reproduced with permission from the American Chemical Society.* (C) TEM section of a mature coccolith within its vesicle. Nucleation is thought to occur on the coencapsulated organic base plate (b, black) (Marsh, 1999). *Reproduced with permission from Springer Science and Business Media.*

Membrane-delimited compartments may also serve as the site of crystal nucleation and growth (Fig. 12.1). This has been reported for the formation of apatite in the matrix vesicles of bone and dentin (Golub, 2009), magnetite in the magnetosomes of magnetotactic bacteria (Komeili, 2012), and calcite coccoliths in Golgi-derived vesicles in coccolithophores (Marsh, 2003; Young & Henriksen, 2003). In matrix vesicles, precipitation is initiated by transport of Ca^{2+} and phosphate through membrane proteins. While there is evidence that the initially precipitated mineral is amorphous, it crystallizes within the matrix vesicle to disordered apatite. It is thought that acidic phospholipids in the membrane promote the nucleation of apatite at the membrane interface. Magnetite nucleation in magnetosomes is also reported to occur on the surface of the encapsulating membrane, defining the orientation of the final crystal (Abreu et al., 2013; Faivre & Schuler, 2008). In contrast, nucleation of coccoliths occurs in a highly controlled fashion in a specific location and orientation on an organic base plate encapsulated in the vesicle. This highlights that cells not only go to great length to control the composition and pH of the solution inside these vesicles but also use tailored surface chemistry of the membrane, integral membrane proteins, and self-assembled scaffolds inside the confined volume to control nucleation and crystal growth. Just how successful the strategy is may become apparent when considering that both highly complex biomineralized elements, such as coccoliths, and mineralization processes of great importance to human health, namely, those mediated by matrix vesicles, start out confined in vesicles, albeit that the vesicle is discarded once it has done its job.

2. MODEL SYSTEMS FOR STUDYING THE INFLUENCE OF CONFINEMENT ON BIOMINERALIZATION

Clearly, privileged environments may play multiple important roles in controlling the nucleation and growth of biominerals. While some of the advantages of compartmentalization are obvious, such as the precise control over size and physiochemical conditions (pH and concentration of ions and biomacromolecules), other factors such as the influence of high surface areas and small volumes are poorly understood. To understand the effect of confinement on biomineralization, a number of model systems have been developed focusing on the precipitation of calcium carbonate.

With decreasing size, the increasing influence of surface energy and entropy are thought to lead to a crossover where more disordered phases become thermodynamically favored (Navrotsky, 2004; Quigley & Rodger, 2008). To identify this threshold, Nudelman and coworkers investigated the stability of nanoparticles of ACC (Nudelman, Sonmezler, Bomans, de With, & Sommerdijk, 2010). Growth and coalescence was inhibited by coating newly formed ACC nanoparticles with a polymer that binds to calcium. Using cryo-transmission electron microscopy (cryo-TEM) to image and probe the stability of the amorphous precursors, a critical size of 100 nm was identified, below which ACC does not crystallize. However, the role of a considerable amount of polymer that interacts extensively with ACC complicates interpretation and generalization. It is not clear, for example, whether the presence of the polymer also affects nucleation and growth of the crystalline polymorphs and whether the polymer-stabilized particles would remain stable in the presence of seed crystals.

The Meldrum group has extensively studied the influence of confinement on precipitation of calcium carbonate at the micron length scale.

Systematic variation in the size of the pores in track-etched membranes (Loste, Park, Warren, & Meldrum, 2004) and the geometry of an annular wedge between cross cylinders (Stephens, Ladden, Meldrum, & Christenson, 2010) suggest that ACC is transiently stabilized by confinement below 5 μm. In both these studies, the stability of the amorphous phase was attributed to limited contact with the bulk solution, which slows dissolution of ACC and reprecipitation of calcite. Even when a carboxyl-terminated self-assembled monolayer was intentionally introduced as a heterogeneous nucleator in picoliter droplets, the lifetime of ACC was enhanced (Stephens, Kim, Evans, Meldrum, & Christenson, 2011). While these systems allow systematic study of the influence of size on ACC stability and crystallization, precipitation is carried out in the presence of multiple interfaces whose influence is difficult to account for. Additionally, the stabilization of ACC by exclusion of water in track-etched membranes or between cross cylinders is a poor model for stabilization in hydrated biological environments.

An alternative method that enables confinement from the nanometer to micron length scale is precipitation inside phospholipid vesicles (liposomes). Unlike precipitation in track-etched membranes, picoliter droplets, or between cross cylinders, precipitation in liposomes occurs in the presence of one well-defined and easily modified interface. The phospholipid bilayer is fundamentally similar to interfaces encountered in biology. Drawing on the extensive research on liposomal drug delivery, liposomes are therefore a versatile platform to study the influence of size, surface chemistry, additives, and supersaturation on biomineral nucleation and growth in confinement. The approach is compatible with the major classes of biominerals, including iron oxides, calcium carbonate, and apatite (Fig. 12.2).

3. PREPARATION OF LIPOSOMES

The general method to prepare liposomes is to rehydrate a thin lipid film. Any dissolved species present during rehydration, for example, Ca^{2+} or the fluorescent dye calcein, will be encapsulated during this step. The size of the liposomes and number of bilayers or lamellarity depend on the method by which the lipid film is rehydrated. The liposome suspension can be further processed to reduce the size, polydispersity, and lamellarity. For precipitation experiments, it is often useful to remove solutes present in the rehydration medium by exchange against an inert and isosmotic medium.

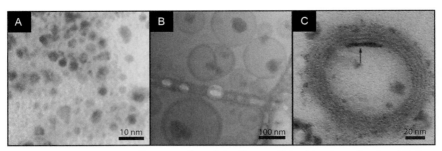

Figure 12.2 TEM images of liposomal models of intracellular biomineralization. (A) Goethite precipitation in phosphatidylcholine liposomes prepared by sonication (Mann & Hannington, 1988). *Reproduced with permission from Elsevier*. (B) Stabilization of amorphous calcium carbonate in phosphatidylcholine liposomes prepared by extrusion (Tester et al., 2011). *Reproduced with permission from The Royal Society of Chemistry*. (C) Apatite crystallization on the inner surface of multilamellar liposomes containing 10 mol% phosphatidic acid (Heywood & Eanes, 1992). *Reproduced with permission from Springer Science and Business Media*. TEM samples were prepared by air-drying (A), plunge freezing (B), and fixation and staining (C).

3.1. Lipid selection

The ability of lipids to self-assemble into supramolecular aggregates is due to their amphiphilic structure (Collier & Messersmith, 2001; Israelachvili, 2011). Phospholipids, which are the major component of cell membranes, are composed of a polar head group and hydrophobic tail, connected by a glycerol backbone (Fig. 12.3). Liposomes can be made from a variety of phospholipids and lipid mixtures, but most work has been done with zwitterionic phosphatidylcholine (PC).

To study precipitation in confinement, lipids may be selected for their potential for interaction with the mineral phase. There have been a number of studies on the influence of acidic phospholipids, capable of binding Ca^{2+}, on the nucleation and growth of calcium phosphate and calcium carbonate. For example, from liposomal models of matrix vesicle mineralization, Eanes and coworkers demonstrate that incorporation of phosphatidic acid (PA) or phosphatidylserine (PS) results in close physical contact between the apatite precipitate and the membrane (Heywood & Eanes, 1992). In contrast, we find that incorporation of PA, PS, or phosphatidylinositol (PI) does not increase interaction of the membrane with calcium carbonate or influence nucleation and growth (Tester et al., 2011). Eanes and coworkers propose that the degree of membrane–mineral interaction is governed by the conformation of the head group, the geometry of the lipids in the membrane, and their electrostatic interaction with the ions. Incorporation of acidic phospholipids may also be used to

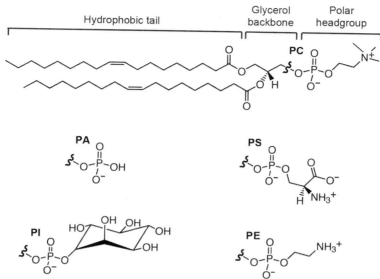

Figure 12.3 Structures of phospholipids. PC, phosphatidylcholine (shown with oleic acid fatty acid tails, dioleoylphosphatidylcholine); PA, phosphatidic acid; PS, phosphatidylserine; PI, phosphatidylinositol; PE, phosphatidylethanolamine.

reduce liposome aggregation and increase encapsulation efficiency (Woodle & Papahadjopoulos, 1989).

The choice of lipid can also be used to tailor liposome stability and membrane fluidity. Among naturally occurring phospholipids, phosphatidylethanolamine (PE) is the only one that does not form closed bilayer vesicles at neutral pH (Woodle & Papahadjopoulos, 1989). It can therefore be used to form pH-sensitive liposomes for triggered release of contents or promote vesicle fusion by stabilizing nonbilayer intermediates. Conversely, cholesterol can be used to decrease the permeability of the phospholipid bilayer to ions or polar molecules. Since liposome formation and processing must be performed above the solid-to-fluid transition temperature of the lipids (T_c), another important consideration is the length and saturation of the acyl chains. To circumvent the challenge of working at elevated temperatures, we choose lipids with an unsaturated acyl chain, with a T_c below room temperature, for instance, dioleoylphosphatidylcholine or soy PC mixtures.

3.2. Preparation of liposomes by extrusion or sonication

Thin films of lipids are prepared by rotary evaporation from solutions in dichloromethane (CH_2Cl_2). The lipid film is rehydrated with an aqueous

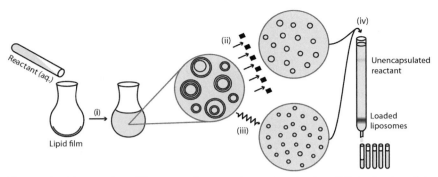

Figure 12.4 Preparation of liposomes by extrusion and sonication. (i) Rehydration of dry lipid film with an aqueous solution of the reactant results in the self-assembly of multilamellar liposomes. The size and polydispersity of the liposomes can be reduced by (ii) extrusion through a track-etched membrane or (iii) sonication. Unencapsulated reactant can be exchanged with an inert isosmotic medium by (iv) size-exclusion or ion exchange chromatography. (For color version of this figure, the reader is referred to the online version of this chapter.)

solution of the reactant and/or other additives and agitated at a temperature above T_c until the film has completely detached from the flask (Fig. 12.4). This results in the separation of the lipid sheets, which close on themselves to form a polydisperse suspension of multilamellar liposomes loaded with solutes present in the rehydration medium. In a typical procedure, a thin film formed from 25 mg of DOPC in 20 mL CH_2Cl_2 in a 50-mL evaporation flask is rehydrated with an aqueous solution of 1 M $CaCl_2$ (5 mL). Rehydration and agitation are performed by gentle spinning of the evaporation flask in the water bath (40 °C) of the rotary evaporator, without vacuum.

The size, number of lamellae, and polydispersity of the liposomes can be reduced by extruding the suspension through a track-etched membrane or by sonication. Extrusion through membranes with pore sizes below 200 nm results in mostly unilamellar liposomes, while larger pore sizes produce polydisperse mixtures of liposomes with varying numbers of lamellae. Alternatively, monodisperse suspensions of unilamellar liposomes can be prepared by sonication in a water bath for several minutes. This disrupts the large multilamellar liposomes, resulting in the formation of small unilamellar liposomes approximately 15–50 nm in diameter. The final size will depend on the sonication time, power, and the volume of the bath (Woodle & Papahadjopoulos, 1989). If using lipids with a T_c above room temperature, extrusion should be performed on a heating block or sonication performed in a heated bath. The size and polydispersity of liposomes prepared by extrusion or sonication can be measured using dynamic light scattering (DLS) (Egelhaaf, Wehrli, Muller, Adrian, & Schurtenberger, 1996).

3.3. Preparation of giant liposomes

Giant liposomes exceeding 1 μm in diameter can be prepared by gentle hydration of the lipid film or by the reverse-phase evaporation method (Fig. 12.5). Both of these methods can be used to encapsulate high ion concentrations, with minimal restriction on the types of lipids that can be used (Walde, Cosentino, Engel, & Stano, 2010).

Of the two methods, we prefer gentle hydration of the lipid film. With this method, the yield and quality of the liposomes will depend on the ability to promote gentle swelling and spontaneous separation the lipid bilayers. This can be facilitated by electrostatic repulsion of negatively charged lipids or by prehydrating the lipid films (Walde et al., 2010). However, in our hands, the best method for the production of liposomes at high concentrations of earth alkaline cations is the agarose-assisted rehydration method developed by Horger and coworkers (Horger, Estes, Capone, & Mayer, 2009).

Here, glass substrates are dip-coated in aqueous solutions of 1 wt% ultra-low-melting agarose, formed by heating agarose gels until liquid. The agarose film is then partially dried at 40 °C using a convection oven or hot plate. The lipid film is deposited on top of the agarose film by evaporation of a small volume of the lipid in CH_2Cl_2 in a desiccator under mild vacuum. Agarose-supported lipid films are then gently rehydrated by submersion in the rehydration medium for at least 1 h. During rehydration, lipid bilayers swell and liposomes bud off into solution. Rehydration requires time and must not be accelerated by agitation, as mechanical disturbance of the lipid film will result in a heterogeneous population of dense lipid aggregates. Progress of rehydration can conveniently be followed by light microscopy.

Preparation of liposomes by reverse-phase evaporation utilizes a chloroform/water two-phase system (Moscho, Orwar, Chiu, Modi, & Zare, 1996). In an evaporation flask, the rehydration medium is layered on top of a chloroform solution of the lipids. A lipid monolayer with the polar head groups in contact with the aqueous phase and the hydrophobic tails immersed in the organic phase forms at the water/chloroform interface. Evaporation of the chloroform at reduced pressure ruptures the lipid monolayer and forces the fragments into the aqueous solution, where they assemble into liposomes. While Michel and coworkers have successfully used the

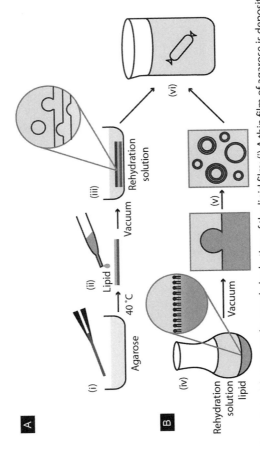

Figure 12.5 (A) Preparation of giant liposomes by gentle hydration of the lipid film: (i) A thin film of agarose is deposited on a glass substrate by dip-coating and drying. (ii) A lipid film is prepared on the agarose-coated substrate by evaporation of the lipid solution. (iii) The lipid film swells during rehydration and liposomes bud off into solution. (B) Preparation of giant liposomes by reverse-phase evaporation: (iv) An aqueous rehydration medium is layered on top of a solution of the lipid in chloroform. (v) On evaporation of chloroform, a lipid monolayer at the chloroform/water interface is forced into the aqueous phase and assembled into giant liposomes. (vi) In the final step, the rehydration medium is exchanged against an inert, isosmotic solution by dialysis. (For color version of this figure, the reader is referred to the online version of this chapter.)

reverse-phase evaporation method to encapsulate calcium ions and alkaline phosphatase (Michel et al., 2004), we find that the liposomes are smaller and have a larger number of lamellae than those formed by gentle hydration (Tester, Whittaker, & Joester, 2013).

3.4. Encapsulation efficiency

Solutes such as inorganic ions and small molecules in the rehydration medium are generally encapsulated in liposomes, but not necessarily at the same concentration. This allows loading liposomes, for instance, with Ca^{2+} for precipitation experiments. Note that for convenience, we omit to mention any counter ions that are in the solution and that are also encapsulated during this process. The encapsulation efficiency depends on the physical and chemical properties of, and interactions between, the phospholipids and the solutes to be encapsulated. Processing of liposomes by extrusion, sonication, or subjection to freeze-thaw cycles can increase encapsulation. However, encapsulation of large macromolecules or nanoparticles typically proceeds at low efficiency (Walde et al., 2010). In giant liposomes, the reverse-phase hydration method results in higher encapsulation efficiency than gentle hydration, but in the case of proteins, the use of organic solvents may result in denaturation (Colletier, Chaize, Winterhalter, & Fournier, 2002). Alternatively, microinjection can be used to introduce macromolecules or nanoparticles after giant liposome formation (Wick, Angelova, Walde, & Luisi, 1996).

3.5. Exchange of ions in the rehydration medium

For nucleation and growth experiments in liposome nano- or microreactors, a membrane-impermeant cation, for instance, Ca^{2+}, is added to the rehydration medium (by adding an appropriate salt) and thus loaded into the liposome. Similarly, small molecules, macromolecules, and even functional enzymes can be encapsulated. However, in many cases, it is important to remove any reactants or additives that did not get encapsulated and are thus present in solution outside the liposomes. At the same time, the encapsulated and exterior solutions should remain isosmotic. This is possible by size-exclusion chromatography (SEC) (Heywood & Eanes, 1992; Tester et al., 2011; Tester, Wu, Weigand, & Joester, 2012), ion exchange chromatography (Mann & Hannington, 1988; Messersmith, Vallabhaneni, & Nguyen, 1998), or dialysis (Tester et al., 2013). The method of choice depends primarily on the size of liposomes and the subsequent methods of characterization.

For liposomes below 1 μm, we find that SEC is the fastest and most efficient method to exchange the rehydration medium against an isosmotic, inert medium. It is also amenable to separation of liposomes prepared at high ionic strength and a wide range of pH values. In SEC, smaller solutes have high retention volumes and elute late, while the comparatively large liposomes elute rapidly. Thus, any small solutes in the rehydration medium can be exchanged against those present in the mobile phase (Fig. 12.4iv). Liposome-containing fractions can be identified from their milky white appearance, if present at high enough concentration, or can be detected by DLS. For the removal of unencapsulated Ca^{2+} from a suspension of $CaCl_2$-loaded liposomes, we use a cross-linked dextran gel column, Sephadex (Sephadex G-25 coarse column: $l=40$ cm, $d=2.5$ cm), equilibrated with an isosmotic aqueous solution of NaCl, and use the same NaCl solution as mobile phase (Tester et al., 2011, 2012). Alternatively, we, as well as others, have used ion exchange chromatography to replace unencapsulated cations with Na^+ (Mann & Hannington, 1988; Messersmith et al., 1998). A drawback of column chromatography is the dilution of the liposome suspension, by band broadening or interaction with the matrix, which may hinder subsequent characterization.

Dialysis is a third method that minimizes liposome dilution and is particularly suited for treating giant liposome suspensions. Similar to SEC, dialysis separates species based on their size. The liposome suspension is placed in a bag made from a porous dialysis membrane with a specific molecular weight cutoff. By placing the dialysis bag in an isosmotic dialysate bath, any solutes below the cutoff size will diffuse out along the concentration gradient, while the dialysate will diffuse in. Dialysis is most effective when the dialysate bath is stirred and repeatedly exchanged to maintain steep concentration gradients. While dialysis is a gentle and straightforward method, effective exchange can take 24 h or longer at room temperature.

4. PRECIPITATION INSIDE LIPOSOMES

To initiate precipitation inside liposomes, the interior must become supersaturated with respect to the reaction product (e.g., calcium carbonate), that is, the product of the activities of the participating ions (Ca^{2+} and CO_3^{2-}) must exceed the solubility product K_{sp}. It is generally useful to encapsulate a membrane-impermeant reactant and increase the concentration of another by passive or facilitated transport from the external medium. The phospholipid bilayer therefore controls the flow of the reactive species. Specific examples for the precipitation of iron oxide, calcium carbonate, and calcium phosphate minerals are given below.

As a model for magnetite formation in magnetotactic bacteria, one of the first reports of precipitation in liposomes was that of iron oxide in PC liposomes by Mann and coworkers (Mann & Hannington, 1988; Mann, Hannington, & Williams, 1986). Briefly, Fe^{2+}- and/or Fe^{3+}-loaded liposomes were prepared by encapsulation from solutions of their salts. Taking advantage of the vastly higher permeability of phospholipid bilayers for anions, precipitation was initiated by passive diffusion of OH^- from the external medium. Despite a very high pH in the outside medium, transport kinetics are slow because of the rate-limiting, charge-balancing efflux of counterions from the liposome interior. These conditions favor the precipitation of stable polymorphs such a goethite (FeO(OH)) over the precipitation of metastable ferrihydrite ($Fe_2O_3 \cdot xH_2O$). Mann and coworkers have also used this method to precipitate silver oxide in Ag-loaded liposomes (Mann & Williams, 1983).

Our work has focused on the influence of liposomal confinement on precipitation of calcium carbonate (Tester et al., 2011, 2012). To precipitate calcium carbonate inside liposomes, ammonium carbonate $(NH_4)_2CO_3$ is added to liposomes loaded with $CaCl_2$. Transport of membrane-impermeant carbonate ions across the membrane occurs by the diffusion of CO_2, which is present in equilibrium (Fig. 12.6). Once CO_2 rehydrates and CO_3^{2-} is formed inside the liposome, rapid precipitation of calcium carbonate is observed.

In liposomes up to 1 μm in size, exactly one calcium carbonate nanoparticle is formed. The particle size scales with the calcium concentration inside and size of the encapsulating liposome (Tester et al., 2011). Regardless of its size, particles consist of ACC instead of one of the more stable crystalline polymorphs of calcium carbonate. We attribute the extended stability of

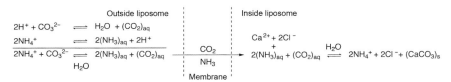

Figure 12.6 Liposome-stabilized ACC is formed by addition of ammonium carbonate and diffusion of CO_2 into Ca-loaded liposomes (Tester et al., 2011, 2013, 2012).

metastable ACC to the suppression of heterogeneous nucleation of the crystalline polymorphs, by exclusion of impurities that act as nucleators (Tester et al., 2013). This method was also used to study precipitation of strontium and barium carbonate. Whereas ACC remains stable for more than a week, amorphous $SrCO_3$ is stabilized for minutes only before strontianite is nucleated, and precipitation of $BaCO_3$ leads to crystalline witherite without formation of an amorphous phase. The increased rate of strontium and barium carbonate crystallization likely results from the combined effects of lower surface energy of the mineral, higher supersaturation, and better wetting of the liposome membrane by crystalline nuclei.

There has been great interest in the precipitation of calcium phosphate inside liposomes, as models for matrix vesicle mediated mineralization of bone-type mineralized tissues. However, the phospholipid membrane is not permeable to phosphate. To overcome this challenge, Eanes and coworkers introduced an ionophore (lasalocid acid, X-537A) into the lipid membrane that binds to calcium ions and facilitates transport across the hydrophobic core of the membrane (Heywood & Eanes, 1992). This allows precipitation of calcium phosphate by the addition of calcium to a suspension of liposomes loaded with phosphate ions. The ionophore can be introduced either during the formation of the lipid films or by adding an ethanolic solution to phosphate-loaded liposomes. Following addition of calcium to the external medium, the first phase to precipitate inside the liposomes is ACP, which quickly converts to crystalline apatite.

An alternative approach reported by Michel and coworkers is to use a membrane-permeant phosphate precursor that is enzymatically cleaved inside the liposome (Michel et al., 2004). Giant liposomes loaded with calcium and alkaline phosphatase were prepared by reverse-phase evaporation and unencapsulated alkaline phosphatase hydrolyzed by proteolytic enzymes. Calcium phosphate precipitation was initiated by the addition of p-nitrophenyl phosphate to the outside medium. The amphiphilic p-nitrophenyl phosphate diffuses across the phospholipid membrane where it is hydrolyzed to p-nitrophenol and inorganic phosphate by encapsulated alkaline phosphatase. The resulting precipitate is a poorly crystalline hydroxyapatite.

5. CHARACTERIZATION METHODS

To minimize artifacts and prevent dissolution or transformation of metastable intermediates, the ideal methods of characterization are those that do not require isolation of the mineral or drying of the liposomes. We have applied several *in situ* imaging, scattering, and spectroscopic techniques to characterize liposomes and encapsulated precipitates (Tester et al., 2011, 2012, 2013). Together, these methods provide information on the structure, morphology, and growth of precipitates from the nanometer to micron length scale (Fig. 12.7). A comprehensive review of *in situ* methods to study the dynamics of biomimetic mineralization is given by Dey, de With, and Sommerdijk (2010).

A central advantage to working with giant liposomes is that encapsulated reactions can be monitored using light microscopy. Precipitation of various polymorphs may be inferred from their characteristic color, in the case of the iron oxides, or morphology, in the case of calcium carbonates. The use of polarized light microscopy further allows the detection of birefringence, characteristic of the development of long-range order. Information on the degree of membrane–mineral interaction can be obtained by confocal microscopy (Tester et al., 2013). The membrane can be fluorescently labeled by incorporation of a commercially available fluorescently conjugated lipid.

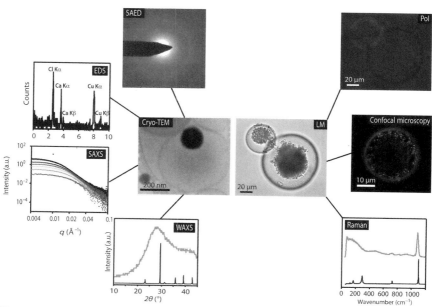

Figure 12.7 Summary of *in situ* methods to characterize the morphology, structure, and growth kinetics of precipitates formed within liposomes. SAED, selected area electron diffraction; EDS, energy-dispersive X-ray spectroscopy; SAXS, small-angle X-ray scattering; WAXS, wide-angle X-ray scattering; LM, light microscopy; Pol, polarized light microscopy. (See color plate.)

The mineral phase can be labeled by coencapsulation with a fluorescent dye that will incorporate into the precipitate. For example, to study the influence of the phospholipid membrane on ACC precipitation, we incorporate 0.5% rhodamine-labeled PE into the membrane and coencapsulate 2 μg/mL of the calcium-binding fluorophore, calcein (Tester et al., 2013). Finally, changes in pH that accompany diffusion of the reactant or mineral precipitation can be monitored by encapsulation of a membrane-impermeant pH-indicator or pH-sensitive fluorophore (Tester et al., 2012).

Liposomes below 1 μm in size must be imaged using transmission electron microscopy. While progress has been made using fixation and heavy-element staining to preserve liposome structure in the high radiation and high vacuum environment of the microscope (Heywood & Eanes, 1992), cryo-electron microscopy of frozen-hydrated liposomes eliminates the artifacts associated with drying and chemical modification (Almgren, Edwards, & Karlsson, 2000). Samples are prepared by deposition of the suspension onto a TEM grid, removal of excess solution, and plunge freezing into liquid ethane. The result is a vitrified film around 100 nm in thickness with liposomes preserved in a frozen-hydrated state that is very similar to their pristine state in suspension. Cryo-frozen liposomes can be imaged without staining due to phase contrast generated by underfocusing the image (Cui et al., 2007). While cryo-TEM is a vast improvement over conventional chemical fixation and staining, it is not without artifacts, primarily due to the formation of crystalline ice during transfer and storage or from high doses of electron irradiation (Friedrich, Frederik, de With, & Sommerdijk, 2010). However, a trained operator using low-dose methods can routinely and rapidly image liposomes up to 200 nm in diameter and characterize precipitate structure by cryo-selected area electron diffraction, elemental composition by cryo-energy dispersive X-ray spectroscopy, and the degree of membrane–mineral interaction by cryo-electron tomography.

In small liposomes, simultaneous small- and wide-angle X-ray scattering (SAXS/WAXS) at synchrotron sources are very well suited to characterize the nucleation and growth of the precipitates *in situ* (Pontoni, Bolze, Dingenouts, Narayanan, & Ballauff, 2003; Tester et al., 2012). The small-angle regime is sensitive to changes in electron density at the nanometer length scale. While it should in principle be possible to investigate nucleation and very early growth, for practical reasons, primarily the limited volume fraction of liposomes, the smallest calcium carbonate particles we can detect are ~20 nm in diameter. However, for larger particles, their time-dependent size and shape can be determined from their scattering intensity as a function of the scattering vector q using analytical evaluation procedures (Peterlik & Fratzl, 2006) or numerical simulation tools (Ilavsky & Jemian, 2009). An upstream WAXS detector allows simultaneous characterization of the long-range order of the precipitates. The availability of synchrotron sources enables detection of low volume fractions of mineral at a millisecond timescale. Radiation damage can be minimized by measurement under continuous flow. This has allowed us to investigate ACC nanoparticle growth kinetics in detail (Tester et al., 2012).

Finally, the structure of the mineral precipitates may also be studied using Raman spectroscopy. Raman spectroscopy is superior to infrared spectroscopy because it tolerates water and can be carried out *in situ*. While this method may in principle be used to study bulk suspensions of liposomes, we find that the volume fraction is generally too low. However, confocal laser Raman microscopy has facilitated the identification of the polymorph of individual precipitates in giant liposomes (Tester et al., 2013). Specifically, we determined the formation of ACC, strontianite, and witherite in giant liposomes by comparison of Raman spectra from precipitates in liposomes with those of reference compounds.

6. CONCLUSIONS

Liposomal encapsulation provides a versatile platform to model biomineralization in privileged environments. To encapsulate a precipitation reaction, liposomes are formed by rehydration of a lipid film with an aqueous solution of a membrane-impermeant reactant. After exchange of the unencapsulated solution, precipitation is initiated by transport of the second reactant into the liposome from the external medium. This is particularly straightforward when the reactant is in equilibrium with a membrane-permeant neutral molecule as, for instance, carbonate ions are in equilibrium with carbon dioxide. Where this is impossible, more sophisticated strategies use an appropriate ionophore or enzyme-mediated cleavage of membrane-permeant precursors. Control over liposome size over three orders of magnitude in diameter (9 orders of magnitude in volume!), lipid composition, and composition of the interior and exterior solutions enables systematic study of the effects of confinement and surface chemistry on nucleation and growth. Initial results highlight that confinement in membrane vesicles greatly suppresses nucleation of crystalline calcium carbonate and to a lesser degree that of other minerals (Tester et al., 2013, 2012). As a consequence, ACC appears greatly stabilized. We envision that liposomal encapsulation strategies in combination with *in situ* characterization methods will be used to improve understanding of biological control over crystal growth and develop bioinspired approaches to material synthesis based on transient stabilization of amorphous precursors. Important capabilities to reach this goal would be the ability to perform experiments under constant supersaturation—this would allow determination of surface energies (Hu et al., 2012)—the incorporation of selective ion transporters in the liposome membrane, immobilization and patterning of liposomes, and controlled fusion of liposomes and their cargoes.

ACKNOWLEDGMENTS
This work was in part supported by the NSF (Grants No. DMR-0805313 and DMR-1106208), by the MRSEC program of the National Science Foundation (DMR-1121262) at the Materials Research Center of Northwestern University, and by the International Institute for Nanotechnology at Northwestern University.

REFERENCES
Abreu, F., Sousa, A. A., Aronova, M. A., Kim, Y., Cox, D., Leapman, R. D., et al. (2013). Cryo-electron tomography of the magnetotactic vibrio Magnetovibrio blakemorei: Insights into the biomineralization of prismatic magnetosomes. *Journal of Structural Biology, 181*, 162–168.
Addadi, L., Joester, D., Nudelman, F., & Weiner, S. (2006). Mollusk shell formation: A source of new concepts for understanding biomineralization processes. *Chemistry (Weinheim an der Bergstrasse, Germany), 12*, 980–987.
Almgren, M., Edwards, K., & Karlsson, G. (2000). Cryo transmission electron microscopy of liposomes and related structures. *Colloids and Surfaces A, Physicochemical and Engineering Aspects, 174*, 3–21.
Beniash, E., Addadi, L., & Weiner, S. (1999). Cellular control over spicule formation in sea urchin embryos: A structural approach. *Journal of Structural Biology, 125*, 50–62.
Bentov, S., Brownlee, C., & Erez, J. (2009). The role of seawater endocytosis in the biomineralization process in calcareous foraminifera. *Proceedings of the National Academy of Sciences of the United States of America, 106*, 21500–21504.
Colletier, J.-P., Chaize, B., Winterhalter, M., & Fournier, D. (2002). Protein encapsulation in liposomes: Efficiency depends on interactions between protein and phospholipid bilayer. *BMC Biotechnology, 2*, 9.
Collier, J. H., & Messersmith, P. B. (2001). Phospholipid strategies in biomineralization and biomaterials research. *Annual Review of Materials Research, 31*, 237–263.
Coradin, T., & Lopez, P. J. (2003). Biogenic silica patterning: Simple chemistry or subtle biology? *ChemBioChem, 4*, 251–259.
Cui, H., Hodgdon, T. K., Kaler, E. W., Abezgauz, L., Danino, D., Lubovsky, M., et al. (2007). Elucidating the assembled structure of amphiphiles in solution via cryogenic transmission electron microscopy. *Soft Matter, 3*, 945–955.
Dey, A., de With, G., & Sommerdijk, N. (2010). In situ techniques in biomimetic mineralization studies of calcium carbonate. *Chemical Society Reviews, 39*, 397–409.
Egelhaaf, S. U., Wehrli, E., Muller, M., Adrian, M., & Schurtenberger, P. (1996). Determination of the size distribution of lecithin liposomes: A comparative study using freeze

fracture, cryoelectron microscopy and dynamic light scattering. *Journal of Microscopy, 184*, 214–228.

Faivre, D., & Schuler, D. (2008). Magnetotactic bacteria and magnetosomes. *Chemical Reviews, 108*, 4875–4898.

Friedrich, H., Frederik, P. M., de With, G., & Sommerdijk, N. (2010). Imaging of self-assembled structures: Interpretation of TEM and cryo-TEM images. *Angewandte Chemie(International Edition), 49*, 7850–7858.

Golub, E. E. (2009). Role of matrix vesicles in biomineralization. *Biochimica et Biophysica Acta, 1790*, 1592–1598.

Heywood, B. R., & Eanes, E. D. (1992). An ultrastructural-study of the effects of acidic phospholipid substitutions on calcium-phosphate precipitation in anionic liposomes. *Calcified Tissue International, 50*, 149–156.

Horger, K. S., Estes, D. J., Capone, R., & Mayer, M. (2009). Films of agarose enable rapid formation of giant liposomes in solutions of physiologic ionic strength. *Journal of the American Chemical Society, 131*, 1810–1819.

Hu, Q., Nielsen, M. H., Freeman, C. L., Hamm, L. M., Tao, J., Lee, J. R. I., et al. (2012). The thermodynamics of calcite nucleation at organic interfaces: Classical vs. non-classical pathways. *Faraday Discussions, 159*, 509–523.

Ilavsky, J., & Jemian, P. R. (2009). Irena: Tool suite for modeling and analysis of small-angle scattering. *Journal of Applied Crystallography, 42*, 347–353.

Israelachvili, J. N. (2011). *Intermolecular and surface forces* (Revised 3rd ed.). San Diego: Academic Press.

Komeili, A. (2012). Molecular mechanisms of compartmentalization and biomineralization in magnetotactic bacteria. *Fems Microbiology Reviews, 36*, 232–255.

Krejci, M. R., Finney, L., Vogt, S., & Joester, D. (2011). Selective sequestration of strontium in desmid green algae by biogenic co-precipitation with barite. *ChemSusChem, 4*, 470–473.

Krejci, M. R., Wasserman, B., Finney, L., McNulty, I., Legnini, D., Vogt, S., et al. (2011). Selectivity in biomineralization of barium and strontium. *Journal of Structural Biology, 176*, 192–202.

Loste, E., Park, R. J., Warren, J., & Meldrum, F. C. (2004). Precipitation of calcium carbonate in confinement. *Advanced Functional Materials, 14*, 1211–1220.

Lowenstam, H. A., & Weiner, S. (1989). *On biomineralization*. New York: Oxford University Press.

Mahamid, J., Sharir, A., Gur, D., Zelzer, E., Addadi, L., & Weiner, S. (2011). Bone mineralization proceeds through intracellular calcium phosphate loaded vesicles: A cryo-electron microscopy study. *Journal of Structural Biology, 174*, 527–535.

Mann, S., & Hannington, J. P. (1988). Formation of iron-oxides in unilamellar vesicles. *Journal of Colloid and Interface Science, 122*, 326–335.

Mann, S., Hannington, J. P., & Williams, R. J. P. (1986). Phospholipid-vesicles as a model system for biomineralization. *Nature, 324*, 565–567.

Mann, S., & Williams, R. J. P. (1983). Precipitation within unilamellar vesicles. 1. Studies of silver(I) oxide formation. *Journal of the Chemical Society, Dalton Transactions, 2*, 311–316.

Marsh, M. E. (1999). Coccolith crystals of Pleurochrysis carterae: Crystallographic faces, organization, and development. *Protoplasma, 207*, 54–66.

Marsh, M. E. (2003). Regulation of $CaCO_3$ formation in coccolithophores. *Comparative Biochemistry and Physiology Part B, Biochemistry and Molecular Biology, 136*, 743–754.

Messersmith, P. B., Vallabhaneni, S., & Nguyen, V. (1998). Preparation of calcium-loaded liposomes and their use in calcium phosphate formation. *Chemistry of Materials, 10*, 109–116.

Michel, M., Winterhalter, M., Darbois, L., Hemmerle, J., Voegel, J. C., Schaaf, P., et al. (2004). Giant liposome microreactors for controlled production of calcium phosphate crystals. *Langmuir, 20*, 6127–6133.

Moscho, A., Orwar, O., Chiu, D. T., Modi, B. P., & Zare, R. N. (1996). Rapid preparation of giant unilamellar vesicles. *Proceedings of the National Academy of Sciences of the United States of America, 93*, 11443–11447.

Navrotsky, A. (2004). Energetic clues to pathways to biomineralization: Precursors, clusters, and nanoparticles. *Proceedings of the National Academy of Sciences of the United States of America, 101*, 12096–12101.

Nudelman, F., Sonmezler, E., Bomans, P. H. H., de With, G., & Sommerdijk, N. (2010). Stabilization of amorphous calcium carbonate by controlling its particle size. *Nanoscale, 2*, 2436–2439.

Peterlik, H., & Fratzl, P. (2006). Small-angle X-ray scattering to characterize nanostructures in inorganic and hybrid materials chemistry. *Monatshefte für Chemie, 137*, 529–543.

Pontoni, D., Bolze, J., Dingenouts, N., Narayanan, T., & Ballauff, M. (2003). Crystallization of calcium carbonate observed in-situ by combined small- and wide-angle X-ray scattering. *The Journal of Physical Chemistry B, 107*, 5123–5125.

Quigley, D., & Rodger, P. M. (2008). Free energy and structure of calcium carbonate nanoparticles during early stages of crystallization. *The Journal of Chemical Physics, 128*, 221101.

Stephens, C. J., Kim, Y. Y., Evans, S. D., Meldrum, F. C., & Christenson, H. K. (2011). Early stages of crystallization of calcium carbonate revealed in picoliter droplets. *Journal of the American Chemical Society, 133*, 5210–5213.

Stephens, C. J., Ladden, S. F., Meldrum, F. C., & Christenson, H. K. (2010). Amorphous calcium carbonate is stabilized in confinement. *Advanced Functional Materials, 20*, 2108–2115.

Tester, C. C., Brock, R. E., Wu, C.-H., Krejci, M. R., Weigand, S., & Joester, D. (2011). In vitro synthesis and stabilization of amorphous calcium carbonate (ACC) nanoparticles within liposomes. *CrystEngComm, 13*, 3975–3978.

Tester, C. C., Whittaker, M. L., & Joester, D. (2013).

Tester, C. C., Wu, C. H., Weigand, S., & Joester, D. (2012). Precipitation of ACC in liposomes—A model for biomineralization in confined volumes. *Faraday Discussions, 159*, 345–356.

Walde, P., Cosentino, K., Engel, H., & Stano, P. (2010). Giant vesicles: Preparations and applications. *ChemBioChem, 11*, 848–865.

Weiner, S., & Addadi, L. (2011). Crystallization pathways in biomineralization. *Annual Review of Materials Research, 41*, 21–40.

Wick, R., Angelova, M. I., Walde, P., & Luisi, P. L. (1996). Microinjection into giant vesicles and light microscopy investigation of enzyme-mediated vesicle transformations. *Chemical Biology, 3*, 105–111.

Woodle, M. C., & Papahadjopoulos, D. (1989). Liposome preparation and size characterization. *Methods in Enzymology, 171*, 193–217.

Young, J. R., & Henriksen, K. (2003). Biomineralization within vesicles: The calcite of coccoliths. In P. M. Dove, J. J. DeYoreo, & S. Weiner (Eds.), *Biomineralization: 54*, (pp. 189–215). Chantilly: Mineralogical Soc Amer.

CHAPTER THIRTEEN

Polymer-Mediated Growth of Crystals and Mesocrystals

Helmut Cölfen[1]
Department of Chemistry, Physical Chemistry, University of Konstanz, Konstanz, Germany
[1]Corresponding author: e-mail address: helmut.coelfen@uni-konstanz.de

Contents

1. Introduction	278
2. Insoluble Polymers	279
2.1 Templating	279
2.2 Incorporation of polymers	285
3. Soluble Polymers	286
3.1 Types of soluble polymers for crystallization control	286
3.2 Nucleation and crystallization inhibition	288
3.3 Polymer-induced liquid precursors	289
3.4 Face-selective adsorption	292
3.5 Selecting between classical and nonclassical growth modes	294
3.6 Mesocrystal formation	297
4. Conclusion	301
References	302

Abstract

Polymers are important additives for the control of mineralization reactions in both biological and bioinspired mineralization. The reason is that they allow for a number of interactions with the growing crystals and even amorphous minerals. These can substantially influence the way the mineral grows on several levels. Already in the prenucleation phase, polymers can control the formation of prenucleation clusters and subsequently the nucleation event. Also, polymers can control whether the further crystallization follows a classical or nonclassical particle-mediated growth path. In this chapter, the main ways in which polymers can be used to control a crystallization reaction will be highlighted. In addition, polymers that are useful for this purpose and the experimental conditions suitable for directing a crystallization reaction into the desired direction through the use of polymers will be described.

1. INTRODUCTION

Biomineralization provides a plethora of examples for highly controlled crystallization events (Lowenstam & Weiner, 1989). In biominerals, control of crystallization becomes obvious at different stages of the formation reaction as well as on different length scales of the often hierarchical biomineral. This crystallization control is exerted by means of biopolymers like proteins or polysaccharides in most cases. Fundamentally, there are two different types of polymers to be distinguished: soluble and insoluble polymers. Insoluble polymers have the role of a scaffold or template for the subsequent mineralization reaction and are also called the insoluble matrix. Typical structural insoluble polymers in biomineralization are collagen, which builds the matrix in bone and tooth (dentin), or chitin, which builds the layered scaffold in nacre. The growth of the organic–inorganic composite material in biomineralization is schematically shown in Fig. 13.1.

The confined reaction space created by the insoluble polymer is subsequently filled with mineral, which often finally crystallizes. This mineralization event is controlled by the soluble polymers. Very often, the soluble polymers are charged and are nucleation or crystallization inhibitors. The inhibition of crystallization until the mineral is shaped is very advantageous for biomineral formation because amorphous precursor material is isotropic and can be brought into any shape allowing for the formation of the fascinating and highly complex biomineral morphologies. Amorphous precursor species have been found in calcium carbonate biominerals like sea urchin spicules (Politi, Arad, Klein, Weiner, & Addadi, 2004) as well as calcium phosphates like zebra fish bones (Mahamid, Sharir, Addadi, & Weiner, 2008). These examples indirectly demonstrate how polymers can be applied to exert control over crystallization reactions. This can be via templating/scaffolding by insoluble polymers or through multiple ways to control a crystallization reaction by soluble polymers that in many cases are charged Song

Figure 13.1 Schematic representation of polymer-mediated biomineral formation. A scaffold is formed by an insoluble polymer creating a confined reaction space. This reaction space is filled with mineral in a second step. Crystallization of the mineral is controlled by soluble polymers. (For color version of this figure, the reader is referred to the online version of this chapter.)

and Coelfen (2011). A comprehensive treatment of all these possibilities is beyond the scope of this chapter, and therefore, the focus will be on the most important possibilities of polymer-mediated crystallization. For the discussed polymer-mediated crystallization reactions, important examples will be presented in more detail. A more detailed discussion of polymer-mediated crystallization can be found elsewhere (Meldrum & Coelfen, 2008; Song & Coelfen, 2011).

2. INSOLUBLE POLYMERS

Insoluble polymers in biomineralization are not only typical structural biopolymers like chitin (nacre) or collagen (bone and teeth) but also polymers that form hydrogels like silk (nacre). Therefore, they can act as templates (chitin or collagen) but can also get occluded into the mineral. For synthetic polymers, there exist a large variety of possibilities to form insoluble polymer matrices, which can serve as template as well as a hydrogel for inclusion into the growing crystal.

2.1. Templating

Polymeric templates for crystallization reactions can be fixed polymer scaffolds, which can be prepared either via polymerization of a monomer inside a mold to obtain a polymer replica or by photocuring polymers that are polymerized by light and can thus be polymerized into any shape using a 3D lithography approach (Sun & Kawata, 2004). A well-known example from the field of bioinspired mineralization is the preparation of a polymeric replica of a sea urchin skeletal plate (Park & Meldrum, 2002). For this, a resin was used, which is typically used to embed samples for electron microscopy. Nine parts LR-white hard-grade electron microscopy resin (Agar Scientific) to one part ethyl acrylate were mixed and used to infiltrate the porous plate by capillary action. Curing in a sealed oxygen-free atmosphere at 60 °C for 24 h yielded the polymer template within the sea urchin plate. Sections of ca. 90 μm with uniform thickness were cut with a microtome, and subsequent dissolution of the sea urchin $CaCO_3$ with diluted hydrochloric acid over 24 h followed by exhaustive washing with water yielded the polymer replica template. $CaCO_3$ was precipitated within this polymer membrane using a U-tube setup as shown in Fig. 13.2. The porous polymer template was placed between the two parts of the U-tube (Fig. 13.2), which were filled with $CaCl_2$ and Na_2CO_3, respectively. After mineralization, the polymer membrane was removed by heating to

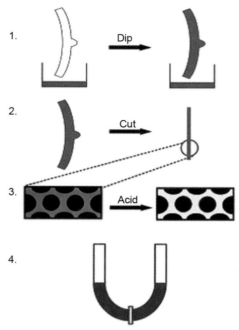

Figure 13.2 Schematic diagram showing the methodology used to grow calcite crystals in a polymer membrane. (1) Sea urchin plate is dipped in monomer solution and cured to a polymer, (2) a thin section is cut with a microtome, and (3) the calcium carbonate is dissolved. (4) Calcium carbonate is then precipitated in the so-formed polymer membrane. *Figure reproduced from Park and Meldrum (2002) with permission of Wiley-VCH.*

500 °C for 45 min or dissolving it in a hot 1:1 mixture of tetrahydrofuran and ethanol (1:1).

With this procedure, remarkable single-crystalline replicas of the initial sea urchin skeletal plate could be achieved when choosing a low reactant concentration (20 mM) and therefore slow crystallization (Fig. 13.3B). Not only were crystal facets reproduced, but also, more impressively, even surfaces curved on the micron scale were obtained. The single crystals were a few hundred micrometers in size. On the other hand, a high reactant concentration failed to reproduce a $CaCO_3$ replica of the complex porous polymer shape and yielded a polycrystalline particle (Fig. 13.3A). The reason is that multiple nucleation events took place at the same time at the higher reactant concentration resulting in the polycrystalline particle (Fig. 13.3A).

On the other hand, the single crystal with complex morphology (Fig. 13.3B) originates from a single nucleation event. However, due to the low reactant concentration, the growth of these complex-shaped

Figure 13.3 Calcium carbonate precipitated in a polymer membrane. (A) Polycrystalline particle precipitated from 400 m*M* reagents and (B) templated single crystal of calcite precipitated from 20 m*M* solutions. *Figure reproduced from Park and Meldrum (2002) with permission of Wiley-VCH.*

single-crystalline particles is limited. It is important to maintain only a few nucleation events inside the template—ideally even only a single nucleation event in order to obtain a single crystal with the complex template shape. When the template surface chemistry was altered in various ways to exhibit negative charges, only polycrystalline precipitates were obtained inside the polymer membrane, because of multiple nucleation events on the negatively charged surfaces. Only the hydrophobic surface, which does not interact with the dissolved ions, is suitable to keep the number of nucleation events low enough to result in the growth of single crystals with the complex template morphology (Wucher, Yue, Kulak, & Meldrum, 2007).

This approach not only is limited to calcium carbonate but also could be extended to other minerals like $SrSO_4$, $PbSO_4$, $PbCO_3$, $NaCl$, and $CuSO_4 \cdot 5H_2O$, and the only requirement to obtain single crystals with the complex shape of the template is that the single-crystal size must exceed the size of the template structures (Yue, Park, Kulak, & Meldrum, 2006).

Not only sea urchin replicas but also in principle every porous polymer membrane can be mineralized by this approach. Examples include the successful mineralization of track etch membranes with single-crystalline $CaCO_3$ where the pore size was important for the pore replication by a single crystal (Loste, Park, Warren, & Meldrum, 2004).

Another template, which was successfully used for the replication with a single crystal, is colloidal crystals, which can be conveniently prepared using monodisperse latexes. An elegant example for this type of templating was reported by Qi and coworkers (Li & Qi, 2008). They used carboxy-functionalized latexes for the preparation of the colloidal crystal template to allow for interaction with calcium ions (Li & Qi, 2008). Quick mixing of a 16 mM $CaCl_2$ and 16 mM Na_2CO_3 solution yielded fast precipitation of amorphous $CaCO_3$ nanoparticles, which were poured onto the template and sucked into the colloidal crystal template with vacuum (Fig. 13.4). Subsequent crystallization to calcite yielded an inverse opal composed of a single crystal of calcite.

Finally, not only static polymer templates but also more dynamic polymeric systems like block copolymer phases can be used for templating crystals with complex shape. Block copolymers consisting of hydrophilic and hydrophobic parts can form a variety of phases in aqueous solution triggered by phase separation depending on polymer concentration, temperature, the block lengths and their length ratios. Especially complex is the triple periodic gyroid morphology (Hamley, 1998). This morphology generated by polystyrene-block-polyisoprene (PS-b-PI) block copolymers was used to template single-crystalline calcite (see Fig. 13.5) (Finnemore et al., 2009). For this, a film was prepared exhibiting the gyroid morphology with continuous PI channels in a PS matrix by the addition of a required amount of PS homopolymer to PS-b-PI solutions with subsequent selective etching of the polyisoprene blocks leaving the polystyrene gyroid morphology as a template (for details, see Finnemore et al. (2009) and schematically Fig. 13.5).

$CaCO_3$ crystallization was carried out by submerging porous PS films, supported on glass, into a $CaCl_2$ aqueous/methanol (20 vol%) solution (20 mL, 10 mM–1 M) to facilitate infiltration of the hydrophobic gyroid phase. The crystallization vessel was placed in a closed desiccator containing a parafilm-sealed, needle-punctured vial of $(NH_4)_2CO_3$ powder to facilitate slow diffusion of CO_2 into the $CaCl_2$ solution. Crystal growth was quenched by the removal of the substrate from the solution and washing

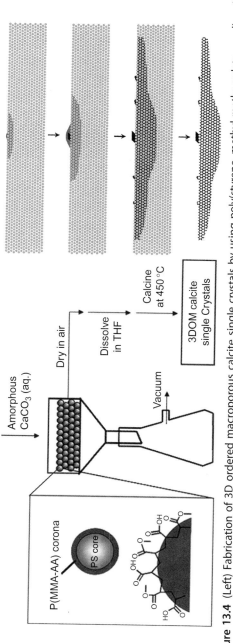

Figure 13.4 (Left) Fabrication of 3D ordered macroporous calcite single crystals by using poly(styrene–methyl methacrylate–acrylic acid) spheres assembled colloidal crystals as template (Li & Qi, 2008). (Right) Formation of calcite single crystals in a colloidal crystal template via amorphous precursors. Infiltration of the template with ACC nanoparticles followed by a calcite nucleation event that leads to crystallization of the ACC phase to calcite with the orientation of the crystal nucleus. Finally, the template was removed by dissolution in THF followed by calcination at 450 °C. *Reproduced from Li and Qi (2008) with permission of Wiley-VCH.*

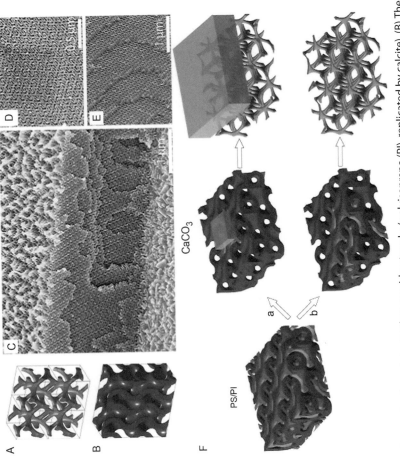

Figure 13.5 (A) The two intertwined, nonintersecting gyroid networks (polyisoprene (PI)), replicated by calcite). (B) The spongelike PS matrix. (C) Cross section of a patterned PS film showing the continuous matrix ((211) face) after PI removal, which exposes the two gyroid networks for crystal infiltration. (D–E) High-magnification images of the (421) and (100) cross-sectional fracture planes of the porous PS film, same magnification. (F) Schematic representation of the crystallization process. (Left to right) Films of the PS/PI copolymer self-assemble into a double-gyroid microphase morphology. After removal of the two PI networks (red and orange) from the PS matrix (blue), calcite crystals nucleate (a) on the surface of or (b) inside the polymer film and grow into the porous networks, leading to a gyroid-patterned single crystal, visible after PS matrix removal. *Image taken from Finnemore et al. (2009) with permission of Wiley-VCH.* (See color plate.)

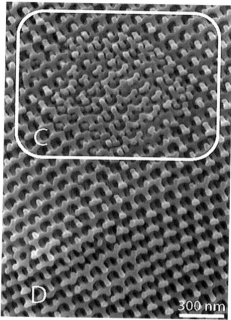

Figure 13.6 Examples of the replication of the full double-gyroid (both networks) and just a single network are shown and are marked C and D, respectively. *Image taken from Finnemore et al. (2009) with permission of Wiley-VCH.*

in a 1:1 water/methanol solution for 1 min. The PS template was removed by heating at 385 °C under oxygen for 2 h.

$CaCl_2$ concentrations >50 mM yielded amorphous precursors, which turned out to be important for a successful infiltration of the template with increasing penetration depth with increasing concentration. The final resulting single calcite crystals reproduced either a single- or a double-gyroid network as shown in Fig. 13.6. However, the size of these nanostructured crystals was limited to the range of microns.

Besides these spectacular gyroid morphologies, also other phases or colloidal objects like block copolymer micelles, vesicles, or any kind of aggregates can be used as templates for single crystals. However, since their size is usually smaller than that of the single crystal, they get occluded into the single crystal (Page, Nassif, Boerner, Antonietti, & Coelfen, 2008) (see Section 2.2).

2.2. Incorporation of polymers

In biominerals, polymers are occluded and it was found that these occluded polymers enhance the mechanical properties (Weiner, Addadi, & Wagner,

2000). This can also be achieved with synthetic polymers. Examples are block copolymer aggregates (Page et al., 2008) or micelles (Kim et al., 2012). But also, an entire polymer network can get occluded into a single crystal if the crystal is grown in a hydrogel. That way, a polymer—single-crystal hybrid material—can be created. One example is the calcite single crystal with an occluded agarose hydrogel (Li, Xin, Muller, & Estroff, 2009).

When calcite was grown in agarose hydrogels (1 w/v%) by the previously described ammonium carbonate gas diffusion technique (see Section 2.1), calcite crystals with the typical rhombohedral morphology were obtained. What is remarkable is that tomographic reconstruction of the agarose network inside the calcite single crystal revealed a 3D random network, which penetrates throughout the crystal as evidenced for the nanometer scale (Li et al., 2009). Since the calcite single-crystal lattice was demonstrated in the areas not occupied by the network fibers, the calcite single crystal must have grown around the agarose fibers without significant disruption or displacement of the fiber network. Otherwise, the fiber network would not be homogenously and randomly distributed throughout the single crystal anymore. This work on inclusion of polymers into single crystals is not an obvious polymer-mediated crystallization but becomes important when it comes to single-crystal–polymer hybrid materials that exhibit a large organic–inorganic interface.

3. SOLUBLE POLYMERS

Soluble polymers are of much interest in terms of polymer-mediated crystallization. Meanwhile, an uncountable number of studies are available on the application of polymers to exert control over a crystallization reactions. However, a considerable number of these studies are descriptive, and it is difficult to draw conclusions from about the precise nature of the polymer interaction with the crystal and the mechanism of polymer-mediated crystallization. Nevertheless, some principles that were revealed will be discussed in the following.

3.1. Types of soluble polymers for crystallization control

A first question when dealing with polymer-mediated crystallization is certainly which type of polymers is especially suitable for crystallization control. A considerable number of biopolymers, which are active in biomineralization of calcium-based biominerals, are charged. Some of these macromolecules like phosphoryns, dentin sialoproteins, asp-rich proteins, or glycopolymers like keratan sulfate contain a very high amount of acidic residues, which are able to interact with calcium. Often, the acidic residues are even present in a block-like arrangement. One example is the acidic binding pentapeptide sequence in statherin, a potent hydroxyapatite growth inhibitor in our saliva. There are two conclusions that can be drawn from these observations: (a) Polyelectrolytes are promising candidates for polymer-mediated crystallization and (b) block copolymer design is advantageous.

While the advantage of polyelectrolytes is obvious due to their electrostatic interaction with ionic crystals, the advantage of a block copolymer design of a potent molecule for polymer-mediated crystallization is less obvious. Similar considerations apply for graft copolymers. The advantage of block copolymers can be demonstrated with the scheme in Fig. 13.7 (Yu & Cölfen, 2004).

Low-molar-mass additives can adsorb well on certain crystal faces and thus influence further crystal growth, but a nanoparticle is not necessarily by adsorption (Fig. 13.7A). On the other hand, unspecific polymer adsorption, which is usually applied for the steric stabilization of nanoparticles, is not face-selective and does not allow control over the direction of further crystal growth (Fig. 13.7B). In a block copolymer, these two features can be combined by designing a block that selectively binds to certain crystal faces, while a second block can be designed to maintain steric nanoparticle stabilization (Fig. 13.7C). While this feature can be advantageous for crystal growth via face-selective polymer adsorption in a classical crystal growth reaction (see Section 3.4), it is clearly advantageous when it comes to non-classical crystallization pathways. In these crystallization reactions, crystallization proceeds via the controlled self-organization of nanoparticles rather

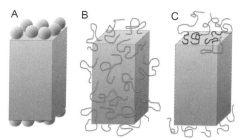

Figure 13.7 Face-selective adsorption of ions or low-molar-mass additives (A), steric particle stabilization by polymers (B), and face-selective adsorption and particle stabilization by a double hydrophilic block copolymer (Cölfen, 2001) (C). (For color version of this figure, the reader is referred to the online version of this chapter.)

than atoms or ions or molecules like in the classical pathway. Here, a block copolymer additive design is of great advantage. While one block can be (ideally) designed as a face-selective binding block, the second block can be designed in a way that it only provides a weak steric stabilization. This allows for a balance of van der Waals attraction and steric repulsion, and therefore, nanoparticles that collide via thermal motion do not immediately stick to each other, nor do they repel each other enough to remain as individual nanoparticles in dispersion. Instead, they stay in close proximity facilitating the formation of superstructures (nanoparticle attraction) with possibilities for positional correction (nanoparticle repulsion). That way, the nanoparticle superstructure can find its energetic minimum of a defect-free superstructure of coaligned nanoparticles. With face-selective polymer adsorption, even directional control over the nanoparticle assembly process can be achieved. Since most of the relevant crystallization reactions take place in aqueous environment, the so-called double hydrophilic block copolymers (Cölfen, 2001) turn out to be especially effective additives for polymer-mediated crystallization reactions—especially nonclassical crystallization reactions.

3.2. Nucleation and crystallization inhibition

Polymers can already influence a crystallization reaction even before nucleation has taken place. The reason is that they can complex ions, which play a role in crystallization, interact with so-called prenucleation clusters (Gebauer, Voelkel, & Coelfen, 2008), and promote or inhibit nucleation. And after nucleation, they can promote or inhibit crystallization, determine whether the further crystallization path is classical or nonclassical, and influence crystal growth in various ways. Prenucleation clusters are a recently discovered species that exists prior to nucleation and thus has to be taken into account if polymers are added for the modification of a crystallization reaction. There are a number of different specific features on how different polymers can already influence the prenucleation phase significantly in a polymer-specific way (Gebauer, Coelfen, Verch, & Antonietti, 2009; Verch, Gebauer, Antonietti, & Colfen, 2011). Prenucleation clusters are treated in a separate chapter in this volume and will not be discussed in further detail here.

However, it is important to mention that one and the same polymer can fulfill multiple roles in a crystallization reaction since there can be different interactions required in the prenucleation, nucleation, and postnucleation phase, which are important for the crystallization reaction.

To further complicate the situation, whether a polymer acts as a nucleator or inhibitor can depend on concentration. In low concentrations, polymers can act as nucleation centers, because they can either adsorb on a surface and act as a center of heterogeneous nucleation or sequester ions by interactions, which also can promote nucleation. At increased concentration however, they can lead to the stabilization of early nucleation species and thus inhibit further growth. This consideration has experimental support. Van der Leeden et al. suggested that low additive concentrations in solution can serve as heterogeneous nucleation centers; however, upon an increase of concentration, the same molecules can inhibit further growth of formed nuclei and nanoparticles (van der Leeden, Kashchiev, & van Rosmalen, 1993). This was experimentally observed for polyelectrolytes and proteins that act not only as inhibitors in solution but also as nucleators when immobilized on a surface (Campbell, Ebrahimpour, Perez, Smesko, & Nancollas, 1989; Ofir, Govrin-Lippman, Garti, & Furedi-Milhofer, 2004; Tsortos, Ohki, Zieba, Baier, & Nancollas, 1996). Therefore, the role of a polymer in the prenucleation, nucleation, and postnucleation phase has to be investigated not only in terms of chemical functionalities, which determine most of the polymer features in crystallization control, but also in terms of the polymer concentration as a nonnegligible variable. What is important to keep in mind is that the influence of a polymer in a polymer-mediated crystallization reaction does not start just after nucleation, but already in the earliest prenucleation stages. This is a reason why polymer-mediated crystallization can lead to precursor species like polymer-induced liquid precursors (PILPs), which can play a significant role in polymer-mediated crystallization (Gower, 2008).

3.3. Polymer-induced liquid precursors

PILPs are a useful precursor phase because they can be molded into any shape (Gower, 2008). Via PILP phases, helical shapes (Fig. 13.8) were generated (Gower & Tirrell, 1998) as well as thin films (Gower & Odom, 2000), for example. In principle, every shape is possible to generate via a PILP phase if a suitable mold is available.

PILP was first found for the $CaCO_3$ system when μg/mL amounts of polyaspartic or polyacrylic acid were added to a $CaCl_2$ solution, which was then transferred into an ammonium carbonate gas diffusion setup (see Section 2.1 for description) (Gower & Tirrell, 1998). The PILP droplets grow up to the micrometer range and can be observed with a conventional

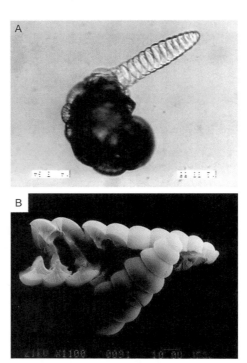

Figure 13.8 Hollow helix synthesized via a PILP phase. (A) is a light micrograph and (B) a scanning electron microscopy image. The helix was fractured by micromanipulation. Bar = 10 μm. *Image reproduced from Gower and Tirrell (1998) with permission of Elsevier.* (For color version of this figure, the reader is referred to the online version of this chapter.)

light microscope. This also allows one to observe the crystallization process. Especially useful in this respect is polarization microscopy, which enables one to detect a crystal via its birefringence. Unfortunately, the gas diffusion method is not very reproducible and the results depend not only on the size of the vessels, surface area, and other parameters but also on the age of the ammonium carbonate itself, which determines its activity to release CO_2. The precise nature of the PILP formation process is still unknown, which could be a result of the difficulty to isolate PILP phases that are very unstable towards crystallization. PILP was meanwhile described for other inorganic systems as well, like calcium phosphate. However, the droplets are very small in that case and cannot be observed under a light microscope.

PILP can also be formed for organic molecular crystals. For these systems, they are much more stable against crystallization compared to their inorganic counterparts. This is likely an effect of the lower lattice energy of organic systems resulting in reduced crystallization tendency. Organic PILP has been described for glutamic acid, lysine, and histidine (Wohlrab, Cölfen, & Antonietti, 2005) as well as an organic azo pigment (Ma et al., 2009). In these cases, the PILP phase could be isolated and its composition analyzed (Wohlrab et al., 2005), showing that a substantial amount of the polymer is included in the PILP phase. What is needed for the generation of the organic PILP phase is a strongly interacting polymer. In case of charged amino acids, this can be a polyelectrolyte with opposite charge. An example is polyethyleneimine, which generates PILP with glutamic acid. Or in case of the azo pigment, a block copolymer was used with one block modified with parts of the pigment molecule, which also maintained a strong interaction between the polymer additive and the crystallizing molecules. When PILP droplets of organic molecules are left to crystallize without further manipulation, they can form porous microspheres as shown in Fig. 13.9. The pigment example also shows that the amount of strongly interacting polymer is very low. In that case, the molar ratio of the polymer to crystallizing molecule was 1:430. If PILPs are used to generate thin films, these

Figure 13.9 SEM images at different magnifications of pigment microspheres of nanoplates obtained in mixed solvents of water and isopropanol (50:50, v/v) via a block copolymer ABABA-acetoacetyl-PEI-b-PEG. The molar ratio of PY181 and ABABA-acetoacetyl-PEI-b-PEG was about 430:1. *Figure reproduced from Ma et al. (2009) with permission of Wiley-VCH.*

films can have a hierarchical structure and can also be mesocrystalline (Jiang, Gong, Volkmer, Gower, & Coelfen, 2011; Jiang et al., 2013). The biggest disadvantage of PILP is that it is only present in a very small region of the phase diagram, as shown in the phase diagram of the glutamic acid–polyethyleneimine system (Jiang, Gower, Volkmer, & Coelfen, 2012). Since this is the solvent-rich region in the phase diagram with only a small amount of the crystallizing component, it is not possible to generate large amounts of PILP to use them in large-scale applications.

3.4. Face-selective adsorption

The most straightforward way to control the morphology of a growing crystal is face-selective adsorption of an additive. The basis for this is Wulff's rule, which was already established in 1901 (Wulff, 1901). This rule defines the shape of a crystal under equilibrium conditions. For all exposed crystal faces, the sum of the products of surface energy and area of the respective crystal face must be minimal. This means that high-energy faces will only have a very small surface area if they are exposed at all, while low-energy faces will become exposed with large surface area. In other words, the morphology of the crystal reflects the surface energies of the exposed faces. Since high-energy faces also grow faster than the low-energy faces, they often vanish upon crystal growth as demonstrated in Fig. 13.10. In this example, the high-energy 101 face grows faster than the low-energy 100 and 001 faces. The dashed line shows that after some time, the area of the 100 and 001 faces became larger (dashed line in Fig. 13.10), and after a second time interval, the 101 face vanished completely (dotted line in Fig. 13.10).

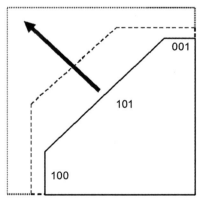

Figure 13.10 Fast-growing high-energy faces (e.g., 101 face) vanish upon crystal growth and the low 100 and 001 crystal faces stay exposed.

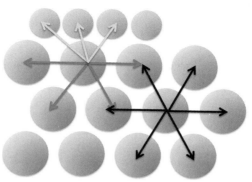

Figure 13.11 Adsorption of an additive (illustrated with spheres in grey) on a crystal surface. The surface energy will be decreased because some bonds can be formed between the surface atoms and the adsorbed molecule (green arrows). This brings the surface atom into a similar situation to a bulk atom that can form bonds with all neighbors (black arrows). (See color plate.)

Therefore, a good handle to change the morphology of a crystal upon growth is to change its surface energy. The surface energy of a crystal face can be lowered upon adsorption of an additive because this attaches to otherwise unsatisfied surface bonds and thus decreases the surface energy (Fig. 13.11). Polymers are especially effective, because one polymer molecule can form multiple bonds to surface atoms.

Especially effective are double hydrophilic block copolymers with one block maintaining water solubility and the second block designed to bind crystal faces selectively. This task is not easy with flexible blocks, but if a rigid, stable block like a ring can be designed, face-selective binding is possible.

An illustrative example for this strategy is the controlled morphosynthesis of gold nanostructures in the presence of a double hydrophilic polyethylenoxide (PEO) block copolymer PEO-*b*-1,4,7,10,13,16-hexaazacycloocatadecan (hexacyclen) ethyleneimine macrocycle (Yu, Colfen, & Mastai, 2004). Gold was synthesized from auric acid by self-reduction in the presence of the polymer in aqueous solution. Very thin and electron-transparent triangles, truncated triangular nanoprisms, and hexagons of gold were produced in the presence of this polymer. The particles were single crystals as confirmed by the selected area electron diffraction pattern. The exposed face is (111). This shows that the hexacyclen block face selectively binds to (111) faces of Au nanoparticles and lowers their surface energy. This leads to the preferential exposure of these faces and in the formation of the observed very thin plates.

The face-selective polymer adsorption onto the (111) faces can be explained by the good geometrical match of the interacting nitrogens in the hexacyclen part to the Au hexagons on the (111) face.

Polar peptides and proteins can also be very effective additives for face-selective binding. They are monodisperse in size and chemical functionality and can form defined secondary structures, which can match the surface structures of selected crystal faces. An α-helical peptide (CBPI) with an array of aspartyl residues was designed to bind to the {1–10} prism faces of calcite (DeOliveira & Laursen, 1997). The CBP1 peptide was added to rhombohedral calcite seed crystals growing from a saturated $Ca(HCO_3)_2$ solution. When CBP1 was added to seed crystals as shown in Fig. 13.12A, and growth was continued, the calcite crystals elongated along the [001] direction (c-axis) with rhombohedral {104} caps (Fig. 13.12B). After washing the crystals with water and replacing the mother solution with fresh saturated $Ca(HCO_3)_2$ solution, a regular rhombohedron formed by a "repair process" occurred with subsequent growth on the putative prism surfaces (Fig. 13.12C). Studded crystals were formed at 25 °C by epitaxial growth perpendicular to each of the six rhombohedral surfaces (Fig. 13.12D and E). After washing these crystals and regrowing them in fresh $Ca(HCO_3)_2$ solution, repair of the nonrhombohedral surfaces was again observed. In each study, a new rhombohedron was formed, and thus, six regular rhombohedra overgrew the original seed (Fig. 13.12F).

Face-selective polymer adsorption depends on the concentration of the polymer. At low polymer concentrations, the polymer can select among several faces for adsorption. As adsorption and desorption take place, an error correction routine can become functional, which leads to the adsorption of the polymer on faces where the interaction energy is the largest. If the polymer concentration is increased above a certain level, nonspecific binding will also take place because there is too much polymer for face selection, since the appropriate crystal faces are already covered by polymer so that nonspecific binding starts to take place.

3.5. Selecting between classical and nonclassical growth modes

Classical crystallization uses atoms, ions, or molecules as building blocks in crystal growth, whereas nonclassical crystallization is particle-mediated. In the latter case, nanoparticles are nucleated first, which are then temporarily stabilized by the polymer to prevent immediate and uncontrolled aggregation. It can therefore be a question of polymer concentration and

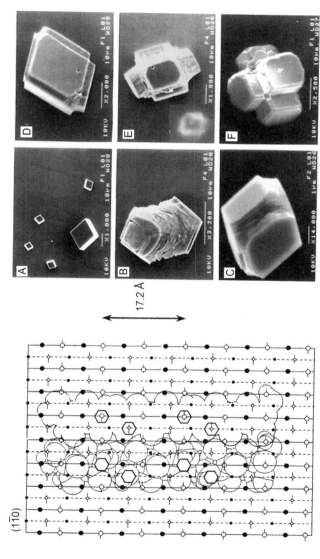

Figure 13.12 (Left) The footprint of two α-helical peptide (CBP1) molecules binding to the (1$\bar{1}$0) prism faces of calcite. The filled circles are Ca^{2+} ions, and open circles are CO_3^{2-} ions. Large circles are ions in the plane of the surface and small circles are 1.28 Å behind this plane. The hexagons indicate that peptide carboxylate ions occupy CO_3^{2-} sites on the corrugated surface. (Right) SEM micrographs showing the effect of CBP1 on the growth of calcite crystals. (A) Calcite seed crystals formed from seed crystals in saturated $Ca(HCO_3)_2$ containing ca. 0.2 mM CBP1. (B) Elongated calcite crystals showing typical rhombohedral morphology. (C) "Repair" and reexpression of rhombohedral surfaces when crystals from (B) are allowed to grow in saturated $Ca(HCO_3)_2$ after removal of CBP1 solution. (D and E) Respective earlier and later stages of growth of calcite crystals from rhombohedral seed crystals at 25 °C in saturated $Ca(HCO_3)_2$ containing ca. 0.2 mM CBP1. (F) "Repair" and reexpression of rhombohedral surfaces when crystals from (E) are allowed to grow in saturated $Ca(HCO_3)_2$ after removal of CBP1. *Reprinted from DeOliveira and Laursen (1997) with permission of the American Chemical Society.*

Figure 13.13 Typical SEM images of calcite mesocrystals obtained on a glass slip by the ammonium carbonate decomposition gas diffusion reaction after 1 day in 1 mL of solution with different concentrations of Ca^{2+} and polystyrene sulfonate: (A) $[Ca^{2+}]=1.25$ mmol/L, [PSS] = 0.1 g/L; (B) $[Ca^{2+}]=1.25$ mmol/L, [PSS] = 0.5 g/L; (C) $[Ca^{2+}]=1.25$ mmol/L, [PSS] = 1.0 g/L; (D) $[Ca^{2+}]=2.5$ mmol/L, [PSS] = 0.1 g/L; (E) $[Ca^{2+}]=2.5$ mmol/L, [PSS] = 0.5 g/L; (F) $[Ca^{2+}]=2.5$ mmol/L, [PSS] = 1.0 g/L; (G) $[Ca^{2+}]=5$ mmol/L, [PSS] = 0.1 g/L; (H) $[Ca^{2+}]=5$ mmol/L, [PSS] = 0.5 g/L; (I) $[Ca^{2+}]=5$ mmol/L, [PSS] = 1.0 g/L. *Reproduced from Wang et al. (2006) with permission of Wiley-VCH.*

supersaturation of the crystallizing species whether a crystallization reaction proceeds along the classical or the nonclassical pathway. This was nicely demonstrated for the system $CaCO_3$ (Fig. 13.13) (Wang, Antonietti, & Cölfen, 2006). At the lowest supersaturation and polymer additive concentration (Fig. 13.13A), the typical calcite rhombohedra are obtained with a slightly truncated triangular corner, which is the high-energy {001} face stabilized by polymer adsorption.

Increasing the supersaturation but leaving the polymer concentration unchanged (Fig. 13.13D) shows a crystal already deviating from the rhombohedral shape but still with the (001) face truncation. At the highest supersaturation and still the same polymer concentration (Fig. 13.13G), a donut-shaped mesocrystal is formed, which is composed of calcite nanoplatelets. This clearly shows that increasing supersaturation shifts the pathway from classical to nonclassical crystallization. This is plausible because increasing

Figure 13.14 Change of the rhombohedral equilibrium $CaCO_3$ morphology with six exposed {104} faces (gray) by lowering the surface energy of the {001} faces (white). The morphology change from rhomboeder to a hexagonal platelet is evident and is maintained just by face-selective interface energy decrease of the two {001} faces. Images modeled with Cerius^2 (Accelrys). *Image reproduced from Meldrum and Coelfen (2008) with permission of the American Chemical Society.*

supersaturation promotes particle nucleation and many nucleation events lead to the formation of many nanoparticles, which can only grow as long as the supersaturation still exists. The result is many nanoparticles. A low supersaturation on the other hand leads only to very few nucleation events. However, the nucleated nanoparticles can then grow to micron size along the classical crystallization pathway.

If on the other hand the polymer concentration is increased, the degree of face-selective adsorption is increased because a larger surface area can be stabilized. This is nicely visible in Fig. 13.13A–C or D–F.

Figure 13.14 shows the effect of face-selective adsorption of an additive onto the high-energy {001} faces in an idealized way for increasing additive concentration from left to right. This is the scenario to be expected for a classical crystallization pathway. For nonclassical crystallization leading to mesocrystals, there is no similar effect with increasing polymer concentration (see Fig. 13.13G–I). In all cases, the polycrystalline character of the mesocrystals is obvious.

3.6. Mesocrystal formation

There are several possibilities for mesocrystal formation. These are shown in Fig. 13.15. For polymer-mediated crystallization, especially possibilities (A) and (F) are of interest.

A typical example for alignment of nanoparticles along a matrix of insoluble molecules (Fig. 13.15A) is the biomineral bone. Here, hydroxyapatite platelets are aligned along a matrix of parallelly oriented collagen molecules Fratzl, Gupta, Paschalis, & Roschger (2004).

Whether or not the confinement in the gap zones in the collagen fibril is already sufficient to orient the nucleating nanoparticles or if noncollagenous

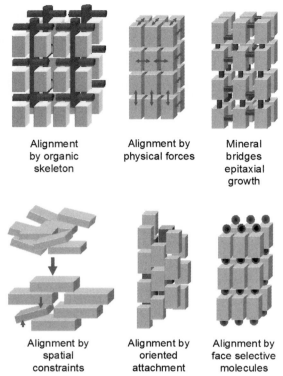

Figure 13.15 Six different possibilities to form a mesocrystal. (For color version of this figure, the reader is referred to the online version of this chapter.)

proteins are needed to initiate and orient the growth of the hydroxyapatite platelets or if the collagen plays an active role in the nucleation and alignment of the nanoplatelets was long discussed. Recent results discuss an active role of the collagen in the mineralization process (Wang et al., 2012). Collagen already controls the apatite structure on the atomic scale as well as size and distribution on larger length scales. To mimic this strategy is not straightforward. One successful attempt was reported for the modification of chitin with peptides with subsequent mineralization of $CaCO_3$ (Kumagai et al., 2012). These peptides are special in the sense that they have a specific binding motif to bind to the oriented chitin matrix and they have a second carboxylic acid function, which is able to bind to $CaCO_3$. This binding peptide leads to the remarkable orientation of calcite crystals along the structured chitin matrix building up a mesocrystal. From the two examples mentioned earlier, it becomes clear that an important requirement is the ordered insoluble structural matrix (collagen, chitin, etc.), but the second requirement is that a forming mineral can orient along this structural matrix. Two possibilities were outlined earlier. The first one is that defined gaps exist as nucleation space for the mineral inside the structural matrix. The second possibility is to use linker molecules to attach forming crystals along an oriented structural matrix. It is clear that such linker molecules are highly specialized and need at least two functions: a binding moiety to the oriented matrix and a binding moiety to the crystal.

Mesocrystal formation can also be triggered by face-selective polymer adsorption onto defined crystal faces (Fig. 13.15F). Face-selective adsorption of a polymer is difficult to achieve. A few selected examples were reported in Section 3.4. However, face-selective polymer adsorption not only can change the morphology of the growing crystal by changing the surface energy of the different crystal faces but also can furthermore trigger the self-assembly of nanocrystals to form a mesocrystal. Such mesocrystals can have very complex shapes as it was demonstrated for helices formed from $BaCO_3$ nanorods that all arranged in a parallel manner to form helical mesocrystalline superstructures (see Fig. 13.16) (Yu, Colfen, Tauer, & Antonietti, 2005). These helices were grown over many days using the slow

Figure 13.16 (Left) Helical nanoparticle superstructures. The helical fibers formed at room temperature with a block copolymer of poly(ethylene oxide) and a phosphonated strongly crystal-interacting block, 1 g/L, starting pH 4, $[BaCl_2] = 10$ mM. (Right) Default orthorhombic morphology of a $BaCO_3$ crystal. Consideration of the different crystal surfaces reveals that only on {110} a favorable completely cationic face termination can exist that is the most likely face for polymer adsorption as compared to all other faces. *Image reproduced from Yu et al. (2005).* (For color version of this figure, the reader is referred to the online version of this chapter.)

ammonium carbonate decomposition gas diffusion method (see Section 2.1 for description).

This remarkable formation of helical mesocrystalline nanoparticle superstructures can be explained on geometrical grounds on basis of face-selective polymer adsorption. The important features of the phosphonated block copolymer are that the polyethylene oxide block maintains water solubility but does not interact with the crystal and the phosphonated block is a strongly interacting block with the cationic {110} surface of the newly nucleated nanoparticles. All other exposed faces are almost neutral. This leads to face-selective polymer adsorption on {110}. The other faces have only little or no polymer adsorbed. If such nanoparticles collide, the {110} faces are sterically stabilized, because the block copolymer has a stiff and sterically demanding phosphonated block. Therefore, a staggered arrangement of aggregating nanoparticles is favorable as shown in Fig. 13.17 (left). In this arrangement, the polymer has enough space to extend into the solution.

If one turns the view by 90° and looks along the long axis of the orthorhombic nanoparticles, it becomes clear that the statistically preferred faces are the four {011} faces. However, in an existing aggregate, two of these faces are not equal for an approaching nanoparticle as symbolized by the orange and yellow half spheres in Fig. 13.17 (center). To fit the nanoparticle at the yellow position, it has to fit to two faces, whereas it only has to fit to one face at the orange position. If the latter is favored, this creates a counterclockwise turn of the aggregate.

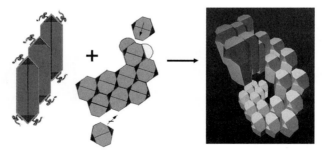

Figure 13.17 Assembly of the orthorhombic nanoparticle units to the chiral mesocrystalline superstructure after polymer adsorption onto {110}. Left of the arrow are two side views of the assembling nanoparticles, and right of the arrow is the resulting superstructure. The sterically demanding polymer adsorbed on the {110} faces forces a staggered arrangement of the BaCO$_3$ nanorods. Turning the view by 90° reveals that for attaching nanoparticles, the {011} faces are not equal and a preferential adsorption side exists. Overlaying these two effects then leads to the formation of a helix structure composed of nanorods shown right of the arrow. (See color plate.)

Together with the staggered arrangement, this creates the formation of a helical mesocrystalline nanoparticle superstructure (Fig. 13.17, right). The handedness of the helix is determined by the direction of the staggered arrangement of the first aggregating nanoparticles. It is equally probable that they arrange into two opposite directions. Consequently, the amount of left- and right-handed helices is equal as was also found experimentally (Yu et al., 2005). This example shows that a complex arrangement of nanoparticles can be coded and triggered by face-selective additive adsorption.

4. CONCLUSION

Polymers are very powerful additives to control crystal growth. They can have a large variety of different functions. One has to distinguish between soluble and insoluble polymers as commonly found in biominerals. The insoluble polymers are called structural matrix and build up a template and scaffold for the subsequent mineralization reaction. They shape the reaction space. The soluble polymers on the other hand are called soluble matrix and influence the crystallization reaction itself. Especially effective for this function are polyelectrolytes and, in the case of true biomineralization, effective proteins that are highly charged. One and the same soluble polymer can have different functions throughout the mineralization reaction. It can, for example, inhibit nucleation and after nucleation face selectively adsorb to control the crystal growth. Also, the polymer additive can control the critical supersaturation and by that way have an influence whether the crystallization reaction proceeds along classical or nonclassical pathways. Therefore, the possibilities to influence a crystallization reaction by polymers are manifold. Polymers can be engineered for face-selective adsorption, and in that way, either the morphology can be controlled in classical crystallization or the aggregation of nanoparticles can be coded in nonclassical crystallization reactions leading to mesocrystals with complex shape. In addition to the crystal morphology, the properties of the crystal are also influenced by the polymer because of the organic–inorganic nature of the hybrid material. Mechanical properties like fracture resistance can especially be improved, because occluded polymers in a crystal hinder crack propagation. Therefore, polymer additives in crystallization reactions are scientifically and technically very interesting, and this research field is very active.

REFERENCES

Campbell, A. A., Ebrahimpour, A., Perez, L., Smesko, S. A., & Nancollas, G. H. (1989). The dual role of poly-electrolytes and proteins as mineralization promoters and inhibitors of calcium-oxalate monohydrate. *Calcified Tissue International, 45*(2), 122–128.

Cölfen, H. (2001). Double-hydrophilic block copolymers: Synthesis and application as novel surfactants and crystal growth modifiers. *Macromolecular Rapid Communications, 22*(4), 219–252.

DeOliveira, D. B., & Laursen, R. A. (1997). Control of calcite crystal morphology by a peptide designed to bind to a specific surface. *Journal of the American Chemical Society, 119*(44), 10627–10631.

Finnemore, A. S., Scherer, M. R. J., Langford, R., Mahajan, S., Ludwigs, S., Meldrum, F. C., et al. (2009). Nanostructured calcite single crystals with gyroid morphologies. *Advanced Materials, 21*(38–39), 3928–3932.

Fratzl, P., Gupta, H. S., Paschalis, E. P., & Roschger, P. (2004). Structure and mechanical quality of the collagen-mineral nano-composite in bone. *Journal of Materials Chemistry, 14*(14), 2115–2123. http://dx.doi.org/10.1039/b402005g.

Gebauer, D., Coelfen, H., Verch, A., & Antonietti, M. (2009). The multiple roles of additives in $CaCO_3$ crystallization: A quantitative case study. *Advanced Materials, 21*(4), 435–439. http://dx.doi.org/10.1002/adma.200801614.

Gebauer, D., Voelkel, A., & Coelfen, H. (2008). Stable prenucleation calcium carbonate clusters. *Science, 322*(5909), 1819–1822. http://dx.doi.org/10.1126/science.1164271.

Gower, L. B. (2008). Biomimetic model systems for investigating the amorphous precursor pathway and its role in biomineralization. *Chemical Reviews, 108*(11), 4551–4627.

Gower, L. B., & Odom, D. J. (2000). Deposition of calcium carbonate films by a polymer-induced liquid-precursor (PILP) process. *Journal of Crystal Growth, 210*(4), 719–734.

Gower, L. A., & Tirrell, D. A. (1998). Calcium carbonate films and helices grown in solutions of poly(aspartate). *Journal of Crystal Growth, 191*(1–2), 153–160.

Hamley, I. W. (1998). *The physics of block copolymers*. Oxford: Oxford Science Publications.

Jiang, Y., Gong, H., Grzywa, M., Volkmer, D., Gower, L., & Coelfen, H. (2013). Microdomain transformations in mosaic mesocrystal thin films. *Advanced Functional Materials, 23*(12), 1547–1555. http://dx.doi.org/10.1002/adfm.201202294.

Jiang, Y., Gong, H., Volkmer, D., Gower, L., & Coelfen, H. (2011). Preparation of hierarchical mesocrystalline DL-lysine center dot HCl-poly(acrylic acid) hybrid thin films. *Advanced Materials, 23*(31), 3548. http://dx.doi.org/10.1002/adma.201101468.

Jiang, Y., Gower, L., Volkmer, D., & Coelfen, H. (2012). The existence region and composition of a polymer-induced liquid precursor phase for DL-glutamic acid crystals. *Physical Chemistry Chemical Physics, 14*(2), 914–919. http://dx.doi.org/10.1039/c1cp21862j.

Kim, Y.-Y., Ganesan, K., Yang, P., Kulak, A. N., Borukhin, S., Pechook, S., et al. (2012). An artificial biomineral formed by incorporation of copolymer micelles in calcite crystals. *Nature Materials, 10*(11), 890–896. http://dx.doi.org/10.1038/nmat3103.

Kumagai, H., Matsunaga, R., Nishimura, T., Yamamoto, Y., Kajiyama, S., Oaki, Y., et al. (2012). $CaCO_3$/chitin hybrids: Recombinant acidic peptides based on a peptide extracted from the exoskeleton of a crayfish controls the structures of the hybrids. *Faraday Discussions, 159*, 483–494. http://dx.doi.org/10.1039/c2fd20057k.

Li, C., & Qi, L. M. (2008). Bioinspired fabrication of 3D ordered macroporous single crystals of calcite from a transient amorphous phase. *Angewandte Chemie, International Edition, 47*, 2388–2393.

Li, H., Xin, H. L., Muller, D. A., & Estroff, L. A. (2009). Visualizing the 3D internal structure of calcite single crystals grown in agarose hydrogels. *Science, 326*(5957), 1244–1247. http://dx.doi.org/10.1126/science.1178583.

Loste, E., Park, R. J., Warren, J., & Meldrum, F. C. (2004). Precipitation of calcium carbonate in confinement. *Advanced Functional Materials, 14*(12), 1211–1220. http://dx.doi.org/10.1002/adfm.200400268.

Lowenstam, H. A., & Weiner, S. (1989). *On biomineralization*. New York: Oxford University Press.

Ma, Y., Mehltretter, G., Plueg, C., Rademacher, N., Schmidt, M. U., & Coelfen, H. (2009). PY181 pigment microspheres of nanoplates synthesized via polymer-induced liquid precursors. *Advanced Functional Materials*, 19(13), 2095–2101. http://dx.doi.org/10.1002/adfm.200900316.

Mahamid, J., Sharir, A., Addadi, L., & Weiner, S. (2008). Amorphous calcium phosphate is a major component of the forming fin bones of zebrafish: Indications for an amorphous precursor phase. *Proceedings of the National Academy of Sciences of the United States of America*, 105(35), 12748–12753. http://dx.doi.org/10.1073/pnas.0803354105.

Meldrum, F. C., & Coelfen, H. (2008). Controlling mineral morphologies and structures in biological and synthetic systems. *Chemical Reviews*, 108(11), 4332–4432. http://dx.doi.org/10.1021/cr8002856.

Ofir, P. B. Y., Govrin-Lippman, R., Garti, N., & Furedi-Milhofer, H. (2004). The influence of polyelectrolytes on the formation and phase transformation of amorphous calcium phosphate. *Crystal Growth & Design*, 4(1), 177–183. http://dx.doi.org/10.1021/cg034148g.

Page, M. G., Nassif, N., Boerner, H. G., Antonietti, M., & Coelfen, H. (2008). Mesoporous calcite by polymer templating. *Crystal Growth & Design*, 8(6), 1792–1794. http://dx.doi.org/10.1021/cg700899s.

Park, R. J., & Meldrum, F. C. (2002). Synthesis of single crystals of calcite with complex morphologies. *Advanced Materials*, 14(16), 1167–1169.

Politi, Y., Arad, T., Klein, E., Weiner, S., & Addadi, L. (2004). Sea urchin spine calcite forms via a transient amorphous calcium carbonate phase. *Science*, 306(5699), 1161–1164. http://dx.doi.org/10.1126/science.1102289.

Song, R.-Q., & Coelfen, H. (2011). Additive controlled crystallization. *CrystEngComm*, 13(5), 1249–1276. http://dx.doi.org/10.1039/c0ce00419g.

Sun, H. B., & Kawata, S. (2004). Two-photon photopolymerization and 3D lithographic microfabrication. *NMR—3D Analysis—Photopolymerization*, 170, 169–273. http://dx.doi.org/10.1007/b94405.

Tsortos, A., Ohki, S., Zieba, A., Baier, R. E., & Nancollas, G. H. (1996). The dual role of fibrinogen as inhibitor and nucleator of calcium phosphate phases: The importance of structure. *Journal of Colloid and Interface Science*, 177(1), 257–262.

van der Leeden, M. C., Kashchiev, D., & van Rosmalen, G. M. (1993). Effect of additives on nucleation rate, crystal growth rate and induction time in precipitation. *Journal of Crystal Growth*, 130(1–2), 221–232.

Verch, A., Gebauer, D., Antonietti, M., & Colfen, H. (2011). How to control the scaling of $CaCO_3$: A "fingerprinting technique" to classify additives. *Physical Chemistry Chemical Physics*, 13(37), 16811–16820. http://dx.doi.org/10.1039/c1cp21328h.

Wang, T. X., Antonietti, M., & Cölfen, H. (2006). Calcite mesocrystals: "Morphing" crystals by a polyelectrolyte. *Chemistry—A European Journal*, 12(22), 5722–5730.

Wang, Y., Azais, T., Robin, M., Vallee, A., Catania, C., Legriel, P., et al. (2012). The predominant role of collagen in the nucleation, growth, structure and orientation of bone apatite. *Nature Materials*, 11(8), 724–733. http://dx.doi.org/10.1038/nmat3362.

Weiner, S., Addadi, L., & Wagner, H. D. (2000). Materials design in biology. *Materials Science & Engineering C-Biomimetic and Supramolecular Systems*, 11(1), 1–8.

Wohlrab, S., Cölfen, H., & Antonietti, M. (2005). Crystalline, porous microspheres made from amino acids using polymer-induced liquid precursor phases. *Angewandte Chemie, International Edition*, 44(26), 4087–4092.

Wucher, B., Yue, W., Kulak, A. N., & Meldrum, F. C. (2007). Designer crystals: Single crystals with complex morphologies. *Chemistry of Materials*, 19(5), 1111–1119. http://dx.doi.org/10.1021/cm0620640.

Wulff, G. (1901). On the question of speed of growth and dissolution of crystal surfaces. *Zeitschrift für Kristallographie und Mineralogie*, 34(5/6), 449–530.

Yu, S. H., & Cölfen, H. (2004). Bio-inspired crystal morphogenesis by hydrophilic polymers. *Journal of Materials Chemistry, 14*, 2124–2147.
Yu, S. H., Colfen, H., & Mastai, Y. (2004). Formation and optical properties of gold nanoparticles synthesized in the presence of double-hydrophilic block copolymers. *Journal of Nanoscience and Nanotechnology, 4*(3), 291–298. http://dx.doi.org/10.1166/jnn.2004.030.
Yu, S. H., Colfen, H., Tauer, K., & Antonietti, M. (2005). Tectonic arrangement of $BaCO_3$ nanocrystals into helices induced by a racemic block copolymer. *Nature Materials, 4*(1), 51–55. http://dx.doi.org/10.1038/nmat1263.
Yue, W., Park, R. J., Kulak, A. N., & Meldrum, F. C. (2006). Macroporous inorganic solids from a biomineral template. *Journal of Crystal Growth, 294*(1), 69–77. http://dx.doi.org/10.1016/j.jcrysgro.2006.05.028.

CHAPTER FOURTEEN

Phage Display for the Discovery of Hydroxyapatite-Associated Peptides

Hyo-Eon Jin[*,†,1], Woo-Jae Chung[*,†,‡,1], Seung-Wuk Lee[*,†,2]

[*]Department of Bioengineering, University of California, Berkeley, California, USA
[†]Physical Biosciences Divisions, Lawrence Berkeley National Laboratory, Berkeley, California, USA
[‡]College of Biotechnology and Bioengineering, Sungkyunkwan University, Suwon, South Korea
[1]These authors contributed equally to this work
[2]Corresponding author: e-mail address: leesw@berkeley.edu

Contents

1. Introduction	306
1.1 Hydroxyapatite-associated proteins	306
1.2 Phage display	307
2. Methods	312
2.1 Materials	312
2.2 Preparation of the target crystals	312
2.3 Phage display for identifying templates for mineralization	313
2.4 Determination of the strongest binding peptides using miniphage library	315
2.5 Alanine-scanning mutation assay	315
2.6 HAP-binding assay using fluorescent dye-conjugated antibody	316
2.7 Solid-phase peptide synthesis	317
2.8 HAP crystal nucleation using peptides	318
2.9 Alignment of the identified peptide with other protein sequences	319
3. Summary	321
References	321

Abstract

In nature, proteins play a critical role in the biomineralization process. Understanding how different peptide or protein sequences selectively interact with the target crystal is of great importance. Identifying such protein structures is one of the critical steps in verifying the molecular mechanisms of biomineralization. One of the promising ways to obtain such information for a particular crystal surface is to screen combinatorial peptide libraries in a high-throughput manner. Among the many combinatorial library screening procedures, phage display is a powerful method to isolate such proteins and peptides. In this chapter, we will describe our established methods to perform phage display with inorganic crystal surfaces. Specifically, we will use hydroxyapatite as a model system for discovery of apatite-associated proteins in bone or tooth biomineralization studies. This model approach can be generalized to other desired crystal surfaces using the same experimental design principles with a little modification of the procedures.

1. INTRODUCTION

In nature, proteins play a critical role in the biomineralization process (Fritz et al., 1994; Mann, 2001; Weiner & Wagner, 1998). It is believed that during the mineralization process, recognition proteins specifically interact with the target surfaces and influence the nucleation and dissolution processes to control the morphology of the target minerals (George et al., 1996; Weiner & Wagner, 1998). Understanding how different peptide or protein sequences selectively interact with the target crystals is of great importance. In order to identify such proteins, the desired protein or mRNA can be extracted and sequenced to determine the primary protein sequences (Belcher et al., 1996; Fritz et al., 1994). Numerous of biochemical assays can then follow to identify active domains of the proteins (Chiu, Li, & Huang, 2010; Pacardo, Sethi, Jones, Naik, & Knecht, 2009). Although conventional approaches to identify these recognition proteins are effective, they are time-consuming and labor-intensive. One of the most promising methods for a particular crystal surface is to screen combinatorial libraries in a high-throughput manner. Among the many combinatorial library screening processes, phage display is a powerful method for isolating recognition peptides (Petrenko, Smith, Gong, & Quinn, 1996; Smith, 1985). In this chapter, we will describe our established methods to perform phage display for inorganic crystal surfaces. Specifically, we will use hydroxyapatite as a model system for discovery of bone- or tooth-associated proteins. This model approach can be generalized to other desired crystal surfaces with a little modification of the procedures using the same experimental design principles.

1.1. Hydroxyapatite-associated proteins

The formation of HAP crystals in the body is thought to occur through templated mineralization of HAP by the surrounding proteins, which include collagen and many HAP-associated proteins (Traub, Arad, &

Weiner, 1989). Although the exact molecular level mechanism of bone mineralization is not understood, it is generally believed that specific proteins associated with mineralization can recognize target surfaces and mineral ions of HAP and control the nucleation of bone and tooth crystals. Once the HAP crystals have fully grown into their programmed shapes and sizes, the proteins inhibit further growth of the crystals and maintain these morphologies throughout a person's lifetime. Proteins are believed to precisely control these processes of crystal growth promotion and inhibition. However, the detailed mechanisms by which mineralization occurs remain elusive.

More than 200 HAP-associated proteins have been identified. These include collagens (Traub et al., 1989; Zhang, Liao, & Cui, 2003), amelogenin (Du, Falini, Fermani, Abbott, & Moradian-Oldak, 2005; Hu et al., 1996; Ishiyama, Mikami, Shimokawa, & Oida, 1998), statherin (Long et al., 1998; Raj, Johnsson, Levine, & Nancollas, 1992), osteocalcin (Ducy et al., 1996; Hoang, Sicheri, Howard, & Yang, 2003), and many acidic phosphoproteins (MacDougall et al., 1997). Although some of the bone-associated proteins, such as osteocalcin and osteonectin, have been analyzed by protein crystallography (Hoang et al., 2003), the molecular mechanisms by which these proteins recognize their target surfaces remain unknown. Proteins change their conformational structure when bound to crystal surfaces. In addition, complex and long encrypted protein sequences vary among all bone- and tooth-related proteins, thus complicating systematic studies of their interactions at the bone interface. Therefore, conventional approaches for studying these proteins have not elucidated their mechanistic role in biomineralization. Recent advancement of genetic engineering of organisms enables us to discover novel functional peptides or proteins in an accelerated manner. Among the many high-throughput screening process of the library techniques, phage display is considered as one of the most powerful techniques to implement in the general laboratory setting, because it generates rapid experimental results and because large combinatorial peptide phage libraries are already commercially available.

1.2. Phage display

Phage display is a combinatorial process used for identifying specific binding peptides through a rapid high-throughput screening process involving phage and bacterial biology (Smith, 1985; Smith & Petrenko, 1997). Various types of phages can be used for this process (see Table 14.1). Due to its commercial availability from New England Biolabs (Ipswich, MA), M13

Table 14.1 Commercially available phage peptide libraries and kits

Company	Library	Library type	Homepages
New England Biolabs	Ph.D.™ Phage Display Peptide Library	Random 7-mer, constrained 7-mer, 12-mer peptide fused to a minor coat protein (pIII) of M13 phage	www.neb.com
Novagen	T7Select® Phage Display System T7Select® cDNA Library	Cloning kit in T7 vectors for library preparation Various human organ cDNA library in T7 vector	www.novagen.com
Dyax	Custom phage display library	M13, lambda, T4 and T7 phages Antibody, peptide, small protein libraries	www.dyax.com
GenScript	Custom antibody phage display	Generation of antibody libraries FASEBA screening technology for expression levels, stabilities, and affinities	www.genscript.com
Creative Biolabs	Custom phage display library	Phage display peptide libraries Antibody libraries including immunized, or naïve, and semisynthetic/synthetic antibody libraries cDNA phage display libraries	www.creative-biolabs.com/

bacteriophage (phage) is the most commonly used phage for this purpose. M13 phage is a bacterial virus composed of a single-stranded DNA encapsulated by various major and minor coat proteins. It has a long-rod filament shape that is approximately 880 nm in length and 6.6 nm in width (Smith & Petrenko, 1997). The viral capsid is mainly composed of 2700 copies of a helically arranged major coat protein, pVIII, and five copies of the pIII and pIX minor coat proteins at the virus ends (Petrenko et al., 1996; Smith & Petrenko, 1997). The insertion of randomized DNA sequences into specific locations of the phage genome creates a highly diverse library of peptides (up to 10^{11} random sequences) that can be displayed on surfaces of the viral particles (Petrenko et al., 1996; Smith & Petrenko, 1997). Various types of phage peptide libraries and library construction kits are also commercially available from a number of biotech companies (Table 14.1).

To select the best peptide sequence that will bind to a given target material, the engineered phage library pool goes through several rounds of a selection process commonly called screening or biopanning (Fig. 14.1).

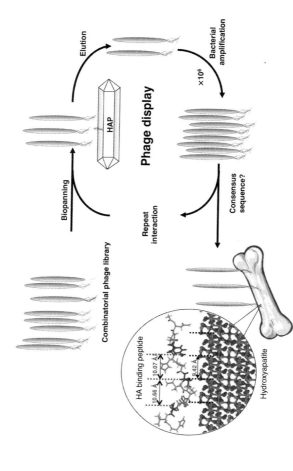

Figure 14.1 Schematic diagram of phage display process for discovering peptide motifs that recognize hydroxyapatite. (See color plate.)

Initially, the entire phage library is allowed to interact with the target. The nonbound phages are washed away, and the bound ones are eluted and amplified through *Escherichia coli* bacterial host infection. This evolutionary approach of selecting the best binding peptide sequences is repeated several times to enrich the eluted phage with those that have the best affinity for the target. Finally, the DNAs of the dominant binding phage are sequenced to identify the peptide inserts that have the greatest affinities for the target. Traditionally, this technique has been used to identify protein epitopes and small molecule antibodies and to study protein–protein interactions (Mullaney & Pallavicini, 2001; Presta, 2005). Recently, phage display has also been utilized to identify peptide sequences that have specificity for a variety of inorganic substances, including semiconductor (Lee, Mao, Flynn, & Belcher, 2002; Whaley, English, Hu, Barbara, & Belcher, 2000), magnetic (Reiss et al., 2004), metallic (Danner & Belasco, 2001), and optically interesting materials (Flynn, Lee, Peelle, & Belcher, 2003; Flynn, Mao, et al., 2003; Lee et al., 2002; Whaley et al., 2000). Moreover, these isolated peptides could induce the crystallization of desired materials by acting as templates (Lee et al., 2002; Mao et al., 2003, 2004; Naik, Stringer, Agarwal, Jones, & Stone, 2002; Sarikaya, Tamerler, Jen, Schulten, & Baneyx, 2003; Sweeney et al., 2004).

Previously, many research groups have used phage display to isolate peptides that interact with polycrystalline and single-crystalline HAP and showed various HAP-specific activities (Gungormus et al., 2008; Segvich, Smith, & Kohn, 2009; Weiger et al., 2010). In our recent work (Chung, Kwon, Song, & Lee, 2011), we also used single-crystal (100) HAP because, based on model bone structures, we hypothesized that the bone crystals mainly expose (100) HAP crystals to interact with collagen type I and collagen-/bone-associated proteins. Using phage display, we identified a 12-residue peptide that bound to single-crystal (100) HAP surfaces under physiological pH conditions (pH 7.5) (Fig. 14.2). This peptide was able to template the nucleation and growth of crystalline HAP mineral in a sequence- and composition-dependent manner (Fig. 14.3). The sequence responsible for the mineralizing activity resembled the tripeptide repeat (Gly-Pro-Hyp; Hyp: hydroxyproline) of type I collagen, a major component of bone extracellular matrix. Using a panel of synthetic peptides, we defined the structural features required for mineralization activity. The results suggest a model for the cooperative noncovalent interaction of the peptide with HAP and support the theory that native collagen has a mineral-templating function *in vivo*.

Identified phage clones at pH 7.5	Aligned amino acid sequences														
CLP12	Q	P	V	H	P	T	I	P	Q	S	V	H			
CLP7C		C	Q	Y	P	T	L	K	S	C					
CLP7			H	A	P	V	Q	P	Q						
Poly12	T	M	G	F	T	A	P	R	F	P	H	Y			
Collagen Type I major repeat	G	P	O	G	P	O	G	P	O	G	P	O	G	P	O
Collagen α I (1301–1315)	Y	P	T	Q	P	S	V	A	Q	K	N	W	Y	I	S
Amelogenin (125–139)	Q	P	Y	Q	P	Q	P	V	Q	P	Q	P	H	Q	P
Statherin (36–50)	G	P	Y	Q	P	V	P	E	Q	P	L	Y	P	Q	P

Figure 14.2 The identified HAP-binding peptide sequences from phage display and corresponding bone- and tooth-associated protein sequences from the protein database. The best binding peptide (CLP12) for the (100) HAP surface at pH 7.5 possessed repeating proline residues at i and $i+3n$ positions (positions 2, 5, and 8) and periodic hydroxylated residues (positions 3, 6, and 10) resembling the major repeat sequence of type I collagen and amelogenin. *Reprinted with permission from Chung et al. (2011). Copyright (2011) American Chemical Society.* (See color plate.)

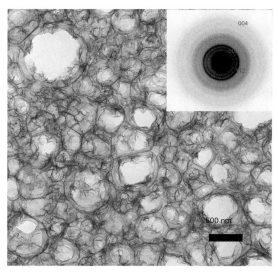

Figure 14.3 Transmission electron micrograph of mineralized HAP crystal templated by CLP12 peptide. *Reprinted with permission from Chung et al. (2011). Copyright (2011) American Chemical Society.* (See color plate.)

We believe that our approach on how to prepare the target crystal surfaces and discover the functional peptides through phage display can be generalized to many crystal surfaces involved in other biomineralization processes by applying similar experimental design principles. Therefore, here we provide the detailed methodology that we use in our laboratory as a model procedure. We will describe (1) target crystal preparation, (2) phage display for the crystal surface, and (3) functional peptide identification.

2. METHODS
2.1. Materials

Ph.D.™-12 and Ph.D.™-C7C Phage Display Peptide Library Kits were purchased from New England Biolabs (Ipswich, MA, USA). Tris–HCl, glycine–HCl, NaCl, BSA, Tween 20, Atto 425-streptavidin, and IPTG/X-gal were purchased from Sigma-Aldrich (St. Louis, MO). Polycrystalline hydroxyapatite was obtained from Alfa Aesar Co. (Milwaukee, WI, USA). For peptide synthesis, preloaded (cysteine and biotinylated lysine) Wang resins were purchased from Novabiochem (San Diego, CA). Fmoc-aminodiethoxy-acetic acid and cleavage reagents trifluoroacetic acid, thioanisole, water, phenol, ethanedithiol, and triisopropyl silane were also obtained from Sigma-Aldrich (St. Louis, MO).

2.2. Preparation of the target crystals

It is critical to prepare well-defined target crystal surfaces for successful discovery of the functional peptides and design of the binding and biomineralization assays. Many researchers have been identifying functional peptides from phage display using polycrystalline HAP as target materials. Although one can obtain interacting peptides this way, it is then very challenging to characterize the interaction between the identified peptides and the target crystalline surfaces. In order to study and optimize the interaction between peptides and HAP surfaces at the molecular level, the target surfaces need to be well defined. The more you know about your crystal targets, the higher chances you have to discover functional peptides. The following is the modified protocol for the synthesis of single-crystal HAP as targets of peptides based on the molten salt synthesis method (Tas, 2001).

2.2.1 Methods
1. Mix potassium sulfate (K_2SO_4) powders with commercial polycrystalline HAP powders at a weight ratio of 1.6 (K_2SO_4/HAP).
2. Place mixture in a clean alumina crucible and heat it in a furnace from room temperature to 1190 °C at a rate of 5 °C/min (~4 h).
3. Hold the temperature at 1190 °C for 3.5 h.
4. Cool the product naturally to room temperature within the furnace.
5. Wash the solidified mass with 90 °C Milli-Q water for three times.
6. Dry single-crystal HAP whiskers in air for 24 h.

7. Place 10 mg of crystals in a microtube. Etch the surface of target HAP single crystals with 1 mL of glycine–HCl buffer (0.2 M, pH 2.2) for 5 min to remove impurities (gently rock at room temperature).
8. Neutralize the mixture by adding 150 µL of pH 9 TBS (1 M Tris–HCl, 150 mM NaCl, autoclaved). Centrifuge the mixture and pipette out the etchant solution.
9. Wash crystals with TBS (50 mM Tris–HCl, pH 7.5, 150 mM NaCl, autoclaved) for five times.
10. Store crystals (~10 mg/mL) in pH 7.5 TBST (TBS containing 0.1%, v/v, Tween 20).

2.3. Phage display for identifying templates for mineralization

We can identify HAP-specific binding peptides by using phage display. The phage display protocol in the succeeding text uses of a mixture of three different commercial libraries.

2.3.1 Methods

1. Mix equal amounts of three different types of commercially available phage library suspensions, 7 mer, 7 mer constrained, and 12 mer (Ph.D.™-7, Ph.D.™-C7C, and Ph.D.™-12, New England Biolabs, MA) to generate a diverse randomized library of more than 6.7×10^9 peptides.
2. Incubate HAP crystal suspension (5 mg/mL) with the combined phage libraries for 1 h at 37 °C.
3. Wash HAP crystal with pH 7.5 TBS 10 times to remove unbound and loosely bound phage.
4. Add 1 mL of 0.2 M glycine–HCl buffer (pH 2.2) to the HAP crystal and incubate the mixture for 10 min. This step elutes phage bound to HAP. Transfer the eluate into a new microcentrifuge tube.
5. Add 150 µL of pH 9 TBS (1 M) to the eluate and mix it gently for neutralization.
6. Incubate 10 mL of LB with *E. coli* strain ER2738 from a plate while vigorously shaking the mixture for 4.5 h at 37 °C prior to the following amplification step.
7. Amplify the rest of eluate by adding it to the 10 mL ER2738 culture with vigorous shaking for 4.5 h at 37 °C.

8. Transfer the culture to a 30-mL centrifuge tube and spin for 10 min at $12{,}000 \times g$ and 4 °C. After transferring the supernatant to a fresh centrifuge tube, respin the supernatant to remove cell pellets.
9. Transfer the upper 80% of the supernatant (7–8 mL) to a fresh centrifuge tube filled with 1.6 mL of 20% PEG in 2.5 M NaCl aqueous solution. Store this phage suspension in the refrigerator overnight.
10. Spin the phage suspension at $12{,}000 \times g$ for 10 min. Discard the supernatant, respin the tube at $12{,}000 \times g$ for 1 min, and pipette out any residual supernatant to isolate phage pellets. Resuspend the phage pellets in 200 μL of TBS.
11. Apply the same procedures to the following rounds of selection with increased amount of surfactant concentration of Tween 20 (0.2%, 0.3%, and 0.4% sequentially from second to fourth rounds).
12. After the elution step of each round, prepare 10- to 10^4-fold serial dilutions of the eluate in LB.
13. Add 10 μL of each phage dilution to each 200 μL of ER2738 culture in a microcentrifuge tube, vortex it briefly and incubate it at room temperature for 5 min.
14. Transfer the infected cells to 40 °C top agar in culture tubes, vortex quickly, and immediately pour culture onto a prewarmed LB/IPTG/X-gal plate (Whaley et al., 2000). Spread top agar evenly by tilting plates. Cool the plates for 10 min to solidify the top layer. Incubate the inverted plates at 37 °C overnight.
15. Pick a well-separated blue plaque from a tittering plate using pipette tip and transfer it to a culture tube filled with 1 mL of diluted ER2738 culture (1/100 in LB). It is recommended to pick about 10–20 plaques out of 100 on the plate.
16. Incubate the tubes with shaking for 4.5 h at 37 °C. Transfer the cultures into microcentrifuge tubes and spin at 10k rpm for 1 min. Transfer the supernatant to a fresh tube and respin. Place the upper supernatant (containing isolated phage) into a fresh tube.
17. For DNA sequencing, place 400 μl of the supernatant into a fresh microcentrifuge tube. Add 160 μl of 20% PEG to 2.5 M NaCl aqueous solution. Invert the tube five times and let it sit for 10 min at room temperature.
18. Spin at 10k rpm for 10 min and discard the supernatant. Respin at 10k rpm for 1 min. Pipette off remaining residual supernatant.
19. Add 100 μl of iodide buffer (10 mM Tris–HCl, pH 8, 1 mM EDTA, 4 M sodium iodide) to the tube and vortex vigorously for 30 s. Add

250 μl of ethanol to the solution and let it sit for 10 min at room temperature.
20. Spin at 10k for 10 min and discard the supernatant. Dry the pellet in the tube in the 50 °C oven for 10 min to remove ethanol.
21. Suspend the single-stranded DNA pellet in 30 μl of DI water. Mix −96 gIII sequencing primer with DNA solution for DNA sequencing.

2.4. Determination of the strongest binding peptides using miniphage library

The dominant binding peptide sequences identified from the phage display are often fast-amplifying peptides biased by the phage biology. In order to determine the strongest HAP-binding peptides from the large subset of phage display results, the phage-bearing HAP-binding peptides need to be collected. Each selected phage sample needs to be separately amplified and diluted into a single minilibrary of HAP-binding phage mixture having 10^6 pfu/μL of each selected phage. The following is the protocol for confirming the strongest binding phage.

2.4.1 Methods
1. Incubate the phage mixture solution with 5 mg of HAP in 1 mL of TBST (50 mM Tris–HCl, 150 mM NaCl, pH 7.5, 0.1%, v/v, Tween 20) for 1 h at 37 °C.
2. Spin the suspension at 10k rpm for 5 min and remove the supernatant. Wash the HAP crystals with TBS 10 times to wash off unbound and weak binding phage. Transfer HAP crystals to a fresh microcentrifuge tube in each washing step.
3. Add 1 mL of 0.2 M glycine–HCl (pH 2.2) to the HAP crystals and incubate for 10 min at 37 °C. Add 150 μL of pH 9 TBS to the mixture for neutralization.
4. Titer the eluate-containing phage on LB X-gal/IPTG agar plates as shown earlier in Section 1.2. Pick and sequence 60–100 blue plaques for determining the strongest binding peptides.

2.5. Alanine-scanning mutation assay

The alanine-scanning technique is a way to determine the contribution of a specific residue to the function (binding affinity) of a peptide. Alanine is used due to its small side chain functional group (methyl). A series of control phage-displaying peptides, which are substituted with alanine(s) at the specific residue(s), can be constructed by genetic engineering of phage DNA. After phage binding to targets is carried out in identical condition, a comparison study can be performed by titering the eluate containing each phage from the target HAP.

2.5.1 Methods
1. Incubate 5 mg of HAP with 1 mL of each control phage solution and the identified phage solution (10^6 pfu/μL) in 1 mL of TBST (50 mM Tris–HCl, 150 mM NaCl, pH 7.5, 0.1%, v/v, Tween 20) for 1 h at 37 °C.
2. Spin the suspension at 10k rpm for 5 min and remove the supernatant. Wash the HAP crystals with TBS 10 times to wash off unbound and weak binding phages. In each washing step, transfer HAP crystals to a fresh microcentrifuge tube.
3. Add 1 mL of 0.2 M glycine–HCl (pH 2.2) to the HAP crystals and incubate the mixture for 10 min at 37 °C. Add 150 μL of pH 9 TBS to the mixture for neutralization.
4. Titer the eluate containing each type of phage on LB X-gal/IPTG agar plates as shown in Section 1.2. Compare the number of blue plaques formed on the plates to determine the respective binding affinity of each type of phage to target HAP.

2.6. HAP-binding assay using fluorescent dye-conjugated antibody

Phage binding to HAP crystals can be detected by labeling the phage bound to the surface with fluorescent dye-conjugated monoclonal pVIII antibody. The extent of the phage binding can be comparatively assayed by observing the fluorescence intensity of the samples using either fluorescence microscopy or FACS (fluorescence-activated cell sorting).

2.6.1 Methods
1. Incubate 4 mg of HAP crystals with $\sim 10^{12}$ pfu/mL of each identified phage or the wild-type phage for 30 min at 37 °C.
2. Spin the suspension at 10k rpm for 5 min and remove the supernatant. Wash the HAP crystals with TBST (50 mM Tris–HCl, 150 mM NaCl, pH 7.5, 0.5%, v/v, Tween 20) 10 times to wash off unbound and weak binding phages. In each washing step, transfer HAP crystals to a fresh microcentrifuge tube.
3. Treat HAP crystals with 1 mg/mL of BSA in TBS (pH 7.5) for 30 min and wash the crystals with TBS 10 times.

4. Label the phage bound to HAP crystals with R-phycoerythrin-conjugated monoclonal pVIII antibody (Amersham Pharmacia Biotech, UK). Acquire fluorescence images (Nikon fluorescence microscope, Japan) and quantify the fluorescence intensity by FACSCalibur flow cytometer (BD Biosciences, CA).
5. Elute the bound phage using 0.2 M glycine–HCl buffer (pH 2.2) and titer the solution for quantification as described earlier in Section 1.2.

2.7. Solid-phase peptide synthesis

After peptide discovery through phage display, it is important to confirm that the same peptide sequences without the phage body structures possess similar or identical function. In order to verify the peptide activity, identified functional peptides from phage display can be prepared by using standard Fmoc solid-phase peptide synthesis. The general procedure of solid-phase peptide synthesis is composed of repeated cycles of coupling/wash/deprotection/wash. The free N-terminal amino group of the peptide attached to a solid-phase with an acid-labile linker is coupled to an Fmoc-protected amino acid building block. The protecting group is then deprotected to reveal a free amino group to which a new amino acid building block can be attached.

2.7.1 Methods

1. Preswell Rink amide resins (1.0 mmol/g, 100–200 mesh, 1% DVB) in NMP for 30 min. Filter off the solvent and deprotect Fmoc group on the resin (200 mg) by adding 3 mL of 3% (v/v) DBU in NMP and shaking the mixture on a rocker for 20 min. Typical washing steps include washing with NMP (3 mL ×3), washing with series of methanol followed by dichloromethane (3 mL each ×3), and washing with NMP (3 mL ×3). Every step is followed by these washing steps.
2. Prepare a mixture solution of Fmoc amino acid, HOBt, and DIC (five equivalents each to the amino group loading on the resin) in NMP solution. The use of concentrations higher than 0.1 M is recommended for each reagent.
3. Perform a coupling step by treating the deprotected resins with the mixture in the preceding text on a rocker for 2 h. Confirm the complete coupling by performing Kaiser ninhydrin test (Bledsoe et al., 2002). The coupling step is repeated until complete coupling is confirmed.

4. Deprotect the Fmoc group of the amino acid coupled to the resin by using 3% (v/v) DBU in NMP as described earlier. Repeat steps 2–4 until the desired peptide is synthesized.
5. Wash resins with dichloromethane three times for more efficient drying of resins prior to peptide cleavage step.
6. Dry resins in the vacuum desiccator for 6 h.
7. Cleave the peptides from resins (200 mg) by treating them with 3 mL of a cleavage mixture of TFA, water, phenol, ethanedithiol, and triisopropylsilane (82.5%:5%:5%:2.5%) for 2 h on a shaker.
8. Collect the solution by filtration into round-bottomed flask. Wash the resins with 1 mL of TFA three times and combine all the filtrates into the flask.
9. Reduce the volume of the combined solution to 1–2 mL using rotary evaporation.
10. Add 10-fold volume of ice-cold diethyl ether to the mixture to precipitate down the peptides.
11. Spin the suspensions at 5k rpm for 2 min to isolate the precipitates. Wash the precipitates with the ice-cold ether (1 mL ×10) and dry in a vacuum desiccator for 6 h.
12. Dissolve the dry solid in DI water to reach a concentration of 5–10 mg/mL. Spin the solution at 10k rpm for 5 min. Filter the supernatant into a fresh tube using syringe filter (0.22 μm pore size).
13. Purify the crude product solution by running preparative HPLC. Collect the desired peptides and freeze them at $-20\,°C$. Perform freeze-drying to obtain lyophilized peptide products.
14. Confirm the desired peptides by analyzing the molecular weight of the product by using liquid chromatography equipped with ESI-MS.

2.8. HAP crystal nucleation using peptides

After the discovery of the functional peptides, the first functional assay that can be utilized is a biomineralization assay. The ability of the peptides to promote nucleation can be characterized by incubating them with the precursor ions of HAP. The resulting peptide-induced mineralization can be observed by transmission electron microscopy (TEM) or scanning electron microscopy (SEM) techniques.

2.8.1 Methods
1. Dissolve peptides in DI water (1.5 mg/mL). Syringe filter the solution to remove any aggregates, which may influence the nucleation process.

Apply 5 μL of aqueous peptide solution to a holey carbon TEM grid for 2 min. Wick the solution away.
2. Prepare the precursor ion solutions by dissolving corresponding salts in DI water and syringe filtering the solution.
3. Treat the peptide-coated TEM grid with 10 μL of 5 mM Na_2HPO_4 on one side and 10 μL of 10 mM $CaCl_2$ on the other side. Wash the TEM grids with 100 μL of DI water three times after 30 min, 1 h, and 2 h. Dry the samples in air and store them in a desiccator.
4. Observe the morphologies of mineralized samples using TEM or SEM.

2.9. Alignment of the identified peptide with other protein sequences

After phage display screening, the identified peptide sequences can be compared with natural bone-associated proteins such as collagen type I, amelogenin, and statherin. Natural bone-associated proteins possess [Pro-(OH)-X]-rich tripeptide patterns, while the identified sequence resembled the tripeptide repeat (Gly-Pro-Hyp) of type I collagen, a major component of bone extracellular matrix (ECM) (Fig. 14.2). Databases for protein information are available on Web sites such as NCBI (http://www.ncbi.nlm.nih.gov) and UniProtKB (http://www.uniprot.org). Web sites for the peptide or protein sequence alignment, such as FASTA (http://fasta.bioch.virginia.edu/fasta), ClustalW (http://www.ch.embnet.org/software/ClustalW), and UniProtKB (http://www.uniprot.org), are also available (Fig. 14.4).

2.9.1 Methods
1. Go to UniProtKB Web site (http://www.uniprot.org) and search your protein of interest for comparison (Fig. 14.4A).
2. Select a box of your protein of interest (Fig. 14.4B). If you want to add other proteins, repeat steps 1 and 2 (Fig. 14.4C).
3. Click "Align" at the bottom to display the result page (Fig. 14.4D).
4. If you want to add additional sequences, write the sequence name as ">," fill in the bottom box with the sequences, and click "Align" (Fig. 14.4D).
5. In addition, you can add "Annotations" and color via "Amino acid properties" on the aligned sequences.

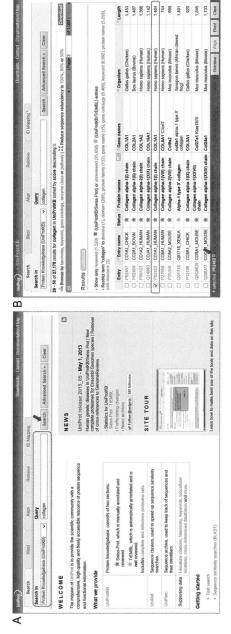

Figure 14.4 Alignment of the peptide and protein sequences. Find a protein of interest in the UniProtKB Web site (http://www.uniprot.org) and align the protein sequences. (For color version of this figure, the reader is referred to the online version of this chapter.)

3. SUMMARY

The methods described here for high-throughput phage peptide library screening against target HAP crystals enable one to identify functional peptide sequences that might be useful in the study of bone- or tooth-associated proteins. The resulting peptide sequences can provide important insights into bone and tooth biomorphogenesis. This approach can be generalized to other desired crystal surfaces using the same experimental design principles with a little modification of the procedures.

REFERENCES

Belcher, A. M., Wu, X. H., Christensen, R. J., Hansma, P. K., Stucky, G. D., & Morse, D. E. (1996). Control of crystal phase switching and orientation by soluble mollusc-shell proteins. *Nature, 381*, 56–58.

Bledsoe, R. K., Montana, V. G., Stanley, T. B., Delves, C. J., Apolito, C. J., McKee, D. D., et al. (2002). Crystal structure of the glucocorticoid receptor ligand binding domain reveals a novel mode of receptor dimerization and coactivator recognition. *Cell, 110*, 93–105.

Chiu, C. Y., Li, Y. J., & Huang, Y. (2010). Size-controlled synthesis of Pd nanocrystals using a specific multifunctional peptide. *Nanoscale, 2*, 927–930.

Chung, W. J., Kwon, K. Y., Song, J., & Lee, S. W. (2011). Evolutionary screening of collagen-like peptides that nucleate hydroxyapatite crystals. *Langmuir, 27*, 7620–7628.

Danner, S., & Belasco, J. G. (2001). T7 phage display: A novel genetic selection system for cloning RNA-binding proteins from cDNA libraries. *Proceedings of the National Academy of Sciences of the United States of America, 98*, 12954–12959.

Du, C., Falini, G., Fermani, S., Abbott, C., & Moradian-Oldak, J. (2005). Supramolecular assembly of amelogenin nanospheres into birefringent microribbons. *Science, 307*, 1450–1454.

Ducy, P., Desbois, C., Boyce, B., Pinero, G., Story, B., Dunstan, C., et al. (1996). Increased bone formation in osteocalcin-deficient mice. *Nature, 382*, 448–452.

Flynn, C. E., Lee, S.-W., Peelle, B. R., & Belcher, A. M. (2003). Viruses as vehicles for growth, organization and assembly of materials. *Acta Materialia, 51*, 5867–5880.

Flynn, C. E., Mao, C. B., Hayhurst, A., Williams, J. L., Georgiou, G., Iverson, B., et al. (2003). Synthesis and organization of nanoscale II-VI semiconductor materials using evolved peptide specificity and viral capsid assembly. *Journal of Materials Chemistry, 13*, 2414–2421.

Fritz, M., Belcher, A. M., Radmacher, M., Walters, D. A., Hansma, P. K., Stucky, G. D., et al. (1994). Flat pearls from biofabrication of organized composites on inorganic substrates. *Nature, 371*, 49–51.

George, A., Bannon, L., Sabsay, B., Dillon, J. W., Malone, J., Veis, A., et al. (1996). The carboxyl-terminal domain of phosphophoryn contains unique extended triplet amino acid repeat sequences forming ordered carboxyl-phosphate interaction ridges that may be essential in the biomineralization process. *Journal of Biological Chemistry, 271*, 32869–32873.

Gungormus, M., Fong, H., Kim, I. W., Evans, J. S., Tamerler, C., & Sarikaya, M. (2008). Regulation of in vitro calcium phosphate mineralization by combinatorially selected hydroxyapatite-binding peptides. *Biomacromolecules, 9*, 966–973.

Hoang, Q. Q., Sicheri, F., Howard, A. J., & Yang, D. S. C. (2003). Bone recognition mechanism of porcine osteocalcin from crystal structure. *Nature, 425*, 977–980.

Hu, C. C., Bartlett, J. D., Zhang, C. H., Qian, Q., Ryu, O. H., & Simmer, J. P. (1996). Cloning, cDNA sequence, and alternative splicing of porcine amelogenin mRNAs. *Journal of Dental Research, 75*, 1735–1741.

Ishiyama, M., Mikami, M., Shimokawa, H., & Oida, S. (1998). Amelogenin protein in tooth germs of the snake Elaphe quadrivirgata, immunohistochemistry, cloning and cDNA sequence. *Archives of Histology and Cytology, 61*, 467–474.

Lee, S.-W., Mao, C., Flynn, C. E., & Belcher, A. M. (2002). Ordering of quantum dots using genetically engineered viruses. *Science, 296*, 892–895.

Long, J. R., Dindot, J. L., Zebroski, H., Kiihne, S., Clark, R. H., Campbell, A. A., et al. (1998). A peptide that inhibits hydroxyapatite growth is in an extended conformation on the crystal surface. *Proceedings of the National Academy of Sciences of the United States of America, 95*, 12083–12087.

MacDougall, M., Simmons, D., Luan, X. H., Nydegger, J., Feng, J., & Gu, T. T. (1997). Dentin phosphoprotein and dentin sialoprotein are cleavage products expressed from a single transcript coded by a gene on human chromosome 4—Dentin phosphoprotein DNA sequence determination. *Journal of Biological Chemistry, 272*, 835–842.

Mann, S. (2001). *Biomineralization: Principles and concepts in bioinorganic materials chemistry*. New York, NY: Oxford University Press.

Mao, C., Flynn, C. E., Hayhurst, A., Sweeney, R., Qi, J., Georgiou, G., et al. (2003). Viral assembly of oriented quantum dot nanowires. *Proceedings of the National Academy of Sciences of the United States of America, 100*, 6946–6951.

Mao, C., Solis, D. J., Reiss, B. D., Kottmann, S. T., Sweeney, R. Y., Hayhurst, A., et al. (2004). Virus-based toolkit for the directed synthesis of magnetic and semiconducting nanowires. *Science, 303*, 213–217.

Mullaney, B. P., & Pallavicini, M. G. (2001). Protein–protein interactions in hematology and phage display. *Experimental Hematology, 29*, 1136–1146.

Naik, R. R., Stringer, S. J., Agarwal, G., Jones, S. E., & Stone, M. O. (2002). Biomimetic synthesis and patterning of silver nanoparticles. *Nature Materials, 1*, 169–172.

Pacardo, D. B., Sethi, M., Jones, S. E., Naik, R. R., & Knecht, M. R. (2009). Biomimetic synthesis of Pd nanocatalysts for the Stille coupling reaction. *ACS Nano, 3*, 1288–1296.

Petrenko, V. A., Smith, G. P., Gong, X., & Quinn, T. (1996). A library of organic landscapes on filamentous phage. *Protein Engineering, 9*, 797–801.

Presta, L. G. (2005). Selection, design, and engineering of therapeutic antibodies. *The Journal of Allergy and Clinical Immunology, 116*, 731–736.

Raj, P. A., Johnsson, M., Levine, M. J., & Nancollas, G. H. (1992). Salivary statherin—Dependence on sequence, charge, hydrogen-bonding potency, and helical conformation for adsorption to hydroxyapatite and inhibition of mineralization. *Journal of Biological Chemistry, 267*, 5968–5976.

Reiss, B. D., Mao, C. B., Solis, D. J., Ryan, K. S., Thomson, T., & Belcher, A. M. (2004). Biological routes to metal alloy ferromagnetic nanostructures. *Nano Letters, 4*, 1127–1132.

Sarikaya, M., Tamerler, C., Jen, A. K. Y., Schulten, K., & Baneyx, F. (2003). Molecular biomimetics: Nanotechnology through biology. *Nature Materials, 2*, 577–585.

Segvich, S. J., Smith, H. C., & Kohn, D. H. (2009). The adsorption of preferential binding peptides to apatite-based materials. *Biomaterials, 30*, 1287–1298.

Smith, G. P. (1985). Filamentous fusion phage—Novel expression vectors that display cloned antigens on the virion surface. *Science, 228*, 1315–1317.

Smith, G. P., & Petrenko, V. A. (1997). Phage display. *Chemistry Review, 97*, 391–410.

Sweeney, R. Y., Mao, C., Gao, X., Burt, J. L., Belcher, A. M., Georgiou, G., et al. (2004). Bacterial biosynthesis of cadmium sulfide nanocrystals. *Chemistry and Biology, 11*, 1553–1559.

Tas, A. C. (2001). Molten salt synthesis of calcium hydroxyapatite whiskers. *Journal of the American Ceramic Society, 84*, 295–300.

Traub, W., Arad, T., & Weiner, S. (1989). Three-dimensional ordered distribution of crystals in turkey tendon collagen fibers. *Proceedings of the National Academy of Sciences of the United States of America, 86*, 9822–9826.

Weiger, M. C., Park, J. J., Roy, M. D., Stafford, C. M., Karim, A., & Becker, M. L. (2010). Quantification of the binding affinity of a specific hydroxyapatite binding peptide. *Biomaterials, 31*, 2955–2963.

Weiner, S., & Wagner, H. D. (1998). The material bone: Structure mechanical function relations. *Annual Review of Materials Science, 28*, 271–298.

Whaley, S. R., English, D. S., Hu, E. L., Barbara, P. F., & Belcher, A. M. (2000). Selection of peptides with semiconductor binding specificity for directed nanocrystal assembly. *Nature, 405*, 665–668.

Zhang, W., Liao, S. S., & Cui, F. Z. (2003). Hierarchical self-assembly of nano-fibrils in mineralized collagen. *Chemistry of Materials, 15*, 3221–3226.

SECTION IV

Protein Structure, Interactions and Function

CHAPTER FIFTEEN

Quantitatively and Kinetically Identifying Binding Motifs of Amelogenin Proteins to Mineral Crystals Through Biochemical and Spectroscopic Assays

Li Zhu[*], Peter Hwang[†], H. Ewa Witkowska[‡], Haichuan Liu[‡], Wu Li[*,1]

[*]Department of Orofacial Sciences, University of California, San Francisco, California, USA
[†]Biochemistry & Biophysics Department, School of Medicine, San Francisco, California, USA
[‡]Sandler–Moore Mass Spectrometry Core Facility, University of California, San Francisco, California, USA
[1]Corresponding author: e-mail address: wu.li@ucsf.edu

Contents

1. Introduction	328
2. Strategy and Rationale	330
3. Experimental Components and Considerations	331
3.1 Syntheses and identification of amelogenin peptides	331
3.2 Synthesis and analysis of bulk HAp crystals	331
3.3 Preparation and identification of single FAp crystal	331
3.4 Expression and purification of recombinant amelogenin	332
3.5 Site-directed mutagenesis of amelogenin mutants with deletions of binding motifs	332
4. Experimental Approaches	332
4.1 Identification of specific binding sequences by peptide screening	332
4.2 Confirmation of the crystal-binding motifs in amelogenin by competitive adsorption assays	335
4.3 Comparison of binding amounts of amelogenin and its mutants on different planes of apatite crystals by enzyme-linked immunosorbent assay	336
4.4 Kinetic analyses of adsorption of amelogenin to HAp crystals	338
5. Data Handling and Processing	340
6. Summary	340
Acknowledgments	341
References	341

Abstract

Tooth enamel is the hardest tissue in vertebrate animals. Consisting of millions of carbonated hydroxyapatite crystals, this highly mineralized tissue develops from a protein matrix in which amelogenin is the predominant component. The enamel matrix proteins are eventually and completely degraded and removed by proteinases to form mineral-enriched tooth enamel. Identification of the apatite-binding motifs in amelogenin is critical for understanding the amelogenin–crystal interactions and amelogenin–proteinases interactions during tooth enamel biomineralization. A stepwise strategy is introduced to kinetically and quantitatively identify the crystal-binding motifs in amelogenin, including a peptide screening assay, a competitive adsorption assay, and a kinetic-binding assay using amelogenin and gene-engineered amelogenin mutants. A modified enzyme-linked immunosorbent assay on crystal surfaces is also applied to compare binding amounts of amelogenin and its mutants on different planes of apatite crystals. We describe the detailed protocols for these assays and provide the considerations for these experiments in this chapter.

1. INTRODUCTION

Tooth enamel has a unique morphological structure and distinctive mechanical properties, making it different from other mineralized tissues in the human body, such as the bone, dentin, and cementum. Tooth enamel is the hardest tissue known in vertebrates because it is composed mainly of numerous hexagonal carbonated hydroxyapatite (HAp) crystals (Ichijo, Yamashita, & Terashima, 1992; Pergolizzi, Anastasi, Santoro, & Trimarchi, 1995). The crystals form enamel rods and interrods, which weave together and orient in different directions (Pergolizzi et al., 1995; Plate & Hohling, 1994). The crystals in tooth enamel are very thin (20–30 nm). In comparison to the thickness, they are extremely long. Many investigators believe that these crystals span the entire breadth of the tooth enamel, a distance up to 2.5 mm (Daculsi, Menanteau, Kerebel, & Mitre, 1984; Leblond & Warshawsky, 1979; Nanci, 2003). This thickness/length ratio is equal to that of a 2000-ft. long rope with a diameter of only 1 in.. The unique shapes and organizations of enamel crystals determine the excellent mechanical properties of tooth enamel and also raise a persistent question as to how these enamel crystals form during tooth development.

Mineralized enamel crystals develop from a layer of enamel protein matrix that is predominately amelogenins (>90%), which themselves form special nanostructures to modulate crystal formation through a biomineralization process that is poorly understood. Amelogenins are gradually and completely removed by enamel proteinases, that is, matrix metalloproteinase-20 (MMP-20) and kallikrein 4 (KLK4) to form highly mineralized mature enamel at the completion of maturation stage. The matured enamel matrix in erupted tooth contains less than 2% of the organic components (Bartlett & Simmer, 1999; Fincham, Moradian-Oldak, & Simmer, 1999). In the secretory, transition, and early maturation stages of tooth enamel development, the interactions between crystal, amelogenin proteins, and proteinases dynamically and delicately control not only the growth rate of the crystal but also its growth direction and morphology.

The hexagonal apatite crystals have different surfaces: the (001) face on the top of the rods and (hk0) faces on their sides (Fig. 15.1). We found that these distinct surfaces exhibit differential interactions with amelogenin proteins (Habelitz et al., 2004), which may lead to subsequently different interactions between adsorbed amelogenin and proteinases, resulting in a directional growth of HAp crystal in the tooth enamel. Therefore, we need to develop a strategy to identify the crystal-binding motifs in amelogenin and quantitatively compare the amelogenin–crystal interactions for the different faces of apatite crystals.

We designed a series of biochemical approaches to analyze amelogenin–apatite interactions. We have identified the apatite-binding sites of amelogenin by monitoring peptide binding using assays and mass

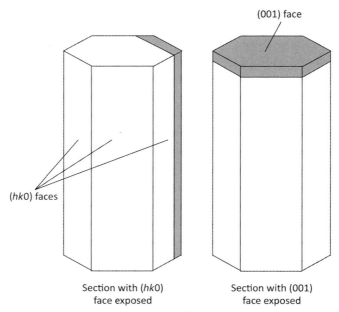

Figure 15.1 FAp single-crystal sections with different faces exposed. Left panel, section with (hk0) crystal face exposed; right panel, section with (001) face exposed. The shaded parts indicate the sections.

spectrometry as the first screening step. The binding affinities of these identified peptides were further confirmed by their ability to compete with the apatite-binding affinities of full-length amelogenin proteins. The binding kinetics of amelogenin–apatite was then investigated to characterize the putative differences in their protein–crystal interactions between wild-type amelogenin and its mutants.

2. STRATEGY AND RATIONALE

To identify the crystal-binding motifs in amelogenin, we have followed a stepwise strategy. First, we used a peptide screening assay to detect the possible binding sequence(s) in amelogenin. The peptides that were preferentially adsorbed on crystals were identified by matrix-assisted laser desorption–ionization time-of-flight mass spectrometry (MALDI-TOF-MS). Second, we preadsorbed the selected peptides onto the crystal surface and then compared their inhibitory effects on the further adsorptions of full-length amelogenin to crystal surfaces. Third, we employed site-directed mutagenesis technique to generate amelogenin mutants carrying deletions within the regions that were indicated to encompass crystal-binding motif(s) and evaluated the impact of the deletions on amelogenin–crystal-binding affinities. The amelogenin mutants containing the different deletions were used to analyze their binding affinity to crystals. For these protein–mineral assays, we utilized two types of targets: bulk HAp crystals (small size and large surface area) and single fluorapatite (FAp) crystals (large size and small surface area) with special planes exposed. In comparison to FAp crystals, HAp powders offer an advantage of much higher surface-binding area per volume and hence a higher detection sensitivity in screening assays. However, assays based on HAp crystals provide only an ensemble view of amelogenin binding onto both (001) and ($hk0$) apatite crystal surfaces. Discrimination between the distinct crystal surfaces was achieved by using FAp crystals. We note that it was previously demonstrated that structures of FAp and HAp crystals are identical, and hence, FAp is a viable substitute of HAp in amelogenin-binding assays (Habelitz et al., 2004; Tarasevich, Lea, Bernt, Engelhard, & Shaw, 2009).

Combination of all the earlier-described studies allowed us to successfully identify apatite crystal-binding motifs at both N- and C-terminal regions of amelogenin.

3. EXPERIMENTAL COMPONENTS AND CONSIDERATIONS

3.1. Syntheses and identification of amelogenin peptides

Two sets of serial peptides—a total of 25 14-mers and 18 20-mers—were designed to cover the whole human amelogenin sequence (175 residues). For each pair of adjacent peptides, a C-terminal portion of the preceding peptide (7 residues in 14-mer peptide set and 10 residues in 20-mer peptide set) overlapped the N-terminal portion of the subsequent peptide. These peptides were commercially synthesized by a solid-phase peptide synthesis method and purified by reversed-phase high-performance liquid chromatography (Deciphergen Biotechnology, Aurora, CO, USA). The identities of peptides were confirmed by MALDI-TOF-MS.

3.2. Synthesis and analysis of bulk HAp crystals

HAp crystals were synthesized as previously described (Featherstone, Mayer, Driessens, Verbeeck, & Heijligers, 1983) and characterized by X-ray diffraction and Fourier transform infrared spectroscopy (Tanimoto et al., 2008). The HAp crystals were sequentially passed through the meshes at sizes of 30 and then 60 μm. The crystals at the size ranging from 30 to 60 μm were collected. The specific surface areas of the apatite particles were measured by the Brunauer–Emmett–Teller method (Brunauer, Emmett, & Teller, 1938) with a Micromeritics TriStar 3000 (Micromeritics Instrument Corp., Norcross, GA, USA). The surface areas for the various synthetic apatites are similar, ranged within 72 and 80 m^2/g.

3.3. Preparation and identification of single FAp crystal

Single FAp crystals were purchased from Earthlight Gems (Kirkland, WA, USA) and their identities were checked by X-ray diffraction. Diffraction data from single-crystal specimens were collected with a Rigaku rotating anode X-ray source and R-axis detector. The crystal form and cell parameters were obtained from analysis of diffraction images, using HKL-2000 or Denzo crystallography software (HKL Research). The determined unit cell parameters, crystal form (trigonal p3), lengths ($a=9.372$, $b=9.372$, and $c=6.883$ Å), and angles (alpha$=90.000°$, beta$=90.000°$, and gamma$=120.000°$) were used to search American Mineralogist Crystal Structure Database (http://rruff.geo.arizona.edu/AMS/amcsd.php) with tolerance at 0.004 Å for *a* and *b* and 0.005 Å for *c*. The results unambiguously confirmed that the identity of this material as genuine is FAp crystal.

The single crystals were sectioned into slides with 100 μm thicknesses by using an IsoMet low-speed diamond saw (Buehler, Lake Bluff, IL, USA) along their *c*-axis or *a*- and *b*-axes to expose their (*hk*0) or (001) faces (Fig. 15.1), respectively. The sections were further shaped into squares at a size of 6 × 6 mm using sandpaper of P200 grit size (ISO/FEPA Grit designation). The sections were further sequentially polished by using sandpapers of the grit sizes of P400 and P800 and P1000 and then by utilizing diamond pastes at the sizes of 0.2 and 0.1 μm.

3.4. Expression and purification of recombinant amelogenin

Wild-type full-length amelogenin (rh174) and its mutants were expressed in *Escherichia coli*. The protein was further purified as previously described (Li, Gao, Yan, & DenBesten, 2003).

3.5. Site-directed mutagenesis of amelogenin mutants with deletions of binding motifs

After identification of possible crystal-binding motifs in amelogenin, pairs of primers were designed to flank the cDNA sequences derived from crystal-binding peptide sequences. A PCR-based mutagenesis was performed using a QuikChange® *Lightning* Site-Directed Mutagenesis Kit (Stratagene Inc., La Jolla, CA, USA) according to the manufacturer's manual. The constructs were sequenced to confirm the correct DNA sequence in frame and the presence of the engineered deletions. The expression, extraction, and purification of the amelogenin mutants followed our previous protocols (Li et al., 2003).

4. EXPERIMENTAL APPROACHES

4.1. Identification of specific binding sequences by peptide screening

The studies employed single FAp sections with either (001) or (*hk*0) plane exposed as binding substrates. The FAp sections with different surfaces exposed were incubated with either a mixture of several different peptides or a single peptide. The adsorbed species were sequentially eluted with 10% and then 50% trifluoroacetic acid (TFA) and analyzed by MALDI-TOF-MS. The following generic protocol describes a typical screening procedure using FAp sections. A similar protocol with minor modifications was also used for studies in which bulk HAp crystal powder served as the substrate.

4.1.1. Prelubricated 1.7 ml microcentrifuge tubes (Corning Incorporated, NY, USA) are used to prevent the adhesion of peptides to the tube, which would interfere with the binding amounts of peptides during the assay.

4.1.2. The ($hk0$) faces of FAp sections with (001) predominantly exposed and the (001) faces of FAp sections with ($hk0$) predominantly exposed are carefully coated by a layer of clear nail polish and air dried. Our preexperiments have confirmed that the amounts of the peptides or proteins adsorbed on the nail polish-coated area are negligible (data not shown).

4.1.3. Each FAp section is placed vertically in the prelubricated microcentrifuge tube to double their surface areas of binding. This position of FAp is used for all the following steps.

4.1.4. The individual peptide or a mixture of different peptides is dissolved in Millipore water at a concentration of 5 μM. Then, 300 μl of peptide solution is added to the tubes to completely immerse the FAp. The peptide–crystal interactions are performed at room temperature for 5 h with gentle shaking on a platform of a horizontally rotating shaker at 50 rotations per minute.

4.1.5. After incubation, the FAp is transferred to a fresh prelubricated microcentrifuge tube and washed three times with Tris buffer (10 mM Tris–HCl, pH 7.4) with gentle shaking, 10 min per wash.

4.1.6. Then, FAp is taken out of the tube and any remaining solution droplets are moved toward one of the corners and carefully removed with a KimWipes tissue.

4.1.7. The FAp is transferred to a new prelubricated tube containing 300 μl of 10% TFA to elute the peptides. After gentle shaking for 30 min, FAp is transferred to another tube for a second step of peptide elution with 50% TFA using the same procedure.

4.1.8. Both 10% and 50% TFA elutes are lyophilized to dryness and resuspended in 10 μl Millipore water.

4.1.9. These eluted peptides are desalted by using a C18 ZipTip (Millipore, Billerica, MA, USA) before MALDI-TOF-MS analyses.

4.1.10. The desalted samples are mixed at 1:1 volume ratio with 5 mg/ml α-cyano-4-hydroxycinnamic acid in 50% acetonitrile/0.1%TFA matrix and spotted onto a MALDI stainless steel target plate.

4.1.11. MALDI-TOF-MS experiments are carried out on a 4800 MALDI TOF/TOF Analyzer (AB Sciex, Foster City, CA, USA) at the UCSF Sandler–Moore Mass Spectrometry Core Facility. The samples are ionized with Nd:YAG laser in the reflector (m/z 800–6000) positive-ion mode.

4.1.12. External optimized calibration utilizes OptiPlate software (AB Sciex, Foster City, CA, USA) and is based on four peptide standards within m/z 1290–3800 (angiotensin I and three ACTH clips: 1–17, 18–30, and 7–38) for reflector data.

4.1.13. Spectra are processed with Data Explorer 4.0 software (AB Sciex) by performing baseline adjustment, noise filtering, and deisotoping.

Figure 15.2 shows an example of mass spectrometric data that allowed us to identify peptide P9 for which the residues lie within N-terminal portion of full-length human amelogenin sequence (h175). The MS data demonstrated that the P9 peptide can bind to the (hk0) face of FAp crystals and can be eluted with 50% TFA. This peptide can also be eluted out by 10% TFA, albeit with much lower abundance (data not shown).

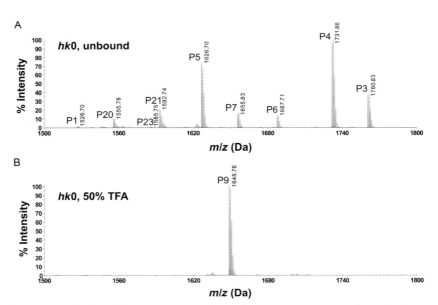

Figure 15.2 MALDI-TOF mass spectra of amelogenin peptides not bound (A) and bound (B) to the (hk0) face of the FAp crystals. Peptide P9 is one of the peptides that was eluted out by 50% TFA from the (hk0) face after binding (B).

4.2. Confirmation of the crystal-binding motifs in amelogenin by competitive adsorption assays

We used a peptide competition strategy to further confirm the amelogenin-binding motifs to apatite crystals by utilizing bulk HAp powders. Following protocol describes the details of this assay. The strategy was to pretreat the crystal surfaces with large access of peptides and then analyze their abilities to inhibit the further binding of full-length amelogenin. The amount of bound amelogenin was measured and compared to the control conditions of protein binding to a "native" crystal surface. The significant decrease in binding amounts after pretreatment with a given peptide would indicate that this peptide likely carries a crystal-binding motif.

4.2.1. The HAp powders are suspended in Tris–HCl buffer (10 mM Tris at pH 7.4) at concentration of 20 mg/ml.

4.2.2. After mixing by pipetting up and down, 5 μl aliquots of HAp suspension (100 μg) are quickly transferred into 1.7 ml prelubricated microcentrifuge tubes.

4.2.3. For the experimental groups, 100 μl of peptide at concentration of 5 μM is added into each tube containing HAp powders. The control is treated with the same amount of Tris–HCl buffer without any peptides. The reactions are incubated at room temperature for 1 h with constant tube vortexing to maintain HAp powder in suspension.

4.2.4. Tubes are centrifuged at $5000 \times g$ for 10 min at room temperature and gel-loading tips are used to carefully remove the supernatants.

4.2.5. The pellets are washed with Tris–HCl buffer three times by centrifugation and resuspension.

4.2.6. After washing, the pallets are resuspended in 100 μl of amelogenin solution (500 μg/ml). Amelogenin is previously prepared by dissolving in Millipore water and its pH is adjusted to 7.4 with 10 mM Tris–HCl at pH 9.0. The undissolved particles are removed by centrifugation at $5000 \times g$ for 10 min.

4.2.7. The reactions are incubated at room temperature with vortexing.

4.2.8. The reactions are centrifuged at $5000 \times g$ for 10 min. The supernatants are transferred to new tubes.

4.2.9. The protein concentrations in supernatants are measured by Bradford protein assay (Bio-Rad, Hercules, CA, USA). The differences from the original concentration (500 μg/ml) will be used to calculate the protein binding amounts.

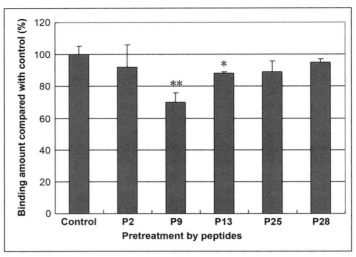

Figure 15.3 Reduced binding amounts of full-length amelogenin in peptide competition assays. The percentages are calculated from the comparisons of binding amounts in samples pretreated by peptides (P2, P9, P13, P25, and P28) to untreated samples. **$P<0.01$, *$P<0.05$.

Using this approach, the binding motifs on specific planes of HAp crystals were further confirmed as shown in Fig. 15.3. The data indicate that the pretreatment of P9 peptide reduces one-third of the binding of full-length amelogenin to HAp crystal powders as compared to unpretreated crystal samples.

4.3. Comparison of binding amounts of amelogenin and its mutants on different planes of apatite crystals by enzyme-linked immunosorbent assay

To confirm the binding domains of amelogenin to different crystal planes, the binding sequences discovered in the previous studies were used to prepare mutated amelogenins with different deletions. All these amelogenin mutants were expressed and purified using the methods as described previously (Le, Gochin, Featherstone, Li, & DenBesten, 2006; Li et al., 2003).

Because the area of the FAp sections is relatively small, the amount of protein bound onto a certain plane is too small to reliably measure using regular protein assays. A system that combines enzyme-linked immunosorbent assay (ELISA) and avidin–biotin complex (ABC) was developed to quantify the amelogenin and its mutants adsorbed on the apatite crystal surfaces. The detailed steps of the experiments are described in the following protocol:

4.3.1. Single FAp sections are prepared as described in Section 3.3 and step 4.1.2. The exposed surfaces are scanned and the images are analyzed by ImageJ software (http://rsb.info.nih.gov/ij/) to determine their surface areas. Each group is analyzed in triplicate for statistical analysis.

4.3.2. FAp is prepared as described in step 4.1.2 and placed vertically in a 1.7-ml prelubricated tube containing 300 µl of amelogenin (500 ng/µl). The protein–crystal interactions are performed at room temperature overnight on a platform of a horizontally rotating shaker at 50 rotations per minute.

4.3.3. After incubation, FAp is transferred to a new tube and unbound proteins are washed away in 1 ml PBS for 10 min (three times) with similar shaking.

4.3.4. One milliliter of blocking buffer (PBS containing 1% BSA) is added to each tube to prevent the nonspecific binding of antibodies and the mixture is incubated at room temperature for 1 h.

4.3.5. After blocking, the FAp sections are washed by PBS with 0.05% Tween-20 (3×10 min) with shaking. Remove the solution by aspiration.

4.3.6. Then, FAp samples are incubated with rabbit amelogenin polyclonal antibody at a dilution of 1:1000 in blocking buffer for 1 h at room temperature on the shaker.

4.3.7. Wash step is repeated as described in step 4.3.5.

4.3.8. Three hundred microliters of biotinylated secondary antibody against rabbit IgG (Dako, Carpinteria, CA, USA) at dilution of 1:1000 in blocking buffer is applied into each tube and incubated for another hour under the same condition as previously mentioned.

4.3.9. Washing step described in step 4.3.5 is repeated.

4.3.10. The FAp is incubated with 300 µl of biotin-conjugated streptavidin–alkaline phosphatase (ABC, Dako), which is diluted at 1:1000 in PBS immediately before use.

4.3.11. After 1 h incubation at room temperature, repeat wash step as step 4.3.5.

4.3.12. During the last washing, the *p*-nitrophenyl phosphate (*p*NPP) substrate solution is freshly prepared by dissolving one *p*NPP tablet (Sigma-Aldrich, St. Louis, MO, USA) and one Tris buffer tablet in 5 ml Millipore water.

4.3.13. Three hundred microliters of $pNPP$ substrate mixture is added to each tube with FAp and incubated in darkness at room temperature for 30 min.

4.3.14. The FAp is quickly removed from the tubes. The reaction solution is mixed until the yellow color developed in the tubes appears homogenous. Two hundred microliters of reaction solution is transferred to each well of 96-well transparent microplate.

4.3.15. The plate is immediately read using absorbance plate reader at 405 nm in a spectrometer (Molecular Devices, CA, USA).

The results in Fig. 15.4 show that (1) amelogenins, both the full-length and its mutant (P9 deletion), bind to (001) and ($hk0$) faces of FAp crystals at different levels; (2) full-length amelogenin binds at a higher level than a mutant carrying P9 deletion; and (3) P9 deletion affects amelogenin binding to the ($hk0$) face of FAp to a larger extent than to the (001) surface.

4.4. Kinetic analyses of adsorption of amelogenin to HAp crystals

The kinetic and equilibrium binding character of amelogenin adsorption onto HAp crystals can be assessed by measuring amelogenin capture as a function of amelogenin concentration and incubation time. The caveat here is that one of the reactants is a solid phase: the exposed crystal surface and surface density of HAp molecules can be used to determine an apparent concentration of HAp in the vortexed suspension.

4.4.1. HAp crystal powder is prepared as described in step 4.2.1. After thoroughly mixing, 2.5 μl of HAp is aliquoted to each 1.7-ml prelubricated tube. The tubes are centrifuged at $1000 \times g$ for 10 min and the supernatant is removed. Please note that the quantity and fineness of HAp crystal powder will significantly affect the measured amelogenin binding rate and response.

4.4.2. Similar to step 4.2.6, amelogenin solution is prepared at three different concentrations (48, 24, and 12 nM). Three is the minimal number of sampling concentrations for generating a binding isotherm. Several more concentrations may be included to improve the accuracy of projected binding saturation.

4.4.3. The HAp pellets are resuspended in amelogenin solution with the three different concentrations. Triplicate samples for each group are used to reduce errors.

4.4.4. The mixture is incubated at room temperature with gentle vortexing. Vortexing speed and consistency are important for

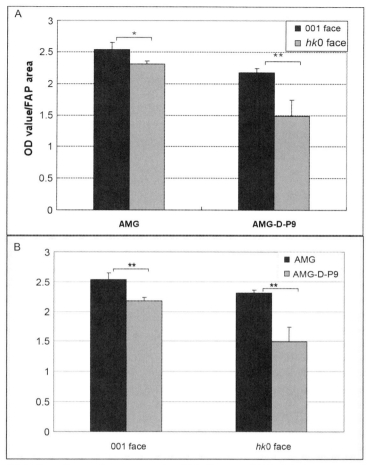

Figure 15.4 The effects of mutagenesis of binding motif P9 deletion on amelogenin mutant adsorptions to different faces of FAp crystals. (A) Comparison of adsorption of amelogenin and P9 deletion mutant between different faces of FAp crystals. (B) Different adsorptions between amelogenin and P9 deletion mutants on either (001) or (hk0) face of FAp crystals.

generating reproducible amelogenin adsorption to the HAp suspension. Therefore, the same vortex apparatus should be used for both test and control groups to ensure reliable comparisons.

4.4.5. The samples collected at four time points (0, 2, 8, and 16 min) are centrifuged at $1000 \times g$ for 10 min.

4.4.6. The supernatant is transferred to a new tube and the protein concentration is measured by Bradford assay.

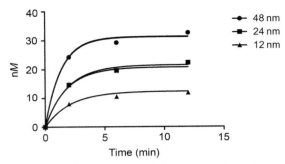

Figure 15.5 Global dynamic fitting curves of full-length amelogenin binding to hydroxyapatite crystals.

4.4.7. The amounts of amelogenin bound to HAp at each time point are calculated by comparing the protein concentrations before and after binding.

4.4.8. The binding kinetics is analyzed by Prism 6 (GraphPad, La Jolla, CA, USA). The nonlinear regression is used for curve fitting the data to the equations of *association kinetics at two or more concentrations of ligand*.

Figure 15.5 shows the kinetic fitting of full-length amelogenin. The kinetic parameters from global fitting are $K_d = 4.968 \times 10^8$ M, $B_{max} = 43.73 \pm 7.028$, $K_{on} = 7.6952 \times 10^{0.06} \pm 1.053 \times 10^6$ s^{-1}, and $K_{off} = 0.3821 \pm 0.048$ s^{-1} M^{-1}.

5. DATA HANDLING AND PROCESSING

All quantitative data are statistically processed for their significance by a student *t*-test and/or ANOVA.

6. SUMMARY

A stepwise protocol was developed to identify the binding of motifs in amelogenin to apatite crystals and to compare the difference in their binding on (001) and (hk0) faces of the apatite crystals. The assays included an overlapping peptide screening, peptide competitive adsorption assay, and ELISA on peptide surfaces using amelogenin protein and its mutants in which binding motifs were deleted. The apatite-binding motifs were identified at both amino- and carboxy-termini of amelogenin using this stepwise strategy.

ACKNOWLEDGMENTS

Supported by NIDCR 2R01DE015821 and 3R01DE015281S1 to W. L. MS Facility acknowledges the support of the Sandler Family and the Gordon and Betty Moore Family Foundations.

REFERENCES

Bartlett, J. D., & Simmer, J. P. (1999). Proteinases in developing dental enamel. *Critical Reviews in Oral Biology and Medicine, 10*, 425–441.
Brunauer, S., Emmett, P. H., & Teller, E. (1938). Adsorption of Gases in Multimolecular Layers. *Journal of the American Chemical Society, 60*, 309–319.
Daculsi, G., Menanteau, J., Kerebel, L. M., & Mitre, D. (1984). Length and shape of enamel crystals. *Calcified Tissue International, 36*, 550–555.
Featherstone, J. D., Mayer, I., Driessens, F. C., Verbeeck, R. M., & Heijligers, H. J. (1983). Synthetic apatites containing Na, Mg, and CO3 and their comparison with tooth enamel mineral. *Calcified Tissue International, 35*, 169–171.
Fincham, A. G., Moradian-Oldak, J., & Simmer, J. P. (1999). The structural biology of the developing dental enamel matrix. *Journal of Structural Biology, 126*, 270–299.
Habelitz, S., Kullar, A., Marshall, S. J., DenBesten, P. K., Balooch, M., Marshall, G. W., et al. (2004). Amelogenin-guided crystal growth on fluoroapatite glass-ceramics. *Journal of Dental Research, 83*, 698–702.
Ichijo, T., Yamashita, Y., & Terashima, T. (1992). Observations on the structural features and characteristics of biological apatite crystals. 2. Observation on the ultrastructure of human enamel crystals. *The Bulletin of Tokyo Medical and Dental University, 39*, 71–80.
Le, T. Q., Gochin, M., Featherstone, J. D., Li, W., & DenBesten, P. K. (2006). Comparative calcium binding of leucine-rich amelogenin peptide and full-length amelogenin. *European Journal of Oral Sciences, 114*(Suppl. 1), 320–326, discussion 327–329, 382.
Leblond, C. P., & Warshawsky, H. (1979). Dynamics of enamel formation in the rat incisor tooth. *Journal of Dental Research, 58*, 950–975.
Li, W., Gao, C., Yan, Y., & DenBesten, P. (2003). X-linked amelogenesis imperfecta may result from decreased formation of tyrosine rich amelogenin peptide (TRAP). *Archives of Oral Biology, 48*, 177–183.
Nanci, A. (2003). *Ten Cate's Oral Histology*. St. Louis: Mosby.
Pergolizzi, S., Anastasi, G., Santoro, G., & Trimarchi, F. (1995). The shape of enamel crystals as seen with high resolution scanning electron microscope. *Italian Journal of Anatomy and Embryology, 100*, 203–209.
Plate, U., & Hohling, H. J. (1994). Morphological and structural studies of early mineral formation in enamel of rat incisors by electron spectroscopic imaging (ESI) and electron spectroscopic diffraction (ESD). *Cell and Tissue Research, 277*, 151–158.
Tanimoto, K., Le, T., Zhu, L., Chen, J., Featherstone, J. D., Li, W., et al. (2008). Effects of fluoride on the interactions between amelogenin and apatite crystals. *Journal of Dental Research, 87*, 39–44.
Tarasevich, B. J., Lea, S., Bernt, W., Engelhard, M., & Shaw, W. J. (2009). Adsorption of amelogenin onto self-assembled and fluoroapatite surfaces. *The Journal of Physical Chemistry, 113*, 1833–1842.

CHAPTER SIXTEEN

Using the RosettaSurface Algorithm to Predict Protein Structure at Mineral Surfaces

Michael S. Pacella[*], Da Chen Emily Koo[†], Robin A. Thottungal[‡], Jeffrey J. Gray[‡,§,1]

[*]Department of Biomedical Engineering, Johns Hopkins University, Baltimore, Maryland, USA
[†]T. C. Jenkins Department of Biophysics, Johns Hopkins University, Baltimore, Maryland, USA
[‡]Program in Molecular Biophysics, Johns Hopkins University, Baltimore, Maryland, USA
[§]Department of Chemical and Biomolecular Engineering, Johns Hopkins University, Baltimore, Maryland, USA
[1]Corresponding author: e-mail address: jgray@jhu.edu

Contents

1. Introduction 344
2. Algorithm Evolution and Prior Results 346
3. Computational Methods 348
    3.1 Overview 348
    3.2 Search algorithm 349
    3.3 Score function 350
4. Protocol Capture 352
    4.1 Running RosettaSurface 352
    4.2 Postprocessing and interpretation of results 353
5. Future Challenges 359
Acknowledgments 360
A. Appendix 360
    A.1 Detailed algorithm description 360
    A.2 Surface parameters 361
    A.3 Formatted PDB and surface vector files 362
    A.4 Fragment files 362
    A.5 Constraint files 363
References 363

Abstract

Determination of protein structure on mineral surfaces is necessary to understand biomineralization processes toward better treatment of biomineralization diseases and design of novel protein-synthesized materials. To date, limited atomic-resolution data have hindered experimental structure determination for proteins on mineral surfaces. Molecular simulation represents a complementary approach. In this chapter, we review

RosettaSurface, a computational structure prediction-based algorithm designed to broadly sample conformational space to identify low-energy structures. We summarize the computational approaches, the published applications, and the new releases of the code in the Rosetta 3 framework. In addition, we provide a protocol capture to demonstrate the practical steps to employ RosettaSurface. As an example, we provide input files and output data analysis for a previously unstudied mineralization protein, osteocalcin. Finally, we summarize ongoing challenges in energy function optimization and conformational searching and suggest that the fusion between experiment and calculation is the best route forward.

1. INTRODUCTION

The interaction of proteins with solid surfaces plays a pivotal role in diverse areas including biomaterials, bioprocessing, nanotechnology, solar energy, materials synthesis, and mineralogy. Treatments for diseases resulting from unregulated biomineralization, such as atherosclerosis, dental calculus, and kidney stone formation, could also be better developed with structural guidance. Protein–surface interactions govern the immune response and biocompatibility of implanted medical devices, the growth and morphology of crystals during biomineralization, and the self-assembly of nanoscale systems (Gray, 2004; Latour, 2008). A better understanding of these interactions would give engineers the ability to design proteins to grow custom materials: a revolutionary, environmentally friendly approach to nanoengineering.

Despite the integral role of protein–surface interactions in these fields, a fundamental understanding of their nature has remained elusive due to limited experimental methods capable of resolving the atomic structure of proteins at solid interfaces (Cohavi et al., 2010, Collier, Vellore, Yancey, Stuart, & Latour, 2012; Goobes, Stayton, & Drobny, 2007). Computational methods offer an alternative approach to explore these interactions at an atomic scale (Latour, 2008).

A wide range of computational methods is currently employed to study protein–surface interactions. In general, these approaches can be broken down into three categories: quantum mechanical (QM)-, molecular dynamics (MD)-, and Monte Carlo (MC)-based docking. QM methods, which treat electronic degrees of freedom (DOFs) explicitly, can be accurate for determining both interaction energies and equilibrium structures of molecules adsorbed on surfaces. However, such methods are limited by poor scaling with the number of atoms in the system and performing geometry optimizations on small peptides using this approach is just beginning to become feasible (Rimola, Aschi, Orlando, & Ugliengo, 2012). Still, QM calculations on smaller systems can guide parameterization of molecular mechanics force fields or score functions (Iori, Di Felice, Molinari, & Corni, 2009; Latour, 2008). Recent QM studies have parameterized the interaction of individual amino acids with hydroxyapatite (HAp) (Corno, Rimola, Bolis, & Ugliengo, 2010) and gold (Iori et al., 2009).

MD methods integrate Newton's laws of motion for each atom in the system and follow the trajectory through time. Theoretically, if MD is run long enough, the time-average of the simulation should approach the thermodynamic ensemble average, allowing for the calculation of thermodynamic properties such as the free energy change of adsorption (Freeman, Harding, Quigley, & Rodger, 2011; Wang, Stuart, & Latour, 2008). Determining the sufficient length of time, however, is difficult and variable from system to system. Whether MD simulations have been run long enough to draw conclusions about the equilibrium behavior of proteins on solid surfaces remains an open question (Latour, 2008), particularly in determining the role of water-mediated protein–surface interactions. The advantage of MD lies in the ability to calculate kinetic properties such as adsorption rate constants from the resulting trajectories. MD also captures correlated motions in a natural way, making it ideal for systems with many correlated DOFs (such as systems containing explicit solvent).

MC-based docking algorithms employ a move set of random perturbations that update explicit DOFs in a molecular system; each move is either accepted or rejected based on the Metropolis criterion. Similarly to MD, if a sufficient number of moves are attempted in MC, the simulation average should converge to the thermodynamic ensemble average. By incorporating minimization into MC-based docking, the ability to sample from a thermodynamic ensemble is lost, but the likelihood of locating a global minimum on the potential energy surface increases dramatically (Li & Scheraga, 1987). Although it lacks the accuracy of QM methods and the kinetic information provided by MD, MC-plus-minimization (MCM) docking is perhaps the best method for locating global minima and determining structural complementarity rapidly. MCM can also perform protein design calculations by expanding the move set to include altering the chemical identity of amino acid side chains, thus searching for a sequence that minimizes the potential energy for a given structure (Kuhlman & Baker, 2000; Kuhlman et al., 2003).

In the field of biomaterials and bioprocessing, the conventional wisdom has been that proteins adsorb to solid surfaces nonspecifically and become denatured upon their adsorption (Gray, 2004; Michaels & Matson, 1985; Ratner & Bryant, 2004). However, in the field of biomineralization, organisms demonstrate exquisite control over crystal growth and morphology using proteins (Belcher et al., 1996; Berman et al., 1993; Weiner & Dove, 2003). With this apparent contradiction in mind, we designed RosettaSurface, a computational tool that simultaneously optimizes the fold, side-chain conformations, and rigid-body orientation of a protein on a solid surface, with the goal of broadly testing the following hypothesis: Does structural recognition analogous to that found in protein–protein complexes exist in the interaction of proteins with crystal surfaces? While QM calculations give accurate energies and equilibrium structures and MD methods provide kinetic data, RosettaSurface employs MCM docking as the proposed route to test potential structural complementarity. Additionally, because of Rosetta's success in predicting protein folding (Raman et al., 2009; Simons, Bonneau, Ruczinski, & Baker, 1999) and protein–protein interfaces (Gray et al., 2003; Sircar, Chaudhury, Kilambi, Berrondo, & Gray, 2010), the algorithms and score functions contained within that framework are logical starting points. RosettaSurface can perform *ab initio* or constraint-based calculations. Including constraints allows users to create a feedback loop between experimental constraint determination and computational refinement based on these constraints.

In this chapter, we present the current version of the RosettaSurface algorithm and the methods it employs. The core RosettaSurface algorithm is available both as a C++ executable and as a modifiable PyRosetta script (Chaudhury, Lyskov, & Gray, 2010). For usage on high-performance computing clusters, we recommend the C++ executable. For users interested in modifying or expanding the existing algorithm, or crafting their own algorithm, PyRosetta scripting is recommended. For both versions, command lines for preprocessing, execution, and postprocessing/analysis are provided.

2. ALGORITHM EVOLUTION AND PRIOR RESULTS

RosettaSurface was developed in stages to address progressively more complex challenges. In the first iteration, the algorithm focused exclusively on sampling rigid-body and side-chain conformational DOFs between a previously refined solution-state structure of the protein of interest and the surface. Restriction to these DOFs precluded the prediction of protein

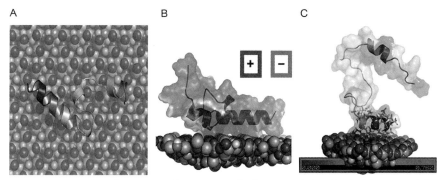

Figure 16.1 Predicted structures from RosettaSurface. Top (A) and side (B) views of the lowest-energy structure of statherin docked onto the [001] face of HAp using the initial version of the algorithm. (C) A representative structure of statherin bound to HAp, colored by the local root-mean-square deviation of 100 structures that best satisfied experimental constraints. *Reprinted from Makrodimitris et al. (2007) and Masica, Ash, et al. (2010) with permission.* (See color plate.)

conformational changes upon adsorption but allowed a thorough coverage of the rigid-body and side-chain conformational space. This algorithm was first applied to the 43-residue salivary protein statherin docking to the [001] plane of HAp (Makrodimitris, Masica, Kim, & Gray, 2007). The algorithm converged on a set of low-energy structures (Fig. 16.1A and B) consistent with previously published solid-state NMR (ssNMR) data and mutagenesis experiments, and it also suggested a helical lysine and arginine molecular recognition motif for the phosphate clusters in HAp.

The next iteration of the algorithm extended sampling to include protein backbone torsional DOFs, not only allowing us to model conformational changes upon adsorption but also increasing the size of conformational space. Again, results were generated for the model statherin/HAp system and benchmarked against experimental ssNMR distance and angle measurements (Masica & Gray, 2009). This algorithm captured an experimentally observed adsorption-induced folding event in statherin's helical binding domain and yielded top structures that matched 15 published ssNMR distance and angle measurements. The algorithm was adapted to incorporate user-provided structural constraints to bias sampling toward conformations that are consistent with existing experimental information. This study was the first to demonstrate that a combination of experimental constraint determination and structure prediction could effectively determine high-resolution protein structures at biomineral interfaces.

The method of establishing a feedback loop between experimental constraint determination and structure prediction was further expanded upon in the next study on statherin/HAp (Masica, Ash, Ndao, Drobny, & Gray, 2010). Areas of structural ambiguity in computational results were targeted for isotopic labeling and ssNMR constraint measurement. Several iterated rounds of experiment and computation yielded a final adsorbed-state structure with minimal ambiguity, shown in Fig. 16.1C. In addition, the ability of predicted structures to comply with the experimental constraints on different HAp faces suggested which faces were most likely to be the binding face.

Since the original algorithm development, RosettaSurface has also been used to study the binding of tissue-nonspecific alkaline phosphatase to HAp (McKee et al., 2011), phosphorylated and nonphosphorylated leucine-rich amelogenin to HAp (Masica, Gray, & Shaw, 2011; Tarasevich et al., 2013), an acidic serine- and aspartate-rich motif of osteopontin to HAp (Addison, Masica, Gray, & McKee, 2010), and other apatite-binding peptides (Addison, Miller, et al., 2010).

3. COMPUTATIONAL METHODS

3.1. Overview

Recently, we incorporated RosettaSurface into the Rosetta 3 software suite (Leaver-Fay et al., 2011), giving object-oriented access to new features such as noncanonical amino acids (Renfrew, Choi, Bonneau, & Kuhlman, 2012), a generalized framework for handling symmetry (DiMaio, Leaver-Fay, Bradley, Baker, & Andre, 2011), variable pH (Kilambi & Gray, 2012), and the scripting interfaces RosettaScripts (Fleishman et al., 2011) and PyRosetta (Chaudhury et al., 2010). Rosetta 3's object-oriented design and Python interface make it possible for users to craft their own MCM docking or folding algorithms and tailor score functions specifically to their needs.

The RosettaSurface algorithm addresses the two principal challenges in structure prediction: (1) sampling of the conformational space available to both the protein and solid surface and (2) scoring of candidate structures (decoys) to identify the global free energy minimum.

Conformational sampling is carried out in two stages: (1) a solution-state stage, where the internal DOFs of the protein are optimized in implicit solvent, independent of the surface, and (2) an adsorbed-state stage, where the internal DOFs of the protein and the rigid-body orientation of the protein relative to the surface are simultaneously optimized on the surface. This process is repeated many times to generate a large ensemble of decoys for both the solution and adsorbed states. Starting from a fresh random fragment assembly process to generate each new decoy avoids oversampling kinetically trapped structures in false minima on the energy landscape and allows for straightforward parallelization.

3.2. Search algorithm

Figure 16.2 presents a flowchart of the RosettaSurface algorithm. The algorithm employs a multiscale MCM protocol based on Rosetta docking (Gray et al., 2003) and folding (Simons et al., 1999) strategies. A key feature in our algorithm is variation in the number of cycles before transitioning from the "solution" to the "adsorbed" stages. In the solution stage, the protein is able to sample both extended and compact structures. Once adsorbed onto the surface, large conformational changes often produce steric clashes between protein and surface atoms, preventing acceptance in a traditional MC

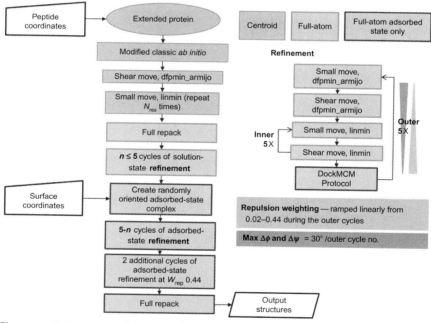

Figure 16.2 Flowchart of the complete RosettaSurface algorithm. Small/shear moves perturb the protein backbone and are described in Bradley, Misura, and Baker (2005); linmin and dfpmin_armijo are energy minimization routines described in the Rosetta user's manual (https://www.rosettacommons.org/manuals). N_{res}, protein length; W_{rep}, repulsive van der Waals weight. (For color version of this figure, the reader is referred to the online version of this chapter.)

algorithm (a computational limitation). In the algorithm, the protein begins the solution stage as a fully extended peptide chain and undergoes several optimization cycles, each resulting in a more compact and "protein-like" structure, containing features that occur frequently in known crystal structures. Varying the number of optimization cycles in solution before transitioning to the adsorbed state ensures that both extended (few optimization cycles in solution) and compact (many optimization cycles in solution) structures are sampled in the adsorbed state. Physically, this feature allows the calculation to capture protein conformational changes upon adsorption (such as unfolding or structural transitions). A detailed description of the algorithm is in the Supporting Information.

An assumption in the search algorithm is that the relative positions of the atoms within the solid surface remain static throughout the simulation and that the surface atoms form a flat plane (no step edges or local defects). These complexities have yet to be incorporated into RosettaSurface.

3.3. Score function

The standard Rosetta score function contains both statistical and physical terms that are optimized to recover known protein crystal structures during folding and docking. However, benchmarking and parameterization are crucial when extending an existing score function or force field into the context of protein–surface interactions (Collier et al., 2012). Due to a lack of experimental benchmarks, very few parameter sets have been developed to specifically describe protein–surface interactions. Additionally, few existing parameter sets designed to simulate protein behavior in solution have been verified to be transferrable to the context of protein–surface interactions (Collier et al., 2012; Vellore, Yancey, Collier, Latour, & Stuart, 2010).

Rosetta's default all-atom score function for protein folding (score12) serves as the starting point for the RosettaSurface score function. To account for electrostatic interactions between protein side chains and charged surface ions, a Coulomb interaction term with a distance-dependent dielectric is added (Kilambi et al., 2012). The final score function is a weighted linear combination dominated by van der Waals (attractive and repulsive), electrostatic (distance-dependent dielectric), hydrogen-bond, and implicit solvent interactions (Table 16.1):

$$E_{total} = W_{vdW} E_{vdW} + W_{elec} E_{elec} + W_{Hbond} E_{Hbond} + W_{solv} E_{solv}$$

Table 16.1 Major score terms used in RosettaSurface

Score term	Weight	Description
fa_atr	0.800	Attractive range of Lennard-Jones 6–12 potential
fa_rep	0.440	Repulsive range of Lennard-Jones 6–12 potential
hack_elec	1.000	Coulomb interactions with distance-dependent dielectric[a]
hbond_lr_bb	1.170	Long-range orientation-dependent backbone–backbone hydrogen bonding[b]
hbond_sr_bb	0.585	Short-range orientation-dependent backbone–backbone hydrogen bonding[b]
hbond_bb_sc	1.170	Orientation-dependent backbone–side-chain hydrogen bonding[b]
hbond_sc	1.100	Orientation-dependent side-chain–side-chain hydrogen bonding[b]
fa_sol	0.650	Implicit solvation model based on EEF1 from Lazaridis and Karplus[c]

[a]Warshel and Russell (1984).
[b]Kortemme, Morozov, and Baker (2003).
[c]Lazaridis and Karplus (1999).

where the weights W_i and the energy functions E_i are defined in Rohl, Strauss, Misura, and Baker, (2004). The statistical score function terms derived from protein structures are only included for intraprotein interactions. Rosetta's accuracy in modeling the intraprotein component of protein–surface interactions is the main advantage of using score12 as a starting point (Leaver-Fay et al., 2013).

When RosettaSurface is run in a constraint-based fashion, the score function is modified to include additional terms that penalize structures deviating from experimental measurements. This penalty ensures that conformational sampling is restricted to structures in agreement with experimental data.

Our studies to date are limited to surfaces of ionic crystals. In this context, electronic delocalization effects are minimal, and assignment of partial charges for electrostatic calculations is a fair approximation. In addition to ignoring electronic effects, we have also employed an implicit solvation model (Lazaridis & Karplus, 1999). Although the reduced DOFs offered by implicit solvation save computation time, hydrogen-bond interactions mediated by explicit water molecules at the solid–liquid interface are lost (Freeman et al.,

2011; Latour, 2008; Sun & Latour, 2006). The precise role of explicit waters in protein–surface interactions remains an area of active research.

4. PROTOCOL CAPTURE

In this section, we provide a step-by-step protocol to generate and postprocess candidate adsorbed protein structures. As an example, we dock osteocalcin to the [100] surface of HAp. Osteocalcin is known to play a role in bone mineralization, but the mechanism of its interaction remains speculative (Ducy et al., 1996). Osteocalcin's solution-state crystal structure is known (PDB ID = 1Q8H) and a model binding structure has been proposed assuming no unfolding upon binding (Hoang, Sicheri, Howard, & Yang, 2003). Since osteocalcin contains three γ-carboxyglutamic acids, it has only become possible to model this protein in the new Rosetta 3 framework with the noncanonical amino acid parameterization (Renfrew et al., 2012).

Input files containing the crystal surface coordinates, the protein sequence, and experimentally determined constraints (if any) are needed to run RosettaSurface. Descriptions of each of these file types are included in the Supporting Information, and the input files for the osteocalcin/HAp example are provided in the `protocol_capture` directory included with Rosetta 3.5 (www.rosettacommons.org).

4.1. Running RosettaSurface

4.1.1 C++ executable

After obtaining and compiling the latest version of the Rosetta source code (see http://www.rosettacommons.org), the RosettaSurface executable can be found in the directory `rosetta_source/bin`. The command line for execution of RosettaSurface is

```
rosetta_source/bin/surface_docking.gccrelease @flags
```

where `gccrelease` may be replaced if other compilers (icc and macos) were used to build Rosetta and `flags` is a plain text file containing a list of all arguments the user wishes to pass to RosettaSurface (Table 16.2). For scientific benchmarking, O (10^5) structures are needed; thus, a high-performance computing cluster is employed, typically using the Condor job distribution system (Thain, Tannenbaum, & Livny, 2005).

4.1.2 PyRosetta script

In addition to the C++ RosettaSurface script, we have written a Python script employing the PyRosetta libraries to allow a user to easily change

Table 16.2 Command-line arguments for the RosettaSurface C++ executable

Arguments	Description
–s filename	PDB file containing protein and surface
–in:file:frag3 filename	Name of 3-mer fragment file
–in:file:frag9 filename	Name of 9-mer fragment file
–in:file:surface_vectors filename	Name of surface vectors file
–database filepath	Path to rosetta_database (default $ROSETTA3_DB)
–nstruct integer	Number of structures to generate
–constraints:cst:file filename	Name of constraints file
–multiple_processes_writing_to_one_directory	*Optional.* Enables parallelization across a computing cluster
–show_simulation_in_pymol integer	*Optional.* Number of seconds between PyMOL updates (enables live viewing of simulation in PyMOL)

and adjust parameters of the protocol to suit specific needs. For example, the number of cycles of refinements may be decreased if sufficient constraint data are available, allowing fewer cycles to obtain a meaningful structure in less time. Examples of parameters that can be changed are listed in Table 16.3.

The PyRosetta scripts can be found in the `apps` directory within PyRosetta and can be run in a similar way to the C++ executable:

```
PyRosetta/apps/surface_docking/surface_docking.py @flags
```

The PyRosetta options differ slightly (Table 16.4). Decoy generation requires 1–15 min, depending on the length of the protein and the distribution of the solution-state and adsorbed-state refinement cycles.

4.2. Postprocessing and interpretation of results

For both the C++ and PyRosetta implementations, the raw data output from RosettaSurface consists of decoys and their corresponding scores. Decoys are output in PDB format in the working directory and are appended with a numerical identification tag. Total scores and individual score terms are appended in each decoy's PDB file and summarized in a score file (.fasc) in the working directory. A breakdown of the hydrogen bonds and the constraint energies is also appended to the end of each PDB file.

Table 16.3 Examples of modifiable PyRosetta algorithm parameters

Parameter	Variable/function name	Default value
Solution-state (full-atom)		
Refinement cycles	sol_cycles	$n \leq 5$ (random)
Outer cycles	sol_outer_cycles	5
Inner cycles	sol_inner_cycles	5
Adsorbed-state		
Refinement cycles	ads_cycles	$5 - n$
Outer cycles	ads_outer_cycles	5
Inner cycles	ads_inner_cycles	5
Other parameters		
Number of small/shear moves before minimization	cen_relax.set_nmoves(n) fa_relax.set_nmoves(n)	6
Maximum angle of perturbation for small/shear moves	cen_relax.set_max_angle(n) fa_relax.set_max_angle(n)	30/outer_cycle

Typically, the top-scoring 1% of decoys are considered for structural analysis. The visualization tool PyMOL (Schrodinger, 2010), which can be interfaced with PyRosetta (Baugh, Lyskov, Weitzner, & Gray, 2011), allows the user to identify interactions that contribute most strongly to the adsorption energy. The script GetTop.sh extracts the top decoys into a separate folder, and the script PostProcessRS.sh generates the secondary structure, contact map, and adsorbed-state surface contact map plots inside that folder. An additional script, extract_scores_and_constraints.py, is used to output the total score and constraint score of each decoy in the directory. A summary of these scripts and their usage is in Table 16.5.

The sample files included with the installation can be found in the directory rosetta_demos/protocol_capture/2013/surface_docking, which includes a README file with additional instructions regarding the environment and paths. A sample flags file is provided to generate ten osteocalcin decoys with a disulfide constraint to bond residues 23 and 29. The commands, run from the sample directory, are

1. surface_docking.py @flags (Use command in §4.1.1 for C++ executable)
   a. Wait until all 10 decoys have been generated (no .in_progress files left)
2. GetTop.sh Ads 4
3. cd TOP4.Ads/
4. PostProcessRS.sh

Table 16.4 Command-line arguments for RosettaSurface Python script

Arguments	Description
-h, --help	Show help message with these descriptions and exit
-s, --start filename	Specify start PDB file containing protein and surface
--database path	Specify Rosetta database path in PyRosetta (default $PYROSETTA_DATABASE)
-n, --nstruct integer	Specify number of structures to generate
-f3, --frag3 filename	*Optional*. Specify 3-mer fragment file. If not specified, *ab initio* and centroid relax will be skipped, which is useful if solution-state structure is known
-f9, --frag9 filename	*Optional*. Specify 9-mer fragment file. If not specified, *ab initio* and centroid relax will be skipped, which is useful if solution-state structure is known
-c, --constraints	*Optional*. Specifies that ssNMR constraints are to be loaded from [startpdbname]_ads##.cst (adsorbed-state) and [startpdbname]_sol##.cst, (solution-state) where ## are integers
--nosmallshear	*Optional*. If included, small/shear refinements are not used. Use if rigid-body docking is desired. If not included, refinements are used (default)
-d, --disulf filename	*Optional*. Specify disulfide constraints file

Table 16.5 Scripts for postprocessing analysis

Script name	Working directory	Usage	Output
GetTop.sh	Main directory with all decoys	GetTop.sh <Ads/Sol> <no. of decoys>	TOP<no. of decoys>.<Ads/Sol> folder containing decoys
PostProcessRS.sh	TOP folders for Ads and Sol	PostProcessRS.sh	SecStruct.png ContactMap.png SurfaceContactMap.png
extract_scores_and_constraints.py	TOP folders for Ads and Sol	extract_scores_and_constraints.py <ads/sol/both>	table_<ads/sol>_scores

In order for this example to run quickly, these commands generate 10 decoys (step 1 in the flags file) and postprocess the four top-scoring decoys. For production runs, and for the plots described in sections 4.2.1–4.2.4, we generated 100,000 structures and analyzed the top-scoring 100.

4.2.1 Secondary structure

A plot is generated to illustrate the distribution of secondary structures at each residue in the set of top-scoring structures. Secondary structure can assess convergence between top-scoring decoys, and comparison of the secondary structure profiles between adsorbed- and solution-state structures can identify adsorption-induced conformational changes.

Output for the case of osteocalcin on HAp is in Fig. 16.3A. The secondary structure distribution of the top adsorbed-state decoys shows a mix of secondary structures at the N-terminus and three major helical structures, matching those observed in the crystal structure (bar above the plot). Residues 1–12 are disordered in the crystal structure (Hoang et al., 2003), so they are excluded, not comparable. The protocol suggests that the solution-state secondary structures are largely retained after adsorption.

4.2.2 Protein–protein contact map

While the secondary structure plot provides local structural information, the contact map provides information about the tertiary structure of the protein. Pairs of residues with at least one atom–atom contact within 8 Å are aggregated over the 100 top-scoring structures and the resulting pairwise frequencies are illustrated in a nonmirrored plot. The x- and y-axes represent the residue numbers and the gradient from white to blue represents minimum to maximum frequencies.

An example of this two-dimensional matrix for the osteocalcin adsorbed-state is shown in Fig. 16.3B. Proximal contacts, such as those found in helices, appear as 4-residue wide stripes along the diagonal, while distal interactions are represented by patches off the diagonal.

Figure 16.3C is a difference plot between the known solution-state structure and the top adsorbed-state decoys, which shows how the tertiary structure changes upon adsorption. (To generate this plot, include a native.pdb file containing the solution-state structure and run the script PlotContactMap.sh.) Contacts that dominate in the adsorbed-state and solution-state decoys are colored blue and red, respectively, with white representing equivalent frequencies. In this example, the first 12 residues are not

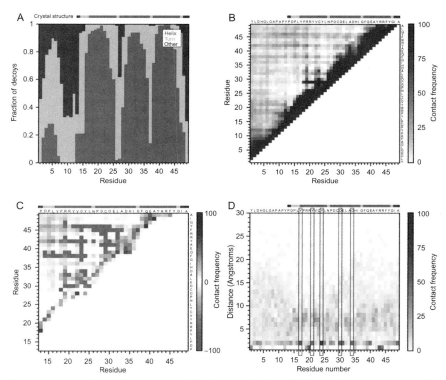

Figure 16.3 Analysis plots generated by RosettaSurface. (A) Secondary structure profile for osteocalcin adsorbed to HAp [100]. (B) Intraprotein contact map for osteocalcin adsorbed to HAp [100]. (C) A difference plot between crystal structure and adsorbed-state decoys. (D) Protein–surface contact map for osteocalcin adsorbed to HAp [100]. γ represents gamma-carboxyglutamic acid. Line above each plot represents the secondary structure of the crystal structure and is colored as in (A). Boxes highlight residues predicted to interact with HAp in Hoang's model. (See color plate.)

comparable and are omitted. The map shows that the prominent interhelical interactions in the solution-state structure are broken during adsorption.

4.2.3 Protein–surface contact map

Similar to the protein–protein contact map, the protein–surface contact map gives information about contacts in the top adsorbed-state decoys. The y-axis is the perpendicular distance of each residue to the surface (in angstroms), and the width of the distance distribution can be used to assess convergence in the models. Residues with ill-defined positions are the most useful protein–surface atom pairs to inform with ssNMR experiments.

Figure 16.3D illustrates the protein–surface contact map for osteocalcin. The three helices exhibit alternating dark spots close to the surface, showing frequent adsorption among the top-scoring structures. Some of the most dominant protein–surface interactions include the three γ-carboxyglutamic acids, two aspartic acids, and two arginine residues and minor contacts with some unmodified glutamic acids. The large spread of distances at both termini shows that a range of conformations are represented in the top-scoring decoys. ssNMR measurements, therefore, would be helpful near the termini to narrow down the candidate structures.

4.2.4 Interpretation of osteocalcin results

Together, the RosettaSurface results in Fig. 16.3 suggest that the adsorbed osteocalcin maintains three helices, adopts an unfolded structure, and interacts with the HAp mineral primarily through three γ-carboxyglutamic acid residues (Gla17, Gla21, and Gla24) of the first helix, two aspartic acid residues (Asp30 and Asp34) of the second helix, and two arginine residues (Arg43 and Arg44) of the third helix. The interacting residues on the first two helices are similar to those proposed by Hoang et al. in their rigid, crystal structure-based, model (Hoang et al., 2003) and both models show symmetrically equivalent calcium atoms coordinated by the three γ-carboxyglutamic acid residues. In fact, the residue contacts proposed by Hoang et al. are all local in sequence, so that many alternate or unfolded conformations can reproduce the same strong contacts with the calcium ions in HAp.

To test whether the unfolded conformations are truly lower in energy, or whether RosettaSurface may have been unable to sample folded binding structures like Hoang et al.'s model, we modified the PyRosetta implementation of RosettaSurface to hold the osteocalcin backbone fixed for rigid docking to the surface. While structures like Hoang et al.'s model were generated during rigid docking (Fig. 16.4B), their energies were higher than those of the lowest-scoring unfolded, adsorbed osteocalcin structures (Fig. 16.4C). It is possible that the unfolded RosettaSurface models are closer to the true structure of adsorbed osteocalcin. However, it is also possible that the balance of interface energy and folding energy is inaccurate; thus, further experimental data are needed to resolve the true structure. The conformations generated by RosettaSurface can be used to suggest sites for ssNMR labeling and measurement. For example, a distance constraint between a side-chain nitrogen atom of Arg43 and a HAp phosphorous atom could provide information on whether or not osteocalcin's third helix interacts with HAp.

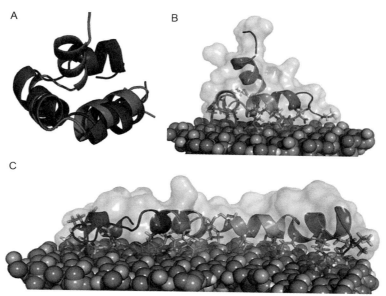

Figure 16.4 Predicted structures for the osteocalcin/HAp system. (A) Predicted solution-state structure of osteocalcin (blue) superimposed with its crystal structure (red). Lowest-energy decoys from rigid (B) and flexible (C) backbone docking. Similar contacts are made between the γ-carboxyglutamic acids and the surface in both models, indicating ambiguity in whether or not osteocalcin adopts extended conformations on HAp. (See color plate.)

5. FUTURE CHALLENGES

The RosettaSurface algorithm provides a flexible platform to explore adsorbed conformations of proteins on mineral surfaces. Because of the paucity of experimental data, it is difficult to validate an energy function (Latour, 2008), and thus, despite the success of the Rosetta energy function for folding, docking, and design, the ranking of adsorbed structures is subject to error. Work is needed to test alternate formulations and parameterizations of the energy function, ideally with additional atomic-resolution structural data on a variety of experimental systems. A second limitation is in RosettaSurface's sampling. Exploring relevant areas of conformational space is particularly difficult for long proteins, since transitional pathways may be hindered by the surface (Masica et al., 2011).

For progress in the study of biomineralization, a combination of computational and experimental approaches is ideal. The weaknesses in energy functions and conformational sampling can be compensated with even sparse experimental data to constrain the search. The constraint-based docking framework employed by RosettaSurface establishes a feedback loop between experimental constraint determination and computational refinement, making it a valuable tool for the study of biomineralization (Masica, Ash, et al., 2010). As the fundamental components of the algorithm improve, design of biomineralization proteins will also improve (Masica, Schrier, Specht, & Gray, 2010; Schrier, Sayeg, & Gray, 2011).

ACKNOWLEDGMENTS

Moon-Young Liza Lee developed early versions of RosettaSurface in PyRosetta. Funding was provided by NSF CAREER grant 0846324 and NIH training grant T32 GM008403 (for RAT).

A. APPENDIX

A.1. Detailed algorithm description

Figure 16.2 in the main text shows the algorithm flowchart. The protein starting structure is an extended chain in implicit solvent, independent from the surface. The protein is initially represented in "centroid mode" where all backbone atoms are explicitly modeled but side chains are represented by a single "pseudoatom" that captures the low-resolution features of the side-chain chemistry.

The extended chain is then rapidly collapsed using a fragment assembly protocol. The fragment assembly protocol follows Simons, Kooperberg, Huang, and Baker (1997), except the radius of gyration score term is removed to allow a greater sampling of extended structures in both the solution and adsorbed stages to capture conformational changes upon adsorption. After each fragment assembly step, the protein backbone is optimized with "shear" and "small" moves coupled with conjugate gradient and line minimization, respectively.

At this point, the protein side chains are converted from "centroid" to "full-atom" mode. In "full-atom" mode, all-atom (including hydrogen) side chains are built from a backbone-dependent rotamer library (Dunbrack & Cohen, 1997) using a simulated annealing protocol (Kuhlman & Baker, 2000). A rotamer library, derived from PDB statistics, represents local minima on the QM energy landscape (Renfrew, Butterfoss, & Kuhlman, 2008).

The protein now undergoes a total of 5 "refinement cycles" with adsorption occurring randomly between cycles 1 and 5. This is done by generating a random number *n* between 1 and 5 and performing *n* cycles of solution stage refinement and 5-n cycles of adsorbed stage refinement. n is varied for each individual decoy, ensuring an approximately equal amount of sampling in both the solution and adsorbed states and yielding a range of compact and extended final structures.

Solution stage refinement includes "outer-" and "inner-loop optimization." During outer-loop optimization, a sequence of backbone perturbation moves are applied, each followed by conjugate-gradient minimization. During inner-loop optimization, the same perturbing moves are applied except each is followed by line minimization along the initial gradient. Separating the refinement stage in this fashion ensures that the least computationally expensive (and least globally perturbing) line minimization calculation is performed in the innermost loop. The outer loop is implemented five times; for each outer loop, the inner loop is repeated five times.

After n cycles of solution stage refinement, an adsorbed-state complex is formed by introducing the solution-refined protein to the surface in a random orientation and bringing the two into contact. The adsorbed protein then undergoes 5-n outer cycles of refinement on the surface. Adsorbed stage refinement is identical to solution stage refinement with the addition of a modified version of RosettaDock's (Gray et al., 2003) high-resolution docking protocol at the end of each cycle. This modified docking protocol consists of a sequence of small rigid-body perturbations, side-chain repacking, and gradient-based minimization in rigid-body space repeated six times; side-chain repacking is combinatorial (full repack), rather than sequential (rotamer trials), every third time. Combinatorial repacking accounts for the possibility of correlated side-chain motions but is more computationally expensive than sequential repacking, which optimizes each side-chain rotamer independently. Hence, the protein undergoes simultaneous backbone, side-chain, and rigid-body optimization on the surface. Finally, an additional 2 cycles of adsorbed stage refinement are performed to arrive at a low-energy structure. With user-input "surface vectors," the protein can be translated to a symmetrically equivalent position at the center of the slab after each round of docking. Because rigid-body docking involves translation in the plane of the surface, an additional "orientation" move is necessary to ensure that the protein is not translated off of the surface slab used for simulation yet remains in a symmetrically equivalent position on the surface.

A.2. Surface parameters

To maintain the residue-centric organization of Rosetta 3, RosettaSurface treats the surface as a sequence of "residues" whose positions remain static throughout the simulation. For ionic surfaces, this implementation is straightforward: each monatomic/polyatomic ion is treated as a distinct residue. The chemical properties of a residue are encoded by its corresponding params file in the Rosetta database. This file includes information such as bond connectivity, equilibrium geometry, atom types, and partial charges. Please see the Rosetta 3.4 manual for a description of each field in a params file.

A.3. Formatted PDB and surface vector files

RosettaSurface reads starting coordinates from a modified PDB format file (www.rcsb.org). The protein should initially be positioned above the surface face the user wishes to dock, with the protein coordinates appearing *after* the coordinates of the surface atoms in the file. The surface atoms should be designated as heteroatoms (HETATM record), belonging to a separate chain from the protein, in the input file with atom and residue designations that match those defined in the params file (see Section A.2).

Coordinates for the surface slab may be produced with CrystalMaker® (CrystalMaker Software Ltd, Oxford, England www.crystalmaker.com), a tool that enables users to prepare surfaces of predefined dimensions for specific crystal faces. The desired length/width of the surface slab will depend on protein size and must accommodate extended conformations without introducing edge effects. A slab thickness of ~10 Å is sufficient, as it exceeds the pairwise interaction cutoff of the Rosetta score function.

Three surface atoms, which define the edges of the surface unit cell, must be specified in a separate surface vectors file. These atoms define a pair of surface unit-cell vectors: translations by any integer linear combination of these vectors are symmetrically equivalent. These vectors can be used to translate the protein to a symmetrically equivalent surface patch at any point during the docking simulation. An example of the required syntax can be seen in the surface vectors file provided in the `protocol_capture` directory.

A.4. Fragment files

The fragment assembly protocol in the low-resolution stage requires pre-generated files that specify candidate backbone conformations, or fragments, for a given 9-mer or 3-mer subsequence. Fragments are most easily obtained using the Robetta server (Kim, Chivian, & Baker, 2004).

A.5. Constraint files

If RosettaSurface is run in a constraint-based fashion, the user must specify the type and parametric form of the constraint in a .cst file (Bowers, Strauss, & Baker, 2000). Three constraint types are currently allowed: atom pair, angle, and dihedral. Ambiguous constraints allow the user to provide a list of several constraints, with the score function only including the contribution of the lowest scoring constraint to the total energy. This type of constraint is needed to incorporate ssNMR distance measurements between protein and surface atoms because there are many symmetrically equivalent copies of the same surface atom in the simulation slab. Common parametric forms include harmonic, circular harmonic, bounded, Gaussian, constant, and linear scaling. An example constraint file for the statherin/HAp system is provided in the `protocol_capture` directory. For a description of constraint file formats, please see the Rosetta 3.4 manual.

In addition to distance and angle constraint files, a file containing residues with disulfide bonds can also be specified to ensure that these bonds are formed. The disulfide file contains residue number pairs, delimited by a space, in separate lines. An example disulfide file is provided in the `protocol_capture` directory.

Multiple ambiguous constraints must be placed in separate files with filenames as follows:

PDB file = [name of input file].pdb
Constraints files
1. [name of input file]_ads01.cst, [name of input file]_ads02.cst etc.
2. [name of input file]_sol01.cst, [name of input file]_sol02.cst etc.

Constraints for the solution and adsorbed states are separated, since protein–surface constraints do not apply to the solution-state refinements.

REFERENCES

Addison, W. N., Masica, D. L., Gray, J. J., & McKee, M. D. (2010). Phosphorylation-dependent inhibition of mineralization by osteopontin ASARM peptides is regulated by PHEX cleavage. *Journal of Bone and Mineral Research, 25*, 695–705.

Addison, W. N., Miller, S. J., Ramaswamy, J., Mansouri, A., Kohn, D. H., & McKee, M. D. (2010). Phosphorylation-dependent mineral-type specificity for apatite-binding peptide sequences. *Biomaterials, 31*, 9422–9430.

Baugh, E. H., Lyskov, S., Weitzner, B. D., & Gray, J. J. (2011). Real-time PyMOL visualization for Rosetta and PyRosetta. *PLos One, 6*, e21931.

Belcher, A. M., Wu, X. H., Christensen, R. J., Hansma, P. K., Stucky, G. D., & Morse, D. E. (1996). Control of crystal phase switching and orientation by soluble mollusc-shell proteins. *Nature, 381*, 56–58.

Berman, A., Hanson, J., Leiserowitz, L., Koetzle, T. F., Weiner, S., & Addadi, L. (1993). Biological-control of crystal texture—A widespread strategy for adapting crystal properties to function. *Science, 259*, 776–779.

Bowers, P. M., Strauss, C. E. M., & Baker, D. (2000). De novo protein structure determination using sparse NMR data. *Journal of Biomolecular NMR, 18*, 311–318.

Bradley, P., Misura, K. M. S., & Baker, D. (2005). Toward high-resolution de novo structure prediction for small proteins. *Science, 309*, 1868–1871.

Chaudhury, S., Lyskov, S., & Gray, J. J. (2010). PyRosetta: A script-based interface for implementing molecular modeling algorithms using Rosetta. *Bioinformatics, 26*, 689–691.

Cohavi, O., Corni, S., De Rienzo, F., Di Felice, R., Gottschalk, K. E., Hoefling, M., et al. (2010). Protein-surface interactions: Challenging experiments and computations. *Journal of Molecular Recognition, 23*, 259–262.

Collier, G., Vellore, N. A., Yancey, J. A., Stuart, S. J., & Latour, R. A. (2012). Comparison between empirical protein force fields for the simulation of the adsorption behavior of structured LK peptides on functionalized surfaces. *Biointerphases, 7*, 24.

Corno, M., Rimola, A., Bolis, V., & Ugliengo, P. (2010). Hydroxyapatite as a key biomaterial: Quantum-mechanical simulation of its surfaces in interaction with biomolecules. *Physical Chemistry Chemical Physics, 12*, 6309–6329.

DiMaio, F., Leaver-Fay, A., Bradley, P., Baker, D., & Andre, I. (2011). Modeling symmetric macromolecular structures in Rosetta3. *PLos One, 6*, e20450.

Ducy, P., Desbois, C., Boyce, B., Pinero, G., Story, B., Dunstan, C., et al. (1996). Increased bone formation in osteocalcin-deficient mice. *Nature, 382*, 448–452.

Dunbrack, R. L., & Cohen, F. E. (1997). Bayesian statistical analysis of protein side-chain rotamer preferences. *Protein Science, 6*, 1661–1681.

Fleishman, S. J., Leaver-Fay, A., Corn, J. E., Strauch, E. M., Khare, S. D., Koga, N., et al. (2011). RosettaScripts: A scripting language interface to the Rosetta macromolecular modeling suite. *PLos One, 6*, e20161.

Freeman, C. L., Harding, J. H., Quigley, D., & Rodger, P. M. (2011). Simulations of ovocleidin-17 binding to calcite surfaces and its implications for eggshell formation. *Journal of Physical Chemistry C, 115*, 8175–8183.

Goobes, G., Stayton, P. S., & Drobny, G. P. (2007). Solid state NMR studies of molecular recognition at protein-mineral interfaces. *Progress in Nuclear Magnetic Resonance Spectroscopy, 50*, 71–85.

Gray, J. J. (2004). The interaction of proteins with solid surfaces. *Current Opinion in Structural Biology, 14*, 110–115.

Gray, J. J., Moughon, S., Wang, C., Schueler-Furman, O., Kuhlman, B., Rohl, C. A., et al. (2003). Protein-protein docking with simultaneous optimization of rigid-body displacement and side-chain conformations. *Journal of Molecular Biology, 331*, 281–299.

Hoang, Q. Q., Sicheri, F., Howard, A. J., & Yang, D. S. C. (2003). Bone recognition mechanism of porcine osteocalcin from crystal structure. *Nature, 425*, 977–980.

Iori, F., Di Felice, R., Molinari, E., & Corni, S. (2009). GolP: An atomistic force-field to describe the interaction of proteins with Au(111) surfaces in water. *Journal of Computational Chemistry, 30*, 1465–1476.

Kilambi, K. P., & Gray, J. J. (2012). Rapid calculation of protein pKa values using Rosetta. *Biophysical Journal, 103*, 587–595.

Kim, D. E., Chivian, D., & Baker, D. (2004). Protein structure prediction and analysis using the Robetta server. *Nucleic Acids Research, 32*, W526–W531.

Kortemme, T., Morozov, A. V., & Baker, D. (2003). An orientation-dependent hydrogen bonding potential improves prediction of specificity and structure for proteins and protein-protein complexes. *Journal of Molecular Biology, 326*, 1239–1259.

Kuhlman, B., & Baker, D. (2000). Native protein sequences are close to optimal for their structures. *Proceedings of the National Academy of Sciences of the United States of America, 97*, 10383–10388.

Kuhlman, B., Dantas, G., Ireton, G. C., Varani, G., Stoddard, B. L., & Baker, D. (2003). Design of a novel globular protein fold with atomic-level accuracy. *Science, 302*, 1364–1368.

Latour, R. A. (2008). Molecular simulation of protein-surface interactions: Benefits, problems, solutions, and future directions. *Biointerphases, 3*, FC2–FC12.

Lazaridis, T., & Karplus, M. (1999). Effective energy function for proteins in solution. *Proteins, 35*, 133–152.

Leaver-Fay, A., O'Meara, M. J., Tyka, M., Jacak, R., Song, Y., Kellogg, E. H., et al. (2013). Scientific benchmarks for guiding macromolecular energy function improvement. *Methods in Enzymology, 523*, 109–143.

Leaver-Fay, A., Tyka, M., Lewis, S. M., Lange, O. F., Thompson, J., Jacak, R., et al. (2011). Rosetta3: An object-oriented software suite for the simulation and design of macromolecules. *Methods in Enzymology, 487*, 545–574, Computer Methods, Pt C.

Li, Z. Q., & Scheraga, H. A. (1987). Monte-Carlo-minimization approach to the multiple-minima problem in protein folding. *Proceedings of the National Academy of Sciences of the United States of America, 84*, 6611–6615.

Makrodimitris, K., Masica, D. L., Kim, E. T., & Gray, J. J. (2007). Structure prediction of protein-solid surface interactions reveals a molecular recognition motif of statherin for hydroxyapatite. *Journal of the American Chemical Society, 129*, 13713–13722.

Masica, D. L., Ash, J. T., Ndao, M., Drobny, G. P., & Gray, J. J. (2010). Toward a structure determination method for biomineral-associated protein using combined solid-state NMR and computational structure prediction. *Structure, 18*, 1678–1687.

Masica, D. L., & Gray, J. J. (2009). Solution- and adsorbed-state structural ensembles predicted for the statherin-hydroxyapatite system. *Biophysical Journal, 96*, 3082–3091.

Masica, D. L., Gray, J. J., & Shaw, W. J. (2011). Partial high-resolution structure of phosphorylated and non-phosphorylated leucine-rich amelogenin protein adsorbed to hydroxyapatite. *The Journal of Physical Chemistry C, Nanomaterials and Interfaces, 115*, 13775–13785.

Masica, D. L., Schrier, S. B., Specht, E. A., & Gray, J. J. (2010). De novo design of peptide-calcite biomineralization systems. *Journal of the American Chemical Society, 132*, 12252–12262.

McKee, M. D., Nakano, Y., Masica, D. L., Gray, J. J., Lemire, I., Heft, R., et al. (2011). Enzyme replacement therapy prevents dental defects in a model of hypophosphatasia. *Journal of Dental Research, 90*, 470–476.

Michaels, A. S., & Matson, S. L. (1985). Membranes in biotechnology—State of the art. *Desalination, 53*, 231–258.

Raman, S., Vernon, R., Thompson, J., Tyka, M., Sadreyev, R., Pei, J. M., et al. (2009). Structure prediction for CASP8 with all-atom refinement using Rosetta. *Proteins, 77*, 89–99.

Ratner, B. D., & Bryant, S. J. (2004). Biomaterials: Where we have been and where we are going. *Annual Review of Biomedical Engineering, 6*, 41–75.

Renfrew, P. D., Butterfoss, G. L., & Kuhlman, B. (2008). Using quantum mechanics to improve estimates of amino acid side chain rotamer energies. *Proteins, 71*, 1637–1646.

Renfrew, P. D., Choi, E. J., Bonneau, R., & Kuhlman, B. (2012). Incorporation of non-canonical amino acids into Rosetta and use in computational protein-peptide interface design. *PLos One, 7*, e32637.

Rimola, A., Aschi, M., Orlando, R., & Ugliengo, P. (2012). Does adsorption at hydroxyapatite surfaces induce peptide folding? Insights from large-scale B3LYP calculations. *Journal of the American Chemical Society, 134*, 10899–10910.

Rohl, C. A., Strauss, C. E., Misura, K. M., & Baker, D. (2004). Protein structure prediction using Rosetta. *Methods in Enzymology, 383*, 66–93.

Schrier, S. B., Sayeg, M. K., & Gray, J. J. (2011). Prediction of calcite morphology from computational and experimental studies of mutations of a de novo-designed peptide. *Langmuir: The ACS Journal of Surfaces and Colloids, 27,* 11520–11527.

Schrodinger, L. L. C. (2010). The PyMOL molecular graphics system, version 1.3r1.

Simons, K. T., Bonneau, R., Ruczinski, I., & Baker, D. (1999). Ab initio protein structure prediction of CASP III targets using ROSETTA. *Proteins, 37,* (Suppl. 3), 171–176.

Simons, K. T., Kooperberg, C., Huang, E., & Baker, D. (1997). Assembly of protein tertiary structures from fragments with similar local sequences using simulated annealing and Bayesian scoring functions. *Journal of Molecular Biology, 268,* 209–225.

Sircar, A., Chaudhury, S., Kilambi, K. P., Berrondo, M., & Gray, J. J. (2010). A generalized approach to sampling backbone conformations with RosettaDock for CAPRI rounds 13-19. *Proteins, 78,* 3115–3123.

Sun, Y., & Latour, R. A. (2006). Comparison of implicit solvent models for the simulation of protein-surface interactions. *Journal of Computational Chemistry, 27,* 1908–1922.

Tarasevich, B. J., Perez-Salas, U., Masica, D. L., Philo, J., Kienzle, P., Krueger, S., et al. (2013). Neutron reflectometry studies of the adsorbed structure of the amelogenin, LRAP. *Journal of Physical Chemistry B, 117,* 3098–3109.

Thain, D., Tannenbaum, T., & Livny, M. (2005). Distributed computing in practice: The Condor experience. *Concurrency and Computation: Practice & Experience, 17,* 323–356.

Vellore, N. A., Yancey, J. A., Collier, G., Latour, R. A., & Stuart, S. J. (2010). Assessment of the transferability of a protein force field for the simulation of peptide-surface interactions. *Langmuir, 26,* 7396–7404.

Wang, F., Stuart, S. J., & Latour, R. A. (2008). Calculation of adsorption free energy for solute-surface interactions using biased replica-exchange molecular dynamics. *Biointerphases, 3,* 9–18.

Warshel, A., & Russell, S. T. (1984). Calculations of electrostatic interactions in biological systems and in solutions. *Quarterly Reviews of Biophysics, 17,* 283–422.

Weiner, S., & Dove, P. M. (2003). An overview of biomineralization processes and the problem of the vital effect. *Biomineralization, 54,* 1–29.

CHAPTER SEVENTEEN

Investigating Protein Function in Biomineralized Tissues Using Molecular Biology Techniques

Christopher E. Killian[*,1], **Fred H. Wilt**[†]

[*]Department of Physics, University of Wisconsin-Madison, Madison, Wisconsin, USA
[†]Department of Molecular and Cell Biology, University of California, Berkeley, California, USA
[1]Corresponding author: e-mail address: cekillian@wisc.edu

Contents

1. Introduction — 368
2. Discovery of Proteins Involved in Biomineralization — 368
   2.1 Generating RNA-seq data from mineralized tissues — 369
   2.2 Sequencing of proteins isolated from mineralized tissues — 376
3. Analysis of Function — 377
   3.1 Loss of function approaches: General comments — 378
   3.2 siRNA loss of function in the mollusk shell — 379
   3.3 Morpholino antisense loss of function in the sea urchin spicule — 381
   3.4 The limitations of loss of function studies — 384
Acknowledgments — 385
References — 385

Abstract

We describe modern molecular biology methods currently used in the study of biomineralization. We focus our descriptions on two areas of biomineralization research in which these methods have been particularly powerful. The first area is the use of modern molecular methods to identify and characterize the so-called occluded matrix proteins present in mineralized tissues. More specifically, we describe the use of RNA-seq and the next generation of DNA sequencers and the use of direct protein sequencing and mass spectrometers as ways of identifying proteins present in mineralized tissues. The second area is the use of molecular methods to examine the function of proteins in biomineralization. RNA interference (RNAi), morpholino antisense, and other methods are described and discussed as ways of elucidating protein function.

1. INTRODUCTION

The modern era of biomineralization research began with the publication in 1989 of the thorough and elegant summaries of the field by Lowenstam and Weiner (1989) and by Simkiss and Wilbur (1989). Since then, powerful technical advances in mineralogy, physical chemistry, biochemistry, and molecular genetics have characterized work in the field.

Traditional genetics—and by this, we mean isolation of variants and study of their inheritance by analysis of progeny—was not an important contributor to biomineralization studies, with the notable exception of the inheritance of certain disease states in humans that displayed abnormalities in teeth and bones. Though of considerable interest, we shall not treat here the use of traditional study of mutants. Some of these studies, such as those of the starmaker mutant found in zebra fish (Söllner et al., 2003) or collagen mutants that cause osteogenesis imperfecta in mammals (Rauch & Glorieux, 2004), combine traditional Mendelian genetics with newer tools of chemistry and materials science to help unravel the role of matrix proteins in formation and function of biomineralized organs.

In this brief chapter, we will describe molecular techniques that are applied to two distinct areas of biomineralization research that have proven to be useful. The first area we examine is the use of modern molecular methods to identify and characterize the so-called occluded matrix proteins—proteins found closely associated with the endo- or exoskeletal elements and dentition of many different genera. See Livingston et al. (2006); Wilt, Killian, and Livingston (2003); Marie, Le Roy, Zanella-Cléon, Becchi, and Marin (2011); and Murdock and Donoghue (2011) for discussions and reviews of these sorts of "matrix" proteins in different organisms.

The second area we examine is the use of molecular, genetic, and cell biology techniques that are increasingly important in functional analyses of the role of organic materials in the formation of, and material properties of, several biominerals (Estroff, 2008). We describe some experimental methods used to examine the function of mineralized tissue matrix proteins.

2. DISCOVERY OF PROTEINS INVOLVED IN BIOMINERALIZATION

There are numerous strategies available to scientists to identify proteins involved in biomineralization. With the advent of high-throughput DNA and protein sequencing, powerful genomic and proteomic techniques are now readily available to be employed for identifying the genes and proteins expressed in mineralized tissues. Experimental strategies for protein identification fall roughly under two general categories. The first strategy examines the mRNA sequences present (also called transcriptomics or RNA-seq) in mineralized tissues (Wang, Gerstein, & Snyder, 2009). With this sort of analysis, a researcher is then able to examine the different proteins encoded by these sequences with molecular, immunological, and cell biological tools. With the advent of the so-called next-generation DNA sequencing technologies, it is not unreasonable to discover nearly all genes expressed by the given tissue (Bai, Zheng, Lin, Wang, & Li, 2013; Mardis, 2011, 2013; Metzker, 2009; Shendure & Ji, 2008). However, one cannot be sure if a given gene has a role in biomineralization just because it is found in the transcriptome of a mineralized tissue, especially if it encodes a unique and previously unidentified protein.

The second experimental approach to identify proteins is to sequence directly the proteins present in mineralized tissues (Bandeira, Clauser, & Pevzner, 2007; Nilsson et al., 2010; Walther & Mann, 2010; Yates, Ruse, & Nakorchevsky, 2009). This direct sequencing of proteins provides some information about the localization of the protein and is particularly powerful when it is complemented with genomic or RNA-seq information (Castellana & Bafna, 2010). Direct protein sequencing can also offer information about posttranslational modifications that are not always accurately inferred from the mRNA sequence (Mann & Jensen, 2003; Witze, Old, Resing, & Ahn, 2007). Once proteins are identified as potentially being involved in biomineralization, a researcher is then able to try to determine their function.

2.1. Generating RNA-seq data from mineralized tissues

Many important discoveries identifying proteins involved in biomineralization have entailed examining mRNA sequences present in mineralized tissues. Researchers synthesize cDNA from mRNA isolated from the mineralizing tissue and then sequence the cDNAs to generate expressed sequence tag (EST) sequence data. An early example of this strategy was provided by Charles Ettensohn's lab, which identified a number of proteins involved in biomineralization in the sea urchin larva by generating a cDNA library from mRNA isolated from isolated primary mesenchyme cells (the cells that synthesize the larva's spicules) (Zhu et al., 2001). They sequenced each clone in the library and identified a number of biomineralization-involved proteins including P16, P19, P58A, and P58B, among others (Adomako-Ankomah & Ettensohn, 2011; Illies, Peeler, Dechtiaruk, & Ettensohn, 2002). They subsequently used *in situ* hybridization, RT-PCR, and morpholino knockout experiments to demonstrate the important role for these proteins in sea urchin biomineralization.

In the not too distant past (at least according to the authors' sense of time), in order to generate a cDNA library, one isolated mRNA, converted it to cDNA, subcloned the cDNAs into a plasmid or phage vector, amplified the libraries in bacteria or bacteria infected with phage, and then plated out the libraries to be screened with radioactive nucleic acid probes or with antibodies (see Benson, Sucov, Stephens, Davidson, & Wilt (1987); George, Killian, & Wilt (1991), as examples). This strategy has been effective in identifying some genes and proteins involved in biomineralization. However, it is a relatively inefficient strategy compared to the most recently developed techniques for identifying interesting genes. With the latest DNA sequencing technologies, one can, in a matter of days, generate a list of nearly all of the genes expressed in the mineralizing tissue of your interest. Instead of subcloning cDNA libraries and growing them in bacteria, mRNA sequence libraries are generated in a test tube and amplified on solid surfaces taking advantage of PCR technologies to amplify the libraries. Degnan's and Jackson's research group was the first to report high-throughput DNA sequencing of cDNAs generated from embryonic mantle cells of the mollusk, *Haliotis asinina*. They were able to identify proteins secreted by mantle cells, many of which were unique proteins (Jackson & Degnan, 2006). Since then, a number of research groups have effectively followed this same strategy (Bai et al., 2013; Joubert et al., 2010; Marin, Le Roy, & Marie, 2012; Pespeni et al., 2013).

The undertaking of RNA sequencing, or the so-called RNA-seq analysis, of mineralized tissues first requires isolating RNA of as high quality as possible. RNA is a labile molecule and ribonucleases are present nearly everywhere in biological tissues. Ribonucleases are quite stable and resistant to inactivation. Care needs to be taken to use procedures that inactivate RNases quickly and minimizes the exposure of RNA to them during isolation. There are a number of effective commercially available reagent kits for the isolation of high-quality RNA. Reagents such as TRIzol and RNAzol have proven to be very effective, and their compositions are based on the reagent described by Chomczynski and Sacchi (1987). These reagents combine the chaotrope, guanidine thiocyanate, with acidic phenol to quickly inactivate RNases, thereby purifying total RNA away from proteins and DNA. Many research groups have successfully used this procedure since it first debuted in 1987 (Chomczynski & Sacchi, 2006), but there are other methods for RNA isolation. A commonly used alternative procedure begins by lysing the tissue and chemically inactivating endogenous RNases with strong chaotropes and then binding RNA to silica resin or membranes, thus avoiding using phenol. Qiagen, Life Technologies/Invitrogen/Ambion, and Millipore, among other manufacturers, sell commercial kits based on this procedure.

Most of the most commonly used RNA-seq procedures and commercial kits are designed for starting with a material that contains between 0.1 and 10 µg of total RNA. The integrity of the isolated RNA samples should be checked. A common procedure for checking the RNA integrity is loading a small portion of the isolated RNA into an Agilent RNA Bioanalyzer 2100 and determining if the RNA has a RIN (RNA integrity number) score of 8 or higher (Schroeder et al., 2006). If an Agilent Bioanalyzer is not available, one can fractionate the RNA on a denaturing 1% agarose–formaldehyde gel. Intact RNA will have distinct 18S and 28S rRNA bands visible with no smearing when the gel is stained for nucleic acids. If contamination of the RNA sample with DNA is a concern, one can treat the sample with RNAse-free DNAse I before use and/or assay for DNA contamination using PCR to amplify intergenic sequences that should not be represented in the isolated RNA sample.

The next step is to synthesize a DNA that is complementary to the isolated mRNAs. This complementary DNA, or cDNA, will ideally represent all the mRNAs present in the tissue. There are a few different procedures available for synthesizing cDNA. mRNA represents 2–5% of total RNA in cells with rRNA representing >80% of the total RNA. One must isolate the mRNA away from the rest of the total RNA.

Depending on which of the high-throughput DNA sequencing procedures is used, it will slightly alter how the cDNA synthesis is handled. No matter which method is used, there are common steps involved in preparation of cDNA for high-throughput sequencing.

The first step is to enrich mRNA in the samples. As mentioned earlier, most of the commercially available RNA-seq kits are designed for use of 0.1–10 µg of total RNA. Some procedures purify the mRNA by taking advantage of the poly(A) tail present on most mRNAs. These procedures bind poly(A)-containing RNA to oligo-dT conjugated to magnetic beads. However, one should be aware that not all mRNAs have poly(A) tails. Also, if the RNA sample is not 100% intact, using oligo-dT selection will skew the mRNA library to sequences at the 3′ end of the mRNA because 5′ portions of the mRNA will have been cleaved off by RNases.

Another strategy one can use to purify mRNA is to degrade or remove rRNA and leave the mRNA intact. The Ribo-Zero kits from Epicenter, the mRNA enrichment kits from Ambion (now Life Technologies), and the RNA-seq kit from Applied Biosystems (now also Life Technologies), as well as other manufacturers, allow the removal of the rRNA from the total RNA in the sample by having rRNA sequences bind to magnetic beads hybridized to the rRNA in the sample. After the removal of the beads, this leaves small RNAs, such as miRNAs, tRNAs, and mRNAs, behind. Recently published studies indicate that using the Ribo-Zero procedures and RNase H procedures for mRNA enrichment from degraded RNA or extremely small quantities of RNA is more effective than using oligo-dT selection of poly(A)-containing mRNA for generating high-quality RNA-seq libraries (Adiconis et al., 2013). Finally, another method to remove rRNA involves mixing isolated total RNA with a mixture of complementary rRNA sequence oligonucleotides and then degrading the rRNA hybrids with RNase H, which specifically digests double-stranded RNA.

Once the mRNA samples have been prepared, the RNAs are fragmented to smaller sizes of between 125 and 700 bases in length depending on the DNA sequencing technology you will be using. RNA-seq procedures from Roche 454 fragment the mRNA sequence after it is converted to cDNA by cleaving the cDNA sequences with a restriction endonuclease, NlaIII, which recognizes a particular 4-base sequence and should randomly cut DNA on average every 256 bases given a random DNA sequence. RNA fragmentation can be done chemically in the presence of divalent cations, or it can be done enzymatically. It is important that this fragmentation step is done effectively and it should be verified by an Agilent Bioanalyzer or by gel electrophoresis. Fragmentation of the mRNA is done because the sequencing reactions of the current technologies are able to sequence relative short stretches of DNA. If the fragments are too long, they will not be sequenced completely. The current DNA sequencing technologies make up for short-run lengths by sequencing several fragments at the same time—from hundreds of thousands to hundreds of millions of fragments. This massively parallel sequencing strategy is designed so that all of the fragments of mRNA will be sequenced. The short sequence reads are aligned later by computer algorithms to obtain the full-length sequence of all of the mRNAs present in the sample.

Once the fragmented mRNA of the appropriate size is obtained, random hexamers are annealed to the mRNA fragments, and cDNA is synthesized with reverse transcriptase—ultimately resulting in small fragment of double-stranded DNA. The next step is to ligate adapters to the cDNA fragments so that each individual fragment can be annealed to a solid surface. Each of the three major next-generation DNA sequencing technologies, Roche 454, Illumina, and Life Technologies' SOLiD, use different custom adapters that work with each system.

With the cDNA fragments coupled with the appropriate adapters ligated at the ends, the next step is to anneal the individual cDNA fragments to a solid surface. The solid surfaces will be either microbeads or a glass slide (a.k.a. flow cell) with oligonucleotides complementary to the appropriate adapters adhered to the solid surface. For all three DNA sequencing technologies that will be described, once the fragments are bound to the surface and physically isolated from the other fragments, DNA fragments are amplified by a polymerase chain reaction. The amplifications of the cDNA fragments are done to boost the sequencing signal for each of the latest technologies. One caveat to keep in mind is that with amplification, not all sequences are amplified equally well. Some are amplified poorly, while others are amplified much more efficiently than most of the other fragments in the library.

There are three so-called next-generation high-throughput DNA sequencing technologies that are most commonly available for scientists today. These technologies are Roche 454 DNA sequencing, Illumina DNA sequencing, and Life Technologies' SOLiD DNA sequencing. All three are optically based, that is, the DNA sequencing reaction for each technology generates light of some sort, and from these reactions, the DNA sequence is determined. Here are brief descriptions of the DNA technologies available today.

2.1.1 Roche 454

Roche 454 DNA sequencing technology is based on pyrophosphate sequencing chemistry. During the preparation for sequencing, each cDNA fragment is attached to microbeads. It is important that each microbead initially binds to only one cDNA fragment for the procedure to work properly. The microbeads with cDNA fragment attached are then placed in an emulsion in which the DNA fragments are amplified millionfold by PCR. After amplification, the beads are released from the emulsion, and the microbeads are each placed in a single well of a picotiter plate where there is an optical detector for each well. Each well is then provided with DNA polymerase, a sequencing primer, as well as sulfurylase and luciferase, and one known deoxyribonucleotide. When a nucleotide is incorporated, pyrophosphate is released and light is emitted. With subsequent rounds of nucleotide incorporation, the sequence of each cDNA fragment in each well is recorded. Typical runs can take up to a day or more to complete with typical read lengths of up to 650–700 bases.

2.1.2 Illumina

Illumina DNA sequencing is based on detection of fluorescent nucleotide incorporation in individual clusters of DNA fragments. To accomplish this, the cDNA fragments are annealed randomly over the surface of a flow cell that has oligonucleotides complementary to the adapters present on the cDNA fragment ends. The annealed DNA fragments are then amplified using bridge PCR to generate millions of copies in the area immediately surrounding where each cDNA fragment initially annealed to the flow cell. Each of these areas of amplified DNA is called a cluster and is addressable by a light sensor. The DNA sequence of each DNA fragment is elucidated by following DNA polymerase incorporation of four uniquely fluorescently labeled nucleotides in each of the clusters. A chemical block at the 3′ end of incorporated nucleotides prevents elongation after each nucleotide is incorporated. This block allows for detection of the incorporated fluorescent nucleotide in each cluster. Subsequent rounds of sequencing occur after the fluorescent tag of the previously incorporated nucleotide is cleaved off and the chemical block at the 3′ end of the last nucleotide incorporated is removed. The next fluorescent nucleotides are then incorporated and recorded in a stepwise fashion. Typical read lengths for Illumina DNA sequencing are up to 100–150 bases long.

2.1.3 SOLiD

Life Technologies' SOLiD DNA sequencing is based on the incorporation of fluorescently labeled oligonucleotides by DNA ligase instead of the incorporation of nucleotides by DNA polymerase as do Roche 454 and Illumina. SOLiD, also sometimes called 2-base encoding, is an acronym for Sequencing by Oligonucleotide Ligation and Detection. Like 454 sequencing, each cDNA fragment is annealed to a glass microbead via the unique adapters. Also similar to 454 sequencing procedure, all of the cDNA fragments are amplified in an oil–aqueous phase emulsion resulting in each microbead having millions of copies of the original cDNA fragment annealed to its surface. After amplification, the microbeads are released from the emulsion and then covalently attached to a glass flow cell. Each microbead attached to the flow cell is then addressable by a light detector.

The SOLiD sequencing reaction is based on rounds of ligation of fluorescently labeled oligonucleotide probes 9 nucleotides in length that have a unique fluorescent tag identifying the first two bases of the probe sequence. For subsequent cycles of ligation to occur, the fluorescent tag of the previously ligated oligonucleotide probe is first removed and part of the oligoprobe is trimmed. The next downstream ligation with another fluorescently labeled probe is ligated and the sequence of the first two nucleotides of this second probe detected. With many cycles of ligation and shifts in the register of which base identities are being read, the sequences of the many thousands to millions of cDNA fragments are determined. The error rate of base identification is very low with this technology because each base is read twice when the whole sequence procedure is finished. The typical base reads for this technology are about 60 bases.

2.1.4 Third-generation DNA sequencing technologies

In addition to the three next-generation technologies for DNA sequencing described earlier, there are the so-called third-generation DNA sequencing technologies just now becoming available. These latest technologies include the Ion Torrent DNA sequencing technology and the Pacific Biosciences RS DNA sequencing technology. Ion Torrent DNA sequencing, owned by Life Technologies, is similar to 454 pyrophosphate sequencing chemistry strategy. But instead of detecting light as the 454 technology does, the ION Torrent sequencer detects H+ ions released as nucleotides are incorporated. Pacific Biosciences RS DNA sequencing technology is based on real-time optical detection of the sequencing of single-DNA molecules. These third-generation technologies offer less expensive and quicker turnaround times. However, they are not as widely used yet. Please see the manufacturers' web sites for more information if you are interested in using them.

No matter which DNA sequencing technology is used, raw base reads from the sequencing for the sequenced fragments are aligned to generate mRNA sequences and identify the proteins encoded by the mRNAs (Gogol-Döring & Chen, 2012; Li & Homer, 2010; Trapnell & Salzberg, 2009). This process is more effective if there is a reference genome with which to map the sequence reads. This is especially true for shorter base reads sequencing data. If the mineralized tissue, which is the subject of investigation, has little or no genomic or transcriptomic sequence information, longer read DNA sequencing technologies, such as Roche 454 coupled with paired reads (sequencing both ends), will make processing of the RNA-seq data more straightforward.

2.2. Sequencing of proteins isolated from mineralized tissues

Another powerful approach in identifying proteins involved in biomineralization is to directly sequence proteins isolated from mineralized tissues. The refinements of mass spectroscopy over the last decade have made protein sequencing a powerful tool for identifying individual proteins from a complex mixture of isolated proteins (Bandeira et al., 2007; Gstaiger & Aebersold, 2009; Nilsson et al., 2010; Steen & Mann, 2004; Walther & Mann, 2010; Yates et al., 2009). There are different types of machine designs available and a review of the different machines is beyond the scope of this chapter.

A common initial step for a proteomic study of the proteins involved in biomineralization is to isolate the proteins embedded within the mineral phase of mineralized tissues. Extensive treatment with bleach to remove any contaminating cells or organic molecules is important. After washing away the bleach, the mineral is usually crushed and washed again with bleach to remove contaminating internal cells and organic material. The ground mineral phase is then washed and the mineral dissolved with acid. Once the mineral is dissolved away, the remaining proteins in solution are dialyzed and concentrated.

The isolated skeletal matrix proteins are then digested with sequence-specific proteases, usually trypsin. Clipping proteins into peptides makes it easier for protein sequencing because full-length proteins are often insoluble and/or do not desorb well in the mass spectrometer (Steen & Mann, 2004). Peptides are usually less problematic as far as solubility.

Before the sample of peptides is sequenced, it is fractionated by some biochemical means. Often, reverse-phase HPLC is used to fractionate the mixture of peptides. As the peptides elute from the column, they are passed onto the mass spectrometer for analysis. Once inside the mass spectrometer, the peptides are ionized and the mass for each is detected. The mass per charge value of each resolved peptide is then compared to a database of these values, and from this, the sequence of the peptide is determined. The probability of the identity is also provided. If the mix of proteins is too complex to resolve all the peptides, a greater amount of fractionation needs to be done on the peptide mixture prior to mass spectrometry. This could involve running samples out on polyacrylamide gels and then cutting portions of the gel with some of the proteins in it, digesting the gel-fractionated proteins with protease and then again loading the sample for analysis.

An advantage of directly sequencing proteins from mineralized tissue is that one is often able to determine posttranslational modifications to proteins that are not predictable from RNA-seq data. Phosphorylation, glycosylation, ubiquitination, and other posttranslation modifications can be detected (Mann & Jensen, 2003; Witze et al., 2007). This sort of information about proteins may prove invaluable in eventually determining function for a particular protein.

With a list of proteins found in a given mineralized tissue, one needs to examine closely what proteins have been detected and keep in mind that there may well be contaminating proteins. This issue has been raised about a study recently published describing the skeletal organic matrix proteome of the stony coral the *Stylophora pistillata*. Drake et al. (2013) identified cytoskeletal matrix proteins as being part of the stony coral skeletal matrix proteome. However, Ramos-Silva, Marin, Kaandorp, and Marie (2013), in a letter responding to the study, pointed out that extensive washing of the finely ground mineral phase of invertebrate hard tissues with concentrated bleach rids these sorts of samples of cytoskeletal proteins. They argue that the cytoskeleton proteins identified as being skeletal matrix protein by Drake et al. (2013) may well be contaminants.

These sorts of concerns are not unique to the Drake et al. (2013) study. There are a number of intracellular proteins that have been identified as being present in the skeletal matrix of sea urchin spines, teeth, and embryonic spicules, for instance (Mann, Poustka, & Mann, 2008a,2008b; Mann, Wilt, & Poustka, 2010). These cellular proteins may well be contaminants. Our pointing out of this issue is not meant to be critical of these findings, but rather cautionary in the interpretation of the findings. These techniques are exquisitely sensitive, and it is prudent to be skeptical if one finds unexpected intracellular proteins in a mineralized tissue matrix sample.

3. ANALYSIS OF FUNCTION

As it has become almost a routine to identify matrix molecules associated with biomineralized structures using modern molecular methods, a vexing problem has developed with the sheer number of them in any given instance. They can be localized, more or less precisely, with respect to the mineral with which they are associated. Amino acid sequences can give some information on likely protein conformation. Probably the most difficult and most interesting issue is to identify the function of any given protein or other macromolecule. The biochemist would attempt a "test tube reconstruction": add all known components together, control the conditions, and reconstruct the biomineral, *in vitro*, by the magic of self-assembly. This has not proven feasible.

A considerable amount of information on function can be garnered by perturbing the expression (including transcription, translation, and/or protein modification) of known matrix genes, a so-called molecular genetic approach. These tests are usually done *in vivo* and serve as a standard against which various hypotheses about function can be measured.

The standard approaches are the following: (1) Change the tissue or place of expression of a given gene, (2) change the time when the gene is expressed, (3) enhance the level of expression (gain of function), and (4) suppress the level of expression (loss of function). The use of misexpression in time or place has not been, to the best of our knowledge, employed in studies of biomineralization. There is, however, no reason why it could not be employed in the developing embryos that undergo biomineralization, such as mollusks, some arthropods, echinoderms, and vertebrates. The experiments would have to employ introduction of molecular vectors whose time and place of expression could be controlled by the experimenter, by judicious choice of promoters and enhancers, or other means.

These experimental approaches are theoretical options at the moment. So, too, the enhancement of expression, by injection of particular mRNAs into an embryo, has not been employed. We shall proceed to describe particular instances of loss of function, the principal tool thus far employed by various workers.

3.1. Loss of function approaches: General comments

There are a number of different strategies for accomplishing a loss of function of a particular gene. One, mentioned at the outset of this chapter, is the use of mutagens to weaken or eliminate gene expression in a genetically tractable organism. This use of standard genetic analysis is still probably the gold standard for understanding function of a component, but cannot be used in most of the organisms currently studied by students of biomineralization because suitable procedures for basic genetic studies in these organisms are not in place.

A second approach, also not employed thus far in the study of biomineralization, is the use of "dominant negatives." This is useful when a protein is involved in cell signaling; a counterfeit ligand is constructed, which binds to the cognate receptor, thereby preventing the interaction with the normal ligand and thus leading to loss of function served by that signaling pathway. This approach has been used, for example, in suppression of *wnt* signaling by flooding cells with counterfeit truncated versions of interacting members of the β-catenin pathway (Marikawa & Elinson, 1999).

A third approach can employ reversible or irreversible inhibitors that interact with the protein in question; this is the kind of approach employed with known enzymes. Carbonic anhydrase, well known to be involved in biomineralization, can be studied by this technique (Supuran, Scozzafava, & Casini, 2003). Or inhibitors of proteases or calcium-channel blockers, both of which are sometimes implicated in biomineralization (Mitsunaga, Makihara, Fujino, & Yasumasu, 1986), can be studied by this method. Inhibition can also be leveraged by comparing gene expression profiles in the presence and absence of a particular inhibitor or other perturbant.

The fourth approach to loss of function uses either knockdown by small interfering RNAs (siRNA) or antisense knockdown of transcription or translation of genes for known matrix proteins.

3.2. siRNA loss of function in the mollusk shell

RNA interference (RNAi) is a relatively recently discovered regulatory pathway that uses double-stranded RNA molecules to selectively destroy specific mRNA molecules. It is part of the normal armamentarium of many plants and animals. Small double-stranded RNA (dsRNA) molecules can be generated from long precursor dsRNA by an endonuclease named Dicer, which cleaves the long dsRNA into short double-stranded fragments—usually about 20 nucleotides long. The short dsRNA is named siRNA (small interfering RNA). The siRNA is unwound into two single-stranded RNAs, one dubbed the "passenger," whereas the other named "guide." The passenger strand is degraded, and the guide strand is incorporated into a multicomponent silencing complex (called RISC). In instances where the guide strand, incorporated into a RISC complex, can anneal by base pairing with a complementary sequence in an mRNA, a nuclease, named Argonaute, embedded in the RISC will cleave the mRNA, thus rendering it inactive. The decreased expression of the gene and its mRNA is thereby accomplished.

The strategy employed in the mollusk shell construction, and in many other instances, takes advantage of this normal cellular mechanism for regulating gene expression. The crucial step is to synthesize, *in vitro*, a long dsRNA corresponding to the presumptive mRNA of a gene suspected to have a role in biomineralization. Suzuki et al. (2009) used this approach. These researchers identified an acidic protein named Pif from the nacre of the shell of the pearl oyster, *Pinctada fucata*, which binds to aragonite. They employed the RNAi strategy to demonstrate that Pif did, indeed, play a crucial role in nacre construction. In conjunction with immunolocalization of Pif and studies of aragonite formation, *in vitro*, they built a strong case for the role of Pif in biomineralization. There are several steps required to carry out this type of experiment; we can illustrate this by looking at the experimental protocols used by Suzuki et al.

3.2.1 Identify a protein of interest, and determine its amino acid sequence

If possible, determine the nucleotide sequence of the mRNA used for synthesis of the protein. The nacre from the pearl oyster was decalcified with acetic acid, and the insoluble matrix was dissolved in the detergent, sodium dodecyl sulfate, and a reducing agent (dithiothreitol), incubated with either calcite or aragonite. The minerals were then washed with small amounts of water, and the eluants examined by acrylamide gel electrophoresis. Almost all the proteins were eluted with water, but in the case of aragonite, one band on the gel, comigrating near 80 kDa, was missing. This putative aragonite-binding protein did not bind well to calcite and was used for subsequent investigation.

3.2.2 Synthesize, in vitro, a long dsRNA molecule corresponding to the mRNA of interest

The next steps used are conventional molecular biology. The 80 kD band was excised from the gel, the protein eluted, and some amino acid sequence obtained by sequencing fragments of the protein. These amino acid sequences are then used to design PCR primers (Apte & Daniel, 2009; Telenius, Carter, Bebb, Ponder, & Tunnacliffe, 1992). Since, in this case, the gene had not been cloned and the exact nucleotide sequence encoding it was not known, it was necessary to use degenerate primers, that is, all four bases are employed for the different codons of each amino acid. PCR was employed to isolate a cDNA from RNA extracted from the mantle. By careful control of annealing conditions during PCR cycles and by sequential use of nested primers, a cDNA representing the Pif protein was obtained from the mantle mRNA. A full-length cDNA was obtained by the RACE PCR strategy (Frohman, 1990).

The cDNA sequence was determined, and primers were designed to produce a long dsRNA corresponding to this mRNA. Both primers included a sequence corresponding to the promoter for T7 RNA polymerase, so that the cDNA produced by PCR could then be used for synthesis of both strands of a dsRNA, *in vitro*. The two RNA strands thus produced were annealed to one another and used in the subsequent experiments. Similar steps were used to produce a control dsRNA, one that would not be expected to produce a knockdown of biomineralization. These workers used an mRNA for GFP (green fluorescent protein) as a control.

3.2.3 Introduce the dsRNA to the biological system being examined
Getting a highly charged, high-molecular-weight polyelectrolyte into cells can be challenging. There are many choices, and one advantage of the exogenous RNAi is that, in some organisms, the process is known to spread in an intact organism from one organ system to another. In the case of the mollusk nacre, the investigators injected different amounts (volume was kept constant; amounts ranged from 5 to 30 micrograms) into the adductor muscle and then waited 7 days to carefully inspect the nacre.

3.2.4 Monitor the effect of the introduced siRNA
It is crucially important to monitor, hopefully by several means, whether the RNAi has actually lowered the concentration of the mRNA of interest. This is often done by quantitative PCR. Primers already used in production of the dsRNA can be used. Suzuki et al. (2009) examined Pif mRNA levels in the RNA extracted from mantle by qPCR and showed a concentration-dependent lowering of mRNA levels by approximately 50%. Examination of the nacre made after RNAi treatment showed that it was disordered and addition of new layers of nacre had been halted. Since control dsRNA had no such effects, they had compiled a strong case for an important and necessary role for Pif in elaboration of nacre.

3.3. Morpholino antisense loss of function in the sea urchin spicule
The use of antisense RNA treatment to selectively inhibit translation of particular proteins has been used to effect "knockdown" of selected genes for several years. The theoretical basis for the loss of function is simple: introduce an RNA sequence that is complementary to a known mRNA into cells

of the organism being studied. The complementary RNA should hybridize specifically to the mRNA of the protein being studied, and the presence of a double-stranded region in the mRNA should and does inhibit its translation (Summerton, 1999).

The devil is in the details. Several different kinds of antisense RNA molecules can be used. A major hurdle has been to use RNA with relatively stable linkages between nucleotide bases, so that the reagent is not degraded by the cell. Another hurdle has been to avoid cytotoxic effects of the reagents used. A major advance was made by the discovery that the use of morpholino-linked pyrimidines and purines produced stable polynucleotides that retained the specificity of a normal phospodiester-linked RNA yet was stable and of low toxicity (Summerton, 1999). This so-called MO (for morpholino) approach has proved invaluable for effecting loss of function analysis in several different organisms and has been especially useful in studying gene regulation in both vertebrate and invertebrate embryos.

This approach was used by us to study the role of a gene and its cognate protein that was known to be present in skeletal spicules of sea urchin embryos. The *SM50* gene was the first gene encoding a known biomineralization matrix protein to be cloned by recombinant DNA technology (Benson et al., 1987; Sucov et al., 1987). Since the genome of the Pacific coast purple sea urchin, *Strongylocentrotus purpuratus*, has been completely sequenced, the genes encoding many known spicule matrix proteins have been identified (Livingston et al., 2006; Mann et al., 2010). However, the function of any of these proteins in biomineralization is not fully understood. We undertook the use of MO approach to gain some insight into the possible function(s) of the *SM50* gene.

Again, there are several steps that need to be employed in this kind of analysis:

3.3.1 Identify the sequence of a protein of potential interest

In this instance, the gene was first identified by cDNA cloning using an antibody directed to all of the spicule matrix proteins. A cDNA library was established in phage, and the phage-infected bacteria that expressed the multitude of proteins in the library were blotted onto nitrocellulose and screened by the use of antibodies specific for isolated spicule matrix proteins (Benson et al., 1987). The SM50 cDNA was isolated and then used to isolate the entire gene from a genomic library. It is a single copy gene with no apparent posttranslational modifications discernable from its sequence (Katoh-Fukui et al., 1991; Killian & Wilt, 1996).

3.3.2 Synthesize an antisense RNA

Prior knowledge of the entire sequence made this step rather simple. Morpholino-linked RNA molecules are synthesized using proprietary reagents and protocols by the Gene Tools, LLC, in Corvallis, OR. It is usually recommended to utilize antisense reagents 22–26 bases in length, complementary to the initial start site for translation, or a sequence very near to this sequence. The use of the design services, included in the price of the order, of Gene Tools is highly recommended. They can also supply control MO reagents that are not expected to produce an effect.

Though not employed in the knockdown of *SM50*, others (e.g., Adomako-Ankomah & Ettensohn, 2011) have also used antisense reagents that are complementary to a splicing site in the pre-mRNA (i.e., nascent transcript). This class of MO reagents thus works in the nucleus and interferes with the production of proper mRNA.

3.3.3 Introduce the antisense MO into the organism

These reagents do not easily permeate the cell. Hence, the MOs are introduced by microinjection into the fertilized egg of the sea urchin embryo. They will, in principle, only act as antisense reagents once the gene is actively being expressed, which in the case of *SM50* in about 18 hours later (at 15 °C) when there are about 500–1000 cells. Microinjection of small volumes (a few picoliters) into eggs or zygotes (100 μm or more) is a widely used procedure, requiring a good inverted microscope, micromanipulators, and devices designed to introduce small volumes via a microinjection pipette with tip diameters of a micron or so (Cheers & Ettensohn, 2004; McMahon et al., 1985). The concentration of the MO can vary widely; for *SM50*, solutions containing 0.5–2.0 mM/L were injected. A fluorescent dye (rhodamine dextran) is included in the injection solution so that the success of the injection can be subsequently monitored using a dissecting microscope equipped with fluorescence filters.

Another popular way to introduce MO reagents is the use of electroporation successfully employed for study of gene expression in ascidian eggs (Zeller, Virata, & Cone, 2006). There are also reagents devised by Gene Tools company that employ temporary low-level disruptions of the phospholipid plasma membrane of cells allowing some permeation of the MO into cells. This is especially useful for cells in tissue culture.

3.3.4 Monitor the effect of the MO

In this instance, knowledge of the biology of the biomineralization system is of importance. The time course of accumulation and its localization was known for the appearance of SM50 in the spicules of the sea urchin embryo. Embryos were examined during development to ensure that overall development was normal, and controls injected with irrelevant MOs were used for comparison. Spicules are visible using ordinary bright field microscopy, so their initial deposition and subsequent elongation and elaboration can be followed by microscopy. The spicules are also intensely birefringent, so polarization microscopy is a useful adjunct. In the case of SM50, it became quite obvious that MO treatment specifically inhibited deposition and elaboration of the spicule, allowing the investigators to conclude that SM50 is indispensable for biomineralization (Wilt, Croker, Killian, and McDonald, 2008).

3.4. The limitations of loss of function studies

It is interesting to consider just what kind of conclusions can be drawn from this approach to study of protein function, and the limitations are not restricted to any particular biological process. The limits of interpretation apply not only to biomineralization but also to other cellular processes as well.

When the reagent, whether it is an siRNA or an MO, interferes with biomineralization, the conclusion that the protein is necessary is probably justified. But that does not allow one to infer what the function is, a limitation of all studies using inhibitors. Supplementary information and probably other experimental approaches are needed. In the case of siRNA, it is clear that Pif is necessary for nacre tablet deposition, narrowing the scope of its action. *In vitro* studies and examination of its sequence lead to hypotheses that it interacts with aragonite and chitin, and so new experiments can be devised.

In the case of SM50, the MO stops biomineralization, when used at high concentrations, from the beginning of biomineralization. But lower concentrations of MO allow the initial calcite granule to be deposited, though later elongation of the spicule rod is severely impaired; does this mean the processes of initial deposition and subsequent elongation are different, at least to some extent? Other information from other approaches will be necessary to get a deeper understanding of SM50 function. Recent work on the effect of SM50 on stabilization of amorphous calcium carbonate (Gong et al., 2012) is a step in this direction.

To generalize, when an inhibitor or agent is used to investigate a biological process and the process is disrupted or stopped, you have only taken the first step in the investigation of the role of this protein in the biological process being investigated.

Future technologies that will likely prove valuable in studies of protein function include those that permit precise editing of gene sequences. The advent of TALENs and CRISPRs as tools to manipulate gene sequences is an exciting step in that direction (Esvelt & Wang, 2013; Gaj, Gersbach, Barbas, & Carlos, 2013; Segal & Meckler, 2013).

ACKNOWLEDGMENTS

C. E. K. works in the laboratory of Dr. Pupa Gilbert (Department of Physics, University of Wisconsin–Madison) and is supported by a DOE grant awarded to Dr. Gilbert (DE-FG02-07ER15899). C. E. K. is also a visiting research associate in the Department of Molecular and Cell Biology at the University of California, Berkeley. F. H. W. receives support from the Committee on Research, University of California, Berkeley.

REFERENCES

Adiconis, X., Borges-Rivera, D., Satija, R., DeLuca, D. S., Busby, M. A., Berlin, A. M., et al. (2013). Comparative analysis of RNA sequencing methods for degraded or low-input samples. *Nature Methods, 10*, 623–629.

Adomako-Ankomah, A., & Ettensohn, C. A. (2011). Novel proteins that mediate skeletogenesis in the sea urchin embryo. *Developmental Biology, 353*, 81–93.

Apte, A., & Daniel, S. (2009). PCR primer design. In C. W. Dieffenbach & G. S. Dveksler (Eds.), *PCR primer: a laboratory manual* (2nd ed., pp. 61–74). New York: Cold Spring Harbor Laboratory Press.

Bai, Z., Zheng, H., Lin, J., Wang, G., & Li, J. (2013). Comparative analysis of the transcriptome in tissues secreting purple and white nacre in the pearl mussel *Hyriopsis cumingii*. *PloS One, 8*, e53617.

Bandeira, N., Clauser, K. R., & Pevzner, P. A. (2007). Shotgun protein sequencing—Assembly of peptide tandem mass spectra from mixtures of modified proteins. *Molecular & Cellular Proteomics, 6*, 1123–1134.

Benson, S., Sucov, H., Stephens, L., Davidson, E., & Wilt, F. (1987). A lineage-specific gene encoding a major matrix protein of the sea urchin embryo spicule. I. Authentication of the cloned gene and its developmental expression. *Developmental Biology, 120*, 499–506.

Castellana, N., & Bafna, V. (2010). Proteogenomics to discover the full coding content of genomes: A computational perspective. *Journal of Proteomics, 73*, 2124–2135.

Cheers, M. S., & Ettensohn, C. A. (2004). Rapid microinjection of fertilized eggs. *Methods in Cell Biology, 74*, 287–310.

Chomczynski, P., & Sacchi, N. (1987). Single-step method of RNA isolation by acid guanidinium thiocyanate-phenol-chloroform extraction. *Analytical Biochemistry, 162*, 156–159.

Chomczynski, P., & Sacchi, N. (2006). The single-step method of RNA isolation by acid guanidinium thiocyanate-phenol-chloroform extraction: Twenty-something years on. *Nature Protocols, 1*, 581–585.

Drake, J. L., Mass, T., Haramaty, L., Zelzion, E., Bhattacharya, D., & Falkowski, P. G. (2013). Proteomic analysis of skeletal organic matrix from the stony coral *Stylophora pistillata*. *Proceedings of the National Academy of Sciences of the United States of America, 110*, 3788–3793.

Estroff, L. A. (2008). Introduction: Biomineralization. *Chemical Reviews, 108*, 4329.

Esvelt, K. M., & Wang, H. H. (2013). Genome-scale engineering for systems and synthetic biology. *Molecular Systems Biology, 9*, 1–17.

Frohman, M. A. (1990). RACE: Rapid amplification of cDNA ends. In M. A. Innis, D. H. Gelfand, J. J. Sninsky, & T. J. White (Eds.), *PCR protocols: A guide to methods and applications* (pp. 28–38). San Diego: Academic Press.

Gaj, T., Gersbach, C. A., Barbas, I., & Carlos, F. (2013). ZFN, TALEN, and CRISPR/Cas-based methods for genome engineering. *Trends in Biotechnology.* http://dx.doi.org/10.1016/j.tibtech.2013.04.004.

George, N. C., Killian, C. E., & Wilt, F. H. (1991). Characterization and expression of a gene encoding a 30.6-kDa *Strongylocentrotus purpuratus* spicule matrix protein. *Developmental Biology, 147*, 334–342.

Gogol-Döring, A., & Chen, W. (2012). An overview of the analysis of next generation sequencing data. In J. Wang, A. C. Tan, & T. Tian (Eds.), *Next generation microarray bioinformatics* (pp. 249–257). New York: Springer.

Gong, Y. U., Killian, C. E., Olson, I. C., Appathurai, N. P., Amasino, A. L., Martin, M. C., et al. (2012). Phase transitions in biogenic amorphous calcium carbonate. *Proceedings of the National Academy of Sciences, 109*, 6088–6093.

Gstaiger, M., & Aebersold, R. (2009). Applying mass spectrometry-based proteomics to genetics, genomics and network biology. *Nature Reviews Genetics, 10*, 617–627.

Illies, M., Peeler, M., Dechtiaruk, A., & Ettensohn, C. (2002). Identification and developmental expression of new biomineralization proteins in the sea urchin *Strongylocentrotus purpuratus*. *Development Genes and Evolution, 212*, 419–431.

Jackson, D. J., & Degnan, B. M. (2006). Expressed sequence tag analysis of genes expressed during development of the tropical abalone *Haliotis asinina*. *Journal of Shellfish Research, 25*, 225–231.

Joubert, C., Piquemal, D., Marie, B., Manchon, L., Pierrat, F., Zanella-Cléon, I., et al. (2010). Transcriptome and proteome analysis of *Pinctada margaritifera* calcifying mantle and shell: Focus on biomineralization. *BMC Genomics, 11*, 613.

Katoh-Fukui, Y., Noce, T., Ueda, T., Fujiwara, Y., Hashimoto, N., Higashinakagawa, T., et al. (1991). The corrected structure of the SM50 *spicule matrix protein of* Strongylocentrotus purpuratus. *Developmental Biology, 145*, 201–202.

Killian, C. E., & Wilt, F. H. (1996). Characterization of the proteins comprising the integral matrix of *Strongylocentrotus purpuratus* embryonic spicules. *Journal of Biological Chemistry, 271*, 9150–9159.

Li, H., & Homer, N. (2010). A survey of sequence alignment algorithms for next-generation sequencing. *Briefings in Bioinformatics, 11*, 473–483.

Livingston, B. T., Killian, C. E., Wilt, F., Cameron, A., Landrum, M. J., Ermolaeva, O., et al. (2006). A genome-wide analysis of biomineralization-related proteins in the sea urchin, *Strongylocentrotus purpuratus*. *Developmental Biology, 300*, 335–348.

Lowenstam, H. A., & Weiner, S. (1989). *On biomineralization*. Oxford: Oxford University Press.

Mann, M., & Jensen, O. N. (2003). Proteomic analysis of post-translational modifications. *Nature Biotechnology, 21*, 255–261.

Mann, K., Poustka, A. J., & Mann, M. (2008a). In-depth, high-accuracy proteomics of sea urchin tooth organic matrix. *Proteome Science, 6*, 33.

Mann, K., Poustka, A. J., & Mann, M. (2008b). The sea urchin (*Strongylocentrotus purpuratus*) test and spine proteomes. *Proteome Science, 6*, 22.

Mann, K., Wilt, F. H., & Poustka, A. (2010). Proteomic analysis of sea urchin (*Strongylocentrotus purpuratus*) spicule matrix. *Proteome Science, 8*, 33.

Mardis, E. R. (2011). A decade's perspective on DNA sequencing technology. *Nature, 470*, 198–203.

Mardis, E. R. (2013). Next-generation sequencing platforms. *Annual Review of Analytical Chemistry, 6*, 287–303.

Marie, B., Le Roy, N., Zanella-Cléon, I., Becchi, M., & Marin, F. (2011). Molecular evolution of mollusc shell proteins: Insights from proteomic analysis of the edible mussel *Mytilus*. *Journal of Molecular Evolution, 72*, 531–546.

Marikawa, Y., & Elinson, R. P. (1999). Relationship of vegetal cortical dorsal factors in the *Xenopus* egg with the wnt/beta-catenin signaling pathway. *Mechanisms of Development, 89*, 93–102.

Marin, F., Le Roy, N., & Marie, B. (2012). The formation and mineralization of mollusk shell. *Frontiers in Bioscience, 4*, 1099–1125.

McMahon, A. P., Flytzanis, C. N., Hough-Evans, B. R., Katula, K. S., Britten, R. J., & Davidson, E. H. (1985). Introduction of cloned DNA into sea urchin egg cytoplasm: Replication and persistence during embryogenesis. *Developmental Biology, 108*, 420–430.

Metzker, M. L. (2009). Sequencing technologies—The next generation. *Nature Reviews Genetics, 11*, 31–46.

Mitsunaga, K., Makihara, R., Fujino, Y., & Yasumasu, I. (1986). Inhibitory effects of ethacrynic acid, furosemide, and nifedipine on the calcification of spicules in cultures of micromeres isolated from sea-urchin eggs. *Differentiation, 30*, 197–204.

Murdock, D., & Donoghue, P. (2011). Evolutionary origins of animal skeletal biomineralization. *Cells, Tissues, Organs, 194*, 98–102.

Nilsson, T., Mann, M., Aebersold, R., Yates, J. R., Bairoch, A., & Bergeron, J. J. (2010). Mass spectrometry in high-throughput proteomics: Ready for the big time. *Nature Methods, 7*, 681–685.

Pespeni, M. H., Sanford, E., Gaylord, B., Hill, T. M., Hosfelt, J. D., Jaris, H. K., et al. (2013). Evolutionary change during experimental ocean acidification. *Proceedings of the National Academy of Sciences of the United States of America, 110*, 6937–6942.

Ramos-Silva, P., Marin, F., Kaandorp, J., & Marie, B. (2013). Biomineralization toolkit: The importance of sample cleaning prior to the characterization of biomineral proteomes. *Proceedings of the National Academy of Sciences of the United States of America, 110*, E2144–E2146.

Rauch, F., & Glorieux, F. H. (2004). Osteogenesis imperfecta. *The Lancet, 363*, 1377–1385.

Schroeder, A., Mueller, O., Stocker, S., Salowsky, R., Leiber, M., Gassmann, M., et al. (2006). The RIN: An RNA integrity number for assigning integrity values to RNA measurements. *BMC Molecular Biology, 7*, 3.

Segal, D. J., & Meckler, J. F. (2013). Genome engineering at the dawn of the golden age. *Annual Review of Genomics and Human Genetics, 14*, 5.1–5.24.

Shendure, J., & Ji, H. (2008). Next-generation DNA sequencing. *Nature Biotechnology, 26*, 1135–1145.

Simkiss, K., & Wilbur, K. M. (1989). *Biomineralization: Cell biology and mineral deposition*. New York: Academic Press.

Söllner, C., Burghammer, M., Busch-Nentwich, E., Berger, J., Schwarz, H., Riekel, C., et al. (2003). Control of crystal size and lattice formation by starmaker in otolith biomineralization. *Science, 302*, 282–286.

Steen, H., & Mann, M. (2004). The ABC's (and XYZ's) of peptide sequencing. *Nature Reviews Molecular Cell Biology, 5*, 699–711.

Sucov, H. M., Benson, S., Robinson, J. J., Britten, R. J., Wilt, F., & Davidson, E. H. (1987). A lineage-specific gene encoding a major matrix protein of the sea urchin embryo spicule. II. Structure of the gene and derived sequence of the protein. *Developmental Biology, 120*, 507–519.

Summerton, J. (1999). Morpholino antisense oligomers: The case for an RNase H-independent structural type. *Biochimica et Biophysica Acta, 1489*, 141–158.

Supuran, C. T., Scozzafava, A., & Casini, A. (2003). Carbonic anhydrase inhibitors. *Medicinal Research Reviews, 23*, 146–189.

Suzuki, M., Saruwatari, K., Kogure, T., Yamamoto, Y., Nishimura, T., Kato, T., et al. (2009). An acidic matrix protein, Pif, is a key macromolecule for nacre formation. *Science*, *325*, 1388–1390.

Telenius, H., Carter, N. P., Bebb, C. E., Ponder, B. A., & Tunnacliffe, A. (1992). Degenerate oligonucleotide-primed PCR: General amplification of target DNA by a single degenerate primer. *Genomics*, *13*, 718–725.

Trapnell, C., & Salzberg, S. L. (2009). How to map billions of short reads onto genomes. *Nature Biotechnology*, *27*, 455–457.

Walther, T. C., & Mann, M. (2010). Mass spectrometry-based proteomics in cell biology. *The Journal of Cell Biology*, *190*, 491–500.

Wang, Z., Gerstein, M., & Snyder, M. (2009). RNA-Seq: A revolutionary tool for transcriptomics. *Nature Reviews Genetics*, *10*, 57–63.

Wilt, F., Croker, L., Killian, C. E., & McDonald, K. (2008). Role of LSM34/SpSM50 proteins in endoskeletal spicule formation in sea urchin embryos. *Invertebrate Biology*, *127*, 452–459.

Wilt, F. H., Killian, C. E., & Livingston, B. T. (2003). Development of calcareous skeletal elements in invertebrates. *Differentiation*, *71*, 237–250.

Witze, E. S., Old, W. M., Resing, K. A., & Ahn, N. G. (2007). Mapping protein post-translational modifications with mass spectrometry. *Nature Methods*, *4*, 798–806.

Yates, J. R., Ruse, C. I., & Nakorchevsky, A. (2009). Proteomics by mass spectrometry: Approaches, advances, and applications. *Annual Review of Biomedical Engineering*, *11*, 49–79.

Zeller, R. W., Virata, M. J., & Cone, A. C. (2006). Predictable mosaic transgene expression in ascidian embryos produced with a simple electroporation device. *Developmental Dynamics*, *235*, 1921–1932.

Zhu, X., Mahairas, G., Illies, M., Cameron, R., Davidson, E. H., & Ettensohn, C. A. (2001). A large-scale analysis of mRNAs expressed by primary mesenchyme cells of the sea urchin embryo. *Development*, *128*, 2615–2627.

SECTION V

Mapping Biomineral Morphology and Ultrastructure

CHAPTER EIGHTEEN

Imaging the Nanostructure of Bone and Dentin Through Small- and Wide-Angle X-Ray Scattering

Silvia Pabisch[*], Wolfgang Wagermaier[*], Thomas Zander[*,†], Chenghao Li[*], Peter Fratzl[*,1]

[*]Max Planck Institute for Colloids and Interfaces, Potsdam, Germany
[†]Helmholtz-Zentrum Geesthacht, Geesthacht, Germany
[1]Corresponding author: e-mail address: peter.fratzl@mpikg.mpg.de

Contents

1. Introduction — 392
2. Experimental Setup — 394
    2.1 Experimental setup of a laboratory SAXS equipment — 395
    2.2 Experimental setup of an X-ray beamline at a synchrotron facility — 395
3. Overview of Parameters — 396
4. Treatment of SAXS Data from Mineralized Tissue — 397
    4.1 Radial and azimuthal intensity profiles — 397
    4.2 Background and transmission — 398
    4.3 ρ-Parameter — 399
    4.4 T-Parameter — 400
    4.5 Analysis of the mineral arrangement with the stack of cards model — 402
5. Treatment of WAXD Data from Mineralized Tissues — 405
    5.1 The texture effect — 405
    5.2 L-Parameter — 407
6. Analysis Tools — 408
7. Combination of Scanning SAXS/WAXD with Other Methods — 409
8. Conclusions — 410
References — 412

Abstract

X-ray scattering is a powerful nondestructive experimental method that is well suited to study biomineralized tissues such as bone. Small-angle X-ray scattering (SAXS) gives information about the size, shape, and predominant orientation of the nanometer-sized mineral particles in the bone. Wide-angle X-ray diffraction (WAXD) allows the characterization of structural parameters, describing size and orientation of the hydroxyapatite crystals.

Furthermore, scanning an area with nano- or micrometer-sized X-ray beams allows one to extend this local information to map large bone or dentin sections. Therefore, this method contributes to obtaining information on several length scales simultaneously. Combining results from scanning SAXS and WAXD with those from other position-sensitive methods such as backscattered electron imaging or X-ray fluorescence spectroscopy of the same bone sections allows the exploration of complex biological processes. The method is described and illustrated by a few examples, including the mapping of a complete tooth and the effect of osteoporosis treatment on the bone mineral.

1. INTRODUCTION

Structural studies of biomineralized tissues generally need information collected over a large range of length scales. Specific techniques are required to relate the functional properties to the hierarchical structure, such as X-ray diffraction methods, in addition to several imaging techniques, such as light microscopy, scanning electron imaging, Raman imaging, or infrared spectroscopy imaging. The combination of those techniques reveals the structural and compositional gradients over several length scales in many biological materials and, in particular, biomineralized tissues (Paris et al., 2011).

The bone and dentin, as heterogeneous materials at all hierarchical levels of organization, require the usage of position-resolved methods at different length scales from the macro- to the nanometer level. Half a century ago, early attempts were made by Eanes and Posner to investigate the composition of the mineral phase in the human bone (Eanes, Termine, & Posner, 1967; Posner, Eanes, Harper, & Zipkin, 1963) by X-ray diffraction, but it took almost further 20 years before measurements on bone continued in several studies using X-ray and neutron diffraction studies to investigate the structure and functions of the bone (Arsenault & Grynpas, 1988; Bacon, Bacon, & Griffiths, 1979; Bonar, Lees, & Mook, 1985; Lees & Prostak, 1988). Small-angle x-ray scattering (SAXS) was used by Fratzl et al. for detailed investigations on calcified tissues like mineralized turkey leg tendon or bone samples of mouse and other animals (Fratzl, Fratzl-Zelman, Klaushofer, Vogl, & Koller, 1991; Fratzl et al., 1992). The thickness of the mineral crystals and their preferred orientation located within the gap zone of the collagen fibrils were derived from these data. In the last decade, the investigation of the bone and its mechanical and structural properties and the effect of fracture and disease on bone quality have multiplied due to new and improved methods applied by several research groups. Deyhle et al. used spatially resolved SAXS combined with microcomputed-tomography (μCT) to study the effect of caries-induced damages on the inorganic and organic components in the human teeth (Deyhle, Bunk, & Müller, 2011). Furthermore, accumulations of key elements in the bone/cartilage interface were exploited by micro-proton-induced X-ray emission and particle-induced gamma-ray emission (Kaaber et al., 2011). *In situ* SAXS and wide-angle X-ray diffraction (WAXD) were used to characterize structural changes of human cortical bone in the submicrometer level and synchrotron X-ray CT and *in situ* fracture-toughness measurements in scanning electron microscope to characterize effects in the micrometer level and therefore reveal age-related changes in mechanical properties (Zimmermann et al., 2011). The method of microfocus *in situ* synchrotron SAXS combined with cantilever bending delivers information about the alterations in fibrillar level bone-material quality affecting macroscopic mechanical competence in metabolic bone diseases (Karunarane et al., 2013).

One can summarize by saying that the hierarchical organization of the bone results in an optimized biomaterial with respect to its specific biological functions (Weiner & Addadi, 1997). The mechanical properties of the bone and its potential fragility depend not only on bone volume, shape, and microarchitecture but also on its material properties (Seeman & Delmas, 2006). The bone is a composite of a collagen-rich organic matrix (Landis, Hodgens, Arena, Song, & McEwen, 1996; Rubin et al., 2003) and plate-shaped inorganic mineral particles (Fratzl, Gupta, Paschalis, & Roschger, 2004; Weiner & Wagner, 1998). The thickness of these mineral platelets is on the order of 2–7 nm with a length of approximately 15–200 nm (Fratzl et al., 2004). Shape, size, and arrangement of the mineral particles and the mineral density distribution have a great influence on the mechanical properties of the bone (Fratzl & Weinkamer, 2007). Especially in the case of a biomaterial such as the bone with its hierarchical structure, an analysis of this material on different length scales at the same time is preferable for receiving information about the local homogeneities and inhomogeneities on the micro- and nanometer level (Gourrier et al., 2007; Rinnerthaler et al., 1999). This can be achieved by scanning SAXS/WAXD measurements, where the sample is moved stepwise in the x- and y-axis plane perpendicular to the direction of the X-ray beam in order to record the scattering and diffraction pattern for every step. On the one hand, information in the micrometer range can be achieved by scanning an area; on the other hand, information in the nanometer range can be obtained by analyzing the pattern at distinct positions in the area.

In this work, an overview about scanning X-ray scattering applied to bone is presented. Different experimental setups and their advantages and disadvantages for specific problems are described and compared. Parameters describing the bones at the nanometer level, such as the thickness, length, and predominant orientation of the mineral particles, are first presented from a basic point of view. These parameters are also discussed in the context of some exemplarily studies, like of the effect of strontium treatment or calcium content related to mineral particle properties in newly formed bone of specimen from different ages obtained by simultaneously measured SAXS, WAXD, and X-ray fluorescence (XRF).

2. EXPERIMENTAL SETUP

In the following two experimental setups to study biomineralized tissues, a laboratory SAXS equipment and a setup at a synchrotron will be described. A general experimental setting is presented in Fig. 18.1 to sketch the scattering of the X-ray beam on the sample. The layout shown follows in both cases basically the same principles. However, a synchrotron has some important advantages. It can provide smaller beam sizes (if necessary, down to the submicrometer range) and consequently a better spatial resolution. Additionally, the beam flux is much higher, thereby decreasing the exposure time for the measurements, as well as the brilliance of synchrotron radiation.

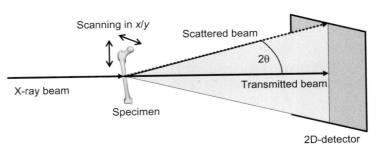

Figure 18.1 Basic principle of an X-ray scattering setup, where the incident beam is scattered on the sample and measured by the 2D detector.

2.1. Experimental setup of a laboratory SAXS equipment

X-ray can be generated in several ways in a common laboratory SAXS equipment. One possibility is that X-rays are generated by a rotating copper anode to produce an X-ray beam with a wave length of $\lambda = 0.154$ nm. Depending on the collimation system and aperture, the beam size can vary between 150 and 500 µm at the sample. For bone samples, a beam size of approximately 150–200 µm is usually appropriate. The thickness of the sample slices should be in the order of the beam size to get a reasonable scattering volume. Different sample-to-detector distances are available to use SAXS (200–1000 mm) and WAXD (<50 mm) setup. A sample-to-detector distance of about 600 mm with a resulting q range between 0.21 and 3.2 nm^{-1} is appropriate for a standard SAXS analysis of the bone. The samples are mounted on a sample holder, which can be moved precisely with a resolution of around 1 µm in the plane perpendicular to the incident beam. This allows scanning SAXS experiments over the entire bone sample. The sample chamber is kept under vacuum during the measurements to minimize scattering in air. To determine the positions of the subsequent SAXS measurements, an X-ray transmission image of the bone using a diode is produced.

2.2. Experimental setup of an X-ray beamline at a synchrotron facility

Synchrotron radiation is produced by cyclic particle accelerators. Electrons are accelerated within the cyclotron, where insertion devices like undulators or wigglers cause the emission of highly intensive, highly collimated, polarized electromagnetic radiation. This allows a wide tunability in energy and wavelength by monochromatization. The advantages of a synchrotron beamline compared to the laboratory equipment are therefore (i) higher resolution of the measurements due to a possible smaller beam size down to a few micrometer or even submicrometer and (ii) the possibility of simultaneous measurements of SAXS and WAXD and possibly other methods such as XRF to measure the calcium content in the bone sample.

For example, the experimental setup of the µ-Spot beamline at BESSY II (Helmholtz-Zentrum Berlin für Materialien und Energie, Berlin, Germany) includes a beam size-defining pinhole, sample stage, an optical microscope with a resolution of 2.2 µm, and a 2D position-sensitive MarMosaic-CCD detector. Two optical systems are available: (i) a MoBC-Multilayer, optimized for fast measurements with high intensity but low resolution of $\Delta E/E = 10^{-2}$, and (ii) single crystals (Ge 1 1 1, Si 1 1 1, and Si 3 1 1) with higher resolution of $\Delta E/E = (1-7) \times 10^{-4}$ but lower intensity. The beam size can be varied from 10 to 100 µm, but in the case of bone, a beam size of approximately 10–30 µm is approved for high-position resolution with a wavelength of $\lambda = 0.082656$ nm. The sample stage consists of an xyz translation stage, which enables scanning measurements of larger bone sections. An optimized q range to determine both SAXS and WAXD concurrently is given at a sample-to-detector distance of about 300 mm. The µ-Spot beamline allows simultaneous SAXS, WAXD, and XRF as mentioned earlier, which includes an additional fluorescence detector with a 100 mm^2 sensitive area and about 167.4 eV energy resolution for the experimental setup. In order to detect the bone sample and define regions of interest, a calcium distribution map of the sample can be created using the XRF signal of calcium. When using a beam size of 30 µm, a thickness of the slices of approximately 30 µm is recommended.

3. OVERVIEW OF PARAMETERS

SAXS and WAXD deliver structural information on different length scales, in bone represented by three important parameters: L-, T-, and ρ-parameter (Fig. 18.2). The mean length of the mineral platelets (L-parameter) can be calculated from the 0 0 2 reflection of the crystallographic planes perpendicular to the c-axis of hydroxyapatite (HAp) (see Fig. 18.2). Information about the 0 0 2 reflection is provided by the WAXD signal. Since the mineral platelets and the fibrils are aligned in parallel to each other, it is also possible to conclude the orientation of the

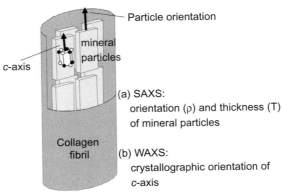

Figure 18.2 Schematic illustration of a collagen fibril and mineral platelets within (not drawn to scale).

fibrils from the 0 0 2 reflection (Fratzl et al., 2004; Landis et al., 1996). Information on the next length scale concerning the size and arrangement of the mineral particles in the bone matrix is obtained by SAXS. In general, two-phase systems can be characterized by SAXS, if (i) the phases differ in their electron density, (ii) the phase transition is characterized by sharp boundaries, (iii) there is uniform electron density within one phase, and (iv) particles are present in a range between 1 and 100 nm. However, some basic information about the system must be known, because by small-angle scattering, it cannot be distinguished whether pores or particles are present in a matrix, because the contrast is due to the square of their electron density difference (Babinet principle). From a SAXS pattern of the bone, one can determine a value for the mean thickness of the mineral (T-parameter) and the orientation and the degree of their alignment in the sample volume (ρ-parameter).

4. TREATMENT OF SAXS DATA FROM MINERALIZED TISSUE

For a correct analysis of the SAXS data, several steps have to be performed and are described in the following sections. This includes the calibration of the data, conversion of the 2D SAXS/WAXD pattern to 1D data, background subtraction, and finally the calculation of the different parameters.

4.1. Radial and azimuthal intensity profiles

The first step is the calibration of the 2D data to determine the exact position of the beam center and the sample-to-detector distance by measuring a calibration standard with well-known lattice parameter. The next step of the SAXS/WAXD data analysis is the conversion of the 2D SAXS/WAXD patterns into 1D data, via summation of the pixel values from the 2D detector. Typically two standard intensity profiles are applied: the radial intensity profile, which is dependent on the length of the scattering vector $\vec{q}$, and the azimuthal intensity profile, which is dependent on the azimuthal angle χ of the 2D data. The radial intensity profile $I(q)$ is obtained by binning all pixels with the same radial distance q to the beam center (Fig. 18.3). In case of the azimuthal intensity profile $I(\chi)$, all pixels with the same azimuthal degree χ are binned (Fig. 18.4). Both formats are needed to determine the structural parameters of the bones. For the calculation of the T-parameter, the radial plot $I(q)$ is used and the azimuthal plot $I(\chi)$ for the ρ-parameter.

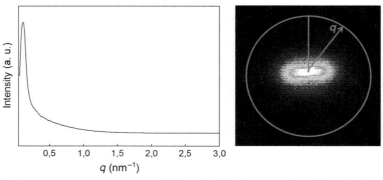

Figure 18.3 Radial plot (left) of the integration in azimuthal direction of a SAXS pattern (right). (See color plate.)

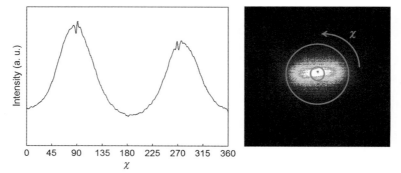

Figure 18.4 Azimuthal plot (left) of the integration in radial direction of a SAXS pattern (right). (For color version of this figure, the reader is referred to the online version of this chapter.)

4.2. Background and transmission

After evaluating the intensity profiles, the background profile, caused by scattering of the experimental setup, has to be subtracted from the sample data. Therefore, a correction of the transmission has to be considered. Additionally, the dark current intensity profile I_{DC} has to be subtracted from the background profile and the sample profiles, as almost all detectors measure intensities even when there is no radiation. One has to divide the sample profile I_{sample} (minus I_{DC}) by the transmission coefficient before subtracting the background profile

$$I_{corr} = \frac{I_{sample} - I_{DC}}{T_s} - (I_{EB} - I_{DC}) \qquad (18.1)$$

where I_{EB} is the empty beam intensity or background intensity. T_s is the ratio between the intensity (meaning the intensity of the primary beam measured with a pin diode) of the incoming beam i_0 and that of the transmitted beam i_r: $T_s = \frac{i_r}{i_0}$. The transmission coefficient has to be calculated for every single measuring point.

With a diode, the transmission coefficient T_s can be easily obtained at each position as the diode is able to directly measure the intensity of the primary beam. Without a diode, the determination of T_s is done by comparing the measured intensity of the sample with unknown T_s with the intensity of a sample with well-known T_s, a standard sample, for example, a thin disk of glassy carbon. Assuming that there is a high scattering intensity of the sample, T_s is written as

$$T_s = \frac{i_{sc} - T_c i_s}{i_c - T_c i_p} \qquad (18.2)$$

where i_{sc} is the intensity of the unknown sample together with the standard sample, i_s the intensity of the unknown sample alone, i_c the intensity of the standard sample alone, i_p is the intensity with no sample, and T_c is T_s of the standard sample. The intensities i_c, i_{sc}, and i_p are determined by integrating the corresponding detector images as a whole. Strong absorption behavior of the standard sample, i.e. very low T_c, simplifies equation 2 to $T_s = i_{sc}/i_c$. If dark current has to be considered, T_s is written as $T_s = (i_{sc} - i_d)/(i_c - i_d)$ with the intensity i_d measured by the detector without beam.

4.3. ρ-Parameter

The ρ-parameter is a measure of the degree of alignment of the mineral platelets. It characterizes the amount of aligned particles in ratio to the irradiated sample volume and gives information about the bone mineral quality. The greater the alignment of the mineral platelets is, the more ordered the minerals within the organic matrix are. The azimuthal plot $I(\chi)$ in Fig. 18.5 shows two peaks separated by $180°$, due to the point symmetry of SAXS pattern. The ρ-parameter is defined as the ratio of the sum of the area under the peaks, A_1 and A_2, to the sum of the total area including the background area B (see Fig. 18.5):

$$\rho = \frac{A_1 + A_2}{A_1 + A_2 + B}. \qquad (18.3)$$

The area under the peaks is proportional to the fraction of aligned particles (HAp platelets), and the area under the constant background is

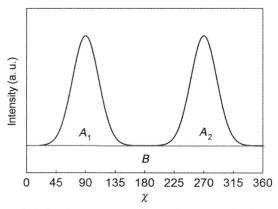

Figure 18.5 Azimuthal plot depicting A_1 and A_2 as the area under the peaks and B as the area under the baseline. (For color version of this figure, the reader is referred to the online version of this chapter.)

proportional to the fraction of randomly oriented mineral particles in the bone. Therefore, the ρ-parameter is the ratio of aligned particles to the total amount of particles in the illuminated volume.

4.4. T-Parameter

For the calculation of the T-parameter, the Porod constant is needed, which was described by Porod in 1951 (Porod, 1951). Porod's law states that for a two-phase system with different electron densities and sharp interfaces, the following equation is valid for large q values:

$$I(q) \xrightarrow{q \to \infty} \frac{P}{q^4} + B. \qquad (18.4)$$

P is the Porod constant and B a constant background. Porod's law concludes that for large q values, the scattering intensity of a biphasic system with distinct interfaces always decreases with q^{-4}. The parameters P and B can be calculated from the so-called Porod plot (Fig. 18.6). By plotting $I \cdot q^4$ versus q^4, the relation should be linear for large q, and it is possible to determine both P and B from a fit of the linear region, where $q_p^{\max}$ has to be chosen such that it is not within the region of Bragg peaks or a region of bad statistics (Fratzl et al., 1991; Rinnerthaler et al., 1999).

The T-parameter is an indicator of the volume-to-surface ratio of the mineral particles, as one definition of the parameter is $T = 4\frac{\Phi(1-\Phi)}{\sigma}$, where Φ is the mineral volume fraction and σ the total surface area of all mineral particles per unit volume. It should be noted that without any additional

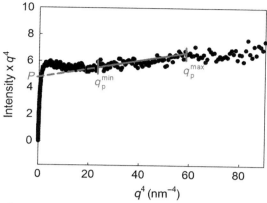

Figure 18.6 Porod plot of the bones showing the Porod region, defined by red lines. The Porod constant P is defined by the intercept of the fit of the Porod region with the y-axis. (For interpretation of the references to color in this figure legend, the reader is referred to the online version of this chapter.)

information about the geometric form or dimension of the minerals, only information about the volume-to-surface ratio can be inferred from the T-parameter. In the case of the mineral particles in the bones, the shape of the particles is supposed to be the thin platelets (Fratzl et al., 2004) where the T-parameter describes the thickness of these platelets.

The T-parameter can be calculated by solving the following equation:

$$T = \frac{4 \cdot J}{\pi \cdot P} = \frac{4}{\pi \cdot P} \int_0^\infty q^2 I^*(q) \, dq \tag{18.5}$$

where I^* is the measured intensity minus the background: $I^* = I(q) - B$. The integral can be solved by determining the area under the curve of the Kratky plot $q^2 I^*$ versus q (Fig. 18.7). One limitation in this procedure comes from the determination of the Porod constant P, as it can only be calculated using the Porod region, a specific part of the scattering curve at high values. Another limitation comes from the experimental restrictions of the accessible q range. The only part of the integral parameter $J = \int_0^\infty q^2 I(q) dq$ that can be measured in this range is $J_{\exp} = \int_{q_{\text{Kratky}}^{\min}}^{q_P^{\min}} q^2 I^*(q) \, dq$. At high q values, the experimental curve can be extrapolated using Porod's law, $J_{\text{Porod}} = \int_{q_P^{\min}}^{\infty} q^2 I^*(q) \, dq = P/q_{\max}$. For low q values (below $q_{\text{Kratky}}^{\min}$), the

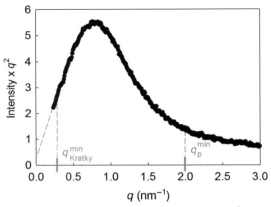

Figure 18.7 Kratky plot of the bone showing the parameters q_{Kratky}^{min} and q_p^{min}. (For color version of this figure, the reader is referred to the online version of this chapter.)

approximation assuming a linear decrease is generally used. The limitation can be overcome by using an alternative approach such as the stack of cards model (Gourrier et al., 2010).

For an even more accurate value for platelike structures, as the T-parameter depends not only on the particle size but also on the mineral volume fraction Φ, an additional parameter W

$$W = \frac{T}{2(1-\Phi)} \qquad (18.6)$$

was defined by Zizak et al. to describe the average smallest dimension of the particles in the bulk of each measurement point (Zizak et al., 2003). This assumption should be taken into account if the volume fraction deviates notably from 50% or changes within the sample. The mineral volume fraction Φ can be measured independently by quantitative backscattered electron imaging (Roschger, Plenk, Klaushofer, & Eschberger, 1995; Zizak et al., 2003) or by µCT as presented in Fig. 18.8 (Märten et al., 2010). However, it should be noted that $T=W$ for $\Phi=0.5$.

4.5. Analysis of the mineral arrangement with the stack of cards model

The T-parameter reveals no information about the shape or arrangement of the mineral particles. However, this information is contained in a rescaled

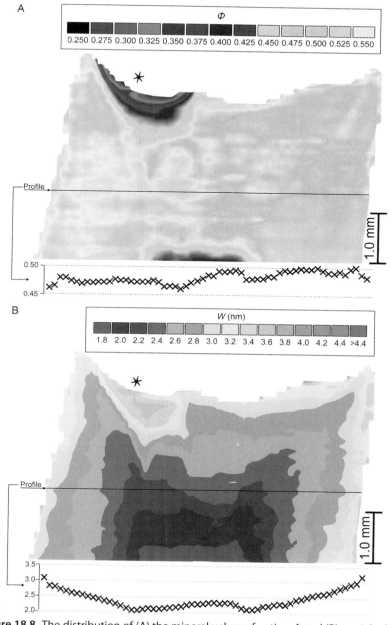

Figure 18.8 The distribution of (A) the mineral volume fraction Φ and (B) particle thickness W of dry dentine were studied by Märten et al. (2010). A typical profile of W-parameter across the data is plotted beneath the map. (See color plate.)

function of the dimensionless parameter $x = qT$ (Fratzl et al., 1996), which is defined as

$$G(x) = x^2 I\left(\frac{x}{T}\right) \Big/ (JT^3) \qquad (18.7)$$

where $\int_0^\infty G(x)\mathrm{d}x = 1$. Different species, bone age, and volume fraction of mineral, all cause systematic changes in G(x) curves, which indicate some systematic differences between tissue structures. Therefore, structure models to describe the G(x) curves and thus the organization of the mineral phase were developed. One model describes the arrangement of the mineral platelets as stacks of cards. It was developed to fit the G(x) curve and provide sets of parameters describing the organization at this structural level (Fratzl et al., 2005). The mineral platelets are assumed to be positioned along the direction normal to the platelets. Therefore, the function G(x) is related to the normalized 1D correlation function $g(\zeta)$ by a Fourier transform (Fratzl et al., 2005):

$$G(x) = \left(\frac{2}{\pi}\right)\int_0^\infty g(\zeta)\cos(\zeta x)\mathrm{d}\zeta. \qquad (18.8)$$

A simple assumption is to describe the correlation function as a damped oscillation:

$$g(\zeta) = A\exp(-\alpha\zeta)\cos(\beta\zeta - \varphi). \qquad (18.9)$$

Here, the period of the oscillation $2\pi/\beta$ gives the typical spacing between the platelets (in units of T). The distance, over which the periodicity in platelet spacing is damped, is described by the exponential decay $(1/\alpha)$. The normalization of G(x) immediately defines the other two constants as $A = 1/\cos\varphi$ and $\tan\varphi = (\alpha - 2)/\beta$, which leads to simple model functions for G(x) and for $g(\zeta)$ of

$$g(\zeta) = \exp(-\alpha\zeta)\left[\cos(\beta\zeta) + \frac{\alpha - 2}{\beta}\sin(\beta\zeta)\right], \qquad (18.10)$$

$$G(x) = \frac{4}{\pi}\frac{x^2 + (\alpha - 1)(\alpha^2 + \beta^2)}{(x^2 + \alpha^2 + \beta^2)^2 + 4\alpha^2\beta^2}. \qquad (18.11)$$

The parameter β describes the degree of spatial correlation between successive plates, which are separated by an average distance $d = T2\pi/\beta$. The parameter α controls the damping of the oscillations and thus provides an indication of the relative extent of the ordering. Smaller values of $2\pi/\alpha$

indicate a reduction in the short-range ordering of the platelets and therefore greater disorder. Consequently, β/α can be seen as a reasonable (scalar) indicator of the degree of ordering at constant T values, as the degree of spatial correlation is modulated by the relative extent of the short-range order.

Using Eq. (18.11) to fit the stack of cards model to the $G(x)$ curves obtained from the experimental data $I(q)$ implies that the parameters α and β will be biased by the approximation at low q in the calculation of $J_{\min}$. Therefore, it is preferable to fit the unbiased experimental data using Eq. (18.6) written as

$$q^2 I(q) = \frac{\pi}{4} P T^2 G(x) \qquad (18.12)$$

using the expression of $G(x)$ given in Eq. (18.11) with $x = qT$ and J given in Eq. (18.5) with $J = TP\pi/4$. Here, T is allowed to vary to account for the unknown value of $J_{\min} = \int_0^{q_{\min}} q^2 I(q) dq$ and P is a fixed parameter as it is calculated independently from the model.

It also follows that the parameter β/α defined previously is insufficient to characterize the degree of ordering as it does not take into account the uncertainty in $J_{\min}$. Instead, a new parameter, κ, is proposed for this purpose and defined as

$$\kappa = \frac{\beta}{\alpha} \frac{T_{\text{ref}}}{T} \qquad (18.13)$$

where T_{ref} is a reference value of T obtained either from a reference sample, from a region of interest of the scan considered to be representative, or even from the mean T value of the scan (Gourrier et al., 2010).

5. TREATMENT OF WAXD DATA FROM MINERALIZED TISSUES

5.1. The texture effect

Mineral crystals in the bone frequently show a preferred orientation. Therefore, the texture effect has to be taken into account during analysis of X-ray scattering data from the bone (Wenk & Heidelbach, 1999). The integral intensity of $I(\chi)$, determined by radial integration of the SAXS/WAXD region, represents the amount of mineral particles in the sample. 3D integration of the data would deliver the most accurate results; however, only 2D information is available, where the texture of the sample influences the results of the measurements and potentially leads to misinterpretation. Information about the HAp content and the amount of mineral particles in the sample are obtained by investigating the HAp (0 0 2)-peak in the WAXD regime and the integral intensity in the SAXS regime. Former investigations were done on HAp powder instead of the bone samples, where no texture effect occurs (Bacon et al., 1979; Termine, Eanes, Greenfield, Nylen, & Harper, 1973). A correction for the influence of texture is needed for the HAp (0 0 2)-peak in the WAXD regime (Lange et al., 2011).

In the case of HAp (0 0 2)-peak in the bone, two coordinate systems that can be converted into each other are defined, one describing the sample system and the other the detector (Fig. 18.9). The sample coordinate system is denoted by $(X,Y,Z) = R(\sin \Psi \cos \Phi, \sin \Psi \sin \Phi, \cos \Psi)$. The angle of any (0 0 2) c-axis of HAp with symmetry direction is defined as Ψ. The angles Ψ and Φ are defined in spherical coordinates and the Z-axis is a unique axis of the cylindrical symmetry. The detector/diffraction coordination system is denoted by $(x,y,z) = R(\sin \varphi \cos(\chi - \chi_0), \sin \varphi \sin(\chi - \chi_0), \cos(\chi - \chi_0))$ with the primary x-ray beam pointing in the z-direction and the area detector in the x–y plane. The direction χ_0 corresponds to the axis of the long bone, and the long bone axis is tilted by an angle μ with respect to the z-axis (primary beam axis). This coordinate rotation can be chosen without restriction of generality around the y-axis, that is, the transformation equation is $(x,y,z) = (X \cos \mu + Z \sin \mu, Y, X \sin \mu + Z \cos \mu)$. According to diffraction theory and the coordinate transformation of Paris and Muller (Paris & Muller, 2003), the equation $\cos(\chi - \chi_0) = (\cos \Psi + \sin \theta \cos \mu)/(\cos \theta \sin \mu)$

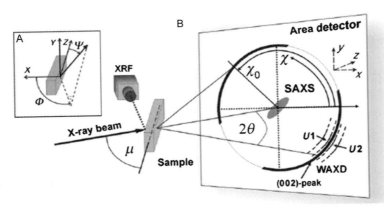

Figure 18.9 Simultaneous measurements of SAXS, WAXD, and XRF. Depiction of (A) sample coordinate system and (B) detector/diffraction coordinate system (Lange et al., 2011).

can be used. In the experiments, the bone sample is perpendicular to the X-ray beam, set $\mu = 90°$, leading to

$$\cos \Psi = \cos\theta \cos(\chi - \chi_0) \quad \text{or} \quad \Psi = \arccos(\cos\theta \cos(\chi - \chi_0)). \quad (18.14)$$

The angle χ_0 can be determined from the WAXD image of the longitudinal bone section and the angle θ is the Bragg angle of the HAp (0 0 2)-peak (Fig. 18.9).

The HAp (0 0 2)-peak obtained by WAXD is azimuthally integrated. The intensity of the (0 0 2)-peak in the WAXD pattern has to be corrected for background scattering, for which reason two regions beside the (0 0 2)-peak, first background U_1 and second background U_2, are integrated as shown in Fig. 18.7. The inner and outer radius for U_1 and U_2 will be defined in units of pixels, so that the difference in the number of pixels between inner and outer radius for both backgrounds is the same. The inner radius of the (0 0 2)-peak integration is equal to the outer radius of the background U_1 and the outer radius of the (0 0 2)-peak integration is equal to the inner radius of the second background U_2 as shown in Fig. 18.7. After the integration of the (0 0 2)-peak, the average of $U_1(\chi)$ and $U_2(\chi)$ is subtracted from the peak intensity (χ) to obtain the background corrected intensity $I(\chi) =$ peak intensity$(\chi) - (U_1(\chi) + U_2(\chi))/2$. The integral intensity of $I(\chi)$ represents the HAp content. In order to correct the orientation influence, the following relation is used to receive the texture corrected integral intensity:

$$\int_{\theta}^{\pi/2} I(\Psi) \sin \Psi \, d\Psi = \cos\theta \int_{0}^{\pi/2} I(\chi' + \chi_0) \sin \chi' \, d\chi' \quad (18.15)$$

where $\chi' = \chi - \chi_0$. Note that the integration in the left-hand term starts at θ and not 0. The integral intensity $I(\Psi)\sin\Psi$ represents the HAp content after texture correction.

A correction of the influence of texture on the SAXS region can be performed as well with software such as Fit2D and is described in detail in the study of Lange et al. (2011).

5.2. L-Parameter

In order to determine the length of the mineral platelets, the (0 0 2)-peak of the HAp crystals is used. This information could also be obtained from the (3 1 0)-peak of HAp powder in the WAXD regime, but not from bone samples (Bacon et al., 1979). Its width in the diffraction pattern is correlated with the average crystal size in the sample. This correlation is described by the Scherrer equation:

$$L = \frac{k\lambda}{B\cos\theta} \qquad (18.16)$$

where L is the crystal length, B is the full width at half maximum of the peak, θ represents the Bragg angle, and k is a constant related to the crystalline shape in the range of 0.87–1.0. Therefore, a decreasing crystal size leads to a broadening of the peaks as the peak width is inversely proportional to the mean crystallite length. The width of the (0 0 2)-peak is fitted by a Gaussian function, where two components contribute to the broadening B, the crystallite length B_L and the instrumental broadening B_I. The calculation of the mean crystallite length is therefore written as

$$B_L = \sqrt{B^2 - B_I^2}. \qquad (18.17)$$

To determine the instrumental broadening contribution on the HAp crystallite, HAp powder is measured, as a calibration standard, on the (0 0 2)-peak width.

6. ANALYSIS TOOLS

AutoFit is an analysis tool for the evaluation of SAXS/WAXD data, custom-made at the Max Planck Institute of Colloids and Interfaces. It contains seven subroutines, where each routine can be selected independently in the menu bar and allows calibrating and integrating the 2D data obtained by the measurements. Automatic routines allow the determination of information of interest, for example, peak fitting and calculation of T- and ρ-parameter by including correction of the background and transmission. Another analysis tool is DPDAK (Directly Programmable Data Analysis Kit). DPDAK is an online open source tool for analyzing large sequences of small-angle scattering data resulting from high-throughput scattering experiments at synchrotron radiation sources, similar to AutoFit, which has been developed in a joint collaboration between the Max Planck Institute of Colloids and Interfaces and DESY (Deutsches Elektronen-Synchrotron, Hamburg, Germany).

7. COMBINATION OF SCANNING SAXS/WAXD WITH OTHER METHODS

Combining scanning SAXS/WAXD with other methods such as XRF analysis was done, for example, in detail by Lange et al. (2011) and Li et al. (2010) and enables one to investigate the bone in even more details such as results about the mineral formation process or the structural organization of the mineral particles in the bone at different length scales.

The storage of strontium during strontium ranelate treatment in newly formed bone was investigated by Li et al. (2010). The combination of scanning X-ray scattering with fluorescence imaging shows the concentration of the strontium in newly formed bone in postmenopausal osteoporotic patients treated with strontium ranelate and the resulting increase of the lattice parameter due to the partial replacement of calcium with strontium (Fig. 18.10). Due to the small beam size of 10 μm, it was possible to differentiate between old and newly formed bones, and therefore, this study

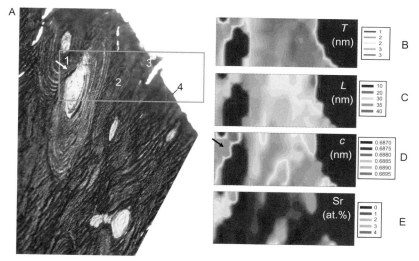

Figure 18.10 Maps of mineral crystal characteristics in osteonal bone from a postmenopausal woman treated for 36 months with strontium ranelate (A). SAXS and XRF analysis derived the mean thickness T (B) and the length L (C) of the mineral crystals. (D) Lattice spacing in the hexagonal direction of apatite shown by the parameter c. (E) Percentage of calcium ions substituted by strontium in the hydroxyapatite crystals. The red lines in the microscopy image (A) indicate regions of increased strontium content in the mineral crystals (Li et al., 2010). (See color plate.)

allowed the exact identification of strontium in the newly formed bone. The mean thickness and length of the mineral platelets were not affected by strontium ranelate treatment. They could prove no indication for a change in human bone tissue quality at the nanoscale after treatment of postmenopausal osteoporotic women with strontium ranelate except in the newly formed bone (Li et al., 2010). To obtain these specific details about strontium treatment, the method of scanning X-ray scattering in combination with XRF played a crucial role.

Lange et al. used SAXS/WAXD combined with XRF simultaneously to reveal the total calcium content in HAp crystals, in particular the mineral development in the bone of fetal and young mice (Lange et al., 2011). The authors showed that a certain fraction of calcium seems not to be part of the HAp crystals in fetal and newborn mice (Fig. 18.11). Furthermore, a complete lack of orientation in the mineral particles at the early fetal stage was revealed via SAXS, whereas 1 day after birth, particles were predominantly aligned parallel to the longitudinal bone axis, with the highest degree of alignment in the midshaft. The mineral particle thickness and length increased with age, while the mineral particles in fetal mice were thicker but much shorter. One can summarize a strong difference in size and orientation of the mineral platelets between fetal and postnatal mice (Lange et al., 2011). Scanning X-ray methods in combination with XRF enabled determination of the development of the HAp content from the fetal to the newborn state (Fig. 18.11).

Future projects are concentrating on combining simultaneous measurements of scanning SAXS/WAXD with further methods such as μCT or Raman spectroscopy. It is essential to understand the temporal and spatial characteristics of regenerated bone. Therefore, the information acquired by 2D imaging techniques such as SAXS/WAXD, XRF, or BEI in combination with 3D imaging techniques such as X-ray computed tomography and their integration can improve the understanding of the spatial organization. Furthermore, the combination of X-ray diffraction with polarized Raman spectroscopy enables the study of hierarchically organized collagen and the crystalline mineral phase at the same time.

8. CONCLUSIONS

Many biological and medical problems in bone research can be tackled by using X-ray scattering methods. X-ray scattering methods give information on the nanoscopic mineral particles, which are—together with

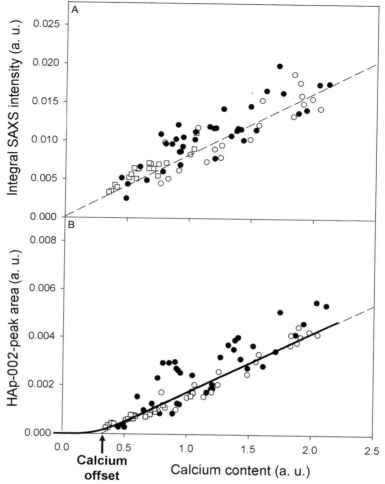

Figure 18.11 Combination of the methods SAXS, WAXD, and XRF data for two representative fetal samples (white symbols) and one postnatal sample (black symbol). (A) Integral SAXS intensity as function of the calcium content. An increase of the amount of mineral particles linearly with the calcium content is notable. (B) HAp (0 0 2)-peak area as function of the calcium content. No linear correlation between HAp and calcium is notable and a calcium offset of the fetal data is revealed via linear regression (Lange et al., 2011).

collagen—the basic building blocks in bone and therefore are fingerprints of bone formation and disease. Quantitative information from the nanometer length scale can be collected over larger areas by scanning X-ray scattering and therefore information at the micrometer level is also available.

Additionally, this allows the comparison of the results with standard procedures such as backscattered electron microscopy or other position-resolved spectroscopic methods. When used in combination with XRF, the total calcium content in HAp crystals can be revealed, in particular the mineral development as a function of age. Further work on combining scanning SAXS/WAXD with other methods such as Raman spectroscopy is in progress. This powerful tool for studies of biomineralized tissues can be provided for a wide range of materials.

REFERENCES

Arsenault, A. L., & Grynpas, M. D. (1988). Crystals in calcified epiphyseal cartilage and cortical bone in the rat. *Calcified Tissue International, 43*, 219–225.

Bacon, G. E., Bacon, P. J., & Griffiths, R. K. (1979). The orientation of apatite crystals in bone. *Journal of Applied Crystallography, 12*, 99–103.

Bonar, L., Lees, S., & Mook, H. (1985). Neutron diffraction studies of collagen in fully mineralized bone. *Journal of Molecular Biology, 181*, 265–270.

Deyhle, H., Bunk, O., & Müller, B. (2011). Nanostructure of healthy and caries-affected human teeth. *Nanomedicine: Nanotechnology, Biology and Medicine, 7*, 694–701.

Eanes, E. D., Termine, J. D., & Posner, A. S. (1967). Amorphous calcium phosphate in skeletal tissues. *Clinical Orthopaedics and Related Research, 53*, 223–235.

Fratzl, P., Fratzl-Zelman, N., Klaushofer, K., Vogl, G., & Koller, K. (1991). Nucleation and growth of mineral crystals in bone studied by small-angle X-ray scattering. *Calcified Tissue International, 48*, 407–413.

Fratzl, P., Groschner, M., Vogl, G., Plenk, H., Eschberger, J., Fratzl-Zelman, N., et al. (1992). Mineral crystals in calcified tissues: A comparative study by SAXS. *Journal of Bone and Mineral Research, 7*, 329–334.

Fratzl, P., Gupta, H. S., Paris, O., Valenta, A., Roschger, P., & Klaushofer, K. (2005). Diffracting "stack of cards"—Some thoughts about small-angle scattering from bone. *Progress in Colloid and Polymer Science, 130*, 33–39.

Fratzl, P., Gupta, H. S., Paschalis, E. P., & Roschger, P. (2004). Structure and mechanical quality of the collagen-mineral nano-composite in bone. *Journal of Materials Chemistry, 14*, 2115–2123.

Fratzl, P., Schreiber, S., Roschger, P., Lafage, M. H., Rodan, G., & Klaushofer, K. (1996). Effects of sodium fluoride and alendronate on the bone mineral in minipigs: A small-angle X-ray scattering and backscattered electron imaging study. *Journal of Bone and Mineral Research, 11*, 248–253.

Fratzl, P., & Weinkamer, R. (2007). Nature's hierarchical materials. *Progress in Materials Science, 52*, 1263–1334.

Gourrier, A., Li, C., Siegel, S., Paris, O., Roschger, P., Klaushofer, K., et al. (2010). Scanning small-angle X-ray scattering analysis of the size and organization of the mineral nanoparticles in fluorotic bone using a stack of cards model. *Journal of Applied Crystallography, 43*, 1385–1392.

Gourrier, A., Wagermaier, W., Burghammer, M., Lammie, D., Gupta, H. S., Fratzl, P., et al. (2007). Scanning X-ray imaging with small-angle scattering contrast. *Journal of Applied Crystallography, 40*, 78–82.

Kaaber, W., Daar, E., Bunk, O., Farquharson, M. J., Laklouk, A., Bailey, M., et al. (2011). Elemental and structural studies at the bone-cartilage interface. *Nuclear Instruments and Methods in Physics Research Section A, 652*, 786–790.

Karunarane, A., Boyde, A., Esapa, C. T., Hiller, J., Terrill, N. J., Brown, S. D., et al. (2013). Symmetrically reduced stiffness and increased extensibility in compression and tension at the mineralized fibrillar level in rachitic bone. *Bone, 52*, 689–698.

Landis, W. J., Hodgens, K. J., Arena, J., Song, M. J., & McEwen, B. F. (1996). Structural relations between collagen and mineral in bone as determined by high voltage electron microscopic tomography. *Microscopy Research and Technique, 33*(2), 192–202.

Lange, C., Li, C., Manjubala, I., Wagermaier, W., Kühnisch, J., Kolanczyk, M., et al. (2011). Fetal and postnatal mouse bone tissue contains more calcium than is present in hydroxyapatite. *Journal of Structural Biology, 176*, 159–167.

Lees, S., & Prostak, K. (1988). The locus of mineral crystals in bone. *Connective Tissue Research, 18*, 41–54.

Li, C., Paris, O., Siegel, S., Roschger, P., Paschalis, E. P., Klaushofer, K., et al. (2010). Strontium is incorporated into mineral crystals only in newly formed bone during strontium ranelate treatment. *Journal of Bone and Mineral Research, 25*, 968–975.

Märten, A., Fratzl, P., Paris, O., & Zaslansky, P. (2010). On the mineral in collagen of human crown dentine. *Biomaterials, 31*, 5479–5490.

Paris, O., Aichmayer, B., Al-Sawalmih, A., Li, C., Siegel, S., & Fratzl, P. (2011). Mapping lattice spacing and composition in biological materials by means of microbeam X-ray diffraction. *Advanced Engineering Materials, 13*, 784–792.

Paris, O., & Muller, M. (2003). Scanning X-ray microdiffraction of complex materials: Diffraction geometry considerations. *Nuclear Instruments and Methods in Physics Research Section B, 200*, 390–396.

Porod, G. (1951). Die Röntgenkleinwinkelstreuung von dichtgepackten kolloiden Systemen. *Colloid & Polymer Science, 124*, 83–114.

Posner, A. S., Eanes, E. D., Harper, R. A., & Zipkin, I. (1963). X-ray diffraction analysis of the effect of fluoride on human bone apatite. *Archives of Oral Biology, 8*, 549–570.

Rinnerthaler, S., Roschger, P., Jakob, H. F., Nader, A., Klaushofer, K., & Fratzl, P. (1999). Scanning small angle X-ray scattering analysis of human bone sections. *Calcified Tissue International, 64*, 422–429.

Roschger, P., Plenk, H., Jr., Klaushofer, K., & Eschberger, J. (1995). A new scanning electron microscopy approach to the quantification of bone mineral density distribution: Backscattered electron image grey-levels correlated to calcium Ka-line intensities. *Scanning Microscopy, 9*, 75–88.

Rubin, M. A., Jasiuk, I., Taylor, J., Rubin, J., Ganey, T., & Apkarian, R. P. (2003). TEM analysis of the nanostructure of normal and osteoporotic human trabecular bone. *Bone, 33*, 207–282.

Seeman, E., & Delmas, P. D. (2006). Mechanisms of disease—Bone quality—The material and structural basis of bone strength and fragility. *New England Journal of Medicine, 254*, 2250–2261.

Termine, J. D., Eanes, E. D., Greenfield, D. J., Nylen, M. U., & Harper, R. A. (1973). Hydrazine-deproteinated bone mineral. *Calcified Tissue Research, 12*, 73–90.

Weiner, S., & Addadi, L. (1997). Design strategies in mineralized biological materials. *Journal of Materials Chemistry, 7*, 689–702.

Weiner, S., & Wagner, H. D. (1998). The material bone: Structure mechanical function relations. *Annual Review of Material Science, 28*, 271–298.

Wenk, H. R., & Heidelbach, F. (1999). Crystal alignment of carbonated apatite in bone and calcified tendon: Results from quantitative texture analysis. *Bone, 24*, 361–369.

Zimmermann, E. A., Schaible, A., Bale, H., Barth, H. D., Tang, S. Y., Reichert, P., et al. (2011). Age-related changes in the plasticity and toughness of human cortical bone at multiple length scales. *Proceedings of the National Academy of Sciences of the United States of America, 108*, 14416–14421.

Zizak, I., Roschger, P., Paris, O., Misof, B. M., Berzlanovich, A., Bernstorff, S., et al. (2003). Characteristics of mineral particles in the human bone/cartilage interface. *Journal of Structural Biology, 141*, 208–217.

CHAPTER NINETEEN

Synchrotron X-Ray Nanomechanical Imaging of Mineralized Fiber Composites

Angelo Karunaratne[*,1], **Nicholas J. Terrill**[†], **Himadri S. Gupta**[*,2]

[*]Queen Mary University of London, School of Engineering and Material Sciences, London, United Kingdom
[†]Diamond Light Source Ltd., Diamond House, Harwell Science and Innovation Campus, Didcot, Oxfordshire, United Kingdom
[1]Present address: The Royal British Legion Centre for Blast Injury Studies, Department of Bioengineering, Imperial College London, London SW7 2AZ, United Kingdom
[2]Corresponding author: e-mail address: h.gupta@qmul.ac.uk

Contents

1. Introduction	416
2. Molecular and Nanoscale Strains and Stresses in a Prototypical Biomineralized Composite	419
2.1 Hierarchical structure of bone: Composition of bone at nanoscale	419
3. Relation between X-Ray Spectra and Nanomechanics	422
3.1 Idealized mineralized fibril and fibril-array structure and SAXS/WAXD pattern	422
3.2 Extractable parameters (definition)	423
4. Sample Preparation for *In Situ* Mechanical Testing with Synchrotron X-Rays	425
4.1 Required Sample geometry for mineralized fibril composites	425
4.2 Sample preparation methods	429
4.3 Mounting samples in mechanical tester	431
5. Mechanical Tester Design	435
5.1 Sample chamber	435
5.2 Tissue strain measurements	436
5.3 Porosity correction for stresses	438
5.4 Sample holders for tensile and bending	439
6. Synchrotron Setup	441
7. *In Situ* Experimental Procedure	442
7.1 Mechanical preconditioning	442
7.2 Locating region of interest	442
7.3 Synchronization between mechanical and SAXS/WAXD measurements	445
7.4 Radiation damage	445
7.5 Continuous or stepwise loading	449
7.6 Methods of scanning: Single point or multiple point scanning	449

8. Data Analysis — 450
   8.1 Data reduction programs — 450
   8.2 Estimated errors associated with sample thickness and scanning — 451
   8.3 Correction for empty cell/empty beam — 452
   8.4 SAXS radial integration — 453
   8.5 WAXD radial integration — 457
   8.6 Calculation of radially averaged intensity profile $I(\chi)$ — 458
9. Case Studies — 461
   9.1 Case study 1: Deformation of mineralized fibrils of rachitic bone — 462
   9.2 Case study 2: Fibril deformation in bending — 463
10. Model Interpretation — 464
11. Summary — 468
Acknowledgments — 469
References — 469

Abstract

In situ synchrotron X-ray scattering and diffraction, in combination with micromechanical testing, can provide quantitative information on the nanoscale mechanics of biomineralized composites, such as bone, nacre, and enamel. Due to the hierarchical architecture of these systems, the methodology for extraction of mechanical parameters at the molecular and supramolecular scale requires special considerations regarding design of mechanical test apparatus, sample preparation and testing, data analysis, and interpretation of X-ray structural information in terms of small-scale mechanics. In this chapter, this methodology is described using as a case study the deformation mechanisms at the fibrillar and mineral particle level in cortical bone. Following a description of the sample preparation, testing, and analysis procedures for bone in general, two applications of the method—to understand fibrillar-level mechanics in tension and bending in a mouse model of rachitic disease—are presented, together with a discussion of how to relate *in situ* scattering and diffraction data acquired during mechanical testing to nanostructural models for deformation of biomineralized composites.

1. INTRODUCTION

Mineral phases such as carbonated apatite or crystalline variants of calcium carbonate play an important mechanical role in biomineralized composites such abalone nacre, bone, and enamel. In these systems, the nature of the mineral structural component comprises amorphous as well as crystalline components (Mahamid, Sharir, Addadi, & Weiner, 2008). The size of the inclusions is typically of the scale of several nanometers; they are usually found in intimate association with protein layers at different structural levels (Fantner et al., 2005) and together with the organic phase form graded composite structures over nanometer, micrometer, and larger length scales (Meyers, Chen, Lin, & Seki, 2008). In the bone, for example, carbonated apatite forms a composite fibril with type I collagen of diameter ~50–200 nm, which forms plywood-like lamellae in the bone with widths ~5–10 μm that in turn form cylindrical osteons at the scale of ~100–200 μm. Due to this well-known hierarchical organization (Weiner & Wagner, 1998), the structural and mechanical properties of the mineral/protein composite at small (submicron) scales are crucial to the mechanical function of the entire organ (Zimmermann et al., 2011). While mechanical test methods at the millimeter and micron scale provide important information on the aggregate stress–strain response, they do not, in isolation, have the ability to determine the nanoscale deformation mechanics and the separate mechanical roles of the mineral and organic phases. These strains in individual mineral platelets and the shearing/plasticity set up in the protein matrix (Gupta et al., 2005) under loading are a measure of the local nanomechanical fields. The local micro-/nanomechanical environment and the changes in this environment across structures such as the shell of abalone or across lamellar deformation mechanisms determine how the material performs its overall function. Therefore, techniques to determine mechanical strains and stresses at submicron and nanometer length scales have been extensively developed in the last few years. These include *in situ* synchrotron X-ray nanomechanical probes (Gupta et al., 2006; Karunaratne, Boyde, et al., 2013; Neil Dong, Almer, & Wang, 2011; Zimmermann et al., 2011) and scanning probe microscopy and nanoindentation of submicron volumes combined with reverse finite element modeling (Tai, Dao, Suresh, Palazoglu, & Ortiz, 2007). It is the purpose of this chapter to provide a description of the methodology underlying the first of these methods—*in situ* synchrotron X-ray scattering/diffraction combined with micromechanical testing—to resolve the nanoscale mechanics of biomineralized composites.

X-ray scattering and diffraction, in conjunction with micromechanical deformation, can in principle resolve the ultrastructural (molecular 0.01–1 nm and supramolecular 1–100 nm)-level strains and (indirectly) stresses in the basic building blocks of biomineralized composites. These building blocks denote (in the bone) the individual mineral platelets (Alexander et al., 2012) and the mineralized collagen fibril (MCF) (Hang & Barber, 2011), and in abalone nacre the single tablets of aragonite in nacre (Bruet et al., 2005), or hypermineralized prisms in enamel (He & Swain, 2007). The wavelength of X-rays is commensurate with interatomic and intermolecular spacings and can hence resolve mechanically induced shifts in the packing arrangements. In mineralized composites, the magnitude, and sign of induced nanoscale deformation, and constitutive relations depend on the coupling between the mineral and organic phases at the smallest level and on the coupling between length scales and are unknown a priori. Biomineralized tissues exhibit crystalline regularity at the molecular level (mineral) and paracrystalline (Arnold et al., 1999) order at the scale of the aggregate of mineral/protein complexes like the MCF. The lattice-level and paracrystalline-level distortions induced by strain and stress can in principle be measured by shifts in the Bragg peaks in the diffraction and scattering spectra. Such applications of X-rays to measure stresses and strains in synthetic materials have a long history (Cullity & Stock, 2001) but have earlier been limited to quasistatic loading conditions, due to the long time ($\sim$ hours) to acquire single spectra with sufficient count statistics from lab sources. When combined with high photon flux at synchrotron sources, the mechanical deformation at the nanoscale can be followed dynamically (Almer & Stock, 2005, 2007; Gupta et al., 2005; Gupta et al., 2006, 2013; Karunaratne et al., 2012; Krauss et al., 2009; Stock & Almer, 2009; Yuan et al., 2011; Zimmermann et al., 2011).

However, the application of synchrotron X-ray nanomechanical imaging to the deformation of biomineralized composites requires special experimental considerations not always found in the testing of synthetic materials. The mineral platelets and inclusions are nanometer scale and also poorly crystalline (Kim et al., 2010), leading to a weak diffraction signal, increased diffuse scattering, and peak broadening (Warren, 1990). The micro- and nanoscale size (volume) of biomineralized tissue units, which are relatively uniform structurally, dictates that the test samples themselves have to be fairly small as well. As a consequence, micromechanical testers often need to be developed for accurate characterization. The relatively low X-ray intensity of the small- and wide-angle diffraction signals requires the high intensity of a synchrotron source for statistically reliable data. The need to resolve the fibrillar-level mechanics on a homogeneous tissue structure requires use of a synchrotron-scale microfocus beam as well (Paris, 2008). Deformation strains in individual components like the mineral platelets may be measurable, but combining the results into a coherent structural model of mechanical deformation requires a structural–mechanical model to be developed concurrently. The structures at the smallest scale (e.g., mineralized fibrils and mineral platelets in the bone) are highly oriented and anisotropic but aggregate into structures such as lamellae with a spread out fiber texture, and the X-ray spectra need to be analyzed in the simplest possible manner consistent with these internal structural complexities. The use of high dosages of X-ray radiation may also damage the mechanical interface between organic and inorganic components in such tissues (Barth, Launey, Macdowell, Ager, & Ritchie, 2010) and needs to be controlled when carrying out combined mechanical and X-ray scattering experiments.

With these considerations in mind, it is the purpose of this chapter to describe (i) the obtainable mechanical information from SAXS/WAXD spectra from hierarchically structured biomineralized tissues (ii) the procedures involved in sample preparation and testing, (iii) data reduction and analysis to obtain mechanical parameters such as fibrillar and molecular strain, and (iv) how to integrate these results into models for the nanoscale mechanical deformation mechanisms. As a specific example to illustrate the procedures involved in the preceding steps, we will consider the deformation mechanisms in uniaxial tension of MCFs in the bone, where considerable work has been carried out (Almer & Stock, 2005, 2007; Gupta et al., 2005, 2006, 2013; Karunaratne et al., 2012; Krauss et al., 2009; Stock & Almer, 2009; Yuan et al., 2011; Zimmermann et al., 2011).

2. MOLECULAR AND NANOSCALE STRAINS AND STRESSES IN A PROTOTYPICAL BIOMINERALIZED COMPOSITE

2.1. Hierarchical structure of bone: Composition of bone at nanoscale

In order to define the mechanical and strain parameters used later, the hierarchical structural model of our model system (MCFs of bone; reviewed in Fratzl & Weinkamer, 2007; Meyers et al., 2008; Weiner & Wagner, 1998) is presented in Fig. 19.1 and briefly discussed here. The MCFs of bone consist of poorly crystalline carbonated apatite (dahllite), type I collagen tropocollagen molecules, and water, which are surrounded by negatively charged noncollagenous proteins. The triple helical tropocollagen molecules contain ~1000 amino acids and are characterized by a glycine–X–Y repeat in amino acid sequence (where X and Y can be proline or hydroxyproline) and a helical pitch of 0.29 nm (Sasaki & Odajima, 1996b). The tropocollagen molecules assemble, together with the apatite mineral, into the MCF, which is between 50 and 200 nm in diameter (Wess, 2005). Carbonated apatite has a hexagonal close-packed (*hcp*) structure where the *c*-axis of the *hcp* lattice

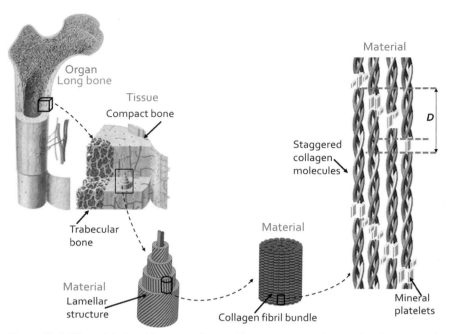

Figure 19.1 Hierarchical architecture of bone. The macroscopic (organ) level consists of two main types: cortical (compact) bone (which will be studied in this chapter) and trabecular or spongy bone. Both bone types have a lamellar structure with lamellae with widths from 5 to 10 μ, which are in turn composed of mineralized collagen fibrils ~50–200 nm in diameter. The internal structure of the mineralized fibril consists of intrafibrillar mineral (shown in a reduced and highly schematic manner described earlier). The fibril is surrounded by extrafibrillar mineral that is not shown for clarity. The triple helical tropocollagen molecules are arranged in a staggered manner, leading to an axial periodicity in electron density known as the D-period. D ~65–67 nm (Fratzl & Weinkamer, 2007; Meyers et al., 2008; Weiner & Wagner, 1998). (See color plate.)

is aligned with the fibril long axis. Mineralization can be both intra- and extrafibrillar (Alexander et al., 2012; Landis, 1996; Rubin, Rubin, & Jasiuk, 2004), but the crystalline registry of the c-axis with the fibril axis is maintained (Wagermaier et al., 2006). Fibrils form into fibril arrays at the scale of 200 nm to 1 μ, which in turn form a plywood-like arrangement (lamellae, which are ~5–10 μ thick) (Weiner, Traub, & Wagner, 1999). The lamellae form into cylindrical osteons ~200 μ in diameter or into trabeculae of similar dimensions. These form, respectively, cortical (compact) and trabecular (spongy) bone, which combine to form the entire organ. In long bones like the femur or tibia, the average orientation of fibrils is along the bone long axis in the cortical shell (Currey, 2002). The material level of bone therefore consists of a two-scale, essentially parallel fibered composite (the MCF), which is arranged into a textured plywood structure at the tissue level. The structure presented will influence the shape of the SAXS/WAXD patterns and the extraction of mechanical parameters from *in situ* SAXS/WAXD spectra, as will now be discussed.

A typical *in situ* X-ray nanomechanical imaging experiment on the fibrillar and mineral platelet deformation in the bone is shown schematically in Fig. 19.2. In brief, a specimen of mineralized tissue, prepared to cross-sectional dimensions of $\sim$100–200 µm $\times$ 200–400 µm, is mounted in a micromechanical tester inserted into a SAXS/WAXD beamline. The sample is deformed while simultaneously measuring SAXS/WAXD spectra using a 2D detector. In the following, the qualitative features of the SAXS and WAXD spectra from the MCFs, their identification with structural properties at the micromolecular, nanomolecular, and molecular scale in the bone, and the definition of nanoscale mechanical deformation parameters and their relation to alterations in the SAXS/WAXD spectra will be presented.

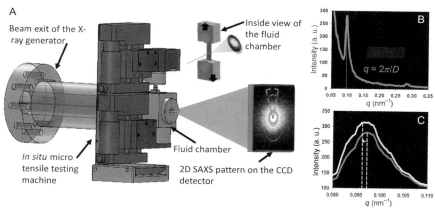

Figure 19.2 *In situ* micromechanics in a synchrotron beamline. Left: Schematic view of the microtensile testing apparatus in a synchrotron SAXS beamline. The sample is immersed in phosphate-buffered saline. Tensile load is applied along the parallel collagen fibril axis as shown in the schematic (black arrows) of MCF. (B) Integrated intensity profile in the radial directions in the angular region (red dash lines) shown in (A), showing the collagen reflections. (C) Shift of the first-order collagen peak position with increasing tissue strain, for tissue strain = 0% (time = 0) (red line) and tissue strain = 0.5% (time = 240 s) (yellow line). (See color plate.)

3. RELATION BETWEEN X-RAY SPECTRA AND NANOMECHANICS

3.1. Idealized mineralized fibril and fibril-array structure and SAXS/WAXD pattern

A well-aligned MCF (as depicted in Fig. 19.3) exhibits a series of parallel layer lines spaced by $2\pi/D$ in the SAXS pattern. The molecular-level reason for the D-stagger is related to the periodicity of hydrophobic and hydrophilic interactions along the collagen amino acid chain leading to a lateral staggered arrangement of tropocollagen chains (for details, see Fratzl, 2008). The width of the layer line in the direction perpendicular to the fibril is proportional to $1/D$, where D is the diameter of the fibril, assuming the fibril to be cylindrical in cross section (Goh et al., 2005). In most natural biomineralized tissue samples except for cases like mineralized tendons (Gupta et al., 2004), the fibrils are not perfectly parallel over the scattering volumes investigated ($\sim 10\ \mu m)^3$ for a typical synchrotron X-ray microfocus setup and larger for lab sources, but have a range of orientations characterized by a fiber orientation distribution $f(\theta)$ where θ is the angle relative to the

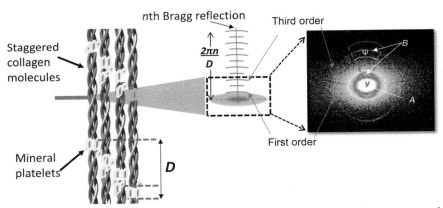

Figure 19.3 Schematic of how SAXS spectra can be linked to structural parameters of the mineralized collagen fibril. Left: X-ray beam incident on the mineralized collagen fibril scatters to form a pattern with two distinct components: the diffuse SAXS scatter from the mineral platelets (ellipse in center) and a series of Bragg diffraction peaks, extended into arced lines due to a combination of fibril orientation distribution due to the lamellar plywood structure and the finite diameter of the fibrils. Right: An experimental SAXS spectrum from murine femoral cortical bone, showing the third-order Bragg peak arising from the meridional D-periodicity and the arcing ψ due to the lamellar architecture. (See color plate.)

vertical axis. As an example, over microscale volumes, the fibril distribution in lamellar bone has been modeled as a five-sublayer plywood structure (Weiner et al., 1999) with a significant fraction of fibrils oriented longitudinally along the bone axis and adjacent sublayers at successively increasing angles to the longitudinal orientation. The angular distribution of fiber orientations leads to the parallel layer lines changing to an arced pattern as shown in Fig. 19.3. This schematic SAXS pattern holds, for example, for the case of the twisted plywood structure (Giraudguille, 1988; Weiner et al., 1999) of lamellar bone with the average fibril orientation along the vertical (as in a longitudinal osteon, Ascenzi, Bonucci, Generali, Ripamonti, & Roveri, 1979).

In most of what follows, we will make a significant approximation in analyzing the SAXS/WAXD pattern as if it were aligned parallel to the loading axis, that is, as in the schematic in Fig. 19.3 instead of the arced experimental data in the same figure. One justification for this approximation is that so far, there is much more modeling and theoretical work for the reader to refer to on the parallel fibered approximation to deformation in nanoscale mineralized fibrillar networks (Bar-On & Wagner, 2012; Buehler, 2007; Gupta et al., 2006; Ji & Gao, 2004b; Yao & Gao, 2007).

For the diffraction pattern from mineral, the WAXD peak of interest is the (0002) c-axis peak that is aligned parallel to the collagen fibril. Shifts in this peak are a measure of strain along the c-axis and also along the fibril. To satisfy the diffraction condition, the (0002) c-axis has to be inclined at half the diffraction angle 2θ ($\sim 18°$ for a X-ray beam wavelength of 0.8857 Å) to the incident X-ray beam. Therefore, the vertical (loading) axis of the micromechanical tester must be inclined to the X-ray beam by this angle, if the test specimen has its main fibril axis along the vertical direction of the tester, and this can be accomplished by a tilt or rotation stage (Gupta et al., 2006). In case the sample has a wide distribution $f(\theta)$ of fibril orientations around the vertical (loading) axis, and the angular width of $f(\theta)$ is much larger than 1/2 2θ, it may be justifiable to omit this last step if it poses problems with mounting the tester, but the width of $f(\theta)$ should ideally be calculated to justify this omission.

3.2. Extractable parameters (definition)

Fibril strain: Fibril strain denotes the percentage change in the axial staggering of tropocollagen molecules along the fibril axis on the application of external load. The axial staggering is denoted by D (Petruska & Hodge, 1964).

The reference state is taken to be the value at zero external stress. For unmineralized tissues with a long region of relatively low stress increase with increased tissue strain, defining the zero point for $\overset{\circ}{D}$ is occasionally difficult. For mineralized tissues that exhibit a rapid increase in stress with applied stress, the reference D value is not problematic to define. Physically, the fibril strain arises from a combination of molecular elongation as well as intermolecular shearing, and the relative contributions to these have been considered for tendon previously (Sasaki & Odajima, 1996a). As will be shown later, determining the collagen molecular strain in mineralized tissues is difficult with X-ray diffraction, and hence in this chapter, we will not attempt to separate the collagen elongation and shearing components, but will consider ε_F as an averaged strain at the fibril level.

Fibril orientation: Fibril orientation is the direction of the long axis of a single fibril. For a group of fibrils with multiple directions (e.g., in a lamella), the mean fibril orientation is the average of the fibril directions, weighted by the probability distribution function of the orientations.

Mineral strain: Mineral crystallite strain ε_M is defined here as the percentage change in the c-axis spacing of the *hcp* crystal lattice of the apatite phase. ε_M can be considered the strain in the MNP along the fibril axis, due to the parallel orientation of the c-axis with the fibril axis.

Mineral orientation: Similar to fibril orientation, mineral orientation denotes the main direction of the c-axis of the mineral platelets. Usually, the fibril orientation and mineral orientation should match closely. However, the degree of orientation may not match as the mineral platelets are smaller than the fibrils and can be relatively widely distributed about the single axis of the fibril even though the mean direction is along the fibril axis. Such phenomena may occur in disorders of mineralization like *osteogenesis imperfecta* (Chang, Shefelbine, & Buehler, 2012).

Types of X-ray nanomechanical experiments: The type of experiment where mechanical information is extracted with SAXS/WAXD to measure mineral or fibril deformation can be, in principle, of three types: *in situ*, scanning, and *in situ* combined with scanning. *In situ* deformation tests are where a single SAXS/WAXD spectrum is taken over a specific scattering volume at successively increasing macroscopic strain or stress levels. Scanning experiments are where a micron-sized X-ray beam is rastered across the sample and SAXS/WAXD spectra are collected at each point. *In situ* scanning experiments are where SAXS/WAXD scans are carried out at successively increasing macroscopic strain or stress levels on the mineralized tissue. The timescales for an *in situ* experiment at high-brilliance synchrotron sources are of the order of minutes or less, as each SAXS/WAXD spectrum can be taken in periods from <1 to 10 s. Timescales for deformation of each sample, even at quasistatic strain rates of 10^{-4} s^{-1}, are of the order of 1–10 min. The combination of SAXS/WAXD measurements with stepwise loading or continuous loading of samples will take of the order of a few minutes, with an additional overhead time needed to locate the sample or to move the sample with respect to the X-ray beam between macroscopic strain increments.

For tissues with negligible viscoelasticity, the SAXS/WAXD experiments may be carried out at lab SAXS/WAXD sources as well. In this case, the timescale for a single SAXS/WAXD spectrum for the sample sizes typically under consideration (scattering volumes $\sim 0.2 \times 0.4 \times 0.2$ mm^3) is about 30 min to 1 h. The total experimental time is accordingly increased to several hours, and such lab-based mineral and fibril strain measurements have been reported, for example, in Tadano, Giri, Sato, Fujisaki, and Todoh (2008).

4. SAMPLE PREPARATION FOR *IN SITU* MECHANICAL TESTING WITH SYNCHROTRON X-RAYS

4.1. Required Sample geometry for mineralized fibril composites

4.1.1 Accessible deformation geometries at fibrillar level

In order to measure and interpret ε_F and ε_M appropriately from SAXS or WAXD measurements, the accessible sample geometries for deformation need to be considered. The uniaxial (tensile or compressive) strain at the scale of the fibril, along the MCF, is determined by percentage changes of the *D*-period. The collagen molecular strain refers to the percentage change in the spacing between amino acid residues along a single chain in the tropocollagen triple helix (Sasaki & Odajima, 1996b). As shown in Fig. 19.4, the Ewald diffraction condition is satisfied for the MCF when it is vertical (or nearly so) with respect to the incident X-ray beam. The fibril orientation in the sample should therefore be in the plane perpendicular to the X-ray beam (considering also the flatness of the Ewald sphere for the low wave vectors used in SAXS). When tensile or compressive strain along the MCF is considered, the simplest geometry is where the fibril is parallel to the loading direction (vertical arrows in Fig. 19.4). The accessible deformation geometries are therefore for tensile/compressive loading along the fibril axis in mineralized collagenous tissues with predominant orientation along the loading direction.

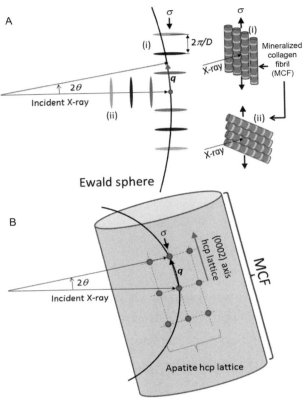

Figure 19.4 Schematic Ewald sphere construction schematic for (A) SAXS and (B) WAXD from mineralized fibrils. (A) SAXS: The sample is mounted perpendicular to the beam. 2θ denotes the Bragg angle and the X-ray is incident from left to right. Due to the low wavelengths considered, the Ewald sphere radius (inversely proportional to the wavelength) is very large and nearly flat as far as the region of interest shown is concerned. The different diffraction orders intersect with the Ewald sphere for a vertically oriented fibril (top, right upper), and deformation along the fibril axis can be captured by shifts in the peak intersections on the Ewald sphere. However, if the fibril is oriented parallel to the beam (peaks shown as horizontally arranged ellipses), then the peaks do not intersect the Ewald sphere and no detectable information on fibril mechanics can be seen from changes in the SAXS spectrum. Lastly, for fibrils oriented perpendicular to the beam but mounted horizontally, compressive stress in the vertical direction can lead to an increase in D-spacing, which will also be detectable. (B) WAXD: For wide-angle X-ray diffraction of the mineral strain in the apatite of the bone, the strain parallel to the c-axis will be detected if the sample is mounted at $1/2\ 2\theta_{(0002)}$, where $2\theta_{(0002)}$ is the Bragg angle for the (0002) reflection. Fibrils mounted vertically or at other angles will not satisfy this condition, and hence, mineral strain along the fibril axis will not be detected. (For color version of this figure, the reader is referred to the online version of this chapter.)

Due to the finite scattering volume from which the SAXS or WAXD signal arises (of the order of 10–100 µm^3 at a minimum and containing at least $\sim 10^7$ fibrils), the signal is the sum of scattering contributions from fibrils at a range of orientations. For samples of cortical bone, fibrolamellar bone, or plexiform bone (Fig. 19.1) tested along the (macroscopic) bone axis, the X-ray scattering will arise from fibrils in the twisted (and rotated) plywood lamellar (thickness $\sim$5–10 µm). The parallel layer lines in the SAXS signal (from a single fibril) will therefore be convoluted with the fibril orientation intensity distribution in the lamella. In plexiform bone and many cortical bone tissues, it is found experimentally that the peak of the orientation distribution is along the bone long axis (and sample long axis), and therefore, when tensile test specimens are loaded along the bone axis, the stress is applied along the predominant fibril orientation direction. As the lamellae are planar units, their fibrils will lie in the plane perpendicular to the X-ray beam when in the geometry shown in Fig. 19.4A, satisfying the Ewald condition, and will contribute to the SAXS signal. The fibril strain ε_F for these MCF aggregates will still approximate to the tensile strain along the main fibril direction, which is chosen as the loading direction. Alternate conditions, for example, when the fibrils are oriented mainly perpendicular to the loading direction in the transverse plane (or at some arbitrary angle in plane), can be used as well, depending on the experiment. For example, a compressive load in the vertical direction in Fig. 19.4A (right lower plot) may result in an elongation of the fibrils in the horizontal direction.

The foregoing description emphasizes that nanoscale MCF deformation of samples with an unknown or arbitrary 3D fibril orientation is not possible without a prior understanding (or measurement) of the fibril orientation, even to an approximate extent, with respect to both the X-ray beam and the loading direction. In certain conditions, for example, where the sample is prepared so that the fibrils are parallel to the beam direction, the Ewald sphere does not intersect the layer lines of the MCFs and no information on MCF deformation can be obtained from SAXS. Alternatively, if the sample contains two groups of fibrils, of which one group is in-plane and the other not, then only the first contribute to the SAXS signal. Using measured changes in the SAXS signal to infer nanoscale strains in this case will therefore be misleading as one structural component will be ignored.

4.1.2 Accessible deformation geometries at mineral platelet level
To measure the strain ε_M in the mineral nanoplatelets (MNPs) in and on the fibril, it is first necessary to assume a specific structural relation of the

crystallographic axes with the nanoscale morphology of the platelet. The average direction of the c-axis ((0002) direction of the hcp crystal structure) is taken parallel to the fibril axis (Wagermaier et al., 2006). Distortions of the hcp lattice along the c-axis therefore measure tensile strain in the MNPs along the fibril axis. To determine strain measurements in other directions perpendicular to the fibril axis is more complicated and needs structural assumptions on how the other crystal axes are oriented with respect to the fibril. Progress in measurement of strains perpendicular to the fibril direction awaits further information on the mineral crystallite orientation in the directions transverse to the fibril axis, and high-resolution TEM (Alexander et al., 2012) may provide the requisite information.

The Ewald sphere diffraction condition for the presence of the (0002) reflection in the WAXD pattern and the relation of the peak position to fibril geometry and the implications for nanoscale mechanical strain in the mineral phase requires a few more comments. For a perfectly aligned system where there is a single orientation direction for the MCF, and all the c-axes of the associated MNPs are perfectly parallel to the MCF long axis, the (0002) peak will appear in the WAXD spectrum only when the sample is inclined at an angle ½ $2\theta_{(0002)}$ to the beam direction, due to the curvature of the Ewald sphere, where $2\theta_{(0002)}$ is the Bragg angle at which the (0002) peak appears (for an incident X-ray wavelength of 0.8789 Å, $2\theta_{(0002)}$ is $\sim 25°$) as shown in Fig. 19.4B. Therefore, in order to measure the tensile strain in the MNPs, the fibrils (and sample) must be both inclined at this angle and deformed in tension or compression along it. However, in biomineralized collagenous tissues where the fibrils are oriented in a range of directions (e.g., in murine cortical bone), the intrinsic spread of fibril orientations in the sample is larger than $2\theta_{(0002)}$, and we have found experimentally that similar results are obtained for mineral strain whether one tilts the sample or not. Lastly, for *in situ* deformation using high-energy X-ray diffraction (Almer & Stock, 2006, 2007; Neil Dong et al., 2011) at energies >50 keV, the radius of the Ewald sphere is very large ($=2\pi/\lambda$ where λ is the wavelength), the curvature is small, and the considerations described earlier can be neglected.

4.1.3 Inaccessible deformation geometries

From the preceding text, it can be seen that at the MCF level, the measurable strain using SAXS is the tensile/compressive strain along the fibril axis. Strains in the transverse direction are not measurable. At the molecular level, the mineral (inorganic) tensile/compressive strain along the fibril axis can be measured similarly. In unmineralized collagen, the lateral spacing of

~1.1–1.5 nm between tropocollagen molecules results in a diffuse equatorial peak that can potentially serve as a probe of lateral strains at the intermolecular level. However, in MCFs, the diffuse scattering from the mineral, in the same equatorial direction, is much more intense and obscures the tropocollagen peak, rendering it unmeasurable for practical purposes. Transverse strains in the MNPs themselves, as discussed earlier, are not possible to interpret without experimentally based structural information of the relation of all the crystallographic axes to the macroscopic platelet morphology and anisotropic orientation of the MNP as a whole. Certain studies (Tadano et al., 2008) interpret the shifts in other crystallographic peaks (such as the (211) plane) in terms of prestrains, but this procedure will not be considered in this chapter and we refer the reader to the original papers for details.

An example of a mineralized fibril geometry, satisfying the constraints mentioned earlier, which is accessible to X-ray nanomechanical imaging using SAXS/WAXD, is bovine fibrolamellar bone (Gupta et al., 2005, 2006). The MCFs are oriented parallel to the loading direction in the average and have a spread of orientations in the scattering volume due to the lamellar plywood structure. The MNPs are oriented parallel to the fibrils, and by inclining the sample by $\frac{1}{2}\, 2\theta_{(0002)}$ to the beam direction, the mineral platelet deformation may be measured as well.

Lastly, it is noted that an implicit assumption of homogeneity of the MCF structure across the thickness of the specimen, in the scattering volume, is made in the experimental configurations shown so far. When two structurally different mineralized tissues (e.g., mineralized cartilage/subchondral bone or osteonal bone/interstitial bone) are present in close proximity, care needs to be taken that the X-ray beam does not progress from one type of tissue at the front of the sample to the other type of tissue at the back. It may be readily seen that thick specimens, where the beam diameter is much smaller than the beam-direction sample thickness, will suffer from this potential problem (the "tunnel" effect) more than thin samples. It is therefore necessary to consider the protocol for sample preparation as well, to which we now turn.

4.2. Sample preparation methods
4.2.1 Sample sectioning for large tissue sections
Coarse sectioning: For large sections of bone $>(10\text{ cm})^3$ in volume (such as the mid-diaphysis of bovine femur), initial sectioning of blocks is carried out with an automatic band saw (e.g., an Exakt cutter, EXAKT Technologies,

Oklahoma City, USA) to regular cuboidal dimensions. For smaller bones, for example, murine bones, it is more convenient to dissect out the whole bone and to prepare it for testing as described in Section 4.3.1. Here, the sample preparation for large vertebrate bone types, such as bovine and sheep bone or antler bone, is described. The initial sectioning should be carried out with a continuous water stream to avoid heat-induced damage on the surface. The target sample tissue is a rectangular sheet of bone isolated, as far as possible, from a single fibrolamellar bone packet. It is convenient to define the coordinate axes of the sample, with the longitudinal (L) direction along the bone long axis, the radial (R) direction from the center of the bone shaft to the periphery, and the tangential direction the remaining orthogonal direction parallel to the tangent (T) to the surface.

Thin sectioning: Following coarse sectioning, a low-speed diamond cutter (e.g., the Buehler IsoMet low-speed sectioning saw from Buehler, Lake Bluff, Illinois) is used to section rectangular sheets of dimensions 200 µm × 5 cm × 5 cm, with the thin direction along the R-axis. Prior to mounting, the sample should be studied under a low-power stereomicroscope to observe the orientation of the fibrolamellar bone packets with respect to the cuboidal surfaces of the sample block. If there is significant misalignment of the bone packets with respect to the block surfaces, triangular sections of Plexiglas should be mounted underneath the bone to compensate for the angular misalignment and to ensure that the saw blade cuts, as far as possible, parallel to the faces of the fibrolamellar bone packets. The angle of the Plexiglas sections should be chosen to be equal to the angle of misalignment of the fibrolamellar bone packets to the surface. A much more accurate and convenient method is by adjusting angles of an oriented goniometer-type attachment to the sample holder (available as accessories from most sectioning firms like Struers GmbH or Buehler) to align the LT-plane parallel to the blade, but not all labs may have access to such an accessory.

With plane parallel mountings available, it is desirable to dismount the sample periodically during sectioning and to check under a stereomicroscope if the section is parallel to the fibrolamellar bone packets. The thin-sectioning step reduces sample dimensions from ∼10 to ∼1–2 cm in the bovine fibrolamellar example here. The sectioning should be carried out under constant irrigation with water or saline, and the speeds should be kept at the lowest setting (typically ∼1 rpm/s for the Buehler IsoMet) to reduce damage or the chance to break off sections of bone. Lastly, for sample holders where the bone sample is held by gravity onto the blade, the additional weights used to increase the rate of cutting should be kept as low as possible to avoid damage or overheating of the bone.

Surface damage: While all sectioning and grinding methods will damage the surface of the tissue to some extent, the depth of the damaged zone is typically of the order of microns. Grinding of very hard pure ceramics like silicon nitride Si_3N_4 induces subsurface damage and stresses that peak at $\sim$5–10 μ below the surface (Sakaida, Tanaka, & Harada, 1997). It is probable that organic/inorganic composites like bone would have a smaller damaged region due to the ability to accommodate surface damage by viscoelastic mechanisms. As the measurements of fibril and mineral strain are averaged across the thickness of the entire sample ($\sim$200 μ in the example here) in the transmission mode SAXS/WAXD measurements, it is expected that the contribution to the X-ray signal from the damaged zone is negligible and will not significantly affect the results. The foregoing implies that for thinner sections (thinner than 10–20 μ), complementary measurements (such as scanning electron microscopy on fractured surfaces of samples viewed transverse to the beam direction) are needed to more carefully control for damage effects.

Methods of storage between stages: Samples may be stored at $-22\,°C$ for up to 8 months before use and at $-80\,°C$ indefinitely without damage to the organic or mineral phase. Thawing should be done in stages, that is, samples should be brought up to $4\,°C$ in a refrigerator prior to testing. Bone tissue may be stored for a week at $4\,°C$ wrapped in PBS-soaked gauze without damage to the material properties. To prevent bacterial action, it is advisable to add an antibacterial agent to the PBS, such as sodium azide (0.1 wt% sodium azide dissolved in PBS is sufficient).

4.3. Mounting samples in mechanical tester

4.3.1 Mounting semimilled sections of bone for tensile tests

For the case of mice bone sample preparation, dissected femurs are skinned, and muscle tissue removed, wrapped in gauze soaked in phosphate-buffered saline, and stored at $-20\,°C$ until mechanical testing (approximately 1 week). For mechanical testing, the distal and proximal ends of the femur are secured in dental cement (Ketac Plus, 3M ESPE, USA), as shown in Fig. 19.5, to grip the samples in the tensile testing device. A custom sample holder for potting the ends of the bones reproducibly is made from acrylonitrile butadiene styrene (ABS) molds. A push-out system is used for removing the specimens after the cement sets (Fig. 19.5). Both bony ends are coated with a bonding agent (OptiBond Solo Plus, Kerr, West Collins)

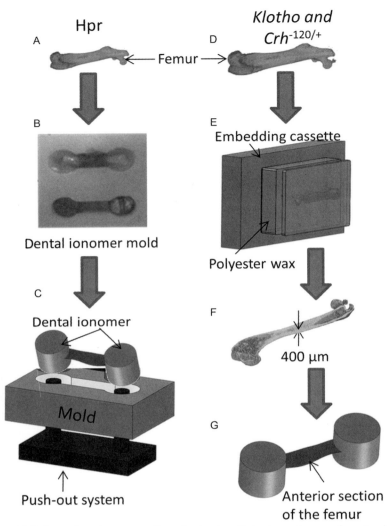

Figure 19.5 Procedure for preparation of microtensile test specimens of murine bone for synchrotron SAXS/WAXD testing. (A and D) Low-power microscope images of femurs from mice suffering from hypophosphatemic rickets (Hpr), premature aging (Klotho), and osteoporosis induced by endogenous production of glucocorticoids (Crh-120/+). (B and C) Production of samples by micromilling: (B) Bones inside the ABS mold and potted at the end with dental cement are micromilled with a Dremel drill (not shown) while mounted inside the mold, for stability, and (C) push-out schematic showing how the potted bone specimens can be removed without damaging the specimen. (E–G) Alternative sample production method where the bone is embedded in low-temperature polyester wax, sectioned with a low-speed diamond saw (not shown) while mounted inside the cassette, leading to a section of cortical bone tissue less than half the width of the bone. (G) The sectioned bone is then embedded inside dental cement molds to form the potted test specimen. (For color version of this figure, the reader is referred to the online version of this chapter.)

before placing inside the ABS molds to increase the adhesion between the bony end surface and the dental cement.

Embedded femurs are machined to form a necked region in the mid-diaphysis using a custom-made micromilling machine. The milling machine consisted of two DC linear-encoder stages (M110.1DG linear stages, Physik Instrumente, United Kingdom) connected to a computer via two motor controllers (C883 Mercury controllers, Physik Instrumente, United Kingdom). A fixed high-speed rotatory milling tool (Dremel 300 series, Dremel Inc., Uxbridge, United Kingdom) with 0.8 mm diameter cutting tool (Dremel Engraving cutter product No. 111, Dremel Inc., Uxbridge, United Kingdom) is used to mill the specimen, which is clamped in a fluid chamber mounted on the movable bidirectional linear stages. Specimens are machined from the posterior quadrants leaving a 200 μm thick anterior quadrant. The width and the thickness of each specimen are measured after milling using a Basler A101f monochrome CCD camera (Basler Vision Technologies, Ahrensburg, Germany). High-resolution (1024 × 768) images of the milled specimens are captured and transferred to the computer and dimensions were measured using ImageJ software (ImageJ, NIH, USA). The typical dimensions of the gauge regions are approximately 0.2 mm (thickness), 1.0 mm (width), and 5.0 mm (length). A selected set of samples are examined by scanning electron microscopy to check whether the machining protocol induced any damage, and no extraneous cracks, notches, or defects in the tissue were found.

4.3.2 Mounting semimilled sections of bone for cantilever tests

Mouse humeri are dissected and transversely sectioned along the mid-diaphysis into two halves (Fig. 19.6) using a diamond saw under constant irrigation with water. The proximal half of the humerus is used, and the head of the humerus is secured in dental cement using a specially made ABS mold system in order to mount the bone sample in the bending tester. The securing process involves applying a dental adhesive to the humeral head and placing it upright into the dental cement within an ABS mold, where the cement was set with a couple of minutes. If UV-curable cements are used, the set period is a few seconds, but a lead shield needs to be placed over the middle region of the sample during curing to avoid changing the material properties of the tissue during UV radiation. Using a diamond wire saw, the bone is further sectioned along the long axis (from the mid-diaphysis to the dental block surface) of the humerus. This procedure generates an elongated cantilever bending strut of ~5 mm length (L), ~200 μm thickness (b), and

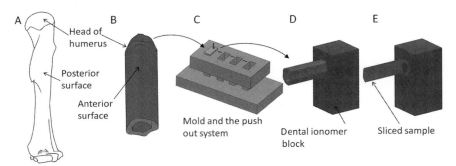

Figure 19.6 Sample preparation for bone used for *in situ* cantilever bending. The whole bone (A) is sectioned in two, (B) mounted in one end on the dental ionomer block (C), and then machined down with a Dremel drill or IsoMet low-speed saw (not shown) to form the required semiflat section for bending experiments. (For color version of this figure, the reader is referred to the online version of this chapter.)

∼500 μm in width (*d*) from the anterior quadrant of the humerus. Samples are wrapped in phosphate-buffered solution-soaked tissue paper and stored at −200 °C until required (approximately 1 week) for mechanical testing.

4.3.3 Mounting whole bones for tensile tests

The same procedure as in (i) the preceding text is followed, but the second milling step is not carried out. SAXS/WAXD measurements of fibril- and mineral strain on whole bones will involve scattering from both the front and back sides of the bone, and give rise to two peaks for the same diffraction order in both SAXS and WAXD. The difference in the sample-to-detector distance will result in split peaks, but the separation is small for SAXS. With a sample-to-detector distance of $L=1$ m (from the front of the specimen), a bone diameter of $d=2$ mm, and X-ray energy of 14 keV, the angular position of the third-order meridional *D*-period is ∼0.227°. The peak position of the signal from the same order, scattering from the back of the specimen, will be at∼0.227° × $(1 - d = 2 \text{ mm}/L = 1000 \text{ mm})$, equivalent to a shift of ∼0.0005° and a percentage change of 0.2%. This change is of the order of the shifts induced by mechanical deformation in fibril peak position (∼0.2–0.5% for bone deformation; e.g., Gupta et al., 2005; Karunaratne et al., 2012; Krauss et al., 2009). Hence, it is important that both sides of the specimen deform equivalently, in order that the measured peak shift (from the sum of the front- and backscattering signals) provides a representative image of fibril strain. For wide-angle X-ray diffraction, where sample-to-detector distances are shorter, the splitting will be larger and use of whole bones is preferably avoided.

5. MECHANICAL TESTER DESIGN
5.1. Sample chamber

Micromechanical testing machine: The mechanical testing of bone together with X-ray diffraction presents several distinct challenges that cannot be attained with conventional mechanical testing rigs (Instron™, Zwick™, or Bose Electroforce™). The device should use two oppositely driven DC linear-encoder stages (M110.1DG, Physik Instrumente, United Kingdom) to stretch the sample, ensuring that the middle of the sample remains at the same position with submicron precision. Thus, when a micron-size X-ray beam irradiates the sample, the initial scattering volume will not shift out of the beam during the tensile deformation. This enables the load-induced nanostructural changes in the same microstructural region to be followed in real time and is critical when combining micromechanics with micron-size X-ray optics in order to ensure changes in SAXS spectra, which are taken from the same scattering volume.

The grips are designed with a circular-shaped hole (6 mm in diameter and 3 mm in depth). Both ends of the femur are secured in dental cement (Filtek™ Supreme XT, 3M ESPE, USA). These dental ionomer ends have a clearance dimension of 5 mm to fit into tensile tester grips. This method of sample gripping avoids any additional forces applied on the bone sample by conventional tightening gripping systems, which may introduce stress concentrations at the grip ends.

The main requirement of the fluid chamber is to keep the bone hydrated in a condition close to the physiological environment of the bone tissue, by immersing the samples in phosphate-buffered saline solution. Furthermore, the chamber should provide a mechanism to create a narrow pathway between the X-ray inlet and outlet to minimize the absorption of X-rays in the saline. To enable this feature, the fluid chamber is designed with tubes protruding into the chamber from the front and back surfaces. These tubes are covered with Ultralene film (SPEX SamplePrep, Metuchen, NJ, USA), which contributes a negligible background X-ray scattering in the SAXS regime.

The compact size of this fluid chamber is very important to reduce the X-ray path length through the liquid (mass attenuation coefficient of the water is 5.529 cm^2/g), so for X-ray energies ~5–15 keV, significant absorption of the X-ray beam will occur when it travels through a few millimeters of water. This feature necessitates a shortening of the total path of the X-ray

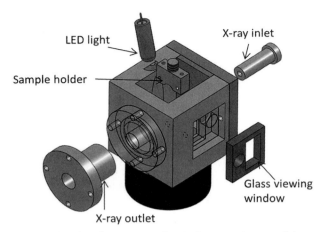

Figure 19.7 Sample chamber for micromechanical testing: image of the sample chamber to contain the test specimen inside the microtensile tester. The inlet and outlet windows, with the tunnels for minimizing water gap, are shown. The glass-viewing window on the right is for imaging the tissue strain on the bone specimens by video extensometry. A LED light is used to illuminate the specimen to highlight the optical markers more clearly. (For color version of this figure, the reader is referred to the online version of this chapter.)

beam in the phosphate-buffered solution to no more than 6 mm. The inlet and outlet windows have the diameters of 6 and 10 mm, respectively. The outlet window is designed with a larger diameter compared to inlet in order to avoid shadowing or blockage of the X-ray scattering. The fluid cell (Fig. 19.7) designed for this micromechanical tester features a viewing window 90° to the X-ray inlet in order to obtain the video extensometry images through the saline solution. A clear (transparent) view through the solution was obtained by fixing a glass slide to the viewing window. The interior of the fluid chamber is illuminated by fixing an LED light inside the fluid chamber as shown in Fig. 19.7, which increases the quality (contrast between marks and bone) of the images captured by the CCD camera.

5.2. Tissue strain measurements

If tissue strain measurements are essential for the experiment, then noncontact video extensometry of displacement of optical markers is the preferred method for tissue-level strain measurements on bone (Gupta et al., 2007).

Noncontact video extensometry: Tissue strain is measured by noncontact video extensometry by imaging the separation of two horizontal lines that are marked on the bone mid-diaphysis surface. Images were captured by a CCD camera (Basler Vision Technologies, Ahrensburg, Germany) viewing the chamber from the side (normal to the X-ray beam). These high-resolution (1024 × 768) images are later analyzed using digital image correlation software. For our work, an in-house LabVIEW/NI Vision-based digital image correlation package was used for tracking optical marker displacement to measure microstrains (Benecke, Kerschnitzki, Fratzl, & Gupta, 2009), but commercial packages like Imaris Bitplane are equally suitable.

The linear region of deformation of bone is identified by noting where the transition to the inelastic regime took place from changes in tangent modulus. For each tissue strain value, the tangent modulus was calculated over a region of 0.1% starting from that tissue strain value. The transition to the inelastic regime was defined as the point where the tangent modulus reduces 10% or less compared to its initial value (Fig. 19.8).

Inhomogeneous microlevel deformation in inelastic zone and tissue strain measurements: The tissue strain measurement method described earlier needs to be refined when considering loading in the inelastic zone for all biomineralized tissues exhibiting plastic deformation, like bone or abalone nacre. Use of two

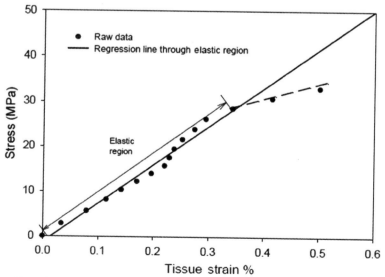

Figure 19.8 Elastic region of deformation. The elastic region of the stress–strain curve is determined experimentally by calculating the point at which the tangent modulus to the stress–strain curve reduces by 10% or more compared to the initial (low-strain) value.

markers spaced by millimeters implies that possible heterogeneities in tissue strain at the scale of 10–100 μ are not resolvable. In tensile loading of bovine fibrolamellar bone, it has been observed that in the elastic regime, the tissue strain is uniform along the loading axis, but when the macroscopic yield strain at ∼0.4% is reached, bands of high deformation with widths ∼100–200 μ appear along the loading axis of the bone (Benecke et al., 2009). For samples that are uniformly shaped along the loading axis, the axial locations of the heterogeneity appear random, although it has micro- and nanostructural origins. Therefore, depending on where the ∼10–100 μm diameter X-ray beam was located along the specimen axis, it may (or may not) coincide with the location of the heterogeneities. As a result, it will not be possible to link eventual measurements of fibril and mineral strain from this scattering volume to tissue strain measurements in the same volume. To overcome this problem, the sample should either

1. be necked in the middle with two optical markers placed very close (<1 mm) on either side of the narrowest region and the X-ray localized to this scattering volume. The disadvantage is that the resolution of the strain will be reduced as the initial gauge length separation is smaller. Further, the ability to scan the sample along the loading axis will be reduced.
2. apply a uniform speckle coating along the entire sample and use spatially resolved regions of interest (Nicolella, Moravits, Gale, Bonewald, & Lankford, 2006) to measure local tissue strain at each location. Possible heterogeneities in tissue strain will be detected with this method, which can be implemented with commercial digital image correlation packages like Imaris (Bitplane AG, Zurich).

5.3. Porosity correction for stresses

The internal porosity of compact bone tissue is mainly due to (i) osteocyte lacunae and canaliculi at the matrix (<10 μm) level and (ii) Haversian systems at scales of 100 μm–1 mm and above. Even when care is taken during sample preparation (Section 4.2) to obtain relatively homogeneous samples that avoid porosity contributions from (ii) mentioned earlier (e.g., plexiform or fibrolamellar bovine bone packets; Gupta et al., 2006) or lamellar bone from mice without Haversian systems (Karunaratne et al., 2012), it is not possible to eliminate material-level porosity contributions (from (i) mentioned earlier) for any mechanical test sample at scales of a few micron and above. The porosity of the sample affects the actual stress on the bone matrix. In an isostrain loading condition (Hull & Clyne, 1996), the stress on the bone matrix will be increased by a factor $1/(1-p)$, where $0 < p < 1$ is the porosity of the material.

Our previous work (e.g., Gupta et al., 2006; Krauss et al., 2009) did not correct for this effect, but more recently (Karunaratne, Bentley, et al., 2013), we have started to use microcomputed X-ray tomography (microCT) and backscattered electron (BSE) imaging in order to measure microstructural porosity. BSE imaging of 2D cross sections of embedded samples after *in situ* SAXS/WAXD testing can be used to estimate the 2D cross-sectional porosity, while microCT can estimate the 3D porosity. While 3D measurements are more useful than 2D, the spatial resolution of lab-based microCT (~15 μm) is lower than that of BSE (~1 μm) and hence underestimates the porosity by not including pores of smaller dimension than ~15 μm. If the sample structure is such that there is no significant variation of porosity along the mechanical loading axis, it can be justified to use the BSE technique for the higher precision obtained.

A simple way to estimate whether there is relative lack of variation is to use microCT for coarse longitudinal scans along the axis of the mechanical load and to study how porosity varies along the longitudinal direction. The porosity fraction for selected regions of interest (100 × 100 pixels each) from BSE and microCT images of the anterior cortex are determined using the plug-in BoneJ (Doube et al., 2010) for the image-processing program ImageJ (Schneider, Rasband, & Eliceiri, 2012). To convert the nominal stress (force/outer cross-sectional area) to tissue stress, the equal strain loading condition should be applied for $\sigma_T = \sigma_N/(1-p)$. We use single-factor analysis of variance (ANOVA) tests to determine if mean values of microCT porosity were statistically different at different regions along the length of the bone in murine femoral tensile test samples prepared as per Section 4.3. No statistically significant ($p > 0.05$) difference across the length was found in representative specimens. Therefore, the higher-resolution (spatial resolution 2–3 μm) BSE measurements of porosity may be used to correct the nominal stress. Such a cross comparison between qBSE and microCT measurements is recommended for the specific system or sample under consideration. In the absence of access to qBSE imaging, microCT measurements (with the implied lower resolution) need to be used to determine microstructural porosity.

5.4. Sample holders for tensile and bending

In situ cantilever bending: The cantilever bending test machine consists of one DC linear-encoder stage, 22N model 31 tension/compression load cell, and

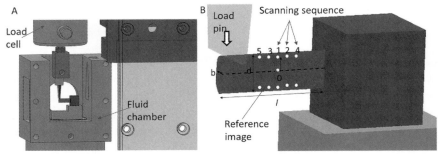

Figure 19.9 *In situ* cantilever bending with synchrotron SAXS/WAXD. (A) Experimental setup for *in situ* cantilever bending combined with microfocus SAXS (BL I22, DLS). The specimen was immersed in a fluid chamber as shown. (B) Higher magnification view of the sample inside the fluid chamber. The anterior quadrant of the humerus is shown intact with mounting block of dental cement. Arrow shows the direction of applied load. White dots denote points where SAXS spectra were taken (middle region of test specimen) for each load step; translation of the vertical mounting stage was used for this purpose. (For color version of this figure, the reader is referred to the online version of this chapter.)

a fluid chamber. As shown in Fig. 19.9, the dental block was inserted into a metal rig and tightened within the fluid chamber of the testing machine.

The loading pin with a sharp edge is set at the end of the exposed length of bone and moved at the constant deflection velocity (1 μm/s) for all samples. Specimens are loaded until fracture; load values are recorded by the load cell, and deflection measurements are obtained by tracking the movement of the linear motor at the point of the contact of load pin and the bone specimen. The bending (flexural) modulus E, bending stress σ, and tissue strain ε_T are calculated from the load deflection data using the following equations (Yuehuei & Robert, 1999):

$$E = \frac{4L^3}{bd^3}\left(\frac{F}{v}\right) \quad (19.B1)$$

$$\sigma = \frac{6FLy}{bd^3} \quad (19.B2)$$

$$\varepsilon = \frac{6FL}{Ebd^2} * y \quad (19.B3)$$

where L (mm) is the length of the bone specimen, b is the thickness, d is the width, F is the load measured from the load cell, v is the deflection movement of the load pin, and y is the distance from the neutral axis to the point where the synchrotron SAXS measurement (for the fibril strain) was carried out for each specimen.

Note that the equations mentioned earlier, based on the Euler–Bernoulli beam theory, are only valid for long slender beams with a large L/d ratio > 6 (Gere & Timoshenko, 1990). In Karunaratne, Boyde, et al. (2013), the L/d ratio of cantilever beam samples is maintained within the specified limit during the sample preparation (average L/d ratio 10.4 ± 2.8).

6. SYNCHROTRON SETUP

Mounting the mechanical stage inside a microfocus synchrotron beamline: Fig. 19.10 shows a typical setup for a microtensile stage mounted on a SAXS/WAXD beamline. Inlet and exit tubes in vacuum or with inert gases like helium can be used to reduce air scattering before and after the sample and increase the peak height relative to background, facilitating fitting and data evaluation. The space along the beam direction can be restricted due to the inlet or exit tubes to restrict air scattering. Mounting and dismounting

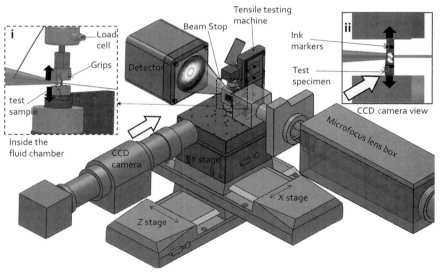

Figure 19.10 The experimental setup for *in situ* microtensile testing combined with microfocus SAXS at beam line I22 (BL I22), diamond light source (DLS). The test sample is immersed inside the fluid chamber and clamped in the grips (inset i). The CCD camera views the sample perpendicular to the incident X-ray beam so as to avoid blocking the beam (inset ii). The 2D CCD detector (Pilatus 6M) is located between 250 mm and ∼1 m from the tensile tester at a wavelength of ∼0.8857 Å (the distance shown is foreshortened for presentation purposes). Samples are replaced by laterally translating the tester out of the beam (Z-stage axis). (See color plate.)

the stage may require a translation of the stage out of the beam path followed by rotating the stage horizontally to facilitate insertion and removal of the specimen.

Detector types: 2D CCD detectors (e.g., Mar SX-165 CCD, Marresearch, Norderstedt, Germany, or Pilatus P3-2M and P2M (Henrich et al., 2009), DECTRIS Ltd, Baden, Switzerland) are the most common form for SAXS/WAXD detection of anisotropic samples at synchrotrons sources. Currently, most detectors have corrections for detector dark current, bad pixels (pixels whose intensity is abnormally high due to electronic defects), and spatial distortions done automatically. Readout times for detectors can range from 10 s (Mar CCD) down to 5 ms for the Pilatus P3 2M where the data are read into a buffer before readout. The readout time is relevant when considering the *time resolution* of the mechanical experiment, especially if strain rates above quasistatic levels ($d\varepsilon_T/dt > 10^{-4}\,\text{s}^{-1}$) are sought. With a readout time of 10 s, reducing exposure times to 1 s or so will not improve the resolution above 10 s. Detectors with much faster readout times, such as the Pilatus3 2M, do not have this limitation.

7. *IN SITU* EXPERIMENTAL PROCEDURE

7.1. Mechanical preconditioning

With the sample preparation described in Section 4.2, slack will exist in the grips of the tissue, leading to a concave upward slope in the stress–strain curve. The slack needs to be eliminated, as it leads to errors in estimation of tissue strain on the abscissa. For bone, loading to a few MPa (e.g., 10 MPa ~ 10–20% of yield stress for bone) followed by relaxation to zero stress is carried out to eliminate this effect. Two to three repeat cycles to ~10–20% of yield stress should be carried out, in general, for any biomineralized composite. If the material is brittle and has no well-defined yield point, 10–20% of ultimate tensile strength, determined on a test sample, should be used as a preloading stress (Fig. 19.11).

7.2. Locating region of interest

Prior to carrying out the *in situ* measurement, the region of interest in the sample needs to be aligned with respect to the X-ray beam. The alignment is especially important when (i) the sample is spatially inhomogeneous in the plane transverse to the beam direction and (ii) changes in the SAXS/WAXD signal between tissue regions are too small to be detected easily in the 2D

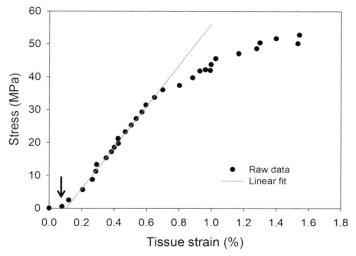

Figure 19.11 Correcting low tissue strains. To eliminate errors in tissue strain measurement from possible slack or reorientation in the grips at very low stresses, a linear extrapolation (red line) from the elastic region back to the abscissa is carried out (vertical arrow indicates intersection point), and the raw stress–strain curve shifted to the left so that the extrapolation passes through the origin. (For interpretation of the references to color in this figure legend, the reader is referred to the online version of this chapter.)

spectra. Examples of such a spatially inhomogeneous and both biologically and biomechanically important tissue are graded mineralized tissues, like the bone–cartilage interface. Further, in such cases, displacements of the sample during mechanical deformation require the initial region to be accurately located in the plane transverse to the beam direction.

Either video or X-ray transmission measurements can be used to locate the region of interest. Video (optical microscopy) is preferred because the sample is not exposed to X-rays during the alignment, minimizing radiation damage. However, many optical microscopes at synchrotron beamlines have long working distance and relatively low power and cannot image microscopic sizes of the sample well. If the sample is located inside a fluid chamber, the difficulty is compounded. The configuration of optical microscopes depends on the beamline available, for example, at the microfocus beamline ID13 at ESRF (ESRF, Grenoble), an inline microscope, and CCD camera system with a hole in the center of the optical sensor enables on-axis imaging of the sample along the beam direction. At the scattering beamline I22 at Diamond Light Source (DLS, Harwell, United Kingdom), a right-angled mirror system enables a microscope to be located perpendicular to the beam direction and to view the sample along the beam axis. If an adequate optical microscope is available, then its motor coordinates and the beam coordinates can be linked.

In absence of an adequate optical microscopy system, X-ray transmission measurements with a photodiode can be used. Exposure times of these diodes are ~0.1 s so the total X-ray dosage is relatively small, comparable to one to two 2D X-ray spectra. For a rectangular-shaped bone specimen oriented perpendicular to the beam in vertical direction, a scan in the horizontal direction is performed to find the midsection. A vertical scan is carried out to find the vertical midpoint. An example of a transmission lines scan on a murine tibia is shown in Fig. 19.12. Transmission through a 200 μ thick section of bone, compared to an empty beam, is typically ~40–60%.

However, if reduction of radiation dosage is a significant issue (e.g., due to a high-intensity microfocus source), it is possible to eliminate the pre-exposure of the sample to radiation when

i. all samples have nearly the same dimension (to within a few tens of microns) and

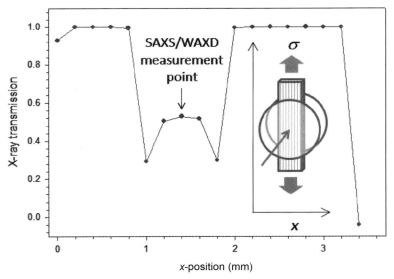

Figure 19.12 X-ray transmission scanning on bone. Image showing X-ray transmission line scan of a murine femoral sample across its width ~0.5 mm. The lower transmission at the sides is due to slight residual curvature in the midcortex of the bone after milling, leading to a higher effective thickness. SAXS and WAXD measurements are carried out in the center of the image (vertical arrow). (For color version of this figure, the reader is referred to the online version of this chapter.)

ii. the sample holder of the micromechanical tester can return reproducibly to given motor positions.

In this case, an initial mapping of a test sample can be made with the X-ray diode to get a 2D transmission map of the sample in the sample holder, and the motor stage y- and z-coordinates (Fig. 19.10) of the region of interest are noted. For the next set of samples, to be used in the actual experimental run, the x- and y-coordinates from the test sample are used without further exposure to the beam.

7.3. Synchronization between mechanical and SAXS/WAXD measurements

The time points for the SAXS/WAXD and mechanical data may be shifted relative to each other if the clocks on the respective control systems differ. A manual synchronization between the two systems is always possible, with accuracy of 1 s. For more accurate synchronization, or to do high strain rate measurements where total experimental time can be ~1–10 s, a synchronization using an on/off signal between the two systems (TTL pulse) should be implemented. The foregoing is not an issue if the control of the tester is the same as the control of the SAXS/WAXD detectors, as would be the case for a dedicated micromechanical tester at the synchrotron beamline. A ~1 s difference between the two control systems will be inevitable in a manual synchronization.

7.4. Radiation damage

Exposure to radiation, for example, gamma irradiation or X-ray radiation, can significantly alter (damage) the structure of the organic (collagen) phase of bone (Barth et al., 2010, 2011; Cornu, Banse, Docquier, Luyckx, & Delloye, 2000). The mechanical effects of this structural deterioration will affect the fibrillar and mineral platelet strain and stress measurements, unless controlled for or minimized. Here, the known effects of irradiation on bone mechanics at the macro- and nanoscale will be summarized, followed by descriptions of how to design experiments where the mechanical effects of the radiation-induced embrittlement are minimized.

Radiation dosage-level calculation: Radiation dosage levels are estimated in grays (or kilograys kGy). As described in Barth et al. (2010), the number of photons/second (n) in the X-ray beam is measured using a calibrated diode or an ionization chamber placed before the sample. The radiation flux density f is given by n/A, where A is the cross-sectional area of the beam. The energy dosage is given by $f \times e \times E_e$ (with E_e the beam energy in eV and $e = 1.6 \times 10^{-19}$ J/eV). With the transmission τ calculated as described in

Section 7.2, the rate at which the X-rays are absorbed by the bone (per unit mass M) is $R = (1-\tau) \times f \times e \times E_e/M$, and hence, the total radiation dose D is Rt, where t is the exposure time. Note that use of a microfocus beam (small A) will significantly increase dose D: going from a 100- to 10-μm diameter beam will increase the dose by a factor of 100 for the same n and t.

Mechanisms for radiation-induced embrittlement: The mechanisms of embrittlement and damage to the organic collagen matrix can be both direct and indirect. Indirectly, in hydrated bone tissue, gamma radiation ionizes water molecules, leading to reactive hydroxyl radical (*OH) formation (Dziedzic-Goclawska, Kaminski, Uhrynowska-Tyszkiewicz, & Stachowicz, 2005) from the water molecules. It has been speculated that these radicals cleave tropocollagen α-helices (Akkus, Belaney, & Das, 2005; Dziedzic-Goclawska et al., 2005; Hamer, Stockley, & Elson, 1999) of collagen. In dry tissues, the effect of radiation on the organic matrix is direct, with the ionizing radiation leading to cleavage or scission of α-helices. The collagen matrix is stabilized (structurally and mechanically) by immature (DHLNL) and mature (pyridinoline or Pyr) cross-links, and irradiation alters the cross-link profile. The fraction of mature cross-links has been shown to reduce with radiation at levels of 60–70 kGy (Barth et al., 2011; Salehpour et al., 1995). Further, an increase of the nonenzymatic AGE cross-links with irradiation has been observed (Barth et al., 2011), which is also observed in the increased brittleness of aging bone (Zimmermann et al., 2011).

Mechanical effects of radiation-induced damage: These radiation-induced molecular structural changes in the organic matrix are associated with significant reductions in the plastic deformation range and work of fracture of bone (Barth et al., 2010, 2011). When samples were exposed to varying levels of radiation prior to the *in situ* experiment, the work to fracture declined from 16.7 (0 kGy) through 5.58 kJ/m^2 (70 kGy) down to 0.22 kJ/m^2 (630 kGy) (Barth et al., 2010). In a subsequent study by the same group (Barth et al., 2011), the mechanical effects of the low-dosage regime (0–70 kGy) were investigated in more detail, combined with synchrotron SAXS/WAXD measurements of the fibril and mineral strain following irradiation. Below 35 kGy, no detectable effect on the work to fracture or strength could be detected. A progressive reduction in work of fracture was observed between 35 and 70 kGy, and above 70 kGy, no postyield deformation was observed. When *in situ* tests combined with SAXS/WAXD measurements were carried out on nonirradiated (0 kGy) versus irradiated (70 kGy), the fibril strain reduced by ∼40%, and the mineral strain by ∼20%.

7.4.1 Methods to minimize effects of radiation on the mechanical outcomes at the nanoscale

i. If the plastic deformation zone is of particular interest, the first option is to reduce radiation damage levels below dosage levels causing reduction in work to fracture in the bone. The safe level has been estimated at 30–35 kGy (Barth et al., 2011). If a single position on the bone is irradiated by the beam, then the number of exposures × exposure time/spectrum needs to be kept sufficiently low that the total radiation dosage does not exceed 30–35 kGy (Zimmermann et al., 2011). In Barth et al. (2011) and Zimmermann et al. (2011), for example, about 25 single SAXS/WAXD exposures can be collected with an exposure time of 0.5 s/frame over the strain range of bone (~2%) during quasistatic testing at ~$10^{-4}\,s^{-1}$ strain rates, similar to those used for other studies (Gupta et al., 2005, 2006). This approach is best when a macroscopic beam (with cross-sectional dimensions at least ~100–200 μ in each direction) is used, that is, where the X-ray beam averages over the entire sample cross section.

ii. The second option to reduce the exposure of a given point in the tissue to radiation is to measure SAXS/WAXD spectra at different locations in the tissue at successively increasing macroscopic strain or stress levels. For example, in uniaxial tension of a long (~5–10 mm) specimen combined with a small (microfocus) X-ray beam (diameter ~10–20 μm), an offset of ~100 μm can be made between each strain and stress increment.

In this way, each location is exposed only once to a radiation dosage. However, with this method, the mechanical and material properties (elastic modulus and work of fracture) of an exposed point will be changed from the native state. From a materials test aspect, the test piece has a progressively increasing number of small zones with different (damaged) material properties after irradiation. As a quantitative model linking fracture strain to the nanostructure of bone does not yet exist, the effect of these holes on the ultimate fracture strains is not clear yet, although it is clear they will reduce the fracture strain and ultimate work of fracture. It is possible, however, to estimate the worst-case effect of these changes on the *elastic* properties quantitatively.

To estimate the effect of material property changes at the exposed point on the elastic properties, we assume a homogeneous sample and a homogeneous strain/stress field along the length of the sample in the elastic zone (Benecke et al., 2009). Let the radiation alter the material properties of the exposed tissue at a site A, changing its elastic modulus, for example, from

E_0 to E_1, over a cross-sectional area of radius R (R is the beam diameter). The elastic strain distribution along the length of the tissue will also have been altered from the original homogeneous strain value, due to the presence of this new elastic "inclusion." For locations far away from the irradiated zone, the strain and stress will have returned to the far-field homogeneous value. The next SAXS/WAXD exposure will be at point B, a distance x from the initial point (at the next higher stress increment), and this procedure is continued (see Section 7.5). Point B should therefore be at a sufficient distance x from A that the elastic strain distribution has reached the original (homogeneous or far-field) value, so that the region B being measured does not have an altered strain and stress distribution due to the irradiated inclusion at A.

To do this, the change in the stress field due to the irradiated inclusion needs to be determined. Analytical expressions for the stress outside of a circular inclusion in a medium under plane tensile stress exist (Deryugin, Lasko, & Schmauder, 2006), and the case for a circular hole (elastic modulus of zero) is a standard example (Barber, 1992). Consider the hole case as a limiting example. In this case, the X-ray has damaged the tissue in the sample to the point of loss of all structural integrity, creating a hole. Along the loading direction, the stress σ_{rr} is (Barber, 1992):

$$\sigma_{rr} = \frac{\sigma}{2}\left(1 - \frac{a^2}{r^2}\right) + \frac{\sigma}{2}\left(1 - \frac{4a^2}{r^2} + \frac{3a^4}{r^4}\right)$$

with a the hole radius and the distance from the center of the hole as r. In this case, the stress will have reached 98% of its far-field value when measured $\sim 10a$ from the center of the hole. Therefore, if the "hole" is considered as the zone of damaged tissue, if subsequent X-ray measurements are displaced by more than $5 \times$ the beam diameter ($=100$–$200\ \mu m$ for a 10–20 μm diameter beam), the stress in the new location will be approximately the stress for a homogeneous sample. More precise estimates can be taken by using the known change in elastic modulus after 70 kGy exposure (Barth et al., 2011). By the principle of superposition, the alterations in stress distribution along the sample for subsequent measurements will be linearly superposed on the original solution. The hole case is an upper bound on the displacement needed, as irradiation at 70 kGy does not change elastic modulus significantly (Barth et al., 2010). Note that a macroscopic beam will not be suited for this purpose as the displacement by $>10 \times$ beam diameter will result in large offsets of 1 mm or greater between SAXS and WAXD spectra and the gauge length of typical test specimens is too short (~5–10 mm) to enable sufficient SAXS/WAXD points (~10 or more) to be acquired.

7.5. Continuous or stepwise loading

Depending on the extent of viscoelasticity in the specimen, stepwise loading or continuous stretching to failure may both be used to determine the nanoscale strain response to macroscopic loads. Loading needs to be combined with scanning (as discussed in Section 7.4) when using a microfocus beam, to minimize radiation damage. Both protocols assume that the nanostructure of the sample is homogeneous across the scanned region, so that fibril and mineral strain results from different spatial locations may be combined in the final analysis. In stepwise loading, the sample is loaded to specified macroscopic stress or strain increments, held for 60–300 s to complete the primary stage of stress relaxation, followed by a SAXS/WAXD measurement (Fig. 19.13). Stepwise loading is generally acceptable for biomineralized tissues like bone that exhibit relatively small ($<10\%$) viscoelastic stress relaxation. Alternatively, in continuous loading, the sample is stretched to failure at a constant rate, and simultaneously, SAXS/WAXD spectra are acquired. In order to expose a different sample location for each SAXS/WAXD spectrum, the motorized stage on which the micromechanical tester is mounted (Fig. 19.10) is translated with a constant velocity in the vertical (y-) direction at the same time. The fibril and mineral strain from each SAXS/WAXD spectrum is averaged over the tissue strain range covered during the time interval for the measurement. Hence, continuous loading may not be appropriate for samples that reach failure strain very quickly at the specified strain rate, and either the strain rate should be lowered or the mode of loading changed to stepwise loading (Fig. 19.13).

7.6. Methods of scanning: Single point or multiple point scanning

The SAXS/WAXD spectra in the *in situ* test mentioned earlier may be acquired in a continuous series (scan) or as a set of single points. In either case, it is advisable to translate the sample by distances larger than ~5–10× the beam diameter in order to reach the homogeneous stress state away from potentially damaged tissue (as discussed in Section 7.4).

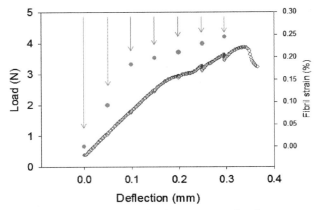

Figure 19.13 Stress and displacement plot for a stepwise loading protocol in bending of murine bone. Scanning SAXS images are acquired at the points where the load has relaxed (arrows); a 30 s relaxation increment was used (Karunaratne, Boyde, et al., 2013). The fibril strain at these points is plotted together with the load. (For color version of this figure, the reader is referred to the online version of this chapter.)

8. DATA ANALYSIS

8.1. Data reduction programs

To extract nanomechanical parameters from the 2D SAXS/WAXD spectra, freely available software such as FIT2D (Hammersley, 1997) can be used. Most commercial SAXS/WAXD systems like Bruker AXS have a system-specific method to integrate and average 2D intensity patterns. Common formats are KLORA (used at ESRF), Nexus ((hdf5) used at DLS), or .TIFF files (used at HASYLAB-DESY, now shut down). Some file formats like BSL/OTOKO contain multiple frames in a single file when a series of SAXS/WAXD spectra are acquired. For these formats, it may be more convenient in these instances to unpack the frames to individual SAXS/WAXD spectra with appropriate numbering. In FIT2D, which will be used for the examples here, the procedure to average intensity from a 2D spectrum $I(q,\chi)$ in either q or χ, to produce $I(q)$ or $I(\chi)$, is via the CAKE or INTEGRATE command (Hammersley, 1997).

Calibration: Calibration in the SAXS regime can be done with the silver behenate (AgBe) standard as well as with dry or wet rat tail tendon (D-period 65 and 67 nm, respectively). The AgBe standard is essential to determine the beam center accurately as the SAXS peaks from the tail tendon form, in two dimensions, only a relatively short arc and do not provide intensity uniformly around a ring, making center determinations more difficult. First, the center is determined using the AgBe standard (in conjunction with the Beam-Centre/Circle Coordinates command of FIT2D). Second, a radial integration (using FIT2D CAKE command) is carried out from 0.27 to 0.31 nm^{-1} with an angular width of 30° oriented along the main fiber direction of the rat tail tendon. The second-, third-, fourth-, and fifth-order meridional peaks are determined, and from the expected D-period of 65 nm (for dry tendon; 67 nm for wet tendon), the sample-to-detector distance L is calculated via

$$q = \frac{2\pi}{D} = \frac{4\pi}{\lambda}\sin\frac{2\theta}{2} \sim \frac{2\pi}{\lambda}2\theta \sim \frac{2\pi}{\lambda}\tan 2\theta = \frac{2\pi}{\lambda}\frac{x}{L}$$

where x is the peak location on the detector relative to the beam center position. The approximation $\sin 2\theta \sim 2\theta$ is valid for SAXS as 2θ is small, $\ll 1°$.

8.2. Estimated errors associated with sample thickness and scanning

Sample thickness: The relation between q and sample-to-detector distance can be used to estimate errors in q arising from finite thickness of the sample. For a specific example, let the X-ray wavelength be $\lambda = 1$ Å and the sample-to-detector distance $L = 250$ mm (measured from the front of the sample). The angular position of the (0002) c-axis peak is $2\theta = 16.8°$. If the sample thickness t is 1 mm, the equivalent Bragg angle 2θ from the scattering from the back face is slightly smaller, and the shift is given by

$$\Delta 2\theta = \frac{xt}{x^2 + L^2}$$

where x is the distance between the (0002) reflection and the beam center on the detector. Inserting numerical values gives $\Delta 2\theta / 2\theta \sim 0.0066\%$. Eventual shifts in mineral peak position (mineral strain) are of the order of 0.1–0.2% (Gupta et al., 2006; Zimmermann et al., 2011). The relative contribution is $0.0066/0.1 = 0.066$, that is, of the order of 5–10% of final mineral strain. While this is small, it is not negligible, so it is advisable to keep sample thicknesses of the order of 1 mm or less.

Sample translation: If, under deformation, the sample moves along the direction of the X-ray beam, a shift in peak position will occur, which is not induced by mechanical forces. The magnitude of the expected shift should therefore be estimated beforehand using the previous relation (where t is now the shift in sample-to-detector distance on lateral translation of the sample), in order to determine the equivalent lateral displacement that would result in this shift.

The foregoing also emphasizes the need for well-fitting sample holders and samples that are uniform in thickness along the axial direction as far as possible. In case shifts in the sample during deformation are unavoidable, it is advisable to deform the specimen in both directions, so that the center of the specimen remains relatively stationary and does not displace along the beam direction. When combining scanning SAXS/WAXD with deformation, the spectra should be measured across a relatively small area ($\sim 500 \times 500$ μm^2) in the center of the deforming zone. These corrections will be necessary for the low strains $< 0.1\%$ typical for hypermineralized structures like nacre, as well as for irregularly shaped biomineralized organs like ossicles of sea urchin.

8.3. Correction for empty cell/empty beam

The SAXS/WAXD spectrum should in principle be corrected for the additional diffuse scattering arising from the air scattering, scattering from slits or other items in the beam path, and other factors independent of the sample. To do this, the SAXS/WAXD spectrum is measured at a location in the cell without the sample (the empty-beam position). A photodiode is used to measure a signal d proportional to the X-ray beam intensity when on the sample and in the empty-beam position, and the corrected SAXS intensity is given by

$$I(q) = I_{raw}(q) - \tau I_{bgr}(q)$$

where the sample transmission $\tau = d_{sample}/d_{bgr}$ is given by the ratio of the diode reading in the sample to the diode reading in the empty-beam position.

Shifts in the third-order collagen D-stagger peak under external strains are relatively unaffected by the subtraction of an empty-beam background correction. The lack of sensitivity to the empty-beam correction is because the linear background term in the model function used to fit the peak (see Section 8.4) includes the diffuse falloff from the empty-beam intensity. In contrast, calculations of parameters from the diffuse SAXS signal, like the Porod length (Fratzl et al., 2000; Rinnerthaler et al., 1999) from the SAXS signal, do require an empty-beam correction. Nevertheless, if a functioning photodiode is available during the experiment, it should be used as described earlier to calculate the corrected $I(q)$ spectrum, in order to compare the

SAXS scattering curves from spectra taken over a period of time (over which the primary beam intensities may possibly vary). Normalization for spectra with different exposure times should be done by a scaling factor inversely proportional to the times for each spectrum.

8.4. SAXS radial integration

Calculation of azimuthally averaged intensity profile I(q): A representative $I(q)$ plot showing the third-order Bragg peak arising from the MCF collagen D-stagger is shown in Fig. 19.14. The third-order peak is the strongest for mineralized collagen after the first order. The first order is not usually used because (i) the scattering of the direct X-ray beam around the beamstop results in a strong background signal at wave vectors <0.10 nm^{-1}, complicating the ability to fit the peak accurately, and (ii) in many practical SAXS/WAXD setups, the finite size of the beamstop around $2\theta = 0°$, combined with the need to block off the divergent primary beam, means that the SAXS intensity in the wave vector range of $q < 0.10$ nm^{-1} is not easily

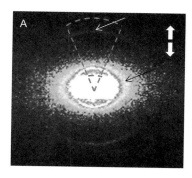
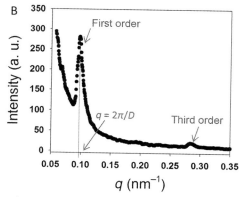

Figure 19.14 SAXS spectra from bone and regions of integration for fibril strain: (A) 2D SAXS/SAXS spectrum from murine cortical bone. Red dashed sector denotes the region rebinned in either q or χ using the FIT2D CAKE command to provide the 1D plot on right. Thick white vertical arrows indicate the direction of applied load. Black arrow shows the region of diffuse scatter arising from the SAXS scattering from mineral platelets (Rinnerthaler et al., 1999). White slanted arrows indicate the third-order meridional reflections arising from the 67 nm collagen D-periodicity; a stronger first-order reflection is visible nearer the apex of the CAKE sector but is obscured by the strong SAXS signal. (B) The azimuthally averaged $I(q)$ plot (from the indicated CAKE sector) from the 2D SAXS image in (A), showing clear diffraction peaks at the first and third orders arising from the collagen meridional D-periodicity. Note that although the third order is much weaker than the first order, the first-order peak has a much higher diffuse SAXS background associated with it. (See color plate.)

accessible. Especially in the case where microfocus X-ray optics is installed, the first order for the collagen reflection is not easily fitted due to divergence of the X-ray beam. Due to the higher-intensity of, the first order, it is easier to fit the peak position but the pixel shift with applied strain will be lower, and this trade-off needs to be considered when deciding whether to use the first or third order to determine MCF strain (Gupta et al., 2006). For the third-order peak, typical limits of integration are from $q_{min} = 0.27$ nm^{-1} to $q_{max} = 0.31$ nm^{-1} with start and end angles of 70° and 110° and the number of radial bins 200, in order to get sufficient points for a reliable fit.

The intensity $I(q)$ is fitted to a Gaussian superposed on a linear background

$$I(q) = a + bq + I_0 \exp\left(\frac{(q-q_0)^2}{2b^2}\right)$$

An example of three representative SAXS profiles at increasing levels of tissue strain is shown in Fig. 19.15. Here, the physically most important parameters are the peak center q_0 and the peak width b. Use of a linear

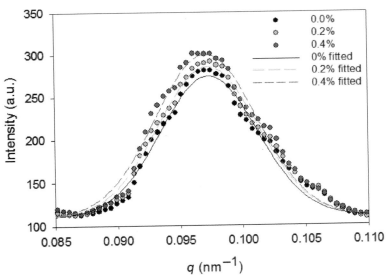

Figure 19.15 Plots of the variation of the $I(q)$ profiles in SAXS (first order of the collagen D-period reflection), after background subtraction, with applied tissue strain (legends). Both experimental data and fit lines are indicated plots for WT femoral mice bone at 0, 0.2%, and 0.4% showing fit lines and experimental data on the same plot. Arrow indicates the direction of peak shift.

background term is justified due to the restricted q-range given earlier, where the full nonlinear character of the diffuse SAXS intensity falloff does not need to be considered. As stated earlier, the nonlinear SAXS background is a greater problem at small wave vectors <0.10 nm^{-1}, and for these cases, model functions like Chebyshev polynomials or inverse fourth-order polynomials should be used (Gupta et al., 2006). In certain configurations (e.g., that at the USAXS beamline ID02 (ESRF, Grenoble)), for first-order reflections, a noticeable asymmetry in the peak *shape* itself is seen by us experimentally (Gupta et al., 2006). To account for this, an asymmetric Gaussian function (exponentially modified Gaussian or EMG, Jeansonne & Foley, 1991) can be used (Gupta et al., 2006). For the fit to the function $I(q)$, the experimentalist should aim to achieve values of R^2 for the fit range of at least 0.8 and p values <0.05. If these levels of fit accuracy are not achieved, then the experimentalist will need to either increase the exposure time, reduce the angular width of the averaging sector for $I(q)$ to avoid contributions from low-intensity tails, or increase the number of radial bins (or points) in the $I(q)$ curve. However, for a given set of experiments that are to be combined and compared to one another, the identical averaging and fitting procedure needs to be followed.

Fibril strain in MCF: The fibril strain ε_F is defined as the percentage increase in collagen axial stagger D with load and is determined from the shift in the peak position q_0 with applied stress. In terms of the collagen axial stagger D, $q_0 = 6\pi/D$ (for the third-order peak position q_0). If the peak position at an external tissue strain level ε_T is $q_0(\varepsilon_T)$, then the fibril strain is given by the relation

$$\varepsilon_F = \frac{q_0(\varepsilon_T = 0) - q_0(\varepsilon_T)}{q_0(\varepsilon_T)}$$

While fibril strain can be calculated directly from the SAXS signal, the stress on the fibril cannot be calculated, without first defining a structural model of load transfer at the fibrillar level, as in, for example, the staggered model (Gupta et al., 2006; Jager & Fratzl, 2000; Ji & Gao, 2004a). These nanoscale fiber composite models relate the stresses and strains in the fibril and interfibrillar phase to each other and to macroscopic loading conditions. These models will be discussed in more detail in Section 10, and here, we note that in certain restricted cases, fibrillar stresses may be calculated. For example, in the parallel fibered geometry of the staggered model in Gupta et al. (2006), Jager and Fratzl (2000), and Ji and Gao (2004a), the averaged stress along the fibril is equal to the macroscopic (external) tensile stress, assuming a negligible interfibrillar volume. With this simplification, the effective elastic modulus of the fibril E_F may be calculated from $E_F = \sigma_T/\varepsilon_F$, where the numerator is determined from the *in situ* testing setup and the denominator from SAXS measurements (Karunaratne, Boyde, et al., 2013).

Assumption of uniaxial vertical loading in MCFs: As previously discussed in Section 4.1.1, the MCF strains can be measured along the fibril axis only (in either tensile or compressive loading). Therefore, in the experiments under consideration here, the fibril axis direction should be along the loading axis, and the SAXS intensity needs to be averaged in a narrow angular sector on the detector centered on this axis. For simplicity, this axis has been taken as vertical here and elsewhere in the text.

We note that variants could include experiments designed to investigate the response of fibrils to off-axis loads. For example, fibrolamellar bone may be sectioned into long strips at angles of 30°, 45°, or 90° to the main bone (and fibril orientation) axis and loaded along the long axis (Seto, Gupta, Zaslansky, Wagner, & Fratzl, 2008). In such a case, the main fibril orientation will be tilted at 30°, 45°, or 90° to the vertical. The angular sector referred to earlier would then be inclined at these angles to the vertical, and the averaging of the SAXS peak will be along this inclined direction. In general, the fibril orientation needs to be determined beforehand by SAXS measurements on unstressed samples. If (within the experimental model assumptions) the tissue has its main fibril axis nearly along the vertical, and the experimental parameters of interest are the uniaxial fibrillar and mineral strains ε_F and ε_M in response to uniaxial macroscopic stress σ, the overall tilt angle of the angular sector can be taken as zero. For simplicity, we will consider this case only in the following, but for experiments designed to investigate anisotropic response, the considerations mentioned earlier need to be kept in mind.

Uniform fibril strain versus heterogeneous fibril strain in MCFs: The shift of the center of the third-order collagen reflection provides a measure of the mean fibril strain, but it is not necessary that the fibril strain is homogeneous in the scattering volume probed by SAXS/WAXD under deformation. The case of antler bone is illustrative in this regard, whereupon crossing the elastic/inelastic transition in macroscopic deformation (also denoted as the yield point), a significant broadening of the third-order peak is observable, that is, an increase in b_{MCF} (Krauss et al., 2009). Physically, such a situation corresponds to a heterogeneous deformation at the MCF level, where some fibrils are deforming to a large extent and others are deforming very little (Krauss et al., 2009). A full treatment of this effect requires a convolution of a distribution of fibril strains, but the following simpler procedure can be used to detect if there is significant heterogeneity during MCF deformation or not.

First, instrumental contributions to the width of the collagen peak need to be subtracted out. A system with perfectly one-dimensional crystalline ordering of tropocollagen molecules in the fibril is needed. The closest experimentally accessible system is the well-ordered, parallel fibered rat tail tendon (RTT) used for SAXS calibration. An azimuthally averaged $I(q)$ for RTT is calculated and the width b_{RTT} of the third-order peak determined as per the fitting procedure described earlier. The true sample width of the MCF peak b_{MCF}^{tr} is $b_{MCF}^{tr} = \sqrt{b_{MCF}^2 - b_{RTT}^2}$, by assuming both the instrumental resolution function and the true MCF peak function are Gaussian (Warren, 1990). By plotting b_{MCF}^{tr} versus macroscopic tissue strain or stress, it can be tested whether heterogeneous fibril deformation occurs to any statistically significant extent in the system under consideration. The physical interpretation of any observed increase in b_{MCF}^{tr} is that of heterogeneous MCF deformation under load, which may occur due to mechanisms like interfibrillar or interlamellar debonding.

8.5. WAXD radial integration

Calculation of azimuthally averaged intensity profile $I(q)$: The procedure for calculating $I(q)$ for mineral scattering is analogous to the SAXS case described earlier. The (0002) crystallographic peak from apatite should be integrated over a range from $\sim$17 to 20 nm^{-1} (Gupta et al., 2006) as the background is low and linear over this range (see Fig. 19.16). The model function is a Gaussian with a linear background. If peak width variations are of interest, peak width correction should be done by comparing to a powder standard of HAP (Lange et al., 2011). Examples of WAXD profiles taken from the same sample at increasing levels of tissue strain are plotted in Fig. 19.17.

Mineral strain in MNP: The mineral strain ε_M is defined as the percentage increase in the c-axis spacing of the hydroxyapatite crystal lattice of the MNP with load and is determined from the shift in the peak position $q_{HAP;0}$ with applied stress. Similar to the procedure for fibril strain, if the peak position at an external tissue strain level ε_T is $q_{HAP;0}(\varepsilon_T)$, then the mineral strain is given by the relation

$$\varepsilon_M = \frac{q_{HAP}(\varepsilon_T = 0)}{q_{HAP}(\varepsilon_T)} - 1$$

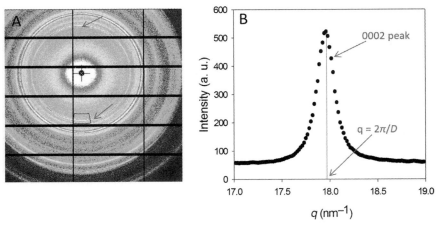

Figure 19.16 WAXD spectra from apatite phase in the bone: (A) 2D WAXD spectrum of murine femoral bone, with the oriented (0002) reflection arising from the apatite mineral indicated by red arrows. The CAKE sector used for azimuthal averaging is shown schematically as an angular sector (black line, lower part of the plot). The black rectangular lines forming a grid on the detector are to be disregarded as they are regions between active detector areas on the Pilatus 6M detector used for measurements. (B) Azimuthally averaged $I(q)$ plot for (0002) order, from the sector shown in (A). Note that the background is nearly constant. (See color plate.)

8.6. Calculation of radially averaged intensity profile $I(\chi)$
8.6.1 SAXS

The angular distribution and predominant fibril orientation of the MCFs are determined from the radially averaged intensity profile of a selected (and relatively strong) Bragg peak arising from the collagen D-stagger. For bone, the third-order peak is an appropriate choice, for similar reasons as described earlier for calculation of the azimuthally averaged intensity $I(q)$: it is stronger than other peaks except for the first order and unlike the first order does not suffer from being partially obscured by the beamstop in lab or synchrotron SAXS/WAXD setups (especially when microfocus endstations are used). However, calculation of the fibril orientation from $I(\chi)$ is complicated by the extraneous SAXS scattering from the mineral phase, which is much larger ($\sim$5–10×) than the X-ray scattering due to the axial D-stagger. The mineral SAXS scattering is anisotropic (Rinnerthaler et al., 1999). It is perpendicular to the main direction of the MCFs, as the MNPs are aligned with their thin direction (thickness) perpendicular to the fibril axis, leading to an elongated signal in the thin

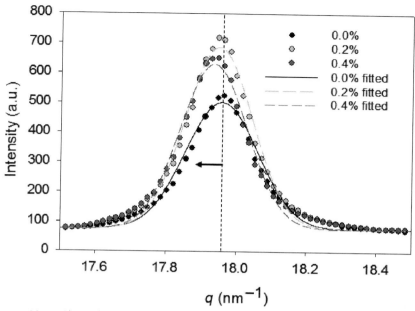

Figure 19.17 Plots of the variation of the $I(q)$ profiles in the WAXD ((0002) reflection from apatite in the bone) with applied tissue strain (legends). Both experimental data and fit lines are indicated plots for WT femoral mice bone at 0, 0.2%, and 0.4% showing fit lines and experimental data on the same plot. Arrow indicates the direction of peak shift.

direction by the principles of SAXS (Fratzl, Jakob, Rinnerthaler, Roschger, & Klaushofer, 1997). It is necessary to subtract out the dominant SAXS signal arising from the mineral.

To do this, the angular intensity distributions in three adjacent annular rings are calculated from the 2D SAXS image, as shown in Fig. 19.18. The wave vector limits for the rings are *inner* (0.27–0.28 nm^{-1}), *middle* (0.28–0.30 nm^{-1}), and *outer* (0.30–0.31 nm^{-1}). Each ring width is equal to 3 × FWHM of the central peak, so that the middle peak contains 99% of the (comparatively low) intensity due to the MCF D-stagger. The corrected $I_{\text{collagen}}(\chi) = I_{\text{middle}}(\chi) - \frac{1}{2}(I_{\text{inner}}(\chi) + I_{\text{outer}}(\chi))$. The second term is essentially an averaged background term due to the SAXS mineral scattering, if the term background is appropriate for a signal much larger than the target. In Fig. 19.18C, a comparison of $I_{\text{collagen}}(\chi)$ and $I_{\text{middle}}(\chi)$ is plotted, showing that while the statistical noise of the peaks corresponding to the MCF alignment is much larger than in the mineral SAXS peaks, two clear peaks are visible. The $I_{\text{collagen}}(\chi)$ is fitted to two Gaussians separated by 180°

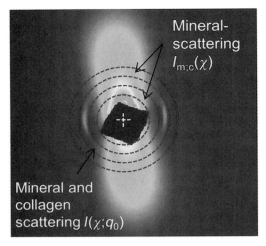
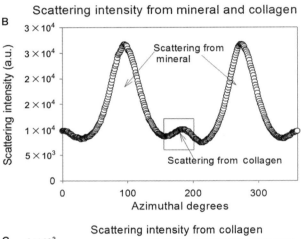
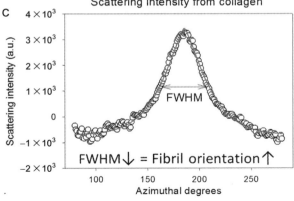

Figure 19.18 Orientation of mineralized collagen fibrils: (A) 2D SAXS plot from murine femoral bone showing the outer, middle, and inner annular rings used to calculate angular intensity distribution $I(\chi)$. (B) Integrated intensity of the middle ring showing the strong peaks arising from the mineral scattering and the much weaker peaks due to collagen meridional D-periodicity. (C) Angular intensity distribution for the peak on the left in (A) after background subtraction of inner and outer rings. Note that the peaks are now clearly visible, but a slight negative offset to the baseline has been introduced during the correction. (See color plate.)

(due to the flatness of the Ewald sphere in the SAXS regime), with zero background term (as the background has been removed in the previous step). R^2 coefficients of at least 0.80 are typical in the case of mouse cortical bone (shown in Fig. 19.18C).

8.6.2 WAXD

To determine the angular distribution for the (0002) MNP peak, a similar procedure is not necessary as there is no analogous interference from diffuse mineral scattering as in the SAXS case. A Gaussian function with a constant background

$$I_{(0002)}(\chi) = I_b + I_0 \exp\left(\frac{(\chi - \chi_0)^2}{2w^2}\right)$$

will suffice.

Analysis of large data sets and use of macros: If a large set of SAXS/WAXD spectra need to be integrated, use of the FIT2D MACRO function is helpful to perform the integrations $I(q)$ and $I(\chi)$ indicated earlier in a standard way. A macro file is a set of commands in FIT2D that can be used to generate $I(q)$ or $I(\chi)$ from a series of commands in a text file as opposed to key clicks. The method to use the MACRO command is described in Hammersley (1997). As the peak shifts in the radial (q-) direction for mineralized tissues are very low prior to macroscopic fracture (~ 0.0006 nm^{-1} for $\varepsilon_Y = 0.4$–0.6% (the typical yield point of bone)), the peak of interest will not shift out of the integration window during deformation. Therefore, the same range of integration (e.g., $q_{min} = 0.28$ nm^{-1} and $q_{max} = 0.30$ nm^{-1} for the third-order collagen peak) can be used across all SAXS/WAXD spectra taken for a single sample across varying stress levels, as well as between similar bone tissue types. For the angular positions, stress-induced shifts in peak position χ_0 are likewise small and the angular integration range can be kept constant, simplifying the construction of the macro. Scripting languages like Perl or Python can be used to construct macro files.

9. CASE STUDIES

Here, two examples of the method described in the preceding text are presented, on the deformation of mineralized fibrils in murine bone.

9.1. Case study 1: Deformation of mineralized fibrils of rachitic bone

MCF strain versus stress in a model of low bone mineralization: Fig. 19.19 shows the variation of macroscopic stress σ_T versus fibril strain ε_F for cortical bone from the femoral mid-diaphyses of a genetically modified murine model for hypophosphatemic rickets (*Hpr*; data from Karunaratne et al., 2012). Only data in the elastic (linear) regime of stress versus macroscopic strain are

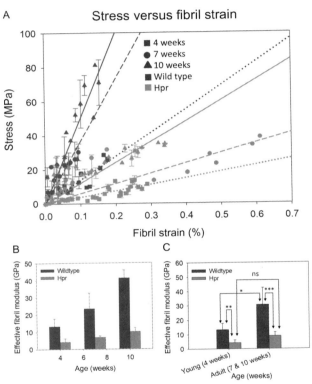

Figure 19.19 Fibrillar-level nanomechanics in rachitic versus wild-type murine femoral cortical bone in tension. (A) Consolidated plots of stress versus fibril strain for a range of rachitic and wild-type specimens (red, rachitic; blue, wild type). Symbols denote different age points: squares (4 weeks), circles (7 weeks), and triangles (10 weeks). (B) Averaged effective fibril moduli of rachitic and wild-type bone plotted as a function of age and genotype. (C) When consolidated into groups of young (4 weeks) and adult (7 and 10 weeks) specimens, they differ significantly both as a function of disease and development. *Data and figure adapted from Karunaratne et al. (2012).* (For interpretation of the references to color in this figure legend, the reader is referred to the online version of this chapter.)

shown in the plot, and stress was corrected for tissue porosity by the $1/(1-p)$ scaling factor described previously. The truncation point between elastic and inelastic regime is carried out as per the procedure stated in Fig. 19.8. While it is clear that there is considerable intragenotype variation, linear regressions between σ_T and ε_F show clear differences between wild type (WT) and *Hpr*. *Hpr* mice bone has much lower gradients $d\sigma_T/d\varepsilon_F$. $d\sigma_T/d\varepsilon_F$ is denoted as effective fibril modulus E_F and the justification for this term will be given in Section 10, following, on *Model Interpretation*. The regression coefficients for the data (R^2 from 0.589 to 0.752) are typical for the degree of linearity expected from individual data sets. To obtain statistically reliable differences between *Hpr* and WT, either the stress may be binned in steps of fibril strain (e.g., ~0.1%) to get a representative stress–fibril strain curve, or the effective fibril modulus may be averaged across genotype and compared (Fig. 19.19B and C). Due to the limited number of samples used in this study ($n=4$ at 3 weeks, $n=3$ at 7 weeks, and $n=2$ at 10 weeks), it is necessary to group the samples into young (3 weeks) and old (7 and 10 weeks) groups to carry out a reliable comparison of the means using a Student's t-test.

9.2. Case study 2: Fibril deformation in bending

MCF strain versus stress in compressive and tensile tissue zones during macroscopic bending of bone: In the second example, a combination of macroscopic beam-bending (Section 5.4) and scanning microfocus SAXS is carried out on *Hpr* bone specimens (Karunaratne, Boyde, et al., 2013). The neutral axis of cantilever-shaped section of bone (100 μm × 200 μm × 10 mm) is taken as the midpoint of the 200 μm span. Note that this assumption is an approximation, because in the bone with inhomogeneous microstructure and mineralization across the cross section, the neutral axis will be shifted upward or downward of the midpoint (Barak, Currey, Weiner, & Shahar, 2009). The bone is bent by cantilever bending in the microgrips shown in Fig. 19.9 and held at specified stress levels (0, 10, 20, and 30 MPa) during the SAXS measurements. At each SAXS measurement interval, the tensile stage is translated by ±200 μm in the vertical direction around the neutral axis, and a SAXS spectrum is first taken in the tensile zone (+200 μm above the neutral axis) followed by a SAXS spectrum taken in the compressive zone (−200 μm below the neutral axis). Stress is calculated from the beam-bending equations (Section 5.4) and fibril strain as described in Section 8.4.

Data from multiple samples are consolidated into a single plot in Fig. 19.20 and plotted with stress σ_T versus fibril strain ε_F. Effective fibril

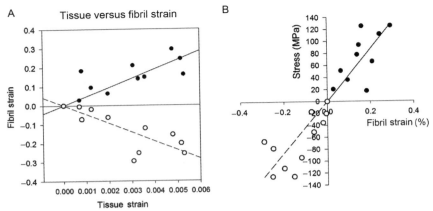

Figure 19.20 Fibrillar-level strain in cantilever bending: example data showing the variation of fibril strain versus tissue strain (A) and stress (B) in cantilever bending. Data from tensile zone are shown as filled circles and from compressive zones as open circles. Data from Karunaratne, Boyde, et al. (2013).

moduli E_F are calculated, as described earlier, for both the compressive and tensile zones, averaged across a specific age and genotype, and plotted with increasing age in Fig. 19.21. Two-way ANOVA tests are carried out to test if E_F varied significantly across age, genotype, and compressive versus tensile locations. Significant differences are found with age (E_F increases with age and tissue development) and disease type (E_F lower in *Hpr* condition), but no significant difference was observed between compressive and tensile zones.

10. MODEL INTERPRETATION

From the *in situ* tensile experiment described earlier, engineering parameters of the biomineralized composite, characteristic of the nanoscale mechanical deformation, including fibril strain ε_F, mineral strain ε_M, ratio of fibril to tissue strain $d\varepsilon_F/d\varepsilon_T$, and effective fibril modulus $E_f = d\sigma_T/d\varepsilon_F$, can be measured. To obtain a physical understanding of the meaning of these parameters in terms of nanomechanical properties, a micro- and nanomechanical structural model of the deformation needs to be developed as well (Fig. 19.22). Several such models have been developed at the molecular and mesoscale for a ranged of biomineralized composites: for example, bone (Bar-On & Wagner, 2012; Garcia, Zysset, Charlebois, & Curnier, 2009; Gupta et al., 2013; Kotha & Guzelsu, 2002), dentin (Xie, Swain, Swadener, Munroe, & Hoffman, 2009), and nacre (Evans et al., 2001).

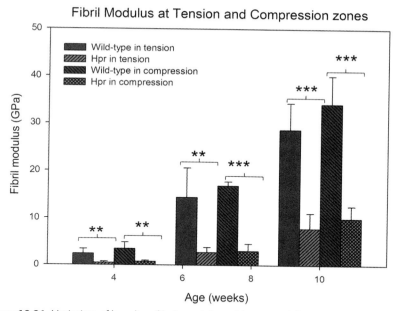

Figure 19.21 Variation of bending fibril modulus with age and disease condition: effective fibril modulus is plotted with age as a function of tension/compression, rachitic (Hpr) or wild type, and development (figure is analogous to Figure 19.19 in tension). Bars denote statistical comparisons between rachitic and wild-type specimens (*significant at $p<0.05$ level; **significant at $p<0.01$ level; and ***significant at $p<0.001$ level). *Data and figure adapted from Karunaratne, Boyde, et al. (2013). (For color version of this figure, the reader is referred to the online version of this chapter.)*

An example for the intrafibrillarly mineralized fibrils of antler bone is shown in Fig. 19.22, where the mineralized fibril array is modeled as a two-level hierarchical composite. In the model, at the molecular scale (within the fibril), collagen layers are interleaved with mineral apatite platelets. Due to the modulus mismatch between collagen and mineral, under tensile loading, the mineral platelets are loaded mainly in tension and the collagen layers in shear (Gupta et al., 2005). A similar consideration holds for the larger structural length scale of mineralized fibrils in an interfibrillar matrix. With these assumptions, force balance between the mineral platelet, collagen layers, mineralized fibril, and interfibrillar matrix can be derived, leading to relationships between nanoscale quantities like ε_M, ε_F, and E_f (determined from the *in situ* SAXS/WAXD experiments of the type described earlier) to macroscopic quantities like tissue strain ε_T and σ_T. In the model proposed in

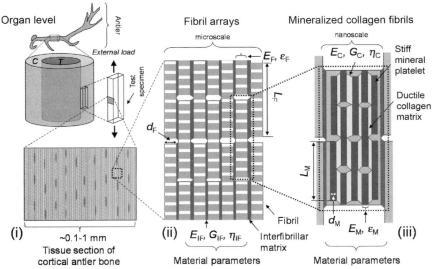

Figure 19.22 Structural model of mineralized collagen fibril deformation in antler bone. (A) Macroscopic, tissue, and lamellar level, showing specimen sectioned out of the cortical shell of antler bone. Ellipses denote osteocytes and other bone cells embedded in the extracellular matrix. (B) Fibril array level showing adjacent mineralized collagen fibrils (striations denote the 67 nm collagen D-periodicity). L_F, length of fibril; d_F, width of fibril; E_F, elastic modulus of fibril; ε_F, strain in the fibril; E_{IF}, tensile modulus of interfibrillar matrix; G_{IF}, shear modulus of interfibrillar matrix; η_{IF}, shear strain in interfibrillar matrix. (C) Intrafibrillar level showing the mineral platelets (dark gray tablets) separated by regions of high shear in the collagen matrix (very light gray, parallelograms). L_M, length of mineral platelet; d_M, width of mineral platelet; E_M, elastic modulus of mineral; ε_M, strain in the mineral; E_C, tensile modulus of collagen; G_C, shear modulus of collagen; η_C, shear strain in collagen. *After Gupta et al. (2013).*

Gupta et al. (2013), for example, the relation between fibrillar and macroscopic tissue strain is

$$\varepsilon_T = \begin{cases} \left(\dfrac{4(1-\Phi_1)}{\rho_1^2 G_{IF}\Phi_1}\left((1-\Phi_2)E_C + \dfrac{\Phi_2 E_M}{k_2}\right) + 1\right)\varepsilon_F & \varepsilon_F \leq \dfrac{k_2\rho_2}{2E_M}\tau^* \\ \left(\dfrac{4(1-\Phi_1)(1-\Phi_2)E_C}{\rho_1^2 G_{IF}\Phi_1} + 1\right)\varepsilon_F + \dfrac{2(1-\Phi_1)\Phi_2\rho_2\tau^*}{\rho_1^2 G_{IF}\Phi_1} & \varepsilon_F > \dfrac{k_2\rho_2}{2E_M}\tau^* \end{cases}$$

where the definitions of the terms are
- Φ_1: fibril volume fraction,
- Φ_2: intrafibrillar mineral volume fraction,

- E_M: mineral (apatite) elastic modulus,
- E_C: collagen elastic modulus,
- G_{IF}: interfibrillar matrix shear modulus,
- ρ_2: mineral platelet aspect ratio,
- ρ_1: fibril aspect ratio,
- k_2: ratio between fibril and mineral strain,
- τ^*: critical shear stress at the collagen/mineral interface at which debonding occurs (leading to plastic deformation).

The constitutive relation between fibril stress and fibril strain is

$$\sigma_F = \begin{cases} \left((1-\Phi_2)E_C + \dfrac{\Phi_2 E_M}{k_2} \right)\varepsilon_F & \varepsilon_F \leq \dfrac{k_2 \rho_2}{2E_M}\tau^* \\ (1-\Phi_2)E_C\varepsilon_F + \dfrac{\Phi_2 \rho_2 \tau^*}{2} & \varepsilon_F > \dfrac{k_2 \rho_2}{2E_M}\tau^* \end{cases}$$

Therefore, if the macroscopic stress–strain curve is known, and the stresses and strains on the fibrils, mineral, and collagen phase known as per the methods described earlier (e.g., in Section 8.4), the nanostructural and mechanical parameters characterizing deformation of the material may be extracted. Alternatively, these models can be used to determine unknown deformation parameters (e.g., collagen shear strain) when other deformation strains are known and the structural parameters of the material can be obtained from experiment (e.g., from high-resolution TEM) (Alexander et al., 2012). Most models (but not all, e.g., Gupta et al., 2013) consider the elastic regime only. Therefore, inelastic deformability is not included, and neither are fracture mechanisms at the nanoscale. It is necessary to confine the comparison, in these models, between experiment and theory to the elastic range.

Spatial averaging of nanomechanical and structural parameters: When comparing the strain and stresses in fibrils, mineral, or other components with experiment, the *spatially averaging* operation inherent in the SAXS/WAXD experiments needs to be kept in mind. The stresses and strains in most of the models cited earlier are not homogeneous spatially. For example, the stress is zero at the ends of the fibril and increases toward the middle (Bar-On & Wagner, 2012; Gupta et al., 2006, 2013). However, the strain measured in the SAXS/WAXD measurements are averaged across these spatially varying parameters, because of the much larger (microscale and above) scattering volume of the SAXS/WAXD beam compared to the nanometer-scale volume of the scattering constituents.

As an example, the scattering volumes in the *in situ* SAXS/WAXD experiments in Gupta et al. (2013) are $\sim$0.0016–0.016 mm^3. These volumes contain a very large number of the nanometer- to micrometer-sized MCFs (10^7–10^8 in Gupta et al., 2013), and the measured X-ray structural information is both averaged across the length of each individual fibril and averaged across all fibrils. The relevance of this point is that the models described earlier provide expressions for strains and stresses as functions of, for example, axial position of the fibril along the tensile axis. As the engineering parameters measured with *in situ* SAXS/WAXD are a volume average, the analytical expressions for moduli, stress, and strain need to be averaged across the length of the fibril before making a comparison.

For uniform tissues of the type considered before (cortical bone from the murine mid-diaphyses), further averaging is not necessary as the stress and strain fields in the models are periodic across the "unit cell"-type repeating units of fibril/interfibrillar matrix/fibril arrangement (Fig. 19.22). However, if the material itself is structurally varying in the architecture of the model across the scanned volume (e.g., if the spacing between mineral platelets, their size, or the fibril diameter or orientation changes), as may be the case near developing tissue interfaces, or at the junctions between lamellae or elsewhere, then the expressions for stress and strain change in the fibril and in other components change as well. This second averaging over the possible variants of the structure across the scattering volume must be considered as well, for a full comparison to a specific model.

11. SUMMARY

High-brilliance synchrotron X-ray radiation, combined with *in situ* micromechanical deformation, can be used to determine strain, stress, and other mechanical parameters at small scales (<100 nm) in nanostructured biomineralized composites. Here, the methodology to extract these nanomechanical parameters at the scale of $\sim$100 nm (using small-angle X-ray diffraction or SAXS) and $\sim$1 nm (using wide-angle X-ray diffraction) has been presented, using as an example the hierarchical deformation patterns in vertebral bone. Issues treated in this chapter include defining appropriate sample geometries for acquiring nanomechanical information, sample preparation and testing methodologies, and analysis and interpretation of the acquired X-ray diffraction patterns in terms of structural models.

ACKNOWLEDGMENTS

The authors gratefully acknowledge collaboration with Rajesh V. Thakker (Oxford Centre for Clinical Endocrinology and Metabolism, Nuffield Department of Medicine, Oxford University), Paul Potter and Chris Esapa (Mammalian Genetics Unit, Medical Research Council Harwell) on deformation mechanics of murine models of bone diseases, and Peter Fratzl (Max Planck Institute of Colloids and Interfaces, Potsdam) on multiscale modeling of deformation mechanisms in the bone. The results presented as examples in this chapter derive largely from these collaborations.

Funding Sources: Medical Research Council, United Kingdom; Diamond Light Source Ltd., Diamond House, Oxfordshire, United Kingdom; School of Engineering and Material Sciences, Queen Mary University of London, London, E1 4NS, United Kingdom; and Engineering and Physical Research Council (EPSRC) United Kingdom, Swindon, United Kingdom.

REFERENCES

Akkus, O., Belaney, R. M., & Das, P. (2005). Free radical scavenging alleviates the biomechanical impairment of gamma radiation sterilized bone tissue. *Journal of Orthopaedic Research, 23*, 838–845.

Alexander, B., Daulton, T. L., Genin, G. M., Lipner, J., Pasteris, J. D., Wopenka, B., et al. (2012). The nanometre-scale physiology of bone: Steric modelling and scanning transmission electron microscopy of collagen-mineral structure. *Journal of the Royal Society Interface, 9*, 1774–1786.

Almer, J. D., & Stock, S. R. (2005). Internal strains and stresses measured in cortical bone via high-energy X-ray diffraction. *Journal of Structural Biology, 152*, 14–27.

Almer, J. D., & Stock, S. R. (2006). Probing the micro-mechanical behavior of bone via high-energy X-rays. In A. Macrander et al. (Ed.) *Fifth international conference on synchrotron radiation materials science Chicago, IL: Proceedings*.

Almer, J. D., & Stock, S. R. (2007). Micromechanical response of mineral and collagen phases in bone. *Journal of Structural Biology, 157*, 365–370.

Arnold, S., Plate, U., Wiesmann, H. P., Stratmann, U., Kohl, H., & Hohling, H. J. (1999). Quantitative electron spectroscopic diffraction analyses of the crystal formation in dentine. *Journal of Microscopy-Oxford, 195*, 58–63.

Ascenzi, A., Bonucci, E., Generali, P., Ripamonti, A., & Roveri, N. (1979). Orientation of apatite in single osteon samples as studied by pole figures. *Calcified Tissue International, 29*, 101–105.

Barak, M. M., Currey, J. D., Weiner, S., & Shahar, R. (2009). Are tensile and compressive Young's moduli of compact bone different? *Journal of the Mechanical Behavior of Biomedical Materials, 2*, 51–60.

Barber, J. R. (1992). *Elasticity (solid mechanics and its applications)*. Berlin, Germany: Springer.

Bar-On, B., & Wagner, H. D. (2012). Effective moduli of multi-scale composites. *Composites Science and Technology, 72*, 566–573.

Barth, H. D., Launey, M. E., Macdowell, A. A., Ager, J. W., & Ritchie, R. O. (2010). On the effect of X-ray irradiation on the deformation and fracture behavior of human cortical bone. *Bone, 46*, 1475–1485.

Barth, H. D., Zimmermann, E. A., Schaible, E., Tang, S. Y., Alliston, T., & Ritchie, R. O. (2011). Characterization of the effects of X-ray irradiation on the hierarchical structure and mechanical properties of human cortical bone. *Biomaterials, 32*, 8892–8904.

Benecke, G., Kerschnitzki, M., Fratzl, P., & Gupta, H. S. (2009). Digital image correlation shows localized deformation bands in inelastic loading of fibrolamellar bone. *Journal of Materials Research, 24*, 421–429.

Bruet, B. J. F., QI, H. J., Boyce, M. C., Panas, R., Tai, K., Frick, L., et al. (2005). Nanoscale morphology and indentation of individual nacre tablets from the gastropod mollusc Trochus niloticus. *Journal of Materials Research, 20,* 2400–2419.

Buehler, M. J. (2007). Molecular nanomechanics of nascent bone: Fibrillar toughening by mineralization. *Nanotechnology, 18,* 295102.

Chang, S. W., Shefelbine, S. J., & Buehler, M. J. (2012). Structural and mechanical differences between collagen homo- and heterotrimers: Relevance for the molecular origin of brittle bone disease. *Biophysical Journal, 102,* 640–648.

Cornu, O., Banse, X., Docquier, P. L., Luyckx, S., & Delloye, C. (2000). Effect of freeze-drying and gamma irradiation on the mechanical properties of human cancellous bone. *Journal of Orthopaedic Research, 18,* 426–431.

Cullity, B. D., & Stock, S. R. (2001). *Elements of X-ray diffraction.* Upper Saddle River, New Jersey: Prentice Hall.

Currey, J. D. (2002). *Bones—Structure and mechanics.* Princeton, NJ: Princeton University Press.

Deryugin, Y., Lasko, G., & Schmauder, S. (2006). Field of stresses in an isotropic plane with circular inclusion under tensile stress. In *CDCM 2006—Conference on damage in composite materials 2006,* Stuttgart, Germany: NDT Net.

Doube, M., Klosowski, M. M., Arganda-Carreras, I., Cordelieres, F. P., Dougherty, R. P., Jackson, J. S., et al. (2010). BoneJ: Free and extensible bone image analysis in ImageJ. *Bone, 47,* 1076–1079.

Dziedzic-Goclawska, A., Kaminski, A., Uhrynowska-Tyszkiewicz, I., & Stachowicz, W. A. (2005). Irradiation as a safety procedure in tissue banking. *Cell and Tissue Banking, 6,* 201–219.

Evans, A. G., Suo, Z., Wang, R. Z., Aksay, I. A., He, M. Y., & Hutchinson, J. W. (2001). Model for the robust mechanical behavior of nacre. *Journal of Materials Research, 16,* 2475–2484.

Fantner, G. E., Hassenkam, T., Kindt, J. H., Weaver, J. C., Birkedal, H., Pechenik, L., et al. (2005). Sacrificial bonds and hidden length dissipate energy as mineralized fibrils separate during bone fracture. *Nature Materials, 4,* 612–616.

Fratzl, P. (2008). *Collagen: Structure and mechanics.* Berlin, Germany: Springer.

Fratzl, P., Jakob, H. F., Rinnerthaler, S., Roschger, P., & Klaushofer, K. (1997). Position-resolved small-angle X-ray scattering of complex biological materials. *Journal of Applied Crystallography, 30,* 765–769.

Fratzl, P., Paris, O., Zizak, I., Lichtenegger, H., Roschger, P., & Klaushofer, K. (2000). Analysis of the hierarchical structure of biological tissues by scanning X-ray scattering using a micro-beam. *Cellular and Molecular Biology, 46,* 993–1004.

FRATZL, P., & WEINKAMER, R. (2007). Nature's hierarchical materials. *Progress in Materials Science, 52,* 1263–1334.

Garcia, D., Zysset, P. K., Charlebois, M., & Curnier, A. (2009). A three-dimensional elastic plastic damage constitutive law for bone tissue. *Biomechanics and Modeling in Mechanobiology, 8,* 149–165.

Gere, J. M., & Timoshenko, S. P. (1990). *Mechanics of materials.* Boston, MA: PWS-KENT Publishing Company.

Giraudguille, M. M. (1988). Twisted plywood architecture of collagen fibrils in human compact-bone osteons. *Calcified Tissue International, 42,* 167–180.

Goh, K. L., Hiller, J., Haston, J. L., Holmes, D. F., Kadler, K. E., Murdoch, A., et al. (2005). Analysis of collagen fibril diameter distribution in connective tissues using small-angle X-ray scattering. *Biochimica Et Biophysica Acta—General Subjects, 1722,* 183–188.

Gupta, H. S., Fratzl, P., Kerschnitzki, M., Benecke, G., Wagermaier, W., & Kirchner, H. O. K. (2007). Evidence for an elementary process in bone plasticity with an activation enthalpy of 1 eV. *Journal of the Royal Society Interface, 4,* 277–282.

Gupta, H. S., Krauss, S., Kerschnitzki, M., Karunaratne, A., Dunlop, J. W., Barber, A. H., et al. (2013). Intrafibrillar plasticity through mineral/collagen sliding is the dominant mechanism for the extreme toughness of antler bone. *The Journal of the Mechanical Behavior of Biomedical Materials,* http://dx.doi.org/10.1016/j.jmbbm.2013.03.020 (in press).

Gupta, H. S., Messmer, P., Roschger, P., Bernstorff, S., Klaushofer, K., & Fratzl, P. (2004). Synchrotron diffraction study of deformation mechanisms in mineralized tendon. *Physical Review Letters, 93,* 158101.

Gupta, H. S., Seto, J., Wagermaier, W., Zaslansky, P., Boesecke, P., & Fratzl, P. (2006). Cooperative deformation of mineral and collagen in bone at the nanoscale. *Proceedings of the National Academy of Sciences of the United States of America, 103,* 17741–17746.

Gupta, H. S., Wagermaier, W., Zickler, G. A., Aroush, D. R. B., Funari, S. S., Roschger, P., et al. (2005). Nanoscale deformation mechanisms in bone. *Nano Letters, 5,* 2108–2111.

Hamer, A. J., Stockley, I., & Elson, R. A. (1999). Changes in allograft bone irradiated at different temperatures. *The Journal of Bone and Joint Surgery British Volume, 81B,* 342–344.

Hammersley, A. P. (1997). *FIT2D: An introduction and overviewESRF Internal Report. ESRF97HA02T.* Grenoble, France: European Synchrotron Radiation Source.

Hang, F., & Barber, A. H. (2011). Nano-mechanical properties of individual mineralized collagen fibrils from bone tissue. *Journal of the Royal Society Interface (The Royal Society), 8,* 500–505.

He, L. H., & Swain, M. V. (2007). Enamel—A "metallic-like" deformable biocomposite. *Journal of Dentistry, 35,* 431–437.

Henrich, B., Bergamaschi, A., Broennimann, C., Dinapoli, C., Eikenberry, E. F., Johnson, I., et al. (2009). PILATUS: A single photon counting pixel detector for X-ray applications. *Nuclear Instruments and Methods in Physics Research A, 607,* 247–249.

Hull, D., & Clyne, T. W. (1996). *An introduction to composite materials.* Cambridge, UK: Cambridge University Press.

Jager, I., & Fratzl, P. (2000). Mineralized collagen fibrils: A mechanical model with a staggered arrangement of mineral particles. *Biophysical Journal, 79,* 1737–1746.

Jeansonne, M. S., & Foley, J. P. (1991). Review of the exponentially modified Gaussian (EMG) function since 1983. *Journal of Chromatographic Science, 29,* 258–266.

Ji, B. H., & Gao, H. J. (2004a). Mechanical properties of nanostructure of biological materials. *Journal of the Mechanics and Physics of Solids, 52,* 1963–1990.

Ji, B. H., & Gao, H. J. (2004b). A study of fracture mechanisms in biological nano-composites via the virtual internal bond model. *Materials Science and Engineering: A Structural Materials Properties Microstructure and Processing, 366,* 96–103.

Karunaratne, A., Boyde, A., Esapa, C. T., Terrill, N. J., Davis, G. R., Brown, S. D. M., Cox, R. D., Bentley, L., Thakker, R. V., Gupta, H. S. (2013). Synchrotron nano-mechanical imaging techniques to understand how altered bone quality increases fracture risk in secondary osteoporosis. Proceedings of the ICBME 2013, IFMBE Proceedings Series, Springer (in press).

Karunaratne, A., Boyde, A., Esapa, C. T., Hiller, J., Terrill, N. J., Brown, S. D. M., et al. (2013). Symmetrically reduced stiffness and increased extensibility in compression and tension at the mineralized fibrillar level in rachitic bone. *Bone, 52,* 689–698.

Karunaratne, A., Esapa, C. R., Hiller, J., Boyde, A., Head, R., Bassett, J. H. D., et al. (2012). Significant deterioration in nanomechanical quality occurs through incomplete extrafibrillar mineralization in rachitic bone: Evidence from in-situ synchrotron X-ray scattering and backscattered electron imaging. *Journal of Bone and Mineral Research, 27,* 876–890.

Kim, Y. K., Gu, L. S., Bryan, T. E., Kim, J. R., Chen, L., Liu, Y., et al. (2010). Mineralisation of reconstituted collagen using polyvinylphosphonic acid/polyacrylic acid templating matrix protein analogues in the presence of calcium, phosphate and hydroxyl ions. *Biomaterials, 31,* 6618–6627.

Kotha, S. P., & Guzelsu, N. (2002). Modeling the tensile mechanical behavior of bone along the longitudinal direction. *Journal of Theoretical Biology, 219*, 269–279.

Krauss, S., Fratzl, P., Seto, J., Currey, J. D., Estevez, J. A., Funari, S. S., et al. (2009). Inhomogeneous fibril stretching in antler starts after macroscopic yielding: Indication for a nanoscale toughening mechanism. *Bone, 44*, 1105–1110.

Landis, W. J. (1996). Mineral characterization in calcifying tissues: Atomic, molecular and macromolecular perspectives. *Connective Tissue Research, 35*, 1–8.

Lange, C., Li, C., Manjubala, I., Wagermaier, W., Kuhnisch, J., Kolanczyk, M., et al. (2011). Fetal and postnatal mouse bone tissue contains more calcium than is present in hydroxyapatite. *Journal of Structural Biology, 176*, 159–167.

Mahamid, J., Sharir, A., Addadi, L., & Weiner, S. (2008). Amorphous calcium phosphate is a major component of the forming fin bones of zebrafish: Indications for an amorphous precursor phase. *Proceedings of the National Academy of Sciences of the United States of America, 105*, 12748–12753.

Meyers, M. A., Chen, P. Y., Lin, A. Y. M., & Seki, Y. (2008). Biological materials: Structure and mechanical properties. *Progress in Materials Science, 53*, 1–206.

Neil Dong, X., Almer, J. D., & Wang, X. (2011). Post-yield nanomechanics of human cortical bone in compression using synchrotron X-ray scattering techniques. *Journal of Biomechanics, 44*, 676–682.

Nicolella, D. P., Moravits, D. E., Gale, A. M., Bonewald, L. F., & Lankford, J. (2006). Osteocyte lacunae tissue strain in cortical bone. *Journal of Biomechanics, 39*, 1735–1743.

Paris, O. (2008). From diffraction to imaging: New avenues in studying hierarchical biological tissues with X-ray microbeams (Review). *Biointerphases, 3*, Fb16–Fb26.

Petruska, J. A., & Hodge, A. J. (1964). A subunit model for the tropocollagen macromolecule. *Proceedings of the National Academy of Sciences of the United States of America, 51*, 6.

Rinnerthaler, S., Roschger, P., Jakob, H. F., Nader, A., Klaushofer, K., & Fratzl, P. (1999). Scanning small angle X-ray scattering analysis of human bone sections. *Calcified Tissue International, 64*, 422–429.

Rubin, M. A., Rubin, J., & Jasiuk, W. (2004). SEM and TEM study of the hierarchical structure of C57BL/6 J and C3H/HeJ mice trabecular bone. *Bone, 35*, 11–20.

Sakaida, Y., Tanaka, K., & Harada, S. (1997). Measurement of residual stress distribution of ground silicon nitride by glancing incidence X-ray diffraction technique. In J. V. Gilfrich & I. Cev Noyan (Eds.), *Advances in X-ray analysis: Proceedings of the forty-fourth annual conference on applications of X-ray analysis*. New York, NY: Plenum Press.

Salehpour, A., Butler, D. L., Proch, E., Schwartz, H. E., Feder, S. M., Doxey, C. M., et al. (1995). Dose-dependent response of gamma irradiation on mechanical properties and related biochemical composition of coat bone-patellar tendon-bone allografts. *Journal of Orthopaedic Research, 13*, 898–906.

Sasaki, N., & Odajima, S. (1996a). Elongation mechanism of collagen fibrils and force-strain relations of tendon at each level of structural hierarchy. *Journal of Biomechanics, 29*, 1131–1136.

Sasaki, N., & Odajima, S. (1996b). Stress-strain curve and Young's modulus of a collagen molecule as determined by the X-ray diffraction technique. *Journal of Biomechanics, 29*, 655–658.

Schneider, C. A., Rasband, W. S., & Eliceiri, K. W. (2012). NIH Image to ImageJ: 25 years of image analysis. *Nature Methods, 9*, 671–675.

Seto, J., Gupta, H. S., Zaslansky, P., Wagner, H. D., & Fratzl, P. (2008). Tough lessons from bone: Extreme mechanical anisotropy at the mesoscale. *Advanced Functional Materials, 18*, 1905–1911.

Stock, S. R., & Almer, J. D. (2009). Strains in bone and tooth via high energy X-ray scattering. *Bone, 44*, S270.

Tadano, S., Giri, B., Sato, T., Fujisaki, K., & Todoh, M. (2008). Estimating nanoscale deformation in bone by X-ray diffraction imaging method. *Journal of Biomechanics, 41,* 945–952.

Tai, K., Dao, M., Suresh, S., Palazoglu, A., & Ortiz, C. (2007). Nanoscale heterogeneity promotes energy dissipation in bone. *Nature Materials, 6,* 454–462.

Wagermaier, W., Gupta, H. S., Gourrier, A., BURGHAMMER, M., ROSCHGER, P., & Fratzl, P. (2006). Spiral twisting of fiber orientation inside bone lamellae. *Biointerphases, 1,* 1–5.

Warren, B. E. (1990). *X-ray diffraction.* New York, NY: Dover.

Weiner, S., Traub, W., & Wagner, H. D. (1999). Lamellar bone: Structure-function relations. *Journal of Structural Biology, 126,* 241–255.

Weiner, S., & Wagner, H. D. (1998). The material bone: Structure mechanical function relations. *Annual Review of Materials Science, 28,* 271–298.

Wess, T. J. (2005). Collagen fibrillar form and function. In D. Parry & J. Squire (Eds.), *Fibrous proteins: Coiled-coils, collagen and elastomers.* Burlington, MA: Elsevier Inc.

Xie, Z. H., Swain, M. V., Swadener, G., Munroe, P., & Hoffman, M. (2009). Effect of microstructure upon elastic behaviour of human tooth enamel. *Journal of Biomechanics, 42,* 1075–1080.

Yao, H. M., & Gao, H. J. (2007). Multi-scale cohesive laws in hierarchical materials. *International Journal of Solids and Structures, 44,* 8177–8193.

Yuan, F., Stock, S. R., Haeffner, D. R., Almer, J. D., Dunand, D. C., & Brinson, L. C. (2011). A new model to simulate the elastic properties of mineralized collagen fibril. *Biomechanics and Modeling in Mechanobiology, 10,* 147–160.

Yuehuei, H., & Robert, A. (1999). *Mechanical testing of bone and the bone-implant interface.* Boca Raton, FL: CRC Press.

Zimmermann, E. A., Schaible, E., Bale, H., Barth, H. D., Tang, S. Y., Reichert, P., et al. (2011). Age-related changes in the plasticity and toughness of human cortical bone at multiple length scales. *Proceedings of the National Academy of Sciences of the United States of America, 109*(35), 14416–14421.

SECTION VI

Mapping Mineral Chemistry

CHAPTER TWENTY

Application of Total X-Ray Scattering Methods and Pair Distribution Function Analysis for Study of Structure of Biominerals

Richard J. Reeder[*,1], F. Marc Michel[†]
[*]Department of Geosciences, Stony Brook University, Stony Brook, New York, USA
[†]Department of Geosciences, Virginia Tech, Blacksburg, Virginia, USA
[1]Corresponding author: e-mail address: rjreeder@stonybrook.edu

Contents

1. Introduction	478
2. Total Scattering Methodology	479
2.1 Experimental geometry and instrument calibration	481
2.2 Sample requirements and *in situ* methods	482
2.3 Data collection times	483
2.4 Background subtraction and data processing	484
3. The PDF	486
4. Extracting Information from the PDF	486
4.1 PDF analysis approaches	486
5. PDF Studies of Biominerals	492
5.1 Ferritin-derived nanocrystalline ferrihydrite	492
5.2 Amorphous calcium carbonate	495
6. Summary	496
Links	498
Acknowledgments	498
References	498

Abstract

Total X-ray scattering and pair distribution function (PDF) analysis, using a high-energy synchrotron source, allow direct study of the short- and intermediate-range structure that distinguish amorphous, structurally disordered, and nanocrystalline biominerals. For such samples in which diffuse scatter is a significant component, care must be taken in the experimental procedures to optimize data quality and extract the useful signal necessary to calculate the PDF. General methods are described for data collection and processing, including commonly used software programs. Methods for analysis and interpretation of PDFs are presented, including direct real-space refinement and reverse Monte Carlo methods. Greater application of PDFs to amorphous and poorly crystallized biominerals will provide new insight into structure, especially over length scales that are not probed by other techniques. The rapid data collection available at synchrotron facilities also allows *in situ* kinetic studies of reactions involving biominerals.

1. INTRODUCTION

Structure is one of the most basic properties of a solid. The arrangement and type of atoms determine the stability, mechanical behavior, and reactivity of a material and usually provide fundamental constraints on reaction pathways. Indeed, structure–function–reactivity relationships have long figured as a central paradigm of structural chemistry. This concept extends equally to surfaces, where structure—rarely a simple truncation of the bulk atomic arrangement—plays a key role in interfacial processes. For minerals, as well as the vast majority of solids, determination of atomic structure has relied mainly on X-ray diffraction (XRD) and to a lesser extent on neutron diffraction, using powders and single crystals.

Yet, understanding structure of biominerals has proved to be more challenging than for most inorganic minerals. Chief among the reasons for this are the conditions under which biominerals typically form. Sites of biomineral formation range from specialized organelles within cells to locations proximal to organisms, commonly in severely constrained volumes under saturation conditions far from equilibrium. Precipitation may be extremely rapid, resulting in metastable phases that are commonly disordered, amorphous, or nanocrystalline. Much attention has focused recently on the role of amorphous biominerals, including amorphous calcium carbonate (ACC) and amorphous calcium phosphate, and their roles as structural components and as precursors for more stable phases (e.g., Addadi, Raz, & Weiner, 2003; Gong et al., 2012; Gower, 2008).

The absence of long-range periodicity of an amorphous phase or its limited spatial scale in a nanocrystalline phase poses special challenges for determining atomic structure. For example, conventional powder XRD of an amorphous phase typically exhibits diffuse, broad maxima, in contrast to the sharp Bragg diffraction lines for well-formed crystals. This has thwarted traditional X-ray structure determination methods, which rely on Bragg intensities and positions and allow determination of the *average* long-range structure. As a result, methods that probe local structure (e.g., EXAFS and NMR) have become well established in structural studies of amorphous and nanocrystalline biominerals. Yet, the broad, diffuse X-ray scattering from such materials contains valuable information about atomic structure, specifically over short to intermediate length scales (up to ~20 Å). A *total scattering* experiment collects both the Bragg and diffuse components and treats these on an equal basis. Fourier transformation of the total scattering, after appropriate normalization for experimental factors and composition, yields a pair distribution function, or PDF. The PDF represents a distribution of interatomic distances, weighted by the scattering power of atom pairs. This real-space representation of structure allows characterization over a complete range of length scales: short, intermediate, and long range.

The total scattering method has not been extensively used within the research communities working on biomineralization. Yet these methods are not new, having been developed during early studies of glasses and liquids (cf. Wright, 1998), revealing intermediate-range order resulting from linkages between coordination polyhedra. However, the present availability of high-energy X-ray sources at synchrotron storage facilities and short-wavelength neutrons at spallation neutron sources has led to significant advances in data collection both in terms of speed and quality, allowing many new applications. The total scattering method is largely similar for X-rays and neutrons; however, the differences in atomic scattering properties between X-rays and neutrons may offer distinct advantages for a particular technique depending on the elements involved (Proffen, 2006). An important limitation for X-rays is the insensitivity to H atoms due to the small scattering cross section (for X-rays). In this chapter, we focus on the X-ray methods, particularly as practiced at synchrotron radiation facilities. Our goal is to provide an introduction to the relevant X-ray scattering techniques and the methods of analysis and to illustrate some of the advantages using amorphous and nanocrystalline biominerals and synthetic analogs that have been studied thus far.

2. TOTAL SCATTERING METHODOLOGY

The goal of the total scattering experiment is to obtain accurate *elastic* Bragg and diffuse scattering intensities from a sample, free of the parasitic scatter that is associated with experimental factors such as the sample holder and its environment. In practice, this normally involves a subtraction of a background dataset from that for the sample, as described in Section 2.4. A primary consideration in the experiment is the range of scattering angles over which data will be collected, more commonly expressed as momentum transfer, $Q = (4\pi\sin\theta)\lambda^{-1}$, where θ is scattering angle and λ is wavelength. A wide Q range is important because the real-space resolution of the PDF can be approximated as $\delta r = \pi/Q_{max}$, where Q_{max} corresponds to the maximum Q value. For X-rays, a high-energy source available at synchrotron storage rings (up to 100 keV) allows data collection over a significantly higher Q range than conventional lab sources (typically 6.4–17.5 keV), as illustrated in Fig. 20.1. The high photon fluxes provided by synchrotron X-ray sources are also desirable since the intensity of the diffuse scattering component can be as much as 8 orders of magnitude less than the Bragg scattering. Third-generation synchrotron facilities, such as the Advanced Photon Source (APS), Argonne National Laboratory, currently provide both high flux and high energies (up to 125 keV) that easily surpass those from even the most intense lab sources. For neutrons, spallation sources are most suitable because of the short wavelengths available (Benmore, 2012). Here, we focus on experimental methods as commonly practiced at synchrotron X-ray experimental beamlines that have been optimized for PDF studies, such as 11-ID-B at the APS.

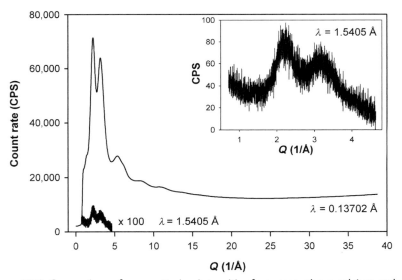

Figure 20.1 Comparison of raw scattering intensities from amorphous calcium carbonate collected using a high-energy synchrotron X-ray source (90.48 keV; $\lambda = 0.13702$ Å) and a Cu Kα lab source (8.05 keV; $\lambda = 1.5405$ Å). Expanded inset shows actual count rate for lab source; this is rescaled ×100 in main window to facilitate comparison.

2.1. Experimental geometry and instrument calibration

The experimental geometry for a total scattering experiment using high-energy synchrotron X-rays is similar to Debye–Scherrer geometry but typically uses a 2D area detector, such as the amorphous silicon-based flat-panel detector system (a-Si) manufactured by General Electric or PerkinElmer (Fig. 20.2). The detector is mounted orthogonal to the incident beam, and the sample-to-detector distance is adjusted to achieve the desired Q range. For example, when using a 40×40 cm a-Si area detector, a sample-to-detector distance of $\sim$12 cm, and an incident X-ray energy of $\sim$58 keV ($\lambda = 0.213$ Å), the maximum usable Q range would be $\sim$25 Å^{-1}, yielding a real-space resolution, $\delta r \approx 0.13$ Å in the PDF. Other geometries including two-circle diffractometers with point counters are also possible (cf. Egami & Billinge, 2003), but these configurations generally do not lend themselves to the rapid data acquisition that is possible with an area detector (Chupas et al., 2003). Calibration is typically performed using common diffraction standards, such as CeO_2 or LaB_6, which are normally available at the experimental station when working

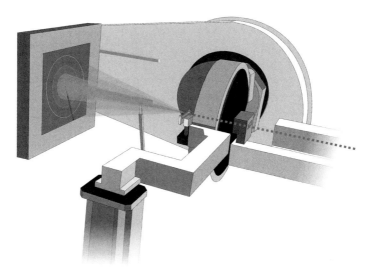

Figure 20.2 Experimental setup with Debye–Scherrer geometry on dedicated PDF beamline 11-ID-B at the APS. Incident high-energy monochromatic beam (dashed line) scatters from the sample before being collected by a tungsten beamstop in place to protect the 40×40 cm area detector positioned to record the scattered intensity.

at a synchrotron beamline. In many instances, the incident X-ray energy of the monochromatic beam is known, and the calibration is used mainly to determine the position of the detector relative to the sample and the incident beam.

2.2. Sample requirements and *in situ* methods

A particular strength of the total X-ray scattering approach is the flexibility of the method in terms of sample state (solid, liquid, and gas), sample matrices, and, in the case of *in situ* experiments, sample environments. There have been many successful applications of the PDF method on polycrystalline, nanocrystalline, and amorphous samples that are in powder (dry) form. Samples, whether environmental, biological, or synthetic, are typically dried and ground to fine, uniform powder prior to analysis. Although sample homogeneity is desirable, it is not a prerequisite. However, analysis of PDF data for polyphasic samples and those containing matrices can suffer from the superposition of the signals. This is true no matter if components are each crystalline, each amorphous, or a combination of the two and is exacerbated when the components are similar in terms of composition or their general structural characteristics. Despite these challenges, there have been a number of successful applications of the total scattering method for polyphasic materials (e.g., Proffen et al., 2005).

The total scattering method is not limited to solids and can also be used to understand structure in liquids and gases. In particular, scattering from liquids containing dissolved complexes or prenucleation clusters could readily be exploited further in studies involving biomineralization. In such cases, the concentration of clusters needs to be sufficiently high to provide acceptable signal, and the best cases will be those in which clusters consist of strongly scattering atoms.

Samples are typically loaded into capillaries made from Kapton (polyimide) or silica glass, as these are devoid of sharp diffraction features, have low scattering cross sections relative to most samples, and are available in a range of diameters. Such characteristics minimize the contribution of the holder to the total scattering and allow this unwanted background signal to be more easily subtracted directly from the data so that the sample scattering can be isolated. A typical Kapton capillary might have an inside diameter of 1 mm and wall thickness of $\sim$25 µm. With the typical incident X-ray beam size less than 1 mm (centered on the capillary), it is evident that very small sample volumes are required for X-ray PDF studies (less than 5 mg). This offers an advantage over neutron experiments, which typically require much larger sample volumes. Environmental cells that allow for liquid or gas flow and/or heating are available at most dedicated PDF experimental stations, allowing time-resolved, *in situ* studies of reactions.

Some recent total scattering experiments have used acoustic levitation to suspend a sample in the X-ray beam without the need for any sample container (Weber, Benmore, Jennings, Wilding, & Parise, 2010). This may be desirable when contact of the sample with the wall of a container might result in reaction, for example, precipitation in the case of an oversaturated liquid.

2.3. Data collection times

How long to collect scattering intensity data for a particular sample is an important consideration. Insufficient counting time, in particular at high Q, can allow unwanted noise due to statistical errors to adversely affect the PDF and possibly lead to misinterpretation.

There are at least three important factors to take into account when deciding on an adequate amount of time to expose a sample, relating to the X-ray source, the sample, and the detector. The flux of X-rays delivered to scattering beamlines with insertion devices at third-generation synchrotron facilities is as much as 4–5 orders of magnitude greater than those produced by most laboratory sources. Data that may have taken days to collect with a lab source (even over a moderate Q range) may require just seconds at a high-flux synchrotron beamline. The flux of X-rays can also vary greatly between different synchrotron scattering beamlines due to differences in beamline source, optics, and, if present, type of insertion device (e.g., undulator or wiggler).

A second factor to consider is the type of detector used for the total scattering experiment. The once-common 1D scintillation and solid-state detectors have now largely been replaced at dedicated PDF synchrotron beamlines by 2D image plate detectors that collect a complete spectrum. Current systems, such as an a-Si flat-panel detector, offer advantages in terms of large detector area and rapid readout times, allowing real-time kinetic studies (Chupas, Chapman, & Lee, 2007). Beamline personnel have expertise in the characteristics of the radiation source and first-hand knowledge of the experimental setup, specifically the type of detector in use, and can generally advise new users to optimize measurement times for their particular experiment.

Finally, the sample under investigation impacts measurement time selection, and the degree of crystallinity, elemental composition, and the overall goal of the total X-ray scattering experiment all must be considered. It is generally desirable to maximize counting statistics and at high Q, in particular, where signal/noise is poorest due to the diminishing coherent scattering signal with increasing Q. However, exposure times will differ dramatically for samples with similar or identical compositions but with structural ordering that varies from crystalline to nanocrystalline to amorphous. Measurement times must be further adjusted if the nature of the sample results in a large scattering component that is considered parasitic and must be treated as background. A sample composed of nanocrystalline Fe oxyhydroxide encapsulated in an organic protein shell and dispersed in water is one such example in which the scattering of the sample holder, protein shell, and water easily exceeds 95% of the total intensity and must be measured separately from the sample so that they can be subtracted. Adequate measurement times of both the sample and these background components are essential to isolate the signal from nanocrystalline portion of the sample in question. Indeed, radiation source, detector technology, and the intrinsic scattering characteristics of a sample, as well as the goals of the total scattering experiment all factor into measurement time selection, and large variability is expected. We offer several examples from our experiences in Section 5 in order to give the reader a general notion of the exposure times used for samples that are crystalline, nanocrystalline, and amorphous.

2.4. Background subtraction and data processing

The conversion of raw total scattering intensity into the total scattering structure function ($S(Q)$ or $f(Q)$), from which the PDF is calculated, requires several steps. Data processing can be done using software that is both widely distributed and freely available for academic use. Here, we mention selected programs often used by our groups but note that this list of software is incomplete and that other packages are available.

Raw scattering images from 2D area detectors are generally corrected for dead time and dark current and then integrated to yield 1D plots of intensity vs. scattering angle (or Q space). If multiple images were collected, they must be averaged or summed. The program FIT2D (Hammersley, Svenson, Hanfland, Fitch, & Häusermann, 1996) is widely used for this, as well as initial calibration of experimental geometry and sample–detector distance.

Successful application of the PDF method relies on accurately removing the parasitic or background scattering from the full scattering intensity. Background in the total scattering experiment originates from a variety of sources including the sample holder, sample matrices (e.g., gases, liquids, and/or solids), and miscellaneous scattering from the instrument. The latter is usually minimized by the addition of shielding to the diffractometer and associated components in order to absorb stray X-rays. This is particularly true in the case of synchrotron-based total scattering since the energies used are relatively high ($>\sim 30$ keV) and flux is large ($>10^{10}$ photons/s). Scattering data for sample and background should be collected under identical experimental conditions to ensure effective removal of background contribution. Additionally, sample matrices, which could include water or some other fluid, can be measured separately and their contribution removed from the data. The subtraction of background is normally done during the processing and normalization of the data to generate the PDF. The program PDFgetX2 (Qiu, Thompson, & Billinge, 2004) is widely used to perform these steps. There may be circumstances when background subtraction is performed postprocessing and in real-space (e.g., Chapman, Chupas, & Kepert, 2005; Harrington et al., 2010).

In addition to background subtraction, corrections must be made for factors that relate to the X-ray source, the sample, and the detector, as described by Egami and Billinge (2003). It is also necessary to account for the composition-dependent, inelastic (Compton) scattering. After background subtraction, correction, and normalization, the X-ray structure function, $S(Q)$, is obtained. This may be Q-weighted to emphasize weaker features at higher Q and then Fourier transformed to obtain the PDF. The entirety of these corrections and calculations is normally performed within specialized data processing programs, such as PDFgetX2.

An important aspect of Fourier transformation is minimization of artifacts. Truncation of scattering data at a finite value Q_{max} introduces termination ripples in the Fourier transform. For samples that are dominated by diffuse scatter, such as amorphous solids, noise can be a significant component at high Q values approaching Q_{max}. Commonly, data for poorly scattering samples must be truncated at a Q lower than Q_{max} to minimize the importance of ripples that result from convolution of noise and truncation error during Fourier transformation. It is possible to introduce a damping function to minimize these effects at high Q, although this broadens features in the PDF and reduces their amplitudes (cf. Egami & Billinge, 2003; Soper & Barney, 2011).

3. THE PDF

The PDF gives the probability of finding two atoms separated by a distance, r, and is often referred to as a "roadmap of interatomic distances." The atomic PDF, $G(r)$, is defined as

$$G(r) = 4\pi r[\rho(r) - \rho_0]$$

where r is radial distance (e.g., from a central atom to a neighboring atom), ρ_0 is the average atomic number density, and $\rho(r)$ is the atomic pair density. The PDF contains the real-space distribution of interatomic distances in a material and is obtained from the Fourier transform of the measurable Bragg and/or diffuse scattering intensities. The relationship of $G(r)$ to the measured scattering pattern (X-rays or neutrons) is through a Fourier transform

$$G(r) = \frac{2}{\pi} \int_{Q=0}^{Q_{max}} Q[S(Q) - 1] \sin(Qr) dQ$$

where $S(Q)$ is the total scattering structure function and contains the measured scattering intensity. The total scattering structure function $S(Q)$ is commonly displayed as the reduced structure function $f(Q)$, or $Q[S(Q) - 1]$. The experimental structure function is related to the coherent part of the total diffracted intensity,

$$S(Q) = \frac{I^{coh}(Q) - \sum c_i |f_i(Q)|^2}{\left|\sum c_i f_i(Q)\right|^2} + 1$$

where $I^{coh}(Q)$ is the corrected measured scattering intensity from the sample. In this equation, c_i and f_i are the atomic concentration and X-ray atomic form factor, respectively, for the atomic species of type i (Egami & Billinge, 2003).

4. EXTRACTING INFORMATION FROM THE PDF

4.1. PDF analysis approaches

4.1.1 Direct information from the PDF

It is important to bear in mind that structural information is represented in one dimension in a PDF, as distances between atom pairs. Yet, that information is invariably interpreted in a 3D framework to be most useful. Owing to the relatively short range over which chemical bonds form, peaks in the 1–3 Å region of a PDF can usually be interpreted as bond lengths, providing direct information about first coordination shells. A Gaussian function can be fit to the peak to get an estimate of the bond length. The integrated area of the peak can be used to calculate average coordination numbers for the pair. The peak width contains information regarding static and dynamic disorder (cf. Egami & Billinge, 2003).

Peaks in the PDF at greater than $\sim$3–4 Å typically correspond to second or higher neighbor distances and may be more difficult to interpret. As noted earlier, pair correlations in the approximate range of 4–20 Å represent intermediate-range structure associated with the connectivity of polyhedral structural units. Peaks in this range can sometimes be interpreted using knowledge of common modes of linkage, for example, corner- and edge-sharing. However, overlap of distances becomes more likely at higher distances, so that development and testing of structure models is commonly a more attractive approach for analysis.

In the PDF from a real sample, we observe that the amplitudes of the peaks decrease with increasing r. This is a result of the instrument resolution and occurs for any sample. However, there are certain instances when the observed attenuation in the amplitudes of peaks with increasing r can be related to the size and, in some cases, morphology of the scattering body. The basic premise in the case of finite-sized nanocrystalline particles is that no atom pairs will be present at distances longer than the maximum dimensions of the sample. For example, given a monodisperse sample of spherical nanocrystalline (i.e., with low structural disorder) particles, the PDF peak amplitudes will decay as r approaches the average dimensions of the coherent scattering domain (CSD), which is essentially equal to the average particle size in the case of a single-domain particle. For amorphous materials, the attenuation of PDF amplitudes with r is more often driven by structural disorder rather than particle size, and the attenuation of signal will increase with increasing disorder. Using the attenuation in the PDF to estimate the range of structural coherence or CSD size is only possible, however, if that size is within the instrumental envelope of the PDF, as determined by instrument resolution. This is a limitation of the spatial coherence of the measurement and is different for each beamline or end station.

To evaluate if the decay of amplitudes with increasing r reflects the structural coherence of the sample or the limitation in the resolution of the instrument, one must measure the instrumental envelope as seen in the PDF. For X-rays, this can be done using the experimental PDF for a strong scattering

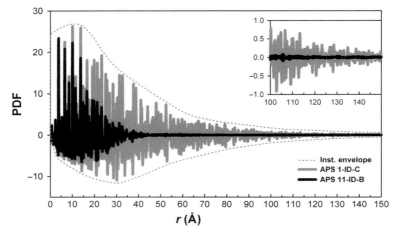

Figure 20.3 Differences in Q resolution for each scattering beamline result in different instrument envelopes as seen in the PDF for identical samples of crystalline CeO_2. The total X-ray scattering intensities were collected to 25 $Å^{-1}$ at APS beamlines 1-ID (gray) and 11-ID-B (black). Higher Q resolution at 1-ID results in a larger instrument envelope and shows PDF correlations that extend to higher r compared with that from 11-ID-B. Inset shows correlations in PDF from APS 1-ID-C are still present beyond 100 Å, while the PDF for 11-ID-B is reduced to statistical noise beyond ~45 Å. Approximate instrument envelope for beamline 1-ID shown as dashed gray line.

material, such as CeO_2, that exhibits long-range structural coherence. The experimental PDFs for CeO_2 attenuate over length scales of nanometers to tens of nanometers when evaluated at two different synchrotron X-ray total scattering beamlines (Fig. 20.3). In both of the cases shown, beyond a certain r, the pair correlations in $G(r)$ disappear and there is only statistical noise that oscillates around zero. The extent and the general shape of the amplitudes for CeO_2 as they decay with increasing r define the instrument envelope for these two experimental setups. As such, it is only possible to estimate the CSD size or range of structural coherence in a material if it is within the instrument envelope of a particular beamline (shaded region, Fig. 20.4). Note that estimating CSD size can be further complicated, for example, in the case of polydisperse samples, materials with a high aspect ratio, or samples that have combined effects of nanometric particle size and structural disorder. A conservative estimated error of ±3 Å (Hall, Zanchet, & Ugarte, 2000) is sometimes reported with CSD sizes reported from the PDF (e.g., Cismasu, Michel, Stebbins, Levard, & Brown, 2012; Toner, Berquó, Michel, Sorenson, & Edwards, 2012).

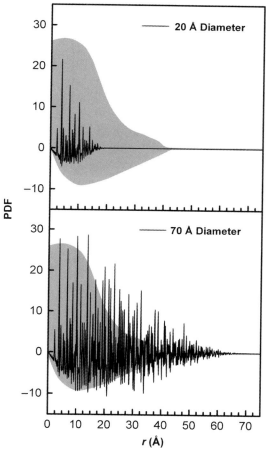

Figure 20.4 Calculated PDFs for spherical CeO_2 nanoparticles with diameters of 20 (black, top pane) and 70 Å (black, bottom pane). The coherent scattering domain size of the 20-Å nanoparticles can be estimated at APS beamline 11-ID-B because the attenuation of correlations in the PDF is within the instrument envelope (shaded region). Correlations for the larger 70-Å nanoparticles extend beyond the instrument envelope and therefore would not be apparent beyond ∼45 Å for data from this same beamline.

4.1.2 Real-space fitting with periodic models

The unit cell approach traditionally used in crystallography may be suitable for interpreting PDFs of samples that exhibit some degree of periodicity. Calculation of the PDF is straightforward for a structure model, which allows least-squares refinement procedures to optimize structure parameters, including unit cell dimensions, atom positions, and atomic displacement parameters, against the experimental PDF. Sometimes called *real-space Rietveld* refinement, this approach has the benefit of using a relatively small number of parameters to define the structure. PDFFIT (Proffen & Billinge, 1999) is one of the most commonly used programs to perform such refinements. This approach can be implemented for clusters and nanosized particles, for example, in the program DISCUS (Neder & Proffen, 2008) and is straightforward for testing proposed structure models that include a space group assignment, unit cell dimensions, and atomic positions and occupancies. Michel et al. (2005) used total X-ray scattering and PDF analysis to evaluate the structure of synthetic iron monosulfide, the so-called amorphous FeS, and showed that it has periodicity over length scales comparable to its particle size (2–3 nm) and also that the PDF could be adequately fit with the model for the known structure of its crystalline counterpart, mackinawite. Michel et al. (2007) used a similar approach to evaluate the structure of synthetic ferrihydrite, a poorly crystalline nanomineral that is ubiquitous in the environment. Here, real-space fitting of the PDFs was used to test previously proposed models and ultimately develop a new single-phase structure model for this phase. Subsequent work on synthetic ferrihydrite by Michel, Barrón, et al. (2010), again using the real-space fitting approach combined with compositional data and analyses of magnetic behavior, suggested that ferrihydrite has inherent structural disorder in the form of Fe vacancies and lattice strain. In samples that exhibit significant disorder or contain other defects (or domains), it may be useful to perform refinements over specific length scales or regions in the PDF (Proffen & Page, 2004). For samples that exhibit no long-range order in the PDF, other nonperiodic approaches have been favored, such as reverse Monte Carlo (RMC).

4.1.3 Reverse Monte Carlo

The RMC method is a computational approach that historically has been applied to glasses and liquids and more recently has also been successfully applied to nanocrystalline solids and disordered crystalline solids. A Monte Carlo type algorithm provides the basis for this method in which random movements are made in atom positions and agreement of the calculated reduced structure function (or PDF) with the experimental function is evaluated. If an atom movement results in an improvement in fit to the experimental function, it is accepted on some statistical basis; movements that result in a poorer fit are rejected. Detailed reviews of the methodology and applications are given by McGreevy (2001) and Keen, Tucker, and Dove

(2005). To achieve an unbiased result, the starting model typically consists of a randomized ensemble of atoms in proportions consistent with the composition of the sample. It is not uncommon for models to contain many thousands of atoms. Hard constraints are typically introduced to avoid unphysical atom separations or to maintain the geometry of an invariant molecule. It is important to recognize that this approach does not yield a unique atomic configuration. Instead, one looks for consistent patterns of coordination that dominate the structure. In this sense, the structure model obtained from RMC simulation, even though constrained by experimental data, is different than the unit cell approach traditionally employed for periodic structures. It has been noted that RMC-derived models tend to be the most disordered configurations that agree with the data. It should be evident why this method has found widespread application in studies of glasses and noncrystalline materials. Goodwin et al. (2010) used RMC to develop a structure model for hydrated ACC. Leading software packages, such as RMCProfile (Tucker, Keen, Dove, Goodwin, & Hui, 2007), now permit simulation using multiple types of structural data, including NMR and EXAFS, simultaneously with total scattering data (both X-ray and neutron). Fits to multiple data types add greater confidence to derived structure models.

4.1.4 Other approaches

There are other approaches for extracting information from total X-ray scattering data and the PDF. Empirical potential structure refinement is a further development of RMC modeling in which the hard constraints that prevent atoms from approaching too closely are replaced by interatomic potentials (Soper, 2005). This method generally provides more constraints on coordination of molecules than RMC does and has been used extensively to model total scattering from liquids. In the case of well-ordered nanoparticles, such as C_{60} (buckyball), it has been shown that there are sufficient numbers of distinct atom-pair distances that the 3D structure can be determined *ab initio* from PDF data alone (Billinge & Levin, 2007). Future work may soon show that unique solutions to more complex structures can also be solved by such methods. Structure models derived from PDF analysis can be tested using molecular dynamics or density functional theoretical approaches to assess the relative stabilities of different model configurations. Singer, Yazaydin, Kirkpatrick, and Bowers (2012) used molecular dynamics to assess a structure model for hydrated ACC. Principal component analysis and linear combination fitting can be used to infer the contributions of different individual components to a multicomponent sample (e.g., Tang et al., 2010).

5. PDF STUDIES OF BIOMINERALS

5.1. Ferritin-derived nanocrystalline ferrihydrite

Ferritin is an omnipresent intracellular protein that mediates iron storage in many living organisms through biomineralization of a ferric (iron) oxyhydroxide particle core in the interior of the protein. The interior diameter of the cavity in ferritin is ~8 nm, and transmission electron microscopy imaging has demonstrated that the core is nanosized (2–6 nm) in both native samples and in apoferritin (APO) that is reconstituted with varying amounts of iron (Chasteen & Harrison, 1999). While it is long recognized that the Fe^{3+} core resembles the nanomineral ferrihydrite, there are a number of challenges that must be overcome in order to probe the structure of the Fe core. Specifically, the nanocrystallinity of the core and presence of significant parasitic scattering from the organic protein shell have impeded prior attempts to fully understand the unperturbed core structure. In addition, ferritin *in vivo* is in solution and, therefore, hydrated. Ideally, the structure of the core would also be evaluated under hydrated conditions in order to avoid undesirable structural changes from heat treatment or other methods of processing and drying.

Michel, Hosein, et al. (2010) measured high-energy total X-ray scattering from samples of APO reconstituted with iron ranging from 500 to 3000 Fe atoms/protein. Total scattering experiments were performed at APS beamline 1-ID, and data were collected at $\lambda = 0.1536$ Å ($E = 80.72$ keV). The beam was focused to a spot size on the sample of 20×20 µm and the flux was an estimated 1×10^{13} photons/s (Shastri et al., 2002). The scattered intensity was collected on an *a*-Si detector (40×40 cm) manufactured by General Electric. Solution and dry (freeze-dried powder) samples and backgrounds were sealed in 1.5 mm and 1.0 mm OD Kapton capillaries, respectively. Optimal counting statistics for dry reconstituted APO were obtained from a total measurement time of 2000 s/sample. Solution samples required additional measurement time of 3200 s each. Scattering intensities of Kapton and Kapton + water + APO were measured independently with exposure times equivalent to the corresponding dry and solution samples, respectively.

For the sample with an average of 2500 Fe atoms/protein and dispersed in water, the profile is dominated by the scattering contributions of the water and the organic protein shell (Fig. 20.5). However, the scattering signal of

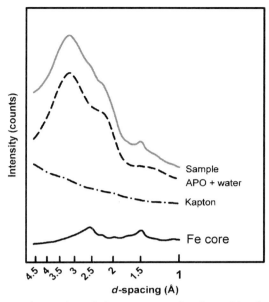

Figure 20.5 Low-angle portion of the raw scattering intensities for apoferritin reconstituted with 2500 Fe atoms/protein (Sample, solid gray) and separate backgrounds of apoferritin in water (APO + water, dashed black) and the capillary holder (Kapton, dash-dot black). Subtraction of the backgrounds from the total signal results in the scattering of the Fe core only (Fe Core, solid black). *Reprinted from Michel, Hosein, et al. (2010), with permission from Elsevier.*

the Fe core can be isolated and identified by careful subtraction of the measured scattering data for the parasitic components (Kapton + water + APO). The Fourier transform of the resulting signal represents the PDF of the Fe core collected *in situ* and comparison to the PDFs for synthetic ferrihydrite derived abiotically show remarkable similarities (Fig. 20.6). PDFs for freeze-dried APO samples reconstituted with amounts of Fe ranging from 500 to 3000 Fe atoms/protein vary in CSD size but have a structure consistent with nanocrystalline ferrihydrite synthesized abiotically (Fig. 20.6). Michel et al. (2007) previously used PDF analysis of ferrihydrite synthesized using common inorganic preparation methods to develop and later refine (Michel, Barrón, et al., 2010) a new single-phase structure model for the abiotic counterpart to the core formed in APO reconstituted with Fe.

Michel, Hosein, et al., (2010) used the attenuation of correlations in the PDFs, described earlier in Section 4.1.1, to estimate CSD sizes for the samples of APO reconstituted with different amounts of Fe. As expected, increasing the Fe content resulted in an increase in the size of the CSD of

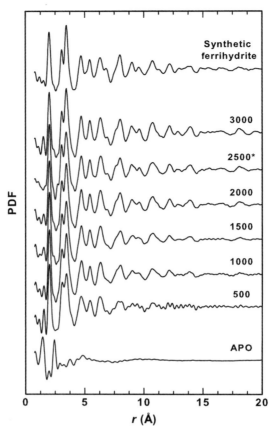

Figure 20.6 PDFs from total X-ray scattering for samples of apoferritin (APO) reconstituted with amounts of Fe ranging from 500 to 3000 Fe atoms/protein. *Sample FeFn 2500 was analyzed in solution while the other reconstituted ferritin samples and the background APO were analyzed as freeze-dried powders. Synthetic ferrihydrite synthesized according to standard abiotic preparation method shown for comparison. *Reprinted from Michel, Hosein, et al. (2010), with permission from Elsevier.*

the core (Fig. 20.7). Interestingly, size analysis using high-resolution transmission electron microscope (TEM) imaging and atomic force microscopy (AFM) for selected samples showed that the physical dimensions of the cores, 2.6–6 nm, were systematically larger than the CSD size estimates from PDF (Fig. 20.7). The differences between the estimates from these methods may be due to the effects of structural disorder in the core and demonstrate how it is not always possible to estimate particle size from the CSD as seen in the PDF.

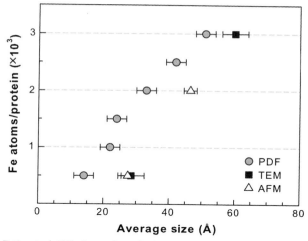

Figure 20.7 Estimated CSD sizes of apoferritin reconstituted with Fe ranging from 500 to 3000 atoms/protein from PDF analysis (gray circles). TEM (black squares) and AFM (open triangles) analyses provide estimates of the physical dimensions (i.e., particle size) for selected samples. *Reprinted from Michel, Hosein, et al. (2010), with permission from Elsevier.*

5.2. Amorphous calcium carbonate

ACC is perhaps one of the most widespread examples of amorphous biominerals, having been the focus of many studies (Addadi et al., 2003). The absence of long-range periodicity was established early using powder XRD, and local structure probes, including Ca K-edge EXAFS, have been used widely to study short-range structure (e.g., Becker et al., 2003; Politi et al., 2006). However, the structural information derived from these methods has largely been limited to length scales less than 4 Å.

Michel et al. (2008) used synchrotron X-ray total scattering and PDF analysis to examine structure in synthetic hydrated ACC. Oscillations in the high-energy total scattering are evident beyond 12 Å^{-1}, far exceeding the Q range typically accessible with a lab source, and with far superior ratio of signal/noise. Proper normalization requires independent knowledge of sample composition, which in many cases can be determined from a chemical analysis. Of particular importance for hydrous samples is the hydrogen content, for which the contribution to the Compton scatter becomes increasingly important at higher Q. In practice, the hydrogen content may be difficult to determine, even in cases where thermogravimetric analysis gives an estimate of H_2O content (assuming H is associated primarily with H_2O). Therefore, it is commonly necessary to try a range of H_2O or H contents, guided by thermogravimetry results, to achieve optimal normalization. Thermogravimetry results for hydrated ACC typically show weight losses in the range of 12–20% before crystallization; if this mass loss is attributed solely to the loss of H_2O, the composition of a hydrated ACC can be estimated as $CaCO_3 \cdot xH_2O$ ($x=0.9$–1.2). The $S(Q)$ and $F(Q)$ obtained are then suitable for Fourier transformation to yield the PDF or $G(r)$ (Fig. 20.8).

The study by Michel et al. (2008) not only provided information on short-range order of the Ca coordination but also revealed intermediate-range order as pair correlations extending up to 12–15 Å. These PDF results served as the basis for RMC modeling to develop a structure model for hydrated ACC (Goodwin et al., 2010). Reeder et al. (2013) used total scattering and PDF analysis to study the hydrated ACC present in gastroliths of the American lobster, finding that the PDF is largely indistinguishable from its synthetic counterpart. Yet ^{1}H NMR revealed differences in the hydrous components between the synthetic and biogenic sample, underscoring the insensitivity of X-ray scattering to H atoms. Radha et al. (2012) have used PDF analysis to characterize a series of amorphous samples spanning the complete hydrated Ca–Mg carbonate system. Pair correlations were found to be limited to 10 Å across the entire range. Ca–O and Mg–O pair correlations are readily distinguished in the PDFs for intermediate compositions, and gradual changes across the system were found.

Our experience at sectors 1-ID and 11-ID-B at the APS, which are optimized for rapid data collection with an a-Si area detector, shows that good quality data with a usable Q range of 20–26 Å^{-1} can be obtained for ACC in well less than 1 min, although longer counting times are desirable when superior signal/noise is needed. The rapid data acquisition possible at such facilities permits *in situ* kinetic studies of reactions. This will allow real-time characterization of structural changes associated with transformation of ACC and other amorphous phases.

6. SUMMARY

X-ray total scattering methods and PDF analysis offer unique advantages for investigations of short- and intermediate-range structure in biominerals, and application to studying amorphous and nanocrystalline phases is certain to expand. The structural information provided by PDF analysis complements studies using local structure techniques, including EXAFS, NMR, and FTIR/Raman spectroscopies. One of the greatest

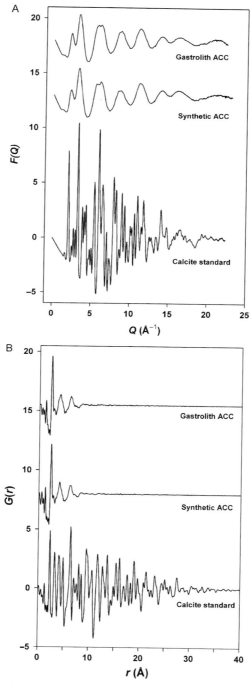

Figure 20.8 Reduced structure functions, $F(Q)$, and pair distribution functions, $G(r)$, for biogenic and synthetic amorphous calcium carbonate samples, compared to a calcite reference sample.

advantages of synchrotron-based PDF analysis is the ability to conduct real-time, *in situ* kinetic studies. Careful attention to the proper techniques of total scattering is essential for successful applications to biomineralization studies. Moreover, new developments in methodology and instrumentation will expand the capabilities, allowing further advances in biomineralization studies.

LINKS

http://www.esrf.eu/computing/scientific/FIT2D/.
http://www.pa.msu.edu/cmp/billinge-group/programs/PDFgetX2/.
http://www.aps.anl.gov/Xray_Science_Division/Structural_Science/Videos/PDFGetX2.html.
http://www.diffpy.org.
http://discus.sourceforge.net.
http://www.rmcprofile.org.

ACKNOWLEDGMENTS

We gratefully acknowledge the assistance provided over many years by the beamline scientists at sectors 1 and 11 at the Advanced Photon Source, Argonne National Laboratory. Use of the Advanced Photon Source, an Office of Science User Facility operated for the U.S. Department of Energy (DOE) Office of Science by Argonne National Laboratory, was supported by the U.S. DOE under Contract No. DE-AC02-06CH11357. RJR acknowledges support from DOE-BES grant DE-FG02-09ER16017.

REFERENCES

Addadi, L., Raz, S., & Weiner, S. (2003). Taking advantage of disorder: Amorphous calcium carbonate and its roles in biomineralization. *Advanced Materials, 15*, 959–970.
Becker, A., Bismayer, U., Epple, M., Fabritius, H., Hasse, B., Shi, J. M., et al. (2003). Structural characterisation of X-ray amorphous calcium carbonate (ACC) in sternal deposits of the crustacea *Porcellio scaber*. *Dalton Transactions, 2003*, 551–555.
Benmore, C. J. (2012). A review of high-energy X-ray diffraction from glasses and liquids. *International Scholarly Research Network Materials Science, 2012*, 852905.
Billinge, S. J. L., & Levin, I. (2007). The problem with determining atomic structure at the nanoscale. *Science, 316*, 561–565.
Chapman, K. W., Chupas, K. J., & Kepert, C. J. (2005). Selective recovery of dynamic guest structure in a nanoporous Prussian Blue through in situ X-ray diffraction: A differential pair distribution function analysis. *Journal of the American Chemical Society, 127*, 11232–11233.
Chasteen, N. D., & Harrison, P. M. (1999). Mineralization in ferritin: An efficient means of iron storage. *Journal of Structural Biology, 126*, 182–194.

Chupas, P. J., Chapman, K. W., & Lee, P. L. (2007). Applications of an amorphous silicon-based area detector for high-resolution, high-sensitivity and fast time-resolved pair distribution function measurements. *Journal of Applied Crystallography, 40*, 463–470.

Chupas, P. J., Qiu, X., Hanson, J. C., Lee, P. L., Grey, C. P., & Billinge, S. J. L. (2003). Rapid-acquisition pair distribution function (RA-PDF) analysis. *Journal of Applied Crystallography, 36*, 1342–1347.

Cismasu, A. C., Michel, F. M., Stebbins, J. F., Levard, C. M., & Brown, G. E., Jr. (2012). Properties of impurity-bearing ferrihydrite I. Effects of Al content and precipitation rate on the structure of 2-line ferrihydrite. *Geochimica et Cosmochimica Acta, 92*, 275–291.

Egami, T., & Billinge, S. J. L. (2003). *Underneath the Bragg Peaks: Structural analysis of complex materials*. Oxford: Pergamon Press.

Gong, Y. U. T., Killian, C. E., Olson, I. C., Appathurai, N. P., Amasino, A. L., Martin, M. C., et al. (2012). Phase transitions in biogenic amorphous calcium carbonate. *Proceedings of the National Academy of Science of the United States of America, 109*, 6088–6093.

Goodwin, A. L., Michel, F. M., Phillips, B. L., Keen, D. A., Dove, M. T., & Reeder, R. J. (2010). Nanoporous structure and medium-range order in synthetic amorphous calcium carbonate. *Chemistry of Materials, 22*, 3197–3205.

Gower, L. B. (2008). Biomimetic systems for investigating the amorphous precursor pathway and its role in biomineralization. *Chemical Reviews, 108*, 4551–4627.

Hall, B. D., Zanchet, D., & Ugarte, D. (2000). Estimating nanoparticle size from diffraction measurements. *Journal of Applied Crystallography, 33*, 1335–1341.

Hammersley, A. P., Svenson, S. O., Hanfland, M., Fitch, A. N., & Häusermann, D. (1996). Two-dimensional detector software: From real detector to idealized image or two-theta scan. *High Pressure Research, 14*, 235–248.

Harrington, R., Hausner, D. B., Bhandarim, N., Strongin, D. R., Chapman, K. W., Chupas, P. J., et al. (2010). Investigation of surface structures by powder diffraction: A differential pair distribution function study on arsenate sorption on ferrihydrite. *Inorganic Chemistry, 49*, 325–330.

Keen, D. A., Tucker, M. G., & Dove, M. T. (2005). Reverse Monte Carlo modelling of crystalline disorder. *Journal of Physics: Condensed Matter, 17*, S15–S22.

McGreevy, R. L. (2001). Reverse Monte Carlo modeling. *Journal of Physics: Condensed Matter, 13*, R877–R913.

Michel, F. M., Antao, S. M., Chupas, P. J., Lee, P. L., Parise, J. B., & Schoonen, M. A. A. (2005). Short- to medium-range atomic order and crystallite size of the initial FeS precipitate from pair distribution function analysis. *Chemistry of Materials, 17*, 6246–6255.

Michel, F. M., Barrón, V., Torrent, J., Morales, M. P., Serna, C. J., Boily, J.-F., et al. (2010). Ordered ferrimagnetic form of ferrihydrite reveals links among structure, composition, and magnetism. *Proceedings of the National Academy of Sciences of the United States of America, 107*, 2787–2792.

Michel, F. M., Ehm, L., Antao, S. M., Lee, P. L., Chupas, P. J., Liu, G., et al. (2007). The structure of ferrihydrite, a nanocrystalline material. *Science, 316*, 1726–1729.

Michel, F. M., Hosein, H.-A., Hausner, D. B., Debnath, S., Parise, J. B., & Strongin, D. R. (2010). Reactivity of ferritin and the structure of ferritin-derived ferrihydrite. *Biochimica et Biophysica Acta, 1800*, 871–885.

Michel, F. M., MacDonald, J., Feng, J., Phillips, B. L., Ehm, L., Tarabrella, C., et al. (2008). Structural characteristics of synthetic amorphous calcium carbonate. *Chemistry of Materials, 20*, 4720–4728.

Neder, R. B., & Proffen, T. (2008). *Diffuse scattering and defect structure simulations*. New York: Oxford University Press.

Politi, Y., Levi-Kalisman, Y., Raz, S., Wilt, F., Addadi, L., Weiner, S., et al. (2006). Structural characterization of the transient amorphous calcium carbonate precursor phase in sea urchin embryos. *Advanced Functional Materials, 16*, 1289–1298.

Proffen, T. (2006). Analysis of disordered materials using total scattering and the atomic pair distribution function. *Reviews in Mineralogy and Geochemistry*, *63*, 255–274.

Proffen, T., & Billinge, S. J. L. (1999). PDFFIT, a program for full profile structural refinement of the atomic pair distribution function. *Journal of Applied Crystallography*, *32*, 572–575.

Proffen, T., & Page, K. L. (2004). Obtaining structural information from the atomic pair distribution function. *Zeitschrift für Kristallographie*, *219*, 130–135.

Proffen, T., Page, K. L., McLain, S. E., Clausen, B., Darling, T. W., TenCate, J. A., et al. (2005). Atomic pair distribution function analysis of materials containing crystalline and amorphous phases. *Zeitschrift für Kristallographie*, *220*, 1002–1008.

Qiu, X., Thompson, J. W., & Billinge, S. J. L. (2004). PDFgetX2: A GUI driven program to obtain the pair distribution function from X-ray powder diffraction data. *Journal of Applied Crystallography*, *37*, 678. http://www.pa.msu.edu/cmp/billinge-group/programs/PDFgetX2.

Radha, A. V., Fernandez-Martinez, A., Hu, Y., Jun, Y.-S., Waychunas, G. A., & Navrotsky, A. (2012). Energetic and structural studies of amorphous $Ca_{1-x}Mg_xCO_3 \cdot nH_2O$. *Geochimica et Cosmochimica Acta*, *90*, 83–95.

Reeder, R. J., Tang, Y., Schmidt, M. P., Kubista, L. M., Cowan, D. F., & Phillips, B. L. (2013). Characterization of structure in biogenic amorphous calcium carbonate: Pair distribution function and nuclear magnetic resonance studies of lobster gastrolith. *Crystal Growth and Design*, *13*, 1905–1914.

Shastri, S. D., Fezzaa, K., Masayekhi, A., Lee, W.-K., Fernandez, P. B., & Lee, P. L. (2002). Cryogenically cooled bent double-Laue monochromator for high-energy undulator X-rays (50-200 keV). *Journal of Synchrotron Radiation*, *9*, 317–322.

Singer, J. W., Yazaydin, A. O., Kirkpatrick, R. J., & Bowers, G. M. (2012). Structure and transformation of amorphous calcium carbonate: A solid-state ^{43}Ca NMR and computational molecular dynamics investigation. *Chemistry of Materials*, *24*, 1828–1836.

Soper, A. K. (2005). Partial structure factors from disordered materials diffraction data: An approach using empirical potential structure refinement. *Physical Review B*, *72*, 104204.

Soper, A. K., & Barney, E. R. (2011). Extracting the pair distribution function from white-beam X-ray total scattering data. *Journal of Applied Crystallography*, *44*, 714–726.

Tang, Y., Michel, F. M., Zhang, L., Harrington, R., Parise, J. B., & Reeder, R. J. (2010). Structural properties of the Cr(III)-Fe(III)-(oxy)hydroxide compositional series: Insights for a nanomaterial "solid solution" *Chemistry of Materials*, *22*, 3589–3598.

Toner, B. M., Berquó, T. S., Michel, F. M., Sorenson, J. V., & Edwards, K. J. (2012). Mineralogy of iron microbial mats from Loihi Seamount. *Frontiers in Microbiology*, *3*(118), 1–18.

Tucker, M. G., Keen, D. A., Dove, M. T., Goodwin, A. L., & Hui, Q. (2007). RMCProfile: Reverse Monte Carlo for polycrystalline materials. *Journal of Physics: Condensed Matter*, *19*, 335218.

Weber, J. K. R., Benmore, C. J., Jennings, G., Wilding, M. C., & Parise, J. B. (2010). Instrumentation for fast in-situ X-ray structure measurements on non-equilibrium liquids. *Nuclear Instruments and Methods in Physics Research A*, *624*, 728–730.

Wright, A. C. (1998). Diffraction studies of glass structure: The first 70 years. *Glass Physics and Chemistry*, *24*, 148–179.

CHAPTER TWENTY-ONE

X-Ray Microdiffraction of Biominerals

Nobumichi Tamura[*], Pupa U.P.A. Gilbert[†,‡,1,2]

[*]Advanced Light Source, Lawrence Berkeley National Laboratory, Berkeley, California, USA
[†]Department of Physics, University of Wisconsin-Madison, Madison, Wisconsin, USA
[‡]Department of Chemistry, University of Wisconsin-Madison, Madison, Wisconsin, USA
[1]Previously publishing as Gelsomina De Stasio
[2]Corresponding author: e-mail address: pupa@physics.wisc.edu

Contents

1. Introduction   501
2. Elements of X-Ray (Micro)Diffraction   504
3. Synchrotron, X-Ray Focusing Optics, and Area Detectors   509
4. Sample Preparation   514
5. Powder X-Ray Microdiffraction   515
6. White-Beam X-Ray Microdiffraction   520
7. Microbeam Small-Angle X-Ray Scattering   522
8. Future Developments   526
References   527

Abstract

Biominerals have complex and heterogeneous architectures, hence diffraction experiments with spatial resolutions between 500 nm and 10 μm are extremely useful to characterize them. X-ray beams in this size range are now routinely produced at many synchrotrons. This chapter provides a review of the different hard X-ray diffraction and scattering techniques, used in conjunction with efficient, state-of-the-art X-ray focusing optics. These include monochromatic X-ray microdiffraction, polychromatic (Laue) X-ray microdiffraction, and microbeam small-angle X-ray scattering. We present some of the most relevant discoveries made in the field of biomineralization using these approaches.

1. INTRODUCTION

The study of biomaterials is a fascinating multidisciplinary field that crosses the boundary between basic and applied sciences. By acquiring an intimate knowledge of the mechanisms fine-tuned by evolution over millions of years of experimentation to produce tools that appear perfectly fitted to their particular functions, new bio-inspired materials with enhanced properties can be devised for use in modern technologies such as bone implants, reef rehabilitation, smart interfacial materials, or hybrid ceramic composites (Munch, Launey, Alsem, & Saiz, 2008; Xia & Jiang, 2008).

Nature often offers microstructural solutions for a materials function that are both superior and more cost-effective than man-made equivalents. For instance, mollusk shell nacre (mother-of-pearl) is a composite material made of 98% aragonite and 2% proteins (Levi-Kalisman, Falini, Addadi, & Weiner, 2001), which is remarkably more fracture-tough than its aragonite mineral component (Currey, 1977), a net gain in mechanical performance that has not been achieved in any man-made composite so far (Gilbert et al., 2008). The spider *Caerostris darwini* is known to produce a silk, which is 10 times stronger than Kevlar (Agnarsson, Kuntner, & Blackledge, 2010), and the surprising robustness of spider web has been very recently demonstrated to originate from the nonlinear behavior of silk (Cranford, Tarakanova, Pugno, & Buehler, 2012). The gecko toe pads have extraordinary adhesive properties that scientists have just started to understand (Autumn et al., 2000, 2002; Huber et al., 2005; Lee, Lee, & Messersmith, 2007). Evolution has led to many different and astute solutions for strong yet flexible dermal armors in animals with performances that no man-made personal body protection can currently provide (Bruet, Song, Boyce, & Ortiz, 2008; Meyers, McKittrick, & Chen, 2013; Yang et al., 2013). The study of the microstructure of biomaterials and of their genesis has therefore great application potentials for the design of new materials with enhanced mechanical robustness (Gao, Ji, Jäger, Arzt, & Fratzl, 2003).

Biominerals constitute a class of biomaterials tuned to provide protection or defense armors, attack weapons, gravity or magnetic sensing, skeletal support to living organisms, etc. (Gilbert, 2012). A common microstructural theme for biominerals is their intrinsic hierarchical and composite nature (Lowenstam & Weiner, 1989). Biominerals as found in bones, teeth, shells, dermal armors, and such are a mixture of mainly inorganic minerals, such as calcium carbonate or calcium phosphate, and a minority of organic molecules such as collagen and other proteins, glycoproteins (Albeck, Weiner, & Addadi, 1996), or polysaccharides such as chitin in crustaceans (Weaver et al., 2012), and mollusks (Falini, Albeck, Weiner, & Addadi, 1996; Levi-Kalisman et al., 2001). The superior mechanical performances of biominerals are explained by the ability of biological systems to carefully control the microstructure (size, shape, crystal orientation, degree of crystallinity, chemical composition, etc.) of the minerals composing them via the action of those organic molecules. Linking biomineral microstructure with their mechanical properties requires the material to be studied at different length scales ranging from the molecular level, at which proteins operate, to the macroscopic scale (Munch et al., 2008; Nalla, Kinney, & Ritchie, 2003; Ortiz & Boyce, 2008; Tai, Dao, Suresh, Palazoglu, & Ortiz, 2007). Generally, a combination of characterization techniques is needed to investigate the multifaceted biomineralization process. Imaging techniques such as scanning electron microscopy, X-ray microscopy, and X-ray tomography provide valuable information about the two- and three-dimensional architecture of biominerals, but they often need to be complemented with diffraction techniques (Lichtenegger, Schöberl, Bartl, Waite, & Stucky, 2002), for assessing the degree of crystallinity, or determining crystal structure, orientation, coorientation, or strain of crystals, and spectroscopic techniques, which provide information on the chemistry of the different constitutive phases (Gilbert, 2014).

Conventional X-ray diffraction using millimeter-size X-ray beams is more than a century old and has been used extensively and successfully to probe the arrangement of atoms in materials (Warren, 1969). X-ray diffraction is useful to determine the crystal structure of an unknown mineral, that is, finding where all atoms are positioned inside the crystallographic unit cell (Giacovazzo, 2002; Zolotoyabko, 2011). It also determines the crystal orientation, that is, how the atomic lattices are oriented with respect to a reference frame. It measures the deformation state of the material, that is, how the atoms are shifted away from their positions due to an applied force (Noyan & Cohen, 1987). Finally, the different crystalline mineral species in a sample can be identified by comparing the position of the experimental diffraction peaks with the positions calculated from literature values (Brindley, 1961). Diffraction is therefore a powerful and information-rich technique to investigate crystalline materials.

For inhomogeneous and hierarchical materials such as biominerals, conventional diffraction would only provide average information on the area illuminated by the beam. As biomineral single-crystalline units have sizes ranging between 10 nm and 10 μm, it is necessary to use small-focus X-ray beams from a synchrotron to obtain most informative diffraction data from biominerals. Micron and submicron spot-size, intense X-ray beams are nowadays routinely produced at most synchrotrons using specialized focusing optics, such as Fresnel zone plates (Di Fabrizio, Romanato, & Gentili, 1999; Lai, Yun, Legnini, & Xiao, 1992), compound refractive lenses (CRLs) (Snigirev, Kohn, Snigireva, Souvorov, & Lengeler, 1998), and Kirkpatrick–Baez (KB) mirrors (Hignette, Cloetens, Rostaing, Bernard, & Morawe, 2005; Ice, Chung, Tischler, Lunt, & Assoufid, 2000; Mimura et al., 2005).

The goal of this chapter is to give an incomplete overview of discoveries made in biominerals using X-ray microdiffraction at various synchrotrons. We will restrict the scope to include only X-ray beams smaller than 10 μm, even though the term microdiffraction has been, and still sometimes is, used to describe diffraction with X-ray beams with spot sizes of a few hundreds of microns, obtained from laboratory sources. The 10 μm limit we impose here selects for synchrotron microdiffraction with state-of-the-art X-ray focusing optics. Nanodiffraction is also possible at very few synchrotrons and indicates X-ray beams below 100 nm (Hignette et al., 2005; Mimura et al., 2005; Riekel, Burghammer, & Davies, 2010). The basics of X-ray diffraction are presented in this section of the chapter. Section 2 introduces synchrotron radiation and explains why it is needed for X-ray microdiffraction and gives an overview of ways to focus X-ray beams. Section 3 describes sample preparation for synchrotron X-ray microdiffraction experiments. Sections 4–6, respectively, deal with monochromatic X-ray microdiffraction, polychromatic (Laue) X-ray microdiffraction, and micro-small-angle X-ray scattering (SAXS), each presenting data from biominerals.

2. ELEMENTS OF X-RAY (MICRO)DIFFRACTION

X-ray diffraction is an X-ray scattering technique that has been used for more than a century to probe interatomic distances in crystalline materials. In a broad sense, X-ray scattering refers to all techniques based on the observation of the scattered intensity by a sample illuminated by an incoming X-ray beam. X-ray diffraction is usually restricted to X-ray scattering by a periodic arrangement of atoms or molecules, that is, a crystal. In the latter case, the constructive interferences between the beams scattered by the atomic planes lead to a structured and discrete distribution of intensities termed "diffraction peaks" or "reflections." Here, we will only consider elastic scattering, that is, scattering in which the energy (wavelength) of the photons remains unchanged after the interaction with the material under study. Diffraction is an elastic process.

Let us consider a set of parallel lattice planes h,k,l from a crystal separated by an interplanar distance d_{hkl} and illuminate it with a perfectly monochromatic X-ray beam of wavelength λ. The incident beam will be reflected by the h,k,l planes in such a way that the incident ($\mathbf{k_i}$) and reflected ($\mathbf{k_r}$) wavevectors are coplanar with the normal to the lattice planes and the angle of incidence θ is equal to the angle of reflection. The scattering vector

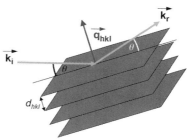

Figure 21.1 Schematic showing Bragg diffraction by a set of parallel atomic planes. In this schematic, θ is both the incidence angle and the reflection angle of the X-ray beam, d_{hkl} is the interplanar distance of atomic planes in the crystal, and λ is the wavelength of the illuminating X-ray beam. A reflection only occurs when all three parameters θ, d_{hkl}, and λ satisfy the Bragg condition. $\mathbf{k_i}$ and $\mathbf{k_r}$ are the incident and reflected wavevectors, respectively, and $\mathbf{q_{hkl}} = \mathbf{k_r} - \mathbf{k_i}$ is the scattering vector normal to the atomic lattice planes. Bold font here indicates vectors. (For color version of this figure, the reader is referred to the online version of this chapter.)

$\mathbf{q_{hkl}} = \mathbf{k_r} - \mathbf{k_i}$, normal to the lattice planes, is pointing to a node of the reciprocal lattice. Furthermore, the reflection will occur only if the Bragg condition is fulfilled, that is, when there are constructive interferences between the reflections from individual atomic planes. The Bragg condition is expressed as:

$$2d_{hkl}\sin\theta = n\lambda$$

where n is an integer number, and the other parameters are described in Fig. 21.1.

The instrument to measure these Bragg reflections is called diffractometer. X-ray diffractometers are typically complex instruments with sample and detector mounted on goniometers with a number of "circles" between 2 and 6, making it possible to precisely rotate the sample to the correct angle satisfying the Bragg condition and position the detector at a location where the reflection can be observed (the replacing of point detectors by large-area detectors has significantly decreased the number of circles used).

Single-crystal X-ray crystallography or, more generally, single-crystal X-ray diffraction refers to the technique that probes atomic arrangement in a single crystal using an X-ray beam and a diffractometer. For unknown crystal structures, a usable set of reflections needs to be systematically collected in order to reconstruct the reciprocal lattice of the crystal, defined as the lattice formed by the end points of all $\mathbf{q_{hkl}}$ scattering vectors. The reciprocal lattice allows in turn to calculate the direct Bravais lattice of the crystal and its unit cell, that is, the smallest set of relative atomic positions in a crystal enabling the reconstruction by translation of the entire lattice of the crystal. The positions of the reflections provide the shape and size of the unit cell, whereas the integrated intensities of the reflections give access to the atomic decoration inside the unit cell, provided the phase information for the structure factors is retrieved (the so-called phase problem of crystallography) (Giacovazzo, 2002). Since the Bragg condition is only rarely and discretely fulfilled, X-ray diffraction of a single crystal with monochromatic X-rays requires data collection at several different angles; thus, the need to rotate the sample under the beam arises (Fig. 21.2). Determining the complete atomic structure of an unknown crystal (a process also called "solving a structure"), as performed on dedicated X-ray crystallography instruments, typically entails collecting hundreds of diffraction patterns. If the crystal structure is already known, the measurement of only a small number of reflections is generally sufficient to establish crystal orientation.

Biominerals are composite and inhomogeneous polycrystalline materials, containing components that are often poorly crystallized. The use of conventional single-crystal diffraction has been therefore of limited use for biomineral samples. An additional problem associated with single-crystal diffraction when combined with a small-size X-ray beam is the sphere of confusion of the diffractometer. All the rotations of the goniometer are concentric only to a limited extent, given by the accuracy of the instrument, which is typically a few microns. Consequently, aligning a (sub)micron-size beam onto a (sub)micron-size sample at the center of rotation of a goniometer is a challenging task. Variations of the single-crystal X-ray diffraction approach using a beam size of a few microns and a limited range of rotations have been employed to study biominerals from mollusks and echinoderms. One such example is described in Section 3 as a case where the single-crystal

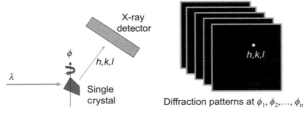

Figure 21.2 In a single-crystal X-ray diffraction approach, the sample must be placed on a rotation stage to be able to collect a usable set of reflections on a two-dimensional area detector. (For color version of this figure, the reader is referred to the online version of this chapter.)

diffraction approach is applied to polycrystalline nacre (DiMasi & Sarikaya, 2004), while the case of a highly textured sea urchin tooth, which behaves nearly as a single crystal, is described in Section 5 (Ma et al., 2009).

An alternate approach to single-crystal diffraction is "powder" diffraction, although the name can be misleading, as the method is applied to a whole range of materials ranging from polycrystalline metals or minerals to "single-crystalline" polymers. When a monochromatic beam illuminates an aggregate of randomly oriented crystallites, the Bragg condition is statistically fulfilled for a number of those crystallites. A particular set of d_{hkl} planes from the ensemble of crystallites in Bragg conditions give rise to a cone of diffraction with semiangle θ_{hkl} (Fig. 21.3). The intensity rings or arcs visible in diffraction patterns from an area detector represent the intersection of diffraction cones of various d_{hkl} values with the plane of the detector. Mathematically, the intersections are "conics" and are called Debye–Scherrer rings. Their shape depends on the position of the detector with respect to the incoming beam (they can be circular when the detector is at $0°$, directly in the path of the incoming beam, elliptical, parabolic, or hyperbolic when the detector is placed at higher angles). Associated with a micron-size incident beam, powder X-ray microdiffraction can be a powerful tool for investigating microstructures of nanocrystalline samples with micron spatial resolution. The use of powder microdiffraction in nacre is discussed in Section 5.

Powder diffraction can be used to measure texture, or strain, and allows for mineral phase identification but generally only provides average information over the large number of grains or crystallites illuminated by the X-ray beam.

To obtain single-crystal information from a sample without the need of rotating it, the Bragg condition can be fulfilled simultaneously for several

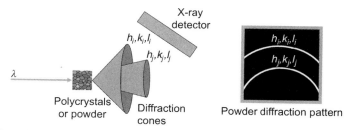

Figure 21.3 Principle of powder X-ray diffraction. Randomly oriented crystallites give rise to diffraction cones of X-ray beam intensities, which intersect the area detector to form Debye–Scherrer rings. (For color version of this figure, the reader is referred to the online version of this chapter.)

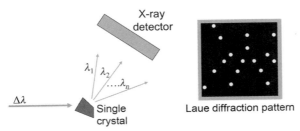

Figure 21.4 With polychromatic (Laue) diffraction, a number of reflections satisfy the Bragg condition simultaneously. (For color version of this figure, the reader is referred to the online version of this chapter.)

reflections when an incident polychromatic X-ray beam is used instead of a monochromatic beam (Fig. 21.4). This technique, called Laue diffraction, has been employed for decades for determining the orientation of crystals with known structure and unit cell.

Although its use for structure solution has been superseded by single-crystal (monochromatic) diffraction, in part because of the difficulty to accurately normalize the intensities of the reflections, Laue diffraction made a comeback for investigating crystallographic transformations in fast processes in macromolecules and other large-unit-cell structures (Moffat, Szebenyi, & Bilderback, 1984). More recently, Laue microdiffraction was used in conjunction with an X-ray microbeam and proved to be very useful for measuring strain distribution in polycrystalline materials (Tamura et al., 2002). Laue microdiffraction in the study of biominerals, including aragonite and calcite mollusk shell layers, is treated in Section 6.

The X-ray diffraction techniques described earlier are generally referred to as wide-angle X-ray scattering (WAXS) by part of the scientific communities, notably polymer and macromolecule scientists, as a contrast to SAXS. Reflections in WAXS patterns are due to the scattering of X-rays by atomic planes and therefore provide measurements of interatomic distances. Because the interatomic spacings in the Bragg equations d_{hkl} are in the few Angstrom regime, the associated reflections appear at high θ angles. Conversely, samples usually studied by SAXS, such as periodic arrangements of nanoparticles, bundles of fibers, or assemblies of proteins with distances between the objects in the nanometer to a few tens of nanometers range, scatter X-rays at very low angles, very close to the transmitted X-ray beam position, typically with $\theta < 3°$. SAXS is shown schematically in Fig. 21.5. Interpreting SAXS patterns provides information on the average size and shape of nanometer-size objects. Micro-SAXS or microbeam small-angle

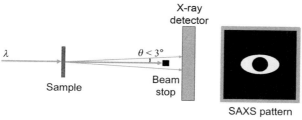

Figure 21.5 Principle of small-angle X-ray scattering. The area detector is placed at 0°. (For color version of this figure, the reader is referred to the online version of this chapter.)

X-ray scattering (μ-SAXS) refers to SAXS using a few micron-size beam and is typically employed to study the arrangement of nanometer- to micrometer-size objects with added spatial resolution and is therefore perfectly applicable to biomaterials such as bones. μ-SAXS of the bone is treated in Section 7.

3. SYNCHROTRON, X-RAY FOCUSING OPTICS, AND AREA DETECTORS

Compared to metals, ceramics, and other materials typically used in industry, biomaterials are made of crystals, which are often smaller and weakly scattering. Poor scattering results from a combination of effects, including small size of coherent domains, chemical composition mostly including light elements, poor degree of crystallinity, and high inhomogeneity. Intense X-ray beams with small spot sizes are therefore essential prerequisite to study biominerals, and both require a synchrotron.

Second- and third-generation synchrotron sources such as the NSLS (National Synchrotron Light Source) on Long Island, near New York; ESRF (European Synchrotron Radiation Facility) in Grenoble, France; the Advanced Photon Source, at Argonne, near Chicago, Illinois; the Advanced Light Source (ALS) at Berkeley, near San Francisco, California; the Stanford Synchrotron Radiation Light Source in Stanford, also near San Francisco, California; the Super Photon Ring 8-GeV (SPRing-8) between Kyoto and Hiroshima, Japan; and Berliner Elektronenspeicherring-Gesellschaft für Synchrotronstrahlung II (BESSY II) in Berlin, Germany, provide highly collimated X-ray beams that are orders of magnitude more intense than X-ray tubes can produce as laboratory sources. Beam collimation and brightness are important characteristics to obtain (sub)micron-size X-ray beams, as they provide high-efficiency diffraction-limited focusing. Another advantage of synchrotron sources over laboratory sources is their energy (wavelength) tunability, allowing for spectroscopic techniques, and those are often used in conjunction with diffraction on the same beamline.

X-ray focusing optics is generally classified into three categories depending on the physical property used to achieve small spot size: they are diffracting, reflective, or refractive optics. Fresnel zone plates (or simply zone plates) are the most popular diffractive optics. Zone plates consist of concentric rings of X-ray-transparent and X-ray-opaque rings (zones) with width and spacing decreasing with increasing radius, in order to obtain constructive interferences at a single focal point. Spot sizes as small as a few nanometers were obtained for soft X-rays below 2 keV (Chao, Harteneck, Liddle, Anderson, & Attwood, 2005). Fabricating hard X-ray zone plates is technically more challenging, as the opaque zones need to be thick to account for the penetrating power of the radiation. However, high-aspect-ratio zones with sufficient quality have been obtained for spot sizes of a few tens of nanometers (Chu et al., 2008).

X-rays achieve total reflection only below a certain critical grazing incidence angle. When an ultrasmooth X-ray-reflecting surface (X-ray mirror) is curved to an elliptical shape, an X-ray beam incoming at a grazing angle below the critical angle will bounce off the mirror and focus in one direction at the image plane of the mirror. A pair of orthogonal elliptical mirrors can be used to focus the beam in both the horizontal and vertical directions. This is the so-called KB configuration (Kirkpatrick & Baez, 1948), which is nowadays routinely used on many synchrotron beamlines. The best KB mirrors achieved spot sizes comparable to hard X-ray zone plates (Mimura et al., 2005), but surface qualities, complexity of the alignment procedure, and nonoptimum beam conditioning often limit the spot size obtained by typical KBs to the 500 nm to 5 μm range. X-ray capillaries are another type of reflective optics in which the X-ray beam is focused inside an ellipsoidally shaped reflecting glass tube (Bilderback, 2003). Despite their lower cost, considerations such as limited working distance and reduced acceptance angle have rendered them less attractive than KB mirrors, and their use outside laboratory sources is now limited.

The X-ray refractive index of materials is slightly lower than 1, so X-ray refractive optics need to be thinner, rather than thicker, in the middle, as opposed to converging lenses used for visible light. Moreover, since the refractive index is only barely below unity, multiple refractive lenses have to be packed in series to achieve substantial focusing, so that X-ray refractive lenses take the form of Compound Refractive Lenses (CRLs) (Schroer et al., 2003; Snigirev et al., 1998). To avoid too much absorption in the lens material, CRLs are typically made of light elements such as beryllium, aluminum, or silicon. They are best suited for hard X-ray beamlines (>>2 keV) as the efficiency drops quickly for soft X-rays due to absorption effects. In addition to the X-ray focusing optics cited earlier, more exotic ones have been developed in recent years: those include X-ray waveguides that use both refraction and reflection (Jark et al., 2001; Pfeiffer, David, Burghammer, Riekel, & Salditt, 2002) and new types of refractive optics: kinoform lenses (Evans-Lutterodt et al., 2007) and prism lenses, consisting of hundreds to thousands of tiny prisms guiding the X-ray beam to a focal point (Jark et al., 2004). A variation of Fresnel zone plates for hard X-rays is multilayer Laue lenses (Kang et al., 2008).

The choice of the optics to use for focusing the X-ray beam depends on the main use of the beamline. The principal advantage of reflective optics with respect to diffractive and refractive ones is their achromaticity, meaning that the position of the focal point is independent of the wavelength, so reflective optics can focus white (polychromatic) radiation and monochromatic radiation. All the other optics are generally used in monochromatic mode only.

A synchrotron beamline specialized in X-ray microbeam diffraction typically has a set of prefocusing and beam conditioning mirrors before the final focusing provided by one of the many optics described earlier. A monochromator would select a specific wavelength from the incoming white light and, in the case of beamlines that can perform Laue diffraction, is removable from the beam so that the beamline can be operated in either monochromatic or white-light mode. Current synchrotrons provide two types of X-ray sources: bending magnets (including superconducting magnets) and insertion devices (undulators and wigglers). Monochromatic diffraction is best done with an undulator source for the beamline, as the flux at harmonic wavelengths is several orders of magnitudes higher than the one from bending magnets. Polychromatic diffraction, instead, works best with a bending magnet or a wiggler as a source for the beamline, because they provide a continuous spectrum over a range of wavelengths (the so-called "white" or more correctly "pink" radiation). Bending magnets are also attractive because they offer wavelength tunability, which is an important capability when diffraction is combined with spectroscopy. Samples are typically placed on a scanning stage, so they can be moved to the focal point of the beam and then raster-scanned under the beam to map different locations within the sample. Some degrees of sample rotations are usually provided as well. X-ray detectors usually consist of a two-

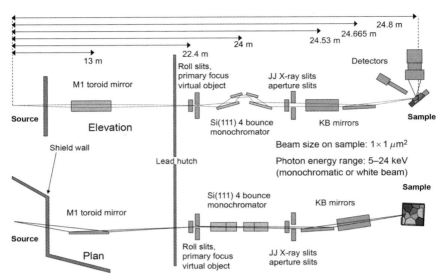

Figure 21.6 Outline of the microdiffraction beamline 12.3.2. at the Advanced Light Source (Kunz et al., 2009). (For color version of this figure, the reader is referred to the online version of this chapter.)

dimensional X-ray area detector to collect the diffraction pattern and a solid-state point detector to collect X-ray fluorescence signal. Optical cameras and lasers are used for positioning the desired area of the specimen onto the beam. One example of an X-ray microdiffraction beamline arrangement is shown in Fig. 21.6.

For X-ray diffraction experiments, area detectors have replaced scintillator point detectors, with the notable exception of high-resolution X-ray diffractometry, which needs higher angular resolution than any area detector can provide. Examples of high-resolution X-ray diffractometry include grazing incidence X-ray scattering, reflectivity measurements, and high-resolution powder diffraction. In all of these, the word high-resolution refers to the angular resolution (not the spatial resolution), which provides the separation of angularly close reflections.

For all other uses, two-dimensional area detectors have now become ubiquitous, as they can capture in a single frame a large cone of the reciprocal space. The most common area detectors for X-rays found on beamlines and laboratories nowadays are charge-coupled device and pixel array detectors, which have replaced films and image plates.

Synchrotron X-ray microbeam beamlines delivering micron- and submicron-size beams were introduced in the late 1990s (Ice, 1997; Iida & Hirano, 1996; MacDowell et al., 2001; Riekel, 2000), but they only gained momentum in the early 2000s (Budai et al., 2003; Do et al., 2004; Evans, Isaacs, Aeppli, Cai, & Lai, 2002; Jo et al., 2010; Rogan, Tamura, Swift, & Ustündag, 2003). Consequently, their use for the study of biominerals is fairly recent. One of the first studies on biominerals was done by DiMasi in 2004 (DiMasi & Sarikaya, 2004) where the authors investigated the orientation of crystals in the prismatic calcite layer and in the nacre aragonite layer of a red abalone (*Haliotis rufescens*) shell. The study was performed using a monochromatic beam of energy 8.5 keV at the X20A beamline of the NSLS at Brookhaven National Laboratory. The X-ray beam was focused with capillary optics to a size of 5–15 μm and data collection was performed using a single-channel scintillator detector. This is a fairly complex operation necessitating tedious sample and detector alignment in order to collect Bragg reflections onto the detector. For each reflection and each crystal orientation, the sample–detector configuration has to be changed in addition to the need to scan the sample position. Nevertheless, DiMasi and Sarikaya were able to map a few crystals and get a first glimpse on the orientation distribution of both calcite and aragonite at the prismatic-nacre boundary of the shell. They found that both calcite and aragonite are less orientationally ordered in a region within 100 μm of the prismatic-nacre boundary. Today's ubiquity of area X-ray detectors has made this type of study more straightforward as will be seen in Section 6 (Fig. 21.7).

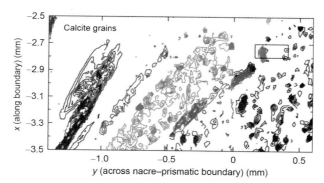

Figure 21.7 Bragg peak intensity map for the calcite (left, $y \leq 0.2$ mm) and aragonite (right, $y \geq 0.2$ mm) grains at and near the prismatic-nacre boundary in a red abalone shell. Each color island indicates a region of cooriented crystals. *Data from DiMasi and Sarikaya (2004).* (See color plate.)

The other pioneering study in biomineralization using X-ray microdiffraction was performed at approximately the same time at the ESRF on the microfocus beamline ID13 using a capillary-focused 5 μm-size X-ray beam, with an area detector. These data are presented in Section 5 (Söllner et al., 2003).

4. SAMPLE PREPARATION

One aspect that makes hard X-ray diffraction techniques quite attractive is that it requires only minimum sample preparation. Surface-sensitive techniques such as electron microscopy using highly interacting particles (electrons) require working in high vacuum. Hard X-rays in contrast are only moderately scattered by air and typically penetrate deep in the sample with little interaction, so that problems such as sample damage due to local absorption and heating of the sample are limited. One of the major limitations of transmission electron microscopy is the requirement that the sample be electron-transparent and therefore not thicker than a few tens of nanometers. Hard X-rays in contrast penetrate most biominerals to a distance of approximately 100 μm (Henke, Gullikson, & Davis, 1993).

Typically, samples for microdiffraction are embedded in epoxy and polished, so the area of interest can easily be identified and analyzed using an optical microscope.

No obvious effects of beam damage are usually detectable on biomineral crystals or their orientations, with the notable exception of amorphous, forming sea urchin spicules, analyzed and exposed to a white-focused X-ray beam within fresh, fixed, dehydrated, but still entire sea urchin embryos. In that particular case, beam damage was observable by eye on the surface of the sample after exposure to white light, as the areas scanned appeared brown or black in color. Unexpectedly, we observed that each spicule instead of being composed in part of single-crystalline calcite and in part of amorphous calcium carbonate, as observed spectroscopically with many different methods (FTIR, EXAFS, and XANES; Beniash, Aizenberg, Addadi, & Weiner, 1997; Gong et al., 2012; Politi et al., 2006, 2008), exhibited multiple crystals of calcite, differently oriented along the spicule. This observation was attributed to beam damage: exposure of the fresh, forming spicule to the intense whitelight 1-μm beam must have induced the crystallization of amorphous calcium carbonate into multiple calcite crystals with random orientations, instead of the expected single-crystal, single-orientation spicule observed for decades with the previously mentioned methods, and also with visible-light microscopy and crossed polarizers in living embryos. Multicrystalline spicules were therefore deemed artifacts of beam damage and were never published nor mentioned before in the literature.

5. POWDER X-RAY MICRODIFFRACTION

Monochromatic beam is preferentially used when the average size of the coherent, that is, cooriented, domains is much smaller than the beam size. This translates into samples with nanocrystalline grains for a micron-sized beam. As mentioned in Section 2, the diffraction patterns are called Debye–Scherrer rings, which are actually conics. When the polycrystalline sample is an aggregate of randomly oriented, sufficiently small crystallites, each ring in the pattern is continuous.

The angular position θ of a ring provides the interplanar atomic spacing d_{hkl} associated with the ring, which is obtained via the Bragg equation. The angular position and the relative intensities of the rings on the area detector are used as the signature of a particular, unique, crystalline species. Powder X-ray diffraction is therefore frequently used as a first tool to identify the crystalline mineral phase or phases present in the sample. The combination of relative intensities of the rings and their angular position is indeed unique to a crystal structure. For instance, calcite gives at moderate angles a strong ring at 3.036 Å (1,0,4) and two weaker ones at 2.495 Å (2,−1,0) and 3.855 Å (1,0,2) and can be easily distinguished from aragonite, which gives strong rings at 3.396 Å (1,1,1), 3.273 Å (1,0,2), and 2.701 Å (2,0,1). For phase identification, it is customary to convert the 2D pattern obtained from an area detector into a 1D pattern called diffractogram, by integrating the intensity along the azimuthal direction. These patterns can then be compared with those in a structure database such as International Center for Diffraction Data that maintains collections of calculated and experimentally determined 1D diffractograms (Fig. 21.8).

Phase identification is not the only information that can be extracted from a Debye–Scherrer ring pattern. Small shifts in the position of the rings relative to their theoretical, unstrained position reflect macrostrain, that is, overall strain that affects all the crystallites in the beam path. In particular, changes in the ellipticity of the rings reflect anisotropic macrostrain and can be extracted using a generalization of the $\sin^2\psi$ technique (Böhm et al., 2004; Noyan & Cohen, 1987; Tamura, 2013). The average crystallite size can be qualitatively assessed by the level of granularity in the rings, along the azimuthal χ direction (Fig. 21.9). Continuous rings

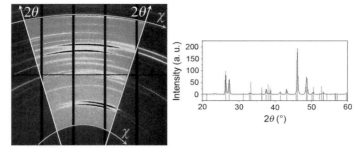

Figure 21.8 Left: X-ray diffraction pattern of Debye–Scherrer rings from a piece of shell from the welk *Urosalpinx cinerea*. The integration area along 2θ and χ is highlighted by inverting the colors, so they appear as complementary to the colors in the rest of the image. This area is the orange sector of an annulus. Right: the resulting 1D diffractograms, obtained from the annular sector on the left, integrated over 2θ and χ. Red lines are the aragonite peak position from a database. All reflections match those of aragonite. *Data courtesy of J. B. Ries*. (See color plate.)

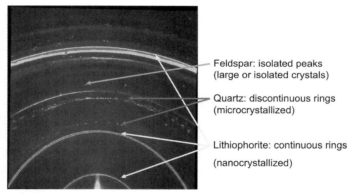

Figure 21.9 This diffraction pattern of a sample of soil shows different type of rings depending on the average grain size of the minerals. *Data courtesy of A. Manceau.* (See color plate.)

are obtained with crystallite grains much smaller than the X-ray beam size, for instance with a micron-size beam, and crystallite sizes on the order of a few tens of nanometers. When the rings show some granularity and reflections from individual grains become recognizable, the average grain size is an order of magnitude smaller than the beam size (i.e., around 100 nm for a 1 μm-size beam). Polycrystals with grain size on the order of or larger than the beam size diffract with isolated spot reflections, if they at all contribute diffraction peaks.

For small crystallite sizes (under 100 nm or so), a quantitative measure can be obtained from the width of the ring (corrected with instrument broadening) using the Scherrer equation:

$$W = \frac{K\lambda}{D\cos\theta}$$

where W is the full width at half maximum of the reflection (line broadening), K is a shape factor (usually set to 0.9 for spherical crystallites), and D the average crystallite size.

Ring width can also be related to microstrain, that is, strain that affects individual crystallites differently. In this case, formulas such as the Stokes–Wilson equation in the succeeding text can be employed for an estimate of the microstrain:

$$W = 2\xi \tan\theta$$

where ξ is the average strain. The contributions of strain and size can be separated using either the Hall–Williamson plot or the more accurate Averbach–Warren method (Noyan & Cohen, 1987; Warren, 1969; Warren & Averbach, 1952; Williamson & Hall, 1953).

When the sample is textured, that is, crystallites have a preferential orientation, the more or less continuous rings are replaced by arcs with visible maxima at certain azimuthal positions (as the example on Fig. 21.8). Analyzing the position of these arcs can reveal the preferred orientation of crystallites. A complete picture is obtained via pole figure analysis. A pole represents a given crystallographic axis in a stereographic projection. For instance, a [001] pole figure shows the distribution, in the sample coordinate system, of all crystallite [001] axes in the beam path. A complete pole figure requires a full angular coverage of all equivalent directions of the crystallites, which can be obtained from a set of diffraction patterns at different ϕ rotation angles of the sample, in a method called "texture analysis" (Heidelbach, Riekel, & Wenk, 1999; Wenk & Grigull, 2003). Pole figures can also be derived from polychromatic microbeam scans as seen in Section 6. In texture analysis, "inverse pole figures" are also often used and can be seen as the opposite of a "pole figure." An inverse pole figure shows the distribution in the crystal reference frame of a selected direction, for example, the normal to the sample surface. Texture analysis using a 1 μm beam has been performed at the ESRF in cortical bone, making it possible to map the distribution of the orientation of the c-axis of the mineral apatite in the sample with high spatial resolution (Wagermaier et al., 2007).

The Rietveld method (Rietveld, 1969) makes it possible to refine lattice parameters and crystal structures from high-resolution X-ray powder data when single crystals of sufficiently large size and good quality are not available (Pokroy, Quintana, Caspi, Berner, & Zolotoyabko, 2004). Rietveld refinement uses a nonlinear least square fit to minimize the differences between the entire set of observed peak intensities in a 1D diffractogram and the peaks calculated from a crystal model. This method has been extended to be used in a wide variety of problems, including the estimation of preferred orientation and microstrains (Lutterotti & Scardi, 1990). The Rietveld method was, for instance, used in combination with texture analysis to study the texture of dinosaur tendon and salmon scales (Lonardelli, Wenk, Lutterotti, & Goodwin, 2005).

Microbeam powder diffraction was used, among other studies, in the study of "starmaker," a protein responsible for the morphogenesis of the aragonite otoliths from the inner ear of the zebrafish (Söllner et al., 2003). In zebrafish, otoliths are biomineral crystallites of aragonite, embedded in a protein framework, used by the fish for gravity sensing, balance, and hearing. Söllner et al. show that as the activity of the starmaker gene is progressively reduced, drastic changes in morphology of the otolith occur: from a round aggregate of aragonite crystallites through a star-like assembly of larger aragonite crystals, and finally to an aggregate of calcite rhombohedral crystals. Synchrotron X-ray powder microdiffraction using a 5 μm beam clearly showed the change in the calcium carbonate polymorph from aragonite to calcite and change in texture as starmaker activity is suppressed (Fig. 21.10).

In the case of highly textured samples, the monochromatic diffraction patterns exhibit an almost single-crystalline quality. One such example of highly cooriented biominerals is given by the grinding tip of the sea urchin tooth (Killian et al., 2009, 2011). Sea urchin teeth are made of low (5–13 mol%) Mg calcite plates and needles, cemented together by a polycrystalline matrix of Mg-rich (40–45 mol%) calcite nanoparticles. Lattice spacings in calcite are a function of Mg content: the larger the Mg content, the smaller the lattice spacings. This can be used to map the Mg concentrations in the sea urchin tooth (Ma et al., 2009). Because of the almost single-crystal quality of the sea urchin tooth, the rings are reduced in the monochromatic beam pattern to broad, isolated spots. Sample rotation was necessary to excite a suitable number of reflections. In that particular experiment, diffraction patterns were obtained using a 10×10 μm^2 monochromatic beam while oscillating the sample by up to 90° in angle. The broadening in the radial direction (2θ) is a function of crystallite size, strain, but also in this case of the gradient in Mg composition. The broadening in

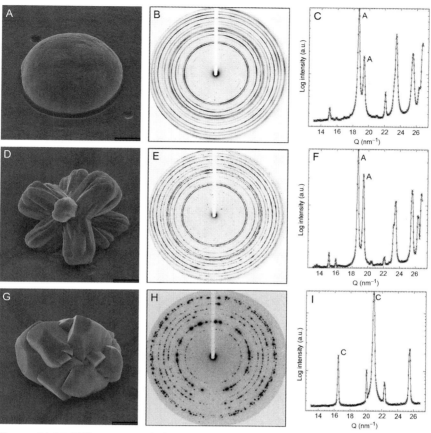

Figure 21.10 Different morphologies and corresponding diffraction patterns of zebrafish otoliths, with decreasing activity of starmaker. The wild-type otolith (A–C) is aragonite, with powder diffraction rings that are more or less continuous (B); thus, the average grain size is at the nanoscale. The diffractogram (C) clearly shows aragonite, with some peak broadening, also indicating nanosized crystallites. With decreasing activity of starmaker, the otolith appears star-shaped (D), the rings are grainier (E) indicating larger grain size, but they still index as aragonite (F) with peaks narrower than in (C). When starmaker is maximally suppressed, the otolith appears as a cluster of rhombohedra (G), the rings appear even spottier (H) indicating micrometric grain size, and the location of the rings indicates that the otolith is entirely composed of calcite (I). Scale bars in are 20 μm. *Data from Söllner et al. (2003).*

the azimuthal direction (χ) is a function of angular spread (AS). Along the azimuthal direction χ, reflections are split into pairs (Fig. 21.11D), indicating the presence of two domains with different orientations.

Monochromatic X-ray microdiffraction was also used to help elucidate the anisotropic mechanical behavior of the ultrathin shell of the pelagic

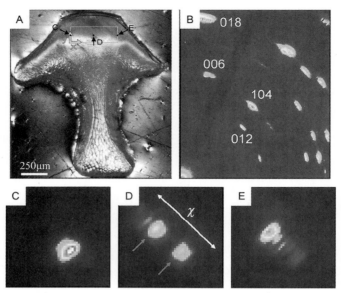

Figure 21.11 Diffraction patterns obtained with a 10×10 μm² monochromatic beam and sample oscillation up to 90°. *Data from Ma et al. (2009).* (See color plate.)

pteropod *Cavolinia uncinata* (Zhang et al., 2011). The shell is made of interlocked helical aragonite nanofibers diffracting almost like a single crystal, with Debye–Scherrer rings reduced to broad spots.

6. WHITE-BEAM X-RAY MICRODIFFRACTION

Monochromatic powder diffraction patterns only provide information averaged over all the crystallites illuminated by the X-ray beam. White-beam Laue diffraction, on the other hand, can provide information on the individual diffracting crystallites. In contrast to monochromatic diffraction, white-beam diffraction does not require sample rotation and a single frame is generally sufficient to characterize the diffracting volume of the sample. In addition, due to crystal imperfections, the wavelength spread of reflections in Laue patterns is typically much larger than the wavelength spread of a fully monochromatized beam, meaning that a white-beam pattern requires a shorter exposure time than a monochromatic-beam one. White-beam diffraction is therefore a fast technique, which can be used for time-resolved experiments or for scanning large areas of the sample. If the incident X-ray beam is small, and a limited number of crystallites are illuminated in a polycrystalline sample, then the indexing of the Laue patterns provides the full orientation in space of the crystallites.

Besides crystallite orientation, Laue patterns also provide information on the deformation state and defect density in the sample. Small deviations of the Laue reflection positions with respect to their "unstrained" positions (theoretical positions calculated from literature values of the undeformed crystal lattice parameters) can be measured to provide information on how the unit cell is deformed and thus the local elastic strain experienced by the sample. Laue patterns in effect provide the full deviatoric elastic strain tensor (Chung & Ice, 1999). Additionally, the shape of the reflections informs on the plastic strain experienced by the sample (Barabash et al., 2001; Chao et al., 2012).

Laue patterns from a polycrystalline sample are generally made of overlapping Laue diffraction patterns generated by all the crystallites in the X-ray beam path. Specifically designed algorithms make it possible to index multiple grains in a single pattern (Tamura, 2013). Figure 21.12 shows an

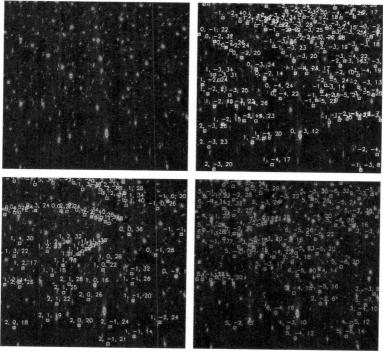

Figure 21.12 The Laue pattern in the top left panel was acquired with a 1 μm X-ray beam from a region of the prismatic calcite layer of the *Haliotis rubra* shell. This pattern shows the composite diffraction of at least three different calcite crystallites, the *h,k,l* indexing of the separate crystallites are highlighted in yellow, orange, and red in other panels. (See color plate.)

example of overlapping Laue patterns taken in the prismatic calcite layer of the shell formed by the gastropod *Haliotis rubra*.

Laue X-ray microdiffraction has been used to study the growth mechanism of nacre in red abalone (*H. rufescens*) shells (Gilbert et al., 2008). In that study, PEEM imaging provided a semiquantitative assessment of misorientations in nacre with a few tens of nanometer resolution and was complemented with Laue X-ray microdiffraction measurements to obtain absolute quantitative values of those misorientations. Abalone nacre is a lamellar arrangement of aragonite layers with regular thickness in which polygonal tablets are stacked together like coins in the direction perpendicular to the layers. Tablets in a stack of coins are frequently cooriented. Since the tablets are thinner than the beam size, and the X-ray beam penetrates deep into the sample (Henke et al., 1993), the resulting Laue pattern is a composite of the contribution of several tablets from many stacks of coins at once. Multigrain indexing provides the orientation of several tablets in a single Laue pattern, with respect to a given coordinate system, and thus provides both the individual contributions and the statistics over multiple tablets as a function of location in the sample. One result of the study is that nacre is gradually ordering over a distance of about 50 μm from the prismatic–nacre boundary. The degree of misorientation is given by the AS of the c-axis of the aragonite tablet normal to the tablet plane. A systematic study on different seashells shows that this ordering differs from species to species of mollusk shells, as shown in Fig. 21.13 (Olson, Blonsky, et al. 2013).

The AS has been found to vary significantly in the prismatic calcite layer of seashells and also to correlate with hardness (Olson, Metzler, et al., 2013). Eight different species of seashells prisms were analyzed, including gastropods (several species of *Haliotis*), bivalves (*Pinctada*, *Atrina*, and *Mytilus*), and one cephalopod (*Nautilus*). Laue X-ray microdiffraction using a 1 μm-size polychromatic beam was used at the ALS and complemented with microindentation hardness measurements. The AS is measured by plotting a pole figure of selected crystallite orientations, obtained by indexing a series of Laue patterns taken during sample mapping (Fig. 21.14). AS turned out to be strongly correlated with hardness, with the *Pinctada* species showing the hardest shell with largest values of AS (Olson, Metzler, et al., 2013).

7. MICROBEAM SMALL-ANGLE X-RAY SCATTERING

SAXS provides information such as average grain size and shape for sample grains in the nanometer (1–1000 nm) size range. In a typical SAXS

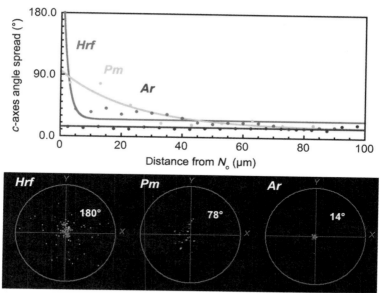

Figure 21.13 Laue X-ray microdiffraction was used to quantitatively assess the angle spread of c-axes in nacre as a function of distance from the first nacre layer N_o. Laue diffraction maps provided area-selected pole figures. The shells represented here are from *Haliotis rufescens* (Hrf), *Pinctada margaritifera* (Pm), and *Atrina rigida* (Ar). Data from Olson, Blonsky, et al. (2013). (See color plate.)

experiment, the sample is mounted in transmission, and X-ray scattering is collected by an area detector, which is placed far downstream from the sample (Fig. 21.5). The distance is necessary to increase spacing and thus angle resolution of the SAXS reflections on the area detector, and a flight tube filled with an inert gas is often necessary to limit air scattering and absorption. A typical SAXS data set is a radial integration of the diffraction pattern, showing the integrated intensity as a function of the magnitude q of the scattering vector:

$$q = 4\pi \sin\theta / \lambda.$$

However, most of the time, SAXS patterns are very complex and extremely difficult to interpret. Monte Carlo simulations are often used to match models of particles size, shape, and spatial distribution with a particular set of SAXS experimental data.

The first example is not a μ-SAXS experiment, as the X-ray beam size used was on the order of 100 μm, but helped to establish a methodology later used for the study of biominerals with smaller beams. The marine

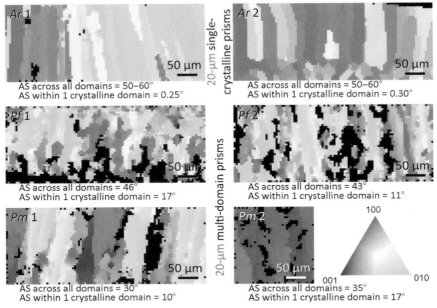

Figure 21.14 Microdiffraction map of the prismatic layer in six shells from three species, with two shells per species: *Atrina rigida* (*Ar1* and *Ar2*), *Pinctada fucata* (*Pf1* and *Pf2*), and *Pinctada margaritifera* (*Pm1* and *Pm2*). The color legend indicates the orientation perpendicular to the image surface. *Data from Olson, Metzler, et al. (2013).* (See color plate.)

bloodworm *Glycera dibranchiata* has jaws with extraordinary resistance to abrasion, far in excess of vertebrate dentin. X-ray diffraction revealed that the mineral constituting the jaw is quite unusual (Lichtenegger et al., 2002): it is a copper-chloride mineral called atacamite [$Cu_2(OH)_3Cl$], which is known to form under extreme environmental conditions, in seawater or in arid hot climates. In fact, the name derives from the Atacama Desert in Chile, the most life-inhospitable place on Earth, as atacamite would dissolve in the presence of freshwater. Moreover, transmission electron microscopy and SAXS indicate that the polycrystalline atacamite is organized into fibers within a protein matrix. SAXS, in particular, was used to determine the orientation and size of the fibers: the average orientation turned out to follow the outer shape of the jaw, interestingly, and the average fiber diameter was determined to be about 80 nm.

Combining μ-SAXS with μ-WAXS is a powerful tool to investigate bone platelets. The micron resolution is needed because of the highly inhomogeneous nature of bone. While μ-WAXS identifies the minerals and

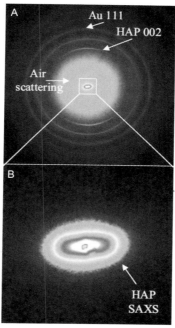

Figure 21.15 Simultaneous WAXS (A) and SAXS (B) obtained on human trabecular bone, acquired at the μ-Spot beamline at BESSY II. The elliptical shape of the SAXS pattern reflects the flat-platelet shape of the hydroxyapatite nanoparticles. *Data from Paris (2008).* (See color plate.)

their crystallographic orientations in the bone, μ-SAXS gives knowledge on the morphology and orientation of the nanoplatelets. Such techniques build on previous usage of SAXS with much larger beam size, on the order of 200 μm (Camacho et al., 1999; Jaschouz et al., 2003). One example of the combined use of μ-SAXS and μ-WAXS is given by the study of the microstructure of human trabecular bone (Paris, 2008). A thin section of trabecular bone is scanned under an X-ray microbeam and both WAXS and SAXS patterns are simultaneously recorded. The WAXS pattern shows texture in the ring of hydroxyapatite, while the SAXS patterns show size, shape, and orientations of the nanoparticles (Fig. 21.15). See Paris (2008) for a review of the combined use of SAXS/WAXS including μ-SAXS/μ-WAXS to study biominerals, such as bone and insect cuticle. A more recent study of the forming bone in zebrafish fin rays observed the transition from amorphous to crystalline bone crystallite using μ-SAXS/μ-WAXS (Mahamid et al., 2010).

8. FUTURE DEVELOPMENTS

X-ray microdiffraction is a powerful tool to investigate the microstructure of biomaterials at the nano-, meso-, and microscales. It makes it possible to map the distribution of crystal orientation, degree of crystallization, mineral species identification, and strain with micron to submicron spatial resolution. However, X-ray microdiffraction, as many other techniques, is by no means a stand-alone method offering all answers at once. X-ray microdiffraction becomes most effective when used in combination with other techniques such as imaging (X-ray microscopy, microtomography, X-PEEM, scanning electron microscopy, EBSD, etc.), X-ray microfluorescence for elemental mapping, micro-XANES for mapping degree of oxidation, micro-EXAFS for mapping local atomic environments, and nano- and microindentation for measuring material hardness.

With the imminent advent of fourth-generation synchrotron sources and X-ray lasers, X-ray microdiffraction will benefit from greater photon flux, higher brightness, and better focusing optics, all of which will decrease the beam spot size on the sample and therefore increase the spatial resolution, eventually leading to microdiffraction being replaced by X-ray nanodiffraction. For polycrystalline materials, nanodiffraction will not really have nanoresolution, as penetrating X-rays would still probe approximately 100 µm deep into the sample, and diffraction patterns would still contain overlapping contribution from multiple crystallites within the path of the beam. 3D X-ray microdiffraction techniques such as differential aperture X-ray microscopy, based on Laue X-ray microdiffraction (Larson, Yang, Ice, Budai, & Tischler, 2002), would provide depth resolution in addition to lateral resolution, to obtain a three-dimensional view of the distribution of crystallites; however, data collection is time-consuming and data reduction a very expensive process. Other technical developments include faster area detectors with greater signal-to-noise ratios. In addition to these technical breakthroughs, data analysis will have to keep up with increasing production rate and complexity of the data, as more demanding experiments are devised. Already today, when used in scanning mode, X-ray microdiffraction generates thousands of diffraction patterns that need to be processed in order to extract meaningful information, and this requires a fair amount of algorithm development and automation. Techniques such as Laue X-ray microdiffraction and SAXS are currently limited by the lack of efficient software and computing power to digest the data and present them in a meaningful format to the users, and this has only recently been acknowledged by all the major facilities around the world.

In conclusion, the powerful and young methods of X-ray microdiffraction in biomineralization, which are just a dozen years old, are expected to have a significantly brighter future and shed light on the plentiful unexplored biomineral structures.

REFERENCES

Agnarsson, I., Kuntner, M., & Blackledge, T. A. (2010). Bioprospecting finds the toughest biological material: Extraordinary silk from a giant riverine orb spider. *PLoS one*, *5*(9), e11234.

Albeck, S., Weiner, S., & Addadi, L. (1996). Polysaccharides of intracrystalline glycoproteins modulate calcite crystal growth in vitro. *Chemistry*, *2*(3), 278–284.

Autumn, K., Liang, Y. A., Hsieh, S. T., Zesch, W., Chan, W. P., Kenny, T. W., et al. (2000). Adhesive force of a single gecko foot-hair. *Nature*, *405*(6787), 681–685.

Autumn, K., Sitti, M., Liang, Y. A., Peattie, A. M., Hansen, W. R., Sponberg, S., et al. (2002). Evidence for van der Waals adhesion in gecko setae. *Proceedings of the National Academy of Sciences of the United States of America*, *99*(19), 12252–12256.

Barabash, R., Ice, G. E., Larson, B. C., Pharr, G. M., Chung, K.-S., & Yang, W. (2001). White microbeam diffraction from distorted crystals. *Applied Physics Letters*, *79*(6), 749.

Beniash, E., Aizenberg, J., Addadi, L., & Weiner, S. (1997). Amorphous calcium carbonate transforms into calcite during sea urchin larval spicule growth. *Proceedings of the Royal Society of London Series B*, *264*(1380), 461–465.

Bilderback, D. H. (2003). Review of capillary x-ray optics from the 2nd international capillary optics meeting. *X-Ray Spectrometry*, *32*(3), 195–207.

Böhm, J., Gruber, P., Spolenak, R., Stierle, A., Wanner, A., & Arzt, E. (2004). Tensile testing of ultrathin polycrystalline films: A synchrotron-based technique. *Review of Scientific Instruments*, *75*(4), 1110.

Brindley, G. W. (1961). The x-ray identification and crystal structures of clay minerals. In G. Brown (Ed.) (2nd ed.). Great Britain: Mineralogical Society, Clay Minerals Group, London, pp. 489–514.

Bruet, B. J. F., Song, J., Boyce, M. C., & Ortiz, C. (2008). Materials design principles of ancient fish armour. *Nature Materials*, *7*(9), 748–756.

Budai, J. D., Yang, W., Tamura, N., Chung, J.-S., Tischler, J. Z., Larson, B. C., et al. (2003). X-ray microdiffraction study of growth modes and crystallographic tilts in oxide films on metal substrates. *Nature Materials*, *2*(7), 487–492.

Camacho, N. P., Rinnerthaler, S., Paschalis, E. P., Mendelsohn, R., Boskey, A., L., & Fratzl, P. (1999). Complementary information on bone ultrastructure from scanning small angle X-ray scattering and Fourier-transform infrared microspectroscopy. *Bone*, *25*(3), 287–293.

Chao, W., Harteneck, B. D., Liddle, J. A., Anderson, E. H., & Attwood, D. T. (2005). Soft X-ray microscopy at a spatial resolution better than 15 nm. *Nature*, *435*(7046), 1210–1213.

Chao, J., Suominen Fuller, M. L., McIntyre, N. S., Carcea, A. G., Newman, R. C., Kunz, M., et al. (2012). The study of stress application and corrosion cracking on Ni–16 Cr–9 Fe (Alloy 600) C-ring samples by polychromatic X-ray microdiffraction. *Acta Materialia*, *60*(3), 781–792.

Chu, Y. S., Yi, J. M., De Carlo, F., Shen, Q., Lee, W.-K., Wu, H. J., et al. (2008). Hard-x-ray microscopy with Fresnel zone plates reaches 40 nm Rayleigh resolution. *Applied Physics Letters*, *92*(10), 103–119.

Chung, J., & Ice, G. (1999). Automated indexing for texture and strain measurement with broad-bandpass x-ray microbeams. *Journal of Applied Physics, 86*(9), 5249–5255.

Cranford, S. W., Tarakanova, A., Pugno, N. M., & Buehler, M. J. (2012). Nonlinear material behaviour of spider silk yields robust webs. *Nature, 482*(7383), 72–76.

Currey, J. D. (1977). Mechanical properties of mother of pearl in tension. *Proceedings of the Royal Society of London Series B, 196,* 443–463.

Di Fabrizio, E., Romanato, F., & Gentili, M. (1999). High-efficiency multilevel zone plates for keV X-rays. *Nature, 401,* 895–898.

DiMasi, E., & Sarikaya, M. (2004). Synchrotron x-ray microbeam diffraction from abalone shell. *Journal of Materials Research, 19*(05), 1471–1476.

Do, D.-H., Evans, P. G., Isaacs, E. D., Kim, D. M., Eom, C. B., & Dufresne, E. M. (2004). Structural visualization of polarization fatigue in epitaxial ferroelectric oxide devices. *Nature Materials, 3*(6), 365–369.

Evans, P. G., Isaacs, E. D., Aeppli, G., Cai, Z., & Lai, B. (2002). X-ray microdiffraction images of antiferromagnetic domain evolution in chromium. *Science, 295*(5557), 1042–1045.

Evans-Lutterodt, K., Stein, A., Ablett, J., Bozovic, N., Taylor, A., & Tennant, D. (2007). Using compound kinoform hard-X-ray lenses to exceed the critical angle limit. *Physical Review Letters, 99*(13), 134801.

Falini, G., Albeck, S., Weiner, S., & Addadi, L. (1996). Control of aragonite or calcite polymorphism by mollusk shell macromolecules. *Science, 271,* 67–69.

Gao, H., Ji, B., Jäger, I., Arzt, E., & Fratzl, P. (2003). Materials become insensitive to flaws at nanoscale: Lessons from nature. *Proceedings of the National Academy of Sciences of the United States of America, 100*(10), 5597–5600.

Giacovazzo, C. (Ed.), (2002). *Fundamentals of crystallography* (2nd ed.). New York: Oxford University Press.

Gilbert, P. U. P. A. (2012). Polarization-dependent imaging contrast (PIC) mapping reveals nanocrystal orientation patterns in carbonate biominerals. *Journal of Electron Spectroscopy and Related Phenomena, 185,* 395–405.

Gilbert, P. U. P. A. (2014). Photoemission spectromicroscopy for the biomineralogist. In E. DiMasi & L. B. Gower (Eds.), *Biomineralization handbook, characterization of biominerals and biomimetic materials.* Boca Raton, FL: CRC Press.

Gilbert, P. U. P. A., Metzler, R. A., Zhou, D., Scholl, A., Doran, A., Young, A., et al. (2008). Gradual ordering in red abalone nacre. *Journal of the American Chemical Society, 130*(51), 17519–17527.

Gong, Y., Killian, C. E., Olson, I. C., Appathurai, N. P., Amasino, A., Martin, M. C., et al. (2012). Phase transitions in biogenic amorphous calcium carbonate. *Proceedings of the National Academy of Sciences of the United States of America, 109*(16), 6088–6093.

Heidelbach, F., Riekel, C., & Wenk, H. R. (1999). Quantitative texture analysis of small domains with synchrotron radiation X-rays. *Journal of Applied Crystallography, 32*(5), 841–849.

Henke, B. L., Gullikson, E. M., & Davis, J. C. (1993). Center for X-ray optics, X-ray attenuation length calculator. http://henke.lbl.gov/optical_constants/atten2.html.

Hignette, O., Cloetens, P., Rostaing, G., Bernard, P., & Morawe, C. (2005). Efficient sub 100 nm focusing of hard x rays. *Review of Scientific Instruments, 76*(6), 063709.

Huber, G., Mantz, H., Spolenak, R., Mecke, K., Jacobs, K., Gorb, S. N., et al. (2005). Evidence for capillarity contributions to gecko adhesion from single spatula nanomechanical measurements. *Proceedings of the National Academy of Sciences of the United States of America, 102*(45), 16293–16296.

Ice, G. E. (1997). Microbeam-forming methods for synchrotron. *X-Ray Spectrometry, 26,* 315–326.

Ice, G. E., Chung, J.-S., Tischler, J. Z., Lunt, A., & Assoufid, L. (2000). Elliptical x-ray microprobe mirrors by differential deposition. *Review of Scientific Instruments, 71*(7), 2635.

Iida, A., & Hirano, K. (1996). Kirkpatrick-Baez optics for a sub-micron synchrotron X-ray microbeam and its applications to X-ray analysis. *Nuclear Instruments and Methods in Physics Research Section B, 114,* 149–153.

Jark, W., Cedola, A., Di Fonzo, S., Fiordelisi, M., Lagomarsino, S., Kovalenko, N. V., et al. (2001). High gain beam compression in new-generation thin-film x-ray waveguides. *Applied Physics Letters, 78*(9), 1192.

Jark, W., Pérennès, F., Matteucci, M., Mancini, L., Montanari, F., Rigon, L., et al. (2004). Focusing X-rays with simple arrays of prism-like structures. *Journal of Synchrotron Radiation, 11*(Pt. 3), 248–253.

Jaschouz, D., Paris, O., Roschger, P., Hwang, H.-S., & Fratzl, P. (2003). Pole figure analysis of mineral nanoparticle orientation in individual trabecula of human vertebral bone. *Journal of Applied Crystallography, 36*(3), 494–498.

Jo, J. Y., Sichel, R. J., Lee, H. N., Nakhmanson, S. M., Dufresne, E. M., & Evans, P. G. (2010). Piezoelectricity in the dielectric component of nanoscale dielectric-ferroelectric superlattices. *Physical Review Letters, 104*(20), 207601.

Kang, H. C., Yan, H., Winarski, R. P., Holt, M. V., Maser, J., Liu, C., et al. (2008). Focusing of hard x-rays to 16 nanometers with a multilayer Laue lens. *Applied Physics Letters, 92*(22), 221114.

Killian, C. E., Metzler, R. A., Gong, Y. U. T., Olson, I. C., Aizenberg, J., Politi, Y., et al. (2009). Mechanism of calcite co-orientation in the sea urchin tooth. *Journal of the American Chemical Society, 131*(51), 18404–18409.

Killian, C. E., Metzler, R. A., Gong, Y. U. T., Churchill, T. H., Olson, I. C., et al. (2011). Self-sharpening mechanism of the sea urchin tooth. *Advanced Functional Materials, 21,* 682–690.

Kirkpatrick, P., & Baez, A. V. (1948). Formation of optical images by x-rays. *Journal of the Optical Society of America, 38*(9), 766–774.

Kunz, M., Tamura, N., Chen, K., MacDowell, A. A., Celestre, R. S., Church, M. M., et al. (2009). A dedicated superbend x-ray microdiffraction beamline for materials, geo-, and environmental sciences at the advanced light source. *The Review of Scientific Instruments, 80*(3), 035108.

Lai, B., Yun, W., Legnini, D., & Xiao, Y. (1992). Hard x-ray phase zone plate fabricated by lithographic techniques. *Applied Physics Letters, 61*(16), 1877–1879.

Larson, B. C., Yang, W., Ice, G. E., Budai, J. D., & Tischler, J. Z. (2002). Three-dimensional X-ray structural microscopy with submicrometre resolution. *Nature, 415*(6874), 887–890.

Lee, H., Lee, B. P., & Messersmith, P. B. (2007). A reversible wet/dry adhesive inspired by mussels and geckos. *Nature, 448*(7151), 338–341.

Levi-Kalisman, Y., Falini, G., Addadi, L., & Weiner, S. (2001). Structure of the nacreous organic matrix of a bivalve mollusk shell examined in the hydrated state using cryo-TEM. *Journal of Structural Biology, 135*(1), 8–17.

Lichtenegger, H. C., Schöberl, T., Bartl, M. H., Waite, H., & Stucky, G. D. (2002). High abrasion resistance with sparse mineralization: Copper biomineral in worm jaws. *Science, 298*(5592), 389–392.

Lonardelli, I., Wenk, H. R., Lutterotti, L., & Goodwin, M. (2005). Texture analysis from synchrotron diffraction images with the Rietveld method: Dinosaur tendon and salmon scale. *Journal of Synchrotron Radiation, 12*(Pt. 3), 354–360.

Lowenstam, H. A., & Weiner, S. (1989). *On biomineralization.* Oxford: Oxford University Press.

Lutterotti, L., & Scardi, P. (1990). Simultaneous structure and size–strain refinement by the Rietveld method. *Journal of Applied Crystallography, 23*(4), 246–252.

Ma, Y., Aichmayer, B., Paris, O., Fratzl, P., Meibom, A., Metzler, R. A., et al. (2009). The grinding tip of the sea urchin tooth exhibits exquisite control over calcite crystal orientation and Mg distribution. *Proceedings of the National Academy of Sciences of the United States of America, 106*(15), 6048–6053.

MacDowell, A. A., Celestre, R. S., Tamura, N., Spolenak, R., Valek, B., Brown, W. L., et al. (2001). Submicron X-ray diffraction. *Nuclear Instruments and Methods in Physics Research Section A, 467–468,* 936–943.

Mahamid, J., Aichmayer, B., Shimoni, E., Ziblat, R., Li, C., Siegel, S., et al. (2010). Mapping amorphous calcium phosphate transformation into crystalline mineral from the cell to the bone in zebrafish fin rays. *Proceedings of the National Academy of Sciences of the United States of America, 107*(14), 6316–6321.

Meyers, M., McKittrick, J., & Chen, P. (2013). Structural biological materials: Critical mechanics-materials connections. *Science, 339,* 773–779.

Mimura, H., Matsuyama, S., Yumoto, H., Hara, H., Yamamura, K., Sano, Y., et al. (2005). Hard X-ray diffraction-limited nanofocusing with Kirkpatrick-Baez mirrors. *Japanese Journal of Applied Physics, 44*(18), L539–L542.

Moffat, K., Szebenyi, D., & Bilderback, D. (1984). X-ray Laue diffraction from protein crystals. *Science, 535,* 1423–1425.

Munch, E., Launey, M., Alsem, D., & Saiz, E. (2008). Tough, bio-inspired hybrid materials. *Science, 322*(5907), 1516–1520.

Nalla, R. K., Kinney, J. H., & Ritchie, R. O. (2003). Mechanistic fracture criteria for the failure of human cortical bone. *Nature Materials, 2*(3), 164–168.

Noyan, I. C., & Cohen, J. B. (1987). *Residual stress measurement by diffraction and interpretation.* New York: Springer-Verlag.

Olson, I. C., Blonsky, A. Z., Tamura, N., Kunz, M., Romao, C. P., White, M. A., et al. (2013). Crystal nucleation and near-epitaxial growth in nacre. *Journal of Structural Biology,* in press.

Olson, I. C., Metzler, R. A., Tamura, N., Kunz, M., Killian, C. E., & Gilbert, P. (2013). Crystal lattice tilting in prismatic calcite. *Journal of Structural Biology, 183,* 180–190.

Ortiz, C., & Boyce, M. (2008). Bioinspired structural materials. *Science, 319,* 1053–1054.

Paris, O. (2008). From diffraction to imaging: New avenues in studying hierarchical biological tissues with x-ray microbeams (Review). *Biointerphases, 3*(2), FB16.

Pfeiffer, F., David, C., Burghammer, M., Riekel, C., & Salditt, T. (2002). Two-dimensional x-ray waveguides and point sources. *Science, 297*(5579), 230–234.

Pokroy, B., Quintana, J. P., Caspi, E. N., Berner, A., & Zolotoyabko, E. (2004). Anisotropic lattice distortions in biogenic aragonite. *Nature Materials, 3*(12), 900–902.

Politi, Y., Levi-Kalisman, Y., Raz, S., Wilt, F., Addadi, L., Weiner, S., et al. (2006). Structural characterization of the transient amorphous calcium carbonate precursor phase in sea urchin embryos. *Advanced Functional Materials, 16*(10), 1289–1298.

Politi, Y., Metzler, R. A., Abrecht, M., Gilbert, B., Wilt, F. H., Sagi, I., et al. (2008). Transformation mechanism of amorphous calcium carbonate into calcite in the sea urchin larval spicule. *Proceedings of the National Academy of Sciences of the United States of America, 105*(45), 17362–17366.

Riekel, C. (2000). New avenues in x-ray microbeam experiments. *Reports on Progress in Physics, 233*(63), 233–262.

Riekel, C., Burghammer, M., & Davies, R. (2010). Progress in micro- and nano-diffraction at the ESRF ID13 beamline. *IOP Conference Series: Materials Science and Engineering, 14,* 012013.

Rietveld, H. M. (1969). A profile refinement method for nuclear and magnetic structures. *Journal of Applied Crystallography, 2*(2), 65–71.

Rogan, R. C., Tamura, N., Swift, G. A., & Üstündag, E. (2003). Direct measurement of triaxial strain fields around ferroelectric domains using X-ray microdiffraction. *Nature Materials, 2*(6), 379–381.

Schroer, C. G., Kuhlmann, M., Hunger, U. T., Günzler, T. F., Kurapova, O., Feste, S., et al. (2003). Nanofocusing parabolic refractive x-ray lenses. *Applied Physics Letters, 82*(9), 1485.

Snigirev, A., Kohn, V., Snigireva, I., Souvorov, A., & Lengeler, B. (1998). Focusing high-energy x rays by compound refractive lenses. *Applied Optics, 37*(4), 653–662.

Söllner, C., Burghammer, M., Busch-Nentwich, E., Berger, J., Schwarz, H., Riekel, C., et al. (2003). Control of crystal size and lattice formation by starmaker in otolith biomineralization. *Science, 302*(5643), 282–286.

Tai, K., Dao, M., Suresh, S., Palazoglu, A., & Ortiz, C. (2007). Nanoscale heterogeneity promotes energy dissipation in bone. *Nature Materials, 6*(6), 454–462.

Tamura, N. (2013). XMAS: A versatile tool for analyzing synchrotron x-ray microdiffraction data. In R. Barabash & G. Ice (Eds.), *Strain and Dislocation Gradients from Diffraction. Spatially-Resolved Local Structure and Defects, World Scientific*. London: Imperial College Press, in press.

Tamura, N., MacDowell, A. A., Celestre, R. S., Padmore, H. A., Valek, B., Bravman, J. C., et al. (2002). High spatial resolution grain orientation and strain mapping in thin films using polychromatic submicron x-ray diffraction. *Applied Physics Letters, 80*(20), 3724.

Wagermaier, W., Gupta, H. S., Gourrier, A., Paris, O., Roschger, P., Burghammer, M., et al. (2007). Scanning texture analysis of lamellar bone using microbeam synchrotron X-ray radiation. *Journal of Applied Crystallography, 40*(1), 115–120.

Warren, B. E. (1969). *X-ray diffraction*. New York: Dover Publications.

Warren, B. E., & Averbach, B. (1952). The separation of cold-work distortion and particle size broadening in X-ray patterns. *Journal of Applied Physics, 23*(4), 497–498.

Weaver, J. C., Milliron, G. W., Miserez, A., Evans-Lutterodt, K., Herrera, S., Gallana, I., et al. (2012). The stomatopod dactyl club: A formidable damage-tolerant biological hammer. *Science, 336*(6086), 1275–1280.

Wenk, H., & Grigull, S. (2003). Synchrotron texture analysis with area detectors. *Journal of Applied Crystallography, 36*, 1040–1049.

Williamson, G., & Hall, W. (1953). X-ray line broadening from filed aluminium and wolfram. *Acta Metallurgica, 1*, 22–31.

Xia, F., & Jiang, L. (2008). Bio-inspired, smart, multiscale interfacial materials. *Advanced Materials, 20*(15), 2842–2858.

Yang, W., Chen, I., Gludovatz, B., Zimmermann, E., Ritchie, R., & Meyers, M. (2013). Natural flexible dermal armor. *Advanced Materials, 25*(1), 31–48.

Zhang, T., Ma, Y., Chen, K., Kunz, M., Tamura, N., Qiang, M., et al. (2011). Structure and mechanical properties of a pteropod shell consisting of interlocked helical aragonite nanofibers. *Angewandte Chemie, 50*(44), 10361–10365.

Zolotoyabko, E. (2011). *Basic concepts of crystallography*. Weinheim: Wiley-VCH, 266 pp.

CHAPTER TWENTY-TWO

FTIR and Raman Studies of Structure and Bonding in Mineral and Organic–Mineral Composites

Jinhui Tao[1]
Physical Sciences Division, Pacific Northwest National Laboratory, Richland, Washington, USA
[1]Corresponding author: e-mail address: jtao@lbl.gov

Contents

1. Introduction 534
2. FTIR and Raman Spectroscopies Overview 535
   2.1 Vibrations in solids 535
3. The Raman Effect 535
   3.1 Infrared absorption 537
   3.2 FTIR and Raman spectroscopies in biomineralization 538
   3.3 Modes and materials for FTIR 542
4. Measuring Biomimetic Crystals and Biominerals 543
   4.1 Example 1: *In situ* FTIR of calcium phosphate transformation in Tris-buffered solution and phosphate-buffered solution 543
   4.2 Example 2: FTIR and Raman spectra of ACC transformation at different stages 547
5. Future Directions 553
Acknowledgments 553
References 553

Abstract

Spectroscopy techniques, such as Fourier transform infrared (FTIR) and Raman, offer methodologies that overlap and expand X-ray diffraction and transmission electron microscopy (TEM) analyses and help gain new insight into mechanisms of biomineralization. FTIR and Raman spectroscopy techniques measure the molecular environment of asymmetrically and symmetrically vibrating bonds, respectively. As such, these techniques have widely been used to gain information on mineral content, phase, and orientation as well as chemical composition of associated organic matrices like collagen, chitin, or lipids. The traditional coupling of optical microscopes to the newer generation FTIR and Raman spectrometers has enabled these analyses to be performed on samples with 0.1–20 µm spatial resolution. Herein, we briefly discuss the basis and protocol for effective measurements using vibrational spectroscopy by taking two systems from our own research as examples.

1. INTRODUCTION

Biomineralization is the process by which living organism manufactures minerals for different functional purposes, such as mechanical stiffening of tissue, magnetic or gravitational sensing, and element storage (Mann, 1988; Sigel, Sigel, & Sigel, 2008; Weiner & Lowenstam, 1989). From the perspective of taxonomic distribution, the most widespread biominerals are calcium carbonate and calcium phosphate, which are firmly associated with organic matrices such as chitin and collagen to form hybrid structure for shells and bones (Arias & Fernández, 2008; Boskey, 1998; Palmer, Newcomb, Kaltz, Spoerke, & Stupp, 2008; Sarikaya, 1999). The structures of these composites have hierarchical levels from the nano- to microscale. Because of the potential application in guiding new composite material design and controllable synthesis, there has been an increasing interest in deciphering the mechanisms of biomineralization over the past decades (Boskey, 1998; Sarikaya, 1999; Sigel et al., 2008).

Typically, research in biomineralization can be divided into two main categories: (a) the study of the ultrastructure, phase, morphology, and chemistry of biominerals *in vivo* at different development stages (Beniash, Aizenberg, Addadi, & Weiner, 1997; Glimcher, 2006; Grynpas, Bonar, & Glimcher, 1984; Lowenstam & Weiner, 1985; Mahamid, Sharir, Addadi, & Weiner, 2008; Politi, Arad, Klein, Weiner, & Addadi, 2004; Politi et al., 2008; Veis, 2003; Weiss, Tuross, Addadi, & Weiner, 2002) and (b) the study the nucleation, growth, and phase transition of minerals in the presence of natural and man-made organic matrices that influence crystal size, morphology, orientation, phase selection, and kinetics, etc. (Cölfen & Mann, 2003; De Yoreo, 2013; De Yoreo & Vekilov, 2003; Elhadj, De Yoreo, Hoyer, & Dove, 2006; Nudelman et al., 2010).

The knowledge of important steps in biomineralization has been substantially improved by the use of modern physiochemical methods. FTIR and Raman spectroscopies can identify virtually any substance through the "fingerprint" of the molecular vibrational absorption or scattering spectrum in the 3–30 μm wavelength region. Chemical bonds like O—H, N—H, C—H, or C=O in organic matrices are found in distinct positions in the FTIR spectra. In the case of mineral phases like calcium carbonate and calcium phosphate, characteristic lattice vibrations (phonons) are additionally excited. These are often quite characteristic for the polymorphic phase and can be used to distinguish different phases, for example, calcite, aragonite, vaterite, calcium carbonate hydrates, or amorphous calcium carbonate (ACC). This chapter is focused on the application of Raman and FTIR in the aspect of phase transformation and organic/inorganic hybrid structure in biomineralization. Its aim is to describe these two methods in a way that enables nonspecialists to understand their potential and to stimulate their application.

2. FTIR AND RAMAN SPECTROSCOPIES OVERVIEW
2.1. Vibrations in solids

At temperatures above absolute zero, all collective vibrations that occur in crystals can be viewed as the superposition of plane waves that virtually propagate to infinity (Kittel, 1996). These plane waves are commonly modeled by quasiparticles called phonons. A normal coordinate $Q = Q_0 \cos(2\pi v_S t)$, which is actually a linear combination of bond lengths and bond angles, is associated with each normal mode. Depending on the dominant term in the normal coordinate, modes can be classified as stretching (v), bending (δ), torsional (τ), librational (pseudorotations/translations), or lattice modes (relative displacement of the unit cells).

For a three-dimensional solid containing N unit cells with p atoms each, $3pN - 6$, different phonons can propagate and their wave vectors all point in a volume of the reciprocal space called the Brillouin zone. There are modes with in-phase oscillations of neighboring atoms (acoustic vibrations) and modes with out-of-phase oscillations (optical vibrations). On the other hand, phonons are referred to as being longitudinal or transverse depending on whether the atoms move parallel or perpendicular to the direction of the wave propagation given by the wave vectors.

3. THE RAMAN EFFECT

The polarization P of the dipoles in solids excited when a laser beam (amplitude E_0 and frequency v_0) interacts with phonons of frequency v_S depends on the polarizability α:

$$P = \alpha E_0 \cos(2\pi v_0 t) \qquad (22.1)$$

where α term can be expressed as a function of the normal vibration coordinates Q; using a Taylor approximation:

$$\alpha = \alpha_0 + \left(\frac{\partial \alpha}{\partial Q}\right)_0 Q + \cdots \qquad (22.2)$$

$$P = \alpha_0 E_0 \cos(2\pi v_0 t) + Q_0 E_0 \left(\frac{\partial \alpha}{\partial Q}\right)_0 \cos(2\pi v_S t) \cos(2\pi v_0 t) + \cdots$$

$$= \alpha_0 E_0 \cos(2\pi v_0 t) + \frac{Q_0 E_0}{2} \left(\frac{\partial \alpha}{\partial Q}\right)_0 [\cos 2\pi (v_0 - v_S) t + \cos 2\pi (v_0 + v_S) t] + \cdots$$

$$(22.3)$$

With the scattered electric field being proportional to P, Eq. (22.3) predicts both quasielastic ($v = v_0$) and inelastic ($v = v_0 \pm v_S$) light scattering (see scheme in Fig. 22.1). The former is called Rayleigh scattering and the latter, which occurs only if vibrations change polarizability $\left(\left(\frac{\partial \alpha}{\partial Q}\right)_0 \neq 0\right)$, is Raman scattering (Gouadec & Colomban, 2007; Lewis & Edwards, 2001; Long, 1977).

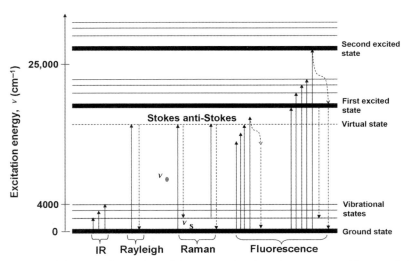

Figure 22.1 Schematic illustration of the energy levels and energy transitions underlying IR, Raman, and fluorescence spectroscopies. The different possibilities of light scattering: Rayleigh scattering (no exchange of energy: incident and scattered photons have the same energy), Stokes scattering (atom or molecule absorbs energy: scattered photon has less energy than the incident photon), and anti-Stokes scattering (atom or molecule loses energy: scattered photon has more energy than the incident photon).

Equation (22.3) reveals the dual sensitivity of Raman spectra to the electrical (α) and mechanical (v_S) properties of the samples. Two types of parameters will therefore influence the spectra:

a. Parameters acting on the "mechanics" like atomic mass, bond strength, or the system geometry (interatomic distances and atomic substitutions) will shift the peak position (the eigenfrequencies of matter vibrations).
b. Parameters acting on the "charge transfer" (ionocovalency, band structure, and electronic insertion) will impact intensity, on the basis of the vibration-induced charge variations occurring at the scale of bonds.

3.1. Infrared absorption

When the frequency of a specific vibration is equal to the frequency of the IR radiation directed on the molecule, the molecule absorbs the radiation. For a molecule of N atoms, among the $3N-6$ (nonlinear molecule) or $3N-5$ (linear molecule) normal modes of vibration, those that produce a net change in the dipole moment may result in an IR activity (Barth & Zscherp, 2002). It should be mentioned that Raman spectroscopy provides complementary information on molecular vibrations. Some vibrational modes of motion are IR-inactive but are Raman-active and vice versa (see scheme in Fig. 22.1).

Calcium carbonate is one of the most common biominerals in marine organisms (Curry & Kohn, 1976). It has three anhydrous polymorphs (calcite, aragonite, and vaterite), two hydrates (calcium carbonate monohydrate and hexahydrate), and amorphous calcium carbonate (ACC). The crystal structure of calcium carbonates, containing two ions of calcium and carbonate, is simple, and the vibrations of carbonate ions in different calcium carbonate structures can be discriminated by FTIR spectra. The absorption bands of carbonate are divided into four areas: symmetric stretch (v_1) at ~ 1080 cm^{-1}, out-of-plane bend (v_2) at ~ 870 cm^{-1}, asymmetric stretch (v_3) at ~ 1400 cm^{-1}, and in-plane bend (v_4) at ~ 700 cm^{-1}. ACC has a broad characteristic v_2 absorption band at ~ 866 cm^{-1} and a split v_3 peak at ~ 1427 and 1477 cm^{-1} and a broad absorption peak of water at ~ 3400 cm^{-1} (Addadi, Raz, & Weiner, 2003). In the FTIR spectra of vaterite, the characteristic absorption bands are located at $v_2 \sim 876$ cm^{-1} and $v_4 \sim 744$ cm^{-1} and a split peak of v_3 at ~ 1439 and 1493 cm^{-1}. Aragonite shows two characteristic absorption bands of $v_2 \sim 856$ cm^{-1} and $v_4 \sim 714$ cm^{-1} coupled with a weak absorption peak at ~ 700 cm^{-1} and also a v_3 absorption band at ~ 1489 cm^{-1}. For calcite, there are two absorption bands at $v_2 \sim 876$ cm^{-1} and v_4

~714 cm^{-1} and an absorption peak at ~1420 cm^{-1}. Table 22.1 shows the FTIR spectra and corresponding assignments of different phases of calcium carbonate (Addadi et al., 2003; Coleyshaw, Crump, & Griffith, 2003; Gueta et al., 2007; Su & Suarez, 1997).

Another important mineral system is calcium phosphate, the main inorganic component of vertebrate bone and tooth (Grynpas et al., 1984). It has potential application in the fields of dentistry, orthopedics, and surgery due to its structural similarity with native mineral of calcified hard tissue (LeGeros, 2008). The crystal structures of calcium phosphate are much more complicated compared to those for calcium carbonate in the respect that there are more polymorphs, nonstoichiometry, and dopant ions such as carbonate. Briefly, an isolated phosphate ion possesses tetrahedral symmetry and four normal internal vibrational modes (v_1, v_2, v_3, and v_4). In the absence of factor group splitting, the v_1 fundamental mode of the isolated phosphate ion representing the symmetric P–O stretching vibration is characterized by a band at 938 cm^{-1}. The triply degenerate asymmetric P–O stretching mode (v_3) is observed at 1017 cm^{-1}. The double-degenerate bending mode (v_2) and the triple-degenerate bending mode (v_4) correspond to the bending vibrations of the O—P—O bond and give rise to bands at 420 and 567 cm^{-1}, respectively (Herzberg, 1945). This study has not considered the impact of the degree of combinations of the various modes of motion, such as stretching or bending, on the fundamental internal vibrational spectrum of calcium phosphates. Table 22.2 indicates the FTIR spectra and corresponding assignments of different species related to calcium phosphate (Berry & Baddiel, 1967; Fowler, Marković, & Brown, 1993; Jillavenkatesa & Condrate, 1998; Klähn et al., 2004; Koutsopoulos, 2002; Taylor, Simkiss, Parker, & Mitchell, 1999; Termine & Lundy, 1974).

3.2. FTIR and Raman spectroscopies in biomineralization

Due to the specificity of vibrational spectra of different phases under FTIR or Raman, these techniques have been widely used in quantitative and qualitative study of phase identification, phase evolution, and phase spatial distribution in both biomimetic and biological minerals. For example, Beniash et al. discovered the first biological example of ACC acting as a transient precursor of calcite in sea urchin larval spicules by FTIR (Beniash et al., 1997). Weiss et al. have used FTIR and Raman spectroscopies to study molluscan larval shells where they could show how the crystalline part of the shell changed with age. ACC was shown to be a precursor for aragonite,

Table 22.1 FTIR spectra and assignments of six polymorphs of calcium carbonate phases and aqueous ions (HCO_3^- and CO_3^{2-}), frequency (cm^{-1})

HCO_3^-	CO_3^{2-}	ACC	Ikaite	Monohydrocalcite	Vaterite	Aragonite	Calcite	Assignment
			3543	3400				ν_{OH} stretch
			3502					ν_{OH} stretch
			3468	3327				ν_{OH} stretch
			3404					ν_{OH} stretch
			3361					ν_{OH} stretch
			3119					ν_{OH} stretch
			3216	3236				ν_{OH} stretch
1668								CO_2 asymmetric stretch
		1637	1673	1700				δ_{HOH} bend
			1644					δ_{HOH} bend
			1616					δ_{HOH} bend
1605								CO_2 asymmetric stretch
		1477	1425	1490	1493	1489		ν_3 asymmetric stretch
	1383	1427	1411	1404	1439		1423	ν_3 asymmetric stretch
1360								CO_2 symmetric stretch
1310								CO_2 symmetric stretch

Continued

Table 22.1 FTIR spectra and assignments of six polymorphs of calcium carbonate phases and aqueous ions (HCO3- and CO32-), frequency (cm^{-1})—cont'd

HCO$_3^-$	CO$_3^{2-}$	ACC	Ikaite	Monohydrocalcite	Vaterite	Aragonite	Calcite	Assignment
1010	1065	1076	1085	1063	1084	1083		v_1 symmetric stretch
843	887	866	876	872	876	856	876	v_2 out-of-plane bend
			743	767	744	714	714	v_4 in-plane bend
			720	712		700		v_4 in-plane bend
			800	588				lattice

Table 22.2 FTIR spectra and assignments of six polymorphs of calcium phosphate phases as well as different aqueous phosphate ions (PO_4^{3-}, HPO_4^{2-}, or $H_2PO_4^-$), frequency (cm^{-1})

$H_2PO_4^-$	HPO_4^{2-}	PO_4^{3-}	ACP	DCPD	β-TCP	OCP	HAP	Assignment
			3400	3535		3400	3572	v_SOH
							2070	Combination
							2000	Combination
				1645		1635–1642		v_2 H_2O bend
				1217, 1200		1295		HPO_4 (6) OH in-plane bend
						1192		HPO_4 (5) OH in-plane bend
1156				1130	1144, 1126	1130, 1119, 1110	1154	v_3 HPO_4 or combination
1077	1080	1017	1052	1064, 1051	1083, 1066, 1044, 1025	1076, 1058, 1036, 1026	1087, 1046, 1032	v_3 PO_4 antisymmetric stretch
944	990	938	945	996–1003, 980–986	990, 970, 945	962	962	v_1 PO_4 symmetric stretch
						913–917		HPO_4 (6) P–OH stretch
879	855			870		861–873		HPO_4 (5) P–OH stretch
				660		621	631	v_LOH or H_2O libration
521	541	567	563	576, 528	609, 594, 555, 544	603, 575, 560, 524	602, 574, 561	v_4 PO_4 antisymmetric bend
394	394	420	435	394	497, 458, 438, 419	466	472, 462	v_2 PO_4 bend

the final biomineral in the shells of marine bivalves *Mercenaria mercenaria* and *Crassostrea gigas* (Weiss et al., 2002). Based on FTIR, Politi et al. showed that sea urchin spine regeneration proceeded via the initial deposition of ACC before the further transformation into calcite of complex shape (Politi et al., 2004). The intensity ratios of the peaks 875/713 and 856/713 cm^{-1} were used to characterize the portion of ACC, when ACC coexists with calcite or aragonite, respectively. The greater intensity ratio of 875/713 or 856/713 cm^{-1} implied the less ACC content in the samples. Hild et al. studied the inhomogeneous chemistry of terrestrial crustacean cuticle cross section by Raman line scans and found that the inner layer was much less crystallized compared to the outer calcitic layer (Hild, Marti, & Ziegler, 2008).

We have mainly used FTIR to monitor the phase transformation of both ACC and ACP in biomimetic and biomineral systems, different from those of Beniash et al. (1997) and Weiss et al. (2002). We decomposed the ACC/calcite hybrid v_2 peak into that of ACC (866 cm^{-1}) and calcite (875 cm^{-1}) and measured the relative amount ratio between ACC and calcite based on two-peak area ratio (Tao, Zhou, Zhang, Xu, & Tang, 2009). The first known literature on quantitatively identifying the degree of crystallinity for calcium phosphate came out at 1966 (Termine & Posner, 1966). A splitting function (SF) of the phosphate v_4 antisymmetric bending frequency at 560–600 cm^{-1} was used as a criterion for the degree of crystallization of calcium phosphate. Typically, A_1 is the area enclosed by the spectrum and the straight line through two maxima at about 600 and 560 cm^{-1}, and A_2 is the area enclosed by spectrum and straight baseline in the range of 450–750 cm^{-1}; the ratio of A_1 to A_2 is defined as SF (Termine & Posner, 1966). The similar method was also adopted by Weiner et al. to evaluate the degree of crystallinity in bone samples (Mahamid et al., 2008; Weiner & Bar-Yosef, 1990).

3.3. Modes and materials for FTIR

FTIR spectra can be collected by transmission modes (KBr pellets, Nujol mulls, disposable cards, liquid cells, and gas cells) or reflection modes (attenuated total reflection (ATR), diffuse reflectance, and specular reflectance). Transmission techniques are classical and will deliver the best-quality spectrum, but they are time-consuming with collection of a spectrum usually taking more than 5 min. Today, more than 70% of new instruments are delivered together with an ATR sampling unit. Collection by the ATR technique does not require special sample preparation. Typical time for collection of a spectrum by ATR together with sample adjustment and crystal cleaning is 2–5 min. The quality of the ATR spectrum is fully sufficient for sample identification, and it should be compared with spectral libraries collected by the same technique. However, the ATR spectrum may have limitations in the case of trace impurities in the sample.

To avoid unwanted peaks in the FTIR spectrum, one has the following possibilities: (a) Use KBr as a sampling material (cells, windows, pellets, etc.). KBr is the only material that has no absorption bands in the wave number region 400–4000 cm^{-1}. An important disadvantage of KBr is its extreme sensitivity to moisture and high solubility in aqueous solution. It is time-consuming to work with, but the spectra have the best quality. (b) Use sampling materials other than KBr and use only the part of the spectral region where the sampling material has no absorption peaks or use other sampling materials and compensate for their peaks—collect the spectrum of the sampling material as a background and subtract it from the sample spectrum. Typical transmission materials include KBr, NaCl, CaF_2, BaF_2, and ZnSe. It should be emphasized that glass or quartz cannot be used due to their own strong absorption in the IR region. Some of the reflection techniques like ATR, diffuse reflectance, or specular reflectance do not need sampling materials. The widely used ATR technique compensates the sampling material peaks automatically by collecting the empty ATR as background. Typical ATR materials include ZnSe, Si, Diamond, and Ge.

4. MEASURING BIOMIMETIC CRYSTALS AND BIOMINERALS

4.1. Example 1: *In situ* FTIR of calcium phosphate transformation in Tris-buffered solution and phosphate-buffered solution

4.1.1 Mineralization reactions

The mineralization reaction is performed in a Tris-buffered saline, consisting of 50 mM Trizma base (Sigma–Aldrich, United States) and 150 mM NaCl (Merck, Germany) in ultrapure water (Millipore, United States) set at pH 7.40 using HCl (25%, Merck). A 10 mM calcium solution and a 10 mM phosphate solution were prepared by dissolving $CaCl_2 \cdot 2H_2O$ (Sigma) and K_2HPO_4 (Merck) in Tris-buffered saline, after which the pH was finely adjusted to 7.40 using 0.1 M NaOH (Merck) or 0.1 M HCl. The reaction was performed at 19 ± 1 °C in a 50 ml beaker, to which

14.7 mL of calcium solution was added. The calcium stock solution was stirred gently at 200 rpm before adding 10.3 mL phosphate solution (the final Ca/P = 1.43). The reaction was followed by stirring for typically 1–4 h.

In order to evaluate the possible influence of Tris on calcium phosphate mineralization, we also studied the mineralization process by using solutions without Tris. Typically, high-purity Na_2HPO_4 (99.95%), KCl (99.999%), NaCl (99.999%), and HCl (1.043 N) from Sigma-Aldrich; $CaCl_2$ (99.99%) from Alfa Aesar; and Milli-Q water with the resistivity of 18.2 MΩ cm were used to prepare stock solutions of $CaCl_2$ (300 mM), Na_2HPO_4 (19 mM), and NaCl (2.23 M). The pH of phosphate solutions was adjusted to 7.400 ± 0.005 using HCl. All the solutions were filtered through filter membranes (pore size 0.22 μm, PVDF, Acrodisc LC, PALL) before use. In the mineralization experiment, we attempted to keep the pH value constant using phosphate as a buffer. $CaCl_2$/NaCl solutions of different concentrations were prepared by dilution from the $CaCl_2$ and NaCl stock solutions mentioned earlier. The supersaturated solutions were made by adding an equal volume of $CaCl_2$/NaCl solution to 19 mM Na_2HPO_4 solution with a fixed ionic strength ($I = 0.0244$ molL^{-1}). During our preliminary studies, we found that the pH of the supersaturated solutions was maintained at 7.400 ± 0.02 at a phosphate concentration of 9.5 mM.

4.1.2 Infrared measurements

In situ IR spectra were recorded on a 1 ml samples from a 25 mL reaction solution using horizontal-attenuated reflection (ATRMaxII™, PIKE Technologies, United States) employing a ZnSe crystal and a FTIR spectrometer (Excalibur™, Varian Inc., United States) equipped with a MCT detector (500 scans and resolution 2 cm^{-1}). The ATR throughput spectra appear different and show different spectral features depending upon the crystal material. One notable feature is the spectral range cutoff (where IR throughput goes to zero) at the long-wavelength end of the spectrum (near 400 cm^{-1}). For the ZnSe crystal plate, the long-wavelength cutoff is at about 520 cm^{-1} and the IR throughput is about 32% at 1000 cm^{-1}. The background spectrum was obtained using Tris-buffered saline in the trough above the ZnSe ATR surface. A cleaned glass coverslip was placed over the solution to reduce the amount of evaporation of the sample on the surface of the crystal. Finally, the Tris-buffered saline background was subtracted from the collected spectra at different time points using Varian Resolutions 4.0 software.

The phases of the calcium phosphates at different stages without Tris were determined by *ex situ* FTIR. The samples were collected by centrifuge at 2700 g for 2 min and the supernatant solution was removed, and then, minerals were rapidly switched from mother solution to ethanol and washed with ethanol for 3×. The resulting samples were then dispersed in ethanol before measurement.

The spectra were obtained in transmission mode. Sample pellets were made by mixing the sample powder with dried spectroscopic grade KBr with mass ratio 1:150 and ground into powder in an agate mortar with a pestle, followed by compacting those into a thin pellet in a stainless steel die of 1 cm inner diameter under a pressure of 1200 psi. The FTIR spectra were collected using a Varian 3100 FTIR spectrometer in the range of 4000–400 cm^{-1} with a resolution of 1 cm^{-1}. Thirty-two spectral scans were conducted on each sample and computer-averaged to produce each FTIR spectrum.

4.1.3 Data analysis

In situ infrared gives us specific information about the chemistry of the phosphate groups in the bulk of the solution. One disadvantage is the interference from PO_4^{3-} stretch vibrations of solute $HPO_4^{2-}/H_2PO_4^-$, which overlap with the calcium phosphate phases. However, as the calcium phosphate mineral has a tendency to precipitate on the ATR crystal, one can distinguish the position of the mineral peaks by analyzing their increase in intensity during precipitation. Due to the strong absorption of water, we are focusing in the region of 800–1300 cm^{-1}.

The IR spectra of clusters ($t=0$ min), dendrites (15 min), and spheres (45 min) (Fig. 22.2A and B) show a similar pattern of a broad signal that is present between the signals of the solute phosphate at 1078 cm^{-1} (P–O stretch $HPO_4^{2-}/H^2PO_4^-$) and 990 cm^{-1} (P–O stretch HPO_4^{2-}) (Habraken et al., 2013; Klähn et al., 2004; Table 22.2). Irrespective of sedimentation effects, this signal is highest for the clusters and decreasing in intensity during the development of the spheres. As from pH measurements, we know that the phosphate that is bound inside the clusters has the same composition as the free phosphate in solution (15% $H_2PO_4^-$ and 85% HPO_4^{2-}), such a broad signal is due to the overlap of the P–O stretches of $HPO_4^{2-}/H_2PO_4^-$ that shift upon cluster formation as well as the increasing P–O–H bend of $HPO_4^{2-}/H_2PO_4^-$ (Klähn et al., 2004). The small signal at around 950 cm^{-1}, only visible with clusters (w) and dendrites (vw), corresponds to a P–O stretch of $H_2PO_4^-$ (Klähn et al., 2004), indicating the presence of $H_2PO_4^-$ inside these structures.

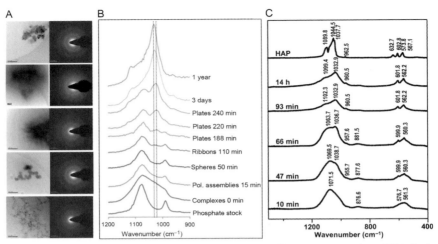

Figure 22.2 (A) Morphology of calcium phosphate at different stages in Cryo-TEM. Scale bar is 200 nm (B) *In situ* ATR–FTIR spectra of the calcium phosphate spectra recorded at different time points in the P–O stretch region. Dotted lines indicate the shift of the increasing PO_4^- stretch vibration from 1021 to 1030 cm^{-1}. (C) FTIR spectra of samples without Tris quenched at times shown in the figure. Solution supersaturation was $\sigma_{HAP} = 3.36$, $\sigma_{OCP} = 1.76$, and $\sigma_{ACP} = 0.12$. Peaks at 524, 877, and 1036 are characteristic of OCP. (For color version of this figure, the reader is referred to the online version of this chapter.)

Proceeding towards the ribbons (110 min) and plates (from 180 min), a distinct signal in the 1020–1030 cm^{-1} region starts to appear that becomes more intense with time and shifts towards 1030 cm^{-1}. This signal corresponds to the P–O stretch of PO_4^{3-}, initially at the value of octacalcium phosphate (OCP) or calcium-deficient apatite (1021 cm^{-1}) (Fowler et al., 1993; Sauer & Wuthier, 1988) and finally shifting to the position of apatite after 3 days to 1 year (Sauer & Wuthier, 1988). Furthermore, the presence of shoulders at 1124/1142 cm^{-1} at 180 min–3 days corresponds to P–O stretch vibrations of HPO_4^{2-} and therefore indicates that the material still has a significant amount of HPO_4^{2-}. The sample after 1 year only has a shoulder at 1115 cm^{-1} and shows a pattern corresponding to apatite that is formed via an OCP precursor (Sauer & Wuthier, 1988). The FTIR of calcium phosphates without Tris also shows a similar transformation pathway in that the initially formed amorphous phase will transform into OCP at 47 min after mixing of calcium and phosphate. The transformation of such a small OCP crystal into apatite happened at 60 min and finished at 93 min (Table 22.2 and Fig. 22.2C).

4.2. Example 2: FTIR and Raman spectra of ACC transformation at different stages

4.2.1 Phase and chemical identification during shell molt of terrestrial crustacean Armadillidium vulgare

The *Armadillidium vulgare* adults (10–14 mm in length) were used in the investigation. The premolt stage was identified by the appearance of white spots at its sternal part (Ziegler, Fabritius, & Hagedorn, 2005). At each molt stage, more than 30 animals were used for examination. The excised parts were washed using triply distilled water for 1–2 s and then rapidly dehydrated in anhydrous methanol for 1 min. The samples were wiped by a piece of filter paper to remove any remnants of soft tissue. The cuticles and sternal deposits were dried under a vacuum condition at room temperature for 24 h and then stored at $-20\ °C$.

4.2.2 Synthesis of ACC

ACC was synthesized according to a method proposed by Wegner et al. (Faatz, Gröhn, & Wegner, 2004; Xu et al., 2008). Typically, 1.25 g NaOH was dissolved in 60 mL water. The NaOH solution was quickly poured into a 240 mL solution containing 1.35 g dimethyl carbonate (DMC) and 0.44 g $CaCl_2 \cdot 2H_2O$. The reaction occurred via a homogeneous manner due to the carbonate released from the hydrolysis of DMC, which was catalyzed by NaOH. The reaction solution was continuously stirred for 3.5 min. The product was separated by filtering through a 0.22 μm membrane and then was washed by anhydrous acetone and ethanol. The solid was dried under a vacuum condition at 25 °C for 48 h.

4.2.3 Controlled crystallization of ACC

Synthesized ACC was dispersed into diethylene glycol (DEG) with a concentration of 3.95 mg/mL. 1.6 mL H_2O, $MgCl_2$ (5 mM), and L-aspartic acid (Asp) (7.5 mM, pH 8.00) aqueous solutions were added into 1.24 mL DEG to study their individual influence, respectively. 0.76 mL ACC–DEG slurry was injected into the reaction solutions and the samples were withdrawn periodically. To examine the switching effect of Asp, 1.36 mL of $MgCl_2$ (5.88 mM) aqueous solution was first added into 0.94 mL DEG. Then, 0.76 mL ACC–DEG slurry was injected into the magnesium solution to obtain the magnesium-stabilized ACC. At 1 h, 0.54 mL Asp solution was added into the slurry, which was prepared by mixing Asp (50 mM, pH 8.00) aqueous solution with DEG at the volume ratio of 0.8. The final concentrations of Asp and Mg^{2+} in aqueous components were 7.5 and

5 mM, respectively. The zero point for the timescale was set at the injection of amino acids. All the experiments were performed at 25 °C.

4.2.4 Infrared and Raman measurements

FTIR spectra of calcium carbonates collected at different time points were collected from 4000 to 400 cm^{-1} in transmission mode using a KBr pellet on a Nexus 670 spectrometer (Nicolet, United States). Sixteen spectral scans were conducted on each sample and computer-averaged to produce each FTIR spectrum.

For Raman spectroscopy measurements, the small pieces of dorsal cuticle or sternal deposit were glued between two aluminum cubes using a double-side carbon tape. Cuticles or sternal deposits were cut sagittally using razor blade to obtain a flat sample surface. The cross section was further polished with a diamond knife in ultramicrotome by successively advancing the sample 10× each by 100 nm.

Raman spectra of biological samples at different molt stages were recorded in backscattering geometry using 514 nm excitation in a LabRAM HR UV Raman spectrometer (Jobin Yvon, France) operated with a spatial resolution of 0.8–1.5 μm and spectral resolution of 4 cm^{-1}. The laser power was set to 10–25 μW and integration times were 30 s to minimize local heating of the sample.

4.2.5 Data analysis

Calcium carbonate polymorph contents have been determined by quantitative analysis of FTIR spectra (Vagenas, Gatsouli, & Kontoyannis, 2003). To calculate of area ratio of component peak 875 cm^{-1} to peak 866 cm^{-1}, the program PeakFit 4.12 (SeaSolve Software Inc.) was used. The spectra in the range from 700 to 1000 cm^{-1} were selected for data processing. The autobaseline subtraction was performed before the peak fitting. Gaussian (amplitude) functions with variable widths were used to fit all the peaks in spectra. The hidden peaks were found by residual method. Typically, a residual was simply the difference in y-value between a data point and the sum of component peaks evaluated at the same data point x-value. By placing peaks in such a way that their total area equaled the area of the data, the hidden peaks could be revealed by these residuals. The amplitude of the threshold value was changed to clear possible noise peaks. The iterations continued until a goodness of fit $R^2 > 0.99$ and the change of $R^2 < 0.001$ between each of the two iterations was obtained (Fig. 22.4C).

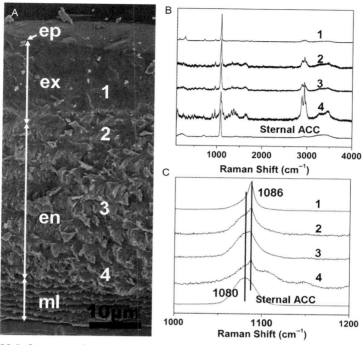

Figure 22.3 Structure, chemistry, and phase distribution in an intermolt cuticle. (A) SEM image of sagittally cleaved fully mineralized cuticle of *Armadillidium vulgare*. The cross section of cuticle showed the thin epicuticle (ep), calcified exocuticle (ex) and endocuticle (en), and membranous layers (ml). 1–4 indicated the positions where Raman spectra were collected. (B) Raman spectra recorded at corresponding sites indicated in A. The spectrum of sternal ACC was used as the contrast. (1) Exocuticle contained well-crystalline calcite and minor ACC. (2) Boundary between exocuticle and endocuticle was composed of ACC and minor calcite. (3) Endocuticle was composed of ACC phase without any detectable calcite. (4) Boundary between endocuticle and membranous layers was composed of ACC and α-chitin. (C) A magnified image of spectra in B.

SEM imaging of sample cross sections was used to study the morphology and phase of different layers in intermolt cuticle (Fig. 22.3A). The distal epicuticle is about 1 μm in thickness. This outermost layer may act as a sensilla and it is not mineralized (Hild et al., 2008). The exocuticle is a 15 μm layer with a distal smooth part and a rough transition part. Within the endocuticle, sublayers of plywood-like stacks of parallel fibers are visible. The membranous layer composes loosely stacked platelike structures. The phase and organic information of mineralized layers is collected under Raman spectroscopy. The bands due to carbonate stretching vibration have their maxima at 1086 and 1078 cm^{-1} for calcite and ACC, respectively

(Hild et al., 2008; Wang, Wallace, De Yoreo, & Dove, 2009). These two phases can also be distinguished by the lattice vibration at 280 cm^{-1} of calcite and a broad band from 100 to 300 cm^{-1} of ACC (Hild et al., 2008; Wang et al., 2009). It is found that the up layer of exocuticle consists of calcite (1 in Fig 22.3B and C). ACC and calcite coexist at the boundary between exocuticle and endocuticle (2 in Fig 22.3B and C). The bands of organic species show their maxima at 1370, 1326, 1104, and 953 (4 in Fig 22.3B and C, Table 22.3), which match well the Raman peaks of α-chitin (Brunner et al., 2009; Ehrlich et al., 2007; Zhang, Geissler, Fischer, Brendler, & Bäucker, 2012). The endocuticle is constructed by ACC and α-chitin (3 in Fig 22.3B and C). The membranous layer contains much higher percentage of α-chitin than other layers (4 in Fig 22.3B and C). The inhomogeneous distribution of inorganic/organic could be the reason for high mechanical property as well as the recycling of minerals during the molt cycles.

We also designed some *in vitro* experiments to evaluate the role of aspartic acid and magnesium during the ACC transformation (Fig. 22.4). The crystallization degree of the calcium carbonate is quantitatively estimated by the area ratio between corresponding component v_2 peaks at 875 cm^{-1} of calcite and 866 cm^{-1} of ACC in FTIR (Fig. 22.4C). This *ex situ* method based upon v_2 symmetry and position is more sensitive than the previous estimation by using v_2:v_4 ratio (Beniash et al., 1997; Politi et al., 2004) so that it can even monitor poorly crystallized calcium carbonate samples. Polarizing microscopy is also used to monitor *in situ* the nucleation of crystals (brilliant domains) in the bulk solutions. The induction time of phase transformation can be estimated by these two complementary methods, and the results match each other well (Fig. 22.4B). The crystallization of ACC in pure water is too fast, so a mixed solvent containing water and diethylene glycol (DEG, a solvent model for hydroxyl rich compound chitin) is used as the dispersing media to slow down the crystallization rates. In all experiments, the water/DEG volume ratio is fixed at 0.8, which is close to the reported water content in cuticle of *Armadillidium vulgare*, 40–60 wt% (Lindqvist, Salminen, & Winston, 1972). In our study, the additives concentrations are represented by their final concentrations in aqueous component. When the ACC contacts water/DEG, the nucleation of calcite occurs within only ~18 min (blue in Fig. 22.4A and D). However, the spontaneous transformation of ACC in the solvent is effectively switched off; no phase transformation is detected even at 40 h in the presence of only 5 mM magnesium (green in Fig. 22.4A and D). Thus, the lifetime of ACC in the solution is significantly increased so that this unstable phase can be temporarily stored.

Table 22.3 Raman spectra and assignments of intermolt cuticle, frequency (cm^{-1})

Raman	Assignment
3459	Disordered hydrogen bond
3268	Ordered hydrogen bond
2965	CH_3 asymmetric stretch
2937	C–H of CH_3 stretch
2884	C–H of ring stretch
1660	Amide I
1619	Amide I
1447	CH_2 bend
1412	CH_2 bend
1370	CH_2 rocking
1326	Amide III
1262	Amide III
1200	Amide III
1147	C–O–C, C–O stretch
1104	C–O–C, C–O stretch
1086	Symmetric stretch of CO_3^{2-} in calcite
1078	Symmetric stretch of CO_3^{2-} in ACC
953	CH_x bend
895	CH_x bend
712	In-plane bend of CO_3^{2-}
500	C–C backbone
456	CCC ring deformation
398	CCC ring deformation
367	CCC ring deformation
320	CCC ring deformation
280	Lattice rotational mode of calcite
156	Lattice translational mode of calcite

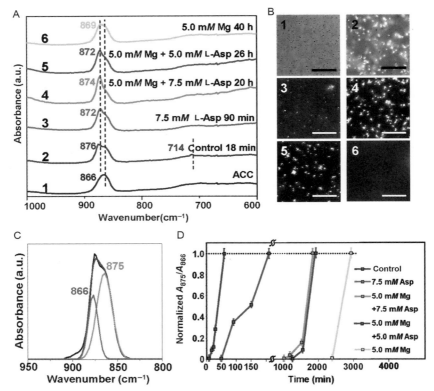

Figure 22.4 Kinetics of phase transformation from ACC to calcite. (A) FTIR spectra of slightly crystallized samples extracted at different timescales under different conditions. (B) The optical images of *in situ* phase evolution investigated by polarizing microscopy (Scale bar is 100 μm) in the solvent. The calcite and ACC phases are expressed as the bright and dark on the image, respectively. The bright densities can reflect the crystallization degrees during the transition visually. (C) Calculation methods for crystallization degree of calcium carbonate. The asymmetric v_2 peak of calcium carbonate can be decomposed into two symmetric component peaks at 875 cm^{-1} (calcite) and 866 cm^{-1} (ACC), respectively. The area ratio between these two peaks was used as a measurement of crystallization degree. (D) Kinetic plot of phase transformation in the absence and presence of magnesium and Asp. (See color plate.)

Asp is also a stabilizer of the ACC phase. The induction time of ACC nucleation is prolonged to 50 min in the presence of 7.5 mM Asp (olive in Fig. 22.4A and D). Although Asp poorly inhibits the crystallization of pure ACC, it can switch on the crystallization of the Mg-stabilized ACC. When the same amount of Asp is added into a prestabilized magnesium–ACC system (the concentration of magnesium is maintained at 5 mM), the magnesium-stabilized ACC becomes unstable and the crystallization can be triggered within 17 h (red in Fig. 22.4A and D). This promotion effect is Asp dose-dependent as the induction time increases to 21 h when the Asp concentration is reduced to 5 mM (purple in Fig. 22.4A and D). The combination of *in situ* polarizing microscopy and FTIR (Fig. Fig. 22.4A and B) demonstrates that the stability of ACC (the transition from ACC to calcite) can be tuned over a wide range by using magnesium together with Asp.

5. FUTURE DIRECTIONS

FTIR and Raman spectroscopies have been proved to be a powerful tool in monitoring the inorganic phase and organic chemical conformation during the biomimetic crystallization and biomineralization in both qualitative and quantitative manner. However, traditional FTIR and Raman spectroscopies face the great challenge of limited spatial resolution even when coupled with optical microscopy with resolution on the magnitude $\sim$1 μm. Recently, FTIR and Raman spectroscopies were combined with near-field microscopy or atomic force microscopy to break through the diffraction limit. The spatial resolution has been increased dramatically to 20–40 nm (Amarie et al., 2012; Bao et al., 2012; Schuck et al., 2013; Weber-Bargioni et al., 2011). These new types of FTIR and Raman spectroscopies could be used to map the organic/inorganic distribution in biominerals and monitor phase transformations and surface organic structures in the nanometer-size range.

ACKNOWLEDGMENTS

The author acknowledge constructive and enlightening discussion with Drs. Xurong Xu, Haihua Pan, and Halei Zhai and Mr. Anhua Cai at Zhejiang University in China. We would like to thank Dr. James De Yoreo for nice discussion and suggestion for phase transformation of ACC and ACP. We also would like to thank Drs. Wouter Habraken and Nico Sommerdijk at Eindhoven University of Technology in the Netherlands for the discussion of calcium phosphate cryo-TEM and FTIR measurements. Much of this work was funded by a grant from the National Institutes of Health (grant no. DK61673).

REFERENCES

Addadi, L., Raz, S., & Weiner, S. (2003). Taking advantage of disorder: Amorphous calcium carbonate and its roles in biomineralization. *Advanced Materials*, *15*, 959–970.

Amarie, S., Zaslansky, P., Kajihara, Y., Griesshaber, E., Schmahl, W. W., & Keilmann, F. (2012). Nano-FTIR chemical mapping of minerals in biological materials. *Beilstein Journal of Nanotechnology*, *3*, 312–323.

Arias, J. L., & Fernández, M. S. (2008). Polysaccharides and proteoglycans in calcium carbonate-based biomineralization. *Chemical Reviews*, *108*(11), 4475–4482.

Bao, W., Melli, M., Caselli, N., Riboli, F., Wiersma, D. S., Staffaroni, M., et al. (2012). Mapping local charge recombination heterogeneity by multidimensional nanospectroscopic imaging. *Science, 338,* 1317–1321.
Barth, A., & Zscherp, C. (2002). What vibrations tell us about proteins. *Quarterly Reviews of Biophysics, 35,* 369–430.
Beniash, E., Aizenberg, J., Addadi, L., & Weiner, S. (1997). Amorphous calcium carbonate transforms into calcite during sea urchin larval spicule growth. *Proceedings of the Royal Society of London B, 264,* 461–465.
Berry, E. E., & Baddiel, C. B. (1967). The infra-red spectrum of dicalcium phosphate dihydrate (brushite). *Spectrochimica Acta, 23A,* 2089–2097.
Boskey, A. L. (1998). Biomineralization: Conflicts, challenges, and opportunities. *Journal of Cellular Biochemistry Supplement, 30–31,* 83–91.
Brunner, E., Ehrlich, H., Schupp, P., Hedrich, R., Hunoldt, S., Kammer, M., et al. (2009). Chitin-based scaffolds are an integral part of the skeleton of the marine demosponge Ianthella basta. *Journal of Structural Biology, 168,* 539–547.
Coleyshaw, E. E., Crump, G., & Griffith, W. P. (2003). Vibrational spectra of the hydrated carbonate minerals ikaite, monohydrocalcite, lansfordite and nesquehonite. *Spectrochimica Acta Part A, 59,* 2231–2239.
Cölfen, H., & Mann, S. (2003). Higher-order organization by mesoscale self-assembly and transformation of hybrid nanostructures. *Angewandte Chemie, International Edition, 42,* 2350–2365.
Curry, J. D., & Kohn, A. J. (1976). Fracture in the crossed-lamellar structure of Conus shells. *Journal of Materials Science, 11,* 1615–1623.
De Yoreo, J. J. (2013). Crystal nucleation: More than one pathway. *Nature Materials, 12,* 284–285.
De Yoreo, J. J., & Vekilov, P. G. (2003). Principles of crystal nucleation and growth. *Reviews in Mineralogy and Geochemistry, 54,* 57–93.
Ehrlich, H., Maldonado, M., Spindler, K. D., Eckert, C., Hanke, T., Born, R., et al. (2007). First evidence of chitin as a component of the skeletal fibers of marine sponges. Part I. Verongidae (demospongia: Porifera). *The Journal of Experimental Zoology, 308B,* 347–356.
Elhadj, S., De Yoreo, J. J., Hoyer, J. R., & Dove, P. M. (2006). Role of molecular charge and hydrophilicity in regulating the kinetics of crystal growth. *Proceedings of the National Academy of Sciences of the United States of America, 103,* 19237–19242.
Faatz, M., Gröhn, F., & Wegner, G. (2004). Amorphous calcium carbonate: Synthesis and potential intermediate in biomineralization. *Advanced Materials, 16,* 996–1000.
Fowler, B. O., Marković, M., & Brown, W. E. (1993). Octacalcium phosphate. 3. Infrared and Raman vibrational spectra. *Chemistry of Materials, 5,* 1417–1423.
Glimcher, M. J. (2006). Bone: Nature of the calcium phosphate crystals and cellular, structural, and physical chemical mechanisms in their formation. *Reviews in Mineralogy and Geochemistry, 64,* 223–282.
Gouadec, G., & Colomban, P. (2007). Raman Spectroscopy of nanomaterials: How spectra relate to disorder, particle size and mechanical properties. *Progress in Crystal Growth and Characterization of Materials, 53,* 1–56.
Grynpas, M. D., Bonar, L. C., & Glimcher, M. J. (1984). Failure to detect an amorphous calcium–phosphate solid phase in bone mineral: A radial distribution function study. *Calcified Tissue International, 36,* 291–302.
Gueta, R., Natan, A., Addadi, L., Weiner, S., Refson, K., & Kronik, L. (2007). Local atomic order and infrared spectra of biogenic calcite. *Angewandte Chemie, International Edition, 46,* 291–294.
Habraken, W. J. E. M., Tao, J., Brylka, L. J., Friedrich, H., Bertinetti, L., Schenk, A. S., et al. (2013). Ion-association complexes unite classical and non-classical theories for the biomimetic nucleation of calcium phosphate. *Nature Communications, 4,* 1507.

Herzberg, G. (1945). *Molecular spectra and molecular structure II. Infrared and Raman spectra of polyatomic molecules*. New York: D. Van Nostrand Company, Inc.

Hild, S., Marti, O., & Ziegler, A. (2008). Spatial distribution of calcite and amorphous calcium carbonate in the cuticle of the terrestrial crustaceans *Porcellio scaber* and *Armadillidium vulgare*. *Journal of Structural Biology, 163*, 100–108.

Jillavenkatesa, A., & Condrate, R. A., Sr. (1998). The infrared and Raman spectra of β- and α-tricalcium phosphate ($Ca_3(PO_4)_2$). *Spectroscopy Letters: An International Journal for Rapid Communication, 31*, 1619–1634.

Kittel, C. (1996). *Introduction to solid state physics* (7th ed.). New York: John Wiley and Sons.

Klähn, M., Mathias, G., Kötting, C., Nonella, M., Schlitter, J., Gerwert, K., et al. (2004). IR spectra of phosphate ions in aqueous solution: Predictions of a DFT/MM approach compared with observations. *The Journal of Physical Chemistry A, 108*, 6186–6194.

Koutsopoulos, S. (2002). Synthesis and characterization of hydroxyapatite crystals: A review study on the analytical methods. *Journal of Biomedical Materials Research, 62*, 600–612.

LeGeros, R. Z. (2008). Calcium phosphate-based osteoinductive materials. *Chemical Reviews, 108*, 4742–4753.

Lewis, I. R., & Edwards, H. G. M. (2001). *Handbook of Raman spectroscopy: From the research laboratory to the process line*. New York: Marcel Dekker Inc.

Lindqvist, O. V., Salminen, I., & Winston, P. W. (1972). Water content and water activity in the cuticle of terrestrial isopods. *The Journal of Experimental Biology, 56*, 49–55.

Long, D. A. (1977). *Raman spectroscopy*. New York: McGraw-Hill.

Lowenstam, H. A., & Weiner, S. (1985). Transformation of amorphous calcium phosphate to crystalline dahllite in the radular teeth of chitons. *Science, 227*, 51–53.

Mahamid, J., Sharir, A., Addadi, L., & Weiner, S. (2008). Amorphous calcium phosphate is a major component of the forming fin bones of zebrafish: Indications for an amorphous precursor phase. *Proceedings of the National Academy of Sciences of the United States of America, 105*, 12748–12753.

Mann, S. (1988). Molecular recognition in biomineralization. *Nature, 332*, 119–124.

Nudelman, F., Pieterse, K., George, A., Bomans, P. H., Friedrich, H., Brylka, L. J., et al. (2010). The role of collagen in bone apatite formation in the presence of hydroxyapatite nucleation inhibitors. *Nature Materials, 9*, 1004–1009.

Palmer, L. C., Newcomb, C. J., Kaltz, S. R., Spoerke, E. D., & Stupp, S. I. (2008). Biomimetic systems for hydroxyapatite mineralization inspired by bone and enamel. *Chemical Reviews, 108*(11), 4754–4783.

Politi, Y., Arad, T., Klein, E., Weiner, S., & Addadi, L. (2004). Sea urchin spine calcite forms via a transient amorphous calcium carbonate phase. *Science, 306*, 1161–1164.

Politi, Y., Metzler, R. A., Abrecht, M., Gilbert, B., Wilt, F. H., Sagi, I., et al. (2008). Transformation mechanism of amorphous calcium carbonate into calcite in the sea urchin larval spicule. *Proceedings of the National Academy of Sciences of the United States of America, 105*, 17362–17366.

Sarikaya, M. (1999). Biomimetics: Materials fabrication through biology. *Proceedings of the National Academy of Sciences of the United States of America, 96*(25), 14183–14185.

Sauer, G. R., & Wuthier, R. E. (1988). Fourier transform infrared characterization of mineral phases formed during induction of mineralization by collagenase-released matrix vesicles in vitro. *The Journal of Biological Chemistry, 263*(17), 13718–13724.

Schuck, P. J., Weber-Bargioni, A., Ashby, P. D., Ogletree, D. F., Schwartzberg, A., & Cabrini, S. (2013). Life beyond diffraction: Opening new routes to materials characterization with next-generation optical near-field approaches. *Advanced Functional Materials, 23*, 2539–2553.

Sigel, A., Sigel, H., & Sigel, R. K. O. (2008). Biomineralization: From nature to application. *Metal ions in life sciences*. Vol. 4. Wiley, ISBN: 978-0-470-03525-2.

Su, C., & Suarez, D. L. (1997). In situ infrared speciation of adsorbed carbonate on aluminum and iron oxides. *Clays and Clay Minerals, 45*, 814–825.

Tao, J., Zhou, D., Zhang, Z., Xu, X., & Tang, R. (2009). Magnesium-aspartate-based crystallization switch inspired from shell molt of crustacean. *Proceedings of the National Academy of Sciences of the United States of America, 106*(52), 22096–22101.

Taylor, M. G., Simkiss, K., Parker, S. F., & Mitchell, P. C. H. (1999). Inelastic neutron scattering studies of synthetic calcium phosphates. *Physical Chemistry Chemical Physics, 1*, 3141–3144.

Termine, J. D., & Lundy, D. R. (1974). Vibrational spectra of some phosphate salts amorphous to X-ray diffraction. *Calcified Tissue Research, 15*, 55–70.

Termine, J. D., & Posner, A. S. (1966). Infra-red Determination of the percentage of crystallinity in apatitic calcium phosphates. *Nature, 211*, 268–270.

Vagenas, N. V., Gatsouli, A., & Kontoyannis, C. G. (2003). Quantitative analysis of synthetic calcium carbonate polymorphs using FT-IR spectroscopy. *Talanta, 59*, 831–836.

Veis, A. (2003). Mineralization in organic matrix frameworks. *Reviews in Mineralogy and Geochemistry, 54*, 249–290.

Wang, D., Wallace, A. F., De Yoreo, J. J., & Dove, P. M. (2009). Carboxylated molecules regulate magnesium content of amorphous calcium carbonates during calcification. *Proceedings of the National Academy of Sciences of the United States of America, 106*, 21511–21516.

Weber-Bargioni, A., Schwartzberg, A., Cornaglia, M., Ismach, A., Urban, J. J., Pang, Y., et al. (2011). Hyperspectral nanoscale imaging on dielectric substrates with coaxial optical antenna scan probes. *Nano Letters, 11*, 1201–1207.

Weiner, S., & Bar-Yosef, O. (1990). States of preservation of bones from prehistoric sites in the Near East: A survey. *Journal of Archaeological Science, 17*, 187–196.

Weiner, S., & Lowenstam, H. A. (1989). *On biomineralization*. Oxford: Oxford University Press, ISBN: 0-19-504977-2.

Weiss, I. M., Tuross, N., Addadi, L., & Weiner, S. (2002). Mollusc larval shell formation: Amorphous calcium carbonate is a precursor phase for aragonite. *The Journal of Experimental Zoology, 293*, 478–491.

Xu, X., Cai, A., Liu, R., Pan, H., Tang, R., & Cho, K. (2008). The roles of water and polyelectrolytes in the phase transformation of amorphous calcium carbonate. *Journal of Crystal Growth, 310*, 3779–3787.

Zhang, K., Geissler, A., Fischer, S., Brendler, E., & Bäucker, E. (2012). Solid-state spectroscopic characterization of α-chitins deacetylated in homogeneous solutions. *The Journal of Physical Chemistry B, 116*, 4584–4592.

Ziegler, A., Fabritius, H., & Hagedorn, M. (2005). Microscopical and functional aspects of calcium-transport and deposition in terrestrial isopods. *Micron, 36*, 137–153.

CHAPTER TWENTY-THREE

A Mixed Flow Reactor Method to Synthesize Amorphous Calcium Carbonate Under Controlled Chemical Conditions

Christina R. Blue, J. Donald Rimstidt, Patricia M. Dove[1]

Department of Geosciences, Virginia Polytechnic Institute and State University, Blacksburg, Virginia, USA
[1]Corresponding author: e-mail address: pdove@vt.edu

Contents

1. Introduction — 558
2. Method — 559
3. Characterization — 562
   3.1 SEM — 562
   3.2 Inductively coupled plasma–optical emission spectrometry — 562
   3.3 Raman spectroscopy — 562
   3.4 Fourier-transform infrared spectroscopy — 564
   3.5 Thermogravimetric analysis — 564
4. Results — 564
References — 567

Abstract

This study describes a new procedure to synthesize amorphous calcium carbonate (ACC) from well-characterized solutions that maintain a constant supersaturation. The method uses a mixed flow reactor to prepare ACC in significant quantities with consistent compositions. The experimental design utilizes a high-precision solution pump that enables the reactant solution to continuously flow through the reactor under constant mixing and allows the precipitation of ACC to reach steady state. As a proof of concept, we produced ACC with controlled Mg contents by regulating the Mg/Ca ratio of the input solution and the carbonate concentration and pH. Our findings show that the Mg/Ca ratio of the reactant solution is the primary control for the Mg content in ACC, as shown in previous studies, but ACC composition is further regulated by the carbonate concentration and pH of the reactant solution. The method offers promise for quantitative studies of ACC composition and properties and for investigating the role of this phase as a reactive precursor to biogenic minerals.

1. INTRODUCTION

Amorphous calcium carbonate (ACC) is now recognized in a wide variety of natural and engineered environments. ACC can be found in some soils and sediments (Jones & Peng, 2012; Sanchez-Roman et al., 2008) and its formation has applications for controlled synthesis in biomedicine and materials sciences (Han & Aizenberg, 2008; Lee et al., 2012). Of particular interest is the ACC that forms during skeletal biomineralization of calcifying organisms (Addadi, Joester, Nudelman, & Weiner, 2006; Politi, Arad, Klein, Weiner, & Addadi, 2004; Tao, Zhou, Zhang, Xu, & Tang, 2009). Characterization studies show that ACC is a truly amorphous phase, with a typical composition of $CaCO_3 \cdot H_2O$, and it exhibits a unique short- and intermediate-range structure (Michel et al., 2008). The realization that ACC can be a reactive precursor to a number of crystalline polymorphs of $CaCO_3$ has motivated an extensive effort to understand the influence of the inorganic and organic factors on ACC composition. For example, the biomineralization community is interested in whether calcification by an ACC pathway affects the Mg content of calcitic biominerals that are used as proxies for seawater temperature (Dissard, Nehrke, Reichary, & Bijma, 2010; Stanley, 2006).

Previous studies of ACC have provided important insights into ACC properties, but a quantitative understanding is not yet established. This is partially because current synthesis procedures are unable to control solution chemistry. For example, a widely used synthesis method involves batch mixing of Na_2CO_3 and $CaCl_2$ solutions, then filtering the precipitate (Koga, Nakagoe, & Tanaka, 1998). However, the closed configuration of this "batch" method inherently causes solution composition to change significantly during synthesis, and the Na_2CO_3 salt increases solution pH to values that are higher (pH 11–13.5) than most natural systems. Another popular approach diffuses $(NH_4)_2CO_3$ into $CaCl_2$ solutions to increase supersaturation and induce ACC formation (Aizenberg, Addadi, Weiner, & Lambert, 1996; Wang, Wallace, De Yoreo, & Dove, 2009). This method produces ACC under highly variable and uncontrolled supersaturation conditions, introduces significant amounts of NH_4^+ into the mineralizing solution, and yields a limited amount of ACC.

To overcome these limitations, we developed a new procedure for ACC synthesis (Fig. 23.1) that uses a mixed flow reactor (MFR) system to maintain constant and controlled solution chemistries. This new procedure for

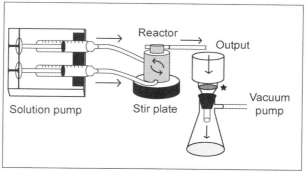

Figure 23.1 Schematic of the mixed flow reactor system. Arrows indicate the direction of flow. Black star denotes the location of the 0.20 μm nylon filter.

producing ACC is adapted from an earlier method that was used for kinetic studies of mineral growth (Saldi, Jordan, Schott, & Oelkers, 2009) and dissolution (Rimstidt & Dove, 1986). The MFR offers a number of advantages over batch and diffusion methods: (1) allows precise control over supersaturation, (2) produces ACC at steady-state conditions, and (3) yields large amounts of ACC with reproducible compositions. We demonstrate this approach by quantifying the factors that influence the Mg content of ACC by producing Mg-ACC from well-characterized solutions under controlled conditions.

2. METHOD

The experimental design (Fig. 23.1) includes a high-precision pump (Harvard Apparatus PHD 2000) fitted with two 100 ml syringes. For this study, one syringe contained Mg-doped $CaCl_2$ solutions with a prescribed Mg/Ca ratio (Table 23.1). A second syringe contained Na_2CO_3 or $NaHCO_3$ solution at a desired concentration (Table 23.2). The syringes were connected by Teflon tubing to opposite sides of the base of a 26-ml reactor unit. These solutions were continuously pumped into the reactor while mixing at a rate of 1000 rpm. As the solutions mixed, the contents of the MFR became supersaturated with respect to ACC, and ACC readily precipitated. The ACC-solution suspension exited the reactor through a single effluent tube located at the top of the reactor (Fig. 23.1) and products were captured on a 0.20-μm nylon filter fitted on a vacuum flask.

Table 23.1 Solution chemistries for variable Mg/Ca ratio experiments

Mg/Ca	0.5:1	1:1	2:1	5:1	8:1
$CaCl_2$ (M)	0.025	0.025	0.01	0.01	0.01
$MgCl_2$ (M)	0.0125	0.025	0.02	0.05	0.08
$NaHCO_3$ (M)	0.10	0.10	0.10	0.10	0.10
Initial pH	9.50	9.50	9.75	9.75	9.75
Steady-state pH	8.80	8.70	9.50	9.30	9.10
$\sigma_{(ACC)}$	4.94	2.43	1.61	1.31	1.02

The average hydraulic residence time (τ) of solution in the reactor was 2 minutes for all experiments, where

$$\tau = \text{flow rate} \, (\text{ml/h}) / \text{volume of reactor} \, (\text{ml})$$

For each experiment, the reactor operated for a minimum of three hydraulic residence times to allow the input solutions and ACC products within the reactor to reach steady-state conditions (Jensen, 2001). All ACC samples were rinsed thoroughly with ethanol and dried in a class II biological safety cabinet for 30 min before storing overnight in a vacuum desiccator. Each sample was weighed the next day to obtain the mass of dry ACC and samples were immediately prepared for their respective mode of analysis. Samples that could not be analyzed immediately were stored at 4 °C for no longer than 3 days.

As a proof of concept, ACC was synthesized from solutions with variable Mg/Ca ratios (Table 23.1) and solutions with different $NaHCO_3$ or Na_2CO_3 concentrations (Table 23.2). The initial pH of the $NaHCO_3$ solutions was adjusted using NaOH, so that the resulting pH was 8.7–9.5 upon mixing with the Mg/Ca solutions (Table 23.1). For the variable saturation experiments (Table 23.2), the initial Mg/Ca ratio was 5:1 and the saturation state (σ_{ACC}) was varied by adjusting carbonate concentration and pH (using NaOH). The saturation state of a solution (σ) with respect to ACC is expressed by:

$$\sigma_{ACC} = \ln\left(\frac{a_{Ca^{2+}} \times a_{CO_3^{2-}}}{K_{sp(ACC)}}\right) \tag{23.1}$$

where a_i is the activity of an ion and $K_{sp\,(ACC)} = 10^{-6.39}$ (25 °C) (Brečević & Nielsen, 1989). (We note that there is some uncertainty in this K_{sp}

Table 23.2 Solution chemistries for constant Mg/Ca ratio (5:1) experiments

Carb. source	Carb. conc. (M)	Initial pH	Steady-state pH	$\sigma_{(ACC)}$
Na_2CO_3	0.03	10.95	10.03	1.11
Na_2CO_3	0.06	11.12	10.13	1.58
Na_2CO_3	0.10	11.24	10.34	1.81
$NaHCO_3$	0.10	9.79	9.31	1.36
$NaHCO_3$	0.10	10.25	9.89	1.70
$NaHCO_3$	0.20	9.14	8.83	1.11
$NaHCO_3$	0.20	9.96	9.75	1.80

value—given that recent studies suggest the solubility of ACC is considerably lower than the reported value (Gebauer, Volkel, & Cölfen, 2008; Hu et al., 2012)—but there have been no further studies on the solubility of ACC, so the Brečević and Nielsen (1989) value is used here.) Activities of ions in solution were calculated using The Geochemist's Workbench© for each solution condition. In this study, saturation states were varied from 1.11 to 1.81, and all experiments were performed at ambient conditions.

The proposed mechanism for ACC particle formation in inorganic environments is the aggregation of aqueous ion pairs into prenucleation clusters (Gebauer et al., 2008), with a stability defined by thermodynamic equilibrium among the solvent, individual ions, and hydrated clusters. This is simplified by the general reaction

$$zCa^{2+} + zCO_3^{2-} = z(CaCO_3)_{aq}, \qquad (23.2)$$

where z is the number of $CaCO_3$ units in a cluster. In the presence of Mg ions, we assume the reaction that is equivalent for the formation of amorphous calcium magnesium carbonate (Wang et al., 2009) is

$$yMg^{2+} + zCa^{2+} + (y+z)(CO_3^{2-}) = \left(Ca_zMg_y(CO_3)_{y+z}\right)_{aq} \qquad (23.3)$$

where y is the number of $MgCO_3$ units in a cluster and the resulting solid solution has y plus z units in the cluster that forms. The equilibrium constant of reaction 2 and absolute values of y and z are unknown, but the ratio of y/z gives the fraction of Mg/Ca in the resulting ACC.

3. CHARACTERIZATION

The synthetic ACC was characterized using a combination of SEM, inductively coupled plasma–optical emission spectrometry (ICP–OES), Raman spectroscopy, FTIR, and thermogravimetric analysis (TGA).

3.1. SEM

Analyses were performed on an FEI Quanta 600 FEG ESEM. Images of the ACC were acquired using secondary electron microscopy. All samples were coated with approximately 10 nm of gold–palladium using a sputter coater before analysis. Optimal viewing parameters included 10 kv and a 3.5 spot size (instrument specific value). Higher voltages destroyed the ACC samples and could not be used. All samples were examined for consistent morphology and particle size variations. SEM analyses showed that all ACC samples exhibited a characteristic spherical morphology and a similar particle size distribution, despite changes in Mg content (Fig. 23.2).

3.2. Inductively coupled plasma–optical emission spectrometry

ICP–OES was performed using a Spectro CirOS VISION. The magnesium and calcium concentrations were determined from calibration curves that were prepared from plasma-grade single-element standards (SPEX CertiPreps). Samples were dissolved in 0.5 N nitric acid for 20 min. Total concentrations of magnesium and calcium (ppm) were corrected for absorbance of excess salt solution by the nylon membrane. A separate experiment was conducted to obtain the levels of calcium and magnesium absorbed by the filter based upon the amount of time the solution was in contact with the membrane, and a calibration table was constructed. Calcium and magnesium standards were analyzed in conjunction with ACC samples to verify the accuracy of the results. Measurements of Mg and Ca content in the ACC samples using ICP–OES showed the method is capable of producing ACC with Mg concentrations of 8–70 mol%.

3.3. Raman spectroscopy

Raman spectra were obtained on a JY Horiba (800 mm) LabRam HR spectrometer equipped with a 632.81 nm He–Ne laser emitting 15 mW power

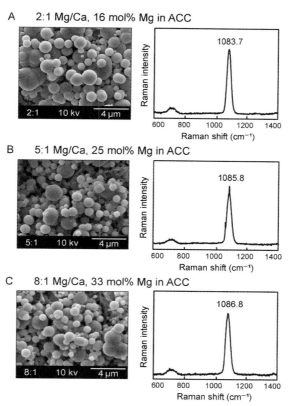

Figure 23.2 SEM images and corresponding Raman spectra of ACC synthesized from (A) 2:1, (B) 5:1, and (C) 8:1 Mg/Ca solutions using the new MFR method. The ACC morphology and particle sizes are similar for all Mg contents. The presence of a broadened carbonate symmetric stretch peak (v_1) in the Raman spectra confirms all products are amorphous. The v_1 peak shifts to higher wave numbers with increases in Mg content.

and focused through an Olympus 50× objective. A small amount of ACC was removed from the nylon filter and placed on a clean glass slide for analysis. Three 180 s spectra accumulations were collected and averaged for each of the samples analyzed and reported in this study. The carbonate peak positions and widths were estimated using a summed Gaussian/Lorentzian peak-fitting routine after baseline corrections were made using Jobin Yvon Labspec version 4 software. Raman spectroscopy confirmed that all products were amorphous; only a broadened v_1 (carbonate symmetric stretch) peak was present in all samples (Fig. 23.2). The peak position shifts to higher wave numbers with Mg content, as reported previously (Wang, Hamm, Bodner, & Dove, 2011).

3.4. Fourier-transform infrared spectroscopy

Fourier-transform infrared spectroscopy (FTIR) was performed on a Varian 670-IR with a Pike GladiATR attachment with a 4 cm^{-1} resolution. Samples were analyzed at a standard 32 scan rate. All results were background ATR corrected using the instrument's software program. Characterization by FTIR corroborated the trend of higher v_1 peak positions with higher Mg concentrations in the ACC for all samples, including those with Mg/Ca ratios in ACC as high as 2.5.

3.5. Thermogravimetric analysis

TGA was performed on a TA Instruments Hi-Res TGA Q5000 Thermogravimetric Analyzer. Samples were heated from 25 °C to 700 °C at a rate of 10 °C min^{-1}. The percentage of water lost during heating was calculated from the difference in the weight percent of the sample between 25 and 250 °C. TGA indicated that all ACC products contained approximately one water molecule per calcium and magnesium carbonate. There was no correlation between the water content and the Mg content of ACC for the conditions of this study.

4. RESULTS

The MFR method produces ACC with consistent compositions that systematically correlate with the solution chemistry conditions without evidence of secondary crystalline phases. At high supersaturation conditions, the amount of ACC approached gram quantities within 1 h of run time, which makes the MFR method ideal for producing a large quantity of samples that can be used for numerous analyses. Experiments conducted with NaHCO$_3$ and regulated input solution Mg/Ca ratios (Table 23.1) produced synthetic ACC with 8–33 mol% Mg (Fig. 23.3). Although all the ACC products contain moderate to very high concentrations of Mg (amorphous calcium and magnesium carbonate), there is no conventional terminology to describe these compositions, so the synthesized materials will simply be referred to as "ACC." Assuming that kinetic effects are negligible, the distribution of Mg between ACC and the solution is expressed by a distribution coefficient (K_D), defined as

$$K_D = \left(\frac{X_{Mg}}{X_{Ca}}\right) / \left(\frac{m_{Mg}}{m_{Ca}}\right) \tag{23.4}$$

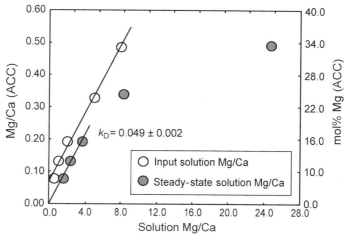

Figure 23.3 ACC composition is dependent upon the Mg/Ca ratio of the reactant solution. The total amount of Mg in ACC increases as the solution Mg/Ca ratio increases, but two different K_D trends exist depending on how the data are plotted (see text for K_D equation). If input solution values are used (open circles), the trend is linear, but if effluent solution values are used (closed circles), the trend is linear only at low Mg/Ca values. Solution Mg/Ca values evolve to higher steady-state values as ACC precipitates in the reactor, and therefore, steady-state Mg/Ca values should be used when calculating K_D. Each data point is an average of three experiments. (For color version of this figure, the reader is referred to the online version of this chapter.)

where (X_{Mg}/X_{Ca}) is the measured Mg/Ca ratio in ACC and (m_{Mg}/m_{Ca}) is the Mg/Ca ratio in the steady-state effluent solution. Equation (23.4) predicts the Mg content of ACC is linearly correlated with solution Mg/Ca, and previous studies (Radha et al., 2012; Wang et al., 2012) have reported a linear correlation when the initial Mg/Ca ratio in solution is >1. Indeed, when the *initial* solution Mg/Ca ratios are plotted versus the Mg/Ca ratio of ACC products (Fig. 23.3), we also find the linear dependence predicted by Eq. (23.4).

To obtain the true partition coefficient, we correlate ACC compositions with their corresponding *steady-state* solution compositions. Figure 23.3 shows two trends with increasing values of solution Mg/Ca that suggest the partitioning of Mg in ACC is more complex than previously assumed. Estimates of K_D obtain 0.049 ± 0.002 for steady-state Mg/Ca ratios of 0–4, but above a solution Mg/Ca ratio of 6, the linear relationship breaks down (Fig. 23.3). This suggests there may be a departure from equilibrium at higher solution Mg/Ca values and that kinetic effects, such as the precipitation rate, may control Mg signatures in ACC. It should also be noted that the K_D values reported here are significantly lower than previous estimates based upon the initial solution Mg/Ca ratio (Radha et al., 2012; Wang et al., 2009), because solution Mg/Ca ratios evolve to higher values at steady state as ACC precipitates.

An additional series of ACC materials were produced using solutions with a constant Mg/Ca ratio of 5:1 while varying saturation state and carbonate source ($NaHCO_3$ and Na_2CO_3). By regulating carbonate solution chemistry, the method produced ACC with 33–70 mol% Mg (Fig. 23.4). Figure 23.4 shows that higher pH and higher carbonate concentrations produce ACC with greater Mg content. Anecdotal evidence suggests these trends are caused in part by the effects of pH and carbonate concentration on precipitation rates. For example, faster precipitation rates cause the Mg/Ca ratio of the steady-state solution to become higher because a larger fraction of Ca^{2+} is consumed during ACC precipitation than Mg^{2+}.

Results from these proof-of-concept experiments reiterate the primary control of solution Mg/Ca ratio on ACC composition (Radha et al., 2012; Wang et al., 2012). By having the ability to independently vary each component of the solution chemistry, we show that ACC composition is

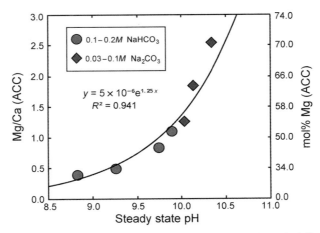

Figure 23.4 ACC synthesized at a constant Mg/Ca ratio (5:1) but with different carbonate concentrations and pH contains variable amounts of Mg, where higher carbonate concentrations and pH promote greater Mg uptake in ACC. The compositions show that carbonate concentrations, and largely pH, provide an additional means of tuning ACC composition. Each data point is an average of three experiments. (For color version of this figure, the reader is referred to the online version of this chapter.)

further regulated by carbonate concentration and pH of the reactant solution. The demonstrated ability of the MFR method to produce ACC with a broad range of desired compositions presents opportunities to study other aspects of ACC formation. These include kinetic versus thermodynamic controls on composition, the incorporation of other trace elements, the factors that regulate isotopic signatures, and the nature of the transformation process from amorphous to crystalline phases. By this approach, it may become possible to decipher and quantify the roles of this phase as a reactive precursor to biogenic minerals.

REFERENCES

Addadi, L., Joester, D., Nudelman, F., & Weiner, S. (2006). Mollusk shell formation: A source of new concepts for understanding biomineralization processes. *Chemistry A European Journal, 12*, 980–987.

Aizenberg, J., Addadi, L., Weiner, S., & Lambert, G. (1996). Stabilization of amorphous calcium carbonate by specialized macromolecules in biological and synthetic precipitates. *Advanced Materials, 8*, 222–226.

Brečević, L., & Nielsen, A. E. (1989). Solubility of amorphous calcium carbonate. *Journal of Crystal Growth, 98*, 504–510.

Dissard, D., Nehrke, G., Reichary, G. J., & Bijma, J. (2010). The impact of salinity on the Mg/Ca and Sr/Ca ratio in the benthic foraminifera ammonia tepida: Results from culture experiments. *Geochimica et Cosmochimica Acta, 74*, 928–940.

Gebauer, D., Volkel, A., & Cölfen, H. (2008). Stable prenucleation calcium carbonate clusters. *Science, 322*, 1819–1822.

Han, T. Y.-J., & Aizenberg, J. (2008). Calcium carbonate storage in amorphous form and its template-induced nucleation. *Chemistry of Materials, 20*, 1064–1068.

Hu, Q., Nielsen, M. H., Freeman, C. L., Hamm, L. M., Tao, J., Lee, J. R. I., et al. (2012). The thermodynamics of calcite nucleation at organic interfaces: Classical vs. non-classical pathways. *Faraday Discussions, 159*, 509–523.

Jensen, J. N. (2001). Approach to steady state in completely mixed flow reactors. *Journal of Environmental Engineering, 127*, 13–18.

Jones, B., & Peng, X. (2012). Amorphous calcium carbonate associated with biofilms in hot spring deposits. *Sedimentary Geology, 269-270*, 58–68.

Koga, N., Nakagoe, Y., & Tanaka, H. (1998). Crystallization of amorphous calcium carbonate. *Thermochimica Acta, 318*, 239–244.

Lee, K., Wagermaier, W., Masic, A., Kommareddy, K. P., Bennet, M., Manjubala, I., et al. (2012). Self-assembly of amorphous calcium carbonate microlens arrays. *Nature Communications, 3*, 725.

Michel, F. M., MacDonald, J., Feng, J., Phillips, B. L., Ehm, L., Tarabrella, C., et al. (2008). Structural characteristics of synthetic amorphous calcium carbonate. *Chemistry of Materials, 20*, 4720–4728.

Politi, Y., Arad, T., Klein, E., Weiner, S., & Addadi, L. (2004). Sea urchin spine calcite forms via a transient amorphous calcium carbonate phase. *Science, 306*, 1161–1164.

Radha, A. V., Fernandez-Martinez, A., Hu, Y., Jun, Y.-S., Waychunas, G. A., & Navrotsky, A. (2012). Energetic and structural studies of amorphous $Ca_{1-x}Mg_xCaCO_3 \cdot nH_2O$ ($0 \leq x \leq 1$). *Geochimica et Cosmochimica Acta, 90*, 83–95.

Rimstidt, J. D., & Dove, P. M. (1986). Mineral/solution reaction rates in a mixed flow reactor: Wollastonite hydrolysis. *Geochimica et Cosmochimica Acta, 50*, 2509–2516.

Saldi, G. D., Jordan, G., Schott, J., & Oelkers, E. H. (2009). Magnesite growth rates as a function of temperature and saturation state. *Geochimica et Cosmochimica Acta, 73*, 5646–5657.

Sanchez-Roman, M., Vasconcelos, C., Schmid, T., Dittrich, M., McKenzie, J. A., Zenobi, R., et al. (2008). Aerobic microbial dolomite at the nanometer scale: Implications for the geologic record. *Geology, 36*, 879–882.

Stanley, S. M. (2006). Influence of seawater chemistry on biomineralization throughout Phanerozoic time: Paleontological and experimental evidence. *Palaeogeography, Palaeoclimatology, Palaeoecology, 232*, 214–236.

Tao, J., Zhou, D., Zhang, Z., Xu, X., & Tang, R. (2009). Magnesium–aspartate-based crystallization switch inspired from shell molt of crustacean. *Proceedings of the National Academy of Sciences of the United States of America, 106*, 22096–22101.

Wang, D., Hamm, L. M., Bodner, R. J., & Dove, P. M. (2011). Raman spectroscopic characterization of the magnesium content in amorphous calcium carbonate. *Journal of Raman Spectroscopy, 43*, 543–548.

Wang, D., Hamm, L. M., Giuffre, A. J., Echigo, T., Rimstidt, J. D., Yoreo, De, et al. (2012). Revisiting geochemical controls on patterns of carbonate deposition through the lens of multiple pathways to mineralization. *Faraday Discussions, 159*, 1–16.

Wang, D., Wallace, A. F., De Yoreo, J. J., & Dove, P. M. (2009). Carboxylated molecules regulate magnesium content of amorphous calcium carbonate. *Proceedings of the National Academy of Sciences of the United States of America, 106*, 21511–21516.

AUTHOR INDEX

Note: Page numbers followed by "*f*" indicate figures and "*t*" indicate tables and "*np*" indicates footnotes.

A

Abbott, C., 307
Abbott, N. L., 216
Abezgauz, L., 272
Ablett, J., 510–511
Abraham, M. J., 79
Abrecht, M., 514–515, 534
Abreu, F., 259–260, 259*f*
Acharya, C., 18
Addadi, L., 28–29, 46–47, 226, 258, 259*f*, 278–279, 285–286, 346, 393–394, 416–417, 478, 495, 502–503, 514–515, 534, 537–542, 550, 558
Addison, W. N., 348
Ade, H., 172–173
Adiconis, X., 372
Adomako-Ankomah, A., 369–370, 383
Adrian, M., 264–265
Aebersold, R., 369, 376
Aeppli, G., 513
Agarwal, G., 308–310
Ager, J. W., 418–419, 445–446, 448–449
Agnarsson, I., 502
Ahmed, I. A. M., 104
Ahn, N. G., 369, 377
Ahner, T. T., 104
Aichmayer, B., 40–41, 392, 506–507, 518–519, 520*f*, 524–525
Aizenberg, J., 166, 212, 214, 215, 216, 217–218, 514–515, 534, 538–542, 550, 558
Akhtar, M. W., 227–229, 241–243, 247
Akkus, O., 446
Aksay, I. A., 464–467
Albeck, S., 502–503
Alexander, B., 417–418, 419–421, 427–428, 467
Ali, A. S., 243–244
Alivisatos, A. P., 155
Allan, N. L., 11
Alliston, T., 445, 446, 447, 448–449
Almer, J. D., 192, 416–418, 419, 428
Almgren, M., 272
Al-Sawalmih, A., 392
Alsem, D., 501–503
Altinoglu, E. L., 37
Amarie, S., 553
Amasino, A. L., 384, 478, 514–515
Amenitsch, H., 97–99
Amjad, Z., 130, 131–132, 135–136
Amoros, J. L., 226–227
Anastasi, G., 328
Andersen, F. A., 26–27, 35*f*
Anderson, E. H., 510
Ando, T., 148
Andre, I., 348
Angelova, M. I., 267
Antao, S. M., 489–490, 492–493
Antonietti, M., 28–29, 46–48, 55–56, 57, 285–286, 288, 290–292, 294–296, 296*f*, 299–300, 299*f*, 301
Apkarian, R. P., 393–394
Apolito, C. J., 317
Appathurai, N. P., 384, 478, 514–515
Apte, A., 380–381
Aquilano, D., 233–234
Aqvist, J., 13
Arad, T., 193, 278–279, 306–307, 534, 538–542, 550, 558
Archer, T., 11, 12*t*
Arena, J., 190–191, 393–394, 396–397
Arganda-Carreras, I., 439
Arias, J. L., 534
Armstrong, M. R., 149
Arnold, R., 181, 216–217
Arnold, S., 47–48, 417–418
Aronova, M. A., 259–260, 259*f*
Aroush, D. R. B., 416–418, 419, 429, 434, 447, 464–467
Arsenault, A. L., 392–393
Arslan, I., 149–150
Artacho, E., 5

Arzt, E., 502, 515–516
Ascenzi, A., 422–423
Aschi, M., 344–345
Ash, J. T., 347f, 348, 359–360
Ashby, P. D., 553
Assoufid, L., 503
Astroem, K. J., 148
Atkinson, G., 14
Attwood, D. T., 510
Atwood, D., 243–244
Auer, M., 193
Autumn, K., 502
Averbach, B., 517
Ayora, C., 85–87, 226
Azais, T., 297–299
Aziz, B., 47–48, 57, 60–61
Azzam, W., 181, 216–217

B

Bacon, G. E., 392–393, 405–406, 407–408
Bacon, P. J., 392–393, 405–406, 407–408
Bacsik, Z., 47–48, 57, 60–61
Baddiel, C. B., 538
Baez, A. V., 510
Bafna, V., 369
Bagus, P. S., 170–172
Bai, Z., 368–369, 370
Baier, R. E., 289
Bailey, J. E., 227–229
Bailey, M., 392–393
Bain, C. D., 217–218
Bairoch, A., 369, 376
Baker, A., 243–244
Baker, D., 345, 346, 348, 349f, 349–351, 351np, 360, 362, 363
Bale, H., 392–393, 416–418, 419, 446, 447, 451
Balköse, D., 227–229, 241, 247
Ballauff, M., 111–115, 112f, 114f, 272–273
Balooch, G., 192
Balooch, M., 192, 329, 330
Bandeira, N., 369, 376
Baneyx, F., 308–310
Banfield, J. F., 46–47, 123–124, 149–151, 153, 155, 251–252
Bannon, L., 306

Bansagi, T., 227–229, 230, 245–246, 247, 248–249
Banse, X., 445
Bao, W., 553
Barabash, R., 521
Barak, M. M., 463
Barbara, P. F., 308–310, 314
Barbas, I., 385
Barber, A. H., 417–418, 419, 464–467, 466f, 468
Barber, J. R., 448–449
Barge, L. M., 250–251
Barnes, G. R. G., 198
Barnes, H. L., 104
Barney, E. R., 485
Bar-On, B., 423, 464–467
Barr, T. L., 243, 247
Barrera, G. D., 11
Barron, T., 11
Barrón, V., 493–494
Barth, A., 537
Barth, H. D., 392–393, 416–419, 445–446, 447, 448–449, 451
Bartl, M. H., 502–503, 523–524
Bartlett, J. D., 307, 328–329
Bar-Yosef, O., 542
Bassett, J. H. D., 417–418, 419, 434, 438–439, 462–463, 462f
Batson, P. E., 173
Bau, H. H., 154, 155–157
Bäucker, E., 549–550
Bäuerlein, E., 226
Baugh, E. H., 354
Baumgartner, J., 39–40, 99–102
Bebb, C. E., 380–381
Becchi, M., 368
Becker, A., 495
Becker, D., 99–102
Becker, M. L., 310
Beckmann, W., 131
Behrens, P., 226
Belaney, R. M., 446
Belasco, J. G., 308–310
Belcher, A. M., 306, 308–310, 314, 346
Benamor, M., 157
Benecke, G., 436–438, 447–448
Benedetti, A., 111

Benedict, J. J., 40
Beniash, E., 131–132, 190–191, 193, 198, 226–227, 258, 514–515, 534, 538–542, 550
Benmore, C. J., 479–480, 483
Bennet, M., 558
Benning, L. G., 39, 99–103, 104, 113–119, 116f, 118f, 212
Benson, S., 370, 382
Bentley, L., 439
Bentov, S., 258
Bergamaschi, A., 442
Berger, J., 368, 514, 518, 519f
Bergeron, J. J., 369, 376
Berlin, A. M., 372
Berman, A., 26, 346
Bernard, P., 503, 504
Berne, B. J., 79–80, 81f
Berner, A., 518
Bernstorff, S., 97–99, 402, 422–423
Bernt, W., 330
Berquó, T. S., 487–488
Berrondo, M., 346
Berry, E. E., 538
Bertinetti, L., 26–28, 33–35, 35f, 37, 38, 39–40, 545
Berzlanovich, A., 402
Besselink, R., 103, 111, 115, 119–120, 121f
Bewernitz, M. A., 40, 51–52
Bhandarim, N., 485
Bhattacharya, D., 377
Bhushan, B., 226
Biebuyck, H. A., 216
Bigi, A., 192
Bijma, J., 558
Bilderback, D. H., 508, 510
Billinge, S. J. L., 85–87, 481–482, 485, 486–487, 489–490, 491–492
Birchall, J. D., 227–229
Bird, D. K., 227–229
Birkedal, H., 416–417
Birse, S., 11, 12t
Bishop, P. N., 192
Bismayer, U., 495
Bittarello, E., 233–234
Black, A. J., 166, 212, 214, 215, 216, 217–218
Blackledge, T. A., 502

Blanco, M., 88
Blank, D. H. A., 103, 111, 119–120, 121f
Blanton, T. N., 97–99
Bledsoe, R. K., 317
Blonsky, A. Z., 522, 523f
Bodner, R. J., 562–563
Boerner, H. G., 285–286
Boesecke, P., 416–418, 419, 423, 429, 438–439, 447, 451, 453–456, 457, 467
Böhm, J., 515–516
Boily, J.-F., 493–494
Bolhuis, P. G., 4–5
Bolis, V., 344–345
Bollinger, J.-C., 57–58
Bolze, J., 111–115, 112f, 114f, 272–273
Bomans, P. H. H., 26–27, 39–40, 60–61, 63, 83, 99–102, 193, 260, 534
Bonar, L. C., 192, 392–393, 534, 538
Bonewald, L. F., 438
Bonneau, R., 346, 348, 349–350, 352
Bonucci, E., 422–423
Börger, L., 53–54
Borges-Rivera, D., 372
Borkiewicz, O., 130
Bormashenko, E., 243, 248
Bormashenko, Y., 243, 248
Born, R., 549–550
Börner, H. G., 57
Borukhin, S., 285–286
Bos, K. J., 192
Bosbach, D., 16
Boskey, A. L., 130, 534
Bostedt, C., 170–172, 181–182, 216–217, 220
Bots, P., 39, 99–102, 103, 113–115, 116f, 212
Bowers, G. M., 87, 491–492
Bowers, P. M., 363
Boyce, A. J., 227–229
Boyce, B., 307, 352
Boyce, M. C., 417–418, 502–503
Boyde, A., 392–393, 416–418, 419, 434, 438–439, 441, 450f, 455–456, 462–463, 462f, 464f, 465f
Boyden, K., 149

Bozovic, N., 510–511
Bradley, P., 348, 349f
Brand, H. E. A., 111
Bras, W., 97, 104
Braun, W., 170–172
Bravman, J. C., 508
Brečević, L., 560–561
Brendler, E., 549–550
Brennan, S., 147–148
Brickmann, J., 83–85
Brindley, G. W., 503
Brinker, C. J., 102–103, 115
Brinson, L. C., 417–418, 419
Brinza, L., 104
Britten, R. J., 382, 383
Brock, R. E., 262–263, 262f, 267–268, 269–271, 269f
Broennimann, C., 442
Broughton, J. Q., 8–9
Brown, G. E. Jr., 487–488
Brown, S. D. M., 392–393, 416–417, 441, 450f, 455–456, 463, 464f, 465f
Brown, W. E., 538, 546
Brown, W. L., 513
Browning, N. D., 149
Brownlee, C., 258
Bruet, B. J. F., 417–418, 502
Brunauer, S., 331
Bruneval, F., 10, 11, 12t
Brunner, E., 549–550
Brunner-Popela, J., 108, 110
Bryan, T. E., 418–419
Bryant, S. J., 346
Brylka, L. J., 26–28, 33–35, 35f, 37, 38, 39–40, 193, 534, 545
Bu, W., 8f
Buck, M., 166–167
Budai, J. D., 513, 526–527
Buehler, M. J., 423, 424, 502
Bunk, O., 392–393
Burghammer, M., 97, 368, 393–394, 419–421, 427–428, 504, 510–511, 514, 517, 518, 519f
Burgot, J., 30–31, 32t
Burt, J. L., 308–310
Busby, M. A., 372
Busch-Nentwich, E., 368, 514, 518, 519f

Busenberg, E., 12t, 16–17
Butler, D. L., 446
Butler, G., 227–229
Butterfoss, G. L., 360

C

Cabrini, S., 553
Cahill, C. L., 104, 130
Cai, A., 547
Cai, Z., 513
Calvo, F., 83
Camacho, N. P., 524–525
Camaioni, D. M., 160–161
Cameron, A., 368, 382
Cameron, R., 369–370
Campbell, A. A., 289, 307
Campbell, G. H., 149
Canals, A., 226
Canham, P. B., 192
Cao, L., 131–132
Capone, R., 265
Carcea, A. G., 521
Carden, A., 192
Carlos, F., 385
Carnerup, A. M., 226–227, 230–232, 233–234, 237–239
Carter, N. P., 380–381
Cartwright, J. H. E., 4, 60–61, 64–66, 230, 241–244, 247, 248
Casana, Y., 149–150
Casati, M. Z., 196
Case, D. A., 7
Caselli, N., 553
Casini, A., 379
Caspi, E. N., 518
Castano-Diez, D., 195
Castellana, N., 369
Castellani, P. P., 191, 201
Castricum, H. L., 103, 104, 111, 115, 119–120, 121f
Catania, C., 297–299
Catti, M., 10–11, 12t
Ceccarelli, M., 74
Cederbaum, L. S., 170–172
Cedola, A., 510–511
Celestre, R. S., 508, 512f, 513
Chabinyc, M. L., 219–220

Author Index

Chaize, B., 267
Chakravarty, S., 227–229, 241–243, 247
Chan, W. P., 502
Chandler, D., 4–5
Chang, S. W., 424
Chao, J., 521
Chao, W., 510
Chapman, J. A., 198, 200
Chapman, K. W., 483, 485
Charlebois, M., 464–467
Chasteen, N. D., 492
Chaturvedi, H. T., 227–229, 241–243, 247
Chaudhury, S., 79, 80–85, 346, 348
Cheatham, T. E., 7
Chebaro, Y., 79
Checa, A. G., 4, 60–61, 64–66
Cheers, M. S., 383
Cheled, S., 26
Chen, I., 502
Chen, J., 331
Chen, K., 512f, 519–520
Chen, L., 418–419
Chen, P. Y., 416–417, 419–421, 420f, 502
Chen, W., 375–376
Chipman, D. M., 160–161
Chipot, C., 13
Chistofferensen, J., 26–27, 35f
Chiu, C. Y., 306
Chiu, D. T., 265–267
Chiu, P.-L., 149–150
Chivian, D., 362
Cho, K., 547
Choi, E. J., 348, 352
Chomczynski, P., 370–371
Chow, L. C., 33–34
Christensen, R. J., 306, 346
Christenson, H. K., 131–132, 260–261
Christoffersen, M. R., 26–27, 35f
Christy, A. G., 226–227, 230–232, 233, 237–239
Chu, Y. S., 510
Chughtai, A., 34–35
Chung, J.-S., 503, 513, 521
Chung, K.-S., 521
Chung, W. J., 310, 311f
Chupas, K. J., 485

Chupas, P. J., 481–482, 483, 485, 489–490, 492–493
Church, M. M., 512f
Cismasu, A. C., 487–488
Claffey, W., 191
Clark, R. H., 307
Clark, S. M., 85
Clausen, B., 482
Clauser, K. R., 369, 376
Cloetens, P., 503, 504
Clyne, T. W., 438–439
Coatman, R. D., 227–229, 228f, 230, 241–243, 248
Coelfen, H., 4, 16–17, 83, 278–279, 285–286, 288, 290–292, 291f, 293f, 297f
Cohavi, O., 344
Cohen, F. E., 360
Cohen, J. B., 503, 515–516, 517
Coleman, M. L., 227–229
Coleman, R. G., 227–229
Coleyshaw, E. E., 537–538
Cölfen, H., 26–27, 28–29, 40, 46–48, 49–52, 53, 54–56, 57, 58–61, 62–66, 99–102, 101f, 103, 131–132, 226–227, 228f, 230–232, 233, 251–252, 287–288, 287f, 290–292, 293–296, 296f, 299–300, 299f, 301, 534, 560–561
Colletier, J.-P., 267
Collier, G., 344, 350
Collier, J. H., 262
Collins, C., 227–229, 230, 241, 243, 247
Colomban, P., 535–536
Colvin, J. D., 149
Conde, M. M., 13
Condrate, R. A. Sr., 538
Cone, A. C., 383
Cooke, B., 77–78
Cooke, D. J., 11, 12t
Cooper, G. J. T., 243, 246
Coradin, T., 258
Cordelieres, F. P., 439
Corn, J. E., 348
Cornaglia, M., 553
Corni, S., 344–345
Corno, M., 344–345
Cornu, O., 445

Cosentino, K., 265, 267
Covington, A. K., 16–17
Cowan, D. F., 85, 496
Cox, D., 259–260, 259f
Craievich, A. F., 96–97, 106, 111
Craig, A. S., 198
Cranford, S. W., 502
Creemer, J. F., 155–157
Croker, L., 383–384
Crommie, M. F., 149–150
Crump, G., 537–538
Csontos, A. A., 149, 160
Cuello, G. J., 85–87
Cui, F. Z., 307
Cui, H., 272
Cui, L., 149–150
Cul, L., 123–124
Cullity, B. D., 417–418
Cunningham, K. M., 38
Curnier, A., 464–467
Currey, J. D., 417–418, 419–421, 434, 439, 456–457, 463, 502
Curry, J. D., 537–538
Cusack, M., 226
Cygan, R. T., 11, 12t

D

Daar, E., 392–393
Daculsi, G., 328
Dahmen, U., 155
Dai, H., 131–132
Daniel, R. M., 227–229
Daniel, S., 380–381
Daniele, P. G., 16–17
Daniels, J. E., 85–87
Danino, D., 272
Danish, E. Y., 16–17
Danner, S., 308–310
Dantas, G., 345
Dao, M., 416–417, 502–503
Darbois, L., 265–267, 270
Darling, T. W., 482
Das, P., 446
Das, S. K., 227–229, 241–243, 247
Dasgupta, S., 10
Daulton, T. L., 417–418, 419–421, 427–428, 467
Davey, R. J., 47–48

David, C., 510–511
Davidson, E. H., 369–370, 382, 383
Davidson, L. E., 104
Davies, R., 504
Davis, G. R., 439
Davis, J. C., 514, 522
Davis, S. A., 226–227
De Carlo, F., 510
de Jonge, N., 149–150, 161
de Leeuw, N. H., 7–8, 8f
De Rienzo, F., 344
De Vivo, M., 74
de With, G., 39–40, 47–48, 60–61, 63, 83, 99–102, 166, 210, 260, 270–271, 272
De Yoreo, J. J., 18, 40, 46–47, 123–124, 130–132, 147–148, 149–151, 153, 155, 157–158, 213–214, 215, 216, 251–252, 534, 549–550, 558, 561, 564–567
Debnath, S., 489–490, 492–493, 493f, 494f, 495f
Dechtiaruk, A., 369–370
Deem, M. W., 74–75, 78
Degnan, B. M., 370
DeHope, W. J., 149
Deisenhofer, J., 196
Delgado-López, J. M., 131–132
Dellago, C., 4–5
Delloye, C., 445
Delmas, P. D., 393–394
DeLuca, D. S., 372
Delves, C. J., 317
DeMartini, B. E., 148
Demeler, B., 53–54
Demichelis, R., 4, 10, 11–13, 12t, 16–17, 18, 58–61, 64–66, 88–89
DenBesten, P. K., 329, 330, 332, 336
Denecke, M. A., 16
Deng, M., 233
DeOliveira, D. B., 294, 295f
Derreumaux, P., 79
Deryugin, Y., 448–449
Desbois, C., 307, 352
Dey, A., 39–40, 47–48, 99–102, 270–271
Deyhle, H., 392–393
Di Fabrizio, E., 503
Di Felice, R., 344–345
Di Fonzo, S., 510–511
di Tommaso, D., 7–8

Dick, B. G. Jr., 7–8
Dick, T., 10
Dillon, J. W., 306
DiMaio, F., 348
DiMasi, E., 14–15, 212, 214, 215, 506–507, 513, 513f
Dinapoli, C., 442
Dindot, J. L., 307
Dingenouts, N., 111–115, 112f, 114f, 272–273
Dissard, D., 558
Dittrich, M., 558
Dixon, D. A., 160–161
Do, D.-H., 513
Docquier, P. L., 445
Doloboff, I. J., 250–251
Donadio, D., 10, 11, 12t
Dong, X., 79
Dongsheng, L., 149–151, 153, 155
Donners, J., 213, 214, 215
Donoghue, P., 368
Doran, A., 502, 522
Döring, W., 99–102
Dorozhkin, S. V., 131–132, 141
Dorozhkina, E. I., 131–132, 141
Doube, M., 439
Double, D. D., 227–229, 228f, 230, 241–243, 248
Dougherty, R. P., 439
Dove, M. T., 10–11, 12t, 85, 87, 490–491, 496
Dove, P. M., 18, 130, 131–132, 147–148, 213–214, 215, 216, 346, 534, 549–550, 558–559, 561, 562–563, 565–566
Dovesi, R., 11–13
Doxey, C. M., 446
Drake, J. L., 377
Drechsler, M., 60–61, 237–238
Driessens, F. C., 331
Drobny, G. P., 344, 347f, 348, 359–360
Du, C., 307
Du, Z., 7–8
Duan, W., 241, 246
Duan, Y., 79
Ducy, P., 307, 352
Duffy, D. M., 11, 12t
Dufresne, E. M., 513
Duggal, T., 243, 247
Dunand, D. C., 417–418, 419

Dunbrack, R. L., 360
Dunin-Barkovskiy, L., 246
Dunlop, J. W., 419, 464–467, 466f, 468
Dunstan, C., 307, 352
Dutrow, G. H., 149–150
Dziedzic-Goclawska, A., 446

E

Eanes, E. D., 26–27, 130, 131–132, 136, 262–263, 262f, 267–268, 270, 272, 392–393, 405–406
Earl, D. J., 74–75, 78
Ebert, T. T., 46–47
Ebrahimpour, A., 289
Echigo, T., 564–565, 566–567
Eckert, C., 549–550
Edwards, H. G. M., 535–536
Edwards, K. J., 272, 487–488
Edwards, S. A., 11, 12t
Eeckhaut, G., 104
Egami, T., 85–87, 481–482, 485, 486–487
Egelhaaf, S. U., 264–265
Egerton, R. F., 155, 160–161
Ehm, L., 85, 489–490, 492–493, 495–496, 558
Ehrlich, H., 549–550
Eiblmeier, J., 226–227, 228f, 229, 230–233, 235–236, 237, 238–240
Eike, D. M., 13–14
Eikenberry, E. F., 442
Elemans, J., 213, 216
Elhadj, S., 18, 534
Eliceiri, K. W., 439
Elinson, R. P., 378–379
Elliott, G. F., 192
Elliott, J. A., 11, 12t
Elson, R. A., 446
Emara, M. M., 14
Emerson, D. J., 77–78
Emmerling, F., 4
Emmett, P. H., 331
Eng, P., 16
Engel, H., 265, 267
Engelhard, M., 330
English, D. S., 308–310, 314
Ensing, B., 74
Eom, C. B., 513
Epple, M., 495
Ercius, P., 149–150

Erez, J., 258
Erickson, B., 192
Erkkilä, H., 111
Erler, R., 104
Ermolaeva, O., 368, 382
Esapa, C. R., 417–418, 419, 434, 438–439, 462–463, 462f
Esapa, C. T., 392–393, 416–417, 439, 441, 450f, 455–456, 463, 464f, 465f
Eschberger, J., 392–393, 402
Escribano, B., 241–244, 247, 248
Estes, D. J., 265
Estevez, J. A., 417–418, 419, 434, 439, 456–457
Estroff, L. A., 181, 210, 212, 214, 215, 285–286, 368
Esvelt, K. M., 385
Ettensohn, C. A., 369–370, 383
Evall, J., 217–218
Evans, A. G., 464–467
Evans, J. E., 149–150
Evans, J. S., 310
Evans, P. G., 513
Evans, S. D., 260–261
Evans-Lutterodt, K., 502–503, 510–511
Eyre, D. R., 190–191

F

Faatz, M., 547
Fabbri, J. D., 183–185
Fabritius, H., 495, 547
Fadley, C. S., 181–182, 220
Faivre, D., 226, 259–260, 259f
Falini, G., 307, 502–503
Falkowski, P. G., 377
Fang, M., 192
Fantner, G. E., 416–417
Farquharson, M. J., 392–393
Featherstone, J. D., 331, 336
Feder, S. M., 446
Feigin, L. A., 106–108, 109
Feng, J., 85, 307, 495–496, 558
Feng, Z., 131–132
Fenter, P., 14–15
Fermani, S., 307
Fernández, M. S., 534
Fernandez, P. B., 492
Fernandez-Martinez, A., 83–87, 86f, 88–89, 496, 564–567
Fernandez-Prini, R., 14
Fernández-Serra, M. V., 5
Ferreira, J. M. F., 131–132
Ferris, F. G., 115
Fery, A., 230–232, 238–239
Feste, S., 510–511
Févotte, G., 38
Fezzaa, K., 492
Fincham, A. G., 328–329
Finnemore, A. S., 282, 284f, 285f
Finney, L., 258
Fiordelisi, M., 510–511
Fischer, S., 549–550
Fitch, A. N., 484
Fleishman, S. J., 348
Flory, P. J., 64–66
Flynn, C. E., 308–310
Flytzanis, C. N., 383
Foley, J. P., 454–455
Folkers, J. P., 217–218
Fong, H., 310
Foti, C., 16–17
Fournet, G., 96–97, 106–107, 108–109
Fournier, D., 267
Fowler, B. O., 538, 546
Fox, M. A., 217–218
Frandsen, C., 46–47, 123–124, 149–151, 153, 155, 251–252
Frantsuzov, P. A., 83
Fratzl, P., 26–27, 39–40, 46–47, 97–102, 190–191, 226, 272–273, 392–394, 396–397, 400–401, 402–404, 403f, 416–418, 419–421, 420f, 422–423, 427–428, 429, 434, 436–439, 447–448, 451, 452–457, 453f, 458–459, 467, 502, 506–507, 518–519, 520f, 524–525
Fratzl-Zelman, N., 392–393, 400
Frederik, P. M., 39–40, 60–61, 63, 83, 99–102, 272
Freeman, C. L., 11, 12t, 83, 130, 147–148, 213–214, 216, 273–274, 345, 351–352, 560–561
Freeman, D. L., 74–75
Freer, A., 226
Freltoft, T., 110, 111, 115, 120
Frenkel, D., 13, 74–75

Frey, S., 180–181
Frick, L., 417–418
Fridriksson, T., 227–229
Friedrich, H., 26–28, 33–35, 35f, 37, 38, 39–40, 193, 272, 534, 545
Friesner, R. A., 79–80, 81f
Fritz, M., 306
Frohman, M. A., 380–381
Fujino, Y., 379
Fujisaki, K., 425, 428–429
Fujiwara, Y., 382
Funari, S. S., 416–418, 419, 429, 434, 439, 447, 456–457, 464–467
Fuoss, P. H., 147–148
Furedi-Milhofer, H., 131–132, 141, 289

G

Gabrielli, C., 157
Gache, N., 57–58
Gaj, T., 385
Gal, J.-Y., 57–58
Gale, A. M., 438
Gale, J. D., 4, 6, 10–15, 12t, 16–17, 18, 58–61, 64–66, 83–87, 86f, 88–89
Gallana, I., 502–503
Ganesan, K., 285–286
Ganey, T., 393–394
Gao, C., 332, 336
Gao, H. J., 423, 455–456, 502
Gao, X., 308–310
García, A. E., 13–14, 78, 79
Garcia, D., 464–467
Garcia-Ruiz, J. M., 226–229, 228f, 230–232, 233–236, 237–239, 240, 241–244, 248–249, 249f, 250–252
Garg, A., 149, 160
Garrett, B. C., 160–161
Garti, N., 131–132, 141, 289
Gassmann, M., 371
Gathercole, L. J., 192
Gatsouli, A., 548
Gaylord, B., 370
Gebauer, D., 4, 10, 11, 12t, 16–17, 18, 26–27, 28–29, 40, 46–48, 49–52, 53, 54–56, 57, 58–61, 62–66, 83, 88–89, 99–102, 101f, 103, 131–132, 288, 560–561
Geiger, B., 193

Geissbuhler, P., 14–15
Geissler, A., 549–550
Geissler, P., 4–5
Geldenhuys, A., 243–244
Generali, P., 422–423
Geney, R., 79, 80–85
Genin, G. M., 417–418, 419–421, 427–428, 467
Gentili, M., 503
George, A., 26–27, 39–40, 130, 193, 306, 534
George, N. C., 370
Georgiou, G., 308–310
Gere, J. M., 441
Gerritsen, J. W., 213, 214, 215
Gersbach, C. A., 385
Gerstein, M., 368–369
Gervasio, F. L., 4–5, 74
Gerwert, K., 538, 545
Geyer, C. J., 74–75
Geyer, W., 180–181
Giacovazzo, C., 503, 505–506
Gianguzza, A., 16–17
Gilbert, B., 514–515, 534
Gilbert, P. U. P. A., 131–132, 502–503, 522, 524f
Gilmer, G. H., 8–9
Gilpin, C. J., 192
Giraudguille, M. M., 422–423
Giri, B., 425, 428–429
Giuffre, A. J., 564–565, 566–567
Gkoumplias, V., 198
Glaab, F., 226–229, 230–233, 235–236, 237–240, 243–244, 248–251, 249f
Glatter, O., 96–99, 106–107, 108, 109, 110
Glazer, L., 26
Glimcher, M. J., 190–191, 534, 538
Glorieux, F. H., 368
Gludovatz, B., 502
Gnanakaran, S., 78, 79
Gochin, M., 336
Goddard, W. A. III., 10, 88
Goetz-Neunhoeffer, F., 131–132
Gogol-Döring, A., 375–376
Goh, K. L., 422–423
Goldberg, H., 28–29
Goldberg, M. C., 38, 196
Goldstein, R. E., 227–229
Golub, E. E., 259–260

Golzhauser, A., 180–181
Gómez-Morales, J., 131–132
Goncalves, P. F., 196
Gong, H., 290–292
Gong, X., 306, 307–308
Gong, Y. U. T., 384, 478, 514–515
Goobes, G., 344
Goodwin, A. L., 85, 87, 490–491, 496
Goodwin, M., 518
Goos, J. A. C. M., 39–40, 60–61, 63, 83
Gorb, S. N., 502
Gossler, B., 237–238
Goto, T., 192
Gottschalk, K. E., 344
Gou, B.-D., 131–132, 136
Gouadec, G., 535–536
Gourrier, A., 226–227, 393–394, 401–402, 405, 419–421, 427–428, 517
Govrin-Lippman, R., 131–132, 141, 289
Gower, L. A., 289–292, 290f
Gower, L. B., 40, 46–47, 51–52, 99–102, 289, 478
Grater, F., 11, 12t
Gray, J. J., 344, 346–347, 347f, 348, 349–351, 354, 359–360, 361
Grazul, J. L., 212, 214, 216
Gready, J. E., 79
Greenfield, D. J., 405–406
Grey, C. P., 481–482
Grey, I. E., 111
Griesshaber, E., 553
Griffith, W. P., 537–538
Griffiths, R. K., 392–393, 405–406, 407–408
Grigull, S., 517
Grogan, J. M., 154, 155–157
Gröhn, F., 547
Groschner, M., 392–393
Gruber, P., 515–516
Grunze, M., 166–167, 170–172, 180–181
Grynpas, M. D., 392–393, 534, 538
Grzywa, M., 290–292
Gstaiger, M., 376
Gu, L. S., 418–419
Gu, T. T., 307
Gueta, R., 537–538
Guillot, B., 7
Guinier, A., 96–97, 106–107, 108–109
Gullikson, E. M., 514, 522

Gunawidjaja, P. N., 47–48, 57, 60–61
Gungormus, M., 310
Günzler, T. F., 510–511
Gupta, H. S., 393–394, 396–397, 400–401, 402–404, 416–418, 419–421, 422–423, 427–428, 429, 434, 436–439, 447–448, 451, 453–456, 457, 464–467, 466f, 468, 517
Gur, D., 258, 259f
Guzelsu, N., 464–467

H

Haasch, R., 219–220
Habelitz, S., 130, 192, 329, 330
Habraken, W. J. E. M., 26–28, 33–35, 35f, 37, 38, 39–40, 545
Haeffner, D. R., 417–418, 419
Hagedorn, M., 547
Hahner, G., 166–167, 170–172
Haimei, Z., 155
Hall, A. J., 227–229
Hall, B. D., 487–488
Hall, W., 517
Hamann, D. R., 212, 214, 216
Hamer, A. J., 446
Hamley, I. W., 282
Hamm, L. M., 130, 147–148, 213–214, 216, 273–274, 560–561, 562–563, 564–565, 566–567
Hammel, M., 107–108
Hammersley, A. P., 450, 461, 484
Han, T. Y.-J., 148–149, 150–151, 157–158, 166–167, 184f, 212, 213, 214, 215, 216, 217–218, 558
Hanc, T. Y., 40
Hanfland, M., 484
Hang, F., 417–418
Hanke, T., 549–550
Hannington, J. P., 262f, 267–268, 269
Hannon, J. B., 149–150, 151
Hansen, W. R., 502
Hansma, P. K., 148, 306, 346
Hansmann, U. H. E., 77–78
Hanson, J. C., 346, 481–482
Hara, H., 503, 504, 510
Harada, S., 431
Harafuji, K., 8–9
Haramaty, L., 377
Harder, E., 10

Harding, J. H., 7–8, 11, 12t, 14–15, 83, 345, 351–352
Härmä, H., 111
Harper, R. A., 392–393, 405–406
Harrington, R., 485, 491–492
Harris, M. J., 10–11, 12t
Harrison, P. M., 492
Harteneck, B. D., 510
Hashimoto, N., 382
Hasse, B., 495
Hassenkam, T., 416–417
Haston, J. L., 422–423
Hatcher, E., 18
Häusermann, D., 484
Hausner, D. B., 485, 489–490, 492–493, 493f, 494f, 495f
Hayhurst, A., 308–310
Haymet, A. D. J., 8–9
Haynes, W. M., 32t
Hazlehurst, T. H., 227–229, 243
He, L. H., 417–418
He, M. Y., 464–467
Head, R., 417–418, 419, 434, 438–439, 462–463, 462f
Heberling, F., 16
Hedges, L. O., 83–87, 86f, 88–89
Hedrich, R., 549–550
Heft, R., 348
Hegner, M., 215
Heidelbach, F., 405–406, 517
Heijligers, H. J., 331
Hellawell, A., 227–229
Hellebusch, D. J., 149–150
Hellwig, C., 170–172
Helm, L., 14
Helveg, S., 155–157
Hemmerle, J., 265–267, 270
Henke, B. L., 514, 522
Henneman, Z. J., 130, 131–132, 141
Henrich, B., 442
Henriksen, K., 259–260
Herden, A., 131
Herrera, S., 502–503
Herzberg, G., 538
Heywood, B. R., 262–263, 262f, 267–268, 270, 272
Higashinakagawa, T., 382
Hignette, O., 503, 504
Hikichi, K., 192

Hild, S., 538–542, 549–550
Hill, T. M., 370
Hiller, J., 392–393, 416–418, 419, 422–423, 434, 438–439, 441, 450f, 455–456, 462–463, 462f, 464f, 465f
Hiltunen, E., 111
Hirano, K., 513
Hoang, Q. Q., 307, 352, 356, 358
Hochrein, O., 83–85
Hodgdon, T. K., 272
Hodge, A. J., 191, 423–424
Hodgens, K. J., 190–191, 393–394, 396–397
Hoefling, M., 344
Hoffman, M., 464–467
Hohling, H. J., 328, 417–418
Holl, M. M. B., 192
Holmes, D. F., 191f, 192, 200, 422–423
Holt, M. V., 510–511
Homer, N., 375–376
Hoppe, E., 243, 247
Horger, K. S., 265
Horn, H., 10
Hornak, V., 79, 80–85
Horvath, D., 230, 243–244, 248
Hosein, H.-A., 489–490, 492–493, 493f, 494f, 495f
Hosfelt, J. D., 370
Hough-Evans, B. R., 383
Hoveling, G. H., 155–157
Howard, A. J., 227–229, 307, 352, 356, 358
Howe, J. M., 149, 160
Howe, M. A., 87
Hoyer, J. R., 18, 131–132, 534
Hsieh, S. T., 502
Hu, C. C., 307
Hu, E. L., 308–310, 314
Hu, J., 131–132
Hu, Q., 40, 130, 147–148, 150–151, 157–158, 213–214, 216, 273–274, 560–561
Hu, Y., 496, 564–567
Huang, E., 360
Huang, J. Y., 131–132, 136, 159–160
Huang, S.-N., 88
Huang, T. C., 97–99
Huang, Y., 306
Huber, G., 502
Huber, K., 36–37
Huenenberger, P. H., 13–14

Hughes, R. A., 192
Hui, Q., 490–491
Hull, D., 438–439
Hull, R., 149–150
Hulsken, B., 213, 216
Hummer, G., 13–14, 77–78
Hunger, S., 104
Hunger, U. T., 510–511
Hunoldt, S., 549–550
Hunter, R. J., 29, 30–31, 36–37
Hura, G. L., 10, 107–108
Huster, D., 192
Hutchinson, J. W., 464–467
Hutter, J., 7
Hwang, H.-S., 524–525
Hyde, S. T., 226–227, 230–232, 233–234, 237–239

I

Iannuzzi, M., 74
Ice, G. E., 503, 513, 521, 526–527
Ichijo, T., 328
Ihli, J., 212
Iida, A., 513
Iijima, M., 131–132
Ilavsky, J., 272–273
Iler, R. K., 101f, 102–103, 115, 228f, 233–234
Illies, M., 369–370
Imai, H., 233–234, 237–238
Iori, F., 344–345
Ireton, G. C., 345
Irmer, G., 38
Irving, T. C., 192, 196–198, 202
Isaacs, E. D., 513
Ishii, H., 174–176
Ishiyama, M., 307
Ismach, A., 553
Ison, H. C. K., 227–229
Israelachvili, J. N., 262
Ito, A., 36–37
Iverson, B., 308–310
Izmailov, A. F., 47–48

J

Jacak, R., 348, 350–351
Jackson, D. J., 370
Jackson, J. S., 439
Jackson, K. A., 8–9
Jackson, R. A., 10–11, 12t
Jacobs, K., 502
Jäger, I., 455–456, 502
Jakob, H. F., 393–394, 400, 452–453, 453f, 458–459
Jalava, J.-H., 111
Jalkanen, J., 141
Jaris, H. K., 370
Jark, W., 510–511
Jarzynski, C., 17
Jaschouz, D., 524–525
Jasiuk, I., 393–394
Jasiuk, W., 419–421
Jeansonne, M. S., 454–455
Jemian, P. R., 272–273
Jen, A. K. Y., 308–310
Jennings, G., 483
Jensen, J. N., 560
Jensen, O. N., 369, 377
Ji, B. H., 423, 455–456, 502
Ji, H., 368–369
Jiang, H. D., 130
Jiang, L., 501–502
Jiang, W., 79
Jiang, Y., 290–292
Jillavenkatesa, A., 538
Jo, J. Y., 513
Joester, D., 258, 262–263, 262f, 265–268, 269–274, 269f, 558
Johnson, I., 442
Johnson, M. A., 160–161
Johnsson, M., 307
Jonah, C. D., 160–161
Jones, B., 115, 558
Jones, S. E., 306, 308–310
Jong Min, Y., 149–150
Jordan, G., 558–559
Joubert, C., 370
Jun, Y.-S., 496, 564–567
Jungjohann, K. L., 149–150
Jungwon, P., 149–150

K

Kaaber, W., 392–393
Kaandorp, J., 377
Kadler, K. E., 191f, 192, 198, 200, 422–423
Kähkönen, H., 111
Kahn, J. L., 147–148

Kajihara, Y., 553
Kajikawa, K., 174–176
Kajiyama, S., 297–299
Kakonyi, G., 104
Kaler, E. W., 272
Kalkan, B., 85
Kaltz, S. R., 534
Kameda, T., 79–80
Kaminski, A., 446
Kamiyama, T., 111
Kammer, M., 549–550
Kang, H. C., 510–511
Kanik, I., 250–251
Kannus, P., 196
Kappl, M., 4
Kaptijn, L., 213, 216
Karim, A., 310
Karim, O. A., 8–9
Karlsson, G., 272
Karplus, M., 351np, 351–352, 351t
Karttunen, M., 141
Karunarane, A., 392–393
Karunaratne, A., 416–418, 419, 434, 438–439, 441, 450f, 455–456, 462–463, 462f, 464–467, 464f, 465f, 466f, 468
Kashchiev, D., 289
Kastenholz, M. A., 13–14
Kato, T., 130, 379–380, 381
Katoh-Fukui, Y., 382
Katsuki, A., 241, 246
Katula, K. S., 383
Kawamura, K., 8–9
Kawano, T., 147–148
Kawata, S., 279–280
Kawska, A., 83–85
Kazanci, M., 26–27
Keen, D. A., 85, 87, 490–491, 496
Keilmann, F., 553
Keim, E. G., 104
Kellermeier, M., 18, 49–51, 54–55, 57, 60–61, 226–229, 228f, 230–234, 235–236, 237–240, 243–244, 248–249, 249f, 250–252
Kelley, D. S., 227–229
Kellogg, E. H., 350–351
Kempter, A., 49–51, 54–55, 57
Kenny, T. W., 502
Kepert, C. J., 485
Kerebel, L. M., 328

Kerisit, S., 14–15, 16
Kerschnitzki, M., 419, 436–438, 447–448, 464–467, 466f, 468
Keyes, T., 4–5
Khare, S. D., 348
Khokhlov, S., 241, 247
Kibalczyc, W., 26–27, 35f, 40
Kienle, L., 60–61, 226–227, 230–233, 235–236, 237, 238–240
Kienzle, P., 348
Kiihne, S., 307
Kilambi, K. P., 346, 348, 350–351
Killian, C. E., 368, 370, 382, 383–384, 478, 514–515, 522, 524f
Kim, B., 79–80, 81f
Kim, D. E., 362
Kim, D. M., 513
Kim, E. T., 216, 346–347, 347f
Kim, I. W., 310
Kim, J. R., 4–5, 418–419
Kim, Y., 41
Kim, Y. K., 418–419
Kim, Y.-Y., 131–132, 226–227, 259–261, 259f, 285–286
Kindt, J. H., 416–417
Kinney, J. H., 502–503
Kinning, D. J., 110, 120–122
Kinzler, M., 170–172
Kirby, N., 111
Kirchner, H. O. K., 436
Kirkpatrick, P., 510
Kirkpatrick, R. J., 87, 491–492
Kirkwood, J. G., 13
Kisielowski, C., 155
Kisker, D. W., 147–148
Kitamura, S., 241, 246
Kitano, Y., 28–29, 213
Kittel, C., 535
Kjems, J. K., 110, 111, 115, 120
Klähn, M., 538, 545
Klamt, A., 83–85
Klaushofer, K., 26–27, 190–191, 392–394, 400, 401–404, 405, 409–410, 409f, 422–423, 452–453, 453f, 458–459
Klein, E., 278–279, 534, 538–542, 550, 558
Klein, M. L., 74
Klepetsanis, P. G., 131–132

Klinowski, J., 227–229, 230, 241, 243, 247
Klivansky, L. M., 148–149, 166–167, 184f, 213, 216, 217–218
Klosowski, M. M., 439
Knecht, M. R., 306
Knott, R. B., 83–85, 111
Ko, J. Y. P., 47–48, 57, 60–61
Koch, M. H. J., 192
Kodera, N., 148
Koetzle, T. F., 346
Kofke, D. A., 78
Koga, N., 348, 558
Kogure, T., 130, 379–380, 381
Kohl, H., 417–418
Kohn, A. J., 537–538
Kohn, D. H., 310, 348
Kohn, V., 503, 510–511
Köktürk, U., 227–229, 241, 247
Kokubo, T., 133
Kolanczyk, M., 405–406, 406f, 407, 409, 410, 411f, 457
Koller, K., 392–393, 400
Komeili, A., 259–260
Kommareddy, K. P., 558
Kone, A., 78
Konhauser, K. O., 115
Königsberger, E., 38
Kontoyannis, C. G., 548
Kooperberg, C., 360
Kortemme, T., 351np
Kotha, S. P., 464–467
Kötting, C., 538, 545
Kottmann, S. T., 308–310
Koutsopoulos, S., 538
Koutsoukos, P. G., 130, 131–132, 135–136
Kovalenko, N. V., 510–511
Kratky, O., 96–99, 106–107, 108, 109, 110
Krauss, S., 417–418, 419, 434, 439, 456–457, 464–467, 466f, 468
Krejci, M. R., 258, 262–263, 262f, 267–268, 269–271, 269f
Kremers, G. J., 149–150
Kriebel, J. K., 181, 210
Kronik, L., 537–538
Kros, A., 41
Krueger, S., 348
Kubista, L. M., 85, 496
Kudryashev, M., 195

Kuhlman, B., 345, 346, 348, 349–350, 352, 360, 361
Kuhlmann, M., 510–511
Kuhnisch, J., 457
Kühnisch, J., 405–406, 406f, 407, 409, 410, 411f
Kulak, A. N., 212, 280–282, 285–286
Kullar, A., 329, 330
Kumagai, H., 297–299
Kumar, A., 216
Kundu, S., 18
Kuntner, M., 502
Kunz, M., 512f, 519–520, 521, 522, 523f, 524f
Kunz, W., 227–229, 228f, 230–232, 233, 238–239, 240, 243–244, 248–249, 249f, 250–251
Kurapova, O., 510–511
Kuzmenko, I., 212, 214, 215
Kwak, S. Y., 212, 214, 215
Kwanpyo, K., 149–150
Kwon, K. Y., 310, 311f

L

Lacmann, R., 131
Ladd, A. J. C., 13
Ladden, S. F., 260–261
Lafage, M. H., 402–404
Laghaei, R., 79
Lagomarsino, S., 510–511
Lai, B., 503, 513
Laibinis, P. E., 217–218
Laio, A., 4–5, 17, 74
Lajoie, G., 141
Laklouk, A., 392–393
Lam, R. S. K., 131–132
Lambert, G., 558
Lammie, D., 393–394
Lamoureux, G., 10
Land, T. A., 147–148
Landis, W. J., 190–191, 393–394, 396–397, 419–421
Landrum, M. J., 368, 382
Lange, C., 405–406, 406f, 407, 409, 410, 411f, 457
Lange, O. F., 348
Langford, R., 282, 284f, 285f
Lankford, J., 438

Lardge, J. S., 11, 12*t*
Larson, B. C., 513, 521, 526–527
Larsson, A. K., 226–227, 230–232, 233–234
Lasko, G., 448–449
Latour, R. A., 344–345, 350, 351–352, 359
Launey, M. E., 418–419, 445–446, 448–449, 501–503
Laursen, R. A., 294, 295*f*
Lausch, A. J., 190–191
Layten, M., 79, 80–85
Lazaridis, T., 351*np*, 351–352, 351*t*
Le Coadou, C., 39–40, 99–102
Le Roy, N., 368, 370
Le, T. Q., 331, 336
Lea, S., 330
Leapman, R. D., 259–260, 259*f*
Leaver-Fay, A., 348, 350–351
Leblond, C. P., 328
Leduc, S., 227–229
Lee, B. P., 502
Lee, H. N., 502, 513
Lee, J. D., 147–148
Lee, J. R. I., 40, 46–47, 123–124, 130, 147–151, 153, 155, 157–158, 166–167, 183–185, 184*f*, 213–214, 215, 216, 217–218, 251–252, 273–274, 560–561
Lee, K., 558
Lee, M. S., 79, 80–85
Lee, P. L., 481–482, 483, 489–490, 492–493
Lee, S.-W., 308–310, 311*f*
Lee, W.-K., 492, 510
Lees, S., 192, 392–393
LeGeros, R. Z., 538
Legnini, D., 258, 503
Legoues, F. K., 149, 158–159
Legriel, P., 297–299
Leiber, M., 371
Leiserowitz, L., 346
Leiterer, J., 4
Leith, A., 190–191
Lemire, I., 348
Lengeler, B., 503, 510–511
Leonardi, L., 191, 201
Les, C. M., 192
Leslie, M., 10–11, 12*t*
Levard, C. M., 487–488
Levi-Kalisman, Y., 495, 502–503, 514–515
Levine, M. J., 307

Lewis, I. R., 535–536
Lewis, S. M., 348
Li, C. H., 40–41, 282, 283*f*, 392, 401–402, 405–406, 406*f*, 407, 409–410, 409*f*, 411*f*, 457, 524–525
Li, D., 46–47, 123–124, 251–252
Li, G., 77–78
Li, H., 212, 214, 215, 285–286, 375–376
Li, J., 368–369, 370
Li, P., 155, 160–161
Li, W., 130, 331, 332, 336
Li, Y. J., 130, 306
Li, Z. Q., 345
Liang, Y. A., 502
Liao, H.-G., 123–124, 149–150
Liao, J.-W., 131–132, 136
Liao, S. S., 307
Liao, Y. Y., 141
Li-Chi, L., 149, 160
Lichtenegger, H. C., 452–453, 502–503, 523–524
Liddle, J. A., 510
Lin, A. Y. M., 416–417, 419–421, 420*f*
Lin, J., 368–369, 370
Lin, S.-T., 88
Linde, A., 196
Lindqvist, O. V., 550
Lipner, J., 417–418, 419–421, 427–428, 467
Liu, B., 153
Liu, C., 510–511
Liu, G., 489–490, 492–493
Liu, J. W., 36–37
Liu, L. J., 47–48, 57, 60–61
Liu, P., 79–80, 81*f*
Liu, R., 547
Liu, X. Y., 130–132
Liu, Y., 148–149, 166–167, 184*f*, 213, 216, 217–218, 418–419
Liu, Z., 74
Livingston, B. T., 368, 382
Livny, M., 352
Lonardelli, I., 518
Long, D. A., 535–536
Long, J. R., 40, 51–52, 307
Lopez, P. J., 258
Lorant, F., 10
Loste, E., 103, 260–261, 282
Loushine, R. J., 131–132

Love, J. C., 181, 210, 219–220
Lowenstam, H. A., 46–47, 130, 258, 278, 368, 502–503, 534
Lu, Y., 191f
Luan, X. H., 307
Lubovsky, M., 272
Lucaveche, C., 196
Ludwigs, S., 282, 284f, 285f
Luisi, P. L., 267
Lundy, D. R., 538
Lunt, A., 503
Lutterotti, L., 518
Lützenkirchen, J., 16
Luyckx, S., 445
Lyskov, S., 346, 348, 354

M

Ma, G. B., 130, 131–132
Ma, W., 131–132
Ma, Y., 46–47, 226–227, 290–292, 291f, 506–507, 518–520, 520f
MacDonald, J., 85, 495–496, 558
MacDougall, M., 307
MacDowell, A. A., 418–419, 445–446, 448–449, 508, 512f, 513
Mächtle, W., 53–54
Mackay, A. L., 230, 241
MacKerell, A. D. Jr., 10
Madura, J., 10
Maginn, E. J., 13–14
Mahairas, G., 369–370
Mahajan, S., 282, 284f, 285f
Mahamid, J., 28–29, 40–41, 46–47, 258, 259f, 278–279, 416–417, 524–525, 534, 542
Maiti, P. K., 88
Makihara, R., 379
Makovicky, E., 113–115
Makrodimitris, K., 346–347, 347f
Malac, M., 155, 160–161
Maldonado, M., 549–550
Malone, J., 306
Manchon, L., 370
Mancini, L., 510–511
Mandelshtam, V. A., 83
Manjubala, I., 405–406, 406f, 407, 409, 410, 411f, 457, 558
Mann, G., 243, 247

Mann, K., 377, 382
Mann, M., 369, 376, 377
Mann, S., 130, 262f, 267–268, 269, 306, 534
Mansouri, A., 348
Mantz, H., 502
Mao, C. B., 308–310
Marchini, M., 191, 192, 201
Mardis, E. R., 368–369
Marie, B., 368, 370, 377
Marikawa, Y., 378–379
Marin, F., 368, 370, 377
Mark, A. E., 78, 79
Marković, M., 538, 546
Marsh, M. E., 259–260, 259f
Marshall, G. W. Jr., 192, 329, 330
Marshall, R., 34–35
Marshall, S. J., 192, 329, 330
Märten, A., 402, 403f
Marti, O., 538–542, 549–550
Martin, M. C., 384, 478, 514–515
Martins, C. I., 147–148
Maruyama, D., 148
Marx, D., 7
Masayekhi, A., 492
Maselko, J., 227–229, 230, 243–244, 248
Maser, J., 510–511
Masic, A., 558
Masica, D. L., 346–347, 347f, 348, 359–360
Mass, T., 377
Massaro, F. R., 233
Massover, W. H., 196
Mastai, Y., 293–294
Mathew, M., 130
Mathias, G., 538, 545
Matson, S. L., 346
Matsudaira, P., 149–150
Matsunaga, R., 297–299
Matsuyama, S., 503, 504, 510
Matteucci, M., 510–511
Mayer, C., 131
Mayer, I., 331
Mayer, M., 265
McDonald, K. L., 193, 383–384
McDowall, A. W., 196
McEwen, B. F., 190–191, 393–394, 396–397
McEwen, L., 190–191
McGreevy, R. L., 87, 123, 490–491

McIntyre, N. S., 521
McKee, D. D., 317
McKee, M. D., 348
McKenzie, J. A., 558
McKittrick, J., 502
McLain, S. E., 482
McLeod, D., 192
McMahan, J. R., 243
McMahon, A. P., 383
McNulty, I., 258
Meadows, R. S., 191f, 192
Mecke, K., 502
Meckler, J. F., 385
Meek, K. M., 192, 198
Meekes, H., 213–214
Mehltretter, G., 290–292, 291f
Meibom, A., 227–229, 506–507, 518–519, 520f
Meldrum, F. C., 41, 46–47, 99–102, 101f, 103, 131–132, 212, 226–227, 260–261, 278–282, 280f, 281f, 284f, 285f, 297f
Melero-García, E., 60–61, 226–227, 228f, 230–236, 237, 238–240
Melli, M., 553
Menanteau, J., 328
Meng, Y., 79
Merbach, A. E., 14
Mersmann, A., 130–131
Messersmith, P. B., 262, 267–268, 502
Messmer, P., 422–423
Metropolis, N., 73–74
Metzker, M. L., 368–369
Metzler, R. A., 131–132, 502, 506–507, 514–515, 518–519, 520f, 522, 524f, 534
Meulenberg, R. W., 148–149, 166–167, 213, 215, 216, 217–218
Meyer, H. J., 12t
Meyer, J. L., 26–27, 131–132, 136
Meyers, M. A., 416–417, 419–421, 420f, 502
Michaels, A. S., 346
Michel, F. M., 85, 87, 487–488, 489–494, 493f, 494f, 495–496, 495f, 558
Michel, M., 265–267, 270
Mikami, M., 111, 307
Miller, A., 192, 196–198, 202

Miller, J. S., 243–244, 362
Miller, S. J., 348
Milliron, G. W., 502–503
Mimura, H., 503, 504, 510
Minesso, A., 111
Mirsaidov, U. M., 149–150
Miserez, A., 502–503
Misof, B. M., 402
Misof, K., 97–99
Misura, K. M. S., 349f, 350–351
Mitchell, P. C. H., 538
Mitre, D., 328
Mitsunaga, K., 379
Mitsutake, A., 83
Miura, T., 233–234, 237–238
Modi, B. P., 265–267
Moffat, K., 508
Mokaya, R., 227–229, 243, 247
Molenbroek, A. M., 155–157
Molinari, E., 344–345
Molinari, J.-F., 11, 12t, 14–15
Montana, V. G., 317
Montanari, F., 510–511
Mook, H., 192, 392–393
Moore, E. W., 57–58
Moore, K. T., 149, 160
Moore, P., 74
Mor, E., 26
Moradian-Oldak, J., 307, 328–329
Morales, M. P., 493–494
Morallon, E., 227–229, 230, 243–244, 248–249, 249f, 250–251
Moravits, D. E., 438
Morawe, C., 503, 504
Moreno, A., 238–239
Morgan, T. T., 37
Morgan, W., 230, 243–244, 248
Morocutti, M., 191, 192, 201
Morozov, A. V., 351np
Morris, N. P., 192
Morse, D. E., 306, 346
Moscho, A., 265–267
Moughon, S., 346, 349–350, 361
Mousseau, N., 79
Mu, Y., 80–83, 82f
Muddana, H. S., 37
Mueller, M., 104
Mueller, O., 371

Mullaney, B. P., 308–310
Müller, B., 392–393
Muller, D. A., 212, 214, 216, 285–286
Müller, F. A., 39–40, 99–102
Müller, M., 97, 264–265, 406–407
Mullin, J. W., 46–47, 130–131
Munch, E., 501–503
Munroe, P., 464–467
Murdoch, A., 422–423
Murdock, D., 368
Murray, T. M., 149, 160
Myerson, A. S., 47–48

N

Nader, A., 393–394, 400, 452–453, 453*f*, 458–459
Naik, R. R., 306, 308–310
Nakagoe, Y., 558
Nakano, Y., 348
Nakata, M., 192
Nakhmanson, S. M., 513
Nakorchevsky, A., 369, 376
Nalla, R. K., 502–503
Nanci, A., 328
Nancollas, G. H., 34–35, 47–48, 130, 131–132, 135–136, 141, 289, 307
Narayanan, T., 111–115, 112*f*, 114*f*, 272–273
Nassif, N., 285–286
Natan, A., 537–538
Nave, C., 192
Navrotsky, A., 260, 496, 564–567
Ndao, M., 347*f*, 348, 359–360
Neder, R. B., 489–490
Nehrke, G., 558
Neil Dong, X., 416–417, 428
Nelson, A. J., 170–172, 216–217
Neubauer, J., 131–132
Newcomb, C. J., 534
Newman, R. C., 521
Nguyen, V., 267–268
Nicolella, D. P., 438
Nielsen, A. E., 131, 560–561
Nielsen, M. H., 40, 46–47, 123–124, 130, 147–151, 153, 155, 157–158, 166–167, 183–185, 184*f*, 213–214, 216, 217–218, 251–252, 273–274, 560–561

Nilsson, J., 148–149, 166–167, 213, 215, 216, 217–218
Nilsson, T., 369, 376
Nishimura, T., 130, 297–299, 379–380, 381
Nisli, G., 227–229, 241, 247
Noce, T., 382
Nociti, F. H. Jr., 196
Nolte, R. J. M., 213, 214, 215, 216
Nonella, M., 538, 545
Novella, M. L., 230, 241–243
Noya, E. G., 13
Noyan, I. C., 503, 515–516, 517
Nudelman, F., 26–27, 39–40, 190–191, 193, 226, 258, 260, 534, 558
Nuzzo, R. G., 181, 210, 217–218
Nydegger, J., 307
Nylen, M. U., 405–406
Nymeyer, H., 78, 79

O

Oaki, Y., 297–299
Odajima, S., 419–421, 423–424, 425
Odom, D. J., 289
Oelkers, E. H., 558–559
Ofir, P. B. Y., 131–132, 141, 289
Ogletree, D. F., 553
Ohara, H., 174–176
Ohki, S., 289
Ohtaki, H., 14
Oida, S., 307
Okamoto, Y., 77–78, 79, 83
Okur, A., 79, 80–85
Old, W. M., 369, 377
Oliver, D. B., 196
Olson, I. C., 384, 478, 514–515, 522, 523*f*, 524*f*
Olson, M. A., 79, 80–85
O'Meara, M. J., 350–351
Ono, Y., 147–148
Onuma, K., 36–37, 131–132
Orgel, J. P. R. O., 192, 196–198, 202
Orlando, R., 344–345
Orme, C. A., 147–148
Orr, B. G., 192
Ortiz, C., 416–417, 502–503
Orwar, O., 265–267
Otalora, F., 226, 230, 241–243

Ouchi, Y., 174–176
Outka, D. A., 169–172
Overhauser, A. W., 7–8
O'Young, J., 141
Özkan, F., 227–229, 241, 247

P

Pacardo, D. B., 306
Padmore, H. A., 508
Paesani, F., 7
Pagano, J. J., 227–229, 230, 244, 245–246, 245f, 247, 248–249
Page, K. L., 482, 489–490
Page, M. G., 285–286
Palazoglu, A., 416–417, 502–503
Pallavicini, M. G., 308–310
Palmer, L. C., 534
Pan, H., 131–132, 547
Panas, R., 417–418
Pang, Y., 553
Panine, P., 111–115, 112f
Pantaleone, J., 230, 248
Papahadjopoulos, D., 262–263, 264–265
Parenty, A. D. C., 243, 246
Paris, O., 190–191, 392, 401–404, 403f, 405, 406–407, 409–410, 409f, 418–419, 452–453, 506–507, 517, 518–519, 520f, 524–525, 525f
Parise, J. B., 104, 483, 489–490, 491–493, 493f, 494f, 495f
Park, J. J., 310
Park, R. J., 260–261, 279–280, 280f, 281–282, 281f
Parker, M. A., 149, 160
Parker, S. C., 7–8, 8f, 14–15, 16
Parker, S. F., 538
Parmar, K., 227–229, 241–243, 247
Parrinello, M., 7, 10, 11, 12t, 17, 74
Parry, D. A. D., 198
Pascal, T. A., 88
Paschalis, E. P., 26–27, 393–394, 396–397, 400–401, 409–410, 409f
Pashley, D. H., 131–132
Pasteris, J. D., 417–418, 419–421, 427–428, 467
Paul, K. E., 219–220

Pavese, A., 10–11, 12t
Peattie, A. M., 502
Pechenik, L., 416–417
Pechook, S., 285–286
Peckauskas, R. A., 27–28
Peckys, D. B., 149–150
Pedersen, J. K., 109, 110, 111
Peeler, M., 369–370
Peelle, B. R., 308–310
Pei, J. M., 346
Peng, B., 111–115, 112f
Peng, X., 558
Penn, R. L., 46–47
Percus, J. K., 120–122
Pérennès, F., 510–511
Perez, L., 289
Perez-Salas, U., 348
Pergolizzi, S., 328
Periole, X., 78, 79
Perrot, H., 157
Pespeni, M. H., 370
Peterlik, H., 272–273
Petrenko, V. A., 306, 307–308
Petruska, J. A., 191, 423–424
Pevzner, P. A., 369, 376
Pfeiffer, F., 510–511
Pharr, G. M., 521
Phillips, B. L., 85, 87, 490–491, 495–496, 558
Philo, J., 348
Phoenix, V. R., 115
Picker, A., 49–51, 54–55, 57
Pierrat, F., 370
Pieterse, K., 26–27, 39–40, 193, 534
Pina, S., 131–132
Pinero, G., 307, 352
Piquemal, D., 370
Piston, D. W., 149–150
Pitera, J. W., 10, 83
Planken, K. L., 47–48
Plate, U., 328, 417–418
Plenk, H. Jr., 392–393, 402
Plimpton, S., 13
Plueg, C., 290–292, 291f
Plummer, L., 12t, 16, 17
Plummer, N. L., 16–17
Pogreb, R., 243, 248
Pohorille, A., 13

Pokroy, B., 518
Politi, Y., 28–29, 46–47, 278–279, 495, 514–515, 534, 538–542, 550, 558
Polte, J., 104
Ponder, B. A., 380–381
Pontoni, D., 111–113, 114f, 272–273
Popp, D., 196
Porod, G., 400
Portale, G., 103, 111, 119–120, 121f
Porter, S., 192
Posner, A. S., 27–28, 392–393, 542
Pouget, E. M., 39–40, 60–61, 63, 83
Poustka, A. J., 377, 382
Pramanik, A., 227–229, 241–243, 247
Pratt, L. R., 13–14
Prenesti, E., 16–17
Presta, L. G., 308–310
Pretzl, M., 230–232, 238–239
Price, G. D., 10–11, 12t
Proch, E., 446
Proffen, T., 479, 482, 489–490
Prostak, K., 392–393
Pugno, N. M., 502
Pusztai, L., 123
Putnam, C. D., 107–108
Pye, C. C., 38

Q

Qi, H. J., 417–418
Qi, J., 308–310
Qi, L. M., 282, 283f
Qian, Q., 307
Qiang, M., 519–520
Qin, Y., 130, 131–132, 141
Qiu, X., 481–482, 485
Quigley, D., 4, 10, 11, 12t, 13–14, 16–17, 18, 58–61, 64–66, 74, 83, 88–89, 260, 345, 351–352
Quinn, T., 306, 307–308
Quintana, J. P., 518

R

Rachel, R., 226–227, 230–233, 235–236, 237, 238–240
Rademacher, N., 290–292, 291f
Rademannm, K., 104
Radha, A. V., 496, 564–567
Radisic, A., 149–150, 151, 158–159

Radmacher, M., 306
Radnai, T., 14
Raiteri, P., 4, 6, 10, 11–15, 12t, 16–17, 18, 58–61, 64–66, 83–87, 86f, 88–89
Raj, P. A., 307
Rakovan, J., 130
Ralston, G., 53–54
Raman, S., 346
Ramaswamy, J., 348
Ramírez-Rodríguez, G. B., 131–132
Ramos-Silva, P., 377
Randel, J. C., 183–185
Rapp, G., 97–99
Rasband, W. S., 439
Ratner, B. D., 346
Rauch, F., 368
Raz, S., 46–47, 478, 495, 514–515, 537–538
Redfern, S., 11, 12t
Reeder, R. J., 85, 87, 490–492, 496
Reedy, M., 196
Refson, K., 537–538
Reichary, G. J., 558
Reichert, P., 392–393, 416–418, 419, 446, 447, 451
Reiss, B. D., 308–310
Renaut, R. W., 115
Renfrew, P. D., 348, 352, 360
Rengstl, D., 228f, 229, 230–232, 233, 239, 240
Resing, K. A., 369, 377
Reuter, M., 149, 158–159
Riboli, F., 553
Ricci, M., 11, 12t, 14–15
Rieckel, C., 97
Rieger, J., 36–37, 251–252
Riekel, C., 368, 504, 510–511, 513, 514, 517, 518, 519f
Riello, P., 111
Rietveld, H. M., 518
Rigon, L., 510–511
Rimola, A., 344–345
Rimstidt, J. D., 558–559, 564–565, 566–567
Rinnerthaler, S., 393–394, 400, 452–453, 453f, 458–459, 524–525
Ripamonti, A., 422–423
Risbud, S. H., 149, 160

Ritchie, C., 243, 246
Ritchie, R. O., 418–419, 445–446, 447, 448–449, 502–503
Robert, A., 440
Robin, M., 297–299
Robins, S. P., 196
Robinson, J. J., 382
Robinson, R. A., 50–51, 55
Rodan, G., 402–404
Rodger, P. M., 10, 11, 12t, 13–14, 16, 17, 74, 83, 260, 345, 351–352
Rodriguez-Blanco, J.-D., 39, 99–102, 103, 113–115, 116f
Rodriguez-Fortea, A., 74
Rogan, R. C., 513
Rohl, A. L., 11, 12t, 83–85
Rohl, C. A., 346, 349–351, 361
Roitberg, A. E., 77–78, 79
Romanato, F., 503
Román-Pérez, G., 5
Roman-Ross, G., 85–87
Romao, C. P., 522, 523f
Roncal-Herrero, T., 39, 99–102, 103, 113–115, 116f
Roschger, P., 393–394, 396–397, 400–404, 405, 409–410, 409f, 416–418, 419–421, 422–423, 427–428, 429, 434, 447, 452–453, 453f, 458–459, 464–467, 517, 524–525
RoseFigura, L., 230, 248
Rosenbluth, A. W., 73–74
Rosenbluth, M. N., 73–74
Ross, F. M., 149–150, 151, 158–159, 161
Rosta, E., 77–78
Rostaing, G., 503, 504
Roszol, L., 245–246
Rouse, S. M., 37
Roux, B., 10, 79
Roveri, N., 192, 422–423
Roy, M. D., 310
Rubin, J., 393–394, 419–421
Rubin, M. A., 393–394, 419–421
Ruczinski, I., 346, 349–350
Rudolph, W. W., 38
Rueggeberg, F. A., 131–132
Ruggeri, A., 191, 192, 201
Ruse, C. I., 369, 376
Russell, M. J., 227–229, 250–251

Russell, S. T., 351np
Ryan, K. S., 308–310
Ryu, O. H., 307

S

Sabanay, I., 193
Sabri Dashti, D., 79
Sabsay, B., 306
Sacchi, N., 370–371
Sader, K., 196
Sadreyev, R., 346
Sagi, I., 514–515, 534
Sainz-Díaz, C. I., 4, 60–61, 64–66, 241–244, 247, 248
Saito, K., 148
Saiz, E., 501–503
Sakaida, Y., 431
Saldi, G. D., 558–559
Salditt, T., 510–511
Salehpour, A., 446
Salje, E. K., 10–11, 12t
Sallum, A. W., 196
Sallum, E. A., 196
Salminen, I., 550
Salowsky, R., 371
Salzberg, S. L., 375–376
Sammartano, S., 16–17
Sanchez-Roman, M., 558
Sand, K. K., 113–115
Sanford, E., 370
Sano, Y., 503, 504, 510
Santisteban-Bailon, R., 228f, 233–236, 238–239
Santoro, G., 328
Sarikaya, M., 308–310, 506–507, 513, 513f, 534
Sarro, P. M., 155–157
Saruwatari, K., 130, 379–380, 381
Sasaki, N., 192, 419–421, 423–424, 425
Satija, R., 372
Sato, T., 425, 428–429
Sauer, G. R., 546
Sayeg, M. K., 359–360
Scardi, P., 518
Scarlett, N. V. Y., 111
Scatchard, G., 59
Schaaf, P., 265–267, 270
Schaible, A., 392–393

Schaible, E., 416–418, 419, 445, 446, 447, 448–449, 451
Scheller, M. K., 170–172
Schenk, A. S., 26–28, 33–35, 35f, 37, 38, 39–40, 41, 545
Scheraga, H. A., 345
Scherer, G. W., 102–103, 115
Scherer, M. R. J., 282, 284f, 285f
Schertel, A., 170–172
Schitter, G., 148
Schlitter, J., 538, 545
Schmahl, W. W., 553
Schmauder, S., 448–449
Schmid, T., 558
Schmidler, S. C., 77–78
Schmidt, C., 251–252
Schmidt, M. P., 85, 496
Schmidt, M. U., 290–292, 291f
Schneider, C. A., 439
Schöberl, T., 502–503, 523–524
Scholl, A., 502, 522
Schoonen, M. A. A., 489–490
Schott, J., 558–559
Schreiber, S., 402–404
Schrier, S. B., 359–360
Schrodinger, L. L. C., 354
Schroeder, A., 371
Schroeder, S. L. M., 47–48
Schroer, C. G., 510–511
Schuck, P. J., 49, 62, 553
Schueler-Furman, O., 346, 349–350, 361
Schüler, D., 226, 259–260, 259f
Schulten, K., 308–310
Schulze-Briese, C., 104
Schupp, P., 549–550
Schurtenberger, P., 264–265
Schuurmann, G., 83–85
Schwartz, H. E., 446
Schwartzberg, A., 553
Schwarz, H., 368, 514, 518, 519f
Scozzafava, A., 379
Scully, C., 192
Sear, R. P., 101f
Searson, P. C., 149–150, 151, 158–159
Seeman, E., 393–394
Segal, D. J., 385
Segvich, S. J., 310
Seki, K., 174–176

Seki, Y., 416–417, 419–421, 420f
Semenza, G., 215
Serna, C. J., 493–494
Seshadri, R., 103
Sethi, M., 306
Seto, J., 57, 226–227, 416–418, 419, 423, 429, 434, 438–439, 447, 451, 453–457, 467
Shahar, R., 463
Shannon, C. E., 196
Sharapov, V. A., 83
Sharir, A., 258, 259f, 278–279, 416–417, 534, 542
Shastri, S. D., 492
Shaw, S., 39, 99–103, 104, 113–119, 116f, 118f
Shaw, W. J., 330, 348, 359
Shechter, A., 26
Shefelbine, S. J., 424
Shen, Q., 510
Shendure, J., 368–369
Sherringham, J. A., 227–229
Shi, J. M., 495
Shiba, K., 131–132
Shim, J., 18
Shimokawa, H., 307
Shimoni, E., 40–41, 524–525
Si, S., 131–132
Sichel, R. J., 513
Sicheri, F., 307, 352, 356, 358
Sidorov, M., 196
Siegel, S., 40–41, 392, 401–402, 405, 409–410, 409f, 524–525
Siegrist, T., 212, 214, 215
Sigel, A., 534
Sigel, H., 534
Sigel, R. K. O., 534
Sigmon, T. W., 149, 160
Simkiss, K., 368, 538
Simmer, J. P., 307, 328–329
Simmons, D., 307
Simons, K. T., 346, 349–350, 360
Sinclair, R., 149, 160
Sindhikara, D. J., 77–78, 79
Singer, J. W., 87, 491–492
Sinha, S. K., 110, 111, 115, 120
Sircar, A., 346
Sitti, M., 502

Sleep, N. H., 227–229
Smesko, S. A., 289
Smit, B., 74–75
Smith, G. P., 306, 307–308
Smith, H. C., 310
Smith, R. K., 155, 243
Smith, W., 13–14
Snigirev, A., 503, 510–511
Snigireva, I., 503, 510–511
Snyder, M., 368–369
Sokolov, S., 104
Soler, J. M., 5
Solis, D. J., 308–310
Söllner, C., 368, 514, 518, 519f
Sommerdijk, N. A. J. M., 39–40, 41, 46–48, 60–61, 63, 83, 99–103, 166, 190–191, 210, 213, 214, 215, 226, 260, 270–271, 272
Sone, E. D., 190–191
Song, J., 310, 311f, 502
Song, K., 79, 80–85
Song, M. J., 190–191, 393–394, 396–397
Song, R.-Q., 278–279, 293f
Song, Y. F., 131–132, 243, 246, 350–351
Sonmezler, E., 260
Sonnenberg, J. L., 7
Soper, A. K., 485, 491–492
Sorensen, C. M., 110, 120
Sorensen, L., 14–15
Sorensen, T. S., 227–229
Sorenson, J. V., 487–488
Sousa, A. A., 259–260, 259f
Souvorov, A., 503, 510–511
Specht, E. A., 359–360
Spijker, P., 11, 12t, 14–15
Spindler, K. D., 549–550
Spoerke, E. D., 534
Spolenak, R., 502, 513, 515–516
Sponberg, S., 502
Squires, G. L., 109
Srajer, G., 14–15
Stachowicz, W. A., 446
Stack, A. G., 6
Staffaroni, M., 553
Stafford, C. M., 310
Stahlberg, H., 195
Stanevsky, O., 243, 248
Stanley, S. M., 558
Stanley, T. B., 317

Stano, P., 265, 267
Starborg, T., 191f
Stawski, T. M., 103, 104, 111, 115, 119–120, 121f
Stayton, P. S., 344
Stebbins, J. F., 487–488
Steen, H., 376
Steijven, E. G. A., 213–214
Stein, A., 510–511
Steinbock, O., 227–229, 230, 244, 245–246, 245f, 247, 248–249
Stellacci, F., 11, 12t, 14–15
Stephens, C. J., 131–132, 260–261
Stephens, L., 370, 382
Stephenson, A. E., 131–132
Stierle, A., 515–516
Stinson, R. H., 196–198, 202
Stipp, S. S. L., 113–115
Stock, S. R., 192, 417–418, 419, 428
Stocker, S., 371
Stockley, I., 446
Stoddard, B. L., 345
Stöhr, J., 169–172, 171f
Stokes, R. H., 50–51, 55
Stone, D. A., 227–229
Stone, M. O., 308–310
Story, B., 307, 352
Stratmann, U., 417–418
Straub, J. E., 4–5
Strauch, E. M., 348
Strauss, C. E. M., 350–351, 363
Streb, C., 243, 246
Stribeck, N., 96–97, 106
Stringer, S. J., 308–310
Strizhak, P., 227–229, 243–244
Strongin, D. R., 485, 489–490, 492–493, 493f, 494f, 495f
Stuart, S. J., 344, 345, 350
Stucky, G. D., 250–251, 306, 346, 502–503, 523–524
Stupp, S. I., 534
Sturchio, N. C., 14–15
Su, C., 537–538
Suarez, D. L., 537–538
Sucov, H. M., 370, 382
Sugita, Y., 77–78, 79, 83
Sullivan, J. P., 159–160
Summerton, J., 381–382

Sumoondur, A., 104
Sun, H. B., 279–280
Sun, J., 226
Sun, L., 33–34
Sun, Y., 351–352
Suo, Z., 464–467
Suominen Fuller, M. L., 521
Supuran, C. T., 379
Suresh, S., 416–417, 502–503
Suzuki, K., 111
Suzuki, M., 130, 379–380, 381
Svenson, S. O., 484
Svergun, D. I., 106–108, 109, 115–119
Swadener, G., 464–467
Swain, M. V., 417–418, 464–467
Sweeney, R. Y., 308–310
Sweeny, P. R., 196–198, 202
Swendsen, R. H., 74–75
Swift, G. A., 513
Swope, W., 10, 83
Szebenyi, D., 508

T

Taavitsainen, V.-M., 111
Tabakovic, A., 37
Tabouillot, T., 37
Tadano, S., 425, 428–429
Tagami, A., 192
Tai, K., 416–418, 502–503
Tainer, J. A., 107–108
Takada, S., 79–80
Takadama, H., 133
Takagi, M., 196
Takagi, S., 130
Takai, E., 148
Talmon, Y., 60–61
Tamerler, C., 308–310
Tamura, N., 508, 512f, 513, 515–516, 519–520, 521–522, 523f, 524f
Tanaka, H., 558
Tanaka, K., 431
Tang, R.-K., 131–132, 136, 542, 547, 558
Tang, S. Y., 392–393, 416–418, 419, 445, 446, 447, 448–449, 451
Tang, Y., 85, 491–492, 496
Taniguchi, M., 192
Tanimoto, K., 331

Tanimoto, Y., 241, 246
Taniwatari, T., 147–148
Tannenbaum, T., 352
Tao, J.-H., 26–28, 33–35, 35f, 37, 38, 39–40, 130, 131–132, 136, 147–148, 213–214, 216, 273–274, 542, 545, 558, 560–561
Tao, Y. T., 217–218
Tarabrella, C., 85, 495–496, 558
Tarakanova, A., 502
Tarasevich, B. J., 330, 348
Tas, A. C., 312
Tauer, K., 299–300, 299f, 301
Tawa, G., 79, 80–85
Tay, F. R., 131–132
Taylor, A., 510–511
Taylor, J., 393–394
Taylor, M. B., 11
Taylor, M. G., 538
Teixeira, J., 110, 111, 115, 120
Telenius, H., 380–381
Teller, A. H., 73–74
Teller, E., 73–74, 331
ten Elshof, J. E., 111, 115
TenCate, J. A., 482
Teng, H. H., 147–148
Tennant, D., 510–511
Tepper, H., 8f, 10
ter Horst, J. H., 47–48
Terada, T., 233–234, 237–238
Terakawa, T., 79–80
Terashima, T., 328
Terfort, A., 181, 216–217
Termine, J. D., 27–28, 130, 392–393, 405–406, 538, 542
Terminello, L. J., 170–172, 181–182, 216–217, 220
Terrill, N. J., 104, 392–393, 416–417, 441, 450f, 455–456, 463, 464f, 465f
Tersoff, J., 149, 158–159
Tester, C. C., 262–263, 262f, 265–268, 269–274, 269f
Thain, D., 352
Thanawala, M. S., 212, 214, 215
Thieme, J., 251–252
Thomas, E. L., 110, 120–122
Thomas, N. L., 227–229, 228f, 230, 241–243, 248

Thompson, D. W., 226
Thompson, E. A., 74–75
Thompson, J. W., 346, 348, 485
Thomson, T., 308–310
Thouvenel-Romans, S., 227–229, 244, 245–246, 245f, 247, 248
Thünermann, A. F., 104
Thurner, P. J., 148
Timon, V., 85–87
Timoshenko, S. P., 441
Tirrell, D. A., 289–290, 290f
Tischler, J. Z., 503, 513, 526–527
Tlili, M. M., 157
Tobler, D. J., 102–103, 115–119, 118f
Toda, A., 148
Todoh, M., 425, 428–429
Todorov, L. V., 147–148
Tokuyasu, K. T., 193
Toledo, S., 196
Tolosa, H., 57–58
Tomson, M. B., 47–48, 130, 131–132, 135–136
Toner, B. M., 487–488
Tong, H., 131–132
Toraya, H., 97–99
Torrent, J., 493–494
Torres, P. M., 131–132
Torrie, G. M., 17, 74
Toth, A., 230, 243–244, 248
Touraud, D., 233–234
Trainor, T. P., 16
Trapnell, C., 375–376
Traub, W., 190–191, 193, 198, 306–307, 419–421, 422–423
Travaille, A. M., 213–214, 215, 216
Travesset, A., 8f
Tremel, W., 4
Tribollet, B., 157
Trimarchi, F., 328
Trinick, J., 196
Tromp, R. M., 149–150, 158–159
Trotter, J. A., 200
Troughton, E. B., 217–218
Tsai, M. M., 149, 160
Tsortos, A., 289
Tsuchiya, T., 8–9, 147–148
Tsuji, T., 131–132
Tucker, M. G., 490–491

Tuckerman, M., 7
Tunnacliffe, A., 380–381
Turner, K. L., 148
Tuross, N., 534, 538–542
Tyka, M., 346, 348, 350–351
Tzaphlidou, M., 198

U

Uechi, I., 241, 246
Ueda, T., 382
Ugarte, D., 487–488
Ugliengo, P., 344–345
Uhrynowska-Tyszkiewicz, I., 446
Ülkü, S., 227–229, 241, 247
Ullmann, S., 155–157
Ulman, A., 166
Ulutan, S., 227–229, 241, 247
Uomi, K., 147–148
Urban, J. J., 553
Uskokovic, V., 130
Ustündag, E., 513

V

Vachette, P., 106–108
Vagenas, N. V., 548
Vaknin, D., 8f
Valek, B., 508, 513
Valenta, A., 402–404
Vallabhaneni, S., 267–268
Valleau, J. P., 17, 74
Vallee, A., 297–299
van Buuren, T., 170–172, 181–182, 216–217, 220
van der Leeden, M. C., 289
van der Spoel, D., 7–8
Van Duin, A. C. T., 10
van Kempen, H., 213–214, 215
Van Kranendonk, M. J., 226–227, 230–232, 237–239
van Maaren, P., 7–8
van Rosmalen, G. M., 289
Vance, A. L., 170–172, 181–182, 216–217, 220
Vanicek, J., 7
Vanommeslaeghe, K., 18
Varani, G., 345
Vasconcelos, C., 558
Vega, C., 13

Veis, A., 130, 190–191, 193, 198, 306, 534
Vekilov, P. G., 46–47, 130–131, 534
Veldhuis, S. A., 103, 104, 111, 119–120, 121*f*
Vellore, N. A., 344, 350
Verbeeck, R. M., 331
Verch, A., 28–29, 47–48, 55–56, 57, 288
Vereecken, P. M., 149–150, 151, 158–159
Verine, H. J., 57–58
Vernon, R., 346
Verwer, P., 213, 216
Viana, J. C., 147–148
Villasuso, R., 226
Virata, M. J., 383
Voegel, J. C., 265–267, 270
Voelkel, A., 4, 16–17, 83, 288
Vogl, G., 392–393, 400
Vogt, S., 258
Voinescu, A. E., 233–234
Voïtchovsky, K., 11, 12*t*, 14–15
Völkel, A., 26–27, 28–29, 47–48, 51–52, 53, 54–55, 58–61, 62–66, 99–102, 103, 131–132, 560–561
Volkmer, D., 290–292
Volmer, M., 99–102
Vorobyov, I., 10
Voter, A., 4–5
Voth, G. A., 7, 8*f*, 10
Vu, H. P., 104

W

Wagermaier, W., 393–394, 405–406, 406*f*, 407, 409, 410, 411*f*, 416–418, 419–421, 423, 427–428, 429, 434, 436, 438–439, 447, 451, 453–456, 457, 464–467, 517, 558
Wagner, H. D., 285–286, 306, 393–394, 416–417, 419–421, 420*f*, 422–423, 456, 464–467
Wagner, P., 215
Waite, H., 502–503, 523–524
Walde, P., 265, 267
Walkush, J., 243–244
Wallace, A. F., 83–87, 86*f*, 88–89, 130, 131–132, 213–214, 215, 216, 549–550, 558, 561, 565–566
Wallace, J. M., 192
Wallqvist, A., 79, 80–85
Walters, D. A., 306

Walther, T. C., 369, 376
Wan, P., 131–132
Wang, B., 131–132
Wang, C. M., 159–160, 346, 349–350, 361
Wang, C.-G., 131–132, 136
Wang, D., 148–149, 166–167, 183–185, 213, 215, 216, 217–218, 549–550, 558, 561, 562–563, 564–567
Wang, F., 345
Wang, G., 368–369, 370
Wang, H. H., 385
Wang, J.-S., 5, 74–75, 77–78, 131–132
Wang, L., 130, 131–132, 141
Wang, M., 104
Wang, R. Z., 464–467
Wang, T. X., 131–132, 294–296, 296*f*
Wang, X., 416–417, 428
Wang, Y. W., 131–132, 297–299
Wang, Z. Q., 130, 131–132, 368–369
Wanner, A., 515–516
Warren, B. E., 418–419, 457, 503, 517
Warren, J., 260–261, 282
Warshawsky, H., 328
Warshel, A., 351*np*
Wasserman, B., 258
Watts, B., 172–173
Waychunas, G., 99–102
Waychunas, G. A., 83–87, 86*f*, 88–89, 496, 564–567
Weaver, A. J., 196
Weaver, J. C., 416–417, 502–503
Weber, A., 99–102
Weber, J. K. R., 483
Weber-Bargioni, A., 553
Wegner, G., 547
Wehrli, E., 264–265
Weigand, S., 262–263, 262*f*, 267–268, 269–274, 269*f*
Weiger, M. C., 310
Weil, S., 26
Weiner, E. R., 38
Weiner, S., 28–29, 46–47, 130, 190–191, 193, 198, 226, 258, 259*f*, 278–279, 285–286, 306–307, 346, 368, 393–394, 416–417, 419–421, 420*f*, 422–423, 463, 478, 495, 502–503, 514–515, 534, 537–542, 550, 558
Weinkamer, R., 393–394, 419–421, 420*f*

Weiss, I. M., 534, 538–542
Weiss, K., 170–172
Weitzner, B. D., 354
Welch, S. A., 46–47
Welham, N. J., 226–227, 230–232, 237–239
Weller, R. N., 131–132
Wenk, H. R., 405–406, 517, 518
Wess, T. J., 192, 196–198, 202, 419–421
Whaley, S. R., 308–310, 314
White, L. M., 250–251
White, M. A., 522, 523f
Whitelam, S., 123–124, 149–150
Whitesides, G. H., 166, 212, 214, 215, 216, 217–218
Whitesides, G. M., 181, 210, 212, 216, 217–218, 219–220
Whittaker, M. L., 192, 265–268, 269–272, 269f, 273–274
Wick, R., 267
Wickstrom, L., 79, 80–85
Wiersma, D. S., 553
Wiesmann, H. P., 417–418
Wilbur, K. M., 368
Wilding, M. C., 483
Will, J., 39–40, 99–102
Willey, T. M., 148–149, 166–167, 170–172, 181–182, 183–185, 184f, 213, 215, 216–218, 220
Williams, J. L., 308–310
Williams, R. J. P., 269
Williamson, G., 517
Williamson, M. J., 149–150
Wilson, R. M., 103
Wilt, F. H., 368, 370, 377, 382, 383–384, 495, 514–515, 534
Winarski, R. P., 510–511
Winkler, B., 10–11, 12t
Winston, P. W., 550
Winterhalter, M., 265–267, 270
Witze, E. S., 369, 377
Wohlrab, S., 290–292
Wolf, S. E., 4
Wolfe, D. B., 219–220
Woll, C., 166–167, 170–172, 181
Wöll, C., 216–217
Wolthers, M., 7–8
Wong, K. F., 7
Wong, P. C. K., 149–150

Woodle, M. C., 262–263, 264–265
Wopenka, B., 417–418, 419–421, 427–428, 467
Wright, A. C., 479
Wright, J. D., 102–103
Wright, K., 11, 12t
Wu, C.-H., 79, 262–263, 262f, 267–268, 269–274, 269f
Wu, H. J., 510
Wu, J.-J., 131–132, 190–191
Wu, K. J., 131–132
Wu, L., 131–132
Wu, W., 130, 131–132
Wu, X. H., 306, 346
Wu, Y., 8f, 10, 97–99
Wucher, B., 280–281
Wulff, G., 292
Wuthier, R. E., 546
Wysocki, L. M., 212, 214, 215

X

Xia, F., 501–502
Xiao, S., 11, 12t
Xiao, Y. Z., 141, 503
Xie, B., 130, 131–132, 141
Xie, Z. H., 464–467
Xin, H. L., 285–286
Xu, H., 131–132
Xu, W., 80–83, 82f, 159–160
Xu, X., 131–132, 542, 547, 558

Y

Yamabi, S., 233–234, 237–238
Yamamoto, A., 131–132
Yamamoto, T., 7
Yamamoto, Y., 130, 174–176, 297–299, 379–380, 381
Yamamura, K., 503, 504, 510
Yamashita, Y., 328
Yan, H., 510–511
Yan, Y., 332, 336
Yancey, J. A., 344, 350
Yang, D. S. C., 307, 352, 356, 358
Yang, L., 153
Yang, P., 285–286
Yang, W., 502, 513, 521, 526–527
Yang, X., 130, 131–132, 141

Yang, Y., 80–83, 82f
Yang, Z., 131–132
Yao, H. M., 423
Yasumasu, I., 379
Yates, J. R., 369, 376
Yazaydin, A. O., 87, 491–492
Yesinowski, J. P., 40
Yevick, G. J., 120–122
Yi, J. M., 510
Yin, H., 243, 246
Young, A., 502, 522
Young, J. R., 259–260
Young-wook, J., 155
Yu, S. H., 287f, 293–294, 299–300, 299f, 301
Yu, X.-Y., 153
Yuan, F., 417–418, 419
Yue, W., 280–282
Yuehuei, H., 440
Yumoto, H., 503, 504, 510
Yun, W., 503
Yvon, J., 548

Z

Zahn, D., 83–85
Zanchet, D., 487–488
Zanella-Cléon, I., 368, 370
Zare, R. N., 265–267
Zaslansky, P., 402, 403f, 416–418, 419, 423, 429, 438–439, 447, 451, 453–456, 457, 467, 553
Zebroski, H., 307
Zeller, R. W., 383
Zelzer, E., 258, 259f
Zelzion, E., 377
Zeng, X., 131–132
Zenobi, R., 558
Zernia, G., 192
Zervakis, M., 198
Zesch, W., 502
Zhang, C. H., 131–132, 307
Zhang, G., 130
Zhang, H. Z., 46–47
Zhang, K., 549–550
Zhang, L. Q., 159–160, 491–492
Zhang, T., 519–520
Zhang, W., 7, 79, 307
Zhang, Z., 131–132, 542, 558
Zharnikov, M., 166–167, 180–181
Zheng, H., 123–124, 149–150, 368–369, 370
Zhong, L., 159–160
Zhong, S., 18
Zhou, D., 131–132, 502, 522, 542, 558
Zhou, W., 230, 241, 247
Zhu, L., 331
Zhu, W., 77–78
Zhu, X., 369–370
Ziblat, R., 40–41, 524–525
Zickler, G. A., 416–418, 419, 429, 434, 447, 464–467
Zieba, A., 289
Ziegler, A., 538–542, 547, 549–550
Zielenkiewicz, A., 40
Zielenkiewicz, W., 40
Zimmermann, E. A., 392–393, 416–418, 419, 445, 446, 447, 448–449, 451, 502
Zipkin, I., 392–393
Zizak, I., 97–99, 402, 452–453
Zolotoyabko, E., 503, 518
Zope, H., 41
Zscherp, C., 537
Zwanzig, R., 13
Zysset, P. K., 464–467

SUBJECT INDEX

Note: Page numbers followed by "*f*" indicate figures and "*t*" indicate tables.

A

ACC. *See* Amorphous calcium carbonate (ACC)
AES. *See* Atomic emission spectroscopy (AES)
AFM. *See* Atomic force microscopy (AFM)
Alanine-scanning mutation assay, 315–316
Amelogenin
 adsorption kinetic analyses, HAp crystals, 338–340, 340*f*
 binding amounts *vs.* mutants, ELISA, 336–338, 339*f*
 binding sequences, peptide screening, 332–334, 334*f*
 confirmation, crystal-binding motifs, 335–336
 data handling and processing, 340
 hexagonal apatite crystals, 329, 329*f*
 identification, crystal-binding motifs, 330
 mineralized enamel crystals, 328–329
 vs. mutants, binding amounts, 336–338, 339*f*
 preparation and identification, single FAp crystal, 331–332
 recombinant, 332
 site-directed mutagenesis, 332
 stepwise protocol, 340
 syntheses and identification, peptides, 331
 synthesis and analysis, bulk HAp crystals, 331
 tooth enamel, 328
Amorphous calcium carbonate (ACC)
 biomineralization, 111–113
 Compton scatter, 435
 concentration, crystalline bulk phase, 27–28
 evolution, 36–37
 formation in biomimetic/biomineralization, 99–102
 FTIR and Raman spectra
  and assignments of intermolt cuticle, frequency, 549–550, 551*t*
 controlled crystallization, 547–548
 in vitro experiments, 550, 552*f*
 intermolt cuticle, 549–550, 549*f*
 IR and measurements, 548
 L-aspartic acid, 552–553
 shell molt, terrestrial crustacean *Armadillidium vulgare*, 547
 synthesis, 547
 individual PDFs, 85–87
 kinetics, 103
 liposome-stabilized, 269, 269*f*
 MFR (*see* Mixed flow reactor (MFR) method)
 and Na_2CO_3 in stopped-flow apparatus, 111–113, 112*f*
 nanoparticles, 260
 pair distribution function (PDF), 85
 PDF analysis, 435
 reverse Monte Carlo (RMC) modeling, 87
 spiculogenesis, sea urchin embryo, 258
 stabilization, 260–261
 supersaturations, 113–115
 synchrotron X-ray total scattering, 435
 system, 99–102
 temperature, 113–115
 thermogravimetry, 435
 time-resolved SAXS patterns, 111–113, 114*f*
 time-resolved WAXS patterns, 115–119, 116*f*
 transformation into microcrystals, 111–113
 weighting factors, 87
Analytical ultracentrifuge (AUC)
 Beckman-Coulter XL-I, 49
 cluster detection and characterization
  determination, s-distributions, 64–66, 66*f*
  hydrodynamic diameter, 64
  hypothetical distributions, 64–66, 65*f*
  Lamm equation, 63

597

Analytical ultracentrifuge (AUC) (*Continued*)
  principle, sedimentation velocity experiments, 61–62, 61f
  Sedfit evaluation routines, 62–63
  Stokes–Einstein equation, 63–64
  sample preparation, measurements, 53–54
Apoferritin (APO), 431–433, 434f
Atomic emission spectroscopy (AES), 247, 250–251
Atomic force microscopy (AFM), 192
Attenuated reflectance (ATR) device, 38
AUC. *See* Analytical ultracentrifuge (AUC)
AutoFit, 408

B

Barium titanate (BaTiO$_3$)
  fractal-like branched oligomeric structures, 120
  growth stages, 119–120
  in situ evolving system, 119–120, 121f
  precursor system, 119–120
  scattering intensity, 120–122
  semiordered agglomerates, 120–122
Biochemical assays. *See* Amelogenin
Biomineralization. *See also* Protein function, biomineralized tissues
  CNT, 99–102
  constraint-based docking framework, 359–360
  definition, 534
  FTIR and Raman spectroscopies, 534–535
  liposomes (*see* Liposomes, intracellular biomineralization)
  organisms, 346
  precipitation, CaCO$_3$, 111–113
  replica exchanges (*see* Replica exchange methods)
  research, 534
  simulation (*see* Force fields (FFs) model)
Biominerals
  PDF analysis
    amorphous calcium carbonate, 433–434
    ferritin-derived nanocrystalline ferrihydrite, 431–433

x-ray microdiffraction (*see* X-ray microdiffraction, biominerals)
Bone
  carbonated apatite, 419–421
  crystalline carbonated apatite, 419–421
  hierarchical architecture, 420f
  SAXS/WAXD patterns, 419–421
  in situ x-ray nanomechanical imaging, 421, 421f
  triple helical tropocollagen molecules, 419–421
Bone and dentin nanaostructure
  AutoFit, 408
  biomaterial, 393–394
  data treatment, WAXD (*see* Wide-angle X-ray diffraction (WAXD))
  experimental setups
    laboratory SAXS equipment, 395
    principle, X-ray scattering, 394, 394f
    X-ray beamline, synchrotron facility, 395–396
  parameters, 396–397, 396f
  position-resolved methods, 392–393
  structural studies, biomineralized tissues, 392
  treatment, SAXS (*see* Small-angle X-ray scattering (SAXS))
Bragg diffraction, 505f
Brillouin zone, 535
Brunauer–Emmett–Teller method, 331

C

Calcium carbonate (CaCO$_3$)
  and ACC (*see* Amorphous calcium carbonate (ACC))
  aqueous systems, 16
  and CNT, 99–102
  crystalline polymorphs, 4
  description, 99–102
  FFs, 10–11, 17
  organic–CaCO$_3$ interaction thermodynamics, 18–19
Calcium phosphate crystallization
  biomineralization, 130
  carbon dioxide, 133–134
  crystal growth, 131
  data handling/processing, 138–141

ex situ and in situ characterization, 131–132
in situ experiments, 135–138
nucleation, 130–131
solution preparation (see Calcium phosphate, solution preparation)
supersaturation, 130–131
Calcium phosphate, solution preparation
carbon dioxide, 133–134
factors, crystal nucleation and growth
foreign particles/bubbles, 135
ionic species ratio, 134
ionic strength (IS), 134
reproducibility, 135
stirring/flow rate, 134
temperature, 134
volume, 134
stock, 132–133
working, 133
CC method. See Constant composition (CC) method
Characterization methods, liposomes
fixation and heavy elements, 272
giant liposomes, 271–272
in situ methods, 270–271, 271f
SAXS/WAXS, 272–273
Classical nucleation theory (CNT), 99–102
Coherent scattering domain (CSD), 429–430
Collagen fibrillar structure
AFM, 192
COL I, 190–191, 192
cryo-electron tomography promise, 203
description, 191, 191f
frozen-hydrated tissue sections, 193
in vitro work, 190–191
image acquisition and processing
cryo-TEM, 195
interpretation, cryo-TEM micrographs, 196
raw/whole micrographs, 196–198, 197f
image alignment, 200–201, 200f, 201f
image selection, 198–199, 199f
measurement, fibril periodicity, 201–202
mineralizing system, 190–191
sample preparation
cryosectioning, 194–195

dental and periodontal tissues, 194
mineralized tissues, 193
section vitrification, 195
soft tissues, 194
solutions, 193–194
scanning EM, 192
spectroscopic methods, 192
Competitive adsorption assays. See Amelogenin
Constant composition (CC) method
crystallization rate, 138
free-drift crystallization processes, 135–136
Le Chatelier's principle, 136
nucleation rate, 139
phosphate ions, 136
protocol, 136–137
reaction order, 139
titrant addition, 135–136
Cryogenic transmission electron microscopy (cryo-TEM)
in situ TEM, 40
investigation, mineralization stages, 39
light microscopy, 39–40
Cryo-scanning electron microscopy (cryo-SEM), 40–41
cryo-SEM. See Cryo-scanning electron microscopy (cryo-SEM)
cryo-TEM. See Cryogenic transmission electron microscopy (cryo-TEM)
Crystallization approaches, SAMs
CO_2-in, 211–213, 211f
fixed composition, 213–214
Kitano method, 213
Crystallization assay
experimental setup, 51–52, 52f
pH titrations, 53
Crystals and mesocrystals
biomineral formation, 278–279
biomineralization, 278
biomineral morphologies, 278–279
confined reaction space, 278–279
fracture resistance, 301
insoluble polymers, 279–286
polymer-mediated biomineral formation, 278, 278f
soluble and insoluble additives, 278–279

Crystals and mesocrystals (*Continued*)
  soluble polymers, 286–301
CSD. *See* Coherent scattering domain (CSD)

D

Data analysis
  data reduction programs
    AgBe standard, 450–451
    calibration, 450–451
    FIT2D, software, 450
    KLORA, 450
    radial integration, 450–451
  empty cell/empty beam, correction, 452–453
  errors estimation
    sample thickness, 451
    sample translation, 451–452
  radially averaged intensity profile calculation (*see* Radially averaged intensity)
  SAXS radial integration
    azimuthally averaged intensity profile, calculation, 453–454
    Chebyshev polynomials, 454–455
    fibril strain in MCF, 455
    representative, 454–455, 454f
    uniaxial vertical loading, 456
    uniform fibril strain *vs.* heterogeneous fibril strain, 456–457
  WAXD radial integration
    apatite phase, bone, 458f
    azimuthally averaged intensity profile, calculation, 457
    mineral strain in MNP, 457–458
Data handling/processing, calcium phosphate
  CC method, 138–139
  pH curve, 139–140
  stopped-flow spectrophotometer, 140–141
Data interpretation, SAXS
  agglomeration and coalescence processes, 110–111
  auxiliary methods, 111
  Coulomb repulsion-attraction phenomena, 109
  interparticle spatial correlations, 109

mathematical functions, 111
pair distribution function (PDF), 107–108
parameters, 106
scattering amplitude, 106–107
shape-independent radius, gyration, 108–109
total scattered intensity, 107
2D CCD detectors, 442
Debye-Scherrer rings, 507
Degrees of freedom (DOFs)
  internal, 348–349
  MC-based docking algorithms, 345
  MD, 345
  protein backbone torsional, 347
Dentin. *See* Bone and dentin nanoostructure
DLS. *See* Dynamic light scattering (DLS)
DNA sequencing
  Illumina, 374
  next-generation high-throughput, 368–369, 373
  RNAs, fragmentation, 372
  Roche 454, 373–374
  SOLiD, 374–375
  third-generation, 375–376
DOFs. *See* Degrees of freedom (DOFs)
DOLLOPs. *See* Dynamically ordered liquid-like oxyanion polymers (DOLLOPs)
Double-stranded RNA (dsRNA), 379
dsRNA. *See* Double-stranded RNA (dsRNA)
Dynamically ordered liquid-like oxyanion polymers (DOLLOPs)
  definition, 60–61
  structural rearrangements, 64–66
Dynamic light scattering (DLS)
  crystal precursors, nucleation, 36–37
  suspension, particles, 37

E

EDS. *See* Energy dispersive X-ray spectroscopy (EDS)
Electron microscopy (EM), 192
ELISA. *See* Enzyme-linked immunosorbent assays (ELISA)
EM. *See* Electron microscopy (EM)
Energy dispersive X-ray spectros- copy (EDS), 153

Enzyme-linked immunosorbent assays (ELISA), 336–338
EXAFS. See Extended X-ray absorption fine structure (EXAFS)
Extended X-ray absorption fine structure (EXAFS), 167–169

F

FAp crystals. See Fluorapatite (FAp) crystals
FEP. See Free-energy perturbation (FEP)
Ferritin-derived nanocrystalline ferrihydrite
  apoferritin (APO), 431–433, 434f
  attenuation of correlations, 434
  CSD sizes, estimation, 436f
  Fourier transform, 433–434
  nanocrystallinity, 431–433
  optimal counting statistics, 433
  scattering experiments, 433
FFs model. See Force fields (FFs) model
Fibril orientation, 424
Fibril strain, 423–424
Fluorapatite (FAp) crystals
  preparation and identification, single, 331–332
  protein-mineral assays, 330
  single-crystal sections, 329, 329f
Force fields (FFs) model
  calcium carbonate (CaCO$_3$), 4
  mineral–aqueous interface (see Mineral(s))
  organics, geochemical systems, 18–19
  potential energy landscape
    description, 5–6
    evaluation, accuracy, 6–7
    sampling, 4–5
  selection and refinement
    ions (see Ions)
    mineral (see Mineral phases, FFs)
    water (see Water model, FFs)
Fourier transform infrared (FTIR) spectroscopy
  ATR device, 38
  characterization, 564
  disadvantage, 38
  flow cell, 37
  modes and materials, 542–543
  and Raman spectra
    ACC transformation, 547–553
    biomineralization, 538–542

  Tris-buffered and phosphate-buffered solutions
    data analysis, 545–546, 546f
    IR measurements, 544–545
    mineralization reactions, 543–544
  vibrations, solids, 535
Free-energy perturbation (FEP)
  Einstein solid, 13
  solvation, 13

G

GLCs. See Graphene liquid cells (GLCs)
Graphene liquid cells (GLCs), 153, 155, 161

H

HAp. See Hydroxyapatite (HAp); Hydroxyapatite-associated proteins (HAp)
HAp crystals. See Hydroxyapatite (HAp) crystals
Hydroxyapatite (HAp)
  osteocalcin, 356, 357f
  RosettaSurface, 348
  structures, 358, 359f
Hydroxyapatite-associated proteins (HAp)
  bone interface, 307
  crystal growth promotion and inhibition, 306–307
  crystal nucleation, peptides, 318–319
  fluorescent dye-conjugated antibody, 316–317
  high-throughput screening process, 307
  molecular level mechanism, 306–307
  peptide phage libraries, 307
  protein crystallography, 307
Hydroxyapatite (HAp) crystals
  vs. FAp, 330
  kinetic analyses, amelogenin adsorption, 338–340, 340f
  synthesis and analysis, bulk, 331

I

ICP-OES. See Inductively coupled plasma-optical emission spectrometry (ICP-OES)
Illumina DNA sequencing, 374
Inductively coupled plasma-optical emission spectrometry (ICP-OES), 562

Infrared (IR) spectroscopy
  absorption
    calcium carbonate, 537–538, 539t
    calcium phosphate, 538, 541t
    frequency, specific vibration, 537
  measurements, 544–545
Inorganic materials
  CaCO$_3$ system, 99–102
  CNT, 99–102
  crystalline materials growth, 99–102, 101f
  evolutionary process, 99
  nucleation and crystallization
    phenomena, 99–102
  SiO$_2$ system, 99–102
In situ crystallization, calcium phosphate
  CC method, 135–137
  pH meter, 137
  stopped-flow spectrophotometer, 137–138
In situ fluid cell TEM
  advantages and disadvantages, 148
  and AFM, 149
  atomic-level resolution imaging, 149–150
  atomic resolution, nanoparticle
    aggregation and attachment, 155, 156f
  battery charging and discharging, 159–160
  calcium and carbonate concentrations, 157–158
  cell fabrication and assembly, 151f, 153–154
  electrochemical nucleation and Cu
    nanoclusters, 158–159, 158f
  electron beam, 155, 160–161
  ex situ experiments, 147–148
  ferrihydrite nanoparticles, 155
  fluid cell design
    copper, platinum and nickel, 151–152
    crystal nucleation, 151
    EDS and electron energy loss
      spectroscopy, 153
    electrochemical and thermal control, 151, 152f
    GLCs, 153
    platinum resistive heater, 151–152
    resistance and temperature, 152–153
    silicon, 150–151, 151f

  van der Waals interaction, 153
  heating control, 160, 161
  lithium-ion batteries, 159–160
  mineralization, 157
  sealed cell-containing reaction solution, 154–155
  Si- and graphene-based liquid cells, 161
  Si$_3$N$_4$ membrane, 155–157
  STM and XPS, 148–149
  time-lapse sequence, 159–160
In situ mechanical testing, synchrotron
    x-rays
  mounting samples, in mechanical tester
    cantilever tests, 433–434
    tensile tests, 431–433
    whole bones, tensile tests, 434
  sample geometry
    in biomineralized collagenous tissues, 428
    bovine fibrolamellar bone, 429
    Ewald diffraction condition, 425, 426f
    mineralized fibril geometry, 429
    mineral nanoplatelets (MNPs), 427–428
    nanoscale MCF deformation, 427
    orientation distribution, 427
    perfectly aligned system, 428
    strain measurements, 427–428
    tensile/compressive strain, 428–429
    thick specimens, 429
  sample preparation methods
    coarse sectioning, 429–430
    storage methods, 431
    surface damage, 431
    thin sectioning, 430
In situ, scattering methods
  BaTiO$_3$, 119–120, 121f
  CaCO$_3$ polymorphs, 103
  and time-resolved structural information, 96
  WAXS, 104
Insoluble polymers
  agarose hydrogels, 286
  biominerals, 285–286
  calcite single-crystal lattice, 286
  crystallization reaction, 301
  mechanical properties, 285–286
  network fibers, 286

Subject Index

organic–inorganic interface, 286
polymeric templates
  bioinspired mineralization, 279–280
  block copolymers, 282
  calcite crystals, membrane, 279–280, 280f
  calcium carbonate precipitation, 280, 281f
  colloidal crystal template, 282
  fabrication, 3D ordered macroporous calcite, 282, 283f
  hydrophobic surface, 280–281
  intertwined, nonintersecting gyroid network, 282, 284f
  monodisperse latexes, 282
  multiple nucleation, 279–280
  phase separation, 282
  polymer replica, 279–280
  replication, full double-gyroid, 285, 285f
  sea urchin replicas, 282
  spectacular gyroid morphologies, 285
  template surface chemistry, 280–281
  triple periodic gyroid morphology, 282
  types, 282
structural matrix, 301
synthetic, 279
tomographic reconstruction, agarose network, 286
Ion potentiometric measurements
  concentration/pH measurements, reaction kinetics (*see* Reaction kinetics)
  determination, bulk species solubility, 33–35
  electrode measurements, Nernst law, 29
  extracted/dried samples, 41
  formation, biological mineral specimens, 26
  mineralization reactions, 27–29
  sampling techniques (*see* Sampling techniques, ion potentiometric measurements)
Ions
  aqueous solution
    case of Ca^{2+}, 13–14
    solvation free energy, 13, 14
  pairing
    experimental techniques, 16–17

    free energy, aqueous solution, 17, 17f
Ion-selective electrode (ISE)
  calibration experiments, 30
  measurements, concentration, 29
  reaction kinetics, 30
ISE. *See* Ion-selective electrode (ISE)

K

Kitano method, 213

L

Liposomes, intracellular biomineralization
  calcium phosphate, 270
  characterization methods (*see* Characterization methods, liposomes)
  formation, liposome-stabilized ACC, 269, 269f
  magnetite formation in magnetotactic bacteria, 269
  membrane-delimited compartments, 259–260, 259f
  membrane-permeant phosphate precursor, 270
  model systems, confinement
    advantages, compartmentalization, 260
    classes, biominerals, 261, 262f
    Meldrum group, 260–261
    surface energy and entropy, 260
  preparation
    encapsulation efficiency, 267
    extrusion/sonication, 263–265, 264f
    giant liposomes, 265–267, 266f
    ions exchange, rehydration medium, 267–268
    lipid selection, 262–263, 263f
  privileged environments, 258
Liquid-liquid separation, 83–85, 86f
Loss of function approaches
  dominant negatives, 378–379
  morpholino (MO) antisense, sea urchin spicule, 381–384
  reversible or irreversible inhibitors, 379
  siRNA (*see* Small interfering RNAs (siRNA))
SM50, 384

M

Magnesium
  ACC, 561
  and calcium concentrations, 562
  carbonate, 564–565
MALDI-TOF-MS. *See* Matrix-assisted laser desorption–ionization time-of-flight mass spectrometry (MALDI-TOF-MS)
Mass spectroscopy
  mix of proteins, 376–377
  refinements, 376
Matrix-assisted laser desorption–ionization time-of-flight mass spectrometry (MALDI-TOF-MS)
  adsorbed species, 332–334
  mass spectra, amelogenin peptides, 334, 334f
  peptides, 330
MC docking. *See* Monte Carlo (MC) docking
MCM docking. *See* MC-plus-minimization (MCM) docking
MC-plus-minimization (MCM) docking, 345, 346, 349–350
MD. *See* Molecular dynamics (MD)
MD simulations. *See* Molecular dynamics (MD) simulations
Mechanical tester design
  sample chamber
    fluid chamber, 435–436
    grips, 435
    micromechanical testing machine, 435, 436f
  in situ cantilever bending, 439–440, 440f
  stresses, porosity correction
    analysis of variance (ANOVA) tests, 439
    backscattered electron (BSE) imaging, 439
    Haversian systems, 438–439
    microcomputed X-ray tomography (microCT), 439
    osteocyte lacunae, 438–439
  tissue strain measurements
    bone deformation, 437
    elastic region of deformation, 437f
    inhomogeneous microlevel deformation, 437–438
    noncontact video extensometry, 436–437
Mesocrystal formation
  alignment, nanoparticles, 297
  chitin matrix, 297–299
  collagen fibril, 297–299
  face-selective polymer adsorption, 299–300
  helical mesocrystalline nanoparticle superstructure, 301
  helical nanoparticle superstructure, 299–300, 299f
  hierarchical structure, human femur bone, 297
  mineralization process, 297–299
  orientation, calcite crystals, 297–299
  orthorhombic nanoparticle units, 300, 300f
  polymer-mediated crystallization, 297
  structural matrix, 297–299
MFR method. *See* Mixed flow reactor (MFR) method
Mineral(s)
  crystallite strain, 424
  mineral–aqueous interface
    ion pairing (*see* Ions)
    solubility, 15–16
    surfaces, 14–15
  orientation, 424
  surfaces (*see* RosettaSurface, protein structure prediction)
Mineral and organic-mineral composites
  biomineralization, 534
  FTIR and Raman spectroscopies (*see* Fourier transform infrared (FTIR) spectroscopy)
  Raman effect (*see* Raman effect)
Mineralization reactions
  gradual increase, concentration, 28–29
  instant (super)saturation, 27–28
  supersaturated solutions, 544
  Tris-buffered saline, 543–544
Mineralized fiber composites, synchrotron x-ray nanomechanical imaging
  biomineralized tissues, 417–418

Subject Index 605

conjunction with micromechanical deformation, 417–418
continuous/stepwise loading, 449
data analysis (see Data analysis)
deformation
　in bending, 463–464
　rachitic bone, 462–463
deformation strains, 418–419
hierarchical organization, 416–417
locating region of interest
　alignment, 442–443
　optical microscopes, configuration, 443–444
　scattering beamline, 443–444
　test sample, mapping, 445
　transmission lines scan, 444, 444f
　X-ray transmission measurements, 443–444
mechanical preconditioning, 442, 443f
mechanical tester design (see Mechanical tester design)
mineralized collagen fibril (MCF), 417–418
mineral phases, 416–417
model interpretation (see Model interpretation)
probes, 416–417
prototypical biomineralized composite (see Bone)
radiation damage (see Radiation damage)
sample preparation (see In situ mechanical testing, synchrotron x-rays)
SAXS/WAXD measurements, 445
scanning methods, 449
synchrotron setup, 441–442, 441f
x-ray spectra and nanomechanics
　extractable parameters, 423–425
　and SAXS/WAXD pattern, 422–423
Mineral phases, FFs
calculation, lattice energy, 10–11
FEP, 13
literature results, calcite properties, 11, 12t
Mineral precipitation
AUC (see Analytical ultracentrifuge (AUC))
Ca^{2+}-ISE, 49–51
conductivity probes, 51

crystallization assay (see Crystallization assay)
multiple-binding (see Multiple-binding model)
nucleation phenomena, 47–48
pH electrodes, 51
titration setup (see Titration)
Mixed flow reactor (MFR) method
ACC particle formation, inorganic environments, 561
advantages, batch and diffusion methods, 558–559
"batch" method, 558
characterization
　FTIR, 564
　ICP-OES, 562
　Raman spectroscopy, 562–563, 563f
　SEM, 562
　TGA, 564
hydraulic residence time, solution, 560
Mg/Ca ratios, 564–565, 565f
partition coefficient, 565–566
pH and carbonate concentrations, 566, 566f
procedure, ACC synthesis, 558–559, 559f
saturation state, solution, 560–561
solution chemistries
　constant Mg/Ca ratio (5:1) experiments, 559, 561t
　variable Mg/Ca ratio experiments, 559, 560t
MO antisense. See Morpholino (MO) antisense
Model interpretation
biomineralized composites, 464–467
deformation parameters, 467
fibril/interfibrillar matrix/fibril arrangement, 468
fibril modulus, variation of bending, 465f
intrafibrillarly mineralized fibrils, 464–467, 466f
nanomechanical and structural parameters, 467
nanoscale quantities, 464–467
Molecular biology techniques. See Protein function, biomineralized tissues
Molecular dynamics (MD)

Molecular dynamics (MD) (*Continued*)
  integration, Newton's laws of motion, 345
  QM calculations, 346
  replica exchange (*see* Replica exchange molecular dynamics (REMD))
  simulations
    atomic, 5
    crystal growth, aqueous solution, 4–5
    FFs, 11
    solubility calculation, 15–16
    steered, 17
Molecular genetic approach, 378
Monte Carlo (MC) docking
  conformational changes, 349–350
  random perturbations, 345
Monte Carlo (MC) sampling
  advantages, 74
  description, 73
  Metropolis method, 73–74
  parallel tempering
    Boltzmann distribution, 75–77
    description, 74–75
    one-dimensional energy landscape, 75–77, 76*f*
    potential energies, 75
    variants, 75
Morpholino (MO) antisense
  identification, protein sequence, 382
  monitor, effect, 383–384
  organism, 383
  RNA treatment, 381–382
  SM50 gene, 382
  synthesization, RNA, 383
Multiple-binding model
  binding scheme, 59–60
  calcium carbonate, 57–58
  DOLLOPs, 60–61
  ion association, 58–59
  temporal profiles, 57
  thermodynamics, cluster formation, 59

N
Near-edge X-ray absorption fine structure (NEXAFS), 167–169
Nernst law, 29
NEXAFS. *See* Near-edge X-ray absorption fine structure (NEXAFS)

Next-generation high-throughput DNA sequencing technologies, 373
NMR spectroscopy. *See* Nuclear magnetic resonance (NMR) spectroscopy
Nuclear magnetic resonance (NMR) spectroscopy, 40
Nucleation, liposomes
  biomineral, 261
  control, 258
  crystal and growth, 259–260, 259*f*
  heterogeneous, 269–270

O
Occluded matrix proteins, 368
Organothiol SAMs
  assessing monolayer quality, 217–218, 218*f*
  biominerals, 210
  contamination, 219–220
  crystallization approaches
    CO_2-in, 211–213
    fixed composition, 213–214
    Kitano method, 213
  film degradation, 220
  monolayer preparation
    carboxyl-terminated SAM monomers, 216
    -COOH, 216
    ethanolic solution, 215
    thiol solutions, 216–217
  myriad potential causes, 221–222
  nucleate calcite, 210
  preparation procedure, 221
  substrate preparation, 214–215
  thiol solution, 220
Osteocalcin
  bone mineralization, 352
  HAp, 356, 357*f*
  interpretation, 358, 359*f*

P
Pair distribution function (PDF) analysis
  analysis approaches
    coherent scattering domain (CSD), 429–430
    conservative estimated error, 430
    decay of amplitudes, evaluation, 430
    peaks, 429

spherical CeO_2 nanoparticles, 426f
structural information, 429
atomic, 428–429
biominerals
   amorphous calcium carbonate,
     433–434
   ferritin-derived nanocrystalline
     ferrihydrite, 431–433
real-space fitting
   DISCUS, 430–431
   PDFFIT, 430–431
   Rietveld refinement, 430–431
   synthetic iron monosulfide, structure,
     430–431
RMC method
   algorithm, 431
   hard constraints, 431
   hydrated ACC, 431
   RMCProfile, 431
roadmap of interatomic distances, 428
total scattering structure function,
  428–429
PCR. *See* Polymerised chain reaction
  (PCR)
Perovskites, 111
Phage display
  alanine-scanning mutation assay, 315–316
  alignment, peptide identification,
   319–320
  binding peptide sequences, 308–310
  bone and tooth biomorphogenesis, 321
  bone- or tooth-associated proteins, 306
  CLP12 peptide, 310, 311f
  dominant binding phage, 308–310
  engineered phage library pool, 307–308
  functional peptide, 311
  HAp-binding assay, fluorescent dye-
   conjugated antibody, 316–317
  HAp-binding peptide sequences, 310,
   311f
  HAp crystal nucleation, peptides,
   318–319
  high-throughput phage peptide library
   screening, 321
  high-throughput screening process,
   307–308
  hydroxyapatite-associated proteins,
   306–307

identification templates, mineralization,
  313–315
library construction kits, 307–308
materials, 312
mineralization activity, 310
mineralization process, 306
miniphage library, 315
peptide libraries and kits, 307–308,
  308t
peptide motifs, hydroxyapatite, 308–310,
  309f
peptide sequence, 307–308
phage genome, 307–308
phage peptide libraries, 307–308
preparation, target crystals, 312–313
screening/biopanning, 307–308
solid-phase peptide synthesis, 317–318
type I collagen, 310
types, 307–308
PH meter
  amorphous calcium phosphate (ACP),
   137
  calcium phosphate solution, pH curves,
   139–140, 139f
  protocol, 137
Phonons, 535
PILPs. *See* Polymer-induced liquid
  precursors (PILPs)
Polychromatic (Laue) diffraction, 507–508,
  508f
Polymer-induced liquid precursors (PILPs)
  $CaCO_3$ system, 289–290
  glutamic acid–polyethyleneimine system,
   290–292
  hollow helix synthesis, 289, 290f
  organic molecular crystals, 290–292
  polarization microscopy, 289–290
Polymerised chain reaction (PCR)
  advantage, 370
  control, annealing conditions, 380–381
  quantitative, 381
Porod's law, 400, 401–402
Potentiometric measurements. *See* Ion
  potentiometric measurements
Potentiometry, 51–53
Prenucleation clusters, 47–48, 83
Prenucleation stage, 27–28
Protein–crystal interactions, 329–330, 337

Protein function, biomineralized tissues
  amino acid sequences, 377–378
  experimental approaches, 378
  generation, RNA-seq data, 369–376
  loss of function approaches (see Loss of function approaches)
  molecular genetic approach, 378
  sequencing, proteins, 376–377
  starmaker mutant, 368
  strategies, protein identification, 368–369
Protein–mineral assays, 330
Protein sequencing
  advantage, 377
  contaminating proteins, 377
  proteomic study, 376
  refinements, mass spectroscopy, 376
Protein–surface interactions
  computational methods, 344–345
  contact map, 357–358
  water-mediated, 345
Proteomics, 368–369, 376
Protocol capture
  input files, 352
  osteocalcin, 352
  postprocessing and interpretation, results
    C++ and PyRosetta implementations, 353
    interpretation, osteocalcin results, 358, 359f
    protein-protein contact map, 356–357
    protein-surface contact map, 357–358
    sample files, 354–356
    scripts, analysis, 354, 355t
    secondary structure, 356, 357f
  Running RosettaSurface, 352–353

Q

Quantum mechanics (QM) simulations, 5

R

Radial distribution function (RDF), 7–8, 8f
Radially averaged intensity
  SAXS
    angular intensity distributions, 459–461
    anisotropic, 458–459
    fibril orientation, calculation, 458–459
    mineralized collagen fibrils, orientation, 460f
    third-order peak, 458–459
  WAXD
    angular distribution, 461
    data sets analysis and macros usage, 461
Radiation damage
  damaged tissue zone, 448–449
  dosage-level calculation, 445–446
  elastic strain distribution, 447–448
  material property changes, effect, 447–448
  mechanical effects, 446
  methods to minimize effects, 447–449
  radiation-induced embrittlement, mechanisms, 446
  stress distribution, alterations, 448–449
Raman effect
  and FTIR (see Fourier transform infrared (FTIR) spectroscopy)
  IR (see Infrared (IR) spectroscopy)
  scattered electric field, 535–536, 536f
  types, parameters, 537
Raman spectroscopy. See Fourier transform infrared (FTIR) spectroscopy
RDF. See Radial distribution function (RDF)
Reaction kinetics
  activity coefficients, 30–31, 32t
  calibration curve, Ca^{2+}-ISE, 30, 31f
  interaction, CO_2
    carbonic acid equilibria, 31–32, 33f
    mineralization reaction, 31–32, 32t
  ion potentiometric measurements, control, 35–36, 35f
  ISEs, 30
REMD. See Replica exchange molecular dynamics (REMD)
Replica exchange methods
  atomistic simulations, 72
  computing resources, 72
  MD (see Replica exchange molecular dynamics (REMD))
  Monte Carlo sampling, 73–77
Replica exchange molecular dynamics (REMD)
  applications
    aggregation-based model, ACC, 85–87
    COSMO, 83–85

Subject Index

hybrid temperature-based REMD approach, 83
nanoparticle growth, stages, 83–85
thermodynamic properties, 87–89
average kinetic energy, 77–78
optimal exchange attempt frequency, 79
simulations, 77–78
system size limitations
alanine peptides, 80–83, 81f
comparison, structural ensembles, 80–83, 82f
exchange probability, 80–83
potential energy, 80, 81f
with solute tempering (REST), 79–80
solvent–solvent interactions, 79
temperatures distributions, 78
RNAi. See RNA interference (RNAi)
RNA interference (RNAi)
definition, 379
exogenous, 381
treatment, 381
RNA-sequence, mineralized tissues
fragmented mRNA, 373
generation, cDNA library, 370
high-throughput DNA sequencing technologies, 373
Illumina DNA sequencing, 374
integrity, 371
mRNA, samples, 371–372
procedures, 372
purification, mRNA, 372
reagents, 370–371
Roche 454, 373–374
SOLiD DNA, 374–375
synthesization, DNA, 371
third-generation DNA, 375–376
Roche 454 DNA sequencing technology, 373–374
RosettaSurface, protein structure prediction
biomaterials and bioprocessing, 346
biomineralization, 359–360
conformational sampling, 348–349
constraint files, 363
description, 360–361
DOFs, 346–347
feedback loop, 348
formatted PDB file, 362
fragment files, 362

HAp, 348
MC-based docking, 345
MD methods, 345
principal challenges, 348
protein-surface interaction, 344
protocol capture (see Protocol capture)
score function (see Score function, RosettaSurface)
search algorithm (see Search algorithm)
ssNMR distance and angle measurements, 347
structures, 346–347, 347f
surface parameters, 361–362
Running RosettaSurface
C++ executable, 352, 353t
PyRosetta script, 352–353, 354t, 355t

S

Samples mounting, in mechanical tester
cantilever tests, 433–434, 434f
tensile tests
acrylonitrile butadiene styrene (ABS) molds, 431–433
micromilling machine, 433
microtensile test specimens, preparation procedure, 431–433, 432f
whole bones, tensile tests, 434
Sampling techniques, ion potentiometric measurements
control, reaction kinetics, 35–36, 35f
cryo-SEM, 40–41
cryo-TEM (see Cryogenic transmission electron microscopy (cryo-TEM))
FTIR/Raman spectroscopy (see Fourier transform infrared (FTIR) spectroscopy)
high-energy radiation, 36
in situ analysis, 40
light scattering/zeta-potential, 36–37
NMR, 40
synchrotron SAXS and WAXS, 38–39
SAXS. See Small-angle X-ray scattering (SAXS)
Score function, RosettaSurface
benchmarking and parameterization, 350
constraint-based fashion, 351
electronic delocalization effects, 351–352

Score function, RosettaSurface (Continued)
score terms, 350–351, 351t
Search algorithm
assumption, 350
flowchart, RosettaSurface algorithm, 349–350, 349f
Sedimentation velocity experiments, 53–54
Self-assembled monolayer (SAM)
angular-dependent XAS measurements, 172, 172f
aromatic ring, 183–185, 184f
characterization, 166–167
crystallization process, 166–167
electronic core-level binding energy, 167–169
EXAFS, 167–169
experiment and data optimization, 178–179
EY and FY detection modes, 173–177
fitting protocol, 183
functional group, 185
incident flux, 177–178
K-edge XAS, 169
linear polarization and energy calibration, 172–173
mineral systems, 166, 167f
NEXAFS, 167–169
organic films, 166
sample preparation and composition, 181–183
soft X-ray beamlines, 167
steric considerations, 183–185
subtle changes, 169
transition dipole moment, 170–172
XAS geometries, 170–172, 171f
X-ray beam, 169–170, 170f
X-ray-induced beam damage, 179–181
Self-assembled monolayers (SAMs).
See Organothiol SAMs
Silica
aqueous and nanoparticulate, 102–103
mass-fractal aggregates, 115
nucleation and growth, 115–119
polymerization, 102–103
precipitation, 115
SAXS patterns, 115–119
SiO_2 system, 102–103

time-resolved SAXS curve, 115–119, 118f
Silica biomorphs
biomimetic approaches, 226
biomimetic self-assembly, inorganic environments, 227–229, 228f
biomineralization and inorganic materials, 226
chemical garden-like structures, 227–229
crystallization, alkaline-earth carbonates, 230
gardens (see Silica gardens)
gels
alkaline sol, 234–235
$BaCl_2$ solution, 235–236
carbonate crystallization, 234–235
custom-designed crystallization cassettes, 234–235, 236f
pH, 233–234
protonation, siliceous species, 233–234
stable and homogeneous silica hydrogel, 233–234
geochemical environments, 227–229
in situ monitoring, growth process, 238–239
microfocus X-ray diffraction, 251–252
morphological and structural ex situ analyses, 236–238
peculiar materials, 227–229
probing concentrations and chemistry, solution, 239–240
in situ analyses, 230
smooth curvatures, 226
solutions
bicarbonate and carbonate ions, 232
crystallization, barium carbonate, 232
experimental procedure, 230–232, 231f
parameters affecting morphogenesis, 233
pH value, 233
precipitates, 232–233
superficial dissolution, carbonate, 232–233
spontaneous compartmentalization, 227–229
Silica gardens
classical preparation method

Subject Index 611

beaker and alkaline silica sol, 241
borates/oxalates, 243
cation determines, 241–243
chemical reaction, 241–243, 242f
custom-designed pellet makers, 243–244
hexacyanoferrates, 243
Mach–Zehnder interferometry, 241–243
metal salt, 241
osmosis, 241–243
rate, salt dissolution, 243–244
isolation and characterization, 246–248
liquid injection method, 244–246
magnetic and electric fields, 246
tracing dynamic diffusion and precipitation processes
AES measurements, 250–251
pH determines, 250–251
physicochemical parameters, 250
salt pellets, 248–250, 249f
self-catalyzed chemical reactors/batteries, 250–251
track time-dependent variations, 248–249
Simulation, biominerals. See Replica exchange methods
SiRNA. See Small interfering RNAs (siRNA)
Site-directed mutagenesis technique, 330
Small-angle X-ray scattering (SAXS)
acquisition accuracy, speed and signal-to-noise levels, 123
agglomerated particles in solution, 2D scattering pattern, 97–99, 98f
background profile and transmission coefficient, 398–399
$BaTiO_3$ precursor system, 119–122
data interpretation (see Data interpretation, SAXS)
data treatment, 106
diffraction pattern, 99
electron microscopy, 123–124
experimental setup, laboratory equipment, 395
formation and growth, ACC (see Amorphous calcium carbonate (ACC))
inorganic materials (see Inorganic materials)
measurements, 97
mineral arrangement, stack of cards model
damped oscillation, 404
function $G(x)$, Fourier transform, 402–404
unbiased experimental data, 405
mineral crystal characteristics, osteonal bone, 409–410, 409f
nanometer-range electronic density variations, 96–97
p-parameter, 399–400, 400f
q-range, 97–99
radial and azimuthal intensity profiles, 397, 398f
scattering intensity, 97
silica (see Silica)
synchrotron-based (see Synchrotron-based SAXS)
T-parameter
Kratky plot, bone, 401–402, 402f
mineral volume fraction imaging, 402, 403f
platelike structures, 402
Porod region, 400, 401f
Porod's law, 400
two-electron density system, 96–97
and WAXD, XRF data, 410, 411f
and WAXS, 38–39
Small interfering RNAs (siRNA)
mollusk shell
dsRNA (see Double-stranded RNA (dsRNA))
monitor, effects, 381
nucleotide sequence, 380
PCR primers, 380–381
RNAi, 379
Pif, 384
SOLiD DNA sequencing, 374–375
Solid-state NMR (ssNMR)
distance measurements, protein and surface atoms, 363
and mutagenesis experiments, 346–347
Soluble polymers
classical and nonclassical growth modes
calcite mesocrystals, glass slip, 294–296, 296f

Soluble polymers (*Continued*)
    classical crystallization pathway, 297
    face-selective adsorption, 297
    particle nucleation, 296–297
    face-selective adsorption
        adsorption, additive, 293, 293*f*
        double hydrophilic block copolymers, 293
        equilibrium conditions, 292
        face-selective adsorption, 292
        face-selective polymer adsorption, 293–294
        fast-growing high-energy faces, 292, 292*f*
        morphosynthesis, gold nanostructures, 293–294
        polar peptides and proteins, 294
        repair process, 294
        surface energy, 292
        water solubility, 293
    low-molar-mass additives, 287–288, 287*f*
    mesocrystal formation, 297–301
    nucleation and crystallization inhibition, 288–289
    PILPs, 289–292
    polymer-mediated crystallization, 286
    types, crystallization control, 286–288
Spectroscopic assays. *See* Amelogenin
SsNMR. *See* Solid-state NMR (ssNMR)
Statherin docking, 346–347, 348
Stopped-flow spectrophotometer
    bromothymol blue (BTB), 137–138
    formation, ACP, 137–138
    protocol, 138
    real-time spectrums, 140
    UV–vis spectrums, pH indicator, 140, 140*f*
Synchrotron-based SAXS
    description, 103–105
    emergent species in size, 103
    growing phases, 103
    in situ and time-resolved measurements, 103
    kinetics, 103
    metal-organic transition phases, 104
    time resolution, 104
    and WAXS, 104

T

TEM. *See* Transmission electron microscopy (TEM)
TEY mode. *See* Total electron yield (TEY) mode
TGA. *See* Thermogravimetric analysis (TGA)
Thermogravimetric analysis (TGA), 564
Third-generation DNA sequencing technologies, 375–376
Time-resolved structural information
    low electron density contrast, 104
    SAXS patterns
        $BaTiO_3$, 119–120, 121*f*
        $CaCl_2$ and Na_2CO_3, 111–113, 112*f*
    scattering methods, 103
    WAXS patterns
        ACC experiment, 113–115, 116*f*
        with electrochemical measurements, 104
Titration
    measurements, 48–49
    quantitative analyses, 49
    treatment, primary data
        calcium carbonate, 55
        development, detected free Ca^{2+}, 55–56, 56*f*
        free ion products, 56–57
        hermodynamic solubility products, 57
        ion-selective electrodes, 54
Total electron yield (TEY) mode
    Auger electrons and reabsorption, 176*f*
    HOPG and MDBA, 168*f*
    invaluable diagnostic, 178–179
    measurements, 175–176
Total scattering methodology
    advanced photon source (APS), 479–480
    Argonne National Laboratory, 479–480
    background subtraction and data processing
        corrections, 485
        data processing, 484
        Fourier transformation, 485
        PDF method, application, 485
        program FIT2D, 484
        total scattering intensity, 484
    Bragg scattering, 479–480

Subject Index

data collection times
  detector type, 483
  factors, 483
  measurement time selection, 484
  synchrotron scattering beamlines, 483
experimental geometry and instrument calibration
  calibration, 481–482
  Debye-Scherrer geometry, 481–482, 481f
high photon fluxes, 479–480
raw scattering intensities, comparison, 480f
sample requirements and *in situ* methods
  acoustic levitation, 483
  flexibility, 482
  homogeneity, 482
  Kapton (polyimide), 482–483
  in liquids and gases, 482
Transmission electron microscopy (TEM)
In situ fluid cell (*see In situ* fluid cell TEM)
tropocollagen, 191
and XRD, 191, 196
Type I collagen (COL I), 190–191

W

Water model, FFs
  biomolecular community, 10
  fixed-charge model, 7
  moving interface simulation, 8–9, 9f
  oxygen–oxygen RDF, 7–8, 8f
  proton transfer events, 10
WAXD. *See* Wide-angle X-ray diffraction (WAXD)
WAXS. *See* Wide-angle X-ray scattering (WAXS)
Wide-angle X-ray diffraction (WAXD)
  L-parameter, 407–408
  mineral crystal characteristics, osteonal bone, 409–410, 409f
  and SAXS, XRF data, 410, 411f
  texture effect
    mineral crystals, 405–406
    orientation influence, 407
    and SAXS, XRF measurements, 406–407, 406f
Wide-angle X-ray scattering (WAXS)
  detectors, 98f, 99

2D patterns, diffraction spots in ACC, 111–113, 114f
3D patterns, pure ACC experiment, 113–115, 116f
and SAXS (*see* Small-angle X-ray scattering (SAXS))
synchrotron-based scattering methods, 103–105
and synchrotron SAXS, 38–39

X

XAS. *See* X-ray absorption spectroscopy (XAS)
X-ray absorption spectroscopy (XAS)
  absorption, 167–169
  EXAFS, 167–169
  geometries, 171f
  K-edge measurements, 169
  monolayer films, 169
  quantitative assignment, orbital orientation, 169–170
  SAM/crystal
    experiment and data optimization, 178–179
    EY and FY detection modes, 173–177
    incident flux, 177–178
    sample preparation and composition, 181–183
    X-ray-induced beam damage, 179–181
  soft X-ray beamlines, 167
X-ray diffraction (XRD)
  atomic structure, determination, 478
  COL I fibrils, 191
  total scattering experiment, 478–479
X-ray microdiffraction, biominerals
  area detectors, 512
  beam collimation and brightness, 509–510
  beam spot size, 526–527
  bending magnets, 511–512
  biomineralization, 527
  crystal structure determination, 503
  3D, 526–527
  diffraction, 503
  elements
    Bragg diffraction, 505f
    Debye-Scherrer rings, 507
    diffraction peaks, 504

X-ray microdiffraction, biominerals
  (*Continued*)
    diffractometer, 505
    intensity rings, 507
    phase problem of crystallography,
      505–506
    polychromatic (Laue) diffraction,
      507–508, 508f
    polycrystalline materials, 506–507
    powder diffraction, 507, 507f
    probes atomic arrangement, 505–506
    single crystalline polymers, 507
    single-crystal X-ray diffraction
      approach, 506f
    small-angle X-ray scattering (SAXS),
      508–509, 509f
    structure solution, 505–506
    wide-angle X-ray scattering (WAXS),
      508–509
  imaging techniques, 502–503
  KB configuration, 510
  microbeam SAXS
    atacamite, 523–524
    data set, 522–523
    WAXS, 524–525, 525f
  microstructural solutions, 502
  microstructure, 526
  monochromatic diffraction, 511–512
  multifaceted biomineralization process,
    502–503
  nacre aragonite layer, study, 513
  polychromatic diffraction, 511–512
  powder
    anisotropic mechanical behavior,
      519–520
    1D diffractograms, 515
    Debye-Scherrer rings, 515
    diffractogram, 515
    microbeampowder diffraction, 518
    monochromatic beam, 515, 520f
  phase identification, 515–516
  polycrystals, 515–516
  Rietveld method, 518
  ring width, 517
  soil sample, 516f
  texture analysis, 517
  zebrafish otoliths, morphologies and
    diffraction patterns, 519f
prismatic calcite layer, study,
  513, 513f
reflection, 510, 511
sample preparation
  hard X-rays, 514
  surface-sensitive techniques, 514
synchrotron sources, 509–510
white-beam
  crystallites, 521–522
  deformation state and defect density,
    521
  Laue diffraction, 520–521
  Laue patterns, 521–522, 521f
  multigrain indexing, 522
  nacre, growth mechanism, 522
X-ray focusing optics, 510
X-ray refractive index, 510–511
X-ray spectra and nanomechanics
  fibril orientation, 424
  fibril strain, 423–424
  mineral crystallite strain, 424
  mineral orientation, 424
  nanomechanical experiments, types,
    424–425
  and SAXS/WAXD pattern
    deformation, 423
    in lamellar bone, 422–423
    mineralized collagen fibril, 422–423,
      422f
    peak of interest, 423
    twisted plywood structure, 422–423
  tissues, negligible viscoelasticity, 425

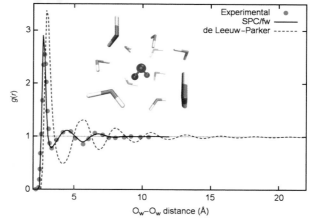

Paolo Raiteri et al., Figure 1.1 Oxygen–oxygen radial distribution function (RDF) of water as measured experimentally (Vaknin, Bu, & Travesset, 2008) and computed according to the de Leeuw and Parker (1998) and SPC/Fw (Wu et al., 2006) FFs. Note the extended range of oscillations out to ∼20 Å for the de Leeuw and Parker model indicative of overstructuring. The inset figure shows a sample configuration of the local coordination environment for a water molecule (blue) by neighboring waters (oxygen in red and hydrogen in white). The integral over $g(r)$ out to the first minimum gives an estimated coordination number for water of 13 for the de Leeuw and Parker model.

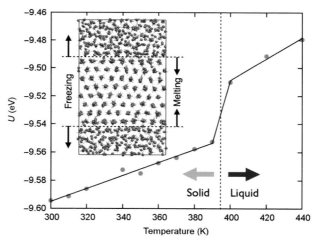

Paolo Raiteri et al., Figure 1.2 Moving interface simulation between ice IX (001) and liquid water. Graph shows the average potential energy per water molecule calculated during a 100 ps simulation as a function of temperature; the first 50 ps were considered as equilibration and discarded. The discontinuity corresponds to the melting point of ice IX. Given this is a metastable phase at these conditions, it therefore represents a lower bound to the melting point. Positive deviations from the potential energy versus temperature curve, such as at 340 K, can be due to defects in the ice structure that are trapped during growth.

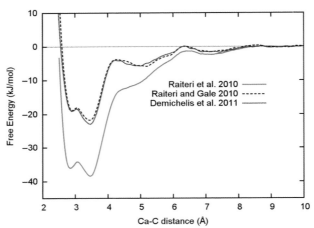

Paolo Raiteri et al., Figure 1.3 Free-energy profiles for the ion pairing of Ca^{2+} and CO_3^{2-} in aqueous solution as a function of the Ca–C distance for three different force fields. The free energy is taken relative to that of the separated ions.

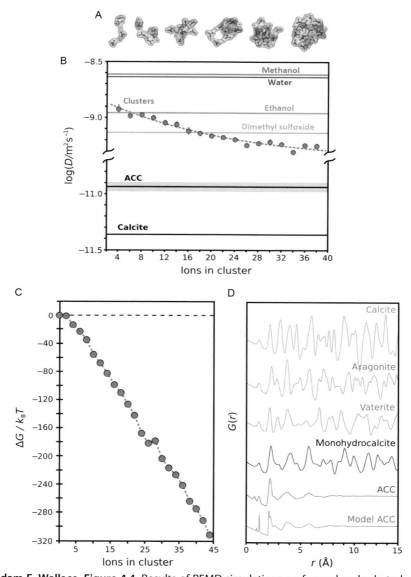

Adam F. Wallace, Figure 4.4 Results of REMD simulations performed on hydrated calcium carbonate clusters. (A) Snapshots taken at various stages of cluster growth. (B) The diffusivity of the calcium ions is plotted as a function of size and compared against calculated values for calcite and hydrous ACC. The significantly greater diffusivity of the ions in the cluster phase along with the similarity with the self-diffusivities of several common solvents suggests that the clusters are in a dense liquid phase. (C) The free energy landscape underlying phase separation and $[Ca^{2+}]=[CO_3^{2-}]=15$ mmol/L. The energy landscape is consistent with liquid–liquid separation and spinodal decomposition. (D) Pair distribution functions for several crystalline and amorphous phases of calcium carbonate as measured by total X-ray scattering. The model ACC structure constructed as per the process described in the text by Wallace et al. (2013) is shown for comparison. *Reproduced with permission from Wallace et al. (2013).*

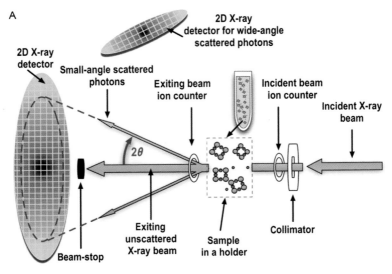
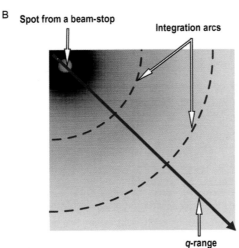

Tomasz M. Stawski and Liane G. Benning, Figure 5.1 (A) Scheme of the typical SAXS/WAXS experiment; (B) example of a scattering pattern from a sample containing agglomerated particles in solution recorded with a 2D SAXS detector. Some of the elements relevant for the data-reduction steps explained in Section 4 are marked in the figure.

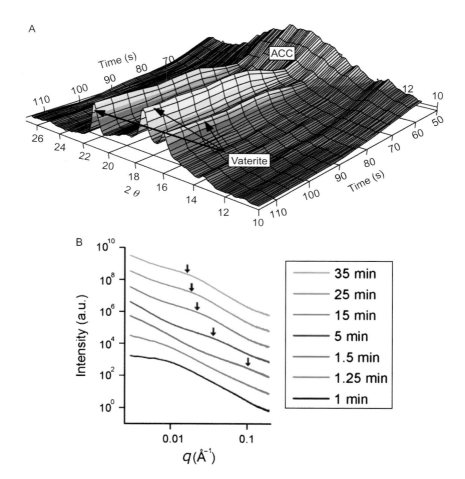

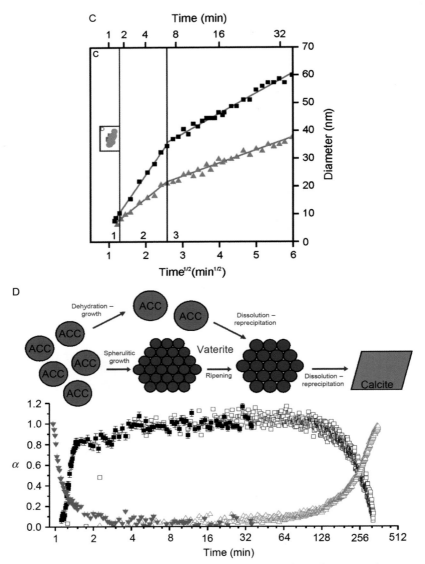

Tomasz M. Stawski and Liane G. Benning, Figure 5.5 (A) Three-dimensional representations of the time-resolved WAXS patterns from the pure ACC experiment (time is plotted on a base 2 log scale for clarity); (B) stacked time series of selected SAXS patterns from the pure ACC experiment, with the legend showing time in minutes and the arrows illustrating the position of the peaks caused by the scattering from the growing vaterite crystallites. (C) ACC nanoparticle and vaterite crystallite sizes derived from the SAXS data versus time (on a $t^{1/2}$ scale) for the pure ACC and an ACC doped with SO_4 experiments. (D) Schematic representation of the multistage ACC → vaterite → calcite crystallization pathway. (E–G) Electron microscope microphotographs of solids quenched throughout a full ACC to calcite transformation reaction. *Printed with permission from Bots et al. (2012). Copyright 2012 American Chemical Society.*

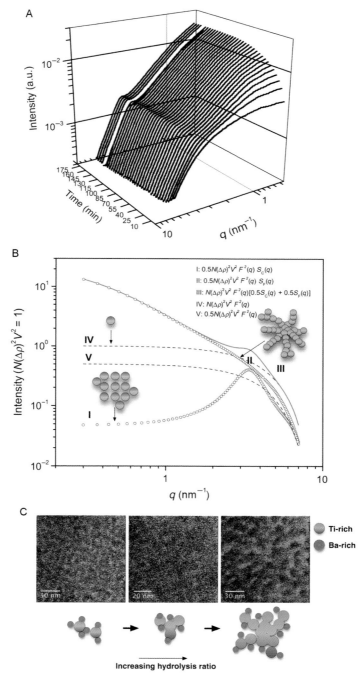

Tomasz M. Stawski and Liane G. Benning, Figure 5.7—See legend on next page

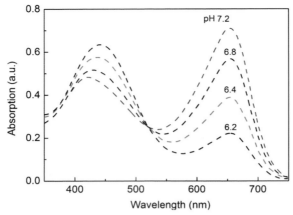

Haihua Pan et al., Figure 6.2 UV–vis spectrums of 30 μM BTB at pH 6.2–7.4. The absorption peak at 418 nm is produced by acid formation and that at 618 nm is produced by base formation.

Tomasz M. Stawski and Liane G. Benning, Figure 5.7 (A) Time-resolved SAXS patterns at 60 °C, showing evolution of barium titanate precursor sols of initial 0.5 mol/dm^3 concentration with hydrolysis ratio [H$_2$O]/[Ti] = 5.6. (B) Simulated SAXS curves using Eq. (5.12), with $N(\Delta\rho)^2 V^2$ set to 1. (i) $I(q)$ of semiordered agglomerates with $r_0 = 0.5$ nm, $v = 0.3$, $2R_{HS} = 1.8$ nm, and $\varepsilon = 0.5$. (ii) $I(q)$ of fractal agglomerates with $r_0 = 0.5$ nm, $D = 1.8$, $R_g = 4.76$ nm, and $\varepsilon = 0.5$. (iii) $I(q)$ of linear combination of structures i and ii. (iv) $I(q)$ of sphere form factor of $r_0 = 0.5$ nm. (v) $I(q)$ of sphere form factor of $r_0 = 0.5$ nm and $\varepsilon = 0.5$. (C) EELS mappings of Ba (red) and Ti (green) of as-dried BTO films with increasing hydrolysis ratios. *Panels (A) and (B) are printed with permission from Stawski et al. (2011a) and panel (C) is printed with permission from Stawski et al. (2011b). Copyright 2011 American Chemical Society.*

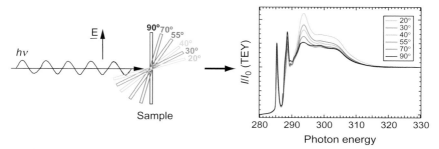

Jonathan R.I. Lee et al., Figure 8.5 Schematic of the geometric configuration for angular-dependent XAS measurements on a planar SAM/crystal sample and an experimental dataset collected at the carbon K-edge for an MDBA SAM on Au(111) using this geometry. Note that the experimental spectra reveal significant angular dependence in the σ^* resonances at ~294 and 304 eV but comparatively little angular dependence in the π^* resonances of the aryl ring (285.4 eV) and carboxyl end-group (~289 eV).

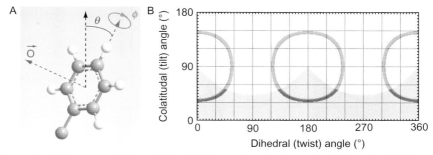

Jonathan R.I. Lee et al., Figure 8.11 (A) Schematic to indicate the tilt (colatitudal, θ) and twist (dihedral, φ) angles that describe the orientation of a phenyl ring. Carbon and hydrogen atoms are denoted by gray and white spheres, respectively. The black arrow is coincident with the surface normal of the metal substrate and the green arrow corresponds to the transition dipole moment, which lies perpendicular to the plane of the ring. (B) A graph indicating the permissible end-group orientations within an MDBA SAM on Au(111) as a subset of all possible orientations. The combinations of colatitudal and tilt angles for the aromatic end-group of MDBA obtained via linear regression analysis are displayed as a red annulus and correspond to the XAS data presented in Fig. 8.2B. Meanwhile, the sterically permissible and forbidden end-group orientations are displayed as gray and white regions, respectively. Thus, overlap between the red annulus and the gray region corresponds to a viable combination of colatitudal and dihedral angles for the MDBA end groups. The blue annulus is a guide for the eye and corresponds to an end-group orientation exactly at the magic angle, ~54.7° (or a random distribution of end-group orientations). Conventions for defining the colatitudal and dihedral angles are available in the literature (Lee et al., 2013).

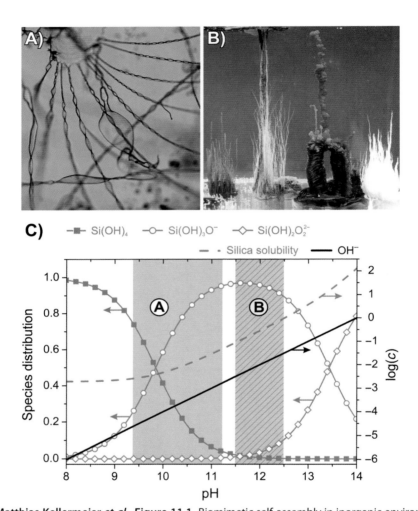

Matthias Kellermeier et al., Figure 11.1 Biomimetic self-assembly in inorganic environments. (A) Silica biomorphs grown by slow crystallization of barium carbonate in a silica gel with a starting pH of 10.5. The dominant morphologies are twisted ribbons and flat sheets, which show optical birefringence when viewed between crossed polarizers. (B) Silica gardens prepared by addition of concentrated silica sol (pH ~ 11.6) to crystals of (from left to right) $CoCl_2$ (blue), $FeCl_2$ (green), $FeCl_3$ (brown-orange), and $CaCl_2$ (white). The precipitates consist of multiple hollow tubes with different thickness and shape. (C) Plot of the fractions of monomeric silica species (symbols), the total solubility of silica (dashed line), and the hydroxide concentration (full line) in solution as a function of pH. The shaded areas represent pH ranges in which the two types of structures were reported to form (Coatman et al., 1980; Eiblmeier, Kellermeier, Rengstl, et al., 2013; Melero-Garcia et al., 2009). At the moderately alkaline pH required for the growth of biomorphs (ca. 9–11), dissolved silica species carry a relatively low number of charges and are prone to condensation reactions (Iler, 1979). This lowers their solubility and leads to the formation of amorphous silica next to carbonate. At the more basic conditions needed for the formation of silica gardens (>11.5), deprotonation of silicate species is enhanced and the solubility of silica increases, so that direct precipitation of SiO_2 is suppressed and the charged silicate anions rather react with the metal cations. In parallel, the higher concentrations of OH^- ions favor the formation of metal hydroxides and oxides. *(A) Reproduced with permission from*

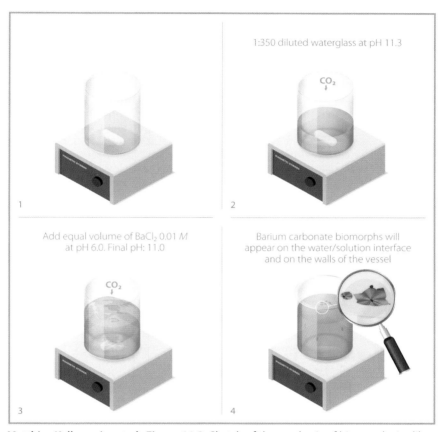

Matthias Kellermeier *et al.*, Figure 11.2 Sketch of the synthesis of biomorphs in dilute silica solutions. See text for explanations.

Matthias Kellermeier et al., Figure 11.6 New preparation method for silica gardens. (A) Image sequence showing the formation of a macroscopic tube from a pellet of $CoCl_2 \cdot 6H_2O$ by controlled dosing of silica sol. Pipette tips indicate that samples can easily be drawn from both the inner solution enclosed by the membrane and the outer reservoir surrounding it. (B) Top view of the tube, demonstrating that it has an open end. (C) Schematic representation of the experimental setup used for characterizing the temporal evolution of the macroscopic silica garden system. Relevant species in the inner and outer solution are indicated together with the devices used for measuring potential differences and pH. *Reproduced with permission from Glaab et al. (2012). Copyright 2012 John Wiley & Sons.*

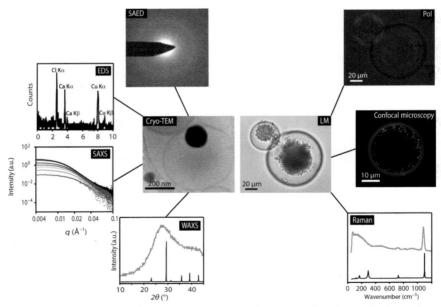

Chantel C. Tester and Derk Joester, Figure 12.7 Summary of *in situ* methods to characterize the morphology, structure, and growth kinetics of precipitates formed within liposomes. SAED, selected area electron diffraction; EDS, energy-dispersive X-ray spectroscopy; SAXS, small-angle X-ray scattering; WAXS, wide-angle X-ray scattering; LM, light microscopy; Pol, polarized light microscopy.

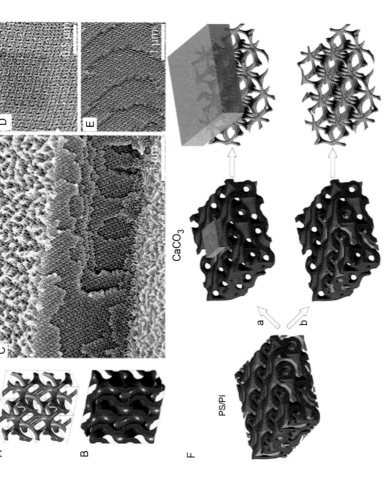

Helmut Cölfen, Figure 13.5 (A) The two intertwined, nonintersecting gyroid networks (polyisoprene (PI), replicated by calcite). (B) The spongelike PS matrix. (C) Cross section of a patterned PS film showing the continuous matrix ((211) face) after PI removal, which exposes the two gyroid networks for crystal infiltration. (D–E) High-magnification images of the (421) and (100) cross-sectional fracture planes of the porous PS film, same magnification. (F) Schematic representation of the crystallization process. (Left to right) Films of the PS/PI copolymer self-assemble into a double-gyroid microphase morphology. After removal of the two PI networks (red and orange) from the PS matrix (blue), calcite crystals nucleate (a) on the surface of or (b) inside the polymer film and grow into the porous networks, leading to a gyroid-patterned single crystal, visible after PS matrix removal. *Image taken from Finnemore et al. (2009) with permission of Wiley-VCH.*

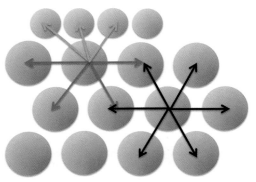

Helmut Cölfen, Figure 13.11 Adsorption of an additive (illustrated with spheres in grey) on a crystal surface. The surface energy will be decreased because some bonds can be formed between the surface atoms and the adsorbed molecule (green arrows). This brings the surface atom into a similar situation to a bulk atom that can form bonds with all neighbors (black arrows).

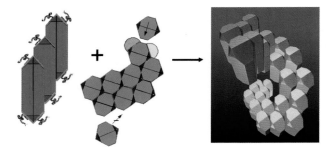

Helmut Cölfen, Figure 13.17 Assembly of the orthorhombic nanoparticle units to the chiral mesocrystalline superstructure after polymer adsorption onto {110}. Left of the arrow are two side views of the assembling nanoparticles, and right of the arrow is the resulting superstructure. The sterically demanding polymer adsorbed on the {110} faces forces a staggered arrangement of the $BaCO_3$ nanorods. Turning the view by 90° reveals that for attaching nanoparticles, the {011} faces are not equal and a preferential adsorption side exists. Overlaying these two effects then leads to the formation of a helix structure composed of nanorods shown right of the arrow.

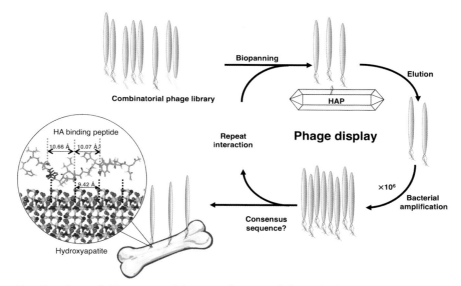

Hyo-Eon Jin et al., Figure 14.1 Schematic diagram of phage display process for discovering peptide motifs that recognize hydroxyapatite.

Identified phage clones at pH 7.5	Aligned amino acid sequences														
CLP12				Q	P	Y	H	P	T	I	P	Q	S	V	H
CLP7C					C	Q	Y	P	T	L	K	S	C		
CLP7						H	A	P	V	Q	P	Q			
Poly12	T	M	G	F	T	A	P	R	F	P	H	Y			
Collagen Type I major repeat				G	P	O	G	P	O	G	P	O	G	P	O
Collagen α I (1301–1315)				Y	P	T	Q	P	S	V	A	Q	K	N	W	Y	I	S
Amelogenin (125–139)				Q	P	Y	Q	P	Q	P	V	Q	P	Q	P	H	Q	P
Statherin (36–50)				G	P	Y	Q	P	V	P	E	Q	P	L	Y	P	Q	P

Hyo-Eon Jin et al., Figure 14.2 The identified HAP-binding peptide sequences from phage display and corresponding bone- and tooth-associated protein sequences from the protein database. The best binding peptide (CLP12) for the (100) HAP surface at pH 7.5 possessed repeating proline residues at i and $i+3n$ positions (positions 2, 5, and 8) and periodic hydroxylated residues (positions 3, 6, and 10) resembling the major repeat sequence of type I collagen and amelogenin. *Reprinted with permission from Chung et al. (2011). Copyright (2011) American Chemical Society.*

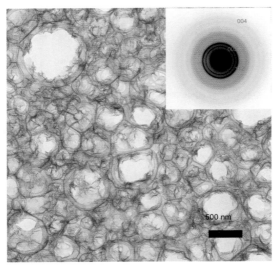

Hyo-Eon Jin et al., Figure 14.3 Transmission electron micrograph of mineralized HAP crystal templated by CLP12 peptide. *Reprinted with permission from Chung et al. (2011). Copyright (2011) American Chemical Society.*

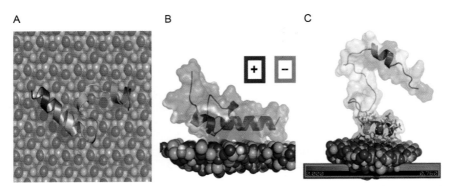

Michael S. Pacella et al., Figure 16.1 Predicted structures from RosettaSurface. Top (A) and side (B) views of the lowest-energy structure of statherin docked onto the [001] face of HAp using the initial version of the algorithm. (C) A representative structure of statherin bound to HAp, colored by the local root-mean-square deviation of 100 structures that best satisfied experimental constraints. *Reprinted from Makrodimitris et al. (2007) and Masica, Ash, et al. (2010) with permission.*

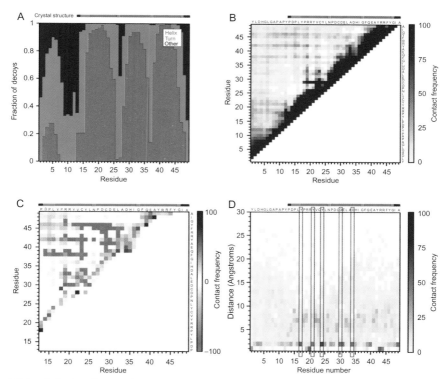

Michael S. Pacella et al., Figure 16.3 Analysis plots generated by RosettaSurface. (A) Secondary structure profile for osteocalcin adsorbed to HAp [100]. (B) Intraprotein contact map for osteocalcin adsorbed to HAp [100]. (C) A difference plot between crystal structure and adsorbed-state decoys. (D) Protein–surface contact map for osteocalcin adsorbed to HAp [100]. γ represents gamma-carboxyglutamic acid. Line above each plot represents the secondary structure of the crystal structure and is colored as in (A). Boxes highlight residues predicted to interact with HAp in Hoang's model.

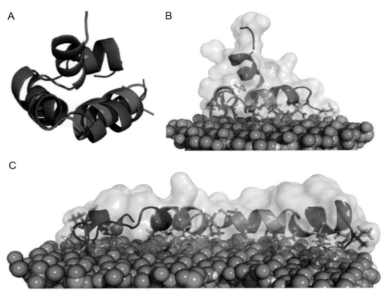

Michael S. Pacella et al., Figure 16.4 Predicted structures for the osteocalcin/HAp system. (A) Predicted solution-state structure of osteocalcin (blue) superimposed with its crystal structure (red). Lowest-energy decoys from rigid (B) and flexible (C) backbone docking. Similar contacts are made between the γ-carboxyglutamic acids and the surface in both models, indicating ambiguity in whether or not osteocalcin adopts extended conformations on HAp.

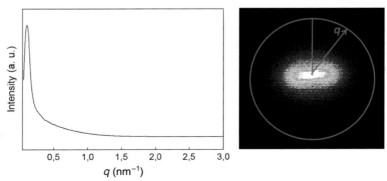

Silvia Pabisch et al., Figure 18.3 Radial plot (left) of the integration in azimuthal direction of a SAXS pattern (right).

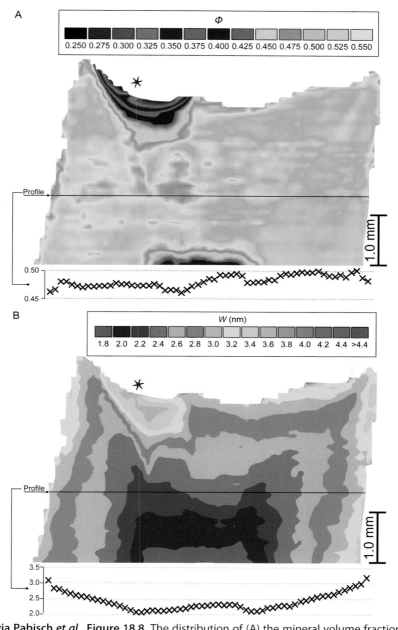

Silvia Pabisch et al., Figure 18.8 The distribution of (A) the mineral volume fraction Φ and (B) particle thickness W of dry dentine were studied by Märten et al. (2005). A typical profile of W-parameter across the data is plotted beneath the map.

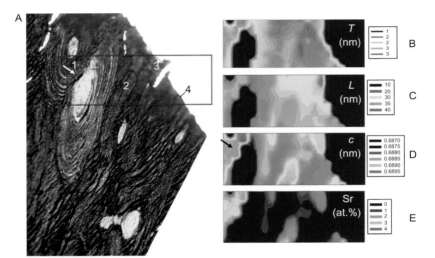

Silvia Pabisch et al., Figure 18.10 Maps of mineral crystal characteristics in osteonal bone from a postmenopausal woman treated for 36 months with strontium ranelate (A). SAXS and XRF analysis derived the mean thickness T (B) and the length L (C) of the mineral crystals. (D) Lattice spacing in the hexagonal direction of apatite shown by the parameter c. (E) Percentage of calcium ions substituted by strontium in the hydroxyapatite crystals. The red lines in the microscopy image (A) indicate regions of increased strontium content in the mineral crystals (Li et al., 2010).

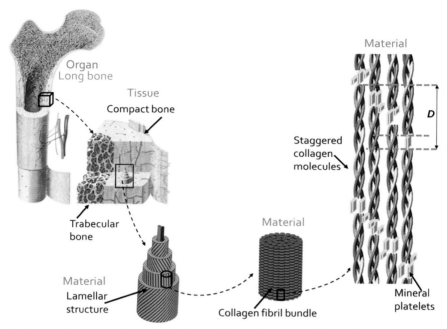

A. Karunaratne et al., Figure 19.1 Hierarchical architecture of bone. The macroscopic (organ) level consists of two main types: cortical (compact) bone (which will be studied in this chapter) and trabecular or spongy bone. Both bone types have a lamellar structure with lamellae with widths from 5 to 10 µ, which are in turn composed of mineralized collagen fibrils ~50–200 nm in diameter. The internal structure of the mineralized fibril consists of intrafibrillar mineral (shown in a reduced and highly schematic manner described earlier). The fibril is surrounded by extrafibrillar mineral that is not shown for clarity. The triple helical tropocollagen molecules are arranged in a staggered manner, leading to an axial periodicity in electron density known as the D-period. $D \sim 65\text{--}67$ nm (Fratzl & Weinkamer, 2007; Meyers et al., 2008; Weiner & Wagner, 1998).

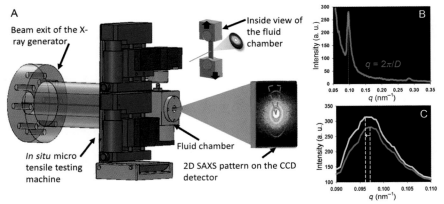

A. Karunaratne et al., Figure 19.2 *In situ* micromechanics in a synchrotron beamline. Left: Schematic view of the microtensile testing apparatus in a synchrotron SAXS beamline. The sample is immersed in phosphate-buffered saline. Tensile load is applied along the parallel collagen fibril axis as shown in the schematic (black arrows) of MCF. (B) Integrated intensity profile in the radial directions in the angular region (red dash lines) shown in (A), showing the collagen reflections. (C) Shift of the first-order collagen peak position with increasing tissue strain, for tissue strain = 0% (time = 0) (red line) and tissue strain = 0.5% (time = 240 s) (yellow line).

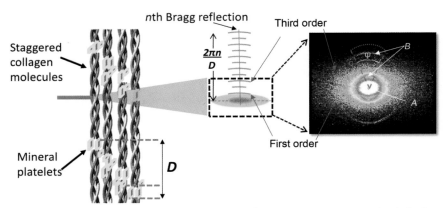

A. Karunaratne et al., Figure 19.3 Schematic of how SAXS spectra can be linked to structural parameters of the mineralized collagen fibril. Left: X-ray beam incident on the mineralized collagen fibril scatters to form a pattern with two distinct components: the diffuse SAXS scatter from the mineral platelets (ellipse in center) and a series of Bragg diffraction peaks, extended into arced lines due to a combination of fibril orientation distribution due to the lamellar plywood structure and the finite diameter of the fibrils. Right: An experimental SAXS spectrum from murine femoral cortical bone, showing the third-order Bragg peak arising from the meridional D-periodicity and the arcing ψ due to the lamellar architecture.

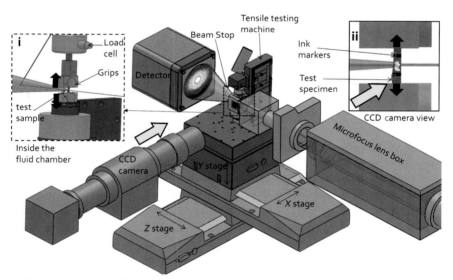

A. Karunaratne et al., Figure 19.10 The experimental setup for *in situ* microtensile testing combined with microfocus SAXS at beam line I22 (BL I22), diamond light source (DLS). The test sample is immersed inside the fluid chamber and clamped in the grips (inset i). The CCD camera views the sample perpendicular to the incident X-ray beam so as to avoid blocking the beam (inset ii). The 2D CCD detector (Pilatus 6M) is located between 250 mm and ~1 m from the tensile tester at a wavelength of ~0.8857 Å (the distance shown is foreshortened for presentation purposes). Samples are replaced by laterally translating the tester out of the beam (Z-stage axis).

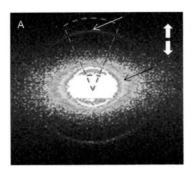
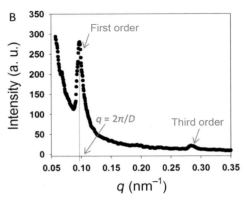

A. Karunaratne et al., Figure 19.14 SAXS spectra from bone and regions of integration for fibril strain: (A) 2D SAXS/SAXS spectrum from murine cortical bone. Red dashed sector denotes the region rebinned in either q or χ using the FIT2D CAKE command to provide the 1D plot on right. Thick white vertical arrows indicate the direction of applied load. Black arrow shows the region of diffuse scatter arising from the SAXS scattering from mineral platelets (Rinnerthaler et al., 1999). White slanted arrows indicate the third-order meridional reflections arising from the 67 nm collagen D-periodicity; a stronger first-order reflection is visible nearer the apex of the CAKE sector but is obscured by the strong SAXS signal. (B) The azimuthally averaged $I(q)$ plot (from the indicated CAKE sector) from the 2D SAXS image in (A), showing clear diffraction peaks at the first and third orders arising from the collagen meridional D-periodicity. Note that although the third order is much weaker than the first order, the first-order peak has a much higher diffuse SAXS background associated with it.

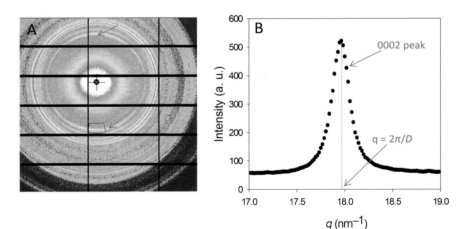

A. Karunaratne et al., Figure 19.16 WAXD spectra from apatite phase in the bone: (A) 2D WAXD spectrum of murine femoral bone, with the oriented (0002) reflection arising from the apatite mineral indicated by red arrows. The CAKE sector used for azimuthal averaging is shown schematically as an angular sector (black line, lower part of the plot). The black rectangular lines forming a grid on the detector are to be disregarded as they are regions between active detector areas on the Pilatus 6M detector used for measurements. (B) Azimuthally averaged $I(q)$ plot for (0002) order, from the sector shown in (A). Note that the background is nearly constant.

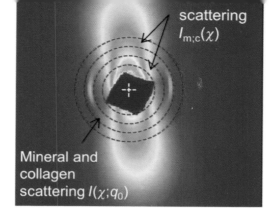
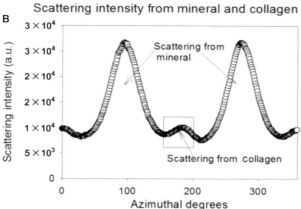
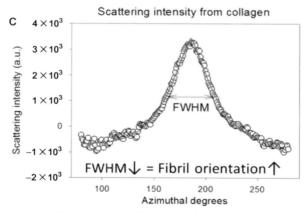

A. Karunaratne et al., **Figure 19.18** Orientation of mineralized collagen fibrils: (A) 2D SAXS plot from murine femoral bone showing the outer, middle, and inner annular rings used to calculate angular intensity distribution $I(\chi)$. (B) Integrated intensity of the middle ring showing the strong peaks arising from the mineral scattering and the much weaker peaks due to collagen meridional D-periodicity. (C) Angular intensity distribution for the peak on the left in (A) after background subtraction of inner and outer rings. Note that the peaks are now clearly visible, but a slight negative offset to the baseline has been introduced during the correction.

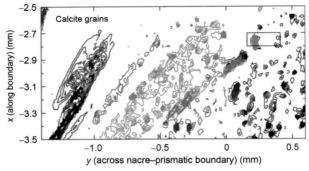

Nobumichi Tamura and Pupa U.P.A. Gilbert, Figure 21.7 Bragg peak intensity map for the calcite (left, $y \leq 0.2$ mm) and aragonite (right, $y \geq 0.2$ mm) grains at and near the prismatic-nacre boundary in a red abalone shell. Each color island indicates a region of cooriented crystals. *Data from DiMasi and Sarikaya (2004).*

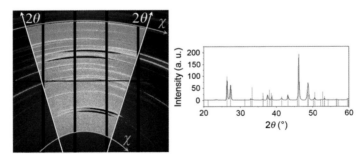

Nobumichi Tamura and Pupa U.P.A. Gilbert, Figure 21.8 Left: X-ray diffraction pattern of Debye–Scherrer rings from a piece of shell from the welk *Urosalpinx cinerea*. The integration area along 2θ and χ is highlighted by inverting the colors, so they appear as complementary to the colors in the rest of the image. This area is the orange sector of an annulus. Right: the resulting 1D diffractograms, obtained from the annular sector on the left, integrated over 2θ and χ. Red lines are the aragonite peak position from a database. All reflections match those of aragonite. *Data courtesy of J. B. Ries.*

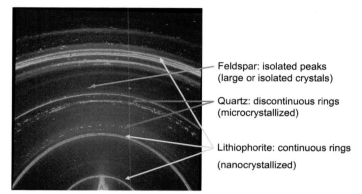

Nobumichi Tamura and Pupa U.P.A. Gilbert, Figure 21.9 This diffraction pattern of a sample of soil shows different type of rings depending on the average grain size of the minerals. *Data courtesy of A. Manceau.*

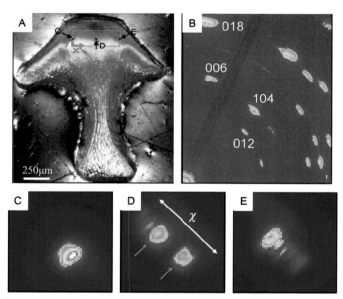

Nobumichi Tamura and Pupa U.P.A. Gilbert, Figure 21.11 Diffraction patterns obtained with a 10×10 μm^2 monochromatic beam and sample oscillation up to 90°. *Data from Ma et al. (2009).*

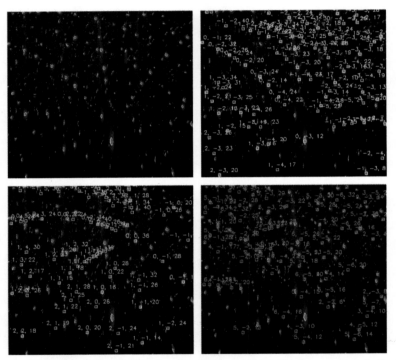

Nobumichi Tamura and Pupa U.P.A. Gilbert, Figure 21.12 The Laue pattern in the top left panel was acquired with a 1 μm X-ray beam from a region of the prismatic calcite layer of the *Haliotis rubra* shell. This pattern shows the composite diffraction of at least three different calcite crystallites, the *h,k,l* indexing of the separate crystallites are highlighted in yellow, orange, and red in other panels.